BIOMATERIALS FROM AQUATIC AND TERRESTRIAL ORGANISMS

BIOMATERIALS FROM AQUATIC AND TERRESTRIAL ORGANISMS

Editors

Milton Fingerman
Rachakonda Nagabhushanam
Department of Ecology and Evolutionary Biology
Tulane University, New Orleans, Louisiana
USA

Science Publishers

Enfield (NH) Jersey Plymouth

SCIENCE PUBLISHERS
An Imprint of Edenbridge Ltd., British Isles.
Post Office Box 699
Enfield, New Hampshire 03748
United States of America

Website: *http://www.scipub.net*

sales@scipub.net (marketing department)
editor@scipub.net (editorial department)
info@scipub.net (for all other enquiries)

ISBN 1-57808-429-6 [10 digits]
978-1-57808-429-6 [13 digits]

Published by Science Publishers, Enfield, NH, USA
An Imprint of Edenbridge Ltd.
Printed in India.

Preface

Organisms, both aquatic and terrestrial, are sources of a wide variety of substances, many of which have already been shown to be bioactive. That is, they have been shown to play a wide variety of physiological and environmental roles. These chemicals include a broad array of proteins, lipids, and polysaccharides. Potentially, many of these natural products will find applications in industry, agriculture, and medicine. In the early phases of the study of these natural products the focus was on their isolation and identification. Now, the emphasis has shifted to testing and the development of new applications to solve medical and environmental problems, among others. These efforts involve collaboration among natural products chemists, oncologists, toxicologists, biochemists, and molecular biologists, to cite some examples of the specialists now involved. We are in an exciting era as potential begins to approach reality.

This volume explores ongoing efforts to develop these natural products into commercially viable materials that will contribute to solving, especially, health and environmental problems world-wide. Among the chapters in this volume are ones that deal with the use of compounds from plants to treat Alzheimer's disease, the antimicrobial activity of terpenes from African plants, antioxidant compounds from plants, antiangiogenic compounds from marine invertebrates, bioactive natural products from marine fungi, and the anti-inflammatory and antiallergenic properties of triterpenoids from plants.

This volume, which presents the most recent information on biomaterials, was written by a highly talented group of scientists who are not only able to communicate very effectively through their writing, but are also responsible for many of the advances that are described herein. We, the editors, have been most fortunate in attracting a highly talented, internationally respected group of investigators to serve as authors. We intentionally set out to present a truly international scope to this volume.

Consequently, appropriate authors from several countries were sought, and to everyone's benefit, our invitations to contribute were accepted.

We take pleasure in thanking the authors for their cooperation and excellent contributions, and for keeping to the publication schedule. The efforts of these individuals made our task much less difficult than it might have been. Also, we especially wish to thank our wives, Maria Esperanza Fingerman and Rachakonda Sarojini, for their constant and undiminishing encouragement and support during the production of this volume. We trust that you, the readers, will agree with us that the efforts of the authors of the chapters in this volume will serve collectively to provide a major thrust toward a better understanding of the good that can come from the ongoing investigations of these interesting biomaterials that have been obtained from both plants and animals, aquatic and terrestrial.

Milton Fingerman
Rachakonda Nagabhushanam

Contents

The Contributors

Ahmed Abdel-Lateff, Pharmacognosy Department, Faculty of Pharmacy, El-Menia University, El-Menia, Egypt

Toshihiro Akihisa, College of Science and Technology, Nihon University, 1-8 Kanda Surugadai, Chiyoda-ku, Tokyo 101-8308, Japan
E-mail: akihisa@chem.cst.nihon-u.ac.jp

Paula B. Andrade, Department of Pharmacognosy, Faculty of Pharmacy, University of Porto, Rua Aníbal Cunha, 164, 4099-030 Porto, Portugal

Shoshana (Malis) Arad, The Institute for Applied Biosciences, Ben-Gurion University of the Negev, Beer-Sheva 84105, Israel
E-mail: arad@bgu.ac.il

Bok Hee Bae, College of Pharmacy, Pusan National University, Keumjeong-ku, Changjeon-dong, Busan 609-735, Korea

Maria de Lourdes Bastos, Department of Toxicology, Faculty of Pharmacy, University of Porto, Rua Aníbal Cunha, 164, 4099-030 Porto, Portugal
E-mail: mlbastos@ff.up.pt

Félix Carvalho, Department of Toxicology, Faculty of Pharmacy, University of Porto, Rua Aníbal Cunha, 164, 4099-030 Porto, Portugal

D. Chuck Dunbar, National Center for Natural Products Research, Research Institute of Pharmaceutical Sciences, School of Pharmacy, University of Mississippi, University, Mississippi 38677, United States of America

Eduarda Fernandes, Department of Physical Chemistry, Faculty of Pharmacy, University of Porto, Rua Aníbal Cunha, 164, 4099-030 Porto, Portugal

Katja M. Fisch, Molecular Biology Laboratory, Department of General Surgery, University Hospital Muenster, Waldeyerstrasse 1, 48149 Muenster, Germany

1

Use of Plants for Management of Alzheimer's Disease

PETER J. HOUGHTON[1] AND MELANIE-JAYNE R. HOWES[2]

[1]*Department of Pharmacy, King's College London, Franklin-Wilkins Building, 150 Stamford Street, London SE1 9NN, UK*

[2]Jodrell Laboratory, The Royal Botanic Gardens, Kew, Richmond, Surrey, TW9 3AB, UK

INTRODUCTION

Alzheimer's disease (AD) is the major disease in a group which is characterised by loss of cognitive function leading to dementia. AD is estimated to account for between 50% and 60% of dementia cases in persons over 65 years of age (Francis *et al.* 1999) and is a progressive, neurodegenerative disease that primarily affects the elderly population. It is a major public health concern in developed countries because of the strains imposed on carers and financial resources by the increasing numbers of sufferers. The main symptoms associated with AD involve cognitive dysfunction, primarily memory loss (Desgranges *et al.* 1998, Förstl *et al.* 1995). Several other features manifest in the later stages of the disease including language deficits, depression, agitation, mood disturbances, and psychosis (McGuffey 1997, Wragg and Jeste 1989).

Although AD as a defined medical condition has only been recognised for about 100 years, the loss of memory and decline in cognitive function associated with advancing years is recorded in many documents from thousands of years ago. In light of the many useful compounds utilised in pharmacy for a whole range of conditions which have been developed from medicinal plants, it is not irrational to suppose that plant species reputed to help memory loss, found in ancient writings, might prove to have activities, and thereby compounds, of value in treating AD. It is only within the last twenty years that methods have been developed to test the reputed claims by scientific research and subsequently determine the structure of the compounds responsible. Since 1990, several compounds with relevant activity, based on natural products, have been introduced into orthodox medicine for the symptomatic treatment of AD and these are discussed below. In addition, plant materials themselves, or derived extracts, are increasingly being used by the general public in the industrially-developed countries for preventative and remedial treatment of mental decline, as well as by naturopaths, herbal practitioners, and the like. In many developing countries such plant materials are the only remedies available to the large majority of the population, and so form an interesting focus of investigation.

The aetiology of AD is still unknown but there is some evidence that various factors appear to increase the risk of developing the disease. Some of these are based on epidemiological evidence and on scientific investigations. The main approach so far to yield therapeutic agents used in the clinic is to elevate levels of acetylcholine (ACh) in the brain, which may be induced by cholinesterase inhibitors (see below). More recently, prevention of glutamate mediated neurotoxicity has been a therapeutic target in AD. The N-methyl-D-aspartate (NMDA) receptor antagonist, memantine, has been introduced in some countries for clinical use in AD patients. However, it should be noted that other avenues are being explored as links between biological activity and reputed usefulness.

The pathological features in the central nervous system (CNS) in AD are senile plaque and neurofibrillary tangle formation, considered by many to be associated with oxidative and inflammatory processes and neurotransmitter disturbances. A consistent example of the latter neuropathological occurrence associated with memory loss is a deficit of ACh, a neurotransmitter which has been demonstrated to be positively connected with cognitive function, and a cholinergic deficit has been correlated with the severity of AD (Bierer *et al.* 1995, Giacobini 1990, Perry *et al.* 1978, Perry 1986). Attempts to restore cholinergic function have been a rational target for drugs used to treat the symptoms of AD and, although cholinergic stimulation has been attempted by some, the prolongation of the availability of ACh released into the neuronal synaptic cleft has been the more fruitful approach, being achieved especially by inhibiting ACh hydrolysis by acetylcholinesterase (AChE), through the use of AChE inhibitors. It should be stressed that such compounds only alleviate some of the cognitive symptoms of the disease and do not achieve any permanent improvement. Since the development of simple *in vitro* procedures to assess AChE inhibition (Marston *et al.* 2002, Rhee *et al.* 2001), several naturally-occurring compounds have been isolated which have this activity.

For several years it was thought that epidemiological evidence indicated that estrogen replacement therapy (ERT) had a preventative role against AD, and estrogen treatment in women with AD was reported to enhance cognitive function (Birge 1997, Fillit *et al.* 1986). However, several more recent studies (Hogervorst *et al.* 2000, Mulnard *et al.* 2000) appear to deny the association between high estrogen levels and reduced incidence of AD, but the estrogen receptor binding properties of some plant extracts have been investigated as possible explanations of their reputation as memory improvers, and there has been some research into this proposed link (Howes *et al.* 2002, Perry *et al.* 2001).

Other reports have indicated that the use of anti-inflammatory drugs may influence the progression of AD. There have been several studies to show that non-steroidal anti-inflammatory drugs (NSAIDs) may reduce the risk of developing AD, and that patients with rheumatoid arthritis, who often use NSAIDs, have a lower incidence of AD (Breitner 1996, Breitner *et al.* 1995, Jenkinson *et al.* 1989, McGeer *et al.* 1990, McGeer *et al.* 1996). Therefore the use of anti-inflammatory compounds may also have potential in AD therapy. Also implicated in the pathology of many diseases including aging processes, atherosclerosis, ischaemic heart disease, and neurodegenerative diseases, including AD, are free radical reactions, which are reported to initiate cell injury (Maxwell 1995, Slater 1984, Spiteller 1993). Consequently, the use of anti-oxidants may slow AD progression and minimise neuronal degeneration.

Numerous plants and their constituents are reputed in traditional practices of medicine to enhance cognitive function and to alleviate other symptoms of AD, including depression (Howes *et al.* 2003a,b). The plant constituents may not only act synergistically with other constituents from the same plant, but may also enhance the activity of compounds from other plant species. This approach has been used in various practices of traditional medicine, including Ayurveda and traditional Chinese medicine (TCM) where a mixture of plants is commonly prescribed. An ethnopharmacological approach has provided leads to identify plants and potential new drugs that are relevant for the treatment of cognitive disorders, including AD, and may aid the discovery of a more varied and efficacious selection of drugs for AD treatment.

This article reviews some of the plants and their active constituents that have been used in traditional practices of medicine for their reputed cognitive-enhancing effects. Plants and their constituents with pharmacological activities relevant to treatment of cognitive disorders, including anticholinesterase (anti-ChE), anti-inflammatory, antioxidant, and estrogenic effects, are discussed.

ACTIVITIES RELEVANT TO THE TREATMENT OF AD

Enhancement of Cholinergic Stimulation

A consistent neuropathological occurrence associated with memory loss is a cholinergic deficit, which has been correlated with the severity of AD (Bierer *et al.* 1995, Giacobini 1990, Perry *et al.* 1978, Perry 1986). Therefore attempts to restore cholinergic function have been a rational target for drugs used to treat the symptoms of AD. Approaches to enhance cholinergic function in AD have been stimulation of cholinergic receptors by cholinomimetic compounds or prolongation of the availability of acetylcholine (ACh) released into the neuronal synaptic cleft. This latter effect is achieved by inhibiting ACh hydrolysis by acetylcholinesterase (AChE), through the use of AChE inhibitors.

This has been the most successful therapeutic approach so far (Francis *et al.* 1999). Over the last decade, some AChE inhibitors have been licensed for clinical use for the symptomatic relief of the early stages of AD, but it should be stressed that their effect is only to slow down the progression of the disease and they do not achieve any permanent improvement. The synthetic drug tacrine (Cognex®) was the first AChE inhibitor to be licensed, but its routine use has been restricted by associated hepatotoxic effects (Hammel *et al.* 1990, Watkins *et al.* 1994). The use of tacrine has been eclipsed by the newer AChE inhibitors such as donepezil (Aricept®), rivastigmine (Exelon®), and galantamine (Reminyl®) (Evans 2002). Factors in their effectiveness include their ability to cross the blood-brain barrier as well as their *in vitro* inhibitory characteristics.

Nicotinic Receptor Stimulation

It is reported that smoking may have protective effects against AD development and that administration of nicotine to AD patients and to healthy (non-AD) elderly people improved cognitive function (Min *et al.* 2001, Sahakian *et al.* 1989, van Duijn and Hofman 1991). Nicotine is also reported to upregulate nicotinic receptors and to increase ACh release (Balfour and Fagerström 1996, Whitehouse and Kalaria 1995). Thus, nicotinic agonists may enhance cholinergic neurotransmission in AD. The cognition enhancing effects of nicotine reported may be due to nicotinic receptor stimulation but nicotine may also protect against AD by other mechanisms. Nicotine has been shown to inhibit β-amyloid formation *in vitro* (Salomon *et al.* 1996), to inhibit the neurotoxic effects of excitatory amino acids (e.g., glutamate), and to enhance the effects of nerve growth factor (NGF) (Whitehouse and Kalaria 1995).

Other compounds, for example lobeline from *Lobelia inflata*, that interact with the nicotinic receptor (Decker *et al.* 1993) may also be exploited to influence cholinergic function in AD. Therefore, nicotine from *Nicotiana tabacum* and other *Nicotiana* spp., or other nicotinic receptor ligands, may have potential in the prevention, delay or treatment of AD.

Anti-Inflammatory Activity

Numerous plants and plant constituents have demonstrated anti-inflammatory properties (Bingöl and Sener 1995, Handa *et al.* 1992, Hoult and Payà 1996, Laughton *et al.* 1991, Welton *et al.* 1986), thus there is potential for more novel anti-inflammatory agents to be identified from plant sources. Novel anti-inflammatory drugs may be useful in AD therapy, with fewer adverse effects than drugs currently available. For example, ferulic acid, which is an anti-oxidant and anti-inflammatory compound derived from plants, ameliorated the β-amyloid-induced reduction in ACh levels in the cortex and the β-amyloid-induced inflammatory responses in the hippocampus in mice, and also improved cognitive function (Yan *et al.* 2001).

Antioxidant Activity

Injury of plant cells, as well as mammalian cells, is associated with the activation of lipoxygenases, which catalyse the formation of hydroperoxides of polyunsaturated fatty acids (PUFAs); a hydroperoxide radical may react with fatty acids to produce dioxoenes, which are regarded as plant defence compounds (Spiteller 1993). The occurrence of oxidative mechanisms in plants may explain why an abundance of antioxidant compounds have been identified in plant tissue. Their antioxidant effects in plants may therefore have relevance in mammals, particularly in disorders involving oxidative stress such as AD.

Estrogenicity and Phytoestrogens

There is considerable evidence to suggest that Estrogen Replacement Therapy (ERT) may have a preventative effect against AD development, and may be useful in slowing disease progression. ERT in women is associated with a reduced risk of developing AD and estrogen treatment in women with AD is reported to enhance cognitive function (Birge 1997, Fillit *et al.* 1986, Honjo *et al.* 1989, Ohkura *et al.* 1994, Schneider *et al.* 1996). However, other studies do not support the use of ERT to assist with the prevention of AD (Hogervorst *et al.* 2000, Mulnard *et al.* 2000). In addition, ERT has been associated with adverse responses including proliferative and oncogenic effects on cells responsive to estrogen. Thus, alternatives to ERT and their potential role in the prevention of AD have been the subject of some investigation.

Some studies indicate that phytoestrogens may alter anxiety, learning and memory *in vivo* (Lephart *et al.* 2002). Phytoestrogens, soya isoflavones in particular, are reported to improve cognitive function, not only in some animal studies but also in some clinical studies, and have been suggested to offer protection against AD development (File *et al.* 2001, Kim *et al.* 2001, Pan *et al.* 2000). One study in student volunteers suggested that a high soya diet (100 mg total isoflavones/day for 10 weeks) may improve short- and long-term memory in both females and males (File *et al.* 2001). Another study showed that twelve weeks consumption of a supplement containing soya isoflavones by postmenopausal women was associated with an improvement in cognitive function (Duffy *et al.* 2003). A longer term study (a six-month, double-blind, randomised, placebo-controlled clinical trial) also showed some cognitive benefits, particularly verbal memory, in postmenopausal women taking supplements containing isoflavones (Kritz-Silverstein *et al.* 2003).

The mode of action of phytoestrogens to explain these observations is unclear, and they may act similarly to ERT or via different mechanisms, perhaps independently of the estrogen receptor. Genistein, an isoflavone from *Glycine max*, has shown a neuroprotective effect against β-amyloid-induced neurotoxicity *in vitro* (Bang *et al.* 2004). Genistein also ameliorated β-amyloid peptide induced hippocampal neuronal apoptosis *in vitro*, which may be explained by various mechanisms such as anti-oxidant effects (Zeng *et al.* 2004). It has been proposed that although phytoestrogens may exert some neuroprotective effects analogous to that of anti-oxidants, they are not functionally equivalent to endogenously active estrogens (Zhao *et al.* 2002).

Soya isoflavones have been subjected to thorough investigation regarding their potential in the prevention of some diseases and their mechanisms of action to explain their pharmacological and clinical effects. Numerous other plants have shown estrogenic activities in some studies, but their physiological significance and any potential relevance in disease prevention remains unclear and requires further study. One species that could be of interest in the search for alternatives to ERT is the TCM plant *Pueria lobata*. *P. lobata* and some of its component isoflavones (e.g., puerarin) have shown estrogenic activity *in vitro* (Boue *et al.* 2003). An *in vivo* study showed that puerarin attenuated the deficits of inhibitory avoidance performance in rats, and these effects were related to increasing cholinergic activity via nicotinic but not muscarinic receptors, activating NMDA receptors, and decreasing serotonergic neuronal activity (Hsieh *et al.* 2002). Treatment for 3 months with *P. lobata* (equivalent to 100 mg isoflavones) in postmenopausal women indicated it may have some potential in improving cognitive function (Woo *et al.* 2003). It is apparent that some phytoestrogens do show potential for use in cognitive related disorders, but more intensive and convincing studies are needed to justify their clinical use.

PLANTS AND PLANT COMPOUNDS OF INTEREST

ALKALOIDAL INHIBITORS OF ACETYLCHOLINESTERASE

Galantamine

Galantamine (**1**) is found in members of the Amaryllidaceae alkaloid, e.g., the Chinese medicinal herb *Lycoris radiata* Herb., and the European *Galanthus nivalis* L. and *Narcissus* spp. (Bores *et al.* 1996). *Galanthus nivalis* was used traditionally in Bulgaria and Turkey for neurological conditions but was developed as a medicine only in the late twentieth century after its cholinesterase inhibitory properties were discovered.

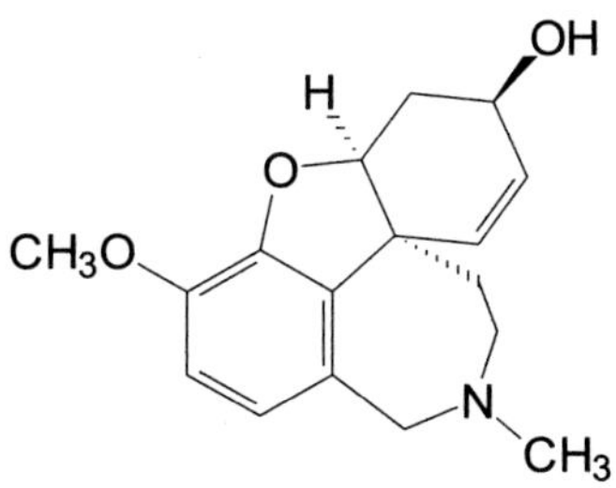

1. Galantamine

Galantamine is licensed in Europe for AD treatment and was well tolerated and significantly improved cognitive function when administered to AD patients, in multi-centre randomised controlled trials (Wilcock *et al.* 2000, Wilkinson and Murray 2001, Marcusson *et al.* 2003). The cognitive benefits appear to be sustained for at least three years, a much longer time than for other drugs of this type (Raskind *et al.* 2004). Galantamine is reported to be more selective for AChE than BuChE, and provides complete oral bioavailability (Bickel *et al.* 1991, Fulton and Benfield 1996, Harvey 1995). Galantamine is also reported to stimulate nicotinic receptors (Pearson 2001, Woodruff-Pak *et al.* 2001), which may also enhance cholinergic function and memory. This additional activity suggests galantamine may have therapeutic advantages over other AChE inhibitors and there is increasing evidence of its value in vascular dementia as well as AD (Small *et al.* 2003). Another alkaloid isolated from *L. radiata*, lycoramine, is also reported to have anti-ChE activity (Irwin and Smith 1960), but it has not been subjected to the vigorous pharmacological and clinical testing to which galantamine has been subjected. It is evident however, that members of the Amaryllidaceae may be a source of other anti-ChE alkaloids that show promise for clinical use.

Huperzine A. *Huperzia serrata* Thunb

The prescription Qian Ceng Ta has been used in TCM to alleviate problems of memory loss (Skolnick 1997) and *Huperzia serrata* (from which Qian Ceng Ta is prepared) is used in TCM for promoting circulation, for fever and inflammation. Huperzine A (**2**), isolated from *H. serrata* (Lycopodiaceae), is a lycopodium alkaloid related to the quinolizidines and it reversibly inhibits AChE *in vitro* and *in vivo* (Bai *et al.* 2000, Laganière *et al.* 1991, Wang *et al.* 1986) and it may also favourably affect other neurotransmitter systems to improve memory (Ou *et al.* 2001). Recent progress on many aspects of huperzine A has been reviewed (Jiang *et al.* 2003).

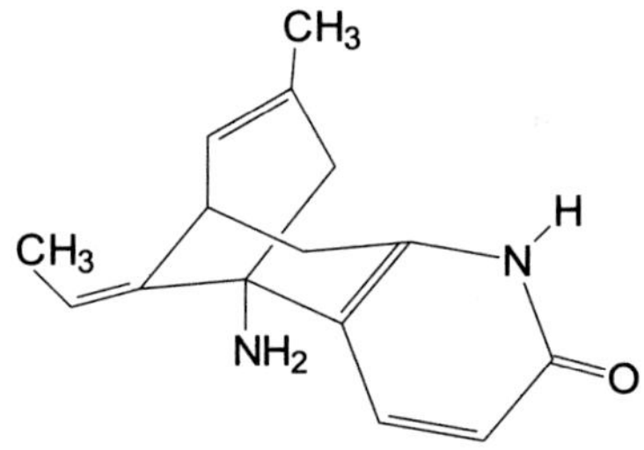

2. Huperzine A

Huperzine A improved memory retention processes in cognitively impaired aged and adult rats (Lu *et al.* 1988) and and in gerbils following ischaemia (Zhou *et al.* 2001). These observations suggest huperzine A has clinical potential in cerebro-vascular disorders, as well as in AD. AChE inhibition may not be the only explanation for the clinical effects observed with huperzine A, as it has also been shown to be neuroprotective. Huperzine A was neuroprotective against β-amyloid peptide fragment 25-53 and free radical-induced cytotoxicity (Xiao *et al.* 2002). Zhou *et al.* (2001) reported that huperzine A attenuates apoptosis by inhibiting the mitochondria-capase pathway, thus inhibiting apoptosis.

Huperzine A (**2**) may also have potential in the attenuation of memory deficits and neuronal damage that occur after ischaemia, so may therefore be beneficial in the treatment of cerebrovascular type dementia (Zhou *et al.* 2001). The indications of possible usefulness in treating AD arising from these *in vitro* and animal studies have been confirmed by recent clinical studies. In a multi-centre, double-blind trial, huperzine A significantly improved memory and behaviour in AD patients, and was reported to be more selective for AChE than butylcholinesterase (BuChE) and was less toxic than the synthetic AChE inhibitors donepezil and tacrine (Small *et al.* 1997). A more recent randomised, placebo-controlled study involved over 200 patients, 100 of whom were given 400 μg huperzine A daily for 12 weeks (Zhang *et al.* 2002). Three types of test, including the ADAS-Cog test and MMSE, were used to assess cognitive function and the test group showed showed significantly higher scores for improvement compared with the placebo group in the various scoring, e.g., the proportion of patients improving by over 4 points in the ADAS-Cog test were 56.1 for the test group and 12.5 for the placebo.

Physostigmine

Physostigma venenosum was used traditionally in Africa as a ritual poison, claimed to determine the guilt or innocence of persons accused of a crime (Evans 2002). Treatment with the indole alkaloid physostigmine, (3) an AChE inhibitor isolated from *P. venenosum*, has improved cognitive function in several *in vivo* studies. For example, physostigmine is reported to protect mice against cognitive impairment caused by oxygen deficit, it improved learning in rats (McCaleb 1990), and antagonised scopolamine-induced impairment of cognitive function in rats (Yoshida and Suzuki 1993). Physostigmine is a short-acting reversible AChE inhibitor and is reported to have shown significant cognitive benefits in both normal and AD patients (Sitaram *et al.* 1978), but clinical use may be limited by its short half-life, which would require multiple daily dosing.

Physostigmine (**3**) has been shown to inhibit both G_1 and G_4 AChE forms, the major AChE isoenzymes present in mammalian CNS (Ogane *et al.* 1992). This effect indicates that physostigmine (**3**) inhibits CNS AChE but, as G_4 AChE levels are reported to be decreased in AD brains (Ogane *et al.* 1992, Younkin *et al.* 1986), there may be clinical implications regarding efficacy. An inhibitor that is more selective for the G_1 AChE form may be more relevant in AD. As well as inhibiting AChE, physostigmine also inhibited butylcholinesterase (BuChE) with similar potency, an enzyme not associated with cholinergic neurotransmission (Otoguro *et al.* 1997). Therefore, adverse effects associated with BuChE inhibition may also occur with physostigmine. However, BuChE has recently been implicated in the aetiology and progression of AD (Greig *et al.* 2001). BuChE inhibition may therefore develop as a therapeutic target in AD. In view of this, physostigmine may have superior clinical efficacy to AChE-selective inhibitors, although the potential clinical advantages of BuChE inhibition require more thorough investigation.

The chemical structure of physostigmine (**3**) has provided a template for the development of rivastigmine (**4**), an AChE inhibitor that is licensed for use in the UK for the symptomatic treatment of mild to moderately severe AD. Rivastigmine, which preferentially inhibits the G_1 form of AChE (Polinsky 1998), improves cognition in AD patients, an effect that has been correlated with the level of AChE inhibition (Agid *et al.* 1998, Grossberg and Desai 2001, Spencer and Noble 1998).

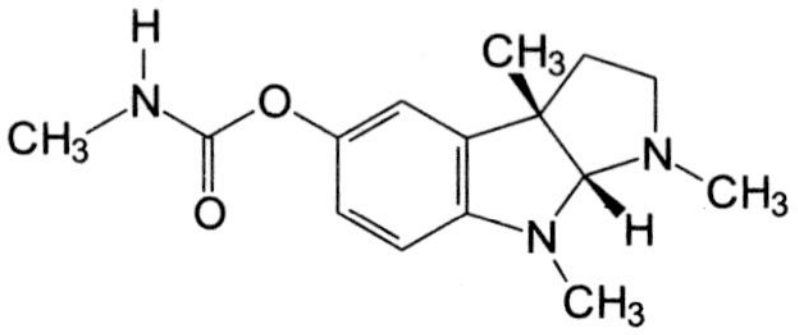

3. Physostigmine

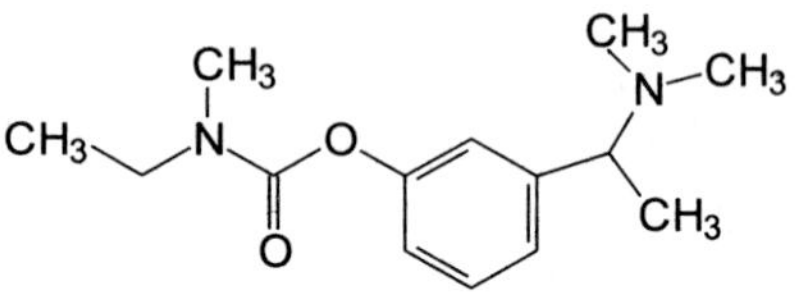

4. Rivastigmine

Rivastigmine is reported to inhibit AChE in the cortex and hippocampus, brain areas involved in cognition (Polinsky 1998). Clinical studies have borne out the usefulness of rivastigmine in mild to moderate AD and recently it has been shown to have positive effects in moderately severe cases (Burns *et al.* 2004).

Plant Extracts and Constituents Displaying Activity Relevant to AD

Bacopa

Bacopa monniera (Scrophulariaceae) is a small plant found extensively in India where it is known as 'Brahmi'. It has a long tradition in Ayurvedic medicine of being useful as a stimulant of the nerves and fatigue of the CNS. Over the last few years it has been investigated scientifically for its effects on the CNS and other relevant biological activities. All relevant studies have been carried out on extracts rather than isolated constituents although the plant is known to contain triterpenoid saponins and these are thought to be associated with the activity.

Clinical studies have so far only been undertaken on healthy volunteers. A double-blind, randomised, placebo-controlled study on patients aged 40-65 years showed that those given a standardised *Bacopa* extract gave a significant improvement in retention of new information but no difference in attention, verbal, and visual short-term memory (Roodenrys *et al.* 2002). Another similar study investigated speed of visual information processing, learning rate and memory consolidation and concluded that the *Bacopa* extract improved higher cognitive processes such as learning and memory (Stough *et al.* 2001). *B. monniera* extract has been shown to possess considerable antioxidant effects and these may contribute to its apparent ability to improve cognitive function (Russo *et al.* 2003).

***Centella asiatica* L.**

One ancient Ayurvedic remedy is *Centella asiatica* (Umbelliferae), which is reputed to restore youth, memory and longevity (Kapoor 1990). For example, an Ayurvedic formulation composed of four herbs, including *C. asiatica*, is used to retard age and prevent dementia, and the herb combined with milk is given to improve memory (Manyam 1999). *C. asiatica* has also been used to treat rheumatic disorders, which suggests it may have anti-inflammatory effects. In TCM *C. asiatica* has been used for various disorders, such as traumatic diseases, and for combating physical and mental exhaustion (Brinkhaus *et al*. 2000, Tang and Eisenbrand 1992).

The pharmacological basis to explain the reputed anti-amnesic effects of *C. asiatica* has been explored experimentally. The essential oil from *C. asiatica* leaf contains monoterpenes, including bornyl acetate (**5**), α-pinene (**6**), β-pinene (**7**) and γ-terpinene (Asakawa *et al*. 1982, Brinkhaus *et al*. 2000), which are reported to inhibit AChE (Miyazawa *et al*. 1997, Perry *et al*. 2000a, Ryan and Byrne 1988). However, monoterpene AChE inhibitors are generally weak compared to the anti-ChE alkaloid, physostigmine (**3**) (Perry *et al*. 2000a).

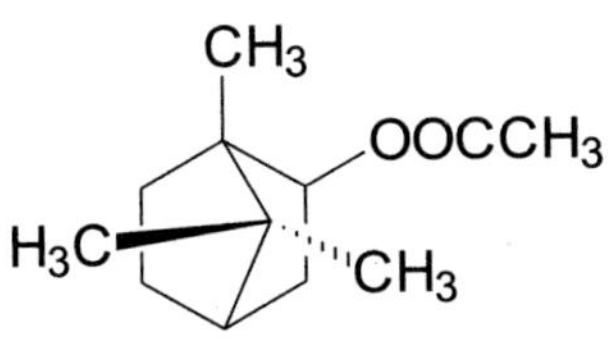

5. Bornyl acetate

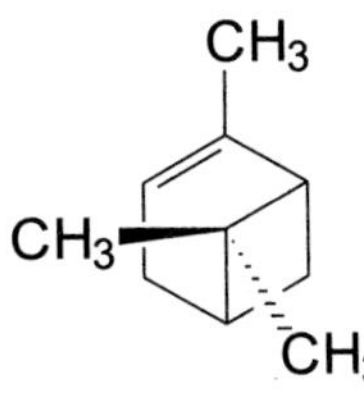

6. α–Pinene

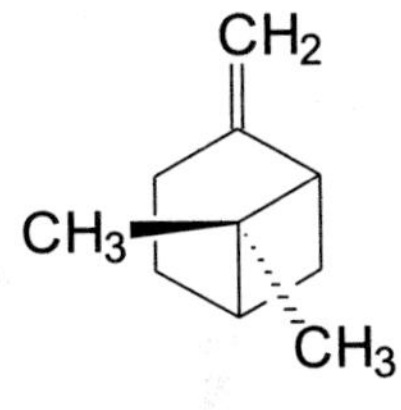

7. β–Pinene

In view of the relatively weak anti-AChE activity of monoterpenes reported to date, it is unlikely that they would be therapeutically effective in cognitive disorders. Although alkaloids are reported to occur in *C. asiatica* (Duke and Ayensu 1985), they have yet to be evaluated for activity against AChE.

Alterations in other neurotransmitter systems have been associated with AD pathology (Advokat and Pellegrin 1992, Fowler *et al*. 1992, Storga *et al*. 1996). Some drugs that modulate neurotransmitter systems have shown some benefits in AD related symptoms (e.g., trazodone improved behavioural symptoms associated with AD) (Lake and Grossberg 1987, Lebert *et al*. 1994, Simpson and Foster 1986). An aqueous extract of *C. asiatica* leaf modulated dopamine, 5-hydroxytryptamine (5-HT), and noradrenaline systems in rat brain and improved learning and memory processes *in vivo* (Nalini *et al*. 1992). Other studies have shown an alcoholic extract of *C. asiatica* to be tranquilising in rats, an activity that was attributed to a triterpene, brahmoside (Kapoor 1990, Sakina and Dandiya 1990). Further studies showed an extract of *C. asiatica* leaf to be sedative, anti-depressant, and potentially cholinomimetic *in vivo* (Sakina and Dandiya 1990). These findings suggest that *C. asiatica* may be appropriate to treat symptoms of depression and anxiety in AD, and that it may also influence cholinergic activity and thus, cognitive function.

Cognitive enhancing effects have been observed in rats following oral administration of an aqueous extract of *C. asiatica*; this effect was associated with an antioxidant mechanism in the CNS (Kumar and Gupta 2002). *C. asiatica* also showed neuroprotective effects on aluminium-induced brain cholinotoxicity and lipid peroxidation *in vivo* (Maheswari 2003, Maheswari and Murali 2004). Limited research has been conducted to identify the compounds responsible for the observed cognitive effects. However, the triterpene asiatic acid from *C. asiatica* and its derivatives have been shown to protect cortical neurons from glutamate-induced excitotoxicity *in vitro* (Lee *et al.* 2000); thus, further research regarding the clinical potential of these compounds may be warranted.

Ginkgo biloba L.

The seeds of *Ginkgo biloba* (Coniferae) have been used in TCM for centuries for the treatment of respiratory disorders; this crude drug is described in the Chinese Pharmacopoeia as an anti-asthmatic and for treatment of polyuria (Kenner and Requena 1996, Pharmacopoeia of the People's Republic of China 2000, Tang and Eisenbrand 1992). *G. biloba* has also been used traditionally in Iran to improve memory loss associated with blood circulation abnormalities (Ross 2001). The use of *G. biloba* leaves for the treatment of circulatory disorders became increasingly popular in the 1960s, particularly in Europe. Numerous investigations have since been conducted regarding the potential of *G. biloba* in the treatment of cognitive disorders, such as AD.

A considerable amount of the research on *G. biloba* has been carried out with the standardised extract EGb 761, which contains flavonoid glycosides, terpene lactones, and various other constituents. There is some evidence (electroencephalographic data) to suggest EGb 761 has a local effect in the CNS (Maurer *et al.* 1997); in addition, this extract has shown various biological activities relevant to the treatment of cognitive dysfunction and Alzheimer's pathology. Favourable effects have been observed on cerebral circulation and neuronal cell metabolism (Heiss and Zeiler 1978, Loffler *et al.* 2001, Tea *et al.* 1987), on the cholinergic system (Kristofiková *et al.* 1992), and the extract showed antioxidant activity (Barth *et al.* 1991, Marcocci *et al.* 1994, Rojas *et al.* 2001) and may also influence caspase-3 activation (Massieu *et al.* 2004). EGb761 was also neuroprotective against β-amyloid- and nitric oxide (NO)-induced toxicity *in vitro* (Bastianetto *et al.* 2000a,b); protection against β-amyloid-induced toxicity has been associated with the flavonoid fraction (CP 205) of EGb 761 (Bastianetto *et al.* 2002).

Apoptosis, a pathogenic mechanism in AD (Emerit *et al.* 2004), could be reduced both *in vitro* and *in vivo* by EGb 761 (Schindowski *et al.* 2001, Yao *et al.* 2001). EGb 761, and a terpene lactone isolated from it (bilobalide), were protective against ischaemia-induced neuronal death and reductions in mitochondrial gene expression *in vivo* (Chandrasekaran *et al.* 2001). In addition, EGb 761 protected mitochondria from β-amyloid-induced toxicity (Eckert *et al.* 2003) and, administration of EGb 761 to vitamin E deficient rats is reported to positively modulate synaptic and mitochondrial plasticity (Bertoni-Freddari *et al.* 2002). In cortical slices, bilobalide reduced both glutamate and aspartate release (Jones *et al.* 2002); a reduction in excitatory amino acid neurotransmitter release may be a rational therapeutic target in AD (Parsons *et al.* 1998).

G. biloba extracts have also been evaluated for their influence on cognitive function in numerous *in vivo* studies. Treatment with *G. biloba* extracts attenuated scopolamine-induced amnesia in rats (Chopin and Briley 1992), enhanced memory retention in both young and old rats (Petkov *et al.* 1993), and improved short-term memory in mice (Stoll *et al.* 1996). A standardised extract of *G. biloba* (containing 24% ginkgo flavone glycosides) inhibited AChE activity *in vitro* and *ex vivo* at doses which correlated with effects against scopolamine-induced deficits in a passive avoidance test in mice (Das *et al.* 2002). Thus, enhancement of cholinergic function may be one mechanism to explain the cognitive effects of *G. biloba*.

Further studies have shown some evidence of clinical efficacy for *G. biloba*. Extracts, including EGb 761, have been associated with modest improvements in cognitive function following administration to both AD and non-AD patients in various studies, including randomised, double-blind, placebo-controlled, multi-centre trials (Hartley *et al.* 2003, Hofferberth 1994, Kanowski *et al.* 1997, Kanowski and Hoerr 2003, Le Bars *et al.* 1997, Le Bars *et al.* 2000, Le Bars *et al.* 2002, Oken *et al.* 1998, Rai *et al.* 1991, Rigney *et al.* 1999). However, a number of the studies assessing the clinical efficacy of *G. biloba* base the results on self-assessment questionnaires; thus the reliability of the results may require confirmation through the use of more objective methods of analysis. However, the numerous investigations *in vitro* and *in vivo* that do show *G. biloba* to display numerous relevant pharmacological activities in relation to AD treatment, do apparently support the results of clinical studies.

The compounds responsible for these pharmacological and clinical observations require further investigation. Activity is perhaps due to different and additive effects, such as cholinergic, anti-inflammatory or vasodilatory effects.

A number of constituents from *G. biloba* leaf have been associated with a vasodilatory action, including some flavonoids (e.g., quercetin (**8**)) and some terpenoids (e.g., bilobalide (**9**) and ginkgolides A (**10**), B (**11**), and C (**12**)) (Kuboto *et al.* 2001, Nishida and Satoh 2004). In addition, ginkgolide B from *G. biloba* antagonised platelet-activating factor (PAF) (Braquet *et al.* 1994) and ginkgetin (**13**), a biflavone from *G. biloba* leaves, inhibited phospholipase A_2; ginkgetin and the biflavone mixture also down-regulated COX-2 expression (Baek *et al.* 1999, Kwak *et al.* 2002).

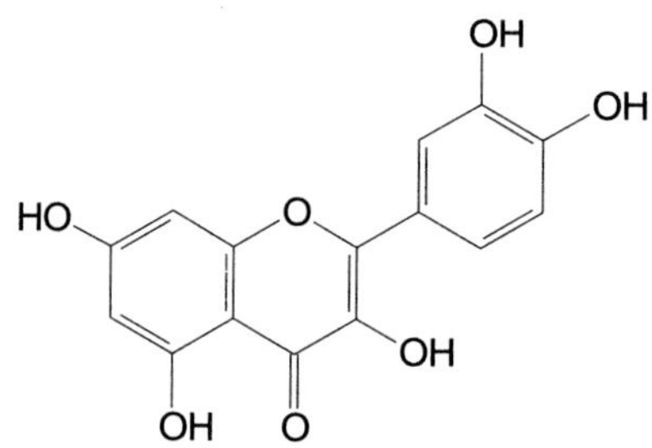

8. Quercetin

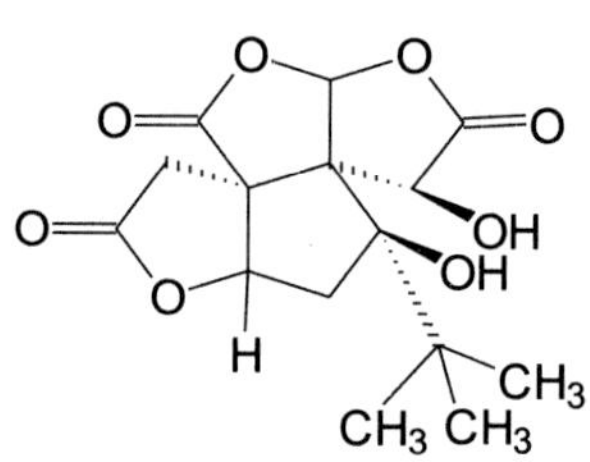

9. Bilobalide

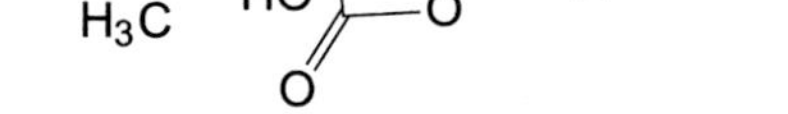

10. R, R' = H Ginkgolide A
11. R = OH, R' = H Ginkgolide B
10. R, R' = OH Ginkgolide C

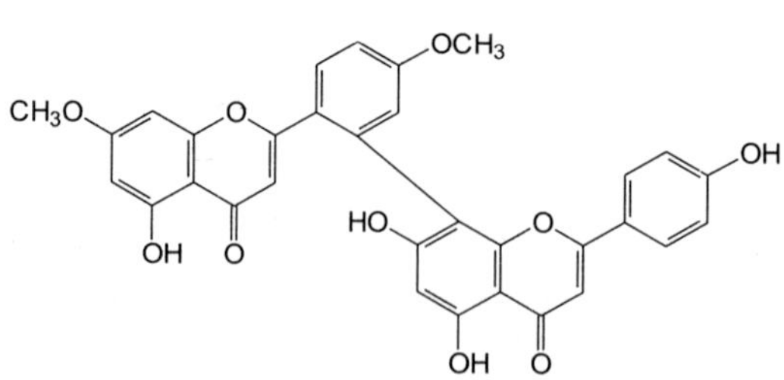

13. Ginkgetin

These observations suggest *G. biloba* may also mediate anti-inflammatory effects. Some evidence also suggests *G. biloba* and some of its flavonoid components, quercetin, kaempferol, and isorhamnetin, may have estrogenic activity (Oh and Chung 2004); however, as with many phytoestrogens, estrogenic activity *in vivo* and the physiological significance require further study.

From evidence to date, it is apparent that *G. biloba* may be of some benefit for the treatment of AD symptoms, but further research is necessary to identify appropriate dosing regimens and the potential for adverse effects with both short- and long-term use. In general, clinical trials have shown oral administration of *G. biloba* to be well-tolerated, with no serious adverse effects (Cieza *et al.* 2003, Maurer *et al.* 1997). Some concerns that have generally been associated with herbal medicines include quality control issues, standardisation of extracts, and potential for interactions with other medicines. For many patients, herbal medicines may be used to complement the effects of conventional medicines, but with unknown interactions. *Ginkgo* supplementation in Alzheimer's patients taking donepezil (5 mg/day) was not reported to have a major impact on the pharmacokinetics and pharmacodynamics of donepezil (Yasui-Furukori *et al.* 2004); however, further studies are needed to investigate the potential interactions of *Ginkgo*, with both other herbal medicines and with drugs.

Melissa officinalis L.

Melissa officinalis (Labiatae) leaf has been used as a medicinal plant for more than 2000 years. In traditional European medicine *M. officinalis* was reputed to be a calming and strengthening remedy and was used to treat migraines, melancholia, neuroses, and hysteria; the plant has been acclaimed for promoting long life and for restoring memory (Bisset 1994, Kenner and Requena 1996, McVicar 1994, Yarnell 1998). In 1751, the herbalist John Hill stated that *M. officinalis* was 'Good for disorders of the head and stomach' (Crellin and Philpott 1990), indicating potential benefits for the use of *Melissa officinalis* for CNS disorders. In Arabic medicine it was used to treat depression (McVicar 1994) and it was also used traditionally in Greek medicine to treat hysteria (Malamas and Marselos 1992). The Commission E Monograph in Germany approves the use of *M. officinalis* for nervous insomnia. In modern alternative medicine, *M. officinalis* essential oil is used in aromatherapy to alleviate depression and insomnia (McVicar 1994).

M. officinalis has been the subject of research regarding its potential as a sedative and anxiolytic, activities that may be appropriate to provide symptomatic relief for some of the behavioural symptoms associated with AD such as agitation. In mice, a hydroalcoholic (30% ethanol) extract of *M. officinalis* leaf was sedative and potentiated barbiturate induced sleep, but the *M. officinalis* essential oil did not demonstrate these sedative effects (Soulimani *et al.* 1991). In one double-blind study *M. officinalis* leaf was reported to alleviate mild anxiety and nervousness, and in combination with *Valeriana officinalis* root, was reported to be as effective as triazolam, and did not cause drowsiness or impair concentration the next day (Yarnell 1998). The results of a double-blind, placebo-controlled trial to investigate the effects of aromatherapy with *M.officinalis* oil in patients with clinically significant agitation associated with dementia indicated that *M. officinalis* oil may be a safe and effective treatment for clinically significant agitation in people with AD (Ballard *et al.* 2002).

More recent studies have focused on the reputed cognitive effects of *M. officinalis*. In one randomised, placebo-controlled, double-blind, balanced crossover study, an improvement in cognitive performance and mood was observed in 20 healthy young participants, following treatment with cholinergically active (determined using *in vitro* studies) dried leaf of *M. officinalis* (Kennedy *et al.* 2003). In another study, an extract of *M. officinalis* was administered to patients with mild to moderate AD for four months (a double-blind, randomised, placebo-controlled trial undertaken in three centres). Results indicted a significantly better outcome on cognitive function than placebo, and it was reported that the *M. officinalis* extract also had a positive effect on agitation (Akhondzadeh *et al.* 2003).

Investigations have been conducted to ascertain a pharmacological basis for the reputed cognitive enhancing effects of *M. officinalis*. Numerous monoterpenes have been identified in the essential oil of *M. officinalis*, including citral (a mixture of the isomers citral a (geranial) and citral b (neral)) (Bisset 1994, Guenther 1949, Sarer and Kökdil 1991, Tittel *et al.* 1982), which weakly inhibit AChE (Ryan and Byrne 1988). Other reported activities of *M. officinalis* leaf extracts and essential oil that may be relevant in AD therapy, and that may explain the reported cognitive improvements, include antioxidant effects (Hohmann *et al.* 1999, Mimica-Dukic *et al.* 2004) and binding to muscarinic and nicotinic receptors *in vitro* (Perry *et al.* 1996, Wake *et al.* 2000), indicating favourable effects on cholinergic function. In addition, some monoterpenes including citral, geraniol and nerol from *M. officinalis* essential oil have been investigated for potential estrogenic effects, with some activity shown in some *in vitro* and receptor binding studies (Howes *et al.* 2002), although the biological significance of these effects is uncertain.

Salvia lavandulaefolia Vahl. and *Salvia officinalis* L.

Research into historical literature has revealed that some activities of sage, particularly its reputation as being good for the memory, may be relevant to AD treatment (Perry *et al.* 1996). Several quotes in 16th century and 17th century English herbals describe its ability to improve the memory (Perry *et al.* 1998, 2003). These ancient reports, together with current usage, which indicates that sage might possess anti-inflammatory properties and also alleviate conditions associated with estrogen imbalance (Barnes *et al.* 2002), suggest that sage might be of possible use in AD. Numerous *in vitro* and *in vivo* tests have been employed to monitor the effects of sage extracts on some of the factors associated with AD, or with a reduction in its incidence.

An ethanolic extract, the steam-distilled oil of *S. officinalis* (Labiatae) and the oil of *S. lavandulaefolia* (Labiatae) have been investigated for anti-ChE activity. It was found that all three samples gave inhibition of AChE at quite low concentrations (Perry *et al.* 1996). This activity was of considerable interest since previous AChE inhibitors were amines, the naturally occurring types being the alkaloids, and these compounds were not known to exist in the *Salvia* species tested.

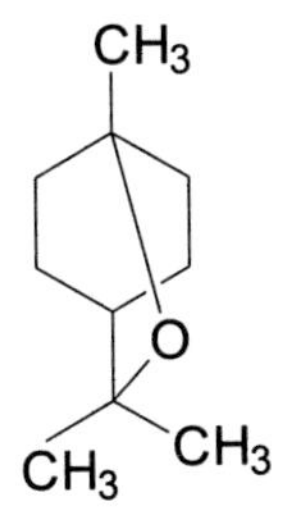

14. 1,8-Cineole

The cholinesterase inhibition shown by the *S. lavandulaefolia* oil was shown to be partly due to the cyclic monoterpenes 1,8-cineole (**14**) and α-pinene (**6**) which were shown to inhibit AChE *in vitro*, with some contribution from other constituents perhaps by acting synergistically (Perry *et al.* 2000a, Savelev *et al.* 2003). However, the monoterpenes were considerably less active, by a factor of at least 10^3, than the alkaloid inhibitors such as physostigmine (**3**) (Perry *et al.* 2000a).

When the antioxidant activity of an aqueous methanolic extract of *S. officinalis* was measured, it was seen to dose-dependently inhibit lipid peroxidation (Hohmann *et al.* 1999). Several antioxidant compounds have been identified in *S. officinalis*, including salvianolic acids I, K and L and various other phenolic compounds (Lu and Foo 2001a,b). Weak antioxidant effects were also shown to be present in an ethanolic extract of *S. lavandulaefolia* (when compared to the standard anti-oxidant 10 μM propyl gallate), and both the water-soluble and chloroform-soluble fractions of this extract gave similar activity (Perry *et al.* 2001).

An ethanol extract of *S. lavandulaefolia* showed only weak inhibition of eicosanoid synthesis, thus indicating a possible anti-inflammatory effect (Perry *et al.* 2001). Ursolic acid has been shown to be an active in vivo anti-inflammatory in *S. officinalis* (Baricevic *et al.* 2001) and some of the essential oil constituents of *S. lavandulaefolia* have also been evaluated for anti-inflammatory effects although only α-pinene (**6**) (comprising 5% of the essential oil) gave significant activity and showed weak selectivity for inhibition of LTB_4 generation (Perry *et al.* 2001), a mechanism which is associated with neurodegeneration so selective inhibition over cyclo-oxygenase (COX) may be relevant therapeutically in AD (Sugaya *et al.* 2000). Dose-dependent oestrogenic activity was present in the ethanolic extract of *S. lavandulaefolia* (Perry *et al.* 2001). The potential oestrogenic activity of the sage extracts and essential oil constituents require further investigation.

These interesting *in vitro* results have stimulated *in vivo* and clinical studies. Oral administration of *S. lavandulaefolia* essential oil to rats decreased striatal AChE activity in both the striatum and the hippocampus compared to the control rats (Perry *et al.* 2002). Thus, it is apparent that following oral administration, one or more constituents of the *S. lavandulaefolia* oil or their metabolites, reach the brain and inhibit AChE in select brain areas, consistent with evidence of inhibition of the brain enzyme *in vitro* (Perry *et al.* 1996).

Clinical studies on human volunteers and even patients with AD have been reported in recent years. In a study by Tildesley *et al.* (2003) the effect of sage in twenty participants using a placebo-controlled, double-blind, balanced, cross-over design was described. There were a number of significant effects on cognition associated with the lowest dose of *Salvia,* including improvements in both immediate and delayed word recall scores. A pilot trial on a small number of eleven patients showing mild to moderate symptoms of AD showed that oral administration of the essential oil of *S. lavandulaefolia* significantly improved cognitive function in one of the three different methods of assessment used (Perry *et al.* 2003). A more thorough multicentre, double-blind, randomised and placebo-controlled trial was carried out in Iran which involved treating 48 patients showing similar symptoms of AD with an extract of *S. officinalis* (Akhonzadeh *et al.* 2003). Patients receiving the *S. officinalis* extract showed significantly better values in measurements of cognitive function and there was no significant difference in side-effects between the treated and placebo groups, the placebo group actually showing a greater incidence of agitation.

Salvia miltiorrhiza Bung

Throughout history *Salvia miltiorrhiza* (Labiatae) has been used for the treatment of a variety of medical conditions. The dried root of *S.*

miltiorrhiza is red in colour, and was therefore used in folk medicine for the management of blood disorders. It is prescribed in TCM to stabilise the heart and calm nerves (Huang 1993). Official indications for the root include treatment of blood circulation disorders, insomnia, neurasthenia, and alleviation of inflammation (Tang and Eisenbrand 1992).

S. miltiorrhiza, or Chinese sage, has been the subject of thorough investigation, and consequently numerous pharmacological activities that may be relevant in CNS disorders, including AD, have been identified. *S. miltiorrhiza* has been employed for the treatment of cerebral vascular disease, and there are several studies to investigate possible mechanisms for the protective effect of *S. miltiorrhiza* against cerebral ischaemia.

S. miltiorrhiza root has been implicated in attenuating dysfunction of vasoactive intestinal peptide (VIP), a neuropeptide distributed within the gastro-intestinal tract and CNS, which may participate in the changes that occur in cerebral ischaemia (Kuang *et al.* 1989). Distribution abnormalities of the neuropeptide substance P have also been associated with some CNS disorders, including AD. Decreased levels of substance P have been suggested as a consequence of neuronal damage following cerebral ischaemia; *S. miltiorrhiza* root has been implicated in protecting neurons from ischaemia (Kuang *et al.* 1991), and so may actively protect against cerebral ischaemia and perhaps other CNS disorders via this mechanism.

Other beneficial effects of *S. miltiorrhiza* root against cerebral ischaemia have also been explored. *S. miltiorrhiza* root may inhibit neuronal cell death by inhibition of presynaptic glutamate release (Kuang and Xiang 1994), and it has been suggested that inhibition of NO formation may also explain CNS protective effects observed with *S. miltiorrhiza* root (Kuang *et al.* 1996a). It should also be considered that the biological function of NO has been suggested to involve the effects of excitatory amino acids, including their effects on brain development, learning, and memory (Moncada *et al.* 1991). This physiological role may aid the explanation of the effects of *S. miltiorrhiza* root on the CNS. Further investigations indicate *S. miltiorrhiza* root may modify ischaemic cell changes by modulating somatostatin, a CNS neuropeptide that has been implicated in learning and memory (Kuang *et al.* 1993).

S. miltiorrhiza root may offer an additional therapeutic approach to management of stroke and ischaemia. Reperfusion to aid recovery of ischaemia can cause further brain damage. During reperfusion, metabolism of free fatty acids from the breakdown of lipid membranes during ischaemia has been proposed to generate oxygen free radicals leading to further brain injury (Traystman *et al.* 1991). *S. miltiorrhiza* root has been shown to offer protection against this process by reducing lipid peroxidation (Kuang *et al.* 1996b, Zhao *et al.* 1996).

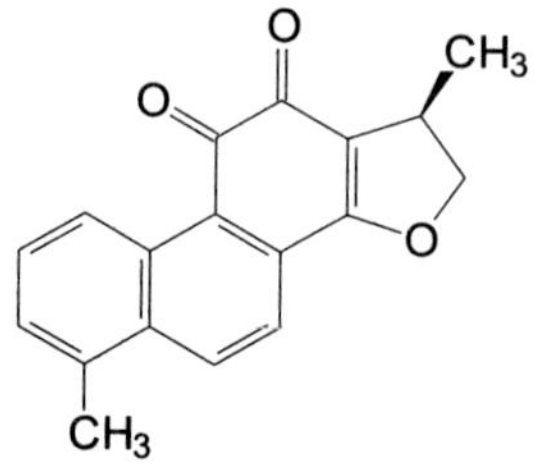

15. Dihydrotanshinone

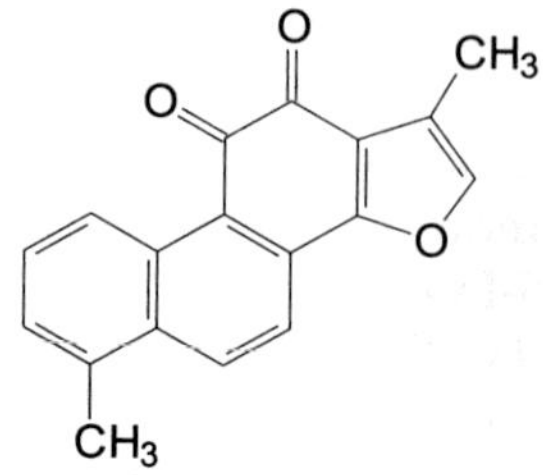

16. Tanshinone I

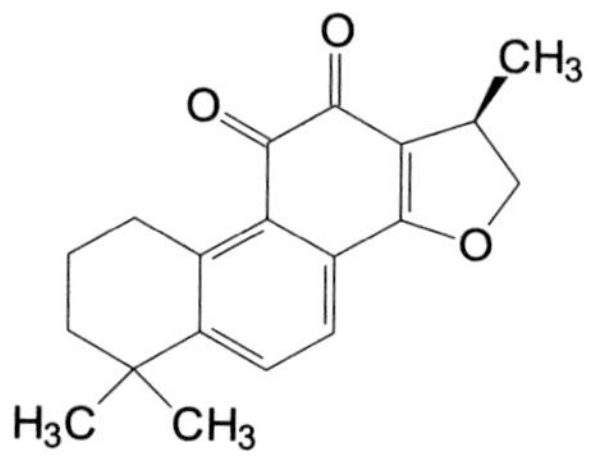

17. Cryptotanshinone

The antioxidant effects of *S. miltiorrhiza* root have been studied and several compounds have been identified with significant antioxidant activity. Several quinones isolated from *S. miltiorrhiza* root have demonstrated an antioxidant effect in lard, with dihydrotanshinone (**15**), tanshinone I (**16**), methylene tanshinquinone and cryptotanshinone (**17**) providing significant antioxidant activity; tanshinone IIa has shown no antioxidant activity (Weng and Gordon 1992).

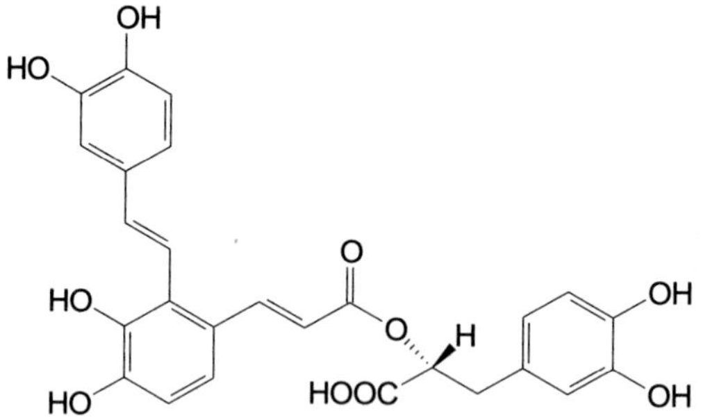

18. Salvianolic acid A

Other components of *S. miltiorrhiza* root have displayed antioxidant effects, including salvianolic acids A (**18**) and B (compounds found to protect against memory impairment induced by cerebral ischaemia in mice (Du *et al.* 2000)) and rosmariquinone (also known as miltirone) (Kang *et al.* 1997, Liu *et al.* 1992). Such antioxidant compounds may be useful in AD therapy.

Tanshinones, e.g., cryptotanshinone (**17**), isolated from *S. miltiorrhiza* root have demonstrated anti-inflammatory activity in mice and were active against 5-LOX in porcine leukocytes, but were not as active as the crude extracts (Chang and But 1986, Paulus and Bauer 2000). The tanshinones are also reported to show weak oestrogenic activity (Chang and But 1986). These activities require further investigation for confirmation, but could be relevant in AD therapy. A recent paper (Ren *et al.* 2004) describes the cholinesterase inhibition of the tanshinones with dihydrotanshinone (**15**) being the most active. Thus the tanshinones show three of the properties which are thought to contribute to a reduced risk of AD and clinical studies on this plant in AD patients would be of great interest.

Withania somnifera **(L.) Dun**

Withania somnifera (Solanaceae) root (ashwagandha) is one of the most highly regarded herbs in Ayurvedic medicine and its use dates back almost 4000 years. It is classed among the rejuvenative tonics ("Rasayanas"), which was emphasised by the Ayurvedic scholar Charaka (10 BC) who wrote about *W. somnifera*, 'One obtains longevity, regains youth, gets a sharp memory and intellect and freedom from diseases, gets a lustrous complexion, and strength of a horse' (Upton 2000).

Numerous studies have been conducted to investigate the cognitive enhancing potential of *W. somnifera*. For example, some steroidal derivatives that have been isolated from *W. somnifera* root are the sitoindosides IX (**19**) and X, which augmented learning acquisition and memory in both young and old rats (Ghosal *et al.* 1989). The mechanisms to explain this effect are unclear, but may involve modulation of cholinergic neurotransmission, as an extract containing the sitoindosides VII-X and withaferin A (**20**) was administered to mice and effects on the neurotransmitter systems in the brain were observed. The results from this study showed that the extract enhanced AChE activity in the lateral septum and globus pallidus and decreased AChE activity in the vertical diagonal band, it also enhanced muscarinic M_1 receptor binding in the lateral and medial septum and in the frontal cortices, and it increased muscarinic M_2 receptor binding sites in cortical regions, but it did not affect γ-aminobutyric acid $(GABA)_A$, benzodiazepine receptor binding, nor NMDA or amino-3-hydroxy-5-methyl-4-isoxazole propionic acid (AMPA) glutamate receptor subtypes (Schliebs *et al.* 1997).

The extract containing the sitoindosides VII-X and withaferin A also reversed the ibotenic acid-induced cognitive deficit and reversed the reduction in cholinergic markers (e.g., ACh, choline acetyltransferase (ChAT)) in rats (Bhattacharya *et al.* 1995). The reputed cognition enhanc-

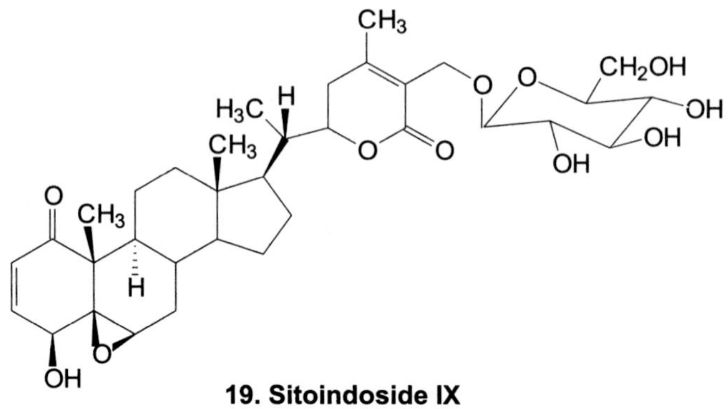

19. Sitoindoside IX

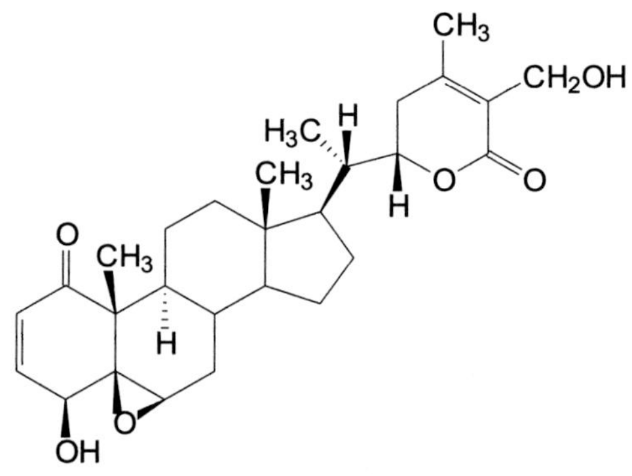

20. Withaferin A

ing effects of *W. somnifera* root may therefore be explained by a preferential action on cholinergic neurotransmission in the cortical and basal forebrain, brain areas involved in cognitive function. These observations indicate that the sitoindosides VII-X and withaferin A could have potential in AD therapy.

A root extract of *W. somnifera* reversed scopolamine-induced disruption of acquisition and attention, and attenuated amnesia following electroconvulsive shock in mice, effects that could be attributed to the root extract having nootropic activity (Dhuley 2001). A methanol extract of *W. somnifera* root dose-dependently promoted dendrite formation in human neuroblastoma cells *in vitro* (Tohda *et al.* 2000) and the compounds that were associated with these effects were some of the withanolide derivatives (Zhao *et al.* 2002b). Further studies have shown that *in vitro*, axons are predominantly extended by withanolide A and dendrites by withanosides IV and VI (Kuboyama *et al.* 2002). The occurrence of these effects in the CNS may promote synaptic formation, which involves neurite outgrowth; thus cholinergic function may be enhanced. A root preparation of *W. somnifera* has also been shown to significantly reduce the number of degenerating cells in the hippocampal brain region of stressed rats, which indicates the root extract may have potential neuroprotective effects in neurodegenerative disorders (Jain *et al.* 2001).

The glycowithanolides showed anxiolytic and antidepressant activities in rats (Bhattacharya *et al.* 2000a), which may be applicable in the symptomatic treatment of AD. An alkaloid extract also had tranquillising effects *in vivo*, and potentiated barbiturate-, ethanol-and urethane-induced hypnosis in mice (Malhotra *et al.* 1965). The pharmacological basis for these observations is unknown, but could reflect the reported GABA-mimetic effect of *W. somnifera* extract (Mehta *et al.* 1991).

Nicotine has been reported to occur in *W. somnifera* root (Kapoor 1990), although some studies have not reported it to be present (Das *et al.* 1964, Schwarting *et al.* 1963). Nicotine has been associated with cognitive enhancement and protection against AD development (Newhouse and Kelton 2000, Sahakian *et al.* 1989, van Duijn and Hofman 1991). If present, nicotine might also contribute to the reputed cognitive effects of *W. somnifera* root.

W. somnifera root and some constituents are also reported to have antioxidant and anti-inflammatory activities, which may also be relevant in AD therapy. Crude extracts are reported to inhibit lipid peroxidation *in vitro* and *in vivo* (Dhuley 1998, Panda *et al.* 1997, Parihar and Hemnani 2003). The root extract and the glycowithanolides (consisting of equimolar concentrations of sitoindosides VII-X and withaferin A (**20**) are hepatoprotective in rats and mice, an effect attributed to the antioxidant activity against hepatic lipid peroxidation (Bhattacharya *et al.* 2000b, Chaurasia *et al.* 2000). The compounds responsible for antioxidant activity include the withanolides (Bhattacharya *et al.* 2001, Mishra *et al.* 2000, Scartezzini and Speroni 2000, Upton, 2000), but other compounds in the root may also be anti-oxidant. The glycowithanolides decreased lipid peroxidation in various tissues including the brain in rodents, and both the glycowithanolides and the sitoindosides (VII-X) enhanced catalase and glutathione peroxidase activities in rat frontal cortex and striatum (Bhattacharya *et al.* 1997, 2001, Chaurasia *et al.* 2000).

A root extract of *W. somnifera* was effective against arthritis in rats (Begum and Sadique 1988) and reduced serum protein levels (e.g., α2-macroglobulin, an indicator of inflammatory conditions) (Anbalagan and Sadique 1981a,b, 1985), and in a double-blind, placebo-controlled, cross-over study, treatment in patients with osteoarthritis resulted in an improvement in symptoms (Kulkarni *et al.* 1991). *W. somnifera* administration to mice treated with a carcinogen reduced IL-1 and TNF-α levels (Dhuley 1997), which may also be relevant in AD, considering the possible involvement of these inflammatory mediators in senile plaque formation and neurodegeneration. *W. somnifera* leaves are also reported to have anti-inflammatory activity (Sudhir *et al.* 1986). The numerous pharmacological activities that have been associated with *W. somnifera* root, indicate that this herb may have multiple beneficial effects in AD patients. This emphasises the advantages that a complex mixture of an extract may have over a pure compound.

CONCLUSIONS

It is apparent that a variety of plants show, or have the potential to show, numerous activities that may be relevant to the treatment of neurodegenerative disorders such as AD. The majority of studies have focused on the anti-ChE alkaloids, such as galantamine (**1**) and physostigmine (**3**). This is perhaps a reflection of the relative success of the use of AChE inhibitors in AD patients, and a lack of understanding of the pathological mechanisms that occur in AD and the subsequent targets for treatment.

Although some plants, such as *Ginkgo biloba* and *Withania somnifera*, have shown beneficial effects on cognitive function, further studies regarding the compounds responsible for activity are necessary, to identify which compounds are responsible for the pharmacological activities observed, or if compounds act synergistically to enhance activity. For many of the plants and compounds that have demonstrated activities relevant to AD therapy, the clinical data is very limited. Clinical efficacy and potential toxicity of active plants and compounds in larger trials requires further assessment, before recommendations concerning their routine use can be identified.

In conclusion, it is apparent that the pharmacological activities of plants often appear to reflect their uses in traditional medicine. The ethnopharmacological approach for selecting plants to investigate for the treatment of a particular disease is a relatively successful method for the identification of plants and compounds that may be exploited for use therapeutically in neurodegenerative and other cognitive disorders.

REFERENCES

Advokat, C., and A.I. Pellegrin. 1992. Excitatory amino acids and memory: Evidence from research on Alzheimer's disease and behavioural pharmacology. *Neurosci. Biobehav. Rev.* 16: 13-24.

Agid Y., B. Dubois, R. Anand, and G. Gharabawi. 1998. Efficacy and tolerability of rivastigmine in patients with dementia of the Alzheimer type. *Curr. Ther. Res. Clin. Exp.* 59: 837-845.

Akhondzadeh, S., M. Noroozian, M. Mohammadi, S. Ohadinia, A.H. Jamshidi, and M. Khani. 2003. *Melissa officinalis* extract in the treatment of patients with mild to moderate Alzheimer's disease: A double blind, randomised, placebo controlled trial. *J. Neurol. Neurosurg. Psychiatry* 74: 863-866.

Akhondzadeh, S., M. Noroozian, M. Mohammadi, S. Ohadinia, A.H. Jamshidi, and M. Knani, 2003. *Salvia officinalis* extract in the treatment of patients with mild to moderate Alzheimer's disease: a double-blind, randomised and placebo-controlled trial. *J. Clin. Pharm. Ther.* 28: 53-59.

Anbalagan, K., and J. Sadique. 1981a. Response of *a*-1 globulins of serum during inflammation. *Curr. Sci.* 50: 88-89.

Anbalagan, K., and J. Sadique. 1981b. Influence of Indian medicine (ashwagandha) on acute-phase reactants in inflammation. *Ind. J. Exp. Biol.* 19: 245-249.

Anbalagan, K., and J. Sadique. 1985. *Withania somnifera* (ashwagandha), a rejuvenating herbal drug which controls *a*-2 macroglobulin synthesis during inflammation. *Int. J. Crude Drug Res.* 23: 177-183.

Asakawa, Y., R. Matsuda, and T. Takemoto. 1982. Monoterpenoids and sesquiterpenoids from *Hydrocotyle* and *Centella* species. *Phytochemistry* 21: 2590-2592.

Baek, S.H., S.S. Yun, T.K. Kwon, J.R. Kim, H.W. Chang, J.Y. Kwak, J.H. Kim, and K.B. Kwun. 1999. The effects of two new antagonists of secretory PLA_2 on TNF, iNOS, and COX-2 expression in activated macrophages. *Shock* 12: 473-478.

Bai, D.L., X.C., Tang and X.C. He. 2000. Huperzine A, a potential therapeutic agent for treatment of Alzheimers disease. *Curr. Med. Chem.* 7: 355-374.

Balfour D.J.K., and K.O. Fagerström. 1996. Pharmacology of nicotine and its therapeutic use in smoking cessation and neurodegenerative disorders. *Pharmacol. Ther.* 72 : 51-81.

Ballard, C.G., J.T. O'Brien, K. Reichelt, and E.K. Perry. 2002. Aromatherapy as a safe and effective treatment for the management of agitation in severe dementia: The results of a double-blind, placebo-controlled trial with *Melissa*. *J. Clin. Psychiatry* 63 : 553-558.

Bang, O.Y., H.S. Hong, D.H. Kim, H. Kim, J.H. Boo, K. Huh, and I. Mook-Jung. 2004. Neuroprotective effect of genistein against beta amyloid-induced neurotoxicity. *Neurobiol. Dis.* 16: 21-28.

Baricevic D., S. Sosa, R. Della Loggia, A. Tubaro, B. Simonovska, A. Krasna, and A. Zupancic. 2001. Topical anti-inflammatory activity of *Salvia officinalis* L. leaves: the relevance of ursolic acid. *J. Ethnopharmacol.* 75 : 125-132.

Barnes J., L.A. Anderson, and J.D. Phillipson. 2002. *Herbal Medicines—A Guide For Health Professionals*, 2nd ed., Pharmaceutical Press, London, UK.

Barth, S.A., G. Inselmann, R. Engemann, and H.T. Heidemann. 1991. Influences of *Ginkgo biloba* on cyclosporin A induced lipid peroxidation in human liver microsomes in comparison to vitamin E, glutathione and N-acetylcysteine. *Biochem. Pharmacol.* 41: 1521-1526.

Bastianetto, S., C. Ramassamy, S. Dore, Y. Christen, J. Poirier, and R. Quirion. 2000a. The *Ginkgo biloba* extract (EGb 761) protects hippocampal neurons against cell death induced by *β*-amyloid. *Eur. J. Neurosci.* 12: 1882-1890.

Bastianetto, S., W.H. Zheng, and R. Quiron. 2000b. The *Ginkgo biloba* extract (EGb 761) protects and rescues hippocampal cells against nitric oxide-induced toxicity: Involvement of its flavonoid constituents and protein kinase C. *J. Neurochem.* 74: 2268-2277.

Bastianetto, S., and R. Quirion. 2002. EGb 761 is a neuroprotective agent against *β*-amyloid toxicity. *Cell. Mol. Biol.* 48: 693-697.

Begum, V., and J. Sadique. 1988. Long term effect of herbal drug *Withania somnifera* on adjuvant induced arthritis in rats. *Ind. J. Exp. Biol.* 26: 887-882.

Bertoni-Freddari, C., P. Fattoretti, U. Caselli, R. Paoloni, and M. Solazzi. 2002. Chronic administration of EGb 761 modulates synaptic and mitochondrial plasticity in adult vitamin E-deficient rats. *Cell. Mol. Biol.* 48: 709-715.

Bhattacharya, S.K., A. Kumar, and S. Ghosal. 1995. Effects of glycowithanolides from *Withania somnifera* on an animal model of Alzheimer's disease and perturbed central cholinergic markers of cognition in rats. *Phytother. Res.* 9: 110-113.

Bhattacharya, S.K., K.S. Satyan, and S. Ghosal. 1997. Anti-oxidant activity of glycowithanolides from *Withania somnifera*. *Ind. J. Exp. Biol.* 35: 236-239.

Bhattacharya, S.K., A. Bhattacharya, K. Sairam, and S. Ghosal. 2000a. Anxiolytic-antidepressant activity of *Withania somnifera* glycowithanolides: An experimental study. *Phytomedicine* 7: 463-469.

Bhattacharya, A., M. Ramanathan, S. Ghosal, and S.K. Bhattacharya. 2000b. Effect of *Withania somnifera* glycowithanolides on iron-induced hepatotoxicity in rats. *Phytother. Res.* 14: 568-570.

Bhattacharya, A., S. Ghosal, and S.K. Bhattacharya. 2001. Anti-oxidant effect of *Withania somnifera* glycowithanolides in chronic footshock stress-induced perturbations of oxidative free radical scavenging enzymes and lipid peroxidation in rat frontal cortex and striatum. *J. Ethnopharmacol.* 74: 1-6.

Bickel U., T. Thomsen, W. Weber, J.P. Fischer, R. Bachus , M. Nitz, and H. Kewitz. 1991. Pharmacokinetics of galanthamine in humans and corresponding cholinesterase inhibition. *Clin. Pharmacol.* 50: 420-428.

Bierer, L.M., V. Haroutunian, S. Gabriel, P.J. Knott, L.S. Carlin, D.P. Purohit, D.P. Perl, J. Schmeidler, P. Kanof, and K.L. Davis. 1995. Neurochemical correlates of dementia severity in Alzheimer's disease: Relative importance of the cholinergic deficits. *J. Neurochem.* 64: 749-760.

Bingöl, F., and B. Şener. 1995. A review of terrestrial plants and marine organisms having anti-inflammatory activity. *Int. J. Pharmacognosy* 33: 81-97.

Birge, S.J. 1997. The role of estrogen in the treatment and prevention of dementia. *Am. J. Med.* 103: 1S-50S.

Bisset, N.G. 1994. *Herbal Drugs and Phytopharmaceuticals,* MedPharm GmbH Scientific Publishers, Stuttgart, Germany.

Bores G.M., F.P. Huger, W. Petko, A.E. Mutlib F. Camacho, D.K. Rush, D.E. Selk, V. Wolf, R.W. Kosley, L. Davis, and H.M. Vargas. 1996. Pharmacological evaluation of novel Alzheimer'' disease therapeutics: acetylcholinesterase inhibitors related to galanthamine. *Am. Soc. Pharmacol. Exp. Therapeut.* 277: 728-738.

Boue, S.M., T.E. Wiese, S. Nehls, M.E. Burow, S. Elliott, C.H. Carter-Wientjes, B.Y. Shih, J.A. McLachlan, and T.E. Cleveland. 2003. Evaluation of the estrogenic effects of legume extracts containing phytoestrogens. *J. Agric. Food Chem.* 51: 2193-2199.

Braquet, P.G., D. Hosford, and M. Koltai. 1994. Ginkgolide B (BN 52021): A natural PAF receptor antagonist. *Neuropsychopharmacology* 10: 693S.

Breitner, J.C.S. 1996. The role of anti-inflammatory drugs in the prevention and treatment of Alzheimer's disease. *Annu. Rev. Med.* 47: 401-411.

Breitner, J.C.S., K.A. Welsh, M.J. Helms, P.C. Gaskell, B.A. Gau, A.D. Roses, M.A. Pericak-Vance, and A.M. Saunders. 1995. Delayed onset of Alzheimer's disease with non-steroidal anti-inflammatory and histamine H_2 blocking drugs. *Neurobiol. Aging* 16: 523-530.

Brinkhaus, B., M. Lindner, D. Schuppan, and E.G. Hahn. 2000. Chemical, pharmacological and clinical profile of the East Asian medicinal plant *Centella asiatica*. *Phytomedicine* 7: 427-448.

Burns, A., R. Spiegel, and P. Quarg. 2004. Efficacy of rivastigmine in subjects with moderately severe Alzheimers disease. *Int. J. Geriatr. Psychiatry* 19: 243-249.

Chandrasekaran, K., Z. Mehrabian, B. Spinnewyn, K. Drieu, and G. Kiskum. 2001. Neuroprotective effects of bilobalide, a component of *Ginkgo biloba* extract (EGb 761), in gerbil global brain ischemia. *Brain Res.* 922: 282-292.

Chang, H. and P.P. But 1986. *Pharmacology and Applications of Chinese Materia Medica,* Vol. II, World Scientific: Singapore.

Chaurasia, S.S., S. Panda, and A. Kar. 2000. *Withania somnifera* root extract in the regulation of lead-induced oxidative damage in male mouse. *Pharmacol. Res.* 41: 663-666.

Chopin, P., and M. Briley. 1992. Effects of four non-cholinergic cognitive enhancers in comparison with tacrine and galanthamine on scopolamine-induced amnesia in rats. *Psychopharmacology* 106: 26-30.

Cieza, A., P. Maier, and E. Poppel. 2003. Effects of *Ginkgo biloba* on mental functioning in healthy volunteers. *Arch. Med. Res.* 34: 373-381.

Crellin, J.K., and Philpott J. 1990. *Herbal Medicine Past and Present* (Vol. 2). *A Reference Guide to Medicinal Plants,* Duke University Press, Durham, North Carolina.

Das, A., G. Shanker, C. Nath, R. Pal, S. Singh, and H.K. Singh. 2002. A comparative study in rodents of standardized extracts of *Bacopa monniera* and *Ginkgo biloba*-Anticholinesterase and cognitive enhancing activities. *Pharmacol. Biochem. Behav.* 73: 893-900.

Das, P.K., C.L. Malhotra, and K. Prasad. 1964. Cardiotonic activity of ashwagandhine and ashwagandhinine, two alkaloids from *Withania ashwagandha* Kaul. *Arch. Int. Pharmacodyn.* 150: 356-362.

Decker, M.W., M.J. Majchrzak, and S.P. Arneric. 1993. Effects of lobeline, a nicotinic receptor agonist, on learning and memory. *Pharmacol. Biochem. Behav.* 45: 571-576.

Desgranges, B., J.-C. Baron, V. de la Sayette, M.-C. Petit-Taboué, K. Benali, B. Landeau, B. Lechevalier, and F. Eustache. 1998. The neural substrates of memory systems impairment in Alzheimer's disease. *Brain* 121: 611-631.

Dhuley, J.N. 1997. Effect of some Indian herbs on macrophage functions in ochratoxin A treated mice. *J. Ethnopharmacol.* 58: 15-20.

Dhuley, J.N. 1998. Effect of ashwagandha on lipid peroxidation in stress-induced animals. *J. Ethnopharmacol.* 60: 173-178.

Dhuley, J.N. 2001. Nootropic-like effect of ashwagandha (*Withania somnifera* L.) in mice. *Phytother. Res.* 15: 524-528.

Du, G.H., Y. Qiu, and J.T. Zhang. 2000. Salvianolic acid B protects the memory functions against transient cerebral ischemia in mice. *J. Asian Natural Products Res.* 2: 145-152.

Duffy, R., H. Wiseman, and S.E. File. 2003. Improved cognitive function in postmenopausal women after 12 weeks of consumption of a soya extract containing isoflavones. *Pharmacol. Biochem. Behav.* 75: 721-729.

Duke, J.A., and E.S. Ayensu. 1985. *Medicinal Plants of China.* Volumes I - II (2nd ed). Reference Publications Inc., Algonac, USA.

Eckert, A., U. Keil, S. Kressmann, K. Schindowski, S. Leutner, S. Leutz, and W.E. Muller. 2003. Effects of EGb 761® *Ginkgo biloba* extract on mitochondrial function and oxidative stress. *Pharmacopsychiatry* 36: S15-S23.

Emerit, J., A. Edeas, and F. Bricaire. 2004. Neurodegenerative diseases and oxidative stress. *Biomed. Pharmacother.* 58: 39-46.

Evans, W.C., 2002. *Trease and Evans' Pharmacognosy,* 15th ed. Saunders, London.

File, S.E., N. Jarrett, E. Fluck, R. Duffy, K. Casey, and H. Wiseman. 2001. Eating soya improves human memory. *Psychopharmacol.* 157: 430-436.

Fillit H., H. Weinreb, I. Cholst, V. Luine, B. McEwan, R. Amador, and J. Zabriskie. 1986. Observations in a preliminary open trial of estradiol therapy for senile dementia-Alzheimer's type. *Psychoneuroendocrinol.* 11: 337-345.

Förstl, H., F. Hentschel, H. Sattel, C. Geiger-Kabisch, C. Besthorn, C. Czech, U. Mönning, and K. Beyreuther. 1995. Age-associated memory impairment and early Alzheimer' disease. *Drug. Res.* 45: 394-397.

Fowler, C.J., C. O'Neill, B. Winblad, and R.T. Cowburn. 1992. Neurotransmitter, receptor and signal transduction disturbances in Alzheimer's disease. *Acta Neurol. Scand.* 85: 59-62.

Francis, P.T., A.M. Palmer, M. Snape, and G.K. Wilcock. 1999. The cholinergic hypothesis of Alzheimer's disease: A review of progress. *J. Neurol. Neurosurg. Psychiatry* 66: 137-147.

Fulton, B., and P. Benfield. 1996. Galantamine *Drugs Aging* 1: 60-65.

Ghosal, S., J. Lal, R. Srivastava, S.K. Bhattacharya, S.N. Upadhyay, A.K. Jaiswal, and U. Chattopadhyay. 1989. Immunomodulatory and CNS effects of sitoindosides IX and X, two new glycowithanolides from *Withania somnifera*. *Phytother. Res.* 3: 201-206.

Giacobini, E. 1990. The cholinergic system in Alzheimer disease. *Prog. Brain Res.* 84: 321-332.

Greig, N.H., T. Utsuki, Q.S. Yu, X.X. Zhu H.W. Holloway, T. Perry, B. Lee, D.K. Ingram, and D.K. Lahiri . 2001. A new therapeutic target in Alzheimer's disease treatment: attention to butyrylcholinesterase. *Curr. Med. Res. Opin.* 17: 159-165.

Grossberg, G., and A. Desai. 2001. Review of rivastigmine and its clinical applications in Alzheimer's disease and related disorders. *Expert Opin. Pharmacother.* 2: 653-666.

Guenther, E. 1949. *The Essential Oils (Volume III): Individual Essential Oils of the Plant Families Rutaceae and Labiatae,* D. Van Nostrand Company, Inc., Toronto, Canada.

Hammel, P., D. Larrey, J. Bernuau, M. Kalafat, E. Fréneaux, G. Babany, C. Degott, G. Feldmann, D. Passayre, and J.P. Benhamou. 1990. Acute hepatitis after tetrahydroaminoacridine administration for Alzheimer's disease. *J. Clin. Gastroenterol.* 12: 329-331.

Handa, S.S., A.S. Chawla, and A.K. Sharma. 1992. Plants with anti-inflammatory activity. *Fitoterapia* 63: 3-31.

Hartley, D.E., L. Heinze, S. Elsabagh, and S.E. File. 2003. Effects on cognition and mood in postmenopausal women of 1-week treatment with *Ginkgo biloba*. *Pharmacol. Biochem. Behav.* 75: 711-720.

Heiss, W.D., and K. Zeiler. 1978. Drug influence on cerebral circulation. *Pharmakotherapie* 1: 137-144.

Hofferberth, B. 1994. The efficacy of EGb 761 in patients with senile dementia of the Alzheimer type: a double-blind, placebo-controlled study on different levels of investigation. *Hum. Psychopharmacol.* 9: 215-222.

Hogervorst, E., J. Williams, M. Budge, W. Riedel, and J. Jolles. 2000. The nature of the effect of female gonadal hormone replacement therapy on cognitive function in post-menopausal women: A meta-analysis. *Neuroscience* 101: 485-512.

Hohmann, J., I. Zupkó, D. Rédei, M. Csányi, G. Falkay, I. Máthé, and G. Janicsák. 1999. Protective effects of the aerial parts of *Salvia officinalis, Melissa officinalis* and *Lavandula angustifolia* and their constituents against enzyme-dependent and enzyme-independent lipid peroxidation. *Planta Med.* 65: 576-578.

Honjo, H., Y. Ogino, K. Naitoh, M. Urabe, J. Kitawaki, J. Yasuda, T. Yamamoto, S. Ishihara, H. Okada, T. Yonezawa, K. Hayashi, and T. Nambara. 1989. *In vivo* effects by estrone sulfate on the central nervous system-senile dementia (Alzheimer's type). *J. Steroid Biochem.* 34: 521-525.

Hoult, J.R.S., and M. Payà. 1996. Pharmacological and biochemical actions of simple coumarins: natural products with therapeutic potential. *Gen. Pharmacol.* 27: 713-722.

Howes, M.-J.R., P.J. Houghton, D.J. Barlow, V.J. Pocock, and S.R. Milligan. 2002. Assessment of estrogenic activity in some common essential oil constituents. *J. Pharm. Pharmacol.* 54: 1521-1528.

Howes, M.-J.R., and P.J Houghton. 2003a. Plants used in Chinese and Indian traditional medicine for improvement of memory and cognitive function. *Pharmacol. Biochem. Behav.* 75: 513-527.

Howes, M.-J.R., N.S.L Perry, and P.J. Houghton. 2003b. Plants with traditional uses and activities, relevant to the management of Alzheimer's disease and other cognitive disorders. *Phytother. Res.* 17: 1-18.

Hsieh, M.T., L.H. Kuo, F.H. Tsai, W.H. Wang, and C.R. Wu. 2002. Effects of puerarin on scopolamine-, mecamylamine-, *p*-chloroamphetamine- and dizocilpine-induced inhibitory avoidance performance impairment in rats. *Planta Med.* 68: 901-905.

Huang, K.C. 1993. *The Pharmacology of Chinese Herbs,* CRC Press, Boca Raton, Florida.

Irwin R.L., and H.J. Smith. 1960. Cholinesterase inhibition by galantamine and lycoramine. *Biochem. Pharmacol.* 3: 147-148.

Jain, S., S.D. Shukla, K. Sharma, and M. Bhatnagar. 2001. Neuroprotective effects of *Withania somnifera* Dunn. in hippocampal sub-regions of female albino rat. *Phytother. Res.* 15: 544-548.

Jenkinson, M.L., M.R. Bliss, A.T. Brain, and D.L. Scott. 1989. Rheumatoid arthritis and senile dementia of the Alzheimer's type. *Br. J. Rheumatol.* 28: 86-88.

Jiang H., X. Luo, and D. Bia. 2003. Progress in clinical, pharmacological, chemical and structural biological studies of huperzine A: a drug of traditional Chinese medicine origin for the treatment of Alzheimer's disease. *Curr. Med. Chem.* 10: 2231- 2252.

Jones, F.A., S.S. Chatterjee, and J.A. Davies. 2002. Effects of bilobalide on amino acid release and electrophysiology of cortical slices. *Amino Acids* 22: 369-379.

Kang, H.S., H.Y. Chung, J.H. Jung, S.S. Kang, and J.S. Choi. 1997. Antioxidant effect of *Salvia miltiorrhiza. Arch. Pharm. Res.* 20: 496-500.

Kanowski, S., W.M. Herrmann, K. Stephan, W. Wierch, and R. Hörr. 1997. Proof of efficacy of the *Ginkgo biloba* special extract EGb 761 in outpatients suffering from mild to moderate primary degenerative dementia of the Alzheimer type or multi-infarct dementia. *Phytomedicine* 4: 3-13.

Kanowski, S., and R. Hoerr. 2003. *Ginkgo biloba* extract EGb 761® in dementia: Intent-to-treat analyses of a 24-week, multi-center, double-blind, placebo-controlled, randomized trial. *Pharmacopsychiatry* 36: 297-303.

Kapoor, L.D. 1990. *Handbook of Ayurvedic Medicinal Plants,* CRC Press, Boca Raton, Florida.

Kennedy, D.O., G. Wake, S. Savelev, N.T.J. Tildesley, E.K. Perry, K.A. Wesnes, and A.B. Scholey. 2003. Modulation of mood and cognitive performance following acute administration of single doses of *Melissa officinalis* (Lemon balm) with human CNS nicotinic and muscarinic receptor-binding properties. *Neuropsychopharmacology* 28: 1871-1881.

Kenner, D., and Y. Requena. 1996. *Botanical Medicine: A European Professional Perspective,* Paradigm Publications, Brookline, Massachusetts.

Kim, H., H. Xia, L. Li, and J. Gewin. 2001. Actions of dietary soya vs. Premarin in mammalian brain. 21st ACS National Meeting, San Diego, California. *Biochem. Soc. Trans.* 29: 216-222.

Kristofiková, Z., O. Benesová, and H. Tejkalová. 1992. Changes of high-affinity choline uptake in the hippocampus of old rats after long-term administration of two nootropic drugs (tacrine and *Ginkgo biloba* extract). *Dementia* 3: 304-307.

Kritz-Silverstein, D., D. Von Muhlen, E. Barrett-Connor, and M.A.B. Bressel. 2003. Isoflavones and cognitive function in older women: the SOy and Postmenopausal Health In Aging (SOPHIA) Study. *Menopause* 10: 196-202.

Kuang, P., W. Wu, J. Liu, F. Zhang, and C. Pu. 1991. The effect of *Radix salviae* miltiorrhizae (RSM) on substance P in cerebral ischemia-animal experiment. *J. Tradit. Chin. Med.* 11: 123-127.

Kuang, P., W. Wu, and K. Zhu. 1993. Evidence for amelioration of cellular damage in ischemic rat brain by *Radix salviae* miltiorrhizae treatment-immunochemistry and histopathology studies. *J. Tradit. Chin. Med.* 13: 38-41.

Kuang P., and J. Xiang. 1994. Effect of *Radix salviae* miltiorrhizae on EAA and IAA during cerebral ischemia in gerbils: a microdialysis study. *J. Tradit. Chin. Med.* 14: 45-50.

Kuang P., Z. Li, F. Zhang, Y. Tao, J. Liu, and W. Wu. 1995. Protective effect of *Radix salviae* miltiorrhizae composita on cerebral ischemia. *J. Tradit. Chin. Med.* 15: 135-140.

Kuang, P., Y. Tao, and Y. Tian. 1996a. Effect of *Radix salviae* miltiorrhizae on nitric oxide in cerebral ischemic-reperfusion injury. *J. Tradit. Chin. Med.* 16: 224-227.

Kuang, P., Y. Tao, and Y. Tian. 1996b. *Radix salviae* miltiorrhizae treatment results in decreased lipid peroxidation in reperfusion injury. *J. Tradit. Chin. Med.* 16: 138-142.

Kuboto, Y., N. Tanaka, K. Umegaki, H. Takenaka, H. Mizuno, K. Nakamura, K. Shinozuka, and M. Kunitomo. 2001. *Ginkgo biloba* extract-induced relaxation of rat aorta is associated with increase in endothelial intracellular calcium level. *Life Sci.* 69: 2327-2336.

Kuboyama, T., C. Tohda, J. Zhao, N. Nakamura, M. Hattori, and K. Komatsu. 2002. Axon- or dendrite-predominant outgrowth induced by constituents from Ashwagandha. *Neuroreport* 13: 1715-1720.

Kulkarni, R.R., P.S. Patki, V.P. Jog, S.G. Gandage, and B. Patwardhan. 1991. Treatment of osteoarthritis with a herbomineral formulation: A double-blind, placebo-controlled, cross-over study. *J. Ethnopharmacol.* 33: 91-95.

Kumar, M.H.V., and Y.K. Gupta. 2002. Effect of different extracts of *Centella asiatica* on cognition and markers of oxidative stress in rats. *J. Ethnopharmacol.* 79: 253-260.

Kwak, W.J., C.K. Han, K.H. Son, H.W. Chang, S.S. Kang, B.K. Park, and H.P. Kim. 2002. Effects of ginkgetin from *Ginkgo biloba* leaves on cyclooxygenases and *in vivo* skin inflammation. *Planta Med.* 68: 316-321.

Laganière, S., J. Corey, X-C. Tang, E. Wülfert, and I. Hanin. 1991. Acute and chronic studies with the anticholinesterase huperzine A: effect on central nervous system cholinergic parameters. *Neuropharmacology* 30: 763-768.

Lake, J.T., and G.T. Grossberg. 1987. Management of psychosis, agitation, and other behavioural problems in Alzheimer's disease. *Psychiatr. Ann.* 17: 371-374.

Laughton M.J., P.J. Evans, M.A. Moroney, J.R.S. Hoult, and B. Halliwell. 1991. Inhibition of mammalian 5-lipoxygenase and cyclo-oxygenase by flavonoids and phenolic dietary additives. *Biochem. Pharmacol.* 42: 1673-1681.

Le Bars, P.L., M.M. Katz, N. Berman, T.M. Itil, A.M. Freedman, and A.F. Schatzberg. 1997. A placebo-controlled, double-blind, randomised trial of an extract of *Ginkgo biloba* for dementia. *JAMA* 278: 1327-1332.

Le Bars, P.L., M. Kieser, and K.Z. Itil. 2000. A 26-week analysis of a double-blind, placebo-controlled trial of the *Ginkgo biloba* extract EGb 761 in dementia. *Dement. Geriatr. Cognit. Disord.* 11: 230-237.

Le Bars, P.L., F.M. Velasco, J.M. Ferguson, E.C. Dessain, M. Kieser, and R. Hoerr. 2002. Influence of the severity of cognitive impairment on the effect of the *Ginkgo biloba* extract EGb 761® in Alzheimer's disease. *Neuropsychobiology* 45: 19-26.

Lebert, F., F. Pasquier, and H. Petit. 1994. Behavioural effects of trazodone in Alzheimer's disease. *J. Clin. Psychiatry* 55: 536-538.

Lee, M.K., S.R. Kim, S.H. Sung, D.Y. Lim, H. Kim, H. Choi, H.K. Park, S.S. Jew, and Y.C. Kim. 2000. Asiatic acid derivatives protect cultured neurons from glutamate-induced excitotoxicity. *Res. Commun. Mol. Pathol. Pharmacol.* 108: 75-86.

Lephart, E.D., T.W. West, K.S. Weber, R.W. Rhees, K.D.R. Setchell, H. Adlercreutze, and T.D. Lund. 2002. Neurobehavioral effects of dietary soy phytoestrogens. *Neurotoxicol. Teratol.* 24: 5-16.

Liu, G., T. Zhang, B. Wang, and Y. Wang. 1992. Protective action of seven natural phenolic compounds against peroxidative damage to biomembranes. *Biochem. Pharmacol.* 43: 147-152.

Loffler, T., S.K. Lee, M. Noldner, S.S. Chatterjee, S. Hoyer, and R. Schliebs. 2001. Effect of *Ginkgo biloba* extract (EGb 761) on glucose metabolism-related markers in streptozotocin-damaged rat brain. *J. Neural Transm.* 108: 1457-1474.

Lu, W-H., J. Shou, and X-C. Tang. 1988. Improving effect of huperzine A in aged rats and adult rats with experimental cognitive impairment. *Acta Pharm. Sinica* 9: 11-15.

Lu, Y.R., and L.Y. Foo . 2001a. Salvianolic acid L, a potent phenolic antioxidant from *Salvia officinalis*. *Tetrahedron Lett.* 42: 8223-8225.

Lu, Y.R., and L.Y. Foo. 2001b. Antioxidant activities of polyphenols from sage (*Salvia officinalis*). *Food Chem.* 75: 197-202.

Maheswari, S.L. 2003. Antioxidant effect of *Centella asiatica* Linn. on aluminium-induced lipid peroxidation in brain regions of rat. *J. Neurochem.* 87 (Suppl. 1): 155.

Maheswari, S.L., and R.V. Murali. 2004. Neuroprotective role of *Centella asiatica* Linn. on aluminium-induced brain cholinotoxicity in Wistar albino rats. *J. Neurochem.* 88 (Suppl. 1): 66.

Malamas, M., and M. Marselos. 1992. The tradition of medicinal plants in Zagori, Epirus (northwestern Greece). *J. Ethnopharmacol.* 37: 197-203.

Malhotra, C.L., V.L. Mehta, P.K. Das, and N.S. Dhalla. 1965. Studies on *Withania* Aswagandha, Kaul. V. The effects of total alkaloids (ashwagandholine) on the central nervous system. *Indian J. Physiol. Pharmacol.* 9: 127-136.

Manyam, B.V. 1999. Dementia in Ayurveda. *J. Altern. Complement. Med.* 5: 81-88.

Marcocci, L., L. Packer, M.-T. Droy-Lefaix, A. Sekaki, and M. Gardès-Albert. 1994. Antioxidant action of *Ginkgo biloba* extract EGb 761. *Methods Enzymol.* 234: 462-475.

Marcusson, J., R. Bullock, S. Gauthier, A. Kurz., and S. Schwalen. 2003. Galantamine demonstrates efficacy and safety in elderly patients with Alzheimers disease. *Alzheimer Dis. Assoc. Disord.* 17 Suppl. 3: S86-91.

Marston, A., J. Kissling, and K. Hostettmann. 2002. A rapid TLC bioautographic method for the detection of acetylcholinesterase and butyrylcholinesterase inhibitors in plants. *Phytochem. Anal.* 13: 51-54.

Massieu, L., J. Moran, and Y. Christen. 2004. Effect of *Ginkgo biloba* (EGb 761) on staurosporine-induced neuronal death and caspase activity in cortical cultured neurons. *Brain Res.* 1002: 76-85.

Maurer, K., R. Ihl, T. Dierks, and L. Frölich. 1997. Clinical efficacy of *Ginkgo biloba* extract EGb 761 in dementia of the Alzheimer type. *J. Psychiat. Res.* 31: 645-655.

Maxwell, S.J. 1995. Prospects for the use of anti-oxidant therapies. *Drugs* 49: 345.

McCaleb, R. 1990. Nature's medicine for memory loss. *HerbalGram.* 23: 15.

McGeer, P.L., E. McGeer, J. Rogers, and J. Sibley. 1990. Anti-inflammatory drugs and Alzheimer's disease. *Lancet* 335: 1037.

McGeer, P.L., M. Schulzer, and E.G. McGeer. 1996. Arthritis and anti-inflammatory agents as possible protective factors for Alzheimer's disease: A review of 17 epidemiologic studies. *Neurology* 47: 425-432.

McGuffey, E.C. 1997. Alzheimer's disease: An overview for the pharmacist. *JAMA* NS37: 347-352.

McVicar, J. 1994. *Jekka's Complete Herb Book,* Kyle Cathie Ltd., London, England.

Mehta, A.K., P. Binkley, S.S. Gandhi, and M.K. Ticku. 1991. Pharmacological effects of *Withania somnifera* root extract on $GABA_A$ receptor complex. *Indian J. Med. Res. B.* 94: 312-325.

Mimica-Dukic, N., B. Bozin, M. Sokovic, and N. Simin. 2004. Antimicrobial and antioxidant activities of *Melissa officinalis* L. (Lamiaceae) essential oil. *J. Agric. Food Chem.* 52: 2485-2489.

Min, S.K., I.W. Moon, R.W. Ko, and H.S. Shin. 2001. Effects of transdermal nicotine on attention and memory in healthy elderly non-smokers. *Psychopharmacol.* 159: 83-88.

Mishra, L., B.B. Singh, and S. Dagenais. 2000. Scientific basis for the therapeutic use of *Withania somnifera*. (ashwagandha): A review. *Altern. Med. Rev.* 5: 334-346.

Miyazawa, M., H. Watanabe, and H. Kameoka. 1997. Inhibition of acetylcholinesterase activity by monoterpenoids with a *p*-menthane skeleton. *J. Agric. Food Chem.* 45: 677-679.

Moncada, S., R.M.J. Palmer, and E.A. Higgs. 1991. Nitric oxide: physiology, pathophysiology, and pharmacology. *Pharmacol. Rev.* 43: 109-142.

Mulnard, R.A., C.W. Cotman, C. Kawas, C.H. van Dyck, M. Sano, R. Doody, E. Koss, E. Pfeiffer, S. Jin, A. Gamst, M. Grundman, R. Thomas, and L.J. Thal. 2000. Estrogen replacement therapy for treatment of mild to moderate Alzheimer disease: A randomised controlled trial. *JAMA* 283: 1007-1015.

Nalini, K., A.R. Aroor, K.S. Karanth, and A. Rao. 1992. Effect of *Centella asiatica* fresh leaf aqueous extract on learning and memory and biogenic amine turnover in albino rats. *Fitoterapia* 63: 232-237.

Newhouse, P.A., and M. Kelton. 2000. Nicotinic systems in central nervous systems disease: Degenerative disorders and beyond. *Pharm. Acta Helv.* 74: 91-101.

Nishida, S., and H. Satoh. 2004. Comparative vasodilating actions among terpenoids and flavonoids contained in *Ginkgo biloba* extract. *Clin. Chim. Acta* 339: 129-133.

Ogane, N., E. Giacobini, and R. Struble. 1992. Differential inhibition of acetylcholinesterase molecular forms in normal and Alzheimer's disease brain. *Brain Res.* 589: 307-312.

Oh, S.M., and K.H. Chung. 2004. Estrogenic activities of *Ginkgo biloba* extracts. *Life Sci.* 74: 1325-1335.

Ohkura, T., K. Isse, K. Akazawa, M. Hamamoto, Y. Yaoi, and N. Hagino. 1994. Low-dose estrogen replacement therapy for Alzheimer disease in women. *Menopause* 1: 125-130.

Oken, B.S., D.M. Storzbach, and J.A. Kaye. 1998. The efficacy of *Ginkgo biloba* on cognitive function in Alzheimer disease. *Arch. Neurol.* 55: 1409-1415.

Otoguro, K., F. Kuno, and S. Ōmura 1997. Arisugacins selective acetylcholinesterase inhibitors of microbial origin. *Pharmacol. Ther.* 76: 45-54.

Ou, L.Y., X.C. Tang, and J.X. Cai . 2001. Effect of huperzine A on working memory in reserpine-or yohimbine-treated monkeys. *Eur. J. Pharmacol.* 433: 151-156.

Pan, Y., M. Anthony, S. Watson, and T.B. Clarkson. 2000. Soy phytoestrogens improve radial arm maze performance in ovariectomised retired breeder rats and do not attenuate benefits of 17β-estradiol treatment. *Menopause* 7: 209-212.

Panda, S., P. Gupta, and A. Kar. 1997. Protective role of ashwagandha in cadmium-induced hepatotoxicity and nephrotoxicity in male mouse. *Curr. Sci.* 72: 546-547.

Parihar, M.S., and T. Hemnani. 2003. Phenolic antioxidants attenuate hippocampal neuronal cell damage against kainic acid induced excitotoxicity. *J. Biosci.* 28: 121-128.

Parsons, C.G., W. Danysz, and G. Quack. 1998. Glutamate in CNS disorders as a target for drug development: An update. *Drug News and Perspectives* 11: 523-569.

Paulus, K., and R. Bauer. 2000. Inhibitory effects from *Salvia miltiorrhiza* on leukotriene biosynthesis. In *Natural Products Research in the New Millenium Abstract Book.* 48th Annual Meeting of the Society for Medicinal Plant Research: Zurich, Switzerland (September 3rd -7th, 2000). P2A/73.

Perry, E.K. 1986. The cholinergic hypothesis: 10 years on. *Br. Med. Bull.* 42: 63-69.

Perry, E.K., A.T. Pickering, W.W. Wang, P.J. Houghton, and N.S.L. Perry. 1998. Medicinal plants and Alzheimer's disease: integrating ethnobotanical and contemporary scientific evidence. *J. Altern. Complement. Med.* 4: 419-428.

Perry, E.K., E. Tomlinson, G. Blessed, K. Bergmann, P. Gibson, and R. Perry. 1978. Correlation of cholinergic abnormalities with senile plaques and mental test scores in senile dementia. *BMJ* 2: 1457-1459.

Perry, N.S.L., G. Court, N. Bidet, J. Court, and E.K. Perry. 1996. European herbs with cholinergic activities: potential in dementia therapy. *Int. J. Geriatr. Psychiatry* 11: 1063-1069.

Perry, N.S.L., P.J. Houghton, A.E. Theobald, P. Jenner, and E.K. Perry. 2000. *In-vitro* inhibition of human erythrocyte acetylcholinesterase by *Salvia lavandulaefolia* essential oil and constituent terpenes. *J. Pharm. Pharmacol.* 52: 895-902.

Perry, N.S.L., P.J Houghton, J. Sampson, S. Hart, M. Lis-Balchin, J.R.S. Hoult, P. Evans, P. Jenner, S. Milligan, and E.K. Perry. 2001. *In vitro* activities of *S. lavandulaefolia* (Spanish sage) relevant to treatment of Alzheimer's disease. *J. Pharm. Pharmacol.* 53: 1347-1356.

Perry, N.S.L., P.J. Houghton, J. Sampson, S. Hart, M. Lis-Balchin, J.R.S. Hoult, P. Evans, P. Jenner, S. Milligan, and E.K. Perry. 2001. *In vitro* activities of *S. lavandulaefolia* (Spanish sage) relevant to treatment of Alzheimer's disease. *J. Pharm. Pharmacol.* 53: 1347-1356.

Perry, N.S.L., P.J. Houghton, P. Jenner, A. Keith, and E.K. Perry. 2002. *Salvia lavandulaefolia* essential oil inhibits cholinesterase *in vivo*. *Phytomedicine* 9: 48-51.

Perry, N.S.L., C. Bollen, E.K. Perry, and C. Ballard. 2003. *Salvia* for dementia therapy: Review of pharmacological activity and pilot tolerability clinical trial. *Pharmacol. Biochem. Behav.* 75: 651-660.

Petkov, V.D., R. Kehayov, S. Belcheva, E. Konstantinova, V.V. Petkov, D. Getova, and V. Markovska. 1993. Memory effects of standardized extracts of *Panax ginseng* (G115), *Ginkgo biloba* (GK501) and their combination Ginkosan (PHL-00701). *Planta Med.* 59: 106-114.

Pharmacopoeia of The People's Republic of China (Vol. I). 2000. The State Pharmacopoeia Commission of the P.R. China, Chemical Industry Press, Beijing, China.

Polinsky, R.J. 1998. Clinical pharmacology of rivastigmine: a new-generation acetylcholinesterase inhibitor for the treatment of Alzheimer's disease. *Clin. Ther.* 20: 634-647.

Rai, G.S., C. Shovlin, and K.A. Wesnes. 1991. A double-blind, placebo controlled study of *Ginkgo biloba* extract (Tanakan) in elderly outpatients with mild to moderate memory impairment. *Curr. Med. Res. Opin.* 12: 350-355.

Raskind, M.A., E.R. Peskind, L. Truyen, P. Kershaw, and C.V. Damaraju, 2004. The cognitive benefits of galantamine are sustained for at least 36 months: a long term-extension trial. *Arch. Neurol.* 61: 252-256.

Ren, Y., P.J. Houghton, R.C. Hider. and M-J.R Howes. 2004. Novel diterpenoid acetylcholinesterase inhibitors from *Salvia miltiorhiza*. *Planta Medica* 70: 201-204.

Rhee, I.K., M. van de Meent, K. Ingkaninan, and R. Verpoorte. 2001. Screening for acetylcholinesterase inhibitors from Amaryllidaceae using silica gel thin-layer chromatography in combination with bioactivity staining. *J. Chromatogr. A.* 915: 217-223.

Rigney. U., S. Kimber, and I. Hindmarch. 1999. The effects of acute doses of standardised *Ginkgo biloba* extract on memory and psychomotor performance in volunteers. *Phytother. Res.* 13: 408-415.

Rojas, P., B. Garduno, C. Rojas, R.M. Vigueras, J. Rojas-Castaneda, C. Rios, and N. Serrano-Garcia. 2001. EGb761 blocks MPP+-induced lipid peroxidation in mouse corpus striatum. *Neurochem. Res.* 26: 1245-1251.

Roodenrys, S., D. Booth., S. Bulzomi, A. Phipps, C. Micallef, and J. Smoker. 2002. Chronic effects of Brahmi (*Bacopa monnieri*) on human memory. *Neuropsychopharmacol.* 27: 279-281.

Ross, I.A. 2001. *Medicinal Plants of the World* (Vol. 2.), *Chemical Constituents, Traditional and Modern Medicinal Uses* Humana Press, Totowa, New Jersey.

Russo, A., A.A. Izzo., F. Borrelli, M. Renis, and A. Vanella, A. 2003. Free radical scavenging capacity and protective effect of *Bacopa monniera* L. on DNA damage. *Phytother. Res.* 17: 870-875.

Ryan, M.F., and O. Byrne. 1988. Plant-insect coevolution and inhibition of acetylcholinesterase. *J. Chem. Ecol.* 14: 1965-1975.

Sahakian, B., G. Jones, R. Levy, J. Gray, and D. Warburton. 1989. The effects of nicotine on attention, information processing, and short term memory in patients with dementia of the Alzheimer type. *Br. J. Psychiatry* 154: 797-800.

Sakina, M.R., and P.C. Dandiya. 1990. A psycho-neuropharmacological profile of *Centella asiatica* extract. *Fitoterapia* 61: 291-296.

Salomon A.R., K.J. Marcinowski, R.P. Friedland, and M.G. Zagorski . 1996. Nicotine inhibits amyloid formation by the *β*-peptide. *Biochemistry* 35: 13568-13578.

Şarer, E., and G. Kökdil. 1991. Constituents of the essential oil from *Melissa officinalis*. *Planta Med.* 57: 89-91.

Savelev, S., E. Okello, N.S.L. Perry, R.M. Wilkins, and E.K. Perry. 2003. Synergistic and antagonistic interactions of anticholinesterase terpenoids in *Salvia lavadulaefolia* essential oil. *Pharmacol. Biochem. Behav.* 75: 661-668.

Scartezzini, P., and E. Speroni. 2000. Review on some plants of Indian traditional medicine with anti-oxidant activity. *J. Ethnopharmacol.* 71: 23-43.

Schindowski, K., S. Leutner, S. Kressmann, A. Eckert, and W.E. Muller. 2001. Age-related increase of oxidative stress-induced apoptosis in mice prevention by *Ginkgo biloba* extract (EGb761). *J. Neural Transm.* 108: 969-978.

Schliebs, R., A. Liebmann, S.K. Bhattacharya, A. Kumar, S. Ghosal, and V. Bigl. 1997. Systemic administration of defined extracts from *Withania somnifera* (Indian ginseng) and shilajit differentially affects cholinergic but not glutamatergic markers in rat brain. *Neurochem. Int.* 30: 181-190.

Schneider, L.S., M.R. Farlow, V.W. Henderson, and J.M. Pogoda. 1996. Effects of estrogen replacement therapy on response to tacrine in patients with Alzheimer's disease. *Neurology* 46: 1580-1584.

Schwarting, A.E., J.M. Bobbitt, A. Rother, K. Atal, K.L. Khanna, J.D. Leary, and W.G. Walter. 1963. The alkaloids of *Withania somnifera*. *Lloydia* 26: 258-273.

Simpson, P.M., and D. Foster. 1986. Improvement in organically disturbed behaviour with trazodone treatment. *J. Clin. Psychiatry* 47: 191-193.

Sitaram, N., H. Weingartner, and J.C. Gillin. 1978. Physostigmine: improvement of long-term memory processes in normal humans. *Science* 201: 272-276.

Skolnick, A.A. 1997. Old herbal Chinese medicine used for fever yields possible new Alzheimer disease therapy. *JAMA* 277: 776.

Slater, T.F. 1984. Free-radical mechanisms in tissue injury. *Biochem. J.* 222: 1-15.

Small, G.W., R.V. Rabins, P.P. Barry, N.S. Buckholtz, S.T. DeKosky, S.H. Ferros, S.I. Finkel, L.P. Gwyther, Z.S. Khachaturian, B.D. Lebowitz, T.D. McRae, J.C. Morris, F. Oakley, L.S. Schneider, J.E. Streim, T. Sunderland, L.A. Teri, and L.E. Tune. 1997. Diagnosis and treatment of Alzheimer's disease and related disorders. *JAMA* 278: 1363-1371.

Small, G., T. Erkinjuntii,, A. Kurz, and S. Lilienfeld. 2003. Galantamine in the treatment of cognitive decline in patients with vascular dementia or Alzheimer's disease with cerebrovascular disease. *CNS Drugs* 17: 905-914.

Soulimani, R., J. Fleurentin, F. Mortier, R. Misslin, G. Derrieu, and J. Pelt. 1991. Neurotropic action of the hydroalcoholic extract of *Melissa officinalis* in the mouse. *Planta Med.* 57: 105-109.

Spencer, C.M., and S. Noble. 1998. Rivastigmine-a review of its use in Alzheimer's disease. *Drugs Aging* 13: 391-411.

Spiteller, G. 1993. Review: On the chemistry of oxidative stress. *J. Lipid Mediators* 7: 199-221.

Stoll, S., K. Scheuer, O. Pohl, and W.E. Müller. 1996. *Ginkgo biloba* extract (EGb 761) independently improves changes in passive avoidance learning and brain membrane fluidity in the ageing mouse. *Pharmacopsychiatry* 29: 144-149.

Storga, D., K. Vrecko, J.G. Birkmayer, and G. Reibnegger. 1996. Monoaminergic neurotransmitters, their precursors and metabolites in brains of Alzheimer patients. *Neurosci. Lett.* 203: 29-32.

Stough, C., J. Lloyd, J. Clarke, L.A. Downey, C.W. Hutchinson, T. Rodgers, and P.J. Nathan. 2001. The chronic effects of an extract of *Bacopa monniera* (Brahmi) on cognitive function in healthy human subjects. *Psychopharmacol*ogy 156: 481-484.

Sudhir, S., R.D. Budhiraja, G.P. Miglani, B. Arora, L.C. Gupta, and K.N. Garg. 1986. Pharmacological studies on leaves of *Withania somnifera*. *Planta Med.* 1: 61-63.

Sugaya, K, T. Uz, V. Kumar, and H. Manev. 2000. New anti-inflammatory treatment strategy in Alzheimer's disease. *Jpn. J. Pharmacol.* 82: 85-94.

Tang, W., and G. Eisenbrand. 1992. *Chinese Drugs of Plant Origin*. Springer-Verlag, Berlin, Heidelberg, Germany.

Tea, S., P. Celsis, M. Clanet, J.-P. Marc-Vergnes, and U. Boeters. 1987. Quantified parameters for the proof of the enhancement of cerebral circulation and metabolism. *Therapiewoche* 37: 2655-2657.

Tildesley, N.T.J., D.O. Kennedy, E.K. Perry, C.G. Ballard, S. Savelev, K.A. Wesnes, and A.B. Scholey. 2003. *Salvia lavandulaefolia* (Spanish Sage) enhances memory in healthy young volunteers. *Pharmacol. Biochem. Behav.* 75: 669-674.

Tittel, G., H. Wagner, and R. Bos. 1982. Chemical composition of the essential oil from *Melissa*. *Planta Med.* 46: 91-98.

Tohda, C., T. Kuboyama, and K. Komatsu. 2000. Dendrite extension by methanol extract of ashwagandha (roots of *Withania somnifera*) in SK-N-SH cells. *Neuroreport* 11: 1981-1985.

Traystman, R.J., J.R. Kirsch, and R.C. Koehler. 1991. Oxygen radical mechanisms of brain injury following ischaemia and reperfusion. *J. Appl. Physiol.* 71: 1185-1195.

Upton, R. 2000. Ashwagandha root. *Withania somnifera*. Analytical, quality control and therapeutic monograph. *American Herbal Pharmacopoeia and Therapeutic Compendium* 1-25.

van Duijn, C.M., and A. Hofman. 1991. Relation between nicotine and Alzheimer's disease. *BMJ* 302: 1491-1494.

Wake, G., J. Court, A. Pickering, R. Lewis, R. Wilkins, and E. Perry. 2000. CNS acetylcholine receptor activity in European medicinal plants traditionally used to improve failing memory. *J. Ethnopharmacol.* 69: 105-114.

Wang, Y.-E, D.-X. Yue, and X-C., Tang. 1986. Anti-cholinesterase activity of huperzine A. *Acta Pharm. Sinica* 7: 110-113.

Watkins, P.B., H.J. Zimmerman, M.J. Knapp, S.I. Gracon, and K.W. Lewis . 1994. Hepatotoxic effects of tacrine administration in patients with Alzheimer's disease. *JAMA* 271: 992-998.

Welton, A.F., L.D. Tobias, C. Fiedler-Nagy, W. Anderson, W. Hope, K. Meyers, and J.W. Coffey. 1986. Effect of flavonoids on arachidonic acid metabolism. *Prog. Clin. Biol. Res.* 213: 231-242.

Weng, X.C., and M.H. Gordon. 1992. Antioxidant activity of quinones extracted from tanshen (*Salvia miltiorrhiza* Bunge). *J. Agric. Food Chem.* 40: 1331-1336.

Whitehouse, P.J., and R.N. Kalaria. 1995. Nicotinic receptors and neurodegenerative dementing diseases: basic research and clinical implications. *Alzheimer Dis. Assoc. Disord.* 9 (Suppl. 2): 3-5.

Wilcock, G.K., S. Lilienfeld, and E. Gaens, on behalf of the Galantamine International-1 Study Group. 2000. Efficacy and safety of galantamine in patients with mild to moderate Alzheimer's disease: multicentre randomised controlled trial. *BMJ* 321: 1445-1449.

Wilkinson, D., and J. Murray. 2001. Galantamine: a randomised, double-blind, dose comparison in patients with Alzheimer's disease. *Int. J. Geriatr. Psychiatry* 16: 852-857.

Woo, J., E. Lau, S.C. Ho, F. Cheng, C. Chan, A.S.Y. Chan, C.J. Haines, T.Y.K. Chan, M. Li, and A. Sham. 2003. Comparison of *Pueraria lobata* with hormone replacement therapy in treating the adverse health consequences of menopause. *Menopause* 10: 352-361.

Woodruff-Pak, D.S., R.W. Vogel, and G.L. Wenk. 2001. Galantamine: effect on nicotinic receptor binding, acetylcholinesterase inhibition, and learning. *Proc. Natl. Acad. Sci., USA* 98: 2089-2094.

Wragg, R.E., and D.V. Jeste. 1989. Overview of depression and psychosis in Alzheimer's disease. *Am. J. Psychiatry* 146: 577-587.

Xiao, X.Q., H.Y. Zhang, and X-C. Tang. 2002. Huperzine A attenuates amyloid β-peptide fragment 25-35-induced apoptosis in rat cortical neurons via inhibiting reactive oxygen species formation and capase-3 activation. *J. Neurosci. Res.* 67: 30-36.

Yan, J.J., J.Y. Cho, H.S. Kim, K.L. Kim, J.S. Jung, S.O. Huh, H.W. Suh, Y.H. Kim, and D.K. Song. 2001. Protection against β-amyloid peptide toxicity *in vivo* with long-term administration of ferulic acid. *Br. J. Pharmacol.* 133: 89-96.

Yao, Z., K. Drieu, and V. Papadopoulos. 2001. The *Ginkgo biloba* extract EGb 761 rescues the PC12 neuronal cells from β-amyloid-induced cell death by inhibiting the formation of β-amyloid-derived diffusible neurotoxic ligands. *Brain Res.* 889: 181-190.

Yarnell, E. 1998. Lemonbalm. *Altern. Complement. Ther.* 4: 417-419.

Yasui-Furukori, N., H. Furukori, A. Kaneda, S. Kaneko, and T. Tateishi. 2004. The effects of *Ginkgo biloba* extracts on the pharmacokinetics and pharmacodynamics of donepezil. *J. Clin. Pharmacol.* 44: 538-542.

Yoshida, S., and N. Suzuki. 1993. Antiamnesic and cholinomimetic side-effects of the cholinesterase inhibitors, physostigmine, tacrine and NIK-247 in rats. *Eur. J. Pharmacol.* 250: 117-124.

Younkin, S.G., B. Goodridge, J. Katz, G. Lockett, D. Nafziger, M.F. Usiak, and L.H. Younkin. 1986. Molecular forms of acetylcholinesterase in Alzheimer's disease. *Fed. Proc.* 45: 2982-2988.

Zeng, H.Y., Q. Chen, and B.L. Zhao. 2004. Genistein ameliorates β-amyloid peptide (25-35)-induced hippocampal neuronal apoptosis. *Free Rad. Biol. Med.* 36: 180-188.

Zhang, K., Y. Bao, P. Wu, R.T. Rosen, and C. Ho 1990. Anti-oxidative components of tanshen (*Salvia miltiorrhiza* Bung). *J. Agric. Food Chem.* 38: 1194-1197.

Zhang, Z., X. Wang, Q. Chen, L. Shu, J. Wang, and G. Shan. 2002 Clinical efficacy and safety of huperzine A in treatment of mild to moderate Alzheimer disease, a placebo-controlled, double-blind, randomised trial. *Chung-Hua I Hsueh Tsa Chih* 82: 941-944.

Zhao, J., N. Nakamura, M. Hattori, T. Kuboyama, C. Tohda, and K. Komatsu. 2002b. Withanolide derivatives from the roots of *Withania somnifera* and their neurite outgrowth activities. *Chem. Pharm. Bull.* 50: 760-765.

Zhou, J., H.Y. Zhang, and X.-C. Tang. 2001. Huperzine A attenuates cognitive deficits and hippocampal neuronal damage after transient ischemia in gerbils. *Neurosci. Lett.* 313: 137-140.

2

Antiviral Activity of Sulfated Polysaccharides of Marine Red Algae

Shoshana (Malis) Arad,* Ariel Ginzberg and Mahmoud Huleihel*

The Institute for Applied Biosciences, Ben-Gurion University of the Negev, Beer-Sheva 84105, Israel

INTRODUCTION

Algal Polysaccharides

Algae are economically important as sources of food, fertilizer, agar, potash, and iodine (Day *et al.* 1999). Production of valuable biochemicals from algae is on the rise. The successful commercial production of β-carotene as a health food from the green microalga *Dunaliella* and the production of phycoerythrin as a diagnostic marker (Phycofluor) from red microalgae are just two examples (Dupre *et al.* 1995, Jin and Melis 2003, Ben-Amotz 2004). The algae-based bio-industry is developing rapidly, the driving force being the trend of the market toward natural—and away from synthetic—products.

Among the algal biochemicals of commercial importance are the sulfated polysaccharides, which can be used as gelling agents, stabilizers, thickeners, and emulsifiers, mainly in food stuffs but also in paints, photographic films, pharmaceuticals, and in tertiary oil recovery (Chen *et al.* 1994, Tache *et al.* 2000, Berteau and Mulloy 2003, Tissot *et al.* 2003a).

One of the features that distinguishes algae from land plants is their characteristic cell wall composed of sulfated polysaccharides. Algal polysaccharides vary in structural and functional properties according the type of organism and the growth conditions (Percival and McDowell

*Corresponding author: S. Arad is to be contacted at: Tel: +972-8-6479069; Fax: +972-8-6479067; Email: arad@bgu.ac.il and M. Huleihel is to be contacted at: Tel: +972-8-6461999; Fax: +972-8-6472970; Email: mahmoudh@bgu.ac.il

1981). In nature, cell-wall polysaccharides fulfill a wide range of different functions, e.g., energy storage, contribution to the structural integrity and mechanical strength of the cell wall, control of osmotic pressure, ion reservoirs, buffer layer that protect the cells against drought, and protection against infective organisms, such as viruses, bacteria, and fungi (Arad 1988, 1999, Kloareg and Quatrano 1988, Lapsin and Pricl 1995, Arad and Richmond, 2004).

To overcome the constant pull of gravity, plant polysaccharides, such as lignin and cellulose, are relatively rigid, but polysaccharides of marine organisms, such as seaweeds, require a more flexible structure to enable them to endure the action of marine currents and waves (Kloareg and Quatrano 1988, Lapsin and Pricl 1995, Goldberg and Kendrick 2004). For marine algae whose natural habitat includes extreme shore environments, the polysaccharides must also be able to withstand drought, salinity, high temperatures, and high solar radiation (Shepherd and Beilby 1999, Tannin-Spitz *et al.* 2004).

The best known of the algal polysaccharides are agar-agar, carrageenan, and alginates, which are extracted from red and brown seaweeds, usually harvested from their natural habitats (Gellenbeck and Chapman 1983, Usov *et al.* 2001). Unfortunately, this source is no longer reliable, due to over-harvesting and ecological damage. Therefore, the search for new—and more stable—sources of sulfated polysaccharides is intensifying. One source of such polysaccharides has been found in bacteria (Baird *et al.* 1983), and another promising new source comprises certain species of red microalgae (Arad 1988, Arad and Weinstein, 2003). In red microalgae, the cells are encapsulated within a sulfated polysaccharide in the form of a gel. During growth in a liquid medium, the external part of the polysaccharide dissolves into the medium and accumulates there (Ramus and Groves 1972, Arad 1988, 1999).

Sulfated polysaccharides from algal sources have been found to possess a variety of biological activities, which may find application in the cosmetic, food and pharmaceutical industries (Tsiapall *et al.* 2001, Matsui *et al.* 2003, Arad and Weinstein 2003, Pardee *et al.* 2004, Talarico *et al.* 2004). Among these activities, this review will focus on antiviral activity of algal polysaccharides.

The antiviral activity of sulfated polysaccharides from marine algae against mumps and influenza B viruses was reported many decades ago (Gerber *et al.* 1958). Later, polysaccharides from extracts of marine red algae were found to be capable of blocking various enveloped viruses, such as *Herpes simplex* virus (HSV) (Burkholderm and Sharma 1969, Deig 1974, Ehresmann e*t al.* 1977, Richards *et al.* 1978, Witvrouw and DeClercq 1997). In addition, algal polysaccharides have been shown to have other biological activities, such as antitumor activity (Zhuang *et al.* 1995, Koyanagi *et al.* 2003), macrophage-stimulation activity (Yoshizawa *et al.*

1995), anti-inflammatory activity (Teixeira and Hellewell 1993, Tissot *et al.* 2003a,b, Matsui *et al.* 2003), antioxidant activity (Tsiapali *et al.* 2001, Tannin-Spitz *et al.* 2004), and anticoagulant activity (Church *et al.* 1989, Pereira *et al.* 1999). Sulfated polysaccharides may also be active in ion-exchange processes in algae: cations like potassium and calcium and heavy metal cations are thought to be selectively absorbed from seawater by algae (Malik *et al.* 2002, Kim *et al.* 2003).

Examples of Sulfated Polysaccharides

Algal polysaccharides are complex sulfated polymers, differing in their chemical compositions and amounts of sulfate, depending on the species (Geresh and Arad 1991, Arad 1999, Lahaye 2001, Berteau and Mulloy 2003, Sakai *et al.* 2003a,b, Arad and Richmond 2004). The complexity of these molecules arises from the large diversity of their monosaccharide constituents, the types of glycoside bonds, and the spherical characteristics of the molecules. As a result, the exact structure of some polysaccharides is not fully known. The biological activity of the polysaccharides is related to their molecular size and structure, linkages of the sugar residues, and the content and position of the sulfate. Representative examples of sulfated polysaccharides are discussed below:

Carrageenans

Carrageenans are sulfated polysaccharides extracted from red seaweeds. To date, 15 types of carrageenan are known, most of which are composed of 4-linked 3, 6-anhydro-α-D-galactose. The most common types: are κ-carrageenan, extracted from *Eucheuma cottonii*; λ-carrageenan, obtained from *Gigartina aciculaire* and *G. pistillata*, two species that grow in close proximity in the sea; and ι-carrageenan, produced by the Philippines seaweed *E. denticulatum* (Lahaye 2001). These polysaccharides are distinguished from one another by the position of the sulfate and by the presence (κ or ι) or absence (λ) of anhydrogalactose: κ-carrageenan is characterized by a repeating unit of β-D-galactopyranose-4-sulfated ($1 \rightarrow 4$)-3, 6 anhydro-α-D-galactopyranose ($1 \rightarrow 3$), and ι-carrageenan is composed of a repeating unit similar to κ-carrageenan with an additional sulfate in position 2 at anhydrogalactose. The repeating unit of λ-carrageenan is β-D-galactopyranose-2-sulfated ($1 \rightarrow 4$) α-D-galactopyranose-2, 6 disulfate ($1 \rightarrow 3$). The degree of sulfation of κ-, ι- and λ-carrageenans is 25%, 30% and 35%, respectively (Malfait *et al.* 1989, Lahaye 2001).

Dextran Sulfate

Dextran is a high-molecular-weight polysaccharide, consisting mainly of $\alpha(1 \rightarrow 6)$ linked D-glucose units, produced by several bacteria. Dextran sulfate was first synthesized in the early 1940s for research purposes (Witvrouw and De Clercq 1997). Later, dextran sulfates of different molecular weights were prepared by partially hydrolyzing dextran and then sulfating the different fractions obtained.

Porphyridium sp. Polysaccharide

The polysaccharides produced by the different species of red microalgae are complex heteropolymers with varying amounts of sulfate (Geresh and Arad 1991). The structure of the polysaccharide produced by the red microalga *Porphyridium* sp. has not been completely elucidated due to its complexity and to the paucity of specific degrading enzymes capable of splitting the complex molecule (Ucko *et al.* 1989, Simon *et al.* 1992, Arad *et al.* 1993). The polysaccharide of *Porphyridium* sp. is composed of 10 sugars. The main monosugars are xylose, glucose, and galactose, with molar ratios of 2.5: 1.0: 1.8, respectively (Geresh and Arad 1991). Glucuronic acid and half-sulfated ester groups are also present, conferring acidic properties on the molecule. The polysaccharide also contains noncovalently bound proteins, the main one being a 66-kDa glycoprotein moiety (Shrestha *et al.* 2004). The polysaccharide was found to contain a basic disaccharide building block (an aldobiuronic acid), 3-*O*-(α-D-glucopyranosyluronic acid)-l-galactopyranose (Geresh *et al.* 1991), which has sulfate groups attached to glucose and galactose in the 6 or 3 position (Lupescu *et al.* 1991). The level of sulfate in this polysaccharide is 7% (based on dry matter).

The molecular mass of *Porphyridium* sp. polysaccharide has been estimated to be $3–5 \times 10^6$ daltons (Geresh and Arad 1991). The polysaccharide is stable over wide range of pH values (2-9), temperatures (30-120 °C), and salinities.

Conformational studies together with X-ray diffraction of the polysaccharide demonstrate an oriented two-fold helical structure with a pitch of diameter 1.6 nm. From intrinsic viscosity measurements, it was estimated that the stiffness of *Porphyridium* sp. polysaccharide chains is in the same range as that of xanthan and DNA (Etheshola *et al.* 1996).

ANTIVIRAL ACTIVITY: FUNCTIONS AND MECHANISMS

General

At one time, it was thought that viruses were so closely associated with their host cells that it would be difficult, if not impossible, to destroy one and not the other. However, with the discovery of the first antiviral drugs in 1950 and their first clinical use in 1962, it became clear that antiviral therapy is indeed possible (Bauer 1985). Following the development of easy assays for detecting viral infections, such as the plaque assay (Rada *et al.* 1960), it became possible to examine large numbers of compounds for their activity against various viruses.

An antiviral compound is usually identified by first gaining an understanding of a virus-specific process (cellular attachment, genome uncoating, viral specific enzymes involved in viral replication processes) and then showing how the compound or the drug inhibits that process. Furthermore, an understanding of the relationship between the molecular structure of the drug and the molecules with which it interacts facilitates the production and examination of related compounds. In this way, it was found that synthetic nucleosides inhibit viral DNA or RNA polymerases (Wood and Geddes 1987, Vasalli and Van Zeijl 2003, Wilson and von Itzstein 2003).

The first reports of the antiviral activity of sulfated polysaccharides from marine algae appeared almost four decades ago. In 1958, algal polysaccharides were found to inhibit the replication of mumps and influenza B viruses (Gerber *et al.* 1958, Schaeffer and Krylov 2000). More recently, polysaccharide fractions from extracts of red algae were shown to exhibit antiviral activity against a variety of viruses, including herpes viruses (Richards *et al.* 1978, Neushul 1990, Ehresmann *et al.* 1977, Serkedjieva 2000, Duarte *et al.* 2001). Sulfated polysaccharides from red algae were also found to inhibit HIV reverse transcriptase (Nakashima *et al.* 1987a,b, Schaeffer and Krylov 2000, Haslin *et al.* 2001). At first, algal polysaccharides were thought to inhibit viral attachment to uninfected cells, rather than inhibiting the virus within the infected cell. Later studies of various sulfated polysaccharides, such as carrageenans, dextran sulfate, and heparin, suggested that these polysaccharides may inhibit DNA and RNA virus infections by acting both outside and inside the infected cells.

Activity against Enveloped DNA Viruses

A large number of studies have reported the antiviral activity of sulfated polysaccharides against various enveloped DNA viruses (Sugawara *et al.* 1989, Andrei *et al.* 1991, Mastromarino e*t al.* 1991, Witvrouw and De Clercq 1997, Huleihel *et al.* 2001, Zhu *et al.* 2003, Ponce *et al.* 2003, Yim *et al.* 2004). In particular, the antiviral activity of red algal polysaccharides has been examined against a number of enveloped DNA viruses, mainly viruses of the herpes family, i.e., *Herpes simplex* virus types 1 and 2 (HSV-1 and HSV-2) mutants resistant to acyclovir, *Varicella zoster* virus (VZV) and cytomegalovirus (CMV) (Gonzalez *et al.* 1987, Neushul 1990, Carlucci *et al.* 1997, 2002, Caceres *et al.* 2000, Duarte *et al.* 2001, Huleihel *et al.* 2001, 2002, Mazumder *et al.* 2002).

Herpes viruses (HSV-1, HSV-2 and VZV) are responsible for severe external infections of the body which makes them suitable for topical treatment with this high molecular weight polysaccharide. For instance, HSV-1 causes recurrent oral and skin lesions and herpes keratitis (an eye infection that may lead to blindness). HSV-2 is the etiological agent in serious genital infections (and is suspected to be one of the causes of cervical cancer) (Vasalli and Van Zeijl 2003, Waggoner-Fountain and Grossman 2004).

It seems that the high molecular weight of the sulfated algal polysaccharides contributes to their antiviral activity: When various samples of dextran sulfate with molecular weights ranging from 1000 to 500,000 were examined for their antiviral activity against different DNA and RNA viruses, it was found that the higher-molecular-weight samples were more effective (Witvrouw and De Clercq 1997).

Studies of various marine red algae (Ehresmann *et al.* 1977, Hatch *et al.* 1979, Richards *et al.* 1978) in a search for anti-herpetic substances culminated in four patents for use of these organisms as botanical agents for the treatment of viral infections. The pharmaceutical industry has, however, been slow in showing an interest in these and other discoveries because of technical difficulties involved in production and handling of these their high molecular-weight carbohydrates.

Polysaccharides obtained from the microalga *Cochlodinium polykrikoides* and from various species of red microalgae (*Porphyridium* sp., *P. aerugineum,* and *Rhodella reticulata*) have been tested *in vitro* for their antiviral activity against different members of the herpes virus family, and some showed significant antiviral activity (Hasui *et al.* 1995, Huleihel *et al.* 2001, 2002). *Porphyridium* sp. polysaccharide showed higher antiviral activity against HSV-1, HSV-2 and VZV than dextran sulfate 500,000 or λ-carrageenan, due to the higher cytotoxicity of the latter two compounds. Treatment of vero cells with 1 $\mu g\ mL^{-1}$ of *Porphyridium* sp. polysaccharide resulted in 50% inhibition of HSV-1 infection. When 1000 $\mu g\ mL^{-1}$ *Porphyridium* sp. polysaccharide was added to these cells no morphologic changes were observed (Huleihel *et al.* 2001). In the above-described experiments, the red microalgae polysaccharides were obtained by extracting and purifying the polysaccharides accumulating in the growth medium, in contrast to the method for obtaining red macroalgae polysaccharides, which are extracted from the thalli of the seaweeds. The former method of preparation seemed to contribute to the low cytotoxicity of the red microalgae polysaccharides relative to other algal polysaccharides, probably as a result of the absence of contaminants from the cells.

The antiviral activity of red algal polysaccharides has also been examined and evaluated *in vivo* in experimental animals. As early as 1979, Hatch *et al.* (1979) examined the ability of polysaccharides extracted from red seaweeds from the coastal areas of northern California to prevent and cure herpes-induced skin lesions in adult mice. The results obtained from this study indicated that these polysaccharides could prevent, but not cure, lesions. The main problems encountered with this animal model related to the nature of the infection and the manner of treatment. The virus in the skin lesions was able to spread to uninfected cells without being exposed to the extracellular environment. Therefore, the high-molecular weight algal polysaccharides, unable to penetrate cell membranes and enter the cells, may have been unable to block the migration of the virus to uninfected cells. In addition, the purification of the polysaccharides was probably insufficient and the method of treatment unsuitable. Neyts and De Clercq (1995) also found that sulfated polyanions, such as dextran sulfate, were able to suppress infection of hairless mice with HSV-2 only when administered at the time of inoculation.

Recently, Huleihel *et al.* (2002) used a different animal model for examining the antiviral activity of red microalgal polysaccharides *in vivo*. Newborn rats were infected with HSV-1, either by scratching a small area of skin with the needle of a syringe containing HSV-1 suspension or by interperitoneal inoculation with an HSV-1 suspension. In a different set of experiments, rabbits were infected by scratching the cornea of the eye with a syringe containing an HSV-1 suspension. In both cases, the infected areas were treated with *Porphyridium* sp. polysaccharide immediately or at various times after inoculation with the virus. The best results (complete prevention of the development of HSV-I infection) were obtained when the animals were treated with the polysaccharide immediately after inoculation. In the rat experiment, a 90% decrease in disease symptoms and a 95% decrease in animal mortability was observed, when the polysaccharide was supplied at a concentration of 100 $\mu g\ mL^{-1}$ postinfection with HSV-1. A higher dosage of polysaccharide (1000 $\mu g\ mL^{-1}$) almost completely diminished symptoms and prevented the death of the animals.

In the rabbit experiment, treatment of the infected eyes with *Porphyridium* sp. polysaccharide three to four days post-infection, when the first symptoms of viral infection in the eyes—keratitis and secretions—appeared, facilitated a complete recovery from the eye infection and prevented death of the infected animals in most cases. The protection provided by Zovirax (acyclovir), as a positive control, was not superior to that conferred by *Porphyridium* sp. polysaccharide (Table 2.1).

The rabbit-eye model for evaluating the antiviral activity of polysaccharides against HSV infection seems to be better than the skin infection model in mice and rats, possibly due to the more precise and more convenient treatment method of administering the polysaccharide in the form of eye drops.

Table 2.1 Effect of *Porphyridium* sp. polysaccharide (*P.* sp.) on the development of HSV-1 infection in rabbit eyes.

Animal group	*100 µg/mL* P. *sp. polysaccharide*					*1,000 µg/mL* P. *sp. polysaccharide*				
	Appearance of sympt.		*Death*		*Infect.*	*Appearance of sympt.*		*Death*		*Infect.*
	Days p.i.	*%*	*Days p.i.*	*%*	*virus, %*	*Days p.i.*	*%*	*Days p.i.*	*%*	*virus, %*
Non-infected, *P.* sp. untreated	—	0	—	0	nt	—	0	—	0	nt
Noninfected, *P.* sp. or Zovirax treated	—	0	—	0	nt	—	0	—	0	nt
Infected, *P.* sp. untreated	4-5	100	10-12	100	100	nt	nt	nt	nt	nt
Infected, *P.* sp. treated	9*	35	—	0	35	14**	13	—	0	13
Infected, Zovirax treated	7-8	40	—	0	40	nt	nt	nt	nt	nt

The number of animals presenting symptoms (sympt.) and subsequently dying was expressed as the percentage of the total number of tested animals. A total of 10 rabbits were used for each treatment. Infective virus [collected on swabs from infected eyes 3 days postinfection (p.i.)] was tested in cell culture. The number of animals containing infected virus was expressed as the percentage of the total number of animals tested. nt = not tested; *Light keratitis with secretions; **Reversible light keratitis

In addition, preliminary studies were performed on the antiviral activity of *Porphyridium* sp. polysaccharide on the development of HSV-I skin lesions in human subjects. The results showed that treatment of herpetic skin lesions by topical application of the polysaccharide twice a day caused a significant reduction in the duration of symptoms and the pain that accompanies such herpetic lesions (Huleihel and Arad, unpublished results).

Mechanism of Action against Enveloped DNA Viruses

A large number of previous reports support the idea that sulfated algal polysaccharides exert their inhibitory action against enveloped viruses, such as herpes viruses, at a very early stage in the viral infection cycle, i.e., virus adsorption onto and/or penetration into the host cell (Sawitzky *et al.* 1990, Neyts *et al.* 1992, Witvrouw and De Clercq 1997, Huleihel *et al.* 2001). Such inhibition could be a result of an interaction between the polysaccharide with all or some of the membrane viral glycoproteins that have defined roles in viral entry into the host cells. For instance, in the case of HSV-1 and HSV-2, the initial interaction of these viruses with the cells is usually by binding of the virion glycoprotein gC or gB to cell surface glycosaminoglycans such as heparin sulfate. The binding affinity of heparin sulfate to the glycoprotein (gC) is structure dependent (Feyzi *et al.* 1997), whereas penetration of the virus requires fusion of the virion envelope with the cell membrane and depends on the action of the viral glycoproteins gB, gD, gH, and gL. Deletion of the genes for any one of these four glycoproteins is lethal and results in production of virion that can bind to (provided gC is present)—but cannot penetrate—the cells (Spear *et al.* 2000, Liu and Thorp 2002). When gC is absent, the heparin sulfate binds to HSV by means of gB (Herold *et al.* 1994).

Although, at present, the receptor-binding site of gC has not been completely elucidated, it is known that there is a cluster of basic amino acids in the vicinity of a very hydrophilic region near the N-terminus of gC, which could be considered as the heparan sulfate-binding domain (Tal-Singer *et al.* 1995, Carlucci *et al.* 1999). This region in the N-terminal portion, shared by other members of the gC family in α-herpes viruses, is perhaps the site with which the algal polysaccharides interact.

The binding of gD to HSV to attain fusion is mediated through several receptors, i.e., the herpes virus entry mediated (HVEM) receptor, belonging to the tumor necrosis factor (TNF) receptor family, and nectin-1 and nectin-2, belonging to the immunoglobulin superfamily (Montgomery *et al.* 1996, Spear *et al.* 2000).

In some cases, such as for xylomannan sulfate F_6 (extracted from the red seaweed *Nothogenia fastigiata*), it was proven that the polysaccharide inhibited HSV-1 and human CMV attachment to the host cell but not their penetration into the cells (Witvrouw and De Clercq 1997). On the other hand, other studies have shown that sulfated polysaccharides (like carrageenan) and other sulfated polyanions prevented viral protein synthesis (Gonzalez *et al.* 1987, Biesert *et al.* 1988). The inhibition of viral protein synthesis could be an indirect effect, resulting from the induced change in cell membrane properties caused by such agents (Biesert *et al.* 1988), if the transmembrane and intracellular signaling processes involved in regulating virus expression are affected.

Sulfated polysaccharides from red microalgae (such as *Porphyridium* sp.) were shown to be capable of blocking various steps in the life cycle of several herpes viruses, such as HSV and VZV (Huleihel and Arad 2001). The latter study showed a direct inhibitory effect of *Porphyridium* sp. polysaccharide on HSV-1, which resulted in significant inhibition of viral infection (Fig. 2.1). It seems likely that this effect is caused by a strong interaction between HSV-1 particles and the *Porphyridium* sp. polysaccharide. The binding was maintained after several dilutions, indicating an irreversible or very strong binding between the virus particles and the polysaccharide. These results may also indicate a weak or reversible interaction between this polysaccharide and the cell membrane (Fig. 1). Therefore, it seems that most of the inhibitory effect of *Porphyridium* sp. polysaccharide on HSV adsorption to host cells is due to its interaction with the virus.

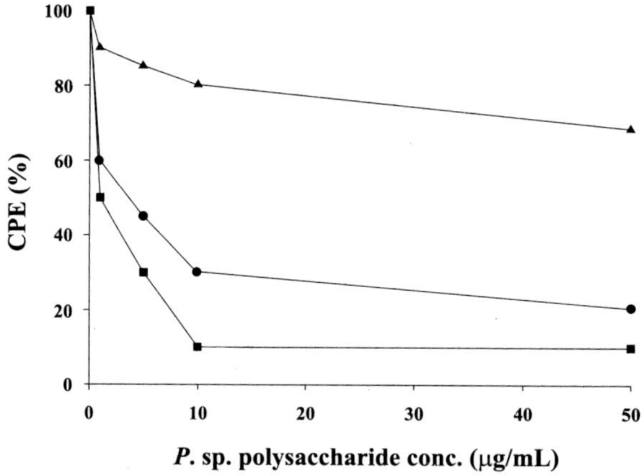

Fig. 2.1 Effect of *Porphyridium* sp. polysaccharide on Vero cell infection with HSV-1. (■) Vero cells infected with 1 multiplicity of infection (moi) of HSV-1 in the presence of various concentrations of *Porphyridium* sp. polysaccharide (■) Cells infected with HSV-1 pretreated with the polysaccharide: Virus particles (5 moi) were preincubated with various concentrations of the polysaccharide at 37°C for 30 min, and then diluted 10^3 times with fresh medium containing 2% of serum before infection. (▲) Cells pretreated with the polysaccharide: Vero cell monolayers were preincubated with various concentrations of *Porphyridium* sp. polysaccharide at 37°C for 2 h, washed × 3 with PBS and infected with 1 moi of HSV-1. Plaque forming units (PFU) were evaluated by the standard plaque assay.

A strong interaction between the negatively charged sulfated polysaccharide and the positively charged enveloped viruses is the most likely scenario. This interaction could depend on the amino acid sequences of the viral envelope glycoproteins that are involved in virus-cell interaction, as mentioned above (Hosoya *et al.* 1991, Carlucci *et al.* 1999). The viruses sensitive to the sulfated polysaccharides and to other polysulfates have a tripeptide segment (Phe-Leu-Gly). This tripeptide segment may be involved, either directly or indirectly, in the inhibitory effects of sulfated polysaccharides (Hosoya *et al.* 1991).

Many previous studies have proven the importance of the sulfate groups in the antiviral activity of algal polysaccharides - the higher the sulfation level, the greater the antiviral activity (Schaeffer and Krylov 2000). For example, the antiherpetic activity of natural carrageenans was directly correlated to the amount of α-D-galactose 2,6-disulfate residues (Carlucci *et al.* 1997). Results from our laboratory showed a significant reduction in the antiviral activity of *Porphyridium* sp. polysaccharide against HSV-1 when the content of sulfate groups was reduced (by depleting sulfate from the growth medium), whereas when *Porphyridium* sp. polysaccharide was chemically over-sulfated, its antiviral activity was significantly increased (Huleihel *et al.* unpublished results). Moreover, among the three species of red microalgae studied, the highest antiviral effect was exhibited by the polysaccharide that had the highest sulfate content, i.e., that of *Porphyridium* sp.

It was also shown that production of new infective viruses during one cycle of replication was strongly inhibited by *Porphyridium* sp. polysaccharide, when the polysaccharide was added immediately post-infection with HSV-1 (Fig. 2.2). It seems therefore that *Porphyridium* sp. polysaccharide is involved in preventing a late step during the virus replication cycle. This premise could also be supported by the results of a study that showed significant inhibition of HSV and VZV infections when *Porphyridium* sp. polysaccharide was added to preinfected cell cultures (Huleihel *et al.* 2001). These findings may suggest a pleiotropic mode of action for sulfated algal polysaccharides during the infection cycle of herpes viruses.

Activity against Enveloped RNA Viruses

Various sulfated polysaccharides from red algae were found to be active against different RNA viruses, such as Sindbis and Semliki forest viruses (togaviruses); vesicular stomatitis virus (VSV) (rhabdovirus); influenza A virus (orthomyxovirus); respiratory syncytial virus (RSV) (paramyxovirus); encephalomyocarditis virus (EMCV) (picornavirus); and human immunodeficiency virus (HIV), human T-cell leukemia virus type 1 (HTLV-1), murine leukemia and sarcoma viruses (MuLV, MuSV)

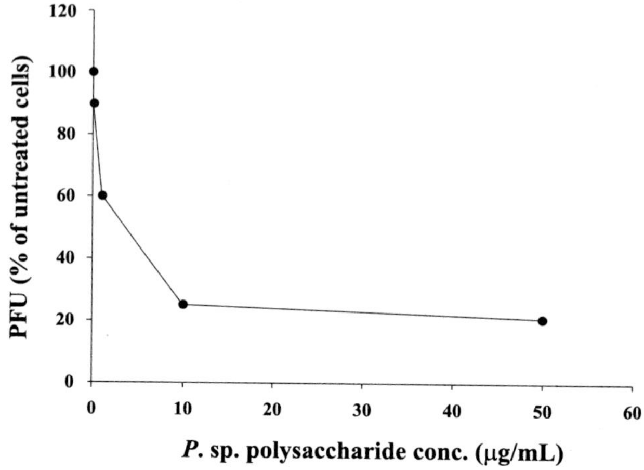

Fig 2.2 Effect of *Porphyridium* sp. polysaccharide on the production of infectious virus in a single replication cycle. Vero cells were infected with 1 moi of HSV-1, and at the end of the infection period they were treated with various concentrations of *Porphyridium* sp. polysaccharide. Twenty-four hours post infection, cells were collected and broken, and infectious virus particles were evaluated by the standard plaque assay on Vero cells. The results are presented as a percent of infected cells not treated with polysaccharide.

(retroviruses) (Witvrouw and De Clercq 1997, Huleihel and Arad 2001, Romanos *et al.* 2002). Many studies have reported the inhibitory effect of algal polysaccharides against HIV replication in cell culture (Nakashima *et al.* 1987a,b, Beress *et al.* 1993, Witvrouw *et al.* 1994, Hasui *et al.* 1995, Haslin *et al.* 2001, Wang and Ng 2001, Tziveleka *et al.* 2003). Most of the research on the anti-HIV activity of marine algae has focused on red and brown macroalgae (Schaeffer and Krylov 2000). Algal polysaccharides have different inhibitory activities against various HIV strains, which may suggest different mechanisms of action (Witvrouw and De Clercq 1997). It was found also that the antiviral activity of algal polysaccharides in general and specifically their activity against HIV in particular, increases with increasing the molecular weight and degree of sulfation (Witvrouw and De Clercq 1997).

Various polysaccharides extracted from marine red algae such as the polysaccharide extracted from the red alga *Schizymenia pacifica*, carrageenans and galactan sulfate extracted from the red seaweed *Agardhiella*, and xylomannan extracted from the red seaweed *Nothogenia fastigiata* demonstrated potent inhibitory activity against HIV replication, as confirmed by assays for inhibition of plaque formation by HIV in MT-4 cells (Baba *et al.* 1990, Damonte *et al.* 1994, Schaeffer and Krylov 2000, Tziveleka *et al.* 2003). Most of these polysaccharides, xylomannan sulfate excepted, were also able to block HIV-induced giant cell (syncytium) formation (Witvrouw and De Clercq 1997). Syncytium formation is induced by the interaction between the gp120 glycoprotein, expressed on the surface of cells infected with HIV, and the CD4 receptor of noninfected $CD4^+$ cells. This giant cell formation may play an important role in the depletion of T4 lymphocytes in AIDS patients (Haseltine 1988). In addition, some of the polysaccharides, such as the polysaccharide extracted from *S. pacifica* and carrageenans, were shown to inhibit the activity of HIV reverse transcriptase (RT) *in vitro* (Nakashima *et al.* 1987a,b, Schaeffer and Krylov 2000, Haslin *et al.* 2001).

Other polysaccharides extracted from marine brown algae, such as *Fucus vesiculous* (Beress *et al.* 1993, Moen and Clark 1993), inhibited both the RT activity (*in vitro*) and the syncytium formation ability of HIV at non-cytotoxic concentrations, although both activities required different doses of the polysaccharides.

Sulfated polysaccharides were also shown to inhibit infection of both CD4+ and CD4– cell lines by HIV, HTLV-1 and the simian retrovirus Mason-Pfizer monkey virus (MPMV), which attaches to receptors other than CD4 (McClure *et al.* 1992, Schaeffer and Krylov 2000). A combination of sulfated polysaccharides, like fucoidan and carrageenan, seems to exhibit better (additive) activity against HIV, a point to be considered for future clinical trials (Bagasra *et al.* 1991). In addition, the anti-HIV proliferation activity of algal polysaccharides was tested in combination with 3′-azido-3′-deoxythymidine (AZT) (Mizumoto *et al.* 1988, Sugawara *et al.* 1989). Fucoidan was found to have synergistic activity with AZT, and it was suggested that this synergism might permit the administration of much lower therapeutic doses of AZT.

In summary, it can be concluded that most of the sulfated polysaccharides that were found to inhibit HIV replication also inhibited syncytium formation, although the structural requirements (molecular weight and sulfation levels) seem to be more stringent for inhibition of syncytium formation than for inhibition of HIV replication (Baba *et al.* 1990, Montefiori e*t al.* 1990, Witvrouw and De Clercq 1997, Haslin *et al.* 2001). In all cases, lower levels of sulfate were required to block HIV replication than the level needed for blocking syncytium formation.

Sulfated polysaccharides from red microalga were also tested for their antiviral activity against retroviruses. *Porphyridium* sp. polysaccharide, was found to significantly inhibit the production of MuLV and the malignant cell transformation of mouse fibroblasts in cell culture by MuSV (Huleihel and Arad 2001). The most effective inhibitory effect of this polysaccharide against both cell transformation and virus production was obtained when the polysaccharide was added to the culture before or at the time of infection. Although addition of the polysaccharide post-infection significantly reduced the number of transformed cells, its effect was less marked than that obtained when the polysaccharide was added before or at the time of infection.

Mechanism of action Against Enveloped RNA Viruses

A large number of reports have shown that polysulfates, including sulfated algal polysaccharides, inhibit HIV replication by preventing the binding of the virions to CD4+ cells (Baba *et al.* 1990, Nakashima *et al.* 1989, Schols *et al.* 1989, Witvrouw *et al.* 1991, Schaffer and Krylor 2000) and consequently the formation of giant cells (syncytium) (Mitsuya *et al.* 1988, Baba *et al.* 1990, Bress *et al.* 1993, Hayashi e*t al.* 1996, Schaffer and Krylor 2000, Haslin *et al.* 2001).

Syncytium formation and virus replication both depend on the interaction between gp120 and CD4 (Baba *et al.* 1990). Syncytium formation, however, depends on gp120 present in cell membrane, while virus replication depends on gp120 located in the viral envelope. Inhibition of giant cell formation was found to be more difficult and required higher concentrations of sulfated polysaccharides than those required for inhibition of virus replication (Baba *et al.* 1990, Schaeffer and Krylov 2000). These results may indicate that it is more difficult to block the gp120 molecules located on the cell surface than on the virus particles.

Some sulfated polysaccharides, such as heparin, dextran sulfate, and glucan sulfate, were found to interact with the positively charged amino acids located in loop V3 of HIV-gp120 (Cardin *et al.* 1989, Callahan *et al.* 1991, Jackson *et al.* 1991, Mbemba *et al.* 1992, Patel *et al.* 1993, Jagodzinski *et al.* 1996). This interaction could prevent HIV from binding and subsequently fusing to the CD4 membrane receptor. In addition, sulfated algal polysaccharides such as carrageenans have been found to inhibit RT activity of retroviruses *in vitro* (Nakashima *et al.* 1987a,b, Moelling *et al.* 1989, Beress *et al.* 1993, Wang and Ng 2001).

Other studies (Pearce-Pratt and Philips 1996) examined and investigated the mechanism of the antiviral activity of sulfated polysaccharides on sexual transmission of HIV-1. Sexual transmission of HIV-1 is cell-mediated process involving initial adhesion of HIV-1-infected mononuclear cells to epithelial cells. Some polysaccharides, such as carrageenan, were found to block both processes—adhesion and viral infection—although at significantly different concentrations. Others, like fucoidan and pentosan polysulfate, had no effect on adhesion, while blocking virus infection. It was thus concluded that sulfated polysaccharides can block HIV-1 infection by mechanisms other than adhesion (Pearce-Pratt and Phillips 1996).

Red microalgal polysaccharides significantly inhibited the production of MuLV and cell transformation by MuSV (Huleihel and Arad 2001). The finding that inhibition of cell transformation by MuSV was reversible after removal of the polysaccharide suggested that red microalgal polysaccharides inhibited a late step after provirus integration into the host genome. It is thus likely that *Porphyridium* sp. polysaccharide affects early steps in virus infection, such as virus adsorption into the host cells, in addition to its effect on a late step after provirus integration. Furthermore, it was shown that there is no direct effect of microalgal polysaccharide on MuSV particles (Huleihel and Arad 2001).

Additional Bioactivities of Algal Sulfated Polysaccharides

Various sulfated polysaccharides from marine algae, such as fucans and carrageenans, were found to have a wide range of activities in biological systems other than their antiviral activities (Itoh *et al.* 1993, Farias *et al.* 2000, Berteau and Mulloy 2003, Tissot *et al.* 2003a,b), i.e., anticoagulant and antithrombotic activity; anti-inflammatory activity; and antitumor activity.

Anticoagulant and Antithrombotic Activity

Anticoagulant and antithrombotic activities are among the most widely studied properties of sulfated polysaccharides. Glycosaminoglycan heparin, an anticoagulant, is an important therapeutic agent for prophylaxis and treatment of thrombosis (Farias *et al.* 2000, Trento *et al.* 2001). Sulfated polysaccharides extracted from marine brown (Church *et al.* 1989, Colliec *et al.* 1991, Nishino *et al.* 1991, Mauray *et al.* 1995, Pereira *et al.* 1999) and red algae (Carlucci et al. 1997, Caceres *et al.* 2000) have also been described as anticoagulants. Unlike heparin, the structural components required for anticoagulant activity of these algal sulfated polysaccharides have not been characterized due to their heterogeneous chemical structures (Pereira *et al.* 1999, Farias *et al.* 2000). Farias *et al.* (2000) suggested that the 2,3-di-*O*-sulfated galactose residues have an "amplifying effect" on the anticoagulant activity of sulfated galactans from red algae. In the case of heparin, a pentasaccharide sequence—with a specific sugar composition and sulfation pattern—is required for its interaction with antithrombin (Lindahl *et al.* 1983, Pereira *et al.* 2002). The anticoagulant effects of sulfated polysaccharides from marine algae depend on an exact pattern of sulfate substitution. A small alteration in the sulfation pattern, from 4-*O*-sulfated to 6-*O*-sulfated galactosamine, leads to almost complete loss of anticoagulant activity in spite of an overall high level of sulfation (Mulloy *et al.* 2000).

Anti-inflammatory Activity

Sulfated polysaccharides, in general, such as heparins and dextran sulfate, and those extracted from algae, such a fucoidan, carrageenan, and red microalgal polysaccharides, were found to have potent anti-inflammatory activity (Teixeira and Hellewell 1993, Angstwurm *et al.* 1995, Dell'Antonio *et al.* 2002, Matsui *et al.* 2003, Tissot *et al.* 2003a,b).

Fucoidan, a sulfated polysaccharide from brown seaweed, was found to inhibit activation of the complement system (Tissot *et al.* 2003), an important part of the immune system that plays an essential part in the host defense against pathogens. Fucoidan was found to be superior to heparin as a complement inhibitor, in that lower doses of fucoidan were required (Maillet *et al.* 1988), and fucoidan exerted its inhibitory action on the earliest steps of the classical pathway of complement activation, before the amplification steps of the cascade and the generation of the pro-inflammatory peptides whose effects are so deleterious (Farkas *et al.* 1998, Winkler 2001). Inhibition of the complement system by fucoidan may explain its anti-inflammatory activity. It is therefore believed that such polysaccharides can be used as therapeutic agents in the management of many inflammatory-tissue injury diseases.

Antitumor Activity

Biological response modifiers that modulate the host biological responses against tumors are under ongoing development for application in cancer therapy. Immunomodulating agents may be capable of potentiating the host immune response without direct cytotoxicity on cancer cells (Franz 1989, Oh *et al.* 1992, Wallace and Morahan 1994).

With a view to developing new antitumor compounds with low toxic potential, various polysaccharides from different biological origins—yeasts, algae, bacteria, higher plants, and fungi—have been investigated for their antitumor and immunomodulating activities (Kraus and Franz 1991, Mizuno *et al.* 1992, Chai *et al.* 1995, Chen *et al.* 1997, Carbonnelle *et al.* 1999). Sulfated polysaccharides from different marine algae have indeed been found to have potent antitumor activity *in vitro* and *in vivo* (animal models) (Mishima *et al.* 1998, Itoh *et al.* 1993, Ren *et al.* 1995, Zhuang *et al.* 1995, Riou *et al.* 1996).

Fucoidan markedly inhibited the growth of Ehrlich ascites carcinoma at a dose of 20 mg/kg per day, with no sign of toxicity in mice. The results of this study demonstrated that fucoidan is an activator of the reticuloendothelial system. Fucoidan enhanced the phagocytosis and chemiluminescence of macrophages (Itoh *et al.* 1993). In another study, it was reported that fucoidan exhibits impressive antitumor and antimetastatic activities in xenograft models (Koyanagi *et al.* 2003). The results of the latter study indicate that the antitumor action of this polysaccharide is due, at least in part, to its antiangiogenic potency and that increasing the number of sulfate groups in the polysaccharide molecule increases the effectiveness of its antiangiogenic and antitumor activities. Angiogenesis, is an important biological event that occurs in a variety of physiological or pathological processes such as tumor progression (Folkman and Shing 1992). Antitumor activity has also been proposed for other algal sulfated polysaccharides, such as funoran, due to its effect on the augmentation of T-helper, T-cytotoxic and NK cells (Ren *et al.* 1995).

SUMMARY

The cells of red algae are encapsulated within a cell wall composed of sulfated polysaccharides that are unique to algae but absent in higher plants. The structural properties of the polysaccharides of red seaweeds and red microalgae vary from species to species in terms of their monosugar compositions, glycoside linkages, glycoproteins, and functional groups. The sulfate content of the polysaccharides also varies among the different species and is influenced by the growth conditions. The sulfated polysaccharides of red algae are characterized by relatively high molecular weights, high viscosities, and acidic properties.

A number of studies have shown that these sulfated polysaccharides have antiviral activities against DNA viruses, such as HSV-1, HSV-2, VZV and CMV, and against RNA viruses, such as VSV, influenza A virus, RSV, EMCV, HIV, HTLV-1, MuLV, and MuSV.

Due to the relatively high molecular weights of the sulfated polysaccharides, their therapeutic applications have been directed mainly to topical applications. Testing of the antiviral activity of the polysaccharides has been conducted in a variety of animal models, and human trials have been undertaken. Treatment of herpetic skin lesions in human volunteers by topical application of the polysaccharide of the red microalga *Porphyridium* sp. caused reduction in the duration of symptoms and the pain that accompanies such herpetic lesions and prevention its further development *In vitro*. The most effective inhibitory effect of the polysaccharide against cell transformation and virus production was obtained when the polysaccharide was added to the culture before or at the time of infection. It seems that the mechanism of action against DNA viruses is different from that of RNA viruses. The anti-RNA virus activity of the polysaccharide was reversed after removal of the polysaccharide supplement from the culture medium. It was thus suggested that for RNA viruses the red microalgal polysaccharide inhibited a late step after provirus integration into the host genome. In contrast, for DNA viruses, the sulfated polysaccharide exerted its effect mainly in the early stages of infection. Such inhibition could be the result of the interaction between the negatively charged polysaccharide with positively charged viral glycoproteins.

In addition to their antiviral activity, the sulfated polysaccharides of marine red algae also possess anti-inflammatory, anticoagulant, antithrombotic, antineoplastic, and anti-cell-proliferative activity. However, for these sulfated polysaccharides to be developed as drugs, further R&D oriented to pharmaceutical applications is necessary.

CONCLUSIONS

Despite the availability of effective drugs against some viruses, such as acyclovir against herpes viruses, the search for new antiviral drugs is still extremely important, due to:

- The evolution of mutant strains resistant to available drugs
- The potential side effects or toxicity of the drugs over long-term use
- The high cost of antiviral drugs
- The lack of drugs effective against certain viruses.

Sulfated polysaccharides from marine algae have enormous potential as possible antiviral drugs, due to:

- The broad spectrum of potent antiviral agents
- Pleiotropic mode of action, which makes development of resistant mutants unlikely
- Lack of side effects and toxicity over long-term use
- Quick and easy preparation of the therapeutic formulation
- Relatively low production costs

In summary, the application of sulfated polysaccharides appears to be a very attractive option for viral therapy, since treatment with drugs that have a pleiotropic mode of action, such as sulfated polysaccharides, is less likely to result in the development of resistant mutants than compounds with a single target during the viral life cycle. Sulfated polysaccharides have also other important bioactivities, such as anti-inflamatory, anti-irritating, immune system enhancement, anti-tumor activity, etc., which could make these polysaccharides much more attractive as drugs for long-term use.

REFERENCES

Andrei, G., R. Snoeck, P. Goubau, J. Desmyter, and E. DeClercq. 1991. Comparative activity of selected antiviral compounds against clinical isolates of human cytomegalovirus. *Eur. J. Microbiol. Infect. Dis.* 19: 1026-1033.

Angstwurm, K., J.R. Weber, A. Segert, W. Burger, M. Weih, D. Freyer, K.M. Einhaupl, and U. Dirnagl. 1995. Fucoidin, a polysaccharide inhibiting leukocyte rolling, attenuates inflammatory responses in experimental pneumococcal meningitis in rats. *Neurosci. Lett.* 191: 1-4.

Arad (Malis), S. 1988. Production of sulfated polysaccharides from red unicellular algae. Pages 76-87 in *Algal Biotechnology*, T. Stadler, J. Million, M.C. Verdus, Y. Karamanos, H. Morvan, and D. Christiaen, eds., Elsevier Applied Science, London.

Arad (Malis), S., G. Keristovsky, B. Simon, Z. Barak, and S. Geresh. 1993. Biodegradation of the sulphated polysaccharide of *Porphyridium* sp. by soil bacteria. *Phytochemistry* 32: 287-290.

Arad (Malis), S. 1999. Polysaccharides of red microalgae. Pages 282-291 in *Chemicals from Microalgae*, Z. Cohen, ed. Taylor & Francis Ltd., London.

Arad (Malis), S., and J. Weinstein. 2003. Novel lubricants from red microalgae: Interplay between genes and products. *Biomedic (Israel)* 1: 32-37.

Arad (Malis), S., and A. Richmond. 2004. Industrial production of microalgal cell-mass and secondary products – species of high potential – *Porphyridium* sp. Pages 289-297 in *Handbook of Microalgal Culture, Biotechnology and Applied Phycology*, A. Richmond, ed. Blackwell Publishing, Oxford, England.

Baba, M., D. Schols, R. Pauwels, H. Nakashima, and E. Declercq. 1990. Sulfated polysaccharides as potent inhibitors of HIV-induced syncytium formation: a new strategy towards AIDS chemotherapy. *J. Acq. Immune Def. Synd.* 3: 493-499.

Bagasra, O., P. Whittle, B. Heins, and R. Pomerantz. 1991. Anti-human immunodeficiency virus type 1 activity of sulfated monosaccharides: comparison with sulfated polysaccharides and other polyions. *J. Infect. Dis.* 164: 1082-1090.

Baird, J.F., P.A. Sandord, and I.W. Cottrell. 1983. Industrial applications of some new microbial polysaccharides. *Bio/Technology* 1: 778-83.

Bauer, D.J. 1985. A history of the discovery and clinical application of antiviral drugs. *Br. Med. Bull.* 41: 309-314.

Ben-Amotz, A. 2004. Industrial production of microalgal cell-mass and secondary products – major industrial species: *Dunaliella*. Pages 273-280 in *Handbook of Microalgal Culture, Biotechnology and Applied Phycology*, A. Richmond, ed., Blackwell Publishing, Oxford, England.

Beress, A., O. Wasserman, S. Tahhan, T. Bruhn, L. Beress, E.N. Kraiselburd, L. Gonzalez, G.E. de Motta, and P.I. Chavez. 1993. A new procedure for the isolation of an anti-HIV compounds (polysaccharides and polyphenols) from the marine alga *Fucus vesiculosus*. *J. Nat. Prod.* 56: 478-488.

Berteau, O., and B. Mulloy. 2003. Sulfated fucans, fresh perspectives: structure, functions and biological properties of sulfated fucans and an overview of enzymes active toward this class of polysaccharide. *Glycobiology* 13: 29-40.

Biesert, L., H. Suhartono, I. Winkler, C. Meichsner, M. Helsberg, G. Hewlett, V. Klimetzek, K. Molling, H.D. Schlumberger, E. Schrinner, H.D. Brede, and H. Rübsamen-Waigmann. 1988. Inhibition of HIV and virus replication by polysulfated polyxylan – HOE/BAU 946, a new antiviral compound. *AIDS* 2: 449-457.

Burkholderm, P.R., and G.M. Sharma. 1969. Antimicrobial agents from the sea. *Lloydia* 32: 466-483.

Caceres, P.J., M.J. Carlucci, E.B. Damonte, B. Matsuhiro, and E.A. Zuniga. 2000. Carrageenans from Chilean samples of *Stenogramme interrupta* (Phyllophoraceae): Structural analysis and biological activity. *Phytochemistry* 53: 81-86.

Callahan, N., M. Mallinson, and M.A. Nocross. 1991. Dextran sulfate blocks antibody binding to the principal neutralizing domain of human immunodeficiency virus type 1 without interfering with gp120-CD4 interactions. *J. Virol*. 65: 1543-1550.

Carbonnelle, D., P. Pondaven, M. Morancais, G. Masse, S. Bosch, C. Jacquot, G. Briand, J.M. Robert, and C. Roussakis. 1999. Antitumor and antiproliferative effects of an aqueous extract from the marine diatom *Hasler ostrearia* (Simonsen) against solid tumors: Lung carcinoma (NSCLC-N6), kidney carcinoma (E39) and melanoma (M96) cell lines. *Anticancer Res.* 19: 621-624.

Huleihel, M., V. Ishanu, J. Tal, and S. Arad (Malis) 2002. Activity of *Porphyridium* sp. polysaccharide against *Herpes simplex* viruses *in vitro* and *in vivo*. *J. Biochem. Biophys. Meth.* 50: 189-200.

Huleihel, M., Z. Vladimir, J. Tal, and S. Arad (Malis) 2001. Antiviral effect of red microalgal polysaccharides on *Herpes simplex* and *Varicella* viruses. *J. Appl. Phycol.* 13: 127-134.

Itoh, H., H. Noda, H. Amano, C. Zhuang, T. Mizuno, and H. Ito. 1993. Antitumor activity and immunological properties of marine algal polysaccharides, especially fucoidan, prepared from *Sargassum thunbergii* of Phaeophyceae. *Anticancer Res.* 13: 2045-2052.

Jackson, R.L., S.J. Busch, and A.D. Cardin. 1991. Glycosaminoglycans: molecular properties, protein interactions, and role in physiological processes. *Physiol. Rev.* 7: 481-539.

Jagodzinski, P.P., J. Wustner, D.K Mieciak, T.J. Wasik, A. Fertala, A.L. Sieron, M. Takahashi, T. Tsuji, T. Mimura, M.S. Fung, M.K. Gorny, M. Kloczewiak M., Y. Kaneko, and D. Kozbor. 1996. Role of the V2, V3 and CD4 binding domains of GP120 in curdlan sulfate neutralization sensitivity of HIV-1 during infection of T lymphocytes. *Virology* 226: 217-227.

Jin, E.S., and A. Melis. 2003. Microalgal biotechnology: Carotenoid production by the green algae *Dunaliella salina*. *Biotechnol. Bioprocess Eng.* 8: 331-337.

Kim, S.K., C.G. Lee, and H.S. Yun. 2003. Heavy metal adsorption characteristics of extracellular polysaccharide produced by *Zoogloea ramigera* grown on various carbon sources. *J. Microbiol. Biotechnol.* 13: 745-750.

Kloareg, B., and R.S. Quatrano. 1988. Structure of the cell wall of marine algae and eco-physiological functions of the matrix polysaccharides. *Oceanog. Mar. Biol. Annu. Rev.* 26: 259-315.

Koyanagi, S., N. Tanigawa, H. Nakagawa, S. Soeda, and H. Shimeno. 2003. Oversulfation of fucoidan enhances its anti-angiogenic and antitumor activities. *Biochem. Pharm.* 65: 173-179.

Kraus, J., and G. Franz. 1991. (1-3) glucans: anti-tumor activity and immunostimulation. Pages 31-444 in *Fungal Cell Wall and Immune Response*, J.P. Latge, and D. Boucias, eds., Springer-Verlag, Berlin.

Kuang, P., and J. Xiang, 1994. Effect of radix *Salviae miltiorrhizae* on EAA and IAA during cerebral ischemia in gerbils: a microdialysis study. *J Chin. Med.* 14: 45-50.

Lahaye, M. 2001. Chemistry and physico-chemistry of phycocolloids. *Cah. Biol. Mar.* 42: 137-157.

Lapsin, R., and S. Pricl. 1995. *Rheology of Industrial Polysaccharides. Theory and Application.* Blackie A & P, London.

Lindahl, U., G. Backstrom, and L. Thunberg. 1983. The antithrombin-binding sequence in heparin. Identification of an essential 6-O-sulfate group. *J. Biol. Chem.* 25: 9826-9830.

Liu, J., and S.C. Thorp. 2002. Cell surface heparan sulfate and its roles in assisting viral infections. *Med. Res. Rev.* 22: 1-25.

Lupescu, N., S. Geresh, S. Arad (Malis) M. Bernstein, and R. Glaser. 1991. Structure of some sulfated sugars isolated after acid hydrolysis of the extracellular polysaccharide of *Porphyridium* sp. unicellular red alga. *Carbohydrate Res.* 210: 349-352.

Maillet, F.M. Petitou, J. Choay, and M.D. Kazatchkine. 1988. Structure-function relationships in the inhibitory effect of heparin on complement activation independency of the anticoagulant and anti-complementary sites on the heparin molecule. *Mol. Immunol.* 25: 917-923.

Malfait T., H. Van Dael, and F. Van Cauwelaert. 1989. Molecular structure of carrageenans and kappa oligomers: a Raman spectroscopic study. *Int. J. Biol. Macromol.* 11: 259-264.

Malik D.J., V. Strelko, M. Streat, and A.M. Puziy. 2002. Characterization of novel modified active carbons and marine algal biomass for the selective adsorption of lead. *Water Res.* 36: 1527-1538.

Mastromarino, P., C. Conti, R. Petruzziello, R. Lapadula, and N. Orsi. 1991. Effect of polyions on the early events of Sindbis virus infection of Vero cells. *Arch. Virol.* 121: 19-27.

Matsui M.S., N. Muizzuddin, S. Arad, and K. Marenus. 2003. Sulfated polysaccharides from red microalgae have anti-inflammatory properties *in vitro* and *in vivo*. *Appl. Biochem. Biotechnol.* 104: 13-22.

Mauray, S., C. Sternberg, J. Theveniaux, J. Millet, C. Sinquin, J. Tapon-Bretaudiere, and A.M. Fisher. 1995. Venous anti-thrombotic and anticoagulant activities of a fucoidan fraction. *Thromb. Haemostasis* 74: 1280-1285.

Mazumder, S., P.K. Ghosal, C.A. Pujol, M.J. Carlucci, E.B. Demonte, and B. Ray 2002. Isolation, chemical investigation and antiviral activity of polysaccharides from *Gracilaria corticata* (Gracilariaceae, Rhodophyta). *Int. J. Biol. Macromol.* 31: 87-95.

Mbemba, E., V. Chams, J.C. Gluckman, D. Klatzmann, and L. Gattegno. 1992. Dextran sulfate suppression of viruses in the HIV family: inhibition of virion binding to CD4+ cells. *Science* 240: 646-649.

McClure, K.O., J.P. Moore, D.F. Blanc, P. Scotting, G.M. Cook, R.J. Keynes, J.N. Weber, D. Davies, and R.A. Weiss. 1992. Investigations into the mechanism by which sulfated polysaccharides inhibit HIV infection *in vitro*. *AIDS Res. Hum. Retroviruses* 8: 19-25.

Mishima, T., J. Murata, M. Toyoshima, H. Fujii, M. Nakajima, T. Hayashi, T. Kato, and I. Saiki. 1998. Inhibition of tumor invasion and metastasis by calcium spirulan (Ca-SP), a novel sulfated polysaccharide derived from a blue-green alga, *Spirulina platensis*. *Clin. Exp. Metastasis* 6: 541-550.

Mitsuya, H., D.J. Looney, S. Kuno, R. Ueno, F. Wong-Staal, and S. Broder. 1988. Dextran sulfate suppression of viruses in the HIV family: inhibition of virion binding to $CD^{+}4$ cells. *Science* 240: 646-649.

Mizumoto, K., I. Sugawara, W. Ito, T. Kodama, M. Hayami, and S. Mori. 1988. Sulfated homopolysaccharides with immunomodulating activities are more potent anti-HTLV-III agents than sulfated heteropolysaccharides. *Jpn J. Exp. Med.* 58: 145-151.

Mizuno, T., M. Ando, R. Sugie, H. Ito, K. Shimura, T. Sumiya, and A. Matsuura. 1992. Antitumor activity of some polysaccharides body and the cultured mycelium of polyporous confluens. *Biosci. Biotech. Biochem.* 56: 34-41.

Moelling, K., T. Schulze, and H. Diringer. 1989. Inhibition of human immunodeficiency virus type 1 Rnase H by sulfated polyanions. *J. Virol.* 63: 5489-5491.

Moen, L.K., and G.F. Clark. 1993. A novel reverse transcriptase inhibitor from *Fucus vesiculosus*. *Int. Conf. AIDS* 9: 145-161.

Montefiori, D.C., W.E. Robinson, A. Modliszewski, J.M. Rowland, S.S. Schuffman, and W.M. Mitchell. 1990. Differential inhibition of HIV-1 cell binding and HIV-1-induced syncytium formation by low-molecular-weight sulfated polysaccharides. *J. Antimic. Chemother.* 25: 313-318.

Montgomery, R.I., M.S. Warner, B.J. Lum, and P.G. Spear. 1996. *Herpes simplex* virus-1 entry into cells mediated by a novel member of the TNF/NGF receptor family. *Cell.* 87: 427-436.

Mulloy, B., P.A. Mourao, and E. Gray. 2000. Structure/function studies of anticoagulant sulfated polysaccharides using NMR. *J. Biotechnol.* 28: 123-135.

Nakashima, H., Y. Kido, N. Kobayashi, Y. Motoki, M. Neushul, and N. Yamamoto. 1987a. Purification and characterization of an avian myeloblastosis and human immunodeficiency virus reverse transcriptase inhibitor: Sulfated polysaccharides extracted from sea algae. *Antimicrob. Agents. Chemother.* 31: 1524-1528.

Nakashima, H., Y. Kido, N. Kobayashi, Y. Motoki, M. Neushul, and N. Yamamoto. 1987b. Antiretroviral activity in a marine red alga: Reverse transcriptase inhibition by an aqueous extract of *Schizymenia pacifica*. *J. Cancer Res. Oncol.* 113: 413-416.

Nakashima, H., O. Yoshida, M. Baba, E. DeClercq, and N. Yamamoto. 1989. Anti-HIV activitiy of dextran sulfate as determined under different experimental conditions. *Antiviral Res.* 11: 233-246.

Neushul, M. 1990. Antiviral carbohydrates from marine red algae. *Hydrobiologia* 205: 99-104.

Neyts J., and E. De Clercq. 1995. Effect of polyanionic compounds on intracutaneous and intravaginal herpes virus infection in mice: impact on the search for vaginal microbicides with anti-HIV activity. *J. Acquir Immune Defic. Syndr. Hum. Retrovirol.* 10: 8-12.

Neyts, J., R. Snoeck, D. Schols, J. Balzarini, J.D. Esko, A. Schepdael, and E. DeClercq. 1992. Sulfated polymers inhibit the interaction of human cytomegalovirus with cell surface heparan sulfate. *Virology* 189: 48-58.

Nishino, T., Y. Aizu, and T. Nagumo. 1991. The influence of sulfate content and molecular weight of a fucan sulfate from the brown seaweed *Ecklonia kurome* on its anti-thrombin activity. *Thromb. Res.* 64: 723-731.

Oh, G.T., S.B. Han, M.H. Kim, M.W. Han, and I.D. Yoo. 1992. Immunostimulating activity of *Phellinus linteus* extracts to B lymphocytes. *Arch. Pharm. Res.* 15: 379-381.

Pardee, K.I., P. Ellis, M. Bouthillier, G.H.N. Towers, and C.J. French. 2004. Plant virus inhibitors from marine algae. *Can. J. Bot.* 82: 304-309.

Patel, M., M. Yanagishita, G. Roderique, Z. D.C. Bou-Habib, T. Oravecz, V.C. Hascall, and M.A. Norcross. 1993. Cell surface heparan sulfate proteoglycan mediates HIV-1 infection of T-cell lines. *AIDS Res. Hum. Retrovir.* 9: 167-174.

Pearce-Pratt, R., and D.M. Phillips. 1996. Sulfated polysaccharides inhibit lymphocyte-to-epithelial transmission of human immunodeficiency virus-1. *Biol. Reprod.* 54: 173-182.

Pearson, V.E. 2001. Galantamine: a new Alzheimer drug with a past life. *Ann. Pharmacother.* 35: 1406-1413.

Pereira, M.S., B. Mulloy, and P.A. Mourao. 1999. Structure and anticoagulant activity of sulfated fucans. *J. Biol. Chem.* 274: 7656-7667.

Pereira, M.S., A.E. Vilela-silva, A.P. Valente, and A.S. Mourao. 2002. A2-sulfated, 3-linked *a*-L-galactan is an anticoagulant polysaccharide. *Carbohydr. Res.* 337: 21-23.

Percival, E., and R.H. McDowell. 1981. Algal walls – composition and biosynthesis. *Encycl. Plant Physiol.* 13: 277-316.

Ponce, N.M., C.A. Pujol, E.B. Damonte, M.L. Flores, and C.A. Stortz. 2003. Fucoidans from the brown seaweed *Adenocystis utricularis*: extraction methods, antiviral activity and structural studies. *Carbohydr. Res.* 338: 53-165.

Rada, B., D. Blaskovic, F. Sorm, and J. Skoda. 1960. The inhibitory effect of 6-azauracil riboside on the multiplication of vaccinia virus. *Experientia* 15: 487-488.

Ramus, J., and S.T. Groves. 1972. Incorporation of sulfate into the capsular polysaccharide of the red alga *Porphyridium*. *J. Cell Biol.* 54: 399-407.

Ren, D.L., J.Z. Wang, H. Noda, H. Amano, and S. Ogawa. 1995. The effects of an algal polysaccharide from *Gloiopeltis tenax* on transplantable tumors and immune activities in mice. *Planta Med.* 61: 120-125.

Richards, J.T., E.R. Kern, L.A. Glasgow, J.C. Overall, E.F. Deign, and M.T. Hatch. 1978. Antiviral activity of extracts from marine algae. *Antimicrob. Agents Chemother.* 14: 24-30.

Riou, D., S. Colliec-Jouault, D.P. Du-Sel, S. Bosch, S. Siavoshian, V. Le-Bert, C. Tomasoni, C. Sinquin, P. Durand, and C. Roussakis. 1996. Antitumor and antiproliferative effects of a fucan extracted from *Ascophyllum nodosum* against a non-small-cell bronchopulmonary carcinoma line. *Anticancer Res.* 16: 1213-1218.

Romanos, M.T.V., M.J. Andrada-Serpa, M.G.M. dos Santos, A.C.F. Ribeiro, Y. Yoneshigue-Valentin, S.S. Costa, and M.D. Wigg. 2002. Inhibitory effect of extracts of Brazilian marine algae on human T-cell lymphotropic virus type 1 (HTLV-1)- induced syncytium formation *in vitro*. *Cancer Invest.* 20: 46-54.

Sakai T., H. Kimura, K. Kojima, K. Shimanaka, K. Ikai, and I. Kato. 2003a. Marine bacterial sulfated fucoglucoronomannan (SFGM) lyase digests brown algal SFGM into trisaccharides. *Mar. Biotechnol.* 5: 70-78.

Sakai T., K. Ishizuka, and I. Kato. 2003b. Isolation and characterization of a fucoidan-degrading marine bacterium. *Mar. Biotechnol.* 5: 409-416.

Sawitzky, D., H. Hampl, and K.P. Habermehl. 1990. Comparison of heparin-sensitive attachment of pseudorabies virus (PRV) and herpes simplex virus type 1 and identification of heparin-binding PRV glycoproteins. *J. Gen. Virol.* 71: 1221-1225.

Schaeffer, D.J., and V.S. Krylov. 2000. Anti-HIV activity of extracts and compounds from alga and cyanobacteria. *Ecotoxicol. Environ. Safety* 45: 208-227.

Schols, D., M. Baba, R. Pauwels, and E. DeClercq. 1989. Flow cytometric method to demonstrate whether anti-HIV-1 agents inhibit virion binding to T4+ cells. *J. AIDS* 2: 10-15.

Serkedjieva, J. 2000. Antiherpes virus effect of the red marine alga *Polysiphonia denudate*. *Z. Naturforsch.* 55: 830-835.

Shepherd, V.A., and M.J. Beilby. 1999. The effect of an extracellular mucilage on the response to osmotic shock in the charophyte alga *Lamprothamnium papulosum*. *J. Membrane. Biol.* 170: 229-242.

Shrestha, R., Y. Weinstein, D. Bar-Zvi, and S. Arad. 2004. A glycoprotein noncovalently associated with cell-wall polysaccharide of the red microalga *Porphyridium* sp. (Rhodophyta). *J. Phycol.* 40: 568-580.

Simon, B., S. Geresh, and S. Arad. 1992. Degradation products of the cell wall polysaccharide of the red microalgae *Porphyridium* sp. (Rhodophyta) by means of enzymatic activity of its predator the dinoflagellate *Gymnodinium* sp. (Gymnodiales). *J. Phycol.* 28: 460-465.

Spear, P., R. Eisenberg, and G. Gray. 2000. Three classes of cell surface receptors for alpha herpes virus entry. *Virology.* 275: 1-8.

Sugawara, I., W. Itoh, S. Kimura, S. Mori, and K. Shimada. 1989. Further characterization of sulfated homopolysaccharides as anti-HIV agents. *Experientia* 45: 996-998.

Tache S., G. Peifer, A.S. Millet, and D.E. Corpet. 2000. Carrageenan gel and aberrant crypt foci in the colon of conventional and human flora-associated rats. *Nutr. Cancer* 37: 193-198.

Talarico, L.B., R.G.M. Zibetti, P.C.S. Faria, L.A. Scolaro, M.E.R. Duarte, M.D. Noseda, C.A. Pujol, and E.B. Damonte. 2004. Anti-herpes simplex virus activity of sulfated galactans from the red seaweeds *Gymnogongrus griffithsiae* and *Cryptonemia crenulata*. *Int. J. Biol. Macromol.* 34: 63-71.

Tal-Singer, R., C. Peng, M. Ponce de Leon, W.R. Abrams, W.B. Banfield, F. Tufaro, G.H. Cohen, and R.J. Eisenberg. 1995. Interaction of herpes simplex virus glycoprotein gC with mammalian cell surface molecules. *J. Virol.* 69: 4471-4483.

Tannin-Spitz, T., M. Bergman, D. van-Moppes, S. Grossman, and S. Arad (Malis). 2004. Antioxidant activity of the Polysaccharide of the red microalga *Porphyridium* sp. *J. Appl. Phycol.* 17: 215-222.

Teixeira, M., and P.G. Hellewell. 1993. Suppression by intradermal administration of heparin of eosinophil accumulation but not oedema formation in inflammatory reactions in guinea-pig skin. *Br. J. Pharmcol.* 110: 1496-1500.

Tissot B., R. Daniel, and C. Place. 2003a. Interaction of the C1 complex of complement with sulfated polysaccharide and DNA probed by single molecule fluorescence microscopy. *Eur. J. Biochem.* 270: 4714-4720.

Tissot, B., B. Montdargent, L. Chevolot, A. Varenne, S. Descroix, P. Gareil, and R. Daniel. 2003b. Interaction of fucoidan with the proteins of the complement classical pathway. *Biochem. Biophys. Acta (BBA) – Proteins & Proteomics* 1651: 5-16.

Trento F., F. Cattaneo, R. Pescador, R. Porta, and L. Ferro. 2001. Antithrombin activity of an algal polysaccharide. *Thrombosis Res.* 102: 457-465.

Tsiapali, E., S. Whaley, J. Kalbfleisch, H.E. Ensley, I.W. Browder, and D.L. Williams. 2001. Glucans exhibit weak antioxidant activity, but stimulate macrophage free radical activity. *Free Radical Biol. Med.* 30: 393-402.

Tziveleka L.A., C. Vagias, and V. Roussis. 2003. Natural products with anti-HIV activity from marine organisms. *Curr. Top. Med. Chem.* 3: 1512-1535.

Ucko, M., E. Cohen, H. Gordin, and S. Arad. 1989. Relationship between the unicellular red alga *Porphyridium* sp. and its predator the dinoflagellate *Gymnodinium* sp. *Appl. and Environ. Microbiol.* 55: 2990-2994.

Usov A.I., G.P. Smirnova, and N.G. Klochkova. 2001. Poplysaccharides of algae: **55**. Polysaccharide composition of several brown algae from Kamchatka. *Russ. J. Bioorg. Chem.* 27: 395-399.

Yoshizawa, Y, A. Ametani, J. Tsunehiro, K. Nomura, M. Itoh, F. Fukui, and S. Kaminogawa. 1995. Macrophage stimulation activity of the polysaccharide fraction from a marine alga (*Porphyra-yezoensis*) - structure-function-relationships and improved solubility. *Biosci. Biotech. Biochem.* 59: 1933-1937.

Vasalli R.J., and M.Van Zeijl. 2003. DNA encapsidation as a target for anti-herpes virus drug therapy. *Antiviral Res.* 59: 73-87.

Waggoner-Fountain L.A., and L.B. Grossman. 2004. *Herpes simplex* virus. *Pediatr. Rev.* 25: 86-93.

Wallace, P.K., and P.S. Morahan. 1994. Role of macrophages in the immunotherapy of Lewis lung peritoneal carcinomatosis. *J. Leukoc. Biol.* 56: 41-51.

Wang, X.H., and T.B. Ng. 2001. Examination of lectins, polysaccharopeptide, polysaccharide, alkaloid, coumarin and trypsin inhibitors for inhibitory activity against human immunodeficiency virus reverse transcriptase and glycohydrolases. *Planta Med.* 67: 669-672.

Wilson J.C., and M., von Itzstein. 2003. Recent strategies in the search for new anti-influenza therapies. *Curr. Drug Targets* 4: 389-408.

Winkler, M. 2001. Ancient proteins and futuristic surgery: the role of complement in discordant xenograft rejection. *Transplant* 33: 3862-3864.

Witvrouw, M., and E. De Clercq. 1997. Sulfated polysaccharides extracted from sea algae as potential antiviral drugs. *Gen. Pharmacol.* 29: 497-511.

Witvrouw, M., J.A. Este, M.G. Mateu, D. Paymen, G. Andrei, R. Snoeck, S. Ikeda, R. Pauels, N.V. Bianchinf, J. Desmyter, and E. Declercq. 1994. Activity of sulfated polysaccharide extracted from the red seaweed *Aghardhiella tenera* against human immunodeficiency virus and other enveloped viruses. *Antiviral Chem. Chemother.* 5: 297-303.

Witvrouw, M., D. Schols, G. Andrei, R. Snoeck, M. Hosoya, R. Pauwels, J. Balzarini, and E. DeClercq. 1991. Antiviral activity of low-MW dextran sulphate (derived from dextran MW1000) compared to dextran sulphate samples of higher MW. *Antiviral Chem. Chemother.* 2: 171-179.

Wood, M.J., and A.M. Geddes. 1987. Antiviral therapy. *Lancet* i: 1189-1192.

Yim, J.H., S.J. Kim, S.H. Ahn, C.K. Lee, K.T. Rhie, and H.K. Lee. 2004. Antiviral effects of sulfated polysaccharide from the marine microalgae *Gyrodinium impudicum* strain KGO3. *Mar. Biotechnol.* 6: 17-25.

Zhao, B., W. Jiang, Y. Zhao, J. Hou, W. Xin, 1996. Scavenging effects of *Salvia miltiorrhiza* on free radicals and its protection for myocardial mitochondrial membranes from ischemia-reperfusion injury. *Biochem. Mol. Biol. Int.* 38: 1171-1182.

Zhu, W., V.E. Ooi, P.K. Chan, and P.J. Ang. 2003. Isolation and characterization of a sulfated polysaccharide from the brown alga *Sargassum patens* and determination of its anti-herpes activity. *Biochem. Cell Biol.* 81: 25-33.

Zhuang, C., H. Itoh, T. Mizuno, and H. Ito. 1995. Antitumor active fucoidan from the brown seaweed, umitoranoo (*Sargassum thunbergii*). *Biosci. Biotech. Biochem.* 59: 563-567.

3

Anti-Inflammatory and Antiallergic Properties of Triterpenoids from Plants

TOSHIHIRO AKIHISA[1] AND KEN YASUKAWA[2]

[1] *College of Science and Technology, Nihon University, 1-8 Kanda Surugadai, Chiyoda-ku, Tokyo 101-8308, Japan*

[2] *College of Pharmacy, Nihon University, 7-7-1 Narashinodai, Funabashi-shi, Chiba 274-8555, Japan*

INTRODUCTION

A large number of different tetracyclic and pentacyclic triterpenoids, together with their derivatives, have been identified in higher plants, fungi, algae and marine animals. In higher plants, oleananes and ursanes, which often occur together, and lupanes occur in a wide range of families. Lanostanes are common in fungi, and cycloartanes in higher plants. Most of the cucurbitanes are found in the family of Cucurbitaceae (Goad and Akihisa 1997). For extraction, separation, and structural elucidation of triterpenoids, refer to the literature (Goad and Akihisa 1997).

Naturally occurring triterpenoids including triterpenoid saponins often exhibit a variety of biological activities such as anti-inflammatory (Liu 1995, Lacaille-Dubois *et al.* 1996, Safayhi and Sailer 1997, Yasukawa and Akihisa, 1997, Rios *et al.* 2000, Rao and Gurfinkel 2000, Akihisa and Yasukawa 2001, De las Heras *et al.* 2003), anti-HIV (De Clercq 2000, Kashiwada *et al.* 2000, Akihisa *et al.* 2001), antitumor-promoting (Liu 1995, Lacaille-Dubois *et al.* 1996, Yasukawa and Akihisa, 1997, Akihisa and Yasukawa 2001), inhibition of DNA polymerases (Akihisa *et al.* 2004, Mizushina *et al.* 2004), and antimycobacterial (Cantrell *et al.* 2001) activities. In the present paper, we summarize the anti-inflammatory activities of various triterpenoids isolated from plants and microorganisms against 12-*O*-tetradecanoylphorbol-13-acetate (TPA)-induced mouse ear edema which are resulted from our own work. In addition, this paper describes selected anti-inflammatory triterpenoids published in the literature mainly from 1998 to 2003.

INFLAMMATION

Inflammation is characterized by redness, heat, pain, edema (swelling), and disturbed tissue function. These symptoms are results of complex pathophysiological processes that include increased blood flow and vascular permeability, activation of humoral and cellular defense mechanisms, sensibilization and activation of nociceptors.

Non-Allergic Inflammation

The various classes of chemical mediators act in concert to amplify inflammation response. The inhibition of prostaglandin (PG) synthesis can give important anti-inflammatory effects and the nonsteroidal anti-inflammatory drugs (NSAIDs, e.g., aspirin, indomethacin), which inhibit the conversion of arachidonic acid (AA) to PGs, are the most prescribed drugs for treatment of inflammatory disease. The PGs not only have direct inflammatory effects, but also act synergistically with other mediators such as bradykinin (BRK). The edema induced by mediators that increase vascular permeability is enhanced by vasodilator PGs. Leukocytes accumulate at sites of inflammation and are believed to contribute to tissue damage by releasing lysosomal enzymes and reactive oxygen species. Another class of metabolites that are also important inflammatory mediators derived from leukocyte AA is the leukotrienes (LTs). LTB_4, in particular, by its potent chemotactic activity, may modulate leukocyte influx to inflammatory sites. A third class of mediator that is linked biosynthetically to the prostanoids and leukotrienes is platelet-activating factor (PAF), a phospholipids that activates most inflammatory cells and has a variety of biological activities related to inflammation (Fig. 3.1) (Croft 1993, Safayhi and Sailer 1997, Rios *et al.* 2000, Fernandez *et al.* 2001).

Allergic Inflammation

The acute early phase of the allergic reaction reflects the actions of mast cell mediators, some of which are preformed and stored in cytoplasmic granules, awaiting release upon allergen provocation, whereas others result from the activation of enzymes. The well-known products of activated mast cells include histamine, serotonin, lipid mediators (PGs, LTs), proteases (tryptase, chymase), chemokines (eotaxin), and cytokines (Fig. 3.1). The activities of allergen-activated mast cells are said to "orchestrate the allergic response". The local effect (e.g., in skin in atopic dermatitis or eczema) include (*a*) enhanced local vascular permeability, leading to leakage of plasma protein, such as fibrinogen (resulting in local deposition of crosslinked fibrin and tissue swelling); (*b*) increased cutaneous blood flow, with intravascular trapping of red cells, owing to arteriolar dilation; (*c*) increased loss of intravascular fluid from postcapillary venules to produce erythema (giving the name of immunoglobulin E (IgE)); and (*d*) other effects such as itching, owing to stimulation of cutaneous sensory nerves by histamine. Local variations on this theme characterize allergic reactions in other allergic diseases, resulting in different signs and symptoms. Chronic asthma has additional complications associated with remodeling of airway tissues, notably smooth muscle hypertrophy and mucus cell hyperplasia, to cause persistent wheezing and congestion (Gould *et al.* 2003).

Mechanistically, the processes of inflammation and carcinogenesis are closely related (Dirsch *et al.* 1997, Keum *et al.* 2003, Place *et al.* 2003). Thus, in multistage carcinogenesis, tumor promotion is closely related

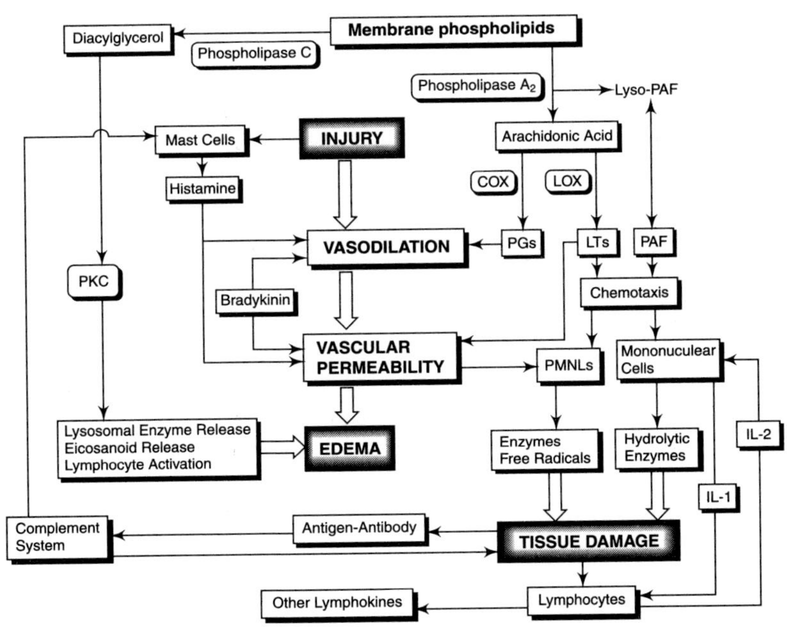

Fig. 3.1 Relationship between lipidic and other mediators and some relevant cellular events (Adapted and modified from Rios *et al.* 2000).

to the biochemical and molecular events as increased activity and/or generation of reactive oxygen species, acute inflammation characterized by skin edema, hyperplasia, increased expression/release of pro-inflammatory cytokines and aberrant up-regulation of pro-inflammatory enzymes including cyclooxygenase-2 (COX-2).

ANTI-INFLAMMATORY DRUGS AND THEIR ACTION

Anti-inflammatory compounds can act on various steps of pathophysiological processes: (*a*) a compound might block the biosynthesis of pro-inflammatory mediators by direct interaction with a key enzyme (e.g., inhibition of COX-2) or by decreasing enzyme expression (e.g., steroidal anti-inflammatory compounds); (*b*) a compound can act by inhibiting the release of performed, stored mediators (e.g., histamine release) or block mediator receptors interaction on target cells (e.g., leukotriene (LT) receptor antagonists); and (*c*) an anti-inflammatory compound may also act by immunostimulation (e.g., maturation of myeloid cells, stimulated phagocytosis) that in turn promotes an increased removal of the insulting signal molecules (organisms or, alternatively, by immunosuppression) that results in a less aggressive inflammatory response to allergen challenge (Safayhi and Sailer 1997, Rios *et al.* 2000).

In general, two major classes of antiphlogistic drugs are available for therapy: NSAIDs which interfere with the COX system (e.g., ibuprofen, aspirin, phenylbutazone and indomethacin), and steroidal anti-inflammatory drugs like cortisone, which have the ability to inhibit both the COX and lipoxygenase (LOX) pathways by interfering with the phospholipase A_2 (PLA_2)-reaction.

RESEARCH ON NEW ANTI-INFLAMMATORY DRUGS

Unfortunately, most of the anti-inflammatory drugs, particularly steroids, besides being effective, can also produce unfavolable side effects such as gastric irritation leading to ulceration, bleeding, altered metabolism, lowered immune system response, and deterioration of the cardiovascular system. Therefore, antiphlogistic agents different from those currently in use are being intensively investigated. Bioassay methods used for screening anti-inflammatory compounds are classified into *in vitro* assays using subcellular structures and *in vivo* assay using animal models. *In vitro* pharmacological studies can serve as an initial screening step in which the presumed effect on a particular aspect of an inflammatory process can be investigated. *In vitro* techniques are reproducible and sensitive, but the usefulness of the results obtained is limited by the fact that they do not reflect what happens *in vivo*. They therefore can not replace the *in vivo* methods, but are a good complement for the conventional procedures. In any case, all potential therapeutics must ultimately be evaluated in *in vivo* models. The most readily available animal models generally reflect acute or subacute reactions because they are simple, reproducible and provide clues to the biochemical and cellular mechanisms involved in the transition from acute to chronic inflammation. The cardinal signs of inflammation are the end points for the measurement of anti-inflammatory drug action (erythema, hyperemia, edema or exudation) (Rios *et al*. 2000).

BIOASSAY METHODS

A number of *in vivo* animal models have been used to assess anti-inflammatory activity. The *in vivo* inflammatory models have the advantage that they measure a biological endpoint or response and reflect the effectiveness of anti-inflammatory compounds, regardless of the possible mechanisms that may be involved. On the other hand, they have the disadvantage of variable biological response and inaccuracies in measurements such as paw volume. In addition, reasonably large amounts of pure compound may be necessary for administration to a large number of animals, in order to construct dose-response curves. Ear edema, paw edema and air pouch models of acute inflammation have been extensively used as *in vivo* animal models. Ear edema can be induced by a great diversity of irritant agents (e.g., phorbol ester, AA and croton oil) and different inflammatory mechanisms have been suggested.

Often, natural products are isolated with only a few milligrams of pure material being available for biological testing. Since isolated rat inflammatory leukocytes or human peripheral leukocytes, in response to stimuli such as calcium ionophore, produce leukotrienes, prostanoids, and PAF, they are good model cells to study the effects of test compounds on the synthesis of the various lipid mediators. Macrophages play a crucial role in modulating the initiation and perpetuation of the inflammatory response. Activation of these cells causes the release of eicosanoids, nitrogen monoxide (NO) and pro-inflammatory cytokines, including tumor necrosis factor (TNF-α) and interleukin (IL)-1β. Cytokines, as well as bacterial lipopolysaccharide (LPS), induce the expression of enzymes such as COX-2, NOS-2 and matrix metalloproteinases in macrophages and other cells, which results in the release of inflammatory mediators and tissue remodeling. This cell system has been widely used to evaluate agents able to block inflammatory mediator production (De las Heras *et al*. 2003).

1. ANIMAL MODELS OF INFLAMMATION

Phorbol Ester-Induced Mouse Ear Edema

The 12-*O*-tetradecanoylphobol-13-acetate (TPA; 4)-induced ear edema test has been well established; and the inflammation induced by this agent is related to the activation of protein kinase C (PKC). The skin inflammation produced by a single application of TPA is characterized by erythema, edema, and polymorphonuclear leukocyte infiltration. Three other phorbol esters, 12-deoxyphorbol-13-phenylacetate (DPP), 12-deoxyphorbol-13- tetradecanoate (DPT), and 12-*O*-hexadecanoyl-16-hydroxyphorbol-13- acetate (HPA), and mezerein, a daphnane ester, also are the PKC activators (Fernandez *et al.* 2001, Surh 2002).

Assay for Inhibition of TPA-Induced Inflammatory Ear Edema in Mice

For screening anti-inflammatory agents, we are using TPA-induced inflammatory ear edema in mice (Fig. 3.2) (Akihisa and Yasukawa 2001). TPA (1μg/ear) dissolved in acetone (20 μL) was applied to the right ear of ICR mice by means of a micropipette. A volume of 10 μL was delivered to both the inner and outer surfaces of the ear. The samples or their vehicles, MeOH-$CHCl_3$-water (2:1:1, 20 μL) or $CHCl_3$ (20 μL), as control, were applied topically about 30 min before TPA treatment. For ear thickness determinations, a pocket thickness gauge with a range of 0 9 mm, graduated at 0.01 mm intervals and modified so that the contact surface area was increased to reduce the tension, was applied to the tip of the ear. The ear thickness was measured before treatment (*a*), and 6 h after TPA treatment (*b* =TPA alone; *b′* =TPA plus sample). The following values were then calculated:
Edema A is induced by TPA alone (*b* – *a*).
Edema B is induced by TPA plus sample (*b′* – *a*).

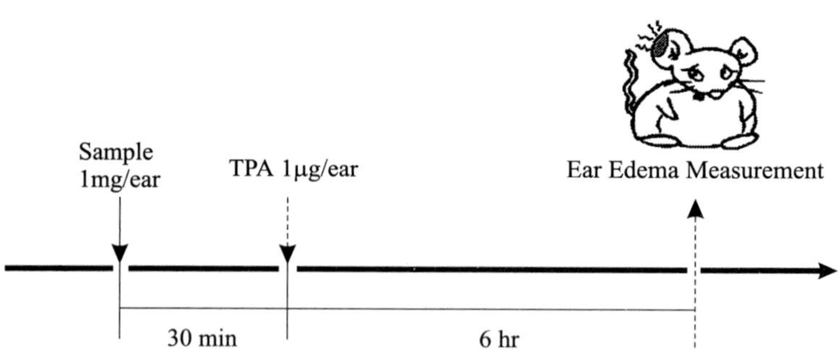

Fig. 3.2 Assay of TPA-induced inflammation in mice.

$$\text{Inhibition ratio (\%)} = \frac{\text{Edema A} - \text{Edema B}}{\text{Edema A}} \times 100$$

Each value was the mean of individual determinations from 4–5 mice. The 50% inhibitory dose (ID_{50}) values were determined by the method of probit-graphic interpolation for four dose levels. This assay is advantageous since it requires only few milligrams of sample, and it can be performed within one day.

Adjuvant-Induced Arthritis

Adjuvant arthritis, a cell mediated chronic inflammatory response, is characterized by severe symmetrical polyarthritis response, and shares many features of human rheumatoid arthritis. It involves most of the joints and associated tissues. Oxygen free radicals and H_2O_2 are closely involved in the pathogenesis of rheumatoid arthritis. Granulocytes, strongly increased in number in this disease, produce large amounts of O_2^- and H_2O_2 during phagocytosis of immune complexes and other materials. Oxygen free radicals are responsible for at least a part of the joint destruction. Adjuvant arthritis in rats has been employed widely as a model for chronic systemic inflammation and possesses many features in common with human rheumatoid arthritis (Duwiejua *et al.* 1993, Geetha *et al.* 1998, Geetha and Varalakshmi 1999).

Arachidonic Acid (AA)-Induced Edema

LTC_4 and PGE_2 are important mediators for ear edema, and the AA-induced mouse ear edema has been applied for screening of LOX inhibitors in the arachidonate pathway. Substances that inhibit this kind of edema could be inhibitors of COX and/or 5-lipoxygenase (5-LOX). Indeed, AA-induced edema has been reported as a model responsive for 5-LOX inhibitors (Inoue *et al.* 1988).

Bradykinin (BRK)-Induced Edema

Plasmatic bradykinin, the most important kinin, is a potent vasodilator and promoter of vascular permeability, and also stimulates the release of histamine from mast cells and the synthesis of prostaglandins (Inoue *et al.* 1993, Rios *et al.* 2000).

Carrageenan (CRG)-Induced Edema

Carrageenan (carrageenin) rat paw edema is a suitable test for evaluating anti-inflammatory drugs which has been frequently used to assess the antiedematous effect of natural products. The biochemical mechanism for the inflammatory reaction induced by CRG in animals is not yet clear. However, chemical mediators such as histamine, serotonin, PGs and kinin are presumed to be involved in the occurrence and development of inflammation. Edema induced by CRG is divided into the first phase in which the histamine and serotonin are involved and the second phase in which PGs and BRK participate through the advance and retreat of swelling (Mukherjee *et al.* 1997, Matsuda *et al.* 1998, Peters *et al.* 1999).

Chronic Dermal Inflammation

In the metabolism of AA, it is well established that LTs are involved in many inflammatory processes, especially in those characterized by a chronic course and damage, like bronchial asthma, arthritis and psoriasis. Repeated doses of TPA cause edema, inflammatory cell infiltration and epidermal hyperplasia. Examination on a system of chronic dermal edema and cellular proliferation caused by repeated application of TPA, to determine their value in skin inflammation, represents an advantage over the topical acute tests (AA and TPA edemas) because a great many skin diseases are essentially chronic and their pharmacological treatment is often applied when the lesion has already occurred (Manez *et al.* 1997).

Ethyl Phenylpropiolate (EPP)-Induced Edema

Topical application of EPP induces ear inflammation which is inhibited only by corticoid-like compounds (dexamethasone) and not by NSAIDs. EPP causes morphological and biochemical responses similar to those induced by TPA, and its action is apparently mediated by a mechanism that does not involve direct interaction with PKC, as occurs with TPA (Recio *et al.* 1995, Giner *et al.* 2000a).

Serotonin (SER)-Induced Edema

The models based on intradermal or subcutaneous injections of serotonin not only provide a good system for measuring the earliest inflammatory phase characterized by a rapid plasma extravasation, but are also utilized as the last step in certain experiments designed to establish whether a drug exerts its anti-inflammatory activity through interactions on glucocorticoid receptor and on DNA protein synthesis (Sugishita *et al.* 1983).

2. *IN VITRO* ANTI-INFLAMMATORY STUDIES

Anticomplement Activity

Complement is one of the major mediators of the inflammatory responses, it was assumed that inhibition of complement activity would be expected to inhibit inflammatory models in which complement activation is involved. The release of anaphylatoxins on the activation of the complement system results in inflammation or allergic reactions for example, asthma, atopy's dermatitis, arthritis systemic lupus erythematous and adult respiratory distress syndrome, by increasing vaslopermeability and edema formation, by releasing lysosomal enzymes and stimulating leukocyte chemotaxis (Kapil and Sharma 1994, Kim *et al.* 1998, Geetha and Varalakshmi, 1999).

Cyclooxygenase-1 (COX-1) and -2 (COX-2)

COX-1 synthesizes prostaglandins that serve noble purposes in housekeeping, such as protecting the stomach from ulcers and regulating renal blood flow. In general, COX-1 is constitutively expressed to fulfill its beneficial housekeeping roles. COX-2, on the other hand, is a key enzyme catalyzing the production of prostaglandins in response to inflammatory stimuli. COX-2 has been recognized as a molecular target of many anti-inflammatory as well as chemopreventive agents (Fernandez *et al.* 2001, Kim *et al.* 2002, Surh 2002).

Cytochrome P450 (CYP)

Cytochrome P450 (CYP) enzymes are involved in the metabolism of endogenous substrates such as fatty acids, cholesterol, and steroids. These enzymes also carry out an important function in the metabolism and ultimate clearance of many structurally distinct zenobiotics such as drugs, carcinogens, and environmental pollutants. It appears that the inhibitor of CYP is, in part, related to its pharmacological activities including anti-inflammatory and anticancer activities (Kim *et al.* 2004).

Hepatoprotective Activity

Experimental hepatitis in animals is induced by different mechanisms. Heavy metals and toxic metabolites of drugs, the latter are radicals and peroxides generated by microsomal enzymes, impair hepatocytes directly whereas the effects of galactosamine and endotoxin (LPS) are promoted by inhibition of anabolic processes and activation of cellular defence mechanisms including immunological responses. The hepatotoxicity test induced by CCl_4 enables evaluation of various anti-inflammatory properties including membrane stabilization, lipid peroxidation lowering, free radical scavenging, and cell regeneration effects. The damaging effects of CCl_4 on the liver are dependent on the peroxidative decomposition of structural membrane lipid, following the cleavage of CCl_4 by the mixed function oxidases to form CCl_3 and highly reactive $OOCCl_3$, which covalently bind to lipid and other cellular macromolecules. As the results of this, hepatic fat accumulation, decreased protein synthesis, glycogenolysis, leakage of cellular enzymes into the blood, and liver necrosis can be manifested (Kweifio-Okai *et al.* 1995, Jeong *et al.* 1997).

Histamine Release

Histamine is a preformed pro-inflammatory mediator stored in mast cells and basophilic granulocytes. Once released, it contacts smooth muscles of bronchi and gut, but powerfully relaxes those of small blood vessels. It promotes edema formation and stimulates gastric acid secretion and sensory nerve endings. Test of the inhibitory activity on histamine release from rat mast cells is a simple *in vitro* assay for anti-inflammatory or anti-allergic drugs, and was proved to be correlated well with the usual *in vivo* assay method (Kohda *et al.* 1985).

Intercellular Adhesion Molecule (ICAM)-1

The adhesion of circulating leukocytes to the endothelium directed by specific cell adhesion molecules is a critical cellular response to inflammation and tissue damage. Endothelial cells express an array of adhesion molecules that control processes such as leukocyte rolling along the endothelium, leukocyte attachment to the vascular wall and transmigration of leukocytes in areas of inflammation. These leukocyte-endothelial interactions require the regulated expression of various adhesion molecules by endothelial cells such as ICAM-1 of which expression is increased in inflamed tissue (Ahn *et al.* 2002, Zhang *et al.* 2003).

IL-1β and TNF-α

Cytokines such as IL-1β and TNF-α are widely believed to act as a serum factor inducing necrosis of transplanted solid tumors in the mouse. These cytokines are important pro-inflammatory mediators. Small amounts of these cytokines play a protective role acting as a host defense factor in immunologic and inflammatory responses. However, the uncontrolled release of cytokines is the basis for the development of inflammatory diseases including asthma, rheumatoid arthritis, and inflammatory bowel disease. Agents inducing the production of TNF-α and IL-1β activate the transcription factors NF-κB and AP-1, and several antioxidants prevent this activation (Allison *et al.* 2001, Fernandez *et al.* 2001, Smolinski and Pestka, 2003).

5-Lipoxygenase (5-LOX)

AA metabolites, especially the products of the LOX pathway, are important mediators of inflammation. LT biosynthesis from AA begins with the 5-lipoxygenase reaction, leading to 5-hydroperoxyeicosatetraenoic acid (5-HPETE), LTA_4 and 5-hydroxyeicosatetraenoic acid (5-HETE) formation. LTB_4 is a very potent chemokinetic and chemotactic agent, which may play an important role in the inflammatory response, whereas LTC_4, D_4 and E_4 are the active components of the slow reacting substance of anaphylaxis (SRS-A), which is one of the dominant mediators in the broncho-constriction symptom of allergic asthma (Inoue *et al.* 1986, Najid *et al.* 1992, Rios *et al.* 2000).

Nuclear Factor-*k*B (NF-*k*B)

Eukaryotic transcription factor nuclear factor-κB (NF-κB), a ubiquitous transcription factor evolutionarily conserved from flies to mammals, is one of the central regulators of an organism's response to various stress signals. NF-κB regulates the transcription of a number of genes involved in immune and inflammatory pathways such as various pro-inflammatory cytokines, adhesion molecules, and apoptosis. Disregulation of NF-κB contributes to a variety of pathological conditions such as septic shock, acute inflammation, viral replication, and some malignancies. Recent studies have demonstrated that NF-κB is involved in regulation of COX-2 and iNOS expression. Several chemopreventive phytochemicals have been shown to inhibit COX-2 and iNOS expression by blocking improper NF-κB activation (Haridas *et al.* 2001, Surh *et al.* 2001, Shishodia *et al.* 2003).

NO Synthesis and Formation of Inducible NO Synthase (iNOS)

There is increasing evidence that nitrogen monoxide (NO) plays a complex role in the modulation of the inflammatory response. Pro-inflammatory properties of NO are attributed to an excessive production of NO by iNOS and are influenced by factors like interaction of NO with other oxidants, duration of NO production and substrate availability. The most conclusive evidence for NO as a mediator of tissue injury has been in arthritis, based on studies in animal models, human osteoarthritis and rheumatoid arthritis. Thus, inhibitors of NO production in macrophages are potential anti-inflammatory drugs (Honda *et al.* 2000, Fernandez *et al.* 2001, Park *et al.* 2004a).

Protein Kinase C (PKC)

Protein kinase C (Ca^{2+-} and phospholipid-dependent protein kinase; PKC) actually constitutes a family of several isozymes that are activated by diacylglycerol (DG). The potent tumor promoter, TPA, binds to, and activates PKC, in competition with DG. PKC stimulation results in phosphorylation of regulatory proteins that affect cell proliferation. There is evidence that carcinogenesis may be suppressed by inhibiting this enzyme. Phorbol tumor promoters such as TPA can replace DG in activating PKC. A number of anti-inflammatory triterpenoids are relatively selective PKA inhibitors, and this suggests that PKA is involved in inflammatory processes but the mechanisms involved are not yet clear. Chemicals that inhibit PKC also inhibit TPA-induced tumor promotion in mouse skin (Hasmeda *et al.* 1999).

ANTI-INFLAMMATORY AND ANTIALLERGIC ACTIVITY OF SELECTED TRITERPENOIDS

1. Influence on Edema in Animal Models

The inhibitory effects of a number of triterpenoids on TPA-induced inflammatory ear edema in mice are shown in Table 3.1. The inhibitory effects of three reference compounds, quercetin (1), a known inhibitor of TPA-induced inflammation in mice, and of two commercially available anti-inflammatory drugs, indomethacin (2) and hydrocortisone (3), were included for comparison. As is evident from Table 3.1, most of the compounds tested exhibited activity almost equivalent to or higher than quercetin (1).

Table 3.1 Inhibitory Effect of Triterpenoids on TPA-induced Inflammation in Mice.

Code	*Compound*	*ID_{50}*[a] *(mg/ear)*	*Source and Occurrence*	*References*
Reference Compound				
1	Quercetin	1.6	Widespread in higher plants	Yasukawa and Akihisa 1997
2	Indometachin	0.3	Commercial	Yasukawa and Akihisa 1997
3	Hydrocortisone	0.03	Commercial	Yasukawa and Akihisa 1997
Oleanane-type triterpenoid				
On1	β-Amyrin	0.4	Widespread in higher plants	Akihisa *et al.* 1996

(*Contd.*)

(*Contd.*)

On2	Maniladiol	0.1	Asteraceae flowers	Yasukawa *et al.* 1996b
On3	Maniladiol 3-*O* -myristate	1	Chrysanthemum flowers	Ukiya *et al.* 2001
On4	Maniladiol 3-*O* -palmitate	0.3	Chrysanthemum flowers	Ukiya *et al.* 2001
On5	Erythrodiol	0.1	Olive; Grape; Asteraceae	Ukiya *et al.* 2001
On6	Deoxoglycyrrhetol di-*O*-PHT 2Na	+	Semi-synthesis	Inoue *et al.* 1989
On8	Longispinogenin	0.2	Asteraceae flowers	Yasukawa *et al.* 1996b
On10	β-Escin	0.3	Horse chestnut seeds	Yasukawa and Akihisa 1997
On11	Verbascosaponin A	+	*Scrophularia auriculata*	Giner *et al.* 2000
On16	11-Methoxyprimulagenin A 3-(Rha 2Glc 2Fuc)	+	*Bupleurum rotundifolium*	Navaro *et al.* 2001
On17	Verbascosaponin	+	*Scrophularia auriculata*	Giner *et al.* 2000
On18	Rotudinoside F	+	*Bupleurum rotundifolium*	Navarro *et al.* 2001
On19	Saikosaponin c	0.4	*Bupleurum falcatum*	Yasukawa *et al.* 1991
On20	Saikosaponin a	0.3	*B. falcatum* roots	Yasukawa *et al.* 1991
On21	Saikosaponin d	0.4	*B. falcatum* roots	Yasukawa *et al.* 1991
On22	Buddlejasaponin I	+	*Scrophularia scorodonia*	Benito *et al.* 1998
On23	Saikosaponin 2	+	*Bupleurum rigidum*	Benito *et al.* 1998
On24	Buddlejasaponin IV (saikosaponin 1)	+	*B. fruticosum* roots; *B. rigidum*	Benito *et al.* 1998
On25	δ-Amyrin	0.3	Theaceae seeds	Akihisa *et al.* 1997b
On26	Heliantriol A_1	0.1	Chrysanthemum flower	Ukiya *et al.* 2001
On27	Germanicol	0.9	Theaceae seeds	Akihisa *et al.* 1997b
On29	Oleana-9(11),12-diene-3β,30-diol di-PHT 2Na	+	Semi-synthesis	Surh 2002
On30	Oleana-11,13(18)-diene-3β,30-diol di-PHT 2Na	+	Semi-synthesis	Surh 2002
On31	Rotudinoside E	+	*Bupleurum rotundifolium*	Navarro *et al.* 2001
On33	Oleanolic acid	0.3	Occurs as glycosides in olive leaves, sugar beet, panax rhizomes, etc.	Yasukawa and Akihisa 1997
On35	Augustic acid	0.09	*Perilla frutescens* leaves	Banno *et al.* 2004
On36	Chikusetsusaponin III	0.5	Panax (*Panax ginseng*)	Yasukawa *et al.* 1991
On37	Chikusetsusaponin IV	1.3	Panax rhizomes	Yasukawa *et al.* 1991
On38	Chikusetsusaponin V (ginsenoside Ro)	1.0	Panax rhizomes	Yasukawa *et al.* 1991
On40	Chrysanthellin A	0.6	Commercial	Yasukawa *et al.* 1991
On41	Rotudinoside C	+	*Bupleurum rotundifolium*	Navarro *et al.* 2001
On42	Echinocystic acid (cochalic acid)	0.2	Sapogenin of *Echinocystis*	Yasukawa *et al.* 1991

[a] + = active.

(*Contd.*)

(*Contd.*)

On43	Echinocystic acid 3-Glc	0.06	*Echinocystis* spp.	Yasukawa *et al.* 1991
On44	Echinocystic acid 3-(Rha 2Glc 2Glc), 28-(Glc 2Glc)	+	*Bupleurum rotundifolium*	Navarro *et al.* 2001
On45	Hederagenin	0.1	Sapogenin from *Clematis, Holboellia, Hedera* spp.	Yasukawa *et al.* 1991
On46	α-Hederin (kalopanaxasaponin A)	0.3	Ivy leaves; *Kalopanax pictus*	Yasukawa *et al.* 1991
On49	Hederacoside C	1.3	Commercial	Yasukawa *et al.* 1991
On50	Chrysanthellin B	0.5	Commercial	Yasukawa *et al.* 1991
On59	Gypsogenin 3-Glc	0.4	*Vaccaria segetalis*	Yasukawa *et al.* 1991
On60	Zanhasaponin A	+	*Zanha africana* root bark	Cuellar *et al.* 1997c
On61	Zanhasaponin B	+	*Z. africana* root bark	Cuellar *et al.* 1997c
On62	Zanhasaponin C	+	*Z. africana* root bark	Cuellar *et al.* 1997c
On63	Tubeimoside I	+	*Bolbostemma paniculatum*	Yu *et al.* 1992
On64	Tubeimoside II	+	*B. paniculatum* stem	Yu *et al.* 2001
On65	Tubeimoside III	+	*B. paniculatum* stem	Ukiya *et al.* 2002
On70	Glycyrrhetinic acid (glycyrrhetic acid)	0.1	Sapogenin from *Glycyrrhiza glabra*	Yasukawa *et al.* 1995
On71	Glycyrrhetinic acid 3-PHT Na	+	Semi-synthesis	Shibata 1990
On72	Glycyrrhizin (glycyrrhizinic acid)	2.5	*Glycyrrhiza glabra* roots	Yasukawa *et al.* 1988
On73	18α-Glycyrrhetinic acid	+	Commercial	Huguet *et al.* 2000
On77	3-Epimaslinic acid	0.1	*Bupleurum fruticescens*	Banno *et al.* 2004
Ursane-type triterpenoid				
Ur1	α-Amyrin	0.2	Widespread in higher plants	Akihisa *et al.* 1996
Ur2	Brein	0.05	Asteraceae flowers	Yasukawa *et al.* 1996b
Ur3	Brein 3-*O*-myristate	0.2	Chrysanthemum flowers	Ukiya *et al.* 2001
Ur4	Brein 3-*O*-palmitate	0.2	Chrysanthemum flowers	Ukiya *et al.* 2001
Ur5	Uvaol	0.1	Olive oil; *Ilex latifolia*; Asteraceae flowers	Yasukawa *et al.* 1996b
Ur6	Ursolic acid	0.1	Wax of apples, pears and other fruits; *Rhododendron* spp.	Banno *et al.* 2004
Ur7	Corosolic acid (2a-hydroxyursolic acid)	0.09	*Perilla frutescens; Prunella vulgaris*	Banno *et al.* 2004

[a] + = active.

(*Contd.*)

(*Contd.*)

Ur8	Pomolic acid	0.12	*Perilla frutescens* leaves	Banno *et al.* 2004
Ur10	Tormentic acid	0.03	*Poterium ancistroides*	Banno *et al.* 2004
Ur11	Dammarolic acid	+	*Terminalia catappa*	Fan *et al.* 2004
Ur14	Asiaticoside	0.6	*Centella asiatica*	Yasukawa *et al.* 1991
Ur15	Maderacoside	0.8	Commercial	Yasukawa *et al.* 1991
Ur16	Hyptadienic acid	0.13	*Perilla frutescens* leaves	Banno *et al.* 2004
Lupane-type triterpenoid				
Lp1	Lupeol	0.6	Widespread in higher plants	Akihisa *et al.* 1996
Lp2	Lupeol acetate	0.6	*Vernonia cinerea*	Yasukawa *et al.* 1995
Lp3	Lupeol palmitate	0.6	*Acacia dealbata*; Stevia	Yasukawa *et al.* 1993
Lp6	3-Epilupeol	0.4	Semi-synthesis	Yasukawa *et al.* 1995
Lp7	Lupenone	1.7	*Acacia dealbata; Selacia beddomei*	Yasukawa *et al.* 1995
Lp8	Calenduladiol	0.2	Asteraceae flowers	Yasukawa *et al.* 1996b
Lp9	Calenduladiol 3-*O*-palmitate	0.3	Chrysanthemum flowers	Ukiya *et al.* 2001
Lp10	Betulin (betulinol)	0.2	Birch bark; *Corylus avellana; Vicia faba*	Yasukawa *et al.* 1995
Lp11	Betulonaldehyde	1.0	Semi-synthesis	Yasukawa *et al.* 1995
Lp12	30-Hydroxylupeol	0.6	*Flourensia heteroleptis*	Yasukawa *et al.* 1995
Lp13	3β-Hydroxylup-20(29)-en-30-al	0.6	Semi-synthesis	Yasukawa *et al.* 1995
Lp14	3β-Acetoxylup-20(29)-en-30-al	0.8	Semi-synthesis	Yasukawa *et al.* 1995
Lp15	Heliantriol B_2	0.05	Asteraceae flowers	Yasukawa *et al.* 1996b
Lp16	Betulinic acid	0.3	*Cornus florida; Zizyphus* sp.	Yasukawa *et al.* 1995
Lp18	Fruticesaponin A	+	*Bupleurum fruticescens* roots	Just *et al.* 1998
Lp19	Fruticesaponin B	+	*B. fruticescens* roots	Just *et al.* 1998
Lp20	Fruticesaponin C	+	*B. fruticescens* roots	Just *et al.* 1998
Lp21	Lupanol	0.2	Semi-synthesis	Yasukawa *et al.* 1995
Multiflorane-type triterpenoid				
Mu2	7-Oxoisomul-tiflorenol	0.2	*Trichosanthes kirilowii* seeds	Akihisa *et al.* 1994a
Mu3	7-Oxodihydro-karounidiol	0.3	*T. kirilowii* seeds	Yasukawa *et al.* 1994
Mu4	Bryonolol	>2.0	*T. kirilowii* seeds	Yasukawa and Akihisa 1997
Mu5	3-Epibryonolol	0.2	*T. kirilowii* seeds	Akihisa *et al.* 1994a
Mu6	7-Oxo-8β-multiflor-9(11)-ene-3a,29-diol diacetate	0.8	*T. kirilowii* seeds	Akihisa *et al.* 1994a
Mu7	Karounidiol	0.4	*T. kirilowii* seeds	Yasukawa *et al.* 1994
Mu8	Karounidiol 3-benzoate	0.2	*T. kirilowii* seeds	Yasukawa *et al.* 1994
Mu9	3-Epikarounidiol	0.6	*T. kirilowii* seeds	Akihisa *et al.* 1994a

[a] + = active.

(*Contd.*)

(*Contd.*)

Taraxastane-type triterpenoid				
Ta1	ψ-Taraxasterol	0.4	Asteraceae	Akihisa *et al.* 1996
Ta2	Faradiol	0.2	Asteraceae flowers	Yasukawa *et al.* 1996b
Ta3	Faradiol 3-*O*-myristate	1.0	Asteraceae flowers	Ukiya *et al.* 2001
Ta4	Faradiol 3-*O*-palmitate	0.9	Asteraceae flowers	Ukiya *et al.* 2001
Ta5	Faradiol α-poxide	0.2	Chrysanthemum flowers	Ukiya *et al.* 2001
Ta6	Heliantriol B0	0.1	Asteraceae flowers	Yasukawa *et al.* 1996b
Ta7	Heliantriol C	0.03	Asteraceae flowers	Yasukawa *et al.* 1996b
Ta8	Heliantriol C 3-*O*-palmitate	0.3	Chrysanthemum flowers	Ukiya *et al.* 2001
Taraxastane-type triterpenoid (continued)				
Ta9	Taraxasterol	0.3	Asteraceae	Akihisa *et al.* 1996
Ta10	Arnidiol	0.1	Asteraceae flowers	Yasukawa *et al.* 1996b
Ta11	Arnidiol 3-*O*-palmitate	1.0	Chrysanthemum flowers	Ukiya *et al.* 2001
Lanostane-type triterpenoid				
La1	24,25-Dihydrol-anosterol	0.2	Solanaceae; Animal fats	Kaminaga *et al.* 1996b
La2	Lanosterol	0.2	Solanaceae seeds; Yeast; Animal fats	Kaminaga *et al.* 1996b
La3	16α-Hydroxy-trametenolic acid	+	*Poria cocos* (Hoelen)	Nukaya *et al.* 1996
La4	16α-Hydroxy-trametenolic acid 3-*O*-acetate	0.02	*P. cocos*	Kaminaga *et al.* 1996b
La5	Tumulosic acid	0.21	*P. cocos*	Kaminaga *et al.* 1996b
La6	Pachymic acid	0.04	*P. cocos* and other fungi	Yasukawa *et al.* 1998a
La7	Polyporenic acid A	+	*Piptoporus betulinus*	Kamo *et al.* 2003
La8	3-*O*-Malonyloxy polyporenic acid A	+	*P. betulinus*	Kamo *et al.* 2003
La9	3-*O*-(3′-hydroxy-3′-methyl-glutaryloxy)-28-Hydroxy polyporenic acid A	+	*P. betulinus*	Kamo *et al.* 2003
La10	3-*O*-(3′-hydroxy-4′-methoxy-carbonyl-3′-methylbutyryloxy) Polyporenic acid A	+	*P. betulinus*	Kamo *et al.* 2003
La11	3-*O*-(3′-hydroxy-3′-methyl-glutaryloxy) Polyporenic acid A	+	*P. betulinus*	Kamo *et al.* 2003

[a] + = active.

(*Contd.*)

(*Contd.*)

La15	Dehydrotra-metenolic acid	0.03	*Poria cocos*	Yasukawa *et al.* 1998a
La16	16α-Hydroxy-dehydro-trametenolic acid	+	*P. cocos*	Nukaya *et al.* 1996
La17	Dehydropachy-mic acid	0.02	*P. cocos*	Yasukawa *et al.* 1998a
La18	Dehydrotumulosic acid	+	*P. cocos* and other fungi	Cuellar *et al.* 1997a
La19	3-Epidehydrotu-mulosic acid	0.09	*P. cocos*	Kaminaga *et al.* 1996b
La20	6α-Hydroxydeh-ydropachymic acid	+	*P. cocos*	Nukaya *et al.* 1996
La21	Dehydroeburi-conic acid	0.03	*P. cocos*	Kaminaga *et al.* 1996b
La22	Polyporenic acid C	0.1	*P. cocos* and *Piptoporus betulinus*	Kaminaga *et al.* 1996b
La23	3β-*p*-Hydroxy-benzoyl-dehy-drotumulosic acid	0.27	*Poria cocos*	Kaminaga *et al.* 1996a
La24	24-Methylenela-nost-9(11)-en-3β-ol	0.4	Theaceae seeds; *Euphorbia kansui* roots	Akihisa *et al.* 1997b
La25	Poricoic acid B	0.02	*Poria cocos*	Yasukawa *et al.* 1998a
La26	Poricoic acid A	0.03	*P. cocos*	Yasukawa *et al.* 1998a
La27	Poricoic acid AM	0.08	*P. cocos*	Kaminaga *et al.* 1996b
La28	Poricoic acid D	0.13	*P. cocos*	Kaminaga *et al.* 1996b
Cycloartane-type triterpenoid				
Cy1	Cycloartenol	0.3	Widespread in higher plants	Akihisa *et al.* 1996
Cy2	Cycloartenol ferulate	0.2	Rice bran; γ-Oryzanol	Yasukawa *et al.* 1998b
Cy3	Cycloartenol *cis*-ferulate	0.8	Rice bran	Akihisa *et al.* 2000
Cy4	Cycloartane-3β, 24*R*,25-triol	1.0	Chrysanthemum flowers	Ukiya *et al.* 2001
Cy5	25-Methoxycy-cloartane-3β, 24*S*-diol	0.3	Chrysanthemum flowers	Ukiya *et al.* 2001
Cy6	24-Methylene-cycloartanol	0.2	Widespread in higher plants	Akihisa *et al.* 1996
Cy7	24-Methylene-cycloartanol ferulate	0.2	Rice bran; γ-Oryzanol	Yasukawa *et al.* 1998b
Cy8	24-Methylene-cycloartanol *cis*-ferulate	0.5	Rice bran	Akihisa *et al.* 2000
Cy9	Astragaloside I	0.6	Astragalus roots	Yasukawa and Akihisa 1997

[a] + = active.

(*Contd.*)

(Contd.)

Dammarane-type triterpenoid				
Da1	Dammaradienol	0.8	Dammar resin; Shea butter	Akihisa *et al.* 1996
Da2	Dammarenediol II	0.3	Chrysanthemum flowers	Ukiya *et al.* 2001
Da3	(24*R*/*S*)-24,25-Epoxy- dammaradienol	0.5	Theaceae seeds	Akihisa et *al.* 1997b
Da4	24-Methylene-dammarenol	0.5	Shea butter	Akihisa *et al.* 1997b
Da7	Ginsenoside R_{b1}	0.4	Panax (*Panax ginseng*) rhizomes	Yasukawa *et al.* 1991
Da8	Ginsenoside R_{b2}	0.6	Panax rhizomes	Yasukawa *et al.* 1991
Da9	Ginsenoside R_d	0.5	Panax rhizomes	Yasukawa *et al.* 1991
Da10	Ginsenoside R_c	0.7	Panax rhizomes	Yasukawa *et al.* 1991
Da11	Ginsenoside R_e	2.3	Panax rhizomes	Yasukawa *et al.* 1991
Da14	Ginsenoside R_{g1}	1.6	Panax rhizomes	Yasukawa *et al.* 1991
Da17	Isoeuphol	0.3	Theaceae seeds	Akihisa *et al.* 1997b
Da18	Isotirucallol	0.3	Theaceae seeds	Akihisa *et al.* 1997b
Cucurbitane-type triterpenoid				
Cb1	24-Dihydro-10α-cucurbitadienol	>2.0	Semi-synthesis	Yasukawa and Akihisa, 1997
Cb2	10α-Cucurbitadienol	>2.0	Cucurbitaceae seeds	Yasukawa and Akihisa, 1997
Cb3	7-Oxo-24-dihydro-10α-cucurbitadienol	0.4	Semi-synthesis	Akihisa *et al.* 1994b
Cb4	7-Oxo-10α-cucurbitadienol	0.7	*Trichosanthes kirilowii* seeds	Akihisa *et al.* 1994b
Cb5	7-Oxo-10α-cucurbitadienol acetate	0.7	Semi-synthesis	Akihisa *et al.* 1994b
Cb7	23,24-Dihydrocucurbitacin B	+	*Cayaponia tayuya*	Recio *et al.* 2004
Cb9	Cucurbitacin R	+	*Bryonia dioica; Cayaponia tayuya*	Recio *et al.* 2004
Cb12	Bryoamaride	0.6	*Bryonia dioica* roots	Ukiya *et al.* 2002
Cb13	Bryonioside G	0.2	*B. dioica* roots	Ukiya *et al.* 2002
Cb14	Bryonioside B	0.2	*B. dioica* roots	Ukiya *et al.* 2002
Cb15	Bryonioside C	0.3	*B. dioica* roots	Ukiya *et al.* 2002
Cb16	Cabenoside D (mogroside I-E_2)	0.6	*B. dioica* roots	Ukiya *et al.* 2002
Cb17	Bryonioside E	0.2	*B. dioica* roots	Ukiya *et al.* 2002
Euphane/Ticulallane-type triterpenoid				
Et1	Butyrospermol	0.6	Theaceae seeds	Akihisa *et al.* 1997b
Et2	(24*R*/*S*)-24,25-Epoxy-butyrospermol	0.5	Theaceae seeds	Akihisa *et al.* 1997b

[a] + = active.

(Contd.)

(*Contd.*)

Et3	Euphol	0.2	*Euphorbia* spp. latex	Yasukawa *et al.* 2000
Et4	D^7-Tirucallol	0.8	Theaceae; *Euphorbia* spp.	Akihisa *et al.* 1996
Et7	Tirucallol	0.4	*Euphorbia* spp. latex	Akihisa *et al.* 1997b
Misceraneous triterpenoid				
Ms1	Taraxerol	0.4	Asteraceae	Akihisa *et al.* 1996
Ms2	Friedelin	0.9	Cork; *Lingnania chungii*	Yasukawa and Akihisa 1997
Ms7	Bacchara-12,21-dien-3β-ol	0.8	Theaceae seeds	Akihisa *et al.* 1997b
Ms8	Helianol	0.1	Asteraceae flowers; Theaceae seeds	Akihisa *et al.* 1996
Ms9	Sasanquol	0.4	*Camellia sasanqua* seeds	Akihisa *et al.* 1998
Ms12	Squalene	>2.0	Fish liver oils; Yeast lipids; Higher plants	Yasukawa and Akihisa 1997

[a] + = active.

Oleanae-Type Triterpenoids

Oleananes (Figs. 3.3 and 3.4) constitute the largest and most important group of triterpenoids. Oleanolic acid (On33) and oleanonic acid (On78) from *Pistacia terabinthus* galls were examined on different inflammation tests (Giner *et al.* 2001). On78 showed activity on the ear edema induced by DPP (40% inhibition at 0.5 mg/ear dose; DPP 4 μg/ear), the dermatitis induced by multiple applications of TPA, and the paw edemas induced by bradykinin (BRK) (61% inhibition at 30 mg/kg dose i.p.; BRK 3 μg/paw) and PLA_2 (84% inhibition at 60 min and 30 mg/kg dose i.p.; PLA_2 1.18 units/paw). The production of LTB_4 from rat peritoneal leukocytes was reduced by On78 with an IC_{50} of 17 μM. Negligible difference was observed in the response of both On33 and On78 to DPP, BRK, and PLA_2, while On78 was more active on the dermatitis by TPA and on the *in vitro* LT formation. It was concluded, that the presence of a ketone at C-3 implies an increase in the inhibitory effects on models related to 5-LOX activity and on associated *in vivo* inflammatory processes. Three oleanane triterpene acids, On33, 3-epioleanolic acid (On76), and 3-epimaslinic acid (On77), from the aerial parts of *Lampaya hieronymi* were evaluated *in vivo* for their ability to inhibit acute inflammation (Alvarez *et al.* 2000). These triterpenoids produced protective effects (ca. 35% inhibition at 80 mg/kg dose i.p.) in CRG-induced paw edema in mouse at 3-5 h after injection of CRG (1.75 mg/paw). The inhibitory effects were almost equivalent with that of phenylbutazone (40% inhibition), and it was suggested that the effect of the triterpenoids may principally relate to inhibition of PG release.

Several natural oleanane-type triterpenoids were assayed on mouse ear edema induced by the PKC activators, mezerein, TPA, two 12-

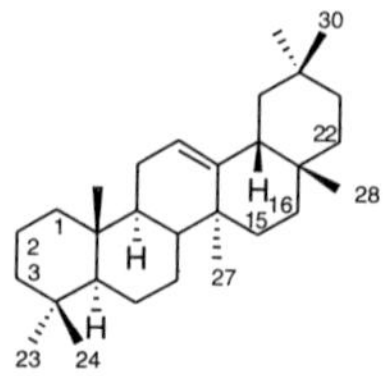

On1 3β-OH
On2 3β,16β-diOH
On3 3β-O-OC$C_{13}H_{27}$, 16β-OH
On4 3β-O-OC$C_{15}H_{31}$, 16β-OH

On5 3β-OH, 28-CH_2OH
On6 3βO-Pht-Na, 30-CH_2O-Pht-Na
On7 3β,22β-diOH, 24-CH_2OH
On8 3β,16β-diOH, 28-CH_2OH
On9 3β,22β-diOH, 24,30-diCH_2OH
On10 3β-O-[Glc-^{2}(Glc-^{4})GlcA], 16α-OH, 21α-O-Tig, 22α-OAc, 24,28-diCH_2OH 22α-OAc, 24,28-diCH_2OH
On11 3β-O-[Rha-^{4}Glc-^{3}(Glc-^{2})Fuc], 11ξ-OMe, 23,28-diCH_2OH

On12 3β-O-[Glc-^{4}Glc-^{3}(Glc-^{2})Fuc], 23,28-diCH_2OH
On13 3β-O-{(Ac-Glc)-^{4}(Ac-Glc)-^{3}[(Ac-Glc)-^{2}](Ac-Fuc)}, 23,28-diCH_2OH
On14 3β-O-[Glc-^{4}Glc-^{3}(Glc-^{2})Fuc], 16β-OH, 23,28-diCH_2OH
On15 3β-O-{(Ac-Glc)-^{4}(Ac-Glc)-^{3}[(Ac-Glc)-^{2}](Ac-Fuc)}, 16β-OH, 23,28-diCH_2OH
On16 3β-O-(Rha-^{2}Glc-^{2}Fuc), 11-OMe,16α-OH, 28-CH_2OH

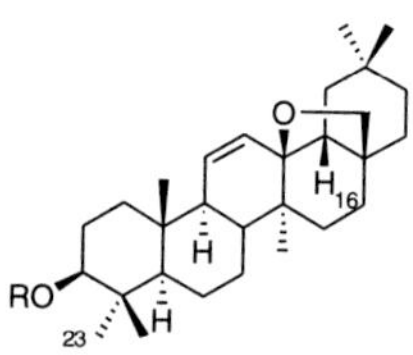

On17 R=[Rha-^{4}Glc-^{4}(Glc-^{2})Fuc], 23-CH_2OH
On18 R=(Rha-^{4}Glc-^{2}Glc-^{2}Fuc), 16α-OH
On19 R=[Glc-^{6}(Rha-^{4})Glc], 16β-OH
On20 R=[Glc-^{3}(6-deoxy-Gal)], 16β-OH, 23-CH_2OH
On21 R=[Glc-^{3}(6-deoxy-Gal)], 16α-OH, 23-CH_2OH
On22 R=[Rha-^{4}Glc-^{3}(Glc-^{2})-6-deoxy-Gal] , 16β-OH, 23-CH_2OH
On23 R={Glc-^{2}[(4-sulfo-Glc)-^{3}]-5-deoxy-Gal}, 16β-OH, 23-CH_2OH
On24 R=[Glc-^{2}(Glc-^{3})6-deoxy-Gal], 16β-OH, 23-CH_2OH

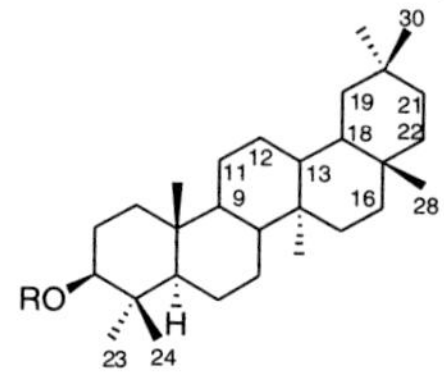

On25 R=H, $\Delta^{13(18)}$
On26 R=H, $\Delta^{13(18)}$, 16β-OH, 28-CH_2OH
On27 R=H, Δ^{18}
On28 R=H, $\Delta^{12,21}$, 24-CH2OH
On29 R=Pht-Na, $\Delta^{9(11),12}$, 30-CH_2O-Pht-Na
On30 R=Pht-Na, $\Delta^{11,13(18)}$, 30-CH_2O-Pht-Na
On31 R=(Rha-^{2}Glc-^{2}Fuc), $\Delta^{11,13(18)}$, 16α-OH, 28-CH_2OH
On32 R=H, $\Delta^{11,13(18)}$, 16α-OH, 23,28-diCH_2OH

Fig. 3.3 Structures of oleannane-type triterpenoids (1) described in this paper.

deoxyphorbol-13-monoesters (DPT and DPP) and bryostatin 1, and by resiniferatoxin, xylene, and AA in order to establish the mode of their anti-inflammatory activity (Huguet *et al.* 2000). The effects on BRK-induced paw edema and on the rat skin inflammation caused by H_2O_2 were also examined. The edema induced by mezerein and DPT was reduced to different extents by the triterpenoids administered epicutaneously (0.5 mg/ear). Against DPT-induced edema, erythrodiol (On5), On33, and hederagenin (On45) were the most effective compounds. Glycyrrhetinic acid (On70), an oleanane possessing a carboxyl group, was active against bryostatin 1-induced edema. Most of the triterpenoids tested were ineffective against the neurogenic inflammation caused by resiniferatoxin and xylene. Some triterpenoids, especially On5, were active against the plantar edema induced by BRK and on the intradermal inflammation induced by H_2O_2. It was concluded that many natural triterpenoids are efficient anti-inflammatory drugs when applied topically against various PKC activators. As most of them are inactive against the inflammation induced by AA and in neurogenic inflammatory models, their effects may depend on *in vivo* inhibition of PKC (Huguet *et al.* 2000).

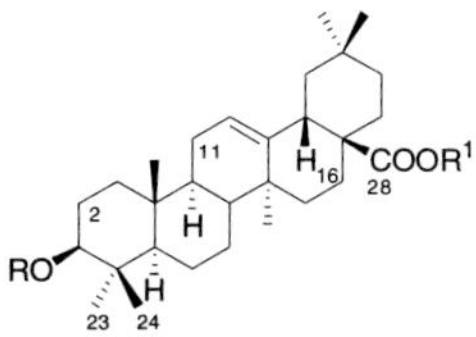

On33 R=R^1=H
On34 R=Caf, R^1=H
On35 R=R^1=H, 2β-OH
On36 R=[Ara-^{6}(Ara-^{4})GlcA], R^1=H
On37 R=GlcA, R^1=Glc
On38 R=(Glc-^{2}GlcA), R^1=Glc
On39 R=[Glc-^{2}(6-Me-GlcA)], R^1=Glc
On40 R=Glc, R^1=(Rha-^{3}Xyl-^{4}Rha-^{2}Xyl)
On41 R=sulfate, R^1=(Glc-^{6}Glc-^{2}Glc-^{2}Glc)
On42 R=R^1=H, 16β-OH
On43 R=Glc, R^1=H, 16β-OH
On44 R=(Rha-^{2}Glc-^{2}Glc), 16β-OH, R^1=(Glc-^{2}Glc)
On45 R=R^1=H, 23-CH$_2$OH
On46 R=(Rha-^{2}Ara), R^1=H, 23-CH$_2$OH
On47 R=(Rha-^{2}Ara),R^1=Me, 23-CH$_2$OH
On48 R=(Xyl-^{3}Rha-^{2}Ara), R^1=H, 23-CH$_2$OH
On49 R=[(6-deoxy-Man)-^{2}Ara], 23-CH$_2$OH, R^1=[(6-deoxy-Man)-^{4}Glc-^{6}Glc]
On50 R=Glc, 23-CH$_2$OH, R^1=(Rha-^{3}Xyl-^{4}Rha-^{2}Xyl)
On51 R=Ara, 23-CH$_2$OH, R^1=[Rha-^{2}(Xyl-^{6})Glc]
On52 R=Glc, 23-CH$_2$OH, R^1= [(6-deoxy-Man)-^{2}(Xyl-^{6})Glc]
On53 R=(Glc-^{4}Rha-^{6}Glc-^{3}Rha-^{2}Ara), R^1=H, 23-CH$_2$OH
On54 R=(Xyl-^{4}Glc-^{4}Glc-^{3}Rha-^{4}Rha-^{2}Ara), R^1=H, 23-CH$_2$OH

On55 R=Ara, 23-CH$_2$OH, R^1=Glc
On56 R=H, 23-CH$_2$OH, R^1=(Glc-^{6}Glc)
On57 R=Ara, 23-CH$_2$OH, R^1=(Glc-^{6}Glc)
On58 R=(Glc-^{3}Rha-^{2}Ara), 23-CH$_2$OH, R^1=(Glc-^{6}Glc)
On59 R=Glc, R^1=H, 23-CHO
On60 R=GlcA, 2β,16α-diOH, 23-COOH, R^1=(Rha-^{2}Rha)
On61 R=GlcA, 2β,16α-diOH, 23-COOH, R^1=(Xyl-^{2}Rha-^{2}Rha)
On62 R=GlcA, 1β,16α-diOH, 23-COOH, R^1=(Xyl-^{3}Xyl-^{2}Rha-^{2}Rha)
On63 R^1→″R=[^{1}Ara^2-^{1}(Xyl-^{3})Rha^4-^{1}HMG^5-^{4}Ara-^{2}Glc], 2β-OH, 23-CH$_2$OH
On64 R^1→″R=[^{1}Ara^2-^{1}(Xyl-^{3})Rha^4-^{1}HMG^5-^{4}Ara-^{2}Glc], 2β,16α-diOH, 23-CH$_2$OH
On65 R^1→″R=[^{1}Ara^2-^{1}(Xyl-^{3})Rha^4-^{1}HMG^5-^{6}Glc-^{2}Glc], 2β-OH, 23-CH$_2$OH
On66 R=Glc-^{3}Glc, R^1=Glc
On67 R=Glc-^{2}Glc, R^1=Glc
On68 R=Rha-^{3}GlcA, R^1=Glc

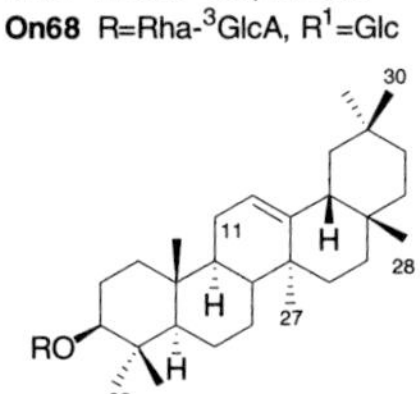

On69 R=(Glc-^{4}Xyl), 2β-OH, 23-CH$_2$OH, 28-COOH, 30-COOMe
On70 R=H, 11-oxo, 30-COOH
On71 R=Pht-Na, 11-oxo, 30-COOH
On72 R=(GlcA-^{2}GlcA), 11-oxo, 30-COOH

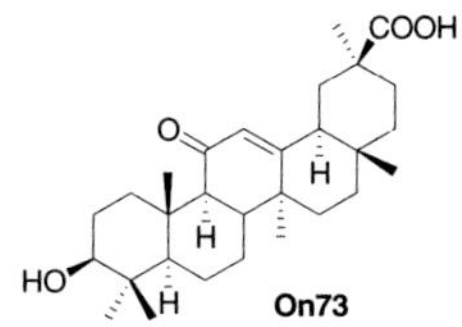

On73

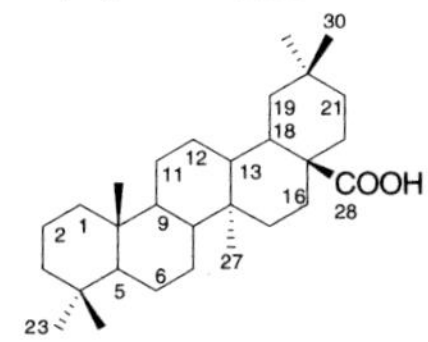

On74 3β-OH, Δ^{18}
On75 3β-O-Caf, Δ^{18}
On76 3α-OH, Δ^{12}
On77 2α,3α-diOH, Δ^{12},
On78 3-oxo, Δ^{12}
On79 3-oxo, Δ^{1}
On80 3,12-dioxo, Δ^{1}
On81 2-CN, 3,12-dioxo, $\Delta^{1,\,9(11)}$
On82 2-CN, 3,12-dioxo, $\Delta^{1,\,9(11)}$, 28-CO-imidazole
On83 2-COOH, 3,12-dioxo, $\Delta^{1,\,9(11)}$, 28-COOMe
On84 3-oxo, $\Delta^{1,\,12}$
On85 2-COOH, 3-oxo, $\Delta^{1,\,12}$
On86 2-CHO, 3-oxo, $\Delta^{1,\,12}$
On87 3β-O-(Xyl-^{2}Fuc-^{6}GlcNAc), 28-COO-[Ara-^{4}(Glc-^{4})Glc], 21β-O-(DHOA$^{1'}$-^{4}Qui-^{6}HMOA)ester, 16α-OH, 23-CH$_2$OH

Fig. 3.4 Structures of oleanane-type triterpenoids (2) dscribed in this paper.

Ursane-Type Triterpenoids

Two ursane-type triterpene acids (Fig. 3.5), ursolic acid (Ur6) and dammarolic acid (Ur11), isolated from the anti-inflammatory $CHCl_3$ extract of *Terminalia catappa* leaves exhibited strong anti-inflammatory activities on TPA-induced ear edema in acute and chronic models (Fan *et al.* 2004). Both ursanes at 0.3 mg/ear dose caused significant reductions (over 50%) of the acute edema, similar to the standard drug indomethacin (57.0%). On chronic inflammation, Ur6 and Ur11 reduced the edema by 96.9 and 94.8%, respectively, at only 0.3 mg/ear. The triterpene acids

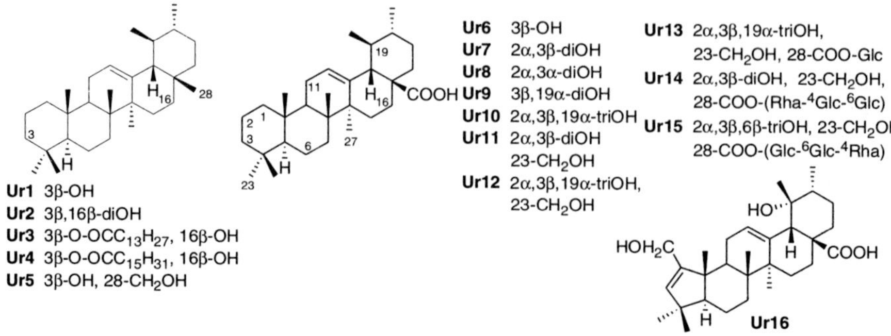

Fig. 3.5 Structures of ursane-type triterpenoids described in this paper.

Ur6 and Ur11, representing 23.8% of the $CHCl_3$ fraction, may be responsible for the anti-inflammatory activity of the extract (Fan *et al.* 2004). The $CHCl_3$ extract of *Salvia officinalis* leaves inhibited dose-dependently the croton oil-induced ear edema in mice (Baricevic *et al.* 2001). Ur6 was isolated from the extract as the main component. The anti-inflammatory effect of Ur6 (ID_{50} 0.14 μM/cm^2) was two fold more potent than that of indomethacin (ID_{50} 0.26 μM/cm^2). The content of Ur6 in sage and sage-based remedies for the topical treatment of inflammatory diseases is proposed as a parameter for quality control purposes (Baricevic *et al.* 2001).

The ether fraction of the resinous bark of *Protium kleinii* (Burseraceae) exhibited antinociceptive activity in several models of nociception in mice (Otuki *et al.* 2001). Brein (Ur2; 10-100 mg/kg, 30 min prior) isolated from the fraction, when given by i.p. route to mice, produced dose-dependent attenuation of both the neurogenic (early) phase and the inflammatory (late) phase of the formalin-induced nociception in mice. The mean ID_{50} values for these effects were 22.7 and 27.2 mg/kg and inhibition effects were 65% and 90% for the first and second phase, respectively. The antinociceptive activity of the ether fraction seems, at least in part, to be related to the presence of Ur2 in the extract.

Oral administration of tormentic acid (Ur10; 30 mg/kg), isolated from the stem bark of the plant *Vochysia divergens*, twice a day for several consecutive days produced time-dependent and pronounced anti-allodynia effect in both ispsilateral and contralateral paws after plantar injection of Complete Freund's Adjuvant (Bortalanza *et al.* 2002). The inhibition observed was 82% and 100%, respectively. Ur10 did not inhibit paw edema formation following adjuvant plantar injection. Ur10 (30 mg/kg, p.o.), given twice a day, inhibited markedly the neuropathic allodynia induced by partial ligation of the sciatic nerve, with inhibition of 91%. The anti-allodynic action of Ur10 was not associated with impairment of the motor activity of the animals, and these results indicate that Ur10 or its derivatives might be of potential interest in the development of new clinically relevant drugs for the management of persistent neuropathic and inflammatory allodynia. The EtOAc- and BuOH fraction and those alkaline hydrolysates of the extract of *Rubus coreanus* fruits showed significant antinociceptive effects as assessed by writhing, hot plate-, and tail flicks tests in mice and rats as well as anti-inflammatory effect in rats with CRG-induced edema (Choi *et al.* 2003). BuOH fraction afforded a large amount of niga-ichigoside F_1 (Ur13) of which hydrolysis in NaOH solution yielded an aglycone 23-hydroxytormentic acid (Ur12). The aglycone Ur12 was much more potent in both antinociceptive and anti-inflammatory tests than the glycoside Ur13. The anti-inflammatory effects of these compounds were further supported by the reduction of CRG-induced lipid peroxidation and hydroxyl radical in serum. These suggest that Ur12 might be an active moiety of Ur13 present in *R. coreanus* (Choi *et al.* 2003).

Lupane-Type Triterpenoids

Lupeol (Lp1) (Fig. 3.6) and lupeol linoleate (Lp4) were investigated for their anti-inflammatory, antinociceptive, antipyretic, and ulcerogenic properties in rats. Lp1 and Lp4 (50 mg/kg dose) showed a reduction in paw swelling by 39 and 58%, respectively, in adjuvant arthritis (Geetha and Varalakshmi 2001). In addition, the effects of Lp1 and Lp4 on the development of complement in adjuvant arthritis in rats were studied and compared with those of indomethacin (Geetha and Varalakshmi 1999). The effect of Lp4 in reducing the foot-pad thickness and complement activity in arthritic rats were even greater than that of Lp1 and indomethacin. Because complement is highly involved in inflammation, the results suggest that the anti-inflammatory activity of triterpenoids may be due to their anticomplement activity. *Leptadenia hastate* (Asclepiadaceae) is widespread in Africa, and the latex of this plant is frequently used in African herbal medicines as an anti-inflammatory and wound healing agent. Lp1, its acetate (Lp2) and palmitate (Lp3) esters isolated from the latex were found to be the main anti-inflammatory constituents in the croton oil-induced ear edema test (Nikiema *et al.* 2001) which proves that the triterpenoids play a pivotal role in the topical anti-inflammatory effect of this latex. Semi-synthetic lupeol eicosapentaenoate (Lp5) (50 mg/kg oral dose) was tested for its inhibitory effect on adjuvant induced arthritis *in vivo* in rats (Latha *et al.* 2001). The increased activities of lysosomal enzymes and glycoproteins associated with decreased collagen in arthritic animals were significantly altered to nearly that of controls. Indomethacin (3 mg/kg), a reference drug, also exerted the influence in almost the same extent with those of Lp5. The beneficial effect of Lp5 proves to be promising as a therapeutic agent in inflammation.

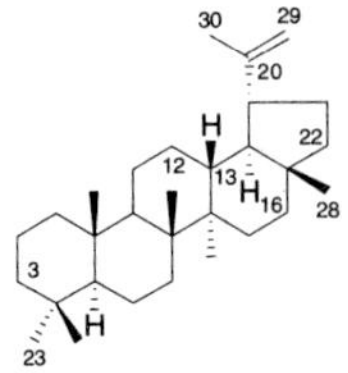

Lp1 3β-OH
Lp2 3β-OAc
Lp3 3β-O-$OCC_{15}H_{31}$
Lp4 3β-O-$OCC_{17}H_{31}$(*n*-6)
Lp5 3β-O-$OCC_{19}H_{29}$(*n*-3)
Lp6 3α-OH
Lp7 3-oxo
Lp8 3β,16β-diOH
Lp9 3β-O-$OCC_{15}H_{31}$, 16β-OH
Lp10 3β-OH, 28-CH_2OH
Lp11 3-oxo, 28-CHO
Lp12 3β-OH, 30-CH_2OH
Lp13 3β-OH, 30-CHO
Lp14 3β-OAc, 30-CHO
Lp15 3β,16β-diOH, 28-CH_2OH
Lp16 3β-OH, 28-COOH
Lp17 3β-O-Caf, 28-COOH
Lp18 3β-[Rha-^{4}(Glc-^{6})Glc], 23,28-diCOOH
Lp19 3β-(Rha-^{4}Glc), 23-COOH, 28-COO-Glc
Lp20 3β-[Rha-^{4}(Glc-^{6})Glc], 23-COOH, 28-COO-Glc
Lp21 3β-OH, 20:29-dihydro

Fig. 3.6 Structures of lupane-type triterpenoids described in this paper.

Several natural lupane-type triterpenoids along with oleananes were evaluated for their anti-inflammatory activity (Huguet *et al.* 2000). Against DPT-induced edema, Lp1, betulin (Lp10), and betulinic acid (Lp16) were the most effective compounds. Lp16, a lupane possessing a carboxyl group was active against bryostatin 1-induced edema.

Lanostane-Type Triterpenoids

The sclerotium of *Poria cocos* Wolf (Polyporaceae) is used in Chinese herbal prescriptions as a diuretic and as a sedative. The MeOH extract of *P. cocos* showed the inhibitory effects on inflammatory mouse ear edema (A Kaminaga *et al.* 1996b). From the active fraction of the extract, several lanostane-type and 3,4-*seco*-lanostane-type triterpene acids have been isolated (Kaminaga *et al.* 1996b, Yasukawa *et al.* 1998a). Upon evaluation of their anti-inflammatory properties, all of them showed marked activity against TPA-induced mouse ear edema, and among which five lanostanes, (Fig. 3.7) 16*α*-hydroxytrametenolic acid 3-*O*-acetate (La4), pachymic acid (La6), dehydrotrametenolic acid (La15), dehydropachymic acid (La17), dehydroeburiconic acid (La21), and two 3,4-*seco*-lanostanes, poricoic acid B (La25) and poricoic acid A (La26), exhibited a strong inhibitory activity with 0.02-0.04 mg/ear of the 50% inhibitory dose (ID_{50}) which was at the same level as hydrocortisone (3; ID_{50} 0.03 mg/ear) (Table 3.1). On AA-induced inflammation test in mice, La15 (ID_{50} 0.8 mg/ear), La17 (1.2 mg/ear), 3*β*-*p*-hydroxybenzoyldehydrotumulosic acid (La23, 1.3 mg/ear), La25 (1.3 mg/ear), and La26 (1.2 mg/ear) exhibited activity almost equivalent with that of indomethacin (ID_{50} 1.0 mg/ear) (Yasukawa *et al.* 1998a).

The hydroalcoholic extract from *P. cocos* has been examined for oral and topical anti-inflammatory activities (Cuellar *et al.* 1997a). It proved to be active against TPA- as well as CRG- and AA-induced acute edemas, TPA chronic inflammation, and oxazolone delayed hypersensitivity in mice. Two lanostane-type carboxylic acids, La6 (ID_{50} 4.7×10^{-3} μmol/ear on acute TPA edema) and dehydrotumulosic acid (La18; ID_{50} 6.8×10^{-4} μmol/ear), have been isolated from the extract.

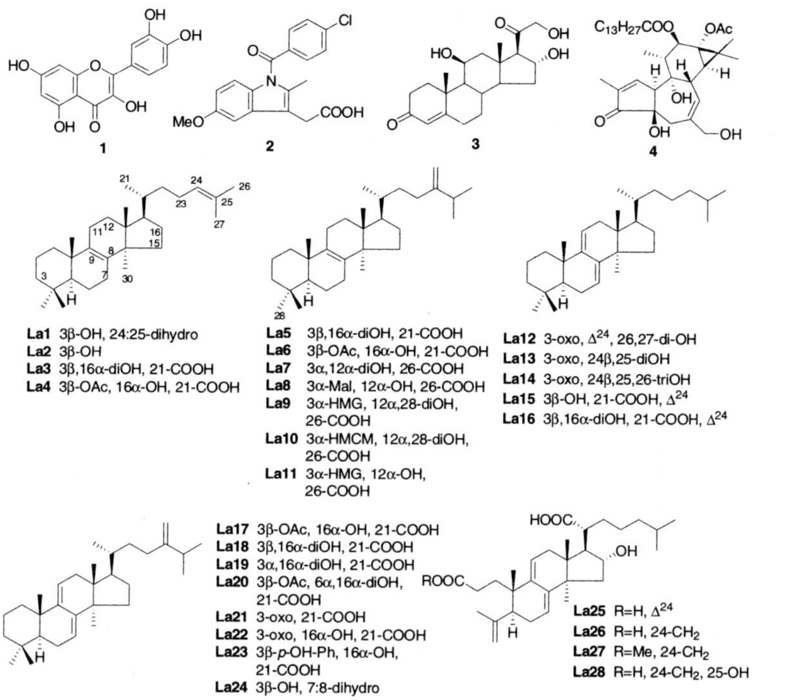

Fig. 3.7 Structures of reference compounds and lanostane-type triterpenoids described in this paper.

All of six lanostane carboxylic acids, 16*α*-hydroxytrametenolic acid (La3), tumulosic acid (La5), La6, 16*α*-hydroxydehydrotrametenolic acid (La16), La18, and 6*α*-hydroxydehydropachymic acid (La20), isolated form the MeOH extract of *P. cocos*, exhibited anti-inflammatory activity against TPA (0.1 μg/ear)-induced mouse ear edema, and among which La20 was found to be the most effective one (65% inhibition at 5 μg/ear dose) (Nukaya *et al.* 1996).

La6 and La18 from *P. cocos* were studied in different models of acute and chronic inflammation. They proved to be active in most of the methods applied. None of them were active against AA-induced ear edema. La18 significantly diminished the mouse ear edema induced by EPP, while La6 was ineffective. When the putative corticoid-like mechanism of both compounds was explored, La6 activity was partially abolished by the glucocorticoid receptor antagonist progesterone, but La18 was not affected. In the phospholipase A_2 (PLA_2)-induced mouse ear edema, both compounds were active (Giner *et al.* 2000b). Six lanostane carboxylic acids isolated from the fruiting bodies of *Piptoporus betulinus* were identified as polyporenic acid A (La7) and their derivatives, La8-La10, and polyporenic acid C (La22) (Kamo *et al.* 2003). All these compounds suppressed the TPA-induced edema on mouse ears by 49-86% with a 0.4 μmol/ear application which was more suppressive than indomethacin (16% inhibition at 0.56 μmol/ear) and glycyrrhetinic acid (On70) (24% inhibition at 0.43 μmol/ear).

Cycloartane-Type Triterpenoids

Inhibitory activity against TPA-induced inflammation in mice was observed in the MeOH extract of rice bran and γ-oryzanol (Fig. 3.8). The active components of rice bran, cycloartenol ferulate (Cy2, ID_{50} 0.2 mg/ear) and 24-methylenecycloartanol ferulate (Cy7, 0.2 mg/ear) inhibited markedly the TPA (1 μg/ear)-induced inflammation (Yasukawa *et al.* 1998b) (Table 3.1).

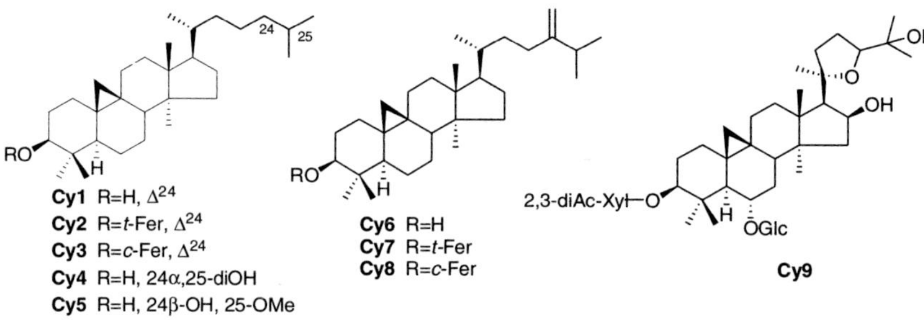

Fig. 3.8 Structures of cycloartane-type triterpenoids described in this paper.

Cucurbitane-Type Triterpenoids

Fractionation of an anti-inflammatory extract from *Cayaponia tayuya* (Cucurbitaceae) roots yielded two active compounds (Fig. 3.9), 23,24-dihydrocucurbitacin B (Cb7) and cucurbitacin R (Cb9). Both compounds showed activity in the following models: TPA-induced acute ear edema (Cb9: 36% inhibition at 4 mg/kg p.o., and 87% inhibition at 0.1 mg/ear topically), CRG-induced mouse paw edema (Cb7: 46% inhibition at 4 mg/kg p.o. and 3 h), PLA_2-induced mouse paw edema (Cb9: 61% inhibition at 3 mg/kg i.p. and 60 min), and serotonin-induced paw edema (Cb7 and Cb9: 73% and 79% inhibition, respectively, at 0.5 mg/kg s.c.) (Recio *et al.* 2004). Cucurbitacin B (Cb6; 0.1 mg/kg, i.p.) isolated from CH_2Cl_2 extract of *Wibrandia ebracteata* (Cucurbitaceae) roots, which are employed in folk medicine in Brazil for the treatment of several diseases, such as rheumatic disease, reduced significantly the levels of PGE_2 by 40.7% (10.4 ng/ml) in the exudate leakage induced by CRG in the pleural cavity of mice (Peters *et al.* 1999).

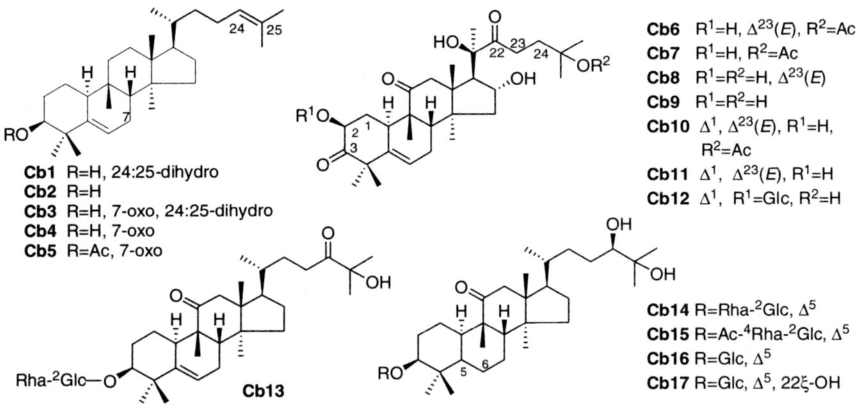

Fig. 3.9 Structures of cucurbitane-type triterpenoids described in this paper.

Dammarane-Type Triterpenoids

The flower of the palmyrah palm (*Borassus flabellifer*) was observed to induce immunosuppression (Fig. 3.10). The 17α-substituted dammarane-type triterpenoid, 17α-dammara-20, 23*E*-dien-3β, 25-diol (Da16), isolated from the flour showed a very promising activity in animal models of immunosuppression and inflammation. It is active in assays for mixed lymphocyte reaction and murine delayed type hypersensitivity, and efficacious in models of acute skin inflammation (oxazolone-induced allergic contact dermatitis) in mouse and domestic pigs upon topical application. Furthermore, efficacy in the mouse model could also be demonstrated after oral administration (Scholz *et al.* 2004).

Euphane/Tirucallane-Type Triterpenoids

Two euphane/tirucallane-type triterpene acids, 3-epiisomasticadienolalic acid (Et9) and isomasticadienonalic acid (Et10) (Fig. 3.11), isolated from *Schinus molle* (Anacardiaceae) fruits, were tested on two models of mice paw inflammation: one of acute inflammation, induced by subcutaneous injection of either PLA_2 or CRG in mice paw, and one of chronic inflammation in the form of eczema, provoked by repeated administration of TPA to the mice ear (Yueqin *et al.* 2003). In the PLA_2-induced paw edema, only Et10 was active (66% inhibition at 60 min and 30 mg/kg dose), whereas both Et9 and Et10 reduced the chronic model of inflammation (39% and 48% of swelling reduction, respectively) and the leukocyte infiltration. An extract of the galls of *Pistacia terebinthus* was proved to be effective against chronic and acute inflammation. Two tirucallanes, masticadienolic acid (Et5) and masticadienonic acid (Et6), and an oleanane, morolic acid (On74), isolated from the galls extract

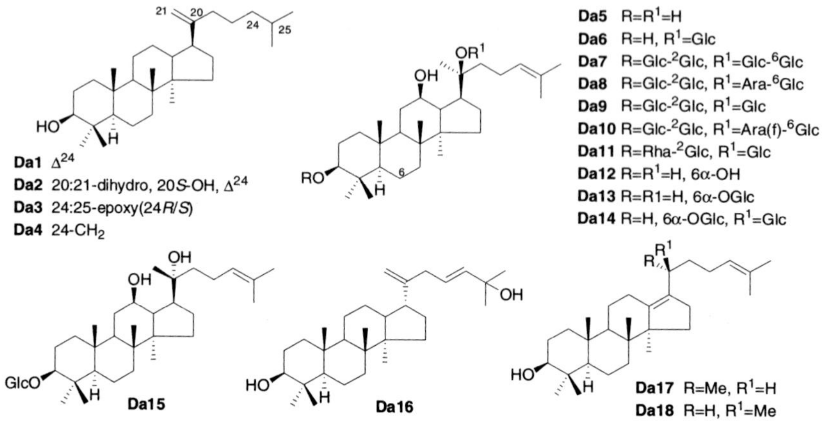

Fig. 3.10 Structures of dammarane-type triterpenoids described in this paper

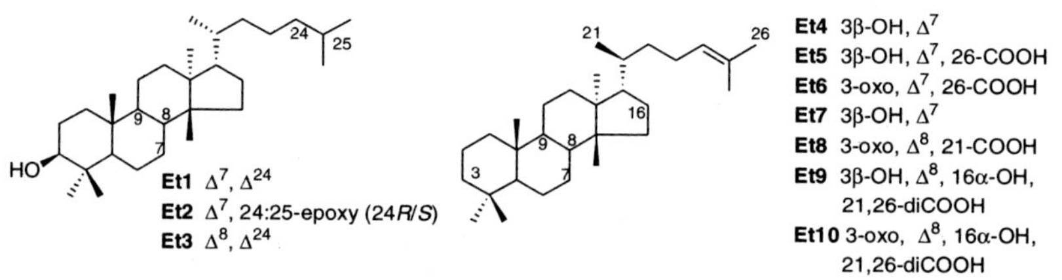

Fig. 3.11 Structures of euphane/tirucallane-type triterpenoids described in this paper.

showed effectiveness on the mouse ear inflammation induced by repeated application of TPA and on the PLA_2-induced paw edema. In addition, they inhibited LTB_4 production in rat polymorphonuclear leucocytes (PMNL) stimulated with calcium ionophore A23187 (Giner *et al.* 2002).

Miscellaneous Triterpenoids

Taraxerol (Ms1) (Figs. 3.12 and 3.13), isolated from *Strobilanthes callosus*, showed a high reduction of edema in the CRG-induced paw edema inflammation model (Singh *et al.* 2002). Myrrhanol A (Ms11), isolated from guggul-gum resin, displayed a potent anti-inflammatory effect on exudative pouch fluid, angiogenesis, and granuloma weights in adjuvant-induced air-pouch granuloma of mice. Its effects were more marked than those of hydrocortisone and Ms11 is a plausible candidate for a potent anti-inflammatory agent (Kimura *et al.* 2001). Antinociceptive and anti-inflammatory effects of limonin (Ms10) isolated from the dried fruits of *Evodia rutaecarpa* were investigated (Matsuda *et al.* 1998). Oral administration of 30 or 100 mg/kg Ms10 significantly decreased the frequency of licking and biting behavior within a unit of time at the late phase without affecting that of the early phase in the formalin test. Ms10

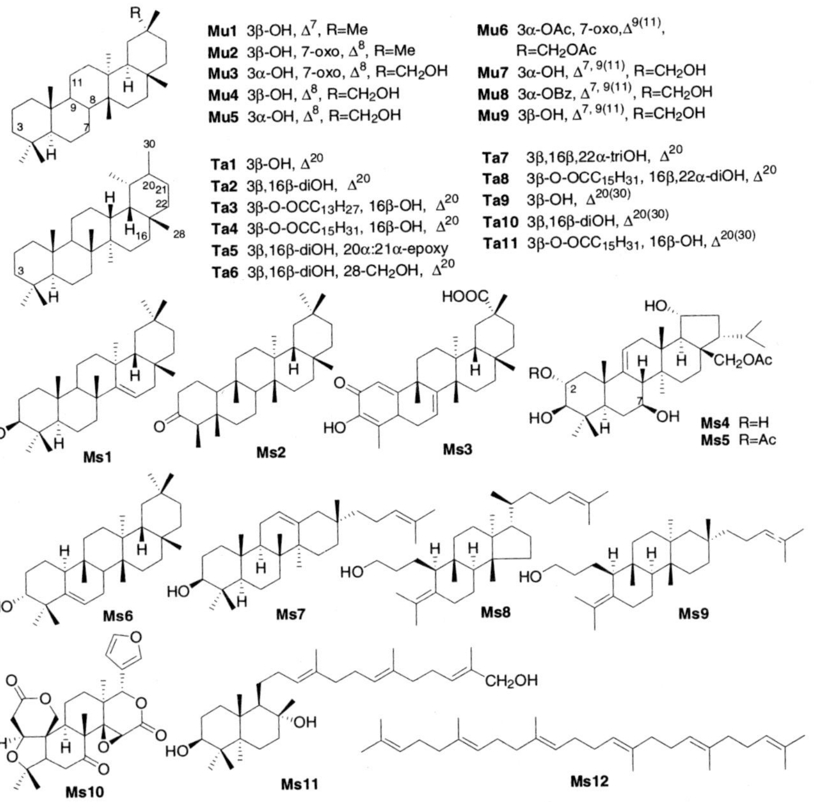

Fig. 3.12 Structures of multiflorane-type taraxastane-type, and other types of triterpenoids described in this paper.

inhibited the rise of vascular permeability induced by acetic acid and the increase of paw edema induced by CRG. Ms10 also showed inhibitory effect on BRK-induced paw edema and AA-induced ear swelling. These results suggest that Ms10 possesses an antinociceptive effect and the effect may be accompanied by an anti-inflammatory action, and that the antinociceptive activity in *Evodiae fructus* is partially attributable to Ms10.

Triterpenoids from Asteraceae

Thirty-six triterpenoids isolated from edible chrysanthemum (*Chrysanthemum morifolium*) flowers and other Asteraceae flowers, viz., 11 monohydroxy tritepenoids, 16 di- and trihydroxy triterpenoids and nine 3-*O*-fatty acid esters of di- and trihydroxy triterpenoids, were evaluated with respect to their anti-inflammatory activity induced by TPA (Akihisa *et al*. 1996, Yasukawa *et al*. 1996a, 1996b, Akihisa *et al*. 1997a, Ukiya *et al*.

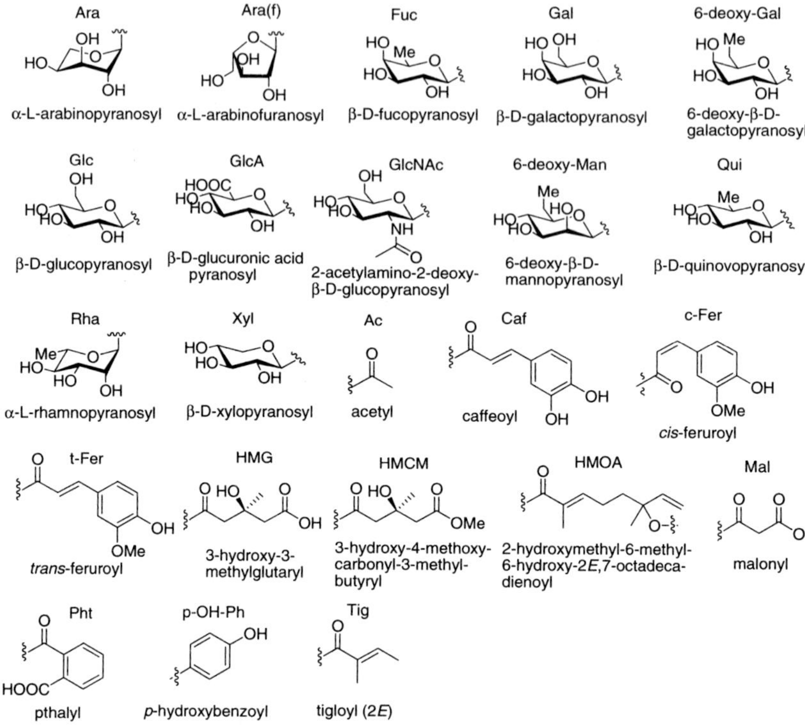

Fig. 3.13 Structures of abbreviated moieties in Figures 3–12 dscribed in this paper.

2001). It was observed that these triterpenoids examined markedly inhibited the inflammation with 0.03–1.0 mg/ear of the 50% inhibitory dose (Table 3.1). There was a close relationship between the hydroxylation of triterpenoids and the inhibitory effects. Di- and trihydroxy triterpenoids always showed higher activity than their corresponding 3*β*-monohydroxy compounds. In the case of Δ^{20}-taraxastenes, faradiol (Ta2; 0.2 mg/ear), showed a higher inhibitory effect than its 3β-monohydroxy homolog, ψ-taraxasterol (Ta1; 0.4 mg/ear). Further hydroxylation of Ta2 at C-22*α* to give heliantriol C (Ta7; 0.03 mg/ear) enhanced the effect markedly. Heliantriol C (Ta7), brein (Ur52; ursane-type), and heliantriol B_2 (Lp15; lupane-type) showed a fairly strong inhibitory effect which was almost comparable with that of hydrocortisone (3). The inhibitory effects of the 3-*O*-fatty acid esters were weaker than their corresponding free hydroxy tritepenoids which could be due to kinetic reasons because highly lipophilic compounds can be trapped in the epidermis, reaching a lower concentration at the underlying action site (Ukiya *et al.* 2001).

Among five triterpenoids isolated from marigold (*Calendula officinalis*) flowers, *β*-amyrin (On1), ψ-taraxasterol (Ta1), faradiol (Ta2), taraxasterol

(Ta9), and lupeol (Lp2), the diol Ta2 was the most active. It showed a dose-dependent effect with a potency that equals that of indomethacin (2) in the topical anti-inflammatory assay with croton oil (Loggia *et al.* 1994). Esterification at C-3 of Ta2 with a fatty acid reduced the activity by more than 50% (Loggia *et al.* 1994) consistent with the observation in the TPA-induced assay described above (Ukiya *et al.* 2001). The anti-inflammatory properties, as determined by croton oil-induced edema of mouse ear, of faradiol-3-*O*-myristate (Ta3) and its 3-*O*-palmitate (Ta4), the main components of lipophilic extracts of *C. officinalis* flowers, were shown to be contribute significantly to the pronounced antiphlogistic activity of the lipophilic extracts of *C. officinalis* flowers (Zitterl-Eglseer *et al.* 1997).

Triterpenoids from Perilla Leaves

Eight triterpene acids, viz., five of the ursane-type, ursolic acid (Ur6), corosolic acid (Ur7), pomolic acid (Ur9), tormentic acid (Ur10), and hyptadienic acid (Ur16), and three of the oleanane-type, oleanolic acid (On33), augustic acid (On35), and 3-epimaslinic acid (On77), among which Ur6 constituted the most predominant triterpene acid, were isolated from EtOH extract of the leaves of perilla (*Perilla frutescens*). These compounds were evaluated for their inhibitory effects on TPA-induced inflammation (1 μg/ear) in mice. All the compounds tested showed a marked anti-inflammatory effects (ID_{50} 0.03–0.3 mg/ear) among which Ur10 exhibited the strongest inhibitory effect (ID_{50} 0.03 mg/ear) which was comparable with that of hydrocortisone (3) (Banno *et al.* 2004).

Oleanane-Type Triterpenoid Saponins

Diverse saponin-rich crude drugs have been used extensively for their anti-inflammatory properties. Glycyrrhetinic acid (On72) is the major bioactive glycoside of licorice root (*Glycyrrhiza radix*) extracts possessing a wide range of pharmacological properties (anti-inflammatory, anti-ulcer, antiallergic, antidote, antioxidant, antitumor, antiviral, etc.). The content of licorice root is 2–24% of the dry weight. The monoammonium salt (glycyram, tussilinar) of On72 is used as an anti-inflammatory and anti-allergic remedy (Baltina 2003). Potent anti-inflammatory agents were found among On72 esters, amides, glycopeptides, and ureides, and some amides of On72 were more potent and less toxic (LD_{50} 5000–8000 mg/kg) anti-inflammatory agents as compared to prednizolone in the CRG-induced inflammation in mice (Baltina 2003).

Scrophularia auriculata is a medicinal plant used in traditional medicine against inflammatory skin diseases. Its aqueous alcohol extract showed a marked effect on both acute and chronic models of inflammation (Cuellar *et al.* 1998). The activity of two oleanane saponins, verbascosaponin A (On11) and verbascosaponin (On17), isolated from *Scrophularia auriculata* have been studied in different models of acute and chronic inflammation (Giner *et al.* 2000c). Both saponins significantly inhibited the mouse paw edema induced by CRG and ear edema induced by single and multiple doses of TPA. On11 (ID_{50} 0.18 μmol/ear) showed a potency twice as high as that of indomethacin (ID_{50} 0.35 μmol/ear) in the acute TPA (2.5 μg/ear) model, and was active after a long latency period against EPP edema, as are glucocorticoids. When the putative corticoid-like mechanism of On11 was studied, it was notably reduced by actinomycin D.

Bupleurum species such as *B. chinense* or *B. falcatum* furnish some crude drugs and extracts commonly used as anti-inflammations in different parts of the world, especially in traditional Chinese medicine. Upon investigation of the MeOH extract of the aerial parts of *B. rotundifolium*, seven oleanane-type saponins were isolated of which five, viz., On16 (ID_{50} 288 nmol/ear), rotundioside F (On18; 99 nmol/ear), rotundioside E (On31; 128 nmol/ear), rotundioside C (On41; 297 nmol/ear), and On44 (248 nmol/ear), were proved to be effective against TPA (2.5 μg/ear)-induced ear edema in mice (Navarro *et al.* 2001). Saponins On31 and On41 were also active on a TPA multiple-dose model of skin chronic inflammation.

The aqueous extract of the roots of *Zanha africana* (Sapindaceae) is used in traditional medicine as a remedy for dysentery. Three oleanane saponins, zanhasaponins A (On60, ID_{50} 14 μg/ear), B (On61, 20 μg/ear), and C (On62, 79 μg/ear), isolated from the MeOH extract of the root bark of *Z. africana*, were active against TPA (2.5 μg/ear)-induced ear edema (Cuellar *et al.* 1997b). These saponins may be considered as responsible, at least partially, for the activity exhibited by the plant extract.

The stem tuber of *Bolbostemma paniculatum* (Cucurbitaceae), a traditional Chinese medicine, was very effective in resolving masses and reducing swellings. Three oleanane saponins, tubeimosides I (On63, 83% inhibition at 0.11 μmol/ear), II (On64, 100% inhibition at 0.11 μmol/ear), and III (On65, 93% inhibition at 0.11 μmol/ear), isolated from the tuber have been shown to inhibit TPA (2 μg/ear)-induced mouse ear edema anti-inflammatory effects (Yu *et al.* 2001). The presence of C-16 hydroxyl group in On64 has been suggested to be responsible for its higher activity than the other two.

In the course of finding new anti-inflammatory natural products, some hederagenin glycosides (100 mg/kg o.d.), viz., loniceroside A (On51), dipsacussaponin B (On53), and dipsacussaponin C (On54), were found to show significant inhibition of ear edema (20-37% inhibition) against AA- and croton oil-induced inflammation, though less potent than indomethacin and hydrocortisone (Kim *et al.* 1999). Three anti-inflammatory saponin components were isolated from the alkaline hydrolysate of a BuOH-soluble portion of *Kalopanax pictus* (Araliaceae) bark extract through an *in vivo* activity-guided fractionation procedure (Li *et al.* 2003). The hydrolysate showed inhibition of adjuvant induced arthritis in rats. After further fractionation, the EtOAc fraction exhibited antiarthritic activity, which resulted in the isolation of α-hederin (On46), α-hederin methyl ester (On47), and kalopanaxsaponin I (On48). All compounds showed inhibition of vascular permeability in mice, but only On47 showed anti-CRG activity in rats and antiarthritic activity in rats and mice. Loniceroside C (On52; 100 mg/kg, p.o.), a hederagenin glycoside, isolated from the aerial parts of *Lonicera japonica* (Caprifoliaceae), showed *in vivo* anti-inflammatory activity (31% inhibition) against mouse ear edema provoked by croton oil (Kwak *et al.* 2003). Aspirin (100 mg/kg, p.o.) exhibited 15% inhibition under the same condition.

Two oleanane saponins On12 (85% inhibition at 0.3 mg/ear dose) and On14 (86% inhibition) isolated from the aerial parts of *Abrus precatorius*, and their acetate derivatives, On13 (95% inhibition) and On14 (95% inhibition), have been tested for anti-inflammatory activity using the croton oil (74 μg/ear) ear model. All the compounds exhibited anti-inflammatory activity but the acetates showed greater inhibition than the parent compounds (Anam 2001).

Lupane-Type Triterpenoid Saponins

Extracts from *Bupleurum fruticescens* (Umbelliferae) were examined for oral and topical anti-inflammatory activities. The BuOH extract proved to be active against CRG and TPA acute edemas and allowed the isolation of three lupane saponins, fruticesaponins A (Lp18), B (Lp19), and C (Lp20) (Just *et al.* 1998). These compounds were studied against CRG, TPA, AA, and EPP acute edemas. Lp19, a bidesmosidic saponin with an unbranched saccharide moiety was the most active in all the tests applied, viz. CRG (1.5 mg): 33% inhibition at 3 h and 100 mg/kg o.d.; TPA (2.5 μg): 71% inhibition at 0.5 mg/ear dose; AA (2 mg): 9% inhibition at 0.3 mg/ear dose; and EPP (1 mg): 32% inhibition at 0.5 mg/ear.

Cucurbitane-Type Triterpenoid Saponins

The roots of *Bryonia dioica* Jacq. (white bryony; Cucurbitaceae), Bryonia Radix, are used as purgative, emmenagogue, and a treatment for gout. Six cucurbitane-type triterpene glycosides isolated from the root extract, viz., bryoamaride (Cb12), bryonioside G (Cb13), bryonioside B (Cb14), bryonioside C (Cb15), cabenoside D (Cb16), and bryonioside E (Cb17), have been evaluated for their inhibitory effects on TPA (1 μg/ear)-induced inflammation. All compounds tested showed marked anti-inflammatory effects with ID_{50} 0.2–0.6 mg/ear which are almost the same with that of a reference drug indomethacin (ID_{50} 0.3 mg/ear) (Ukiya *et al.* 2002).

2. INHIBITION OF INFLAMMATORY MEDIATORS

Oleanane- and Ursane-Type Triterpenoids

Ursolic acid (Ur6), isolated from the hexane extract of *Plantago major*, showed a significant COX-2 inhibitory effect, directly on the enzyme activity, with an IC_{50} value of 130 μM and a COX-2/COX-1 selectivity ratio of 0.6 (Ringbom *et al.* 1998). The structural isomer oleanolic acid (On33) was found to be less active than Ur6 (IC_{50} 295 μM), but showed a similar selectivity ratio (0.8). No significant inhibition on COX-2 or COX-1 was observed by glycyrrhetinic acid (On70). The direct inhibitory effect of Ur6 and On33 on COX-2 catalyzed prostaglandin biosynthesis increased with preincubation, indicating a time-dependent inhibition, while the effect on COX-1 was found to be independent of preincubation time (Ringbom *et al.* 1998). Ur6 was revealed to suppress tumor necrosis factor (TNF)-induced expression of cyclin D1, COX-2, and matrix metalloproteinase-9 expression, which are NF-κB-regulated genes involved in initiation, promotion, and metastasis of tumors (Shishodia *et al.* 2003). This is consistent with that mentioned above (Ringbom *et al.* 1998) demonstrating that Ur6 blocks expression of COX-2 which correlates to both anti-inflammatory and chemopreventive properties. Ur6 inhibits expression of inducible NO synthase (iNOS) (Suh *et al.* 1998), and iNOS is another gene known to be regulated by NF-κB (Shishodia *et al.* 2003).

On33 and Ur6 were tested for interactions with the COX and 5-LOX pathways of AA metabolism in calcium ionophore-stimulated mouse peritoneal macrophages and human platelets (Diaz *et al.* 2000). On33 (IC_{50} 23.51 μM) and Ur6 (60.91 μM) showed a significant effect on PGE_2-release, with less inhibition percentages than reference drug indomethacin (IC_{50} 0.95 μM). In the LTC_4-assay, only On33 showed a significant effect (IC_{50} 16.79 μM). In addition, in the assay on thromboxane B_2 (TXB_2)-release induced by calcium ionophore in human platelets, Ur6 (IC_{50} 50.21 μM) showed a significant effect although with less potency than the reference drug ibuprofen (IC_{50} 1.27 μM). Ur6 was investigated for the expression of COX-2 in TPA-treated human mammary and oral epithelial cells (Subbaramaniah *et al.* 2000). Treatment with Ur6 suppressed TPA-mediated induction of COX-2 protein and synthesis of PGE_2. Ur6 also suppressed the induction of COX-2 mRNA by TPA. Ur6 inhibited TPA-mediated activation of PKC, extracellular signal-regulated kinase 1/2, c-Jun *N*-terminal kinase, and p38 mitogen-activated protein kinases. These data are important for understanding the anti-inflammatory properties of Ur6. On33 and Ur6 were tested furthermore for their ability to modulate the activities of several cytochrome P450 (CYP) enzymes using human liver microsomes (Kim *et al.* 2004). On33 was revealed to have an inhibitory effect on CYP1A2 and CYP3A4, whereas Ur6 inhibited CYP2C19. It is thus likely that either consumption of herbal medicine containing On33 and Ur6, or administration of On33 or Ur6 themselves, can cause drug interactions in humans when co-administered with drugs that are metabolized primarily by CYPs. In addition, it appears that the inhibitory effect of On33 on CYP1A2 is, in part, related to its pharmacological activities including anti-inflammatory and anticancer activities.

More than eighty novel triterpenoids have been synthesized from On33 and Ur6, and have been tested for their ability to suppress the *de novo* formation of two enzymes, iNOS and COX-2 (Suh *et al.* 1998). Two synthetic oleananes, 3, 11-dioxoolean-1,12-dien-28-oic acid (On86) and 3,12-dioxoolean-1-en-28-oic acid (On80), were found to be highly active. Then, new derivatives with electron-withdrawing substituents at the C-2 position of 3-oxoolean-1-en-28-oic acid (On79) were synthesized (Honda *et al.* 1997, Suh *et al.* 1998). Among them, 2-cyano-3,12-dioxooleane-1,9(11)-dien-28-oic acid (CDDO) (On81) was 400 times potent than previous compounds prepared as an inhibitor of iNOS (IC_{50}, 0.4 nM). The potency of On81 was similar to that of dexamethasone, although On81 does not act through the glucocorticoid receptor (Honda *et al.* 1998, Suh *et al.* 1999). On81 suppressed the abilities of various inflammatory cytokines, such as IFN-γ, IL-1, and TNF-α, to induce *de novo* formation of the enzymes iNOS and COX-2 in mouse peritoneal macrophages, rat brain microglia, and human colon fibroblasts. These activities have been found at concentrations ranging from 10^{-6} to 10^{-9} M in cell cultures (Suh *et al.* 1999). Subsequently, novel oleanane triterpenoids with modified rings A and C were synthesized, and among them, methyl 2-carboxy-3,12-dioxooleana-1,9(11)-dien-28-oate (On83) showed similar high inhibitory activity (IC_{50} 0.8 nM) to On81 against production of NO induced by IFN-γ in mouse macrophages (Honda *et al.* 1999). Nineteen novel olean- and urs-12-ene trieterpenoids with a 1-en-3-one functionality having a substituent at C-2 in ring A have been newly synthesized (Honda *et al.* 2000). Among them, 3-oxooleana-1,12-diene derivatives with carboxyl, methoxycarbonyl, and nitrite groups at C-2 showed high activity. In particular, 2-carboxy-3-oxooleana-1,12-dien-28-oic acid (On85) had the highest activity (IC_{50} 0.07 μM) in this group of triterpenodis. The potency of On85 was similar to that of hydrocortisone (IC_{50} 0.01 μM), although On85 does not act through the glucocorticoid receptor (Honda *et al.* 2000b). Overall, the combination of modified rings A and C increases the potency by about 10,000 times compared with the lead compound 3-oxooleana-1,12-dien-28-oic acid (On84) (IC_{50} 1 μM level). The selected oleanane triterpenoid, CDDO (On81), was found to be a potent, multifunctional agent in various *in vitro* assays and to show anti-inflammatory activity against thioglycolate-IFN-γ-induced mouse peritonitis (Honda *et al.* 2000a). Recently, it was shown that 1[2-cyano-3,12-dioxooleana-1,9(11)-dien-28-oyl]imidazole (CDDO-Im) (On82), a novel synthetic triterpenoid, is more potent than its parent compound, On81, both *in vitro* and *in vivo* (Place *et al.* 2003). *In vitro*, On82 is a potent inhibitor of *de novo* iNOS expression in primary mouse macrophages.

Glycyrrhetinic acid (On70) is formed from glycyrrhizin (On72) in the intestine by intestinal bacteria. On70 was investigated for its effects on the release of NO and at the level of iNOS gene expression in mouse macrophage (Jeong and Kim 2002). On70 elicited a dose-dependent increase in NO production and in the level of iNOS mRNA. Transient expression assays with NF-κB binding sites linked to the luciferase gene revealed that the increased level of iNOS mRNA, induced by On70, was mediated by the NF-κB transcription factor complex. Thus, On70 stimulates NO production and is able to up-regulate iNOS expression through NF-κB transactivation in macrophages.

The caffeoyl derivatives of oleanolic acid (On34) and morolic acid (On75) isolated from evening primrose oil showed radical scavenging activity. These compounds exhibited potent *in vitro* inhibitory effects on COX-1 and COX-2 and on human leukocyte elastase (Humberger *et al.* 2002). It thus appeared that these lipophilic caffeoyl derivatives could be of importance for the quality of evening primrose oil.

The inhibitory effects of oleanane-type triterpenoids from fabaceous plants on the TNF-α-induced expression of cell adhesion molecules on THP-1 human monocytic leukemia cells were examined (Ahn *et al.* 2002). Abrisapogenol E (On9), and soyasapogenols B (On7) and C (On28) significantly inhibited adhesion molecule (ICAM-1) expression (39.5% inhibition at 30 μM) in a cell-based ELISA. Moreover, these triterpenoids showed the same activity as dexamethasone. It was concluded that the activity of oleanane saponins and sapogenins against ICAM-1 expression are dependent upon the position of the hydroxyl group, and in particular upon the status of the C-21 and C-24 positions and of the glycosyl group at C-3 position.

The anticomplementary activity of ginseng saponins and their degradation products obtained by chemical treatment of Korean red ginseng saponins was investigated (Kim *et al.* 1998). The total saponin and its major components showed anticomplement activity through the classical pathway, and among which three oleananes, oleanolic acid (On33; IC_{50} 69 μM), ginsenoside Ro (On38, Chikusetsusaponin V; IC_{50} 58 μM), and 6′-*O*-methyl-ginsenoside Ro (On39; IC_{50} 10 μM), showed strong anticompliment activity.

The effect of saikogenin D (On32) on PGE_2 production and intracellular Ca^{2+} concentration ($[Ca^{2+}]_i$) in C6 rat glioma cells was examined (Kodama *et al.* 2003). On32 (1-20 μM) inhibited PGE_2 production induced by the Ca^{2+} ionophore A23187 in a concentration-dependent manner with the IC_{50} of about 3 μM. On32 did not affect the conversion of AA into PGE_2 in microsomal preparations. On the other hand, On32 elevated $[Ca^{2+}]_i$ in a concentration-dependent manner (10-100 μM) with the EC_{50} value of about 35 μM in the presence or absence of extracellular Ca^{2+}. These results suggest that On32 possesses a dual effect: an inhibition of A23187-induced PGE_2 production without a direct inhibition of COX activity; and an elevation of $[Ca^{2+}]_i$ that is attributed to Ca^{2+} release from intracellular stores.

Ursolic acid (Ur6), corosolic acid (Ur7) , and 3-epicorosolic acid (Ur8) were isolated from the herb extract of *Prunella vulgaris* (Labiatae) (Ryu *et al.* 2000). Ur8 was demonstrated significant inhibition on the release of *β*-hexosaminidase from the cultured RBL-2H3 cells in a dose dependent manner (IC_{50} 57 μM). On the other hand, Ur6 (IC_{50} 17 μM) and Ur7 (IC_{50} 27 μM) exhibited strong inhibitory activity against the production of NO from cultured LPS-stimulated murine macrophage RAW 264.7 cells (Ryu *et al.* 2000).

Lupane-Type Triterpenoids

The anti-infammatory triterpenoid lupeol (Lp1) was evaluated in *in vitro* assays (Fernandez *et al.* 2001). Lp1 significantly reduced PGE_2 production from A23187-stimulated macrophages, but failed to affect LTC_4 release. It was a weak inhibitor of NO release, but dose-dependently suppressed PGE_2. Cytokine production (TNF-*α* and IL-1*β*) was inhibited in the range 10-100 μM in LPS-treated macrophage. Thus, Lp1 is able to prevent the production of some pro-inflammatory mediators, which likely contributed to its anti-inflammatory activity. Betulinic acid caffeate (Lp2717) isolated from evening primrose oil showed radical scavenging activity. This exhibited potent *in vitro* inhibitory effects on COX-1 and COX-2 and on human leukocyte elastase (Humberger *et al.* 2002).

Lanostane-Type Triterpenoids

The fruiting body of *Ganoderma lucidum* (Polyporaceae) is well-known in oriental medicine, and has been used clinically in East Asia as both a tonic and a sedative. It is also used for the treatment of chronic bronchitis, cancer, arteriosclerosis, hypertension, and muscular dystrophy. Twelve lanostane-type triterpenoids were isolated from the *G. lucidum* spores, and among which, ganoderiol F (La12; IC_{50} 4.8 μM), ganodermanondiol (La13; IC_{50} 41.7 μM), and ganodermanontriol (La14; IC_{50} 17.2 μM) were found to exhibit a strong anti-complement activity against the classical pathway (CP) of the complement system (Min *et al.* 2001).

Cucurbitane-Type Triterpenoids

Bioassay-guided purification of an extract of *Cucurbita andreana* fruits yielded cucurbitacins B (Cb6), D (Cb8), E (Cb10), and I (Cb11) (Jayaprakasam *et al.* 2003). These cucurbitacins were evaluated for their inhibitory effects on COX-1 and COX-2 enzymes and on lipid peroxidation. Inhibitory potentials against COX-2 enzymes were Cb6 (32%), Cb8 (29%), Cb10 (35%), and Cb11 (27%) at 100 μg/mL dose. These compounds however, did not inhibit the COX-1 enzyme at this concentration. Ibuprofen (2.1 μg/mL), a reference drug, exhibited 53% COX-1 and 53% COX-2 inhibitory activities. Cb6 and Cb11 inhibited lipid peroxidation by 59 and 23%, respectively, at 100 μg/mL. The effects of 23,24-dihydrocucurbitacin D (Cb7) on macrophage NO generation were examined (Park *et al.* 2004a). Cb7 (<80 μM) effectively abolished NO generation from macrophages activated with lipopolysaccharide (LPS) and INF-γ. Cb7 decreased the levels of protein and mRNA for iNOS, and potently blocked NF-κB activation, a process necessary for transcriptional activation of iNOS. These results suggested that Cb7 inhibited NO generation by blocking NF-κB activation and iNOS gene transcription. Because NF-κB activation is necessary not only for NO generation but also for many inflammatory processes, Cb7 and its derivative could be developed as anti-inflammatory drugs.

Dammarane-Type Triterpenoids

(20*S*)-Protopanaxatriol (Da12), one of the major metabolites of ginsenosides by human intestinal bacteria, has been found to block the increase in LPS-induced iNOS and COX-2 expressions through inactivation of NF-κB by preventing I-κBα phosphorylation and degradation (Oh *et al.* 2004). Thus, Da12 may be a useful agent for chemoprevention of inflammatory diseases or cancer.

Euphae/Tirucallane-Type Triterpenoids

Commercially available extracts from *Boswellia serrata* resin used as anti-inflammatory drugs or phytonutrients show paradoxical concentration-dependent potentiating and inhibitory actions on 5-LOX product synthesis in stimulated polymorphonuclear cells (PMNs) (Boden *et al.* 2001). 3-Oxotirucallic a cid (Et8), a stimulating constituent, was shown to act as inhibitor of the 5-LOX in the cell-free assay, initiate 5-LOX activation by a mitogen-activated protein kinase (MEK)-inhibitor sensitive mechanism and potentiate stimulated product syntheses in intact cells. Since Et8 contributes significantly to the overall biological effects of *B. serrata* resin extracts, special precaution for standardization is recommended when using *B. serrata* preparations as drugs or dietary supplements.

Other Types of Triterpenoids

The MeOH extract of the wax gourd, the fruits of *Benincasa hispida* (Cucurbitaceae), was found to show inhibitory activity on histamine release from rat exudates cells induced by antigen-antibody reaction. Through bioassay-guided separation, several compounds were isolated, and among the active principles, multiflorenol (Mu1; 34% inhibition at 1 nM) and alnusenol (Ms6; 16% inhibition at 1 nM), were found to potently inhibit the histamine release (Yoshizumi *et al.* 1998). The aqueous acetone extract from the roots of a Chinese herbal medicine, *Rubia yunnanensis*, showed potent inhibitory effect on NO production in LPS-activated macrophages (Morikawa *et al.* 2003). Among the triterpenoids isolated from the extract, rubianol-d (Ms4) and rubianol-e (Ms5) showed potent inhibitory activity. The MeOH extract of the roots of *Celastrus orbiculatus* (Celasturaceae) showed potent inhibition of NF-κB activation in RAW264.7 cells transfected with NF-κB reporter construct. Celastrol (Ms3), a component of the extract, exhibited strong activity in both NF-κB activation (IC_{50} 0.27 μM) and NO production (IC_{50} 0.23 μM) (Jin *et al.* 2002). Ms3 is further proved to have effects of suppression of TNF-α and IL-1β production in human monocytes and by human lung macrophages, and suppression of induced NO production by macrophage lineage cells and by endothelial cells (Allison *et al.* 2001).

Oleanane-Type Tritepenoid Saponins

Anticomplement activity of hederagenin (On45) and related saponins isolated from *Dipsacus asper* was investigated *in vitro*. HN saponin F (On55; IC_{50} 3.7 $\times$ 10^{-5} M) was the most potent followed by On58 (7.9 $\times$ 10^{-5} M), On57 (9.9 $\times$ 10^{-5} M), dipsacussaponin A (On56; 1.1 $\times$ 10^{-4} M), and On45 (1.6 $\times$ 10^{-4} M) on the CP of complement system (Oh *et al.*

1999). Three oleanolic acid bisdesmosides, On66, On67, and On68, isolated from the MeOH extract of *Tiarella polyphylla* were shown to possess anticomplement activities (Park *et al.* 1999).

The effects of various hederagenin monodesmosides isolated from the stem bark of *Kalopanax pictus* have been evaluated on LPS-induced NO, PGE_2, and TNF-α release by the macrophage cell line RAW 264.7. Among the tested monodesmosides, kalopanaxsaponin A (α-hederin, On46) was the most potent inhibitor of NO production, and it also significantly decreased PGE_2 and TNF-α release. In consistent with these observations, the expression level of iNOS and COX-2 enzyme was inhibited by On46 in a concentration-dependent manner. Thus, it was suggested that On46-mediated inhibition of iNOS, COX-2 expression, and TNF-α release may be one of the mechanisms responsible for the anti-inflammatory effects of the stem bark of *K. pictus* (Kim *et al.* 2002).

Esculentoside A (On69), isolated from the roots of the Chinese herb *Phytolacca esculenta*, is reported to possess potent anti-inflammatory effects in acute and chronic experimental models. *In vitro* experiments, it was demonstrated that On69 (0.1–10 μmol/l) significantly reduced the release of TNF from the peritoneal macrophages derived from mice pretreated with thioglycolate. IL-1 and IL-6 secretion was also obviously inhibited in a concentration-dependent manner by On69 from 0.01 to 10 μmol/l. Inhibition of inflammatory cytokine production may contribute to the anti-inflammatory effects of On69 (Ju *et al.* 1998).

In screening for *in vitro* effects of anti-inflammatory saikosaponins, buddlejasaponin I (On22), saikosaponin 1 (buddlejasaponin IV, On24) and saikosaponin 2 (On23), on cellular systems generating COX and LOX metabolites, all of the saponins exhibited a significant effect. The action is more marked on LOX metabolite LTC_4. These support the inhibition of AA metabolism as one of the biochemical mechanisms that might be the rationale for the putative antiphlogistic activity of these saikosaponins (Benito *et al.* 1998). Saikosaponin a (On20), isolated from the extract of *Bupleurum falcatum*, at more than 1 mg/kg (i.v.) significantly inhibited the passive cutaneous anaphylaxis (PCA) reaction in rats in a dose-dependent manner, attaining a maximum inhibition of ca. 60% with 10 mg/kg. On20 at 3 and 10 mg/kg also suppressed asthmatic broncho-constriction in sensitized guinea pigs. On20 possesses a weak inhibitory activity on histamine-induced tracheal contraction in guinea-pigs and on histamine release induced by A23187 in rat mast cells. These indicate that On20 has an inhibitory activity against allergic asthma which seems to derive from both antagonism of the histamine action and inhibition of allergic mediators (Park *et al.* 2002).

Avicin G (elliptoside A, On87) was isolated from the Australian desert tree *Acacia victoriae* (Fabaceae), and this was found to be a potent inhibitor of TNF-induced NF-κB in human T cell leukemia. On87 reduced the expression of NF-κB-regulated proteins such as iNOS and COX-2, those form an essential component of an inflammatory cascade critical for the repair of oxidative damage. On87 may prove important for reducing both oxidative and nitrosative cellular stress (Haridas *et al.* 2001).

Hepatoprotective Oleanane-Type Triterpenoid Saponins

Oleanane-type tritepenoid saponins are known to exhibit hepatoprotective activity. These saponins are characterized by possessing a oleanene-glucuronide structure with olean-12-ene type skeleton, a C-28 methyl group, and a glucuronic acid moiety linked at C-3 of the aglycone. Glycyrrhizin (On72) is a representative one, and a preparation of this saponin has been widely used as a clinical antihepatitis drug (Safayhi and Sailer 1997). Other well known hepatoprotective saponins are saikosaponins, and soyasaponins (Kinjo *et al.* 1999b, Miyao *et al.* 1998, Kinjo *et al.* 1998, Ikeda *et al.* 1998, Udayama *et al.* 1998, Kinjo *et al.* 1999c).

Dammarane-Type Triterpenoid Saponins

When ginsenoside R_c (Da10) was anaerobically incubated with human fecal microflora, all specimens metabolized Da10 to compound K (Da6) and (20*S*)-protopanaxadiol (Da5). The inhibitory activity on the release of β-hexosaminidase and NO production were tested for Da10 and its metabolites, and Da6 exhibited the most potent activity (β-hexoaminidase release inhibition: IC_{50} 24 μM; NO production inhibition: ID_{50} 53 μM) (Bae *et al.* 2002). Antiallergic property of Da6 was further substantiated by the strong inhibitory activity of histamine release from rat peritoneal mast cells by compound 48/80 (IC_{50} 13 μM) and of IgE-induced passive PCA reaction in mice (98% inhibition at 25 mg/kg dose p.o.; 99% inhibition at 25 mg/kg dose i.p.) (Choo *et al.* 2003). The antiallergic activities of ginsenosides, which were isolated from acid-treated ginseng (*Panax ginseng*, Araliaceae), and their metabolites by human intestinal bacteria were measured (Park *et al.* 2003). Ginsenoside Rh_2 (Da15), a main metabolite, had the most potent inhibitory activity (IC_{50} 30 μM) on β-hexosaminidase release from RBL-2H3 cells and in the PCA reaction. The inhibitory activity of Da15 was more potent than that of disodium cromoglycate (IC_{50} 0.5 mM), a commercial antiallergic drug. Da15 showed membrane stabilizing action upon differential scanning calorimetry and inhibited NO (IC_{50} 32 μM) and PGE_2 (IC_{50} 8 μM) in LPS-stimulated RAW cells. These suggest that Da15 can exhibit antiallergic activity originating from cell membrane-stabilizing activity and anti-inflammatory activity by the inhibition of NO and PGE_2 production. Ginsenoside Rb_1 (Da7) from ginseng possesses anti-inflammatory properties, and this was proved to inhibit the production of LPS-induced pro-inflammatory cytokines, IL-6 and TNF-α in culture (Smolinski *et al.* 2003). Ginsenoside Rh_1 (Da13) also was evaluated for its antiallergic properties (Park *et al.* 2004b). Da13 potently inhibited histamine release from rat peritoneal mast cells and the IgE-mediated PCA reaction (87% inhibition at 25 mg/kg) in mice. Da13 was also found to have a membrane-stabilizing action as revealed by differential scanning calorimetry. Futhermore, it inhibited iNOS and COX-2 protein expression, and activated NF-κB in nuclear fraction. The antiallergic action of Da13 may originate from its cell membrane-stabilizing and anti-inflammatory activities, and can improve the inflammation caused by allergies (Park *et al.* 2004b).

SUMMARY

Triterpenoids are widely distributed in plants, and in many cases are the principles responsible for the anit-inflammatory and anti-allergic effects. Many of these compounds are active in different *in vivo* experimental models such as ear edema induced by phorbol esters, ethylphenylpropiolate, and AA, hind paw edema induced by CRG, serotonin, and PLA_2, adjuvant arthritis, and experimental models of allergy. Other effects have been studied *in vitro*, and some triterpenoids are active against inflammatory enzymes like 5-LOX, elastase, and PLA_2. Others inhibit histamine and interleukin release, metabolism of endogeneous corticoids, and complement and protein-kinase activities. In the future, more mechanistic-oriented basic research is needed to elucidate the mechanism of action. Studies of derivatives of naturally occurring triterpenoids are also necessary to elucidate the structure-activity relationship and to guide the development of novel anti-inflammatory drugs.

List of Abbreviations

AA = Arachidonic acid
BRK = Bradykinin
COX-2 = Cyclooxygenase-2
CRG = Carrageenan
CP = Classical pathway
CYP = Cytochrome P450
DPP = 12-Deoxyphorbol-13-phenylacetate

DPT = 12-Deoxyphorbol-13-tetradecanoate
EPP = Ethyl phenylpropiolate
HPA = 12-*O*-Hexadecanoyl-16-hydroxyphorbol-13-acetate
IFN-γ = Interferon-gamma
IgE = Immunoglobulin E
IL-1 = Interleukin-1
iNOS = Inducible NO synthase
LOX = Lipoxygenase
LTs = Leukotrienes
NF-κB = Nuclear factor-κB
NO = Nitrogen monoxide
NSAIDs = Nonsteroidal anti-inflammatory drugs
PAF = Platelet-activating factor
PCM = Passive cutaneous anaphylaxis
PG = Prostaglandin
PKC = Protein kinase C
PLA_2 = Phospholipase A_2
PMNL = Polymorphonuclear leucocytes
TNF-α = Tumor necrosis factor
TPA = 12-*O*-Tetradecanoylphorbol-13-acetate

REFERENCES

Akihisa, T., K. Yasukawa, Y. Kimura, M. Takido, W.C.M.C. Kokke, and T. Tamura. 1994a. Five D:C-friedo-oleanane triterpenes from the seeds of *Trichosanthes kirilowii* Maxim. and their anti-inflammatory effects. *Chem. Pharm. Bull.* 42: 1103-1105.

Akihisa, T., K. Yasukawa, Y. Kimura, M. Takido, W.C.M.C. Kokke, and T. Tamura. 1994b. 7-Oxo-10*α*-cucurbitadienol from the seeds of *Trichosanthes kirilowii* and its anti-inflammatory effect. *Phytochemistry* 36: 153-157.

Akihisa, T., K. Yasukawa, H. Oinuma, Y. Kasahara, S. Yamanouchi, M. Takido, K. Kumaki, and T. Tamura, 1996. Triterpene alcohols from flowers of Compositae and their anti-inflammatory effects. *Phytochemistry* 43: 1255-1260.

Akihisa, T., K. Yasukawa, and Y. Kasahara. 1997a. Triterpenoids from the flowers of Compositae and their anti-inflammatory effects. *Current Topics Phytochem.* 1: 137-144.

Akihisa, T., K. Yasukawa, Y. Kimura, S. Takase, S. Yamanouchi, and T. Tamura. 1997b. Triterpene alcohols from camellia and sasanqua oils and their anti-inflammatory effects. *Chem. Pharm. Bull.* 45: 2016-2023.

Akihisa, T., K. Yasukawa, Y. Kimura, S. Yamanouchi, and T. Tamura. 1998. Sasanquol, a 3,4-*seco*-triterpene alcohol from sasanqua oil, and its anti-inflammatory effect. *Phytochemistry* 48: 301-305.

Akihisa, T., K. Yasukawa, M. Yamaura, M. Ukiya, Y. Kimura, N. Shimizu, and K. Arai. 2000. Triterpene alcohol and sterol ferulates from rice bran and their anti-inflammatory effects. *J. Agric. Food Chem.* 58: 2313-2319.

Akihisa, T., and K. Yasukawa. 2001. Antitumor-promoting and anti-inflammatory activities of triterpenoids and sterols from plants and fungi. Pages 43-87 in *Studies in Nat. Prod. Chem., Vol. 25, Bioactive Nat. Prod. (Part F)*, Atta-ur-Rahman, ed., Elsevier Science B. V., Amsterdam.

Akihisa, T., J. Ogihara, J. Kato, K. Yasukawa, M. Ukiya, S. Yamanouchi, and K. Oishi. 2001. Inhibitory effects of triterpenoids and sterols on human immunodeficiency virus-1 reverse transcriptase. *Lipids* 36: 507-512.

Akihisa, T., Y. Mizushina, M. Ukiya, M. Oshikubo, S. Kondo, Y. Kimura, T. Suzuki, and T. Tai. 2004. Dehydrotrametenonic acid and dehydroeburiconic acid from *Poria cocos* and their inhibitory effects on eukaryotic DNA polymerase α and β. *Biosci. Biotechnol. Biochem.* 68: 448-450.

Akihisa, T., and K. Yasukawa. 2001. Antitumor-promoting and anti-inflammatory activities of triterpenoids and sterols from plants and fungi. Pages 43-87 in *Studies in Nat. Prod. Chem., Vol.* 25, *Bioactive Nat. Prod. (Part F)*, Atta-ur-Rahman, ed., Elsevier Science B. V., Amsterdam.

Allison, A.C., R. Cacabelos, V.R.M. Lombardi, X.A. Alvarez, and C. Vigo. 2001. Celastrol, a potent antioxidant and anti-inflammatory drug, as a possible treatment for Alzheimer's disease. *Prog. Neuro-Psychopharmacol. Biol. Psychiatry* 25: 1341-1357.

Alvarez, M.E., A.E. Rotelli, L.E. Pelzer, J.R. Saad, and O. Giordano. 2000. Phytochemical study and anti-inflammatory properties of *Lamopaya hieronymi* Schum. Ex. Moldenke. *Farmaco* 55: 502-505.

Ahn, K.-S., J.-H. Kim, S.-R. Oh, B.-S. Min, J. Kinjo, and H.-K. Lee. 2002. Effects of oleanane-type triterpenoids from fabaceous plants on the expression of ICAM-1. *Biol. Pharm. Bull.* 25: 1105-1107.

Anam, E.M. 2001. Anti-inflammatory activity of compounds isolated from the aerial parts of *Abrus precatorius* (Fabaceae). *Phytomedicine* 8: 24-27.

Bae, E.-A., M.-K. Choo, E.-K. Park, S.-Y. Park, H.-Y. Shin, and D.-H. Kim. 2002. Metabolism of ginsenoside Rc by human intestinal bacteria and its related antiallergic activity. *Biol. Pharm. Bull.* 25: 743-747.

Baltina, L.A. 2003. Chemical modification of glycyrrhizic acid as a route to new bioactive compounds for medicine. *Curr. Med. Chem.* 10: 155-171.

Banno, N., T. Akihisa, H. Tokuda, K. Yasukawa, H. Higashihara, M. Ukiya, K. Watanabe, Y. Kimura, J. Hasegawa, and H. Nishino. 2004. Triterpene acids from the leaves of *Perilla frutescens* and their anti-inflammatory and antitumor-promoting effects. *Biosci. Biotechnol. Biochem.* 68: 85-90.

Baricevic, D., S. Sosa, R.D. Loggia, A. Tubaro, B. Simonovska, A. Krasna, and A. Zupancic. 2001. Topical anti-inflammatory activity of *Salvia offinalis* L. leaves: the relevance of ursolic acid. *J. Ethnopharmacol.* 75: 125-132.

Benito, P.B., M.J.A. Martinez, A.M.S. Sen, A.S. Gomze, L.F. Matellano, S.S. Cotrearas, and A.M.D. Lanza. 1998. *In vivo* and *in vitro* antiinflammatory activity of saikosaponins. *Life Sci.* 63: 1147-1156.

Boden, S.E., S. Schweizer, T. Bertsche, M. Duefer, G. Drews, and H. Safayhi. 2001. Stimulation of leukotriene synthesis in intact polymorphonuclear cells by the 5-lipoxygenase inhibitor 3-oxo-tirucallic acid. *Mol. Pharmacol.* 60: 267-273.

Bortalanza, L.B., J. Ferreira, S.C. Hess, F. Delle Monache, R.A. Yunes, J.B. Calixto. 2002. Anti-allodynic action of the tormentic acid, a triterpene isolated from plant, against neuropathic and inflammatory persistent pain in mice. *Eur J. Pharmacol.* 453: 203-208.

Cantrell, C.L., S.G. Franzblau, and N.H. Fischer, N. H. 2001. Antimycobacterial plant terpenoids. *Planta Med.* 67: 685-694.

Choi, J., K.-T. Lee, S.-Y. Yun, C.-D. Ko, H.-J. Jung, and H.-H. Park. 2003. Antinociceptive and antiinflammatory effects of niga-ichinoside F_1 and 23-hydroxytormentic acid obtained from *Rubus coreanus*. *Biol. Pharm. Bull.* 26: 1436-1441.

Choo, M.K., E.K. Park, M.J. Han, and D.H. Kim. 2003. Antiallergic activity of ginseng and its ginsenosides. *Planta Med.* 69: 518-522.

Croft, K.D. 1993. Assessment of anti-inflammatory activity of natural products. Pages 265-278 in *Bioactive Natural Products - Detection, Isolation, and Structural Determination*, S. M. Colegate, and R.J. Molyneux, eds., CRC Press, Inc., Boca Raton, Florida.

Cuellar, M.J., R.M. Giner, M.C. Recio, M.J. Just, S. Manez, and J.L. Rios. 1997a. Effect of the basidiomycete *Poria cocos* on experimental dermatitis and other inflammatory conditions. *Chem. Pharm. Bull.* 45: 492-494.

Cuellar, M.J., R.M. Giner, M.C. Recio, M.J. Just, S. Manez, J.L. Rios, A.R. Bilia, J. Msonthi, and K. Hostettmann. 1997b. Three new oleanane saponins from *Zanha africana*. *J. Nat. Prod.* 60: 191-194.

Cuellar, M.J., R.M. Giner, M.C. Recio, M.J. Just, S. Manez, M. Cerda, K. Hostettman, J.-L. Rios. 1997c. Zanhasaponins A and B, antiphospholipase A_2 saponins from an antiinflammatory extract of *Zanha africana* root bark. *J. Nat. Prod.* 60: 1158-1160.

Cuellar, M.J., R.M. Giner, M.C. Resio, M.J. Justa, S. Manez, M. Cerda, and J.L. Rios. 1998. Screening of anti-inflammatory medicinal plants used in traditional medicine against skin diseases. *Phytother. Res.* 12: 18-23.

De Clercq, E. 2000. Current lead natural products for the chemotherapy of human immunodeficiency virus (HIV) infection. *Med. Res. Rev.* 20: 323-349.

De las Heras, B., B. Rodriguez, L. Bosca, and A.M. Villar. 2003. Terpenoids: sources, structure elucidation and therapeutic potential in inflammation. *Curr. Topics Med. Chem.* 3: 171-185.

Diaz, A.M., M.J. Abad, L. Fernandez, C. Recuero, L. Villaescusa, A.M. Silvan, and P. Bermejo. 2000. In vitro anti-inflammatory activity of iridoids and triterpenoid compounds isolated from *Phillyrea latifolia* L. *Biol. Pharm. Bull.* 23: 1307-1313.

Dirsch, V.M., A.K. Kiemer, H. Wagner, and A.M. Vollmar. 1997. The triterpenoid quinonemethide pristimerin inhibits induction of unducible nitric oxide synthase in murine macrophages. *Eur. J. Pharmacol.* 336: 211-217.

Duwiejua, M., I.J. Zeitlin, P.G. Waterman, J. Chapman, G.J. Mhango, and G. J. Provan. 1993. Anti-inflammatory activity of resins from some species of the plant family Burseraceae. *Planta Med.* 59: 12-16.

Fan, Y.M., L.Z. Xu, J. Gao, Y. Wang, X.H. Tang, X.N. Zhao, Z.X. Zhang. 2004. Phytochemical and anti-inflammatory studies on *Terminalia catappa*. *Fitoterapia* 75: 253-260.

Fernandez, M.A., B. de las Heras, M.D. Garcia, M.T. Saenz, and A. Villar. 2001. New insights into the mechanism of action of the anti-inflammatory triterpene lupeol. *J. Pharm. Pharmacol.* 53: 1533-1539.

Geetha, T., P. Varalakshmi, and R.M. Latha. 1998. Effect of triterpenes from *Crataeva nurvala* stem bark on lipid peroxidation in adjuvant induced arthritis in rats. *Pharmacol. Res.* 37: 191-195.

Geetha, T. and P. Varalakshmi. 1999. Anticomplement activity of triterpenes from *Crataea nurvala* stem bark in adjuvant arthritis in rats. *Gen. Pharmacol.* 32: 495-497.

Geetha, T., and P. Varalakshmi. 2001. Anti-inflammatory activity of lupeol and lupeol linoleate in rats. *J. Ethnopharmacol.* 76: 77-80.

Giner, E.-M., S. Manez, M.-C. Recio, R.-M. Giner, M. Cerda-Nicolas, and J.-L. Rios. 2000a. *In vivo* studies on the anti-inflammatory activity of pachymic and dehydrotumulosic acids. *Planta Med.* 66: 221-227.

Giner, E.-M., M.-L. Villalba, M.-C. Recio, S. Manez, M. Cerda-Nicolas, and J.-L. Rio. 2000b. Anti-inflammatory glycoterpenoids from *Scrophularia auriculata*. *Eur. J. Pharmacol.* 389: 243-252.

Giner, E.-M., M.-L. Villalba, M.-C. Recio, S. Manez, M. Cerda-Nicolas, and J.-L. Rio. 2000c. Anti-inflammatory glycoterpenoids from *Scrophularia auriculata*. *Eur. J. Pharmacol.* 389: 243-252.

Giner, E.-M., S. Manez, M.C. Recio, R.M. Giner, J.M. Prieto, M. Cerda-Nicolas, and J.L. Rios. 2001. Oleanonic acid, a 3-oxotriterpene from Pistacia, inhibits leukotriene synthesis and has anti-inflammatory activity. *Eur. J. Pharmacol.* 428: 137-143.

Giner, E.-M., S. Manez, R.M. Giner, M.C. Recio, J.M. Prieto, M. Cerda-Nicolas, and J.L. Rios. 2002. Anti-inflammatory triterpenes from *Pistacia terebinthus* galls. *Planta Med.* 68: 311-315.

Goad, L.J. and T. Akihisa. 1997. *Analysis of Sterols*, Blackie Academic & Professional, London.

Gould, H.J., B.J. Sutton, A.J. Beavil, R.L. Beavil, N. McCloskey, H.A. Coker, D. Fear, and L. Smurthwaite. 2003. The biology of IgE and the bass of allergic disease. *Ann. Rev. Immunol.* 21: 579-628.

Haridas, V., C.J. Arntzen, and J.U. Gutterman. 2001. Avicins, a family of triterpenoid saponins from *Acacia victoriae* (Bentham), inhibit activation of nuclear factor-*k*B by inhibiting both its nuclear localization and ability to bind DNA. *Proc. Natl. Acad. Sci. U.S.A.* 98: 11557-11562.

Hasmeda, M.G. Kweifio-Okai, T. Macrides, and G.M. Poyla. 1999. Selective inhibition of eukaryote protein kinases by anti-inflammatory triterpenoids. *Planta Med.* 65: 14-18.

Honda, T., H.J. Finlay, G.W. Gribble, N. Suh, and M.B. Sporn. 1997. New enone derivatives of oleanolic acid and ursolic acid as inhibitors of nitric oxide production in mouse macrophages. *Bioorg. Med. Chem. Lett.* 7: 1623-1628.

Honda, T., B.V. Rounds, G.W. Gribbe, N. Suh, Y. Wang, and M.B. Sporn. 1998. Design and synthesis of 2-cyano-3,12-dioooxoolean-1,9-dien-28-oic acid, a novel and highly active inhibitor of nitric oxide production in mouse macrophages. *Bioorg. Med. Chem. Let.* 8: 2711-2714.

Honda, T., B.V. Rounds, L. Bore, F.G. Favaloro, Jr., G.W. Gribble, N. Suh, Y. Wang, and M. B. Sporn. 1999. Novel synthetic oleanane triteprenoids; a series of highly active inhibitors of nitric oxide production in mouse macrophages. *Bioorg. Med. Chem. Lett.* 9: 3429-3434.

Honda, T., B.V. Rounds, L. Bore, H.J. Finlay, F.G. Favaloro, Jr., N. Suh, Y. Wang, M.B. Sporn, and G.W. Gribble. 2000a. Synthetic oleanane and ursane triterpenoids with modified rings A and C: a series of highly active inhibitors of nitric oxide production in mouse macrophages. *J. Med. Chem.* 43: 4233-4246.

Honda, T., G.W. Gribble, N. Suh, H.J. Finlay, R.V. Rouncs, L. Bore, F.G. Favaloro, Jr., Y. Wang, and M.B. Sporn. 2000b. Novel synthetic oleanane and ursane triterpenoids with various enone functionalities in ring A as inhibitors of nitric oxide production in mouse macrophages. *J. Med. Chem.* 43: 1866-1877.

Huguet, A.-I., M.C. Recio, S. Manez, R.-M. Giner, and J.-L. Rios. 2000. Effect of triterpenoids on the inflammation induced by protein kinase C activators, neuronally acting irritants and other agents. *Eur. J. Pharmacol.* 410: 69-81.

Humberger, M., U. Riese, H. Graf, M.F. Melzig, S. Ciesielski, D. Baumann, K. Dittmann, and C. Wegner. 2002. Constituents in evening primrose oil with radical scavenging, cyclooxygenase, and neutrophil elastase inhibitory activities. *J. Agric. Food Chem.* 50: 5533-5538.

Ikeda, T., M. Udayama, M. Okawa, T. Arao, J. Kinjo, and T. Nohara. 1998. Partial hydrolysis of soyasaponin I and the hepatoprotective effects of the hydrolytic products. Study of the structure-hepatoprotective relationship of soyasapogenol B analogs. *Chem. Pharm. Bull.* 46: 359-361.

Inoue, H., H. Saito, Y. Koshihara, and S. Murota. 1986. Inhibitory effect of glycyrrhetinic acid derivatives on lipoxygenase and prostaglandin synthetase. *Chem. Pharm. Bull.* 34: 897-901.

Inoue, H., T. Mori, S. Shibata, and Y. Koshihara. 1988. Inhibitory effect of glycyrrhetinic acid derivatives on arachidonic acid-induced mouse ear oedema. *J. Pharm. Pharmacol.* 40: 272-277.

Inoue, H., T. Mori, S. Shibata, and Y. Koshihara. 1989. Modulation by glycyrrhetinic acid derivatives of TPA-induced mouse ear oedema. *Br. J. Pharmacol.* 96: 204-210.

Inoue, H., K. Inoue, T. Takeuchi, N. Nagata, and S. Shibata. 1993. Inhibition of rat acute inflammatory paw oedema by dihemiphthalate of glycyrrhetinic acid derivatives: comparison with glycyrrhetinic acid. *J. Pharm. Pharmacol.* 45: 1067-1071.

Jayaprakasam, B., N.P. Seeram, and M.G. Nair. 2003. Anticancer and anti-inflammatory activities of cucurbitacins from *Cucurbita andreana*. *Cancer Lett.* 189: 11-16.

Jeong, H.G., and J.Y. Kim. 2002. Induction of inducible nitric oxide synthase expression by 18β-glycyrrhetinic acid in macrophages. *FEBS Lett.* 513: 208-212.

Jeong, T.C., H.J. Kim, J.I. Park, C.S. Ha, J.D. Park, S.I. Kim, and J.K. Roh. 1997. Protective effects of red ginseng saponins against carbon tetrachloride-induced hepatotoxicity in Sprague Dawley rats. *Planta Med.* 63: 136-140.

Jin, H.Z., B.Y. Hwang, H.S. Kim, J.H. Lee, Y.H. Kim, and J.J. Lee. 2002. Antiinflammatory constituents of *Celastrus orbiculatus* inhibit the NF-kappaB activation and NO production. *J. Nat. Prod.* 65: 89-91.

Ju, D.W., Q.Y. Zheng, X. Cao, J. Fang, and H.B. Wang. 1998. Esculentoside A inhibits tumor necrosis factor, interleukin-1, and interleukin-6 production induced by lipopolysaccharide in mice. *Pharmacology* 56: 187-195.

Just, M.J., M.C. Recio, R.M. Giner, M.J. Cuellar, S. Manez, A.R. Billia, and J.-L. Rios. 1998. Anti-inflammatory activity of unusual lupane saponins from *Bupleurum fruticescens*. *Planta Med.* 64: 404-407.

Kaminaga, Y., K. Yasukawa, H. Kanno, T. Tai, Y. Nunoura, and M. Takido. 1996a. Inhibitory effects of lanostane-type triterpene acids, the components of *Poria cocos*, on tumor promotion by 12-*O*-tetradecanoylphorbol-13-acetate in two-stage carcinogenesis in mouse skin. *Oncology* 53: 382-385.

Kaminaga, T., K. Yasuakwa, M. Takido, T. Tai, and Y. Nunoura, 1996b. Inhibitory effect of *Poria cocos* on 12-*O*-tetradecanoylphorbol-13-acetate-induced ear oedema and tumour promotion in mouse skin. *Phytother. Res.* 10: 581-584.

Kamo T., M. Asanoma, H. Shibata, and M. Hirota. 2003. Anti-inflammatory lanostane-type triterpene acids from *Piptopurus betulinus*. *J. Nat. Prod.* 66: 1104-1106.

Kapil, A., and S. Sharma. 1994. Anti-complement activity of oleanolic acid: an inhibitor of C_3-convertase of the classical complement pathway. *J. Pharm. Pharmacol.* 46: 922-923.

Kashiwada, Y., T. Nagao, A. Hashimoto, Y. Ikeshiro, H. Okabe, L.M. Cosentino, and K.-H. Lee. 2000, Anti-AIDS agents 38. Anti-HIV activity of 3-*O*-acetyl ursolic acid derivatives. *J. Nat. Prod.* 63: 1619-1622.

Keum, Y.-S., S.S. Han, K.-S. Chun, K.-K. Park, J.-H. Park, S.K. Lee, and Y.-J. Surh. 2003. Inhibitory effects of the ginsenoside Rg_3 on phorbol ester-induced cyclooxygenase-2 expression, NF-*k*B activation and tumor promotion. *Mutation Res.* 523-524: 75-85.

Kim, D.S., S.R. Oh, I.S. Lee, K.Y. Jung, J.D. Park, S.I. Kim, and H.-K. Lee. 1998. Anticomplementary activity of ginseng saponins and their degradation products. *Phytochemistry* 47: 397-399.

Kim, S.Y., K.H. Son, H.W. Chang, S.S. Kang, and H.P. Kim. 1999. Inhibition of mouse ear edema by steroidal and triterpenoid saponins. *Arch. Pharm. Res.* 22: 313-316.

Kim, Y.-K., R.-G. Kim, S.-J. Park, J.-H. Ha, J.-W. Choi, H.-J. Park, and K.-T. Lee. 2002. *In vitro* anti-inflammatory activity of kalopanaxsaponin A isolated from *Kalopanax pictus* in murine macrophage RAW 264.7 cells. *Biol. Pharm. Bull.* 25: 472-476.

Kim, K.-A., J.-S. Lee, H.-J. Park, J.-W. Kim, C.-J. Kim, I.-S. Shim, N.-J. Kim, S.-H. Han, and S. Lim. 2004. Inhibition of cytochrome P450 activities by oleanolic acid and ursolic acid in human liver microsomes. *Life Sci.* 74: 2769-2779.

Kimura, I., M. Yoshikawa, S. Kobayashi, Y. Sugihara, M. Suzuki, H. Oominami, T. Murakami, H. Matsuda, and V.V. Doiphode. 2001. New triterpenes, myrrhanol A and myrrhanone A, from guggul-gum resins, and their potent anti-inflammatory effect on adjuvant-induced air-pouch granuloma of mice. *Bioorg. Med. Chem. Lett.* 11: 985-989.

Kinjo, J., M. Imagire, M. Udayama, T. Arao, and T. Nohara. 1998. Structure-hepatoprotective relationship study of soyasaponins I-IV having soyasapogenol B as aglycone. *Planta Med.* 64: 233-236.

Kinjo, J., M. Udayama, M. Okawa, and T. Nohara. 1999a. Study of structure-hepatoprotective relationship of oleanane-type triterpenoidal glucuronides obtained from several fabaceous plants on rat primary hepatocyte cultures. *Biol. Pharm. Bull.* 22: 203-206.

Kinjo, J., M. Okawa, M., M. Udayama, Y. Sohno, T. Hirakawa, Y. Shii, and T. Nohara. 1999b. Hepatoprotective and hepatotoxic actions of oleanane acid-type triterpenoidal glucuronides on rat primary hepatocyte cultures. *Chem. Pharm. Bull.* 47: 290-292.

Kinjo, J., K. Aoki, M. Okawa, Y. Shii, T. Hirakawa, T. Nohara, Y. Nakajima, T. Yamazaki, T. Hosono, M. Someya, Y. Niiho, and T. Kurashige. 1999c. HPLC profile analysis of hepatoprotective oleanane-glucuronides in Puerariae Flos. *Chem. Pharm. Bull.* 47: 708-710.

Kodama, Y., X. Lu, C. Tsuchiya, Y. Ohizumi, M. Yoshida, and N. Nakahata. 2003. Dual effect of saikogenin D: in vitro inhibition of prostaglandin E_2 production and elevation of intracellular free Ca^{2+} concentration in C6 rat glioma cells. *Planta Med.* 69: 765-767.

Kohda, H., W. Tokumoto, K. Sakamoto, M. Fujii, M.Y. Hirai, K. Yamasaki, Y. Komoda, H. Nakamura, S. Ishihara, and M. Uchida. 1985. The biologically active constituents of *Ganoderma lucidum* (Fr.) Karst. Histamine release-inhibitory triterpenes. *Chem. Pharm. Bull.* 33: 1367-1374.

Kwak, W.J., C.K. Han, H.W. Chang, H.P. Kim, S.S. Kang, and K.H. Son. 2003. Loniceroside C, an antiinflammatory saponin from *Lonicera japonica*. *Chem. Pharm. Bull.* 51: 333-335.

Kweifio-Okai, G., F.D. Munk, T.A. Macrides, P. Smith, and B.A. Rumble. 1995. Antiarthritic mechanisms of lupeol triterpenes. *Drug. Dev. Res.* 36: 20-24.

Latha, R.M., M. Lenin, M. Rasool, and P. Varalakshmi. 2001. A novel derivative pentacyclic triterpene and w3 fatty acid [lupeol-EPA] in relation to lysosomal enzymes glycoproteins and collagen in adjuvant induced arthritis in rats. *Prostagl. Leukotri. Essent. Fatty Acids* 64: 81-85.

Lacaille-Dubois, M.A., and H. Wagner. 1996. A review of the biological and pharmacological activities of saponins. *Phytomedicine* 2: 363-386.

Li, D.W., J.E. Hyun, C.S. Jeong, Y.S. Kim, and E.B. Lee. 2003. Antiinflammatory activity of α-hederin methyl ester from the alkaline hydrolysate of the butanol fraction of *Kalopanax pictus* bard extract. *Biol. Pharm. Bull.* 26: 429-433.

Liu, J. 1995. Pharmacology of oleanolic acid and ursolic acid. *J. Ethnopharmacol.* 49: 57-68.

Loggia, R.D., A. Tubaro, S. Sosa, H. Becker, St. Saar, and O. Isaac. 1994. The role of triterpenoids in the topical anti-inflammatory activity of *Calendula officinalis* flowers. *Planta Med.* 60: 516-520.

Manez, S., M.C. Recio, R.M. Giner, and J.-L. Rios. 1997. Effect of selected triterpenoids on chronic dermal inflammation. *Eur. J. Pharmacol.* 334: 103-105.

Matsuda, H., M. Yoshikawa, M. Iinuma, and M. Kubo. 1998. Antinociceptive and anti-inflammatory activities of limonin isolated from the fruits of *Evodia rutaecarpa* var. *bodinieri*. *Planta Med.* 64: 339-342.

Min, B.S., J.J. Gao, M. Hattori, H.K. Lee, and Y.H. Kim. 2001. Anticomplement activity of terpenoids from the spores of *Ganoderma lucidum*. *Planta Med.* 67: 811-814.

Miyao, H., T. Arao, M. Udayama, J. Kinjo, and T. Nohara. 1998. Kaikasaponin III and soyasaponin I, major triterpene saponins of *Abrus cantoniensis*, act on GOT and GPT: Influence on transaminase elevation of rat liver cells concomitantly exposed to CCl_4 for one hour. *Planta Med.* 64: 5-7.

Mizushina, Y., T. Akihisa, M. Ukiya, C. Murakami, I. Kuriyama, X. Xu, H. Yoshida, and K. Sakaguchi. 2004. A novel DNA topoisomerase inhibitor: dehydroebriconic acid, one of the lanostane-type triterpene acids from *Poria cocos*. *Cancer Sci.* 95: 354-360.

Morikawa, T., J. Tao, S. Ando, H. Matsuda, and M. Yoshikawa. 2003. Absolute stereostructures of new arborinane-type triterpenoids and inhibitors of nitric oxide production from *Rubia yunnanensis*. *J. Nat. Prod.* 66: 638-645.

Mukherjee, P.K., K. Saha, J. Das, M. Pal, and B.P. Saha. 1997. Studies on the anti-inflammatory activity of rhizomes of *Nelumbo nucifera*. *Planta Med.* 63: 367-369.

Najid, A., A. Simon, J. Cook, H. Chable-Rabinovitch, C. Delage, A.J. Chulia, and M. Rigaud. 1992. Characterization of ursolic acid as a lipoxygenase and cyclooxygenase inhibitor using macrophages, platelets and differentiated HL60 leukemic cells. *FEBS Lett.* 299: 213-217.

Navarro, P., R.M. Giner, M.C. Recio, S. Manez, M. Cerda-Nicolas, and J.-L. Rios. 2001. *In vivo* anti-inflammatory activity of saponins from *Bupleurum rotundifolium*. *Life Sci.* 68: 1199-1206.

Nikiema, J.B., R. Vanhaelen-Fastre, M. Vanhaelen, J. Fontaine, C. De Graef, and M. Heenen. 2001. Effects of antiinflammatory triterpenes isolated from *Leptadenia hastata* latex on keratinocyte proliferation. *Phytother Res.* 15: 131-134.

Nukaya, H., H. Yamashiro, H. Fukazawa, H. Ishida, and K. Tsuji. 1996. Isolation of inhibitors of TPA-induced mouse ear edema from hoelen, *Poria cocos*. *Chem. Pharm. Bull.* 44: 847-849.

Oh, S.R., K.Y. Jung, K.H. Son, S.H. Park, I.S. Lee, K.S. Ahn, and H.K. Lee. 1999. *In vitro* anticomplementary activity of hederagenin saponins isolated form roots of *Dipsacus asper*. *Arch. Pharm. Res.* 22: 317-319.

Oh, G.S., H.O. Pae, B.M. Choi, E.A. Seo, D.H. Kim, M.K. Shin, J.D. Kim, J.B. Kim, and H.T. Chung. 2004. 20(*S*)-Protopanaxatriol, one of ginsenoside metabolites, inhibits inducible nitric oxide synthase and cylooxygenase-2 expressions through inactivation of nuclear factor-*k*B in RAW 264.7 macrophage stimulated with lipoplysaccharide. *Cancer Lett.* 205: 23-29.

Otuki, M.F., F.V. Lima, A. Malheiros, V. Cechniel-Filho, F.D. Monache, R.A. Yunes, and J.B. Calixto. 2001. Evaluation of the antinociceptive action caused by ether fraction and a triterpene isolated form resin of *Protium kleinii*. *Life Sci.* 69: 2225-2236.

Park, S.-H., S.R. Oh, K.Y. Jung, I.S. Lee, K.S. Ahn, J.-G. Kim, J.J. Lee, and H.-K. Lee. 1999. Anticomplement activities of oleanolic acid monodesmosides and bisdesmosides isolated from *Tiarella polyphylla*. *Arch. Pharm. Res.* 22: 428-431.

Park, K.H., J. Park, D. Koh, and Y. Lim. 2002. Effect of saikosaponin-A, a triterpenoid glycoside, isolated from *Bupleurum falcatum* on experimental allergic asthma. *Phytother. Res.* 16: 359-363.

Park, E.-K., M.-K. Choo, E.-J. Kim, M. J. Han, and D.-H. Kim. 2003. Antiallergic activity of ginsenoside Rh_2. *Biol. Pharm. Bull.* 26: 1581-1584.

Park, C.S., H. Lim, K.J. Han, S.H. Baek, H.O. Sohn, D.W. Lee, Y.G. Kim, H.Y. Yun, K.J. Baek, and N.S. Kwon. 2004a. Inhibition of nitric oxide generation by 23,24-dihydrocucurbitacin D in mouse peritoneal macrophages. *J. Pharmacol. Exp. Ther.* 309: 705-710.

Park, E.K., M.K. Choo, M.J. Han, and D.H. Kim. 2004b. Ginsenoside Rh_1 possesses antiallergic and anti-inflammatory activities. *Int. Arch. Allergy Immunol.* 133: 113-120.

Peters, R.R., T.F. Saleh, M. Lora, C. Patry, A.J. de Brum-Fernandes, M.R. Farias, and R.M. Ribeiro-do-Valle. 1999. Anti-inflammatory effects of the products from *Wibrandia ebracteata* on carrageenan-induced pleurisy in mice. *Life Sci.* 64: 2429-2437.

Place, A.E., N. Suh, C.R. Williams, R. Risingsong, T. Honda, Y. Honda, G.W. Gribble, L.M. Leesnitzer, J.B. Stimmel, T.M. Willson, E. Rosen, and M.B. Sporn. 2003. The novel synthetic triterpenoid, CDDO-imidazolide, inhibits inflammatory response and tumor growth *in vivo*. *Clin. Cancer Res.* 9: 2798-2806.

Rao, A.V., and D.M. Gurfinkel. 2000. The bioactivity of saponins: triterpenoid and steroidal glycoside. *Drug Metabol. Drug Interact.* 17: 211-235.

Recio, M.C., R.M. Giner, S. Manez, J. Gueho, H.R. Julien, K. Hostettman, and J.L. Rios. 1995. Investigations on the steroidal anti-inflammatory activity of triterpenoids from *Diospyros leucomelas*. *Planta Med.* 61: 9-12.

Recio, M.C., M. Pieto, M. Bonucelli, C. Orsi, S. Manez, R.M. Giner, M. Cerda-Nicolas, and J.-L. Rios. 2004. Anti-inflammatory activity of two cucurbitacins isolated from *Cayaponia tayuya* roots. *Planta Med.* 70: 414-420.

Ringbom, T., L. Segura, Y. Noreen, P. Perera, and L. Bohlin. 1998. Ursolic acid from Plantago major, a selective inhibitor of mouse ear by ursolic acid and 4,4-dimethylcholestane derivatives. *J. Nat. Prod.* 61: 1212-1215.

Rios, J.L., M.C. Recio, S. Manez, and R.M. Giner, R.M. 2000. Natural triterpenoids as anti-inflammatory agents. Pages 93-143 in *Studies in Nat. Prod. Chem., Vol.* 22, *Bioactive Nat. Prod.* (*Part C*), Atta-ur-Rahman, ed., Elsevier Science B.V., Amsterdam.

Ryu, S.Y., M.H. Oak, S.K. Yoon, D.I. Cho, G.S. Yoo, T.S. Kim, and K.M. Kim. 2000. Anti-allergic and anti-inflammatory triterpenes from the herb of *Prunella vulgaris*. *Planta Med.* 66: 358-60.

Safayhi, H., and E.-R. Sailer. 1997. Anti-inflammatory actions of pentacyclic triterpenes. *Planta Med.* 63: 487-493.

Scholz, D., K. Baumann, M. Grassberger, B. Wolff-Winniski, G. Rihs, H. Walter, and J.G. Meingassner. 2004. Synthesis of dammarane-type tritepenoids witn anti-inflammatory activity *in vivo*. *Bioorg. Med. Chem. Lett.* 14: 2983-2986.

Shibata, S. 1990. Biological activities of oleanane-type triterpenes (I). *Minophagen Medical Reviw* 35: 1-13.

Shishodia, S., S. Majumdar, S. Banerjee, and B.B. Agarwal. 2003. Ursolic acid inhibits nuclear factor-kappaB activation induced by carcinogenic agents through suppression of IkappaBalpha kinase and p65 phosphorylation: correlation with down-regulation of cyclooxygenase 2, matrix metalloproteinase 9, and cyclin D1. *Cancer Res.* 63: 4375-4383.

Singh, B., P.M. Sahu, and M.K. Sharma. 2002. Anti-inflammatory and antimicrobial activities of triterpenoids from *Strobilanthes callosus* nees. *Phytomedicine* 9: 355-359.

Smolinski, A.T., and J.J. Pestka. 2003. Modulation of lipopolysaccharide-induced proinflammatory cytokine production in vitro and in vivo by the herbal constituents apigenin (chamomile), ginsenoside Rb_1 (ginseng) and parthenolide (feverfew). *Food Chem. Toxicol.* 41: 1381-1390.

Subbaramaniah, K., P. Michaluart, M.B. Sporn, and A.J. Dannenberg. 2000. Ursolic acid inhibits cyclooxygenase-2 transcription in human mammary epithelial cells. *Cancer Res.* 60: 2399-2404.

Sugishita, E., S. Amagaya, and Y. Ogihara. 1983. Studies on the mechanism of anti-inflammatory activities of papyriogenin A and paryriogenin C. *J. Pharm.Dyn* 6: 287-294.

Suh, N.T. Honda, H.J. Finlay, A. Barchowsky, C. Williams, N.E. Benoit, Q.-W. Xie, C. Nathan, G.W. Gribble, and M.B. Sporn. 1998. Novel triterpenoids suppress inducible nitric oxide synthase (iNOS) and inducible cyclooxygenase (COX-2) in mouse macrophages. *Cancer Res.* 58: 717-723.

Suh, N., Y. Wang, T. Honda, G.W. Gribble, E. Dmitrovsky, W. F. Hickey, R.A. Maue, A.E. Place, D.M. Porter, M.J. Spinella, C.R. Williams, G. Wu, A.J. Dannenberg, K.C. Flanders, J.J. Letterio, D.J. Mangelsdorf, C.F. Nathan, L. Ngyyen, W.W. Porter, R.F. Ren, A.B. Roberts, N.S. Roche, K. Subbramaniah, and M.B. Sporn, 1999. A novel synthetic oleanane triterpenoid, 2-cyano-3,12-dioxoolean-1,9-dien-28-oic acid, with potent differentiating, antiproliferative, and anti-inflammatory activity. *Cancer Res.* 59: 336-341.

Surh, Y.-J., K.-S. Chun, H.-H. Cha, S.S. Han, Y.-S. Keum, K.-K. Park, and S.S. Lee. 2001. Molecular mechanisms underlying chemopreventive activities of anti-inflammatory phytochemicals: down-regulation of COX-2 and iNOS through suppression of NF-*k*B activation. *Mutation Res*. 480-481: 243-268.

Surh, Y.-J. 2002. Anti-tumor promoting potential of selected spice ingredients with antioxidative and anti-inflammatory activities: a short review. *Food Chem. Toxicol*. 40: 1091-1097.

Udayama, M., M. Ohkawa, N. Yoshida, J. Kinjo, and T. Nohara. 1998. Structures of three new oleanane glucuronides isolated from *Lathyrus palustris* var. *pilosus* and hepatoprotective activity. *Chem. Pharm. Bull*. 46: 1412-1415.

Ukiya, M.; T. Akihisa, K. Yasukawa, Y. Kasahara, Y. Kimura, K. Koike, T. Nikaido, and M. Takido. 2001. Constituents of Compositae plants. 2. Triterpene diols, tiols, and their 3-*O*-fatty acid esters from edible chrysanthemum flower extract and their anti-inflammatory effects. *J. Agric. Food Chem*. 49: 3187-3197.

Ukiya, M., T. Akihisa, K. Yasukawa, H. Tokuda, M. Toriumi, K. Koike, Y. Kimura, T. Nikaido, W. Aoi, H. Nishino, and M. Takido. 2002. Anti-inflammatory and anti-tumor-promoting effects of cucurbitane glycosides from the roots of *Bryonia dioica*. *J. Nat. Prod*. 65: 179-183.

Yasukawa, K., M. Takido, M. Takeuchi, and S. Nakagawa. 1988. Inhibitory effect of glycyrrhizin and caffeine on two-stage carcinogenesis in mice. *Yakugaku Zasshi* 108: 794-796.

Yasukawa, K., M. Takido, T. Matsumoto, M. Takeuchi, and S. Nakagawa. 1991. Sterol and triterpene derivatives from plants inhibit the effects of a tumor promoter, and sitosterol and betulinic acid inhibit tumor formation in mouse skin in two-stage carcinogenesis. *Oncology* 48: 72-76.

Yasukawa, K., A. Yamaguchi, J. Arita, S. Sakurai, A. Ikeda, and M. Takido. 1993. Inhibitory effect of edible plant extracts on 12-*O*-tetradecanoylphorbol-13-acetate-induced ear edema in mice. *Phytother. Res*. 7: 185-189.

Yasukawa, K., T. Akihisa, T. Tamura, and M. Takido. 1994. Inhibitory effect of karounidiol on 12-*O*-tetradecanoylphorbol-13-acetate-induced tumor promotion. *Biol. Pharm. Bull*. 17: 460-462.

Yasukawa, K., S.Y. Yu, S. Yamanouchi, M. Takido, T. Akihisa, and T. Tamura. 1995. Some lupine-type triterpene inhibit tumor promotion by 12-*O*-tetradecanoylphorobol-13-acetate in two-stage carcinogenesis in mouse skin. *Phytomedicine* 1: 309-313.

Yasukawa, K., T. Akihisa, H. Oinuma, T. Kaminaga, H. Kanno, Y. Kasahara, T. Tamura, K. Kumaki, S. Yamanouchi, and M. Takido. 1996a. Inhibitory effect of taraxastane-type triterpenes on tumor promotion by 12-*O*-tetradecanoylphorbol-13-acetate in two-stage carcinogenesis in mouse skin. *Oncology* 53: 341-344.

Yasukawa, K., T. Akihisa, H. Oinuma, Y. Kasahara, Y. Kimura, S. Yamanouchi, K. Kumaki, T. Tamura, and M. Takido. 1996b. Inhibitory effect of di- and trihydroxy triterpenes from the flowers of Compositae on 12-*O*-tetradecanoylphorobl-13-acetate-induced inflammation in mice. *Biol. Pharm. Bull*. 19: 1329-1331.

Yasukawa, K., and T. Akihisa. 1997. Anti-inflammatory and antitumor-promoting activities of sterols, triterpene alcohols and their derivatives. *Recent Res. Devel. Oil Chem*. 1: 115-125.

Yasukawa, K., T. Kaminaga, S. Kitanaka, T. Tai, Y. Nunoura, S. Natori, and M. Takido. 1998a. 3β-*p*-Hydroxybenzoyldehydrotumulosic acid from Poria cocos, and its anti-inflammatory effects. *Phytochemistry* 48: 1357-1360.

Yasukawa, K., T. Akihisa, Y. Kimura, T. Tamura, and M. Takido, 1998b. Inhibitory effect of cycloartenol ferulate, a component of rice bran, on tumor promotion in two-stage carcinogenesis in mouse skin. *Biol. Pharm. Bull*. 21: 1072-1076.

Yasukawa, K., T. Akihisa, Z. Yoshida, and M. Takido. 2000. Inhibitory effect of euphol, a triterpene alcohol from the roots of *Euphorbia kansui*, on tumour promotion by 12-*O*-tetradecanoylphorbol-13-acetate in two-stage carcinogenesis in mouse skin. *J. Pharm. Pharmacol.* 52: 119-124.

Yoshizumi, S., T. Murakami, M. Kadoya, H. Matsuda, J. Yamahara, and M. Yoshikawa. 1998. Medicinal foodstuffs. XI. Histamine release inhibitors from wax gourd, the fruits of *Benincasa hispida* Gogn. *Yakugaku Zasshi* 118: 188-192.

Yu, L., R. Ma, Y. Wang, H. Nishino, J. Takayasu, W. He, M. Chang, J. Zhen, W. Liu, and S. Fan. 1992. Potent anti-tumorigenic effect of tubeimoside 1 isolated form the bulb of *Bobostemma paniculatum* (Maxim) Franquet. *Int. J. Cancer* 50: 635-638.

Yu, T.X., R.D. Ma, and L.J. Yu. 2001. Structure-activity relationship of tubeimosides in anti-inflammatory, antitumor, and antitumor-promoting effects. *Acta Pharmacol. Sin.* 22: 463-468.

Yueqin, Z., M.C. Recio, S. Manez, R.M. Giner, M. Cerda-Nicolas, and J.L. Rios. 2003. Isolation of two triterpenoids and a biflavanone with anti-inflammatory activity from *Schinus molle* fruits. *Planta Med.* 69: 893-898.

Zhang, W.-J., P. Hufnagl, B.R. Binder, and J. Wojta. 2003. Antiinflammatory activity of astragaloside IV is mediated by inhibition of NF-*k*B activation and adhesion molecule expression. *Thromb. Haemost.* 90: 904-914.

Zitterl-Eglseer, K., S. Sosa, J. Jurenitsch, M. Schubert-Zsilavecz, R.D. Loggia, A. Tubaro, M. Bertoldi, and C. Franz. 1997. Anti-oedematous activities of the main triterpendiol esters of marigold (*Calendula officinalis* L.). *J. Ethnopharmacol.* 57: 139-144.

4

Antioxidant Compounds Extracted from Several Plant Materials

Rosa M. Seabra[1], Paula B. Andrade[1], Patrícia Valentão[1], Eduarda Fernandes[2], Félix Carvalho[3] and Maria de Lourdes Bastos[3*]

[1]*REQUIMTE, Department of Pharmacognosy, Faculty of Pharmacy, University of Porto, Rua Aníbal Cunha, 164, 4099-030 Porto, Portugal*

[2]*REQUIMTE, Department of Physical Chemistry, Faculty of Pharmacy, University of Porto, Rua Aníbal Cunha, 164, 4099-030 Porto, Portugal*

[3]*REQUIMTE, Department of Toxicology, Faculty of Pharmacy, University of Porto, Rua Aníbal Cunha, 164, 4099-030 Porto, Portugal*

INTRODUCTION

Nowadays there is considerable evidence that the antioxidants contained in fruits, vegetables, and beverages play an important role in the maintenance of health and in prevention of disease. These antioxidants are believed to play a very important role in the body defence system against the various reactive oxidant species that are generated during several physiological and pathological processes (Cantuti-Castelvetri *et al.* 2000). Moreover, when added to foods, antioxidants minimize rancidity, retard the formation of toxic oxidation products, maintain nutritional quality, and increase shelf-life (Jadhav *et al.* 1995). The present review aims to provide a glimpse at the importance of the antioxidants of plant origin and the methodologies used to evaluate their activity, an insight into some relevant antioxidant chemical families of plant origin, and finally into sources of antioxidants for the food industry.

A broad definition of an antioxidant may be "any substance that, when present at low concentrations compared to those of an oxidizable substrate, significantly delays or prevents oxidation of that substrate", the term "oxidizable substrate" being almost everything found in foods and in living tissues including proteins, lipids, carbohydrates and DNA

*Corresponding Author

(Halliwell 1995). A compound might exert antioxidant actions *in vivo* or in food by inhibiting the generation of reactive species, or by directly scavenging free radicals. Additionally, *in vivo* an antioxidant might act indirectly by raising the levels of endogenous antioxidant defences (e.g., by upregulating the expression of the genes encoding superoxide dismutase, catalase or glutathione peroxidase) (Halliwell 1995).

The scientific community has an increasing need to evaluate the antioxidant capacity of physiological fluids, foods, beverages, and natural products. As explained below in more detail, this need is derived from the proven evidence of the importance of antioxidants to scavenge the reactive oxygen/nitrogen species (ROS/RNS), which, when produced in excess, may be involved in pathological conditions related to the impairment of the oxidative/antioxidative balance in favour of the former, leading to oxidative stress. Examples of ROS include the free radicals superoxide ($O_2^{\bullet -}$), peroxyl ($ROO^{\bullet}$), alkoxyl ($RO^{\bullet}$), hydroxyl ($HO^{\bullet}$), the non radical species singlet oxygen ($^1O_2$), hydrogen peroxide (H_2O_2), hypochlorous acid (HOCl), and lipid hydroperoxides. The RNS of primary concern are nitric oxide ($^{\bullet}NO$) and peroxynitrite ($ONOO^-$; a product of the combination of $O_2^{\bullet -}$ and $^{\bullet}NO$). Other potentially important RNS include nitrogen dioxide radical ($^{\bullet}NO_2$), nitrosonium (NO^+), and nitronium (NO_2^+) ions.

ROS and RNS are continuously generated in small amounts by normal cellular processes. Exposure to environmental insults such as infectious agents, pollution, UV light, cigarette smoke, and radiation, have also been shown to increase free radical formation (Caporaso 2003, Cooke *et al.* 2003). The endogenously produced ROS and RNS are essential for life, being involved in several biological functions, namely energy production, phagocytosis, regulation of cell growth and cellular signalling, synthesis of biologically important compounds, and xenobiotic metabolism (Suzuki *et al.* 1997, Babior 2000, Ali and Mann 2004, Azzi *et al.* 2004). However, when produced in excess, or in case of antioxidants depletion, these reactive species may also be extremely deleterious, causing oxidative stress by oxidatively damaging biomolecules, like lipids in cell membranes, proteins in tissues or enzymes, carbohydrates, and DNA, which cause membrane damage, protein modification (including enzymes), and DNA damage. Also important, oxidative stress has been implicated in the aetiology of several diseases and aging (Halliwell and Gutteridge 1999, Vendemiale *et al.* 1999, Beckman and Ames 1998, Feng *et al.* 2001, van Dam 2002, Embola *et al.* 2002). Thus, in a normal cellular milieu, ROS and RNS are essential for life while in cases of antioxidant exhaustion they may become detrimental.

The human body has several mechanisms for defence against free radicals and other reactive oxygen species composed of an array of enzymatic and nonenzymatic endogenous antioxidant defences (Sies 1993, Fridovich 1997). One important line of defence is a system of enzymes, including glutathione peroxidases, superoxide dismutases, and catalase, which decrease the concentration of the most harmful oxidants. A second line of defence is small-molecular-weight compounds which act as antioxidants; that is, they react with oxidising chemicals, reducing their capacity to cause deleterious effects. Some, such as glutathione, ubiquinol, and uric acid, are produced by normal metabolism. Other small-molecular-weight antioxidants are found in the diet, the best known being vitamin E, vitamin C, and carotenoids (Sies 1993, Fridovich 1997). The various defences are complementary to one another by acting on different oxidants or in different cellular compartments. Also, in addition to their individual effects, antioxidants work in concert with one another by a series of oxidation-reduction (redox) reactions to quench reactive oxidant species. Redox buffering systems of the common antioxidants are all related in a stepwise, sequential and synergistic recycling process, which provides ongoing neutralization of free radicals (Packer 1992). A typical example of synergism between antioxidants is glutathione regenerating ascorbate and ascorbate regenerating α-tocopherol (Chan 1993, Beyer 1994, Fornai *et al.* 1999). However, despite these defence antioxidants (able either to suppress free radical formation and chain initiation or to scavenge free radical and chain propagation), some ROS and RNS still escape to cause damage. Thus, the body antioxidant system is provided also with repair antioxidants (able to repair damage, and based on proteases, lipases, transferases, and DNA repair enzymes) (Pietta 2000).

In situations of increased free radical generation, the reinforcement of endogenous antioxidants may be particularly important in diminishing the cumulative effects of oxidatively damaged molecules. Indeed, a large amount of evidence has accumulated over the years that has confirmed the putative positive benefits of specific antioxidants as well as of diets rich in fruits and vegetables in altering, reversing, or forestalling certain types of cancer and cardiovascular diseases as well as neuronal/behavioral decrements, with varying degrees of success (Cantuti-Castelvetri *et al.* 2000). Certain diets are particularly rich in antioxidants such as vitamins A, C, E, carotenoids, and bioflavonoids (such as flavones, tannins, anthocyanins, and quercetin), as well as in several essential minerals including selenium, copper, manganese, and zinc, which are involved in the structure or catalytic activity of antioxidant enzymes (Dodge 1985, Oomah and Mazza 1999, Ottley 2000, Waine 2001). Besides these antioxidants, other substances in plants might account for at least part of the health benefits associated with vegetable and fruit consumption. Over the past decade evidence has been accumulated that plant polyphenols are an important class of defence antioxidants. These compounds are widespread virtually in all plant foods, often at high levels, and include phenolic derivatives, flavonoids, tannins, and lignans (Pietta 2000). Thus, it might be important to examine the impact of the antioxidants contained in different foods and other sources of plant origin in order to ascertain whether the course of oxidative-stress related diseases can be altered.

On the other hand, the growing interest in the substitution of synthetic food antioxidants by natural ones has fostered research on vegetable sources and on the screening of raw materials to identify new antioxidants (Moure *et al.* 2001). In fact, the oxidative deterioration of fats and oils in foods is responsible for rancid odours and flavours, with a consequent decrease in nutritional quality and safety caused by the formation of secondary, potentially toxic compounds (Gertz *et al.* 2000, Yanishlieva *et al.* 2001). Many of these secondary oxidation products are highly reactive themselves and may initiate the oxidative chain process *in vivo* (Yanishlieva *et al.* 2001). For these reasons a high oxidative stability of lipids is very important for human health and it is also economically important. Thus, the addition of antioxidants is required to preserve flavour and colours and to avoid vitamin destruction. Furthermore, oxidation reactions are not an exclusive concern for the food industry; antioxidants are widely believed to be necessary to prevent deterioration of other oxidizable goods, such as cosmetics and pharmaceuticals (Irache *et al.* 1992, Aruoma 2003a), and plastics (Scott 1988).

CHARACTERIZATION OF ANTIOXIDANTS

The antioxidant activity (capacity or potential) is widely used as a useful parameter to characterize the redox properties of different plant materials. This activity is associated with compounds capable of protecting a biological system against the potentially harmful effect of processes or reactions that cause excessive oxidation, involving pro-oxidant reactive species. The scientific literature is endowed with a vast number of different approaches for the evaluation and demonstration of antioxidant activity, the characterization of antioxidants being the topic of several excellent reviews (Halliwell 1995, Arnao *et al.* 1999, Niki and Noguchi 2000, Arnao *et al.* 2001, Aruoma 1996, Aruoma 1999, Arnao 2000, Aruoma 2003b, Valgimigli *et al.* 2001, Antolovich *et al.* 2002, Kohen and Nyska 2002, Sánchez-Moreno 2002), and the citations therein. Thus, only succinct remarks are hereby presented.

As mentioned above, a compound might exert antioxidant actions by directly scavenging pro-oxidant reactive species (e.g., free radicals and other ROS and RNS, metal ions), by inhibiting their generation, and/or by up-regulating antioxidant defences, and repairing damage. Based on these action mechanisms, several methodologies have been implemented for evaluating antioxidant activity. As represented by several studies performed by our group both with plant extracts and isolated compounds, the screening methodologies are usually carried out using simple *in vitro* non-cellular systems, (Valentão *et al.* 2001, 2002a, 2002b, 2002c, 2003), but the confirmation of the antioxidant potential should be performed using *in vitro* systems, namely isolated cells (Fernandes *et al.* 1995, 2003, Lima *et al.* 2004, Valentão *et al.* 2004a) and ultimately in *in vivo* studies (Pires das Neves *et al.* 2004, Valentão *et al.* 2004b). The direct scavenging of radicals can be followed using the electron spin resonance technique (EPR), which allows the detection of relatively stable radicals. A variation of this technique is the spin trapping method, in which a highly reactive radical reacts with a trap molecule to produce a stable radical product, which can be evaluated (Valgimigli *et al.* 2001). Other trapping procedures allow a radical to react with a detector molecule to yield a stable product that can then be evaluated using a variety of techniques. Known examples are the hydroxylation of salicylic acid for detection of $HO^{\cdot}$, the cytochrome c reduction assay for detection of $O_2^{\cdot-}$, and detection of ROS and RNS by colored, fluorescent, and chemiluminiscent end-product compounds (Fernandes *et al.* 1995, Arnao *et al.* 1999, Antolovich *et al.* 2002, Kohen and Nyska 2002, Fernandes *et al.* 2003). Voltammetric measurements are also useful to measure electron transfer between molecules and evaluate oxidation/reduction potentials of various redox-active compounds. These methodologies can provide information concerning thermodynamic, kinetic, and analytical features of the tested compounds (Kohen and Nyska 2002). However, in the more usual approach for antioxidant characterization, various indirect measurements are used to assess the effectiveness of an antioxidant in preventing oxidative damage to a suitable substrate. In general, the oxidation is carried out in the presence of pro-oxidant systems and the extent of suppression of the rate of oxidation and duration of inhibition produced by the antioxidant are measured. Essentially, the antioxidant activity involves the interaction with metals, radicals, or other types of pro-oxidant ROS and RNS (Halliwell 1995, Arnao *et al.* 1999, Niki and Noguchi 2000, Aruoma 2003b). The assessment of antioxidant activity should always be carefully performed in order to avoid possible interferences in the assays and to obtain correct interpretations of the data. Thus, as often pointed out, there is no single method, which correctly evaluates the total antioxidant capacity (Niki and Noguchi 2000). It is important to study relevant reactive species and sources generating such species; the choice will depend on whether effects *in vivo* or effects in the food matrix are being considered. An antioxidant may protect lipids against oxidative damage whilst accelerating damage to other biological molecules (Aruoma *et al.* 1997, Antolovich *et al.* 2002, Sánchez-Moreno 2002). Antioxidant efficacy may be demonstrated in one system but to fail to protect, or even sometimes cause damage, in others. This is of particular importance as certain antioxidant inhibitors of lipid peroxidation may not protect other targets (such as DNA and protein) against damage, and sometimes can even aggravate such damage (Halliwell 1995, Niki and Noguchi 2000, Aruoma 2003b). Measurements of lipid peroxidation using microsomes, phospholipid liposomes, arachidonic acid, and other lipid model systems (e.g., bulk oil and emulsified oil systems) should be the first line of tests to establish the potential antioxidant action of dietary antioxidant compounds. The interfacial characteristics of the antioxidants (in part determined by their partition coefficients) are equally critical (Aruoma 2003b). The extent to which oxidation of fatty acids and their esters occurs in foods depends on the chemical structure of the fatty acid, the nature of food processing and the temperature at which the foods are stored and/or cooked, and the minor constituent antioxidants, thus being essential to test the antioxidants in different assay systems (Aruoma 2003b).

During *in vitro* testing, it is essential to examine the action of a compound under physiological pH values, and over a concentration range that is expected to be achieved in plasma during its therapeutic use unless there is good reason to suspect that it concentrates at a particular reservoir/target site *in vivo*. These "screening" approaches can be used to rule out direct antioxidant activity *in vivo*: a compound that is not very effective *in vitro* will not be any better *in vivo* (Halliwell 1995, Aruoma 2003b).

When an antioxidant scavenges active radicals, the radical is transformed to a nonradical, stable product while the antioxidant may itself give rise to damaging radical species. The fate of this antioxidant-derived radical is also important in determining its antioxidant efficacy (Niki and Noguchi 2000).

In vitro, the antioxidant activity is often measured for isolated compounds. This is necessary to evaluate its own capacity and to study the structure-reactivity relationship (Niki and Noguchi 2000). However, even though individual measurement of the antioxidant activity of all components in a sample is possible, this can be time consuming and expensive. In addition, there may be synergism between antioxidants and examining one in isolation may not accurately reflect their combined action

(Antolovich *et al.* 2002). Thus, both types of measurements (individually and on the complex mixture) should be performed in order to have a complete understanding of the antioxidant potential of the material under study.

In vivo, the direct detection of the production of ROS and RNS is difficult, as these species are short-lived and highly reactive. In some cases, trapping procedures are used *in vivo* for the detection of ROS and RNS (Floyd *et al.* 1984, Carvalho *et al.* 2001, Liu *et al.* 2004). However, ongoing oxidative damage is generally analyzed by the measurement of secondary products, including derivatives of amino acids, nucleic acids, and lipid peroxidation (Kohen and Nyska 2002).

In designing antioxidant drugs and/or supplements for therapeutic purposes, it is also important to consider the physiological state of the subject consuming the antioxidant. The bio-efficacy of drugs is determined by the degree of absorption, metabolism, distribution, and subsequent elimination, which may be altered by several disease states. In any case, the antioxidant activity needs to be supported by testing in the actual substrate and conditions of interest, since the drugs must become available at a physiologically relevant concentration for the intended target (e.g., if one is interested in the therapeutic strategies to prevent progressive neuronal loss based on antioxidant activity, the antioxidant must be able to cross the blood brain barrier and occur at the respective brain region for neuroprotection). How much of the consumed antioxidant drugs and/or supplement becomes available to the patient on consumption, depends on the chemical nature of the molecule, the nature of the food components, interactions of components within the food matrix, the mix of diet, the health status, and other drugs taken concomitantly. The presence of competitive enzymes and antioxidants in addition to pro-oxidants may also profoundly affect the *in vivo* activity of test antioxidants. In order to evaluate the antioxidant capacity, these factors should be separately assessed and considered (Aruoma 1996, Niki and Noguchi 2000, Antolovich *et al.* 2002, Aruoma 2003b). One of the major difficulties in characterizing the antioxidant activity *in vivo* comes from the fact that it varies depending on the oxidative stress related disease and the pro-oxidant agents involved in that disease, which is usually not completely identified. Another important point is that, *in vivo*, the antioxidant probably does not exert its effect by itself alone but acts in combination with other antioxidants, including endogenous compounds. Such interaction may affect the final antioxidant capacity (Niki and Noguchi 2000).

It is now clear that both oxidants and antioxidants can regulate a multitude of different cellular functions, signal transduction pathways, and gene expression. The mechanism by which the transcription of specific eukaryotic genes is redox regulated is complex. However, recent findings suggest that redox-sensitive transcription factors play an essential role in this process (Arrigo 1999). Although the mechanism by which oxidative stress and/or variation in the intracellular redox state modulate gene expression still remains far from being completely understood, the implementation of screening assays in this area is clearly warranted. Accordingly, the well-known antioxidant compound silibinin has been shown to strongly suppress photocarcinogenesis by an effect that involves the inhibition of DNA synthesis and cell proliferation and induction of apoptosis. These antitumor effects of silibinin were accompanied by an increased accumulation of p53 protein and an induction of CDK inhibitors with a decrease in CDK-cyclin levels. Furthermore, silibinin activated MAPK but inhibited the Akt pathway and decreased survival level with a concomitant increase in caspase-3 activation (Mallikarjuna *et al.* 2004). In another study, silibinin proved to be an immune-response modifier *in vivo*, inhibiting the intrahepatic expression of tumor necrosis factor, interferon-gamma, interleukin (IL)-4, IL-2, and iNOS, and augmenting synthesis of IL-10 (Schumann *et al.* 2003). Another paradigmatic study, using vitamin E, has shown that this antioxidant also functions in the control of the expression of several genes involved in responses to oxidative stress and may regulate signal transduction pathways (Niki and Noguchi 2000). On the other hand, the activation of transcriptional factors, such as nuclear factor kB (NF-kB) and c-myc, is modulated by oxidative stress and participates in the up-regulation of endogenous antioxidants like the enzymes superoxide dismutase, catalase or glutathione peroxidase (Arrigo 1999, Kohen and Nyska 2002, Aruoma 2003b). Thus, the cell signalling response to antioxidants and/or oxidative stress may be variable, and can be up-regulated as well as down-regulated, and different responses to dose or magnitude to antioxidant/prooxidant stimuli can be demonstrated. Consequently, the effect from supplementation with an antioxidant is difficult to predict, and ultimately must be thoroughly assessed, with an important focus on the molecular mechanism involved (Poulsen *et al.* 2000).

PLANT METABOLITES WITH ANTIOXIDANT ACTIVITY

Chlorophylls

Photosynthesis, an essential process of life in all green plants, is also a source of oxidizing species. Chlorophylls are the main photosensitisers in plants and the electron transport system is considered to be the major source of singlet oxygen and superoxide radical (Arora *et al.* 2002). On the other hand, plants are also exposed to high doses of sunlight, which is known to contain energetic short-wavelength ultraviolet photons that are potentially detrimental because of their capacity to interact with many cellular molecules. One of the possible fates of electronic energy taken up by a UV-absorbing molecule is the transference of that energy to molecular oxygen, giving rise to singlet oxygen (Larson 1988).

For these reasons, plants were forced, in the course of evolution, to develop several chemical defences against oxidizing species. Therefore, it is not surprising that so many plant molecules have the ability to control, one way or another, oxidative stress.

The term *chlorophyll* was first used by Pelletier and Cavantou, in 1818, to describe the complex pigment responsible for the green colour of plants. Several years later it was demonstrated that this complex pigment was a mixture of two green pigments and several yellow ones, now known to be carotenoids. The two green pigments are referred as chlorophyll *a* and chlorophyll *b* (Fig. 4.1).

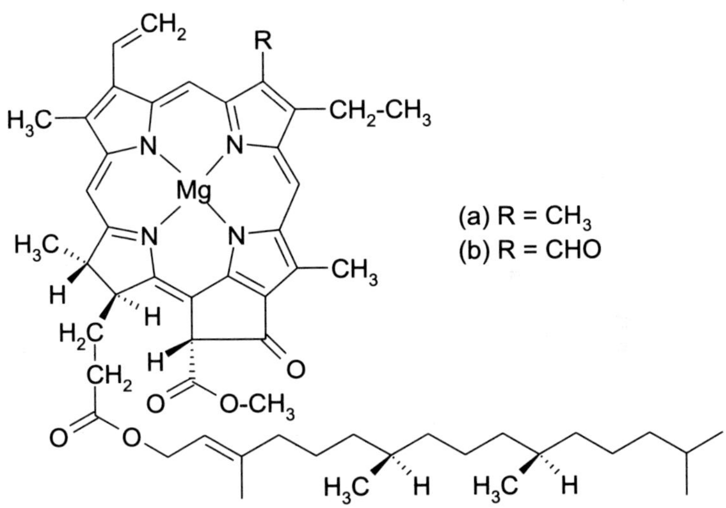

Fig. 4.1 Structures of chlorophylls *a* and *b*.

As referred to above, chlorophylls are photosensitive molecules, which means that they are easily excited by light; the excess energy is, afterwards, transferred to other molecules and, therefore, chlorophyll is a pro-oxidant when exposed to light (Usuki *et al.* 1984, Endo *et al.* 1984). However, it was found that, in the dark, both chlorophylls, but chlorophyll *a* with higher intensity, were able to retard the oxidative deterioration of triglycerides in rapeseed and soybean oil, although the molecular mechanism of such antioxidant action remains unclear (Larson 1988, Rajalakshmi and Narasimhan 1995).

More recently, chlorophyllin (water soluble form of chlorophyll obtained by alkaline hydrolysis) showed the ability to suppress the nitric oxide production by cells stimulated by lipopolysaccharide (Cho 2000). Park *et al.* (2003) verified that chlorophyllin and some chemically related compounds were able to scavenge superoxide radicals generated by 12-O-tetradenaoylphorbol-13-acetate and hydroxyl radicals generated by the Fenton reaction.

Chlorophyllins have also demonstrated antimutagenic activity (Terwel and van der Hoeven 1985, Ong *et al.* 1986). The mechanism of antimutagenecity of chlorophyllyn is still unknown, but it was suggested to be attributed to the scavenging of free radicals (Ong *et al.* 1986).

Carotenoids

The carotenoids are a class of compounds, mainly tetraterpenoids (C40), characterized by the presence of a series of conjugated double bonds that are responsible for their bright yellow to red colours (Britton 1976). These compounds are synthesized by the union of eight isoprenic units, in a way that the structure is reversed at the centre. The linear and symmetric backbone can be modified by several reactions, leading to a large diversity of structures (Britton 1976). It is usual to classify the carotenoids in two groups: the carotenes (if devoid of oxygen, like *β*-carotene) and the xanthophylls (when endowed with oxygen, like lutein). They can maintain a linear structure, like lycopene or, more frequently, acquire two cyclic structures (mostly of iononic type) at both ends of the linear portion of the molecule (Fig. 4.2).

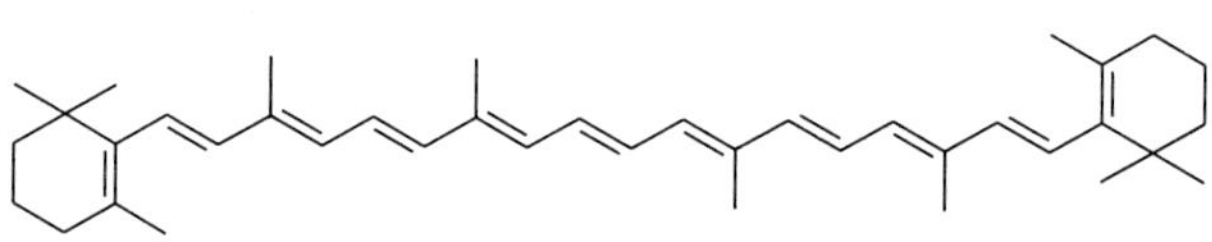

β-Carotene

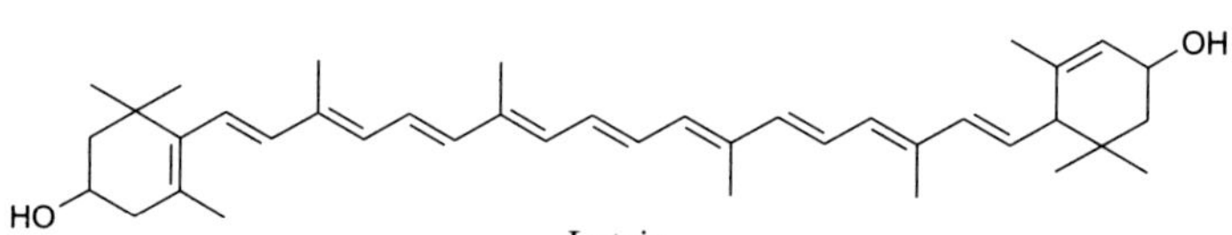

Lutein

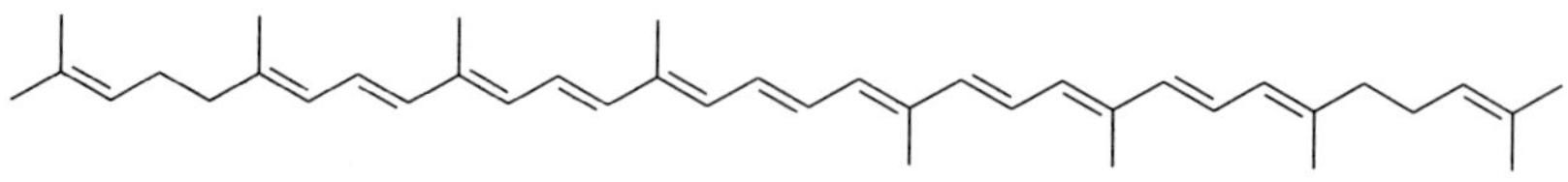

Lycopene

Fig. 4.2 Representative structures of carotenoids.

They are actually present in all photosynthetic organisms—at least in small amounts—where they have two main functions: (i) to act as "antenna pigments" gathering wavelengths of light that are not absorbed by chlorophylls and (ii) protecting chlorophylls from the effects of oxidation. Chloroplasts usually contain the same profile of carotenoids, consisting lutein (≈45%), β-carotene (25–30%), violaxantin (15%), and neoxanthin (15%) (Britton 1991).

In plants, carotenoids are responsible for many of the red (tomato and pepper), orange (carrot and citrus fruits), and yellow (autumn plant leaves, sunflower, narcissus) colours. Roots (except for carrots) and seeds are usually poor in carotenoids. In fruits, they are usually concentrated in the epicarp and are responsible or co-responsible for the colours of citrus, olive fruits, mangoes, and others.

Animals appear to be unable to synthesize carotenoids. However, once obtained from the diet, these compounds may be further chemically modified and incorporated in cellular structures; therefore, carotenoids are also responsible for the colours of some insects, fishes (e.g., salmon and rainbow trout), birds (flamingos), and crustaceans. Asthaxanthin, equinenone, and canthaxanthin are characteristically animal carotenoids.

The colouring properties of carotenoids justify their use as natural food colourings and as additives for aquaculture.

Another property associated with carotenoids, for more than 70 years, is their ability to be converted into vitamin A. More specifically, only the molecules in which a non-substituted β-ionone group is present have such a capacity (Britton 1995). β-Carotene is, doubtless, the main compound responsible for the synthesis of vitamin A in humans. It is present in several fruits and vegetables, such as carrots, spinach, peaches, apricots, and others. Vitamin A can also be formed from α-carotene (found in carrots, pumpkins, and peppers) and cryptoxanthin (found in oranges, tangerines, papayas, and nectarines).

More recently, the attention of the scientific community has been attracted by the antioxidant activities of carotenoids. Several experimental and epidemiological studies have revealed that this group of compounds is able to protect against certain types of cancers (Heber and Lu 2002) and other diseases, like erythropoietic protoporphyria (Mathews-Roth 1993) and age-related macular degeneration (Landrum and Bone 2001). Evidence suggests that this protection is related, at least in part, to their antioxidant properties.

Carotenoids exert their antioxidant activity by two different mechanisms: by singlet oxygen quenching and by radical scavenging. The singlet oxygen quenching property of carotenoids is due to the presence of the conjugated polyene system and is performed, predominantly, by a physical route, a process that implies the transfer of energy from singlet oxygen to carotenoids; in this process the oxygen returns to its triplet ground state and an excited triplet carotenoid is formed (Yanishlieva-Maslarova 2001). The energy of this new form of carotenoids is dissipated through rotational and vibrational interactions between the carotenoids and the surrounding molecules, that is, as heat. Since the carotenoids remain intact during physical quenching of singlet oxygen, they can be reused several times in such quenching cycles (Stahl and Sies 2003). This quenching capacity is responsible for the effectiveness of carotenoids in preventing damage in photobiological systems. Experiments were carried out on dealing with this ability of carotenoids, in order to evaluate and compare the effectiveness of some carotenoids: several singlet oxygen generator systems, in homogeneous solutions or in various types of liposomes, were used but, obviously, the results depend on the experimental conditions used (Di Mascio *et al.* 1989, Di Mascio *et al.* 1990, Conn *et al.* 1991).

The radical scavenging property of carotenoids is also due to the polyene system and can occur, basically, in two ways, depending on the radical to be scavenged: (i) the carotenoid can give an electron to the radical and is transformed into a cationic radical; this is what happens when carotenoids react with a nitrogen dioxide radical ($^{\bullet}NO_2$) (Mortensen *et al.* 1997, Everet *et al.* 1996); (ii) on the other hand, on reacting with other radicals, like the mercaptoethanol thiyl radical (Mortensen *et al.* 1997), or a thiyl radical (Everet *et al.* 1996) an adduct is formed with an unpaired electron, located on the carotenoid portion, being stabilized by resonance:

$$HOCH_2CH_2\text{-}S. + \text{Carotenoid} \rightarrow [HOCH_2CH_2S\text{-Carotenoid}]^{\bullet}$$

In some cases the adduct can decompose forming an anion and the carotenoid radical cation (Mortensen *et al.* 1997).

Peroxyl radicals are also scavenged *via* adduct formation (Fig. 4.3). If the oxygen concentration is low, a second peroxyl radical can be added to produce a non-radical species, one that is probably unable to exert prooxidative effects (Belitz and Grosch 1999).

Several works have been published referring to the parameters influencing the radical scavenging properties of carotenoids, such as the partial pressure of oxygen, carotenoid concentration, and the nature of the radical to be scavenged; but, under certain experimental conditions, a pro-oxidant effect was observed (Burton and Ingold 1984, Jorgensen and Skibsted 1993, Belitz and Grosch 1999).

Obviously the effectiveness of carotenoids is also influenced by their structure. Factors like the number of conjugated double bonds, the presence of oxo and/or hydroxyl groups, and even the geometry of the

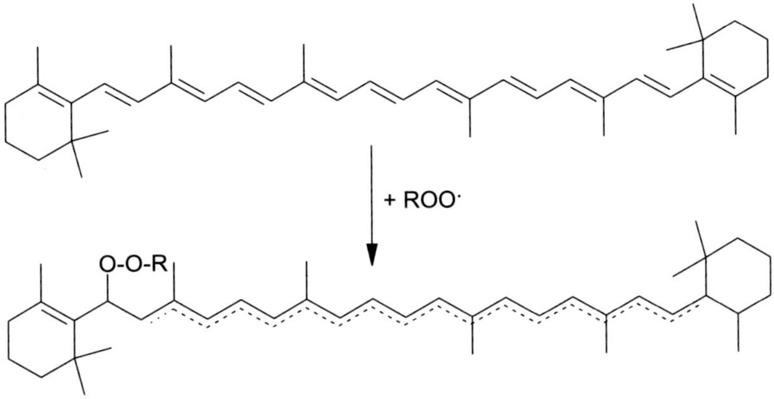

Fig. 4.3 Adduct formation with peroxyl radicals.

molecule, affect both the polarity of the carotenoid and its scavenging capacity (Goto *et al.* 2001, Terao 1989, Miller *et al.* 1996).

Cooperative effects of carotenoids with other antioxidants are also described and were reviewed by Stahl and Sies (2003). At least in some *in vitro* studies, β-carotene was able to regenerate tocopherol from the tocopheryl radical; the resulting carotenoid radical cation may subsequently be repaired by ascorbic acid. In comparison to the individual antioxidants, and as to what concerns the scavenging of reactive nitrogen species, α-tocopherol, ascorbic acid, and β-carotene exhibited cooperative synergistic effects (Böhm *et al.* 1998a). In some assays, a mixture of carotenoids was revealed to be more effective than individual compounds, showing that a synergistic effect can also work among them (Stahl *et al.* 1998).

Tocopherols

Tocopherols (*lato sensu*) are a group of eight closely related phenolic compounds consisting of a benzochromane system, synthesized by the shikimate pathway, linked to an isoprenoid side chain. This side chain is saturated in the case of tocols (usually called tocopherols, *stricto sensu*) or unsaturated in the case of tocotrienols. In each of these groups there are four compounds according to the nature of the substituents on the benzene ring (Fig. 4.4).

α-Tocopherol, the first member identified, was recognized to have vitamin properties and was named vitamin E; therefore, for several years (even till now in some of the literature) the two terms were taken as synonymous. Nowadays the designation of vitamin E is extended to all compounds of the group, although they do not have exactly the same properties: indeed α-tocopherol is the one with more pronounced vita-

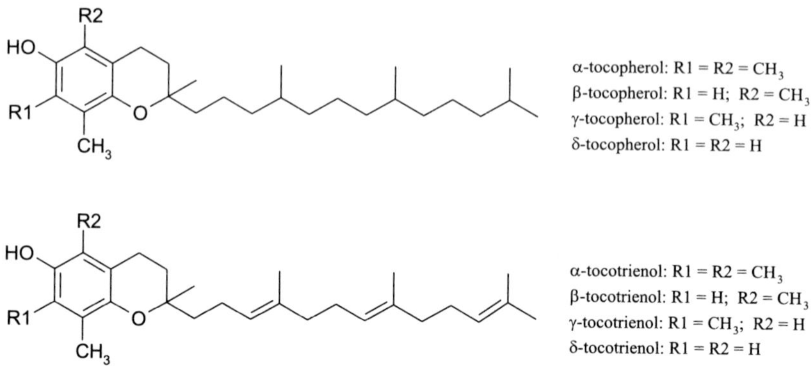

Fig. 4.4 Representative structures of tocopherols.

min E properties but other differences have been noticed among the members of the group (Brigelius-Flohé *et al.* 2002).

In plants they are present within the chloroplasts and, therefore, in the leaves of most plants; since they are fat-soluble they are stored, predominantly, in plant oils. Principal food sources, in decreasing order of magnitude, are: wheat germ, barley, palm oil, rice bran, soya bean oil, safflower oil, oat, peanut oil, cocoa butter, and olive oil. One must have in mind that the eight compounds are not found in all of the above mentioned foods and that they are not present in the same proportion.

In the human body, the most abundant of the eight tocopherols and, probably, the most physiologically important, is *α*-tocopherol, also the most studied of the components of the group. Due to its lipophilic properties it constitutes part and parcel of the biomembranes and lipid storage organelles. One supposes that *α*-tocopherol is oriented with the aromatic ring in close association with the polar fraction of the phospholipid and with the phytyl chain associated with the fatty acids in the hydrophobic inner regions of the membrane bilayer.

Due to their phenolic nature tocopherols behave as antioxidants just in the same way as other phenols do: they give a hydrogen atom to the peroxyl radical and are transformed into a less reactive radical. The unpaired electron of the tocopheroxyl radical is stabilized by delocalization but needs to be reduced back to tocopherol; otherwise it can react with lipids to generate lipid radicals, thereby promoting the chain reaction instead of breaking it (Brigelius-Flohé *et al.* 2002). The privileged position of the tocopherol molecule in membranes allows the tocopheroxyl radical to be converted back to *α*-tocopherol in redox cycle reactions involving ascorbate in the aqueous phase (Niki 1987b, Bast *et al.* 1991). In fact, *α*-tocoferol is considered to be the major chain breaking antioxidant in membranes: it is an efficient scavenger of lipid peroxyl radicals as they diffuse to the aqueous phase.

According to some researchers, tocopherols are also able to quench singlet oxygen, just in the way carotenoids do, either when chlorophyll is used as photosensitizer (Jung *et al.* 1991, Fryer 1992) or with other singlet oxygen generators (Fukuzawa *et al.* 1998a,b). However, this activity was not confirmed in cell membranes (Schafer *et al.* 2002). Some references to an action on hydroperoxides are also described (Mäkinen *et al.* 2000, Shimpuku *et al.* 2000).

The relative strength of various members of the tocopherol family has been studied giving different results dependent on the experimental conditions used (Jung *et al.* 1991, Huang *et al.* 1994, 1995, Lampi *et al.* 1999, Mäkinen *et al.* 2000).

Vitamin E is believed to be involved in sexual reproduction and in some neurological and immune functions (Wang and Quinn 1999, Feki *et al.* 2001). Due to its antioxidative properties it is supposed to prevent problems such as cardiovascular disease, cancer, chronic inflammation and neurological disorders. The state of the art concerning vitamin E and the mechanisms underlying its functions were recently analysed by a group of European scientists. According to them, the ability of tocopherols to undergo redox reactions is probably pivotal in their actions; however, tocopherols may have roles not related to this specific ability (Brigelius-Flohé *et al.* 2002).

Phospholipids

Phospholipids and phospholipid concentrates have been reported to possess antioxidant capacity. Tsaliki *et al.* (1999) have evaluated the antioxidant activity of lupin seed flour and derivatives by measuring the coupled oxidation of carotene and linoleic acid. They found that their pronounced antioxidant effect correlated with the presence of phenolic compounds and phospholipids at high concentrations. Polyenyphosphatidylcholine, a mixture of polyunsaturated phosphatidylcholines extracted from soybeans, also exhibited antioxidant effects by protecting against CCl_4-induced hepatic and alcohol-induced pancreatic oxidative stress in rats (Aleynik *et al.* 1997, 1999). Dilinoleoylphosphatidylcholine, the main component of polyenyphosphatidylcholine, is known to be responsible for the protective effects obtained with that mixture in several experimental models (Navder *et al.* 2000, Navder and Lieber 2002). It appears paradoxical that dilinoleoylphosphatidylcholine displays antioxidant properties despite being rich in unsaturated fatty acids, which constitute the main substrates for lipid peroxidation. The exact mechanism by which this phospholipid acts as an antioxidant is unknown, but it seems to be specific, since its antioxidant ability is not shared by phospholipids containing either saturated fatty acids of the same chain length as linoleate in both positions 1 and 2 or a single linoleate bound to either position 1 or 2 of the glycerol molecule (Navder *et al.* 2000). It was suggested by Aleynik *et al.* (1999) that the mechanism whereby the polyunsaturated phosphatidylcholines act as antioxidants could be due to its great capacity to repair the membranes damaged by the oxidative stress, through their high bioavailability and rapid incorporation into these membranes.

Phospholipids are one of the most important natural antioxidants in oils (Yanishlieva and Marinova 2001). Part of the higher stability of crude oils relative to processed ones is due to the presence of phosphorous-containing compounds. A positive correlation was found by Ramadan *et al.* (2003) between the radical scavenging activity, towards galvinoxyl and 1,1-diphenyl-2-picrylhydrazyl (DPPH) radicals, of black cumin, coriander, and niger seed oils and the total phospholipid content. In this study coriander seed oil and its unsaponifiable fraction exhibited the strongest activity. The phospholipid pattern was similar among the three seed oils, wherein phosphatidylcholine followed by phosphatidyle-thanolamine (Fig. 4.5) were the major components. It was also observed that the scavenging activity was highly correlated with

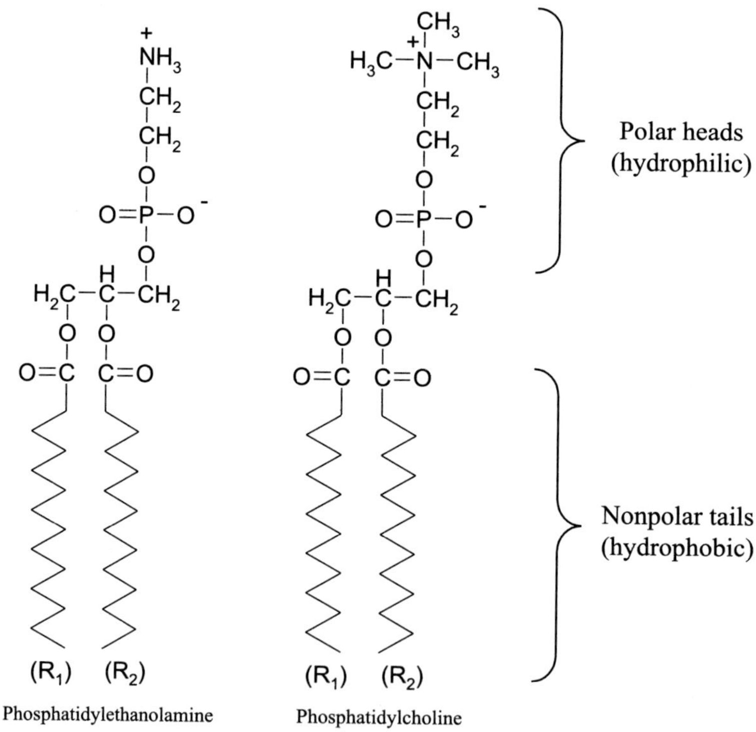

Fig. 4.5 General structures of phosphatidylethanolamine and phosphatidylcholine.

the degree of fatty acid saturation. The same conclusion was found when the oxidative stability of the three referred to crude seed oils were compared with that of the correspondent stripped oils, by measuring the formation of oxidative products after storage under accelerated oxidation conditions (Ramadan and Mörsel 2004). A direct correlation between the concentration of polar lipids and the resistance to autoxidation was observed, indicating that crude polar lipids, especially phospholipids, are potent antioxidant compounds.

The ability of phospholipids to stabilize lipids is affected by the chain length and degree of saturation of the fatty acids on their molecules: those with longer chain length and containing more saturated fatty acid are the most effective antioxidants. Though the exact mechanism of action of phospholipids is still not fully established, their antioxidant activity may be explained by the (i) synergism with tocopherols, in which the amine group can facilitate hydrogen or electron donation to tocopherols, (ii) chelation of metals by phosphate groups, (iii) formation of antioxidant Maillard-type products between phospholipids and oxidation products, and the (iv) action as an oxygen barrier between oil/air interfaces. Furthermore, the emulsifying action of phospholipids can play an important role by increasing the contact between other antioxidants present in the water phase and the oxidizing fat (Ramadan *et al.* 2003).

Organic Acids

Some organic acids are also effective antioxidants. Ascorbic acid (which, together with dehydroascorbic acid, constitutes what we call vitamin C) is, most probably, the most widely distributed water soluble antioxidant in vegetables (Naidu 2003). The richest foods in ascorbic acid, by decreasing order of magnitude, are: rose hips, parsley, black currant, pepper, Brussels sprouts, and broccoli. Although also present in the majority of animals, it is not synthesised by humans who have to obtain it from their diet.

Although typically hydrosoluble, ascorbic acid can work in heterogeneous systems: in milk it plays a major role in the oxidation stability of lipids and riboflavin (Jung *et al.* 1998, Lee *et al.* 1998). Ascorbic acid is also a major effective water soluble antioxidant in plasma: it can protect against lipid peroxidation by trapping the peroxyl radical in the aqueous phase before it can get into the lipid membrane of lipoproteins (Niki 1987a).

When involved in radical scavenging it gives a single electron to the free radical species thus originating the less reactive ascorbyl free radical. This ascorbyl radical can be reduced back to ascorbic acid or oxidized to dehydroascorbic acid, which in turn can be reduced back to ascorbic acid by reducing agents such as glutathione (Fig. 4.6).

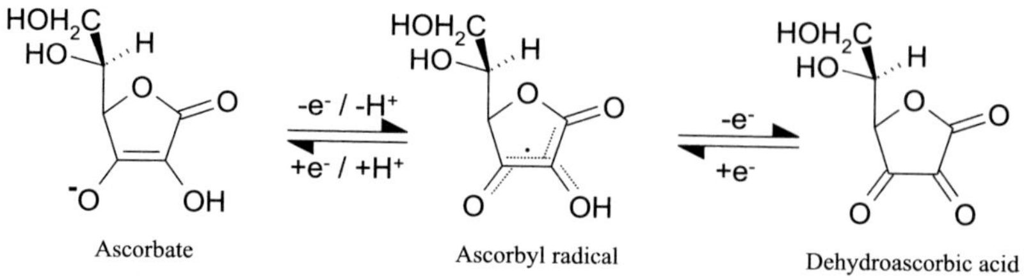

Fig. 4.6 Oxidation of ascorbate to give the ascorbyl radical and dehydroascorbic acid (adapted from Carr and Frei 1999).

Ascorbic acid can behave as a scavenger/neutralizer of a wide range of oxygen and nitrogen reactive species such as: hydroxyl radical, alkoxyl radicals, peroxyl radicals, superoxide radical/hydroperoxyl radical, hypochlorous acid, ozone, singlet oxygen, nitrogen dioxide radical, and peroxynitrite/peroxynitrous acid (Carr and Frei 1999).

Besides this direct action on oxidizing species, ascorbic acid amplifies the action of other antioxidants since it can regenerate α-tocopheroxyl, urate, and the β-carotene radical cation from their radical species (Naidu 2003, Niki 1987b, Niki *et al.* 1995). Therefore, we can speak of synergism of several antioxidants. In liposomal and biological membranes in aqueous dispersions, vitamin C works in association with vitamin E: vitamin C scavenges radicals in the aqueous region whereas vitamin E scavenges radicals within the membrane. This interaction between the two vitamins can take place not only in homogeneous solutions but also in liposomal membrane systems, where the two vitamins reside separately outside and within the membranes respectively (Niki 1987b, Niki *et al.* 1995). Synergisms were also observed in other models (Böhm 1998a,b).

The antioxidant properties of ascorbic acid are recognised as the most probable basis for the action that it seems to exert on atherosclerosis (Frei 1997, Martin and Frei 1997) and cancer prevention (Campbell *et al.* 1999, Naidu 2003).

Recently some concern has appeared in the literature because in *in vitro* experiments, ascorbic acid, when in low concentrations (0.2 M), is able to catalytically reduce active metal ions and therefore contribute to oxidative damage through the production of hydroxyl and alkoxyl radicals (Bast *et al.* 1991, Buettner and Jurkiewicz 1996, Proteggente *et al.* 2000). However this does not necessarily imply that ascorbic acid can be a pro-oxidant *in vivo* and many controversies are still unresolved in this area (Carr and Frei 1999, Chen *et al.* 2000, Halliwell 1996).

Other organic acids can also work as antioxidants. Oxalic acid is the simplest dicarboxylic acid. The most striking chemical impact of this-compound is its strong chelating ability for multivalent cations. Other carboxylic acids such as tartaric, malic, citric, succinic, and hydroxy-glutaric behave as antioxidants because they also have the ability to chelate metals. They are, therefore, classified as "preventive" or synergistic.

Amino Acids

Amino acids, and consequently peptides and proteins, are a main target for oxidizing species in biological systems (Stadtman and Levine 2003). Although they obey the general formula $RCH(NH_2)COOH$ the nature of the R substituent is different enough to allow us to say that this group presents heterogeneity. The preferential loci for oxidation are the amino group (when free), the imidazol cycle (in histidine), or the sulfur containing groups like those of cysteine and methionine. The nature and the length of the rest of the molecule may also influence the degree of oxidation. The amino acids more susceptible to oxidation are histidine, cysteine, methionine, lysine, and tryptophan (Jadhav *et al.* 1995). Given this scenario, it is not surprising that some amino acids were shown to be efficient antioxidants in experimental models.

Because of the above mentioned heterogeneity, no generalizations can be made, but several mechanisms have been ascribed to the antioxidant effect of amino acids: (i) they act as synergists of primary antioxidants by regenerating their oxidized forms (Martinéz-Tomé *et al.* 2001, Pokorny and Korczak 2001); (ii) they chelate trace metals, an ability that seems to be particularly strong for histidine containing peptides (Chen *et al.* 1998); in heterogeneous systems, amino acids transport chelated metals to the water phase, where they are innocuous as catalysts, since heavy metals behave as pro-oxidants mainly in their free state in the fat phase (Martinéz-Tomé *et al.* 2001, Wu *et al.* 2003); (iii) they reduce lipid hydroperoxides in a non-radical way; in this reaction the amino acids are oxidized to form imines, sulphides, thiosulphinates, or sulphoxides, depending on their structure (Pokorny and Korczak 2001); (iv) they (or at least histidine-containing peptides like carnosine and anserine) are able to quench singlet oxygen (Egorov *et al.* 1992, 1997, Chen *et al.* 1998); and (v) they are also able to scavenge hypochlorous acid (Martínez-Tomé *et al.* 2001) and free radicals, such as peroxyl radicals (Kohen *et al.* 1988), hydroxyl radical (Chen *et al.* 1998, Martínez-Tomé *et al.* 2001), and DPPH (Wu *et al.* 2003).

An interesting finding from Wu *et al.* (2003) was that, at least in the four experimental methods used by this group, the peptides showed higher antioxidant activities than simple amino acids.

References to pro-oxidant activity of amino acids are also found in the literature (Yong and Karel 1978, Searle and Wilson 1983, Martínez-Tomé *et al.* 2001). This pro-oxidant activity is highly dependent on the experimental conditions used during the assay, the presence of metals and the pH of the media (Martínez-Tomé *et al.* 2001).

During food roasting and baking, free amino acids (or free amino groups of amino acids, peptides, or proteins) react with reducing sugars in a non-enzymatic way, giving rise to what are called Maillard's reaction products. This usually is a complex reaction, involving several steps of which the entire mechanism, though not fully understood, giving rise to coloured macromolecules that have shown powerful antioxidant properties (Pokorny and Schmidt 2001). The final Maillard's reaction products are typical metal chelators, but several intermediate products have been shown to be free radical scavengers or hydroperoxides reducers. The simpler compound, Nα-(1-deoxy-D-fructos-1-yl)-L-arginine, present in aged garlic extract, has a high antioxidant activity, and its hydrogen peroxide reducing effect was comparable to that of ascorbic acid (Ryu *et al.* 2001).

Phenolic Compounds

Phenolic compounds form one of the largest and most ubiquitous groups of plant metabolites. Although a large variety of plant phenols exists, most compounds arise from a common origin: the amino acids phenylalanine or tyrosine. These amino acids are deaminated to cinnamic acids, which enter the phenylpropanoid pathway. A key step in this biosynthetic route is the introduction of one or more hydroxyl groups into the phenyl ring. As a result, these compounds are derived from a common building block in their carbon skeleton: the phenylpropanoid unit C_6-C_3. Biosynthesis according to this pathway produces the large variety of plant phenols: cinnamic acids (C_6-C_3), benzoic acids (C_6-C_1), flavonoids (C_6-C_3-C_6), proanthocyanidins ((C_6-C_3-C_6)$_n$), coumarins (C_6-C_3), stilbenes (C_6-C_2-C_6), lignans (C_6-C_3-C_3-C_6), and lignins ((C_6-C_3)$_n$) (Fig. 4.7).

In past years great attention has been focused on the antioxidant properties of these compounds. Previous studies found that there was a direct relationship between antioxidant activity and total phenolic content in selected herbs, vegetables, and fruits. Phenolic compounds were shown to provide a major contribution to antioxidant activity (Velioglu *et al.* 1998, Zheng and Wang 2001). Recently, Cai *et al.* (2004) characterized the antioxidant activity and phenolic composition of 112 traditional Chinese medicinal plants and a positive relationship was found between antioxidant capacity and phenolic content. The major types of phenolic compounds included all the classes referred to above.

Phenolics are able to act as antioxidants in a number of ways. Phenolic hydroxyl groups are good hydrogen donors: hydrogen-donating antioxidants can react with reactive oxygen and nitrogen species (Valentão *et al.* 2001, 2002a,b,c, 2003, Heim *et al.* 2002, Payá *et al.* 1992, Choi *et al.* 2002) in a termination reaction, which breaks the cycle of generation of

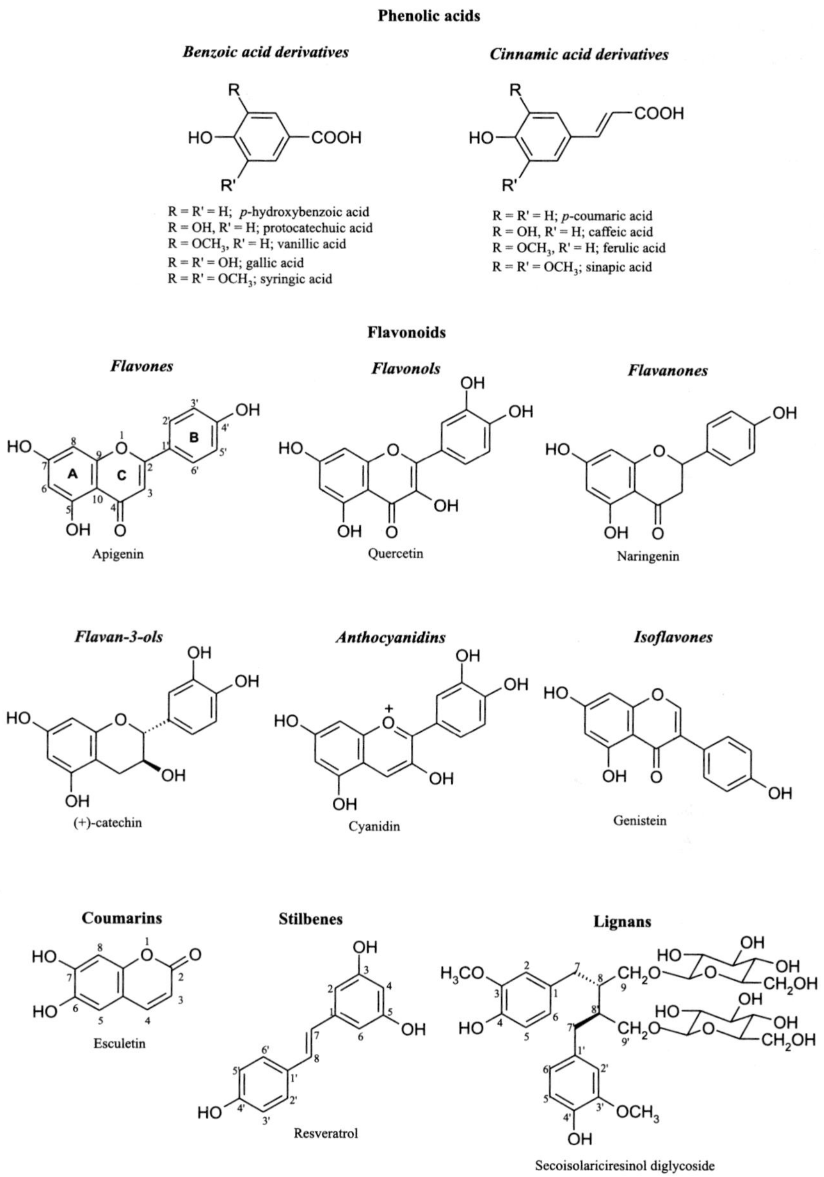

Fig. 4.7 Representative structures of selected classes of dietary polyphenols.

new radicals. Following interaction with the initial reactive species, a radical form of the antioxidant is produced, having a much greater chemical stability than the initial radical. The interaction of the hydroxyl groups of phenolics with the π-electrons of the benzene ring gives the molecules special properties, most notably the ability to generate free radicals where the radical is stabilized by delocalization. The formation of these relatively long-lived radicals is able to modify radical-mediated oxidation processes (Parr and Bolwell 2000).

The antioxidant capacity of phenolic compounds is also attributed to their ability to chelate metal ions involved in the production of free radicals (Rice-Evans *et al.* 1995, Decker 1997, Brown *et al.* 1998, Croft 1998, Yang *et al.* 2001). However, phenolics can act as pro-oxidants by chelating metals in a manner that maintains or increases their catalytic activity or by reducing metals, thus increasing their ability to form free radicals (Decker 1997, Croft 1998, Yang *et al.* 2001).

Phenolic structures often have the potential to strongly interact with proteins, due to their hydrophobic benzenoid rings and the hydrogen-bonding potential of the phenolic hydroxyl groups. This gives phenolics the ability to act as antioxidants also by virtue of their capacity to inhibit some enzymes involved in radical generation, such as various cytochrome P450 isoforms, lipoxygenases, cyclo-oxygenase, and xanthine oxidase (Laughton *et al.* 1991, Cos *et al.* 1998, Parr and Bolwell 2000).

Additionally, synergistic effects of phenolics with other antioxidants, namely ascorbic acid, β-carotene, and α-tocopherol (Croft 1998, Liao and Yin 2000), and regulation of intracellular glutathione levels have also been described (Myhrstad *et al.* 2002).

The antioxidant capacity of a phenolic compound is dependent upon its partitioning ability between lipid and aqueous media. The interaction of an agent with biomembranes, or the uptake of an agent into the membranes, is strongly related to its lipophilicity, which is expressed as the partition coefficient. Higher partition coefficients provide deeper interaction with the membranes, but can lead to a lower antioxidant effect when water-soluble oxidants, such as Fe^{2+} or Cu^{+}, are involved, or when free radicals are present in aqueous media (Foti *et al.* 1996, Brown *et al.* 1998, Liao and Yin 2000).

Phenolics are able to bind proteins and, consequently, interaction of phenolic compounds with proteins affects their antioxidant efficacy. Arts *et al.* (2002) observed that the antioxidant capacity of flavonoids was lower in the presence of proteins. These findings indicate that phenolics do not reach their optimum scavenging capacity in matrixes containing proteins.

Benzoic and cinnamic acids (Fig. 4.7) account for approximately one third of the phenolic compounds in our diet. The most representative cinnamic acid is caffeic acid, which occurs in fruits, vegetables and coffee mainly as an ester with quinic acid (chlorogenic acid or 5-caffeoylquinic acid) (Yang *et al.* 2001, Tapiero *et al.* 2002).

The antioxidant activity of phenolic acids is related to the number and position of hydroxyl groups in the molecule. The antioxidant efficiency of monophenols is strongly enhanced by the introduction of a second hydroxyl group in the *ortho* or *para* positions, and is increased by one or two methoxy substitutions in position *ortho* to the hydroxyl group (Laranjinha *et al.* 1994, Foti *et al.* 1996, Rice-Evans *et al.* 1996, Natella *et al.* 1999, Fukumoto and Mazza 2000).

Cinnamic acid derivatives exhibit greater antioxidant capacity than their benzoic counterparts. This fact is linked to the presence of the propenoic side chain, instead of the carboxylic group of benzoic derivatives: the electron withdrawing potential of the single carboxyl functional group in benzoic acids has a negative influence on the H-donating abilities of hydroxyl benzoates (Rice-Evans *et al.* 1996). The conjugated double bond in the side chain of cinnamic acids could have a stabilizing effect by resonance on the phenoxyl radical, thus enhancing the antioxidant activity of the aromatic ring (Laranjinha *et al.* 1994, Natella *et al.* 1999). Substitutions by electron-donating alkyl or methoxy groups in the *ortho* position allow increased stabilization of the resulting aryloxyl radical through electron delocalization after hydrogen donation by the hydroxyl group (Rice-Evans *et al.* 1996).

Tyrosol (4-hydroxyphenylethanol) and hydroxytyrosol (3,4-dihydroxyphenylethanol) (Fig. 4.8), are two phenolic alcohols which distribution in nature is not so wide as the previously mentioned phenolic acids, but have been the subject of numerous studies. These two compounds, together with caffeic acid and the oleuropein derivatives (discussed at the "Iridoids and secoiridoids" section), constitute the polar fraction of virgin olive oil, and are responsible for its stability and resistance to auto-oxidation, contributing as well to its characteristic flavour and taste (Puerta *et al.* 2001). Their health benefits (in which the antioxidant activity is included) were reviewed by Tuck and Hayball (2002) and by Visioli *et al.* (2002). The antioxidant properties of phenolic fraction of olive oil have also been the object of studies (Visioli *et al.* 1998, Visioli and Galli 1998a).

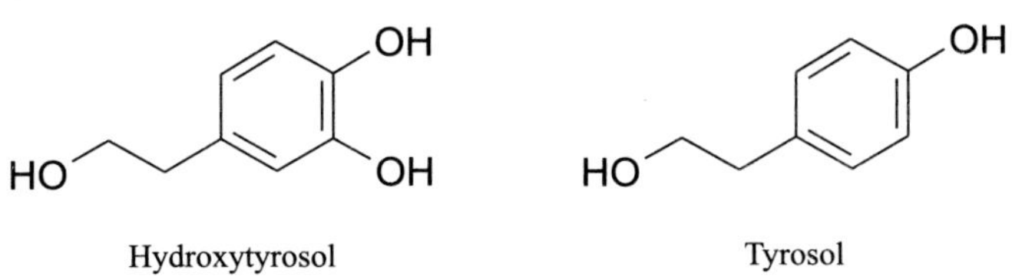

Fig. 4.8 Structures of tyrosol and hydroxytyrosol.

Hydroxytyrosol potently and dose-dependently inhibited copper sulphate-induced oxidation of LDL (Visioli *et al.* 1995, Aruoma *et al.* 1998). This compound was demonstrated to be a strong metal-chelator, a free radical scavenger of DPPH (Visioli *et al.* 1995, Visioli and Galli 1998b) and a potent scavenger of superoxide radical and hypochlorous acid (Visioli *et al.* 1998). Furthermore, hydroxytyrosol has been proven to scavenge peroxynitrite. Hydroxytyrosol showed prooxidant activities, but these are 40-fold weaker than those of the reducing agent ascorbate (Deiana *et al.* 1999).

The radical scavenging properties of hydroxytyrosol and its capacity to inhibit fatty acid peroxidation rates make it a candidate as a substitute for synthetic antioxidants used in common food and for a protective effect on human cells. It also has the advantage of being both hydro- and liposoluble, thus useful in emulsions or systems containing both water and oil phases (Visioli *et al.* 1999).

Flavonoids are the largest class of secondary plant phenolics and comprise an extensive group of naturally occurring substances found in all parts of the plant. They are important components in the human diet, although generally considered as non-nutrients. Flavonoid intake can range between 50 and 800 mg/day, accounting for two-thirds of the total phenols, depending on the consumption of vegetables and fruit, and on specific beverages, such as red wine, tea, and unfiltered beer. Other significant sources of flavonoids are different medicinal plants and related phytomedicines (Pietta, 2000, Tapiero *et al.* 2002).

The basic flavonoid structure is the flavan nucleus, consisting of two aromatic rings linked by a three-carbon aliphatic chain, which normally has been condensed to form a pyran ring. Flavonoids can be classified into several classes according to the oxidation and pattern of substitution of their central C ring (Fig. 4.7), in which there are different arrangements of their hydroxyl, methoxyl, and glycosidic side groups, and in their conjugation between the A and B rings. Flavonoids generally occur in plants as glycosylated derivatives (Pietta 2000, Heim *et al.* 2002). The main flavan-3-ols are catechins, which are the most abundant compounds in tea (Wheeler and Wheeler 2004). Quercetin is the main flavonol in human diet and is particularly abundant in onions. The main sources of flavanones are in citrus fruits. Soybeans are the only significant source of isoflavones, which are considered as phytoestrogens, due to their estrogenic properties. Flavones are less common and are found in sweet red pepper and celery. Anthocyanins are the pigments present in blue to red fruits (Yang *et al.* 2001, Tapiero *et al.* 2002).

The antioxidant activity of flavonoids depends upon the arrangement of functional groups around the nuclear structure. In fact, the spatial arrangement of substituents has, perhaps, a higher importance for antioxidant activity than the flavan backbone alone.

The B ring hydroxyl configuration is the most significant determinant for the scavenging of both reactive oxygen and nitrogen species. A 3′, 4′-catechol structure in the B ring enhances scavenging activity

(Fukumoto and Mazza 2000, Burda and Oleszek 2001) since it has better electron-donating properties and is a radical target: oxidation of a flavonoid occurs on the B ring when the catechol is present, yielding the fairly stable *ortho*-semiquinone radical, through facilitating electron delocalization, which may react with a second radical, acquiring a quinone structure (Fig. 4.9) (Rice-Evans *et al.* 1996, Pietta 2000, Heim *et al.* 2002, Kähkönen and Heinonen 2003).

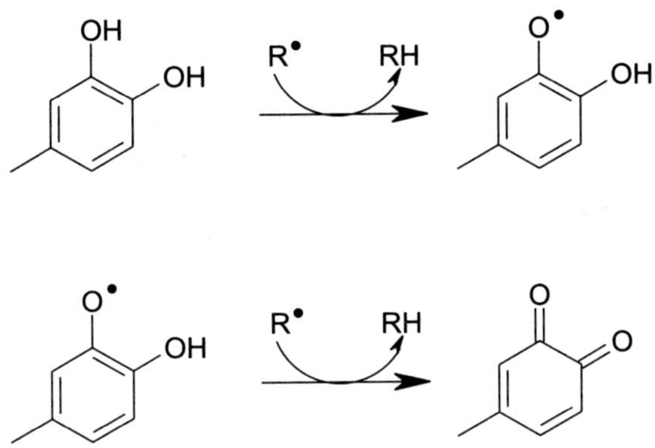

Fig. 4.9 Free radical scavenging by flavonoids.

Free radical scavenging by flavonoids is highly dependent on the presence of a free 3-hydroxyl group (Hopia and Heinonen 1999, Burda and Oleszek 2001), which is thought to increase the stability of the flavonoid radical. Thus, flavonols and flavones containing a catechol group in ring B are highly active, with flavonols more potent than the corresponding flavones because of the presence of the 3-hydroxyl group (Pietta 2000). The torsion angle of the B ring, with respect to the rest of the molecule, strongly influences scavenging ability: flavonols and flavanols with a 3-OH are planar, while the flavones and flavanones, lacking this feature, are slightly twisted. Planarity allows conjugation, electron dislocation and an increase in flavonoid phenoxyl radical stability, removal of a 3-OH abrogates coplanarity and conjugation, thereby compromising scavenging capacity (Heim *et al.* 2002). B ring hydroxyl radicals form hydrogen bonds with the heterocycle and the A ring. Eliminating this hydrogen bond effects a minor twist of the B ring compromising the electron delocalization capacity. Due to this intramolecular hydrogen bonding, the influence of a 3-OH is potentiated by the presence of a 3′, 4′-catechol, which explains the potent antioxidant activity of flavan-3-ols and flavonols that possess the latter feature (Rice-Evans *et al.* 1996). Compared to the B ring hydroxylation pattern, the impact of the A ring substitution on antioxidant activity is of questionable significance.

However, a 5-OH may contribute to the scavenging potential by increasing resonance in the presence of a 2, 3 double bond and 3-OH and 4-oxo groups (Rice-Evans *et al.* 1996, Heim *et al.* 2002).

In the case of isoflavones, the location of B ring at 3-position of the heterocyclic ring greatly affects radical scavenging capacity. The single 4′-OH group is required and the methoxylation of this group diminishes the potency (Pietta 2000).

The differences in antioxidant activity between polyhydroxylated and polymethoxylated flavonoids are most likely due to differences in both hydrophobicity and molecular planarity. Although the ratio of methoxyl to hydroxyl substituents does not necessarily predict the scavenging ability of a flavonoid, the B ring is particularly sensitive to the position of methoxy group. Steric obstruction of 3′, 4′-catechol structure by 4′-*O*-methylation significantly compromises antioxidant ability. Multiple A ring methoxyl groups also reverse the positive effect of a B ring catechol (Heim *et al.* 2002).

The 2, 3 double bond in conjugation with the 4-carbonyl group in the C ring is responsible for electron delocalization of the B ring, resulting in an increased antioxidant activity (Foti *et al.* 1996, Rice-Evans *et al.* 1996, Fukumoto and Mazza 2000). Apparently, the 2, 3 double bond has a little contribution to the hydrogen-donating ability without the diphenolic structure in the B ring. Conjugation between the A and B rings allows a resonance effect of the aromatic nucleus that lends stability to the flavonoid radical and is therefore critical in optimizing the phenoxyl radical-stabilizing effect of a 3′, 4′-catechol (Rice-Evans *et al.* 1996).

The glycosylation of flavonoids reduces their antioxidant activity when compared to the corresponding aglycones (Rice-Evans *et al.* 1996, Hopia and Heinonen 1999, Fukumoto and Mazza 2000, Burda and Oleszek 2001). The different activities among glycosides can be partly explained by different solubilities and by differences in oxidizability of glycosides (Hopia and Heinonen 1999, Fukumoto and Mazza 2000). Aside from occupying free hydroxyl groups necessary for hydrogen abstraction and radical scavenging, any substituent sugar is capable of diminishing the coplanarity of the B ring relatively to the whole structure of the flavonoid and the ability to delocalize electrons. On the other hand, glycosylation lends hydrophilicity and alters the access of the antioxidant to the lipid phase (Heim *et al.* 2002). However, Kähkönen and Heinonen (2003) have shown that, depending on the anthocyanidin, different glycosylation patterns either enhanced or diminished the antioxidant power. For the most part, the activities of the glycosides and the aglycons did not differ remarkably in emulsion. In LDL the aglycons exhibited, in general, greater activity, while the glycosides were more effective in bulk oil.

The chelating properties of flavonoids also contribute to their antioxidant activity. The proposed sites for attachment of transition metal ions are the catechol moiety in ring B, the ketol structure, 4-oxo, 3-hydroxyl, in the heterocyclic ring, or 4-oxo, 5-hydroxyl, in ring C (Fig. 4.10) (Rice-Evans *et al.* 1996, Pietta 2000).

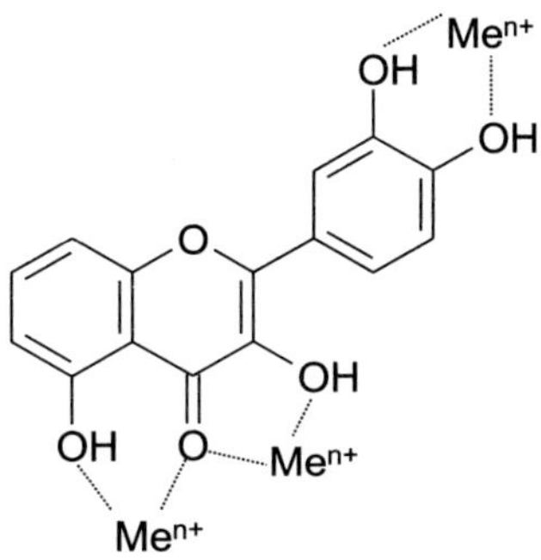

Fig. 4.10 Flavonoids binding sites for trace metals.

Tannins are substances that are able to combine with proteins of animal hide preventing their putrefaction and converting them into leather. They are polyphenolic compounds that can be grouped into two classes: the hydrolysable and the condensed ones: hydrolysable tannins are esters of gallic acid (or their oxidation derivatives) with a polyalcohol (usually glucose), while condensed tannins (now preferentially called procyanidins) are di- and oligomeric derivatives of catechins and catechin-gallate esters (Fig. 4.11) (Bors and Michel 1999).

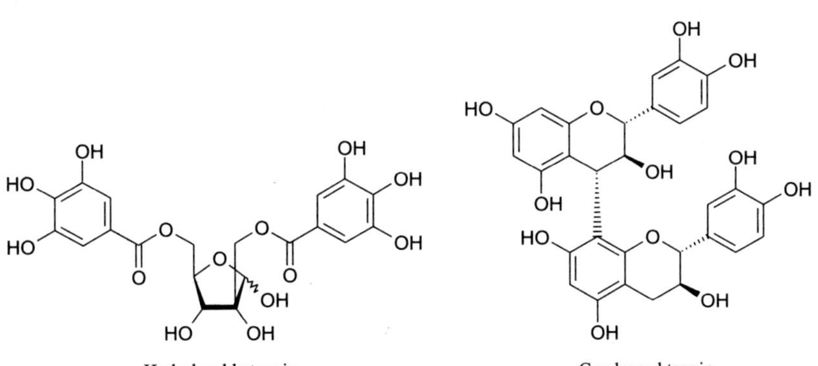

Fig. 4.11 Representative structures of hydrolyzable and condensed tannins.

Given their relationship to phenolic acids and flavonoids, their antioxidant properties are not a surprise: they exert their antioxidant activity by scavenging free radicals, chelating trace metals and by binding proteins with suppression of their enzymatic activity (Hässig *et al.* 1999). Yokozawa *et al.* (1998) have shown that the scavenging activity of tannins increases with an increase in the number of galloyl groups and molecular weight and in the presence of an *ortho*-hydroxyl structure: the hydroxyl groups are responsible for the chelating and radical scavenging properties of these compounds.

Coumarins are derivatives of benzo-α-pyrone (Fig. 4.7) and are common in plants both in the free state or as glycosides. Several studies confirm the ability of coumarins to act as antioxidants by inhibiting lipid peroxidation and by scavenging superoxide and hydroxyl radicals, hypochlorous acid and DPPH (Payá *et al.* 1992, Foti *et al.* 1996, Kaneko *et al.* 2003). Inhibition of 5-lipoxygenase and cyclo-oxigenase has also been described (Hoult and Payá 1996). Recently, Zhang and Wang (2004) demonstrated that coumarins without the catechol moiety have few structural factors beneficial to reduce O-H bond dissociation enthalpy and, consequently, are weaker antioxidants than the counterparts containing a catechol group. Loss of the dihydroxyl function by replacement of one hydroxyl group by a methoxyl group or by a glycosyl moiety is detrimental to the antioxidant activity (Payá *et al.* 1992, Foti *et al.* 1996, Kaneko *et al.* 2003). Additionally, the 1,2-pyrone is a weaker electron-withdrawing group than the 1,4-pyrone present in flavonoids, having little influence on the antioxidant activity of coumarins.

Stilbenes contain two phenyl compounds connected by a 2-carbon methylene bridge. They occur in nature in a restricted number of families, most of them acting in plants as phytoalexins. Resveratrol (3,5,4'-trihydroxy-trans-stilbene) (Fig. 4.7) is the most extensively studied compound and it is present in grapes and peanuts (Yang *et al.* 2001). Resveratrol and its analogues have been shown to behave as chain-breaking antioxidants by trapping the propagating peroxyl radicals involved in the peroxidation of linoleic acid in micelles. Additionally, these compounds exhibit a synergistic antioxidant effect with α-tocopherol, since they are able to regenerate it by reducing the α-tocopheroxyl radical (Fang *et al.* 2002). Matsuda *et al.* (2001) also observed the antioxidant capacity of these compounds against *tert*-butylhydroperoxide-induced lipid peroxidation of the erythrocyte membrane and the scavenging effect for DPPH and superoxide radicals.

Several structural features contribute to the antioxidant activity of stilbenes. Molecules bearing *ortho*-dihydroxyl functionality possess higher antioxidant capacity, because the phenoxyl radical formed is more stable due to the intramolecular hydrogen bonding interaction. On the other hand, it is easier to further oxidize the phenoxyl radical and/or *ortho*-semiquinone radical anion to form the final *ortho*-quinone. The

4′-hydroxyl group also enhances the activity, since it can stabilize the semi-quinone radical anion intermediate by resonance through the *trans* double bond. The galloyl moiety enhances activity, while glycosyl reduces it. In addition, stilbenes with both 3-hydroxyl and 4′-methoxyl groups have been shown to inhibit xanthine oxidase (Matsuda *et al.* 2001, Fang *et al.* 2002). The antioxidant activity also correlates with the electrochemical behaviour of the molecule, exhibiting higher activity relative to those with lower oxidation potentials and reversible cyclic voltamnograms (Fang *et al.* 2002).

Lignans are secondary plant constituents recognized as phytoestrogens. They can be characterized as oxidatively coupled products of two or more phenylpropanoids. Flaxseed is the richest source of plant lignans. It is characterized by the presence of secoisolariciresinol diglycoside (Fig. 4.7), which is metabolized by human colonic microflora to enterodiol and enterolactone. Both secoisolariciresinol diglycoside and its mammalian metabolites revealed the ability to inhibit lipid peroxidation, to scavenge hydroxyl radicals and to have a metal ion chelation capacity (Prasad 1997, Kitts *et al.* 1999). An aromatic methoxy group seemed to be important for the antioxidant activity (Niemeyer and Metzler 2003).

Sesame, an important oilseed from *Sesamum indicum,* is one of the oldest oilseeds known to humans. Sesame lignans are naturally occurring antioxidant components present in sesame oil, with sesamin and sesamolin (Fig. 4.12) being the predominant lignan compounds. Sesaminol (Fig. 4.12), a minor sesame lignan, results from the acid-catalysed rearrangement of sesamolin during the bleaching of crude sesame oil, also formed during the roasting process of sesame seeds and during frying (Kochhar 2000, Dachtler *et al.* 2003). Sesamol (Fig. 4.12) is a minor phenolic compound resulting from the transformation of sesamolin during the bleaching process of sesame oil (Chang *et al.* 2002). The high resistance of sesame oil to oxidative deterioration is attributed to the presence of endogenous antioxidants, that include the compounds referred to and γ-tocopherol (Kang *et al.* 2000, Chang *et al.* 2002). Dachtler *et al.* (2003) have evaluated the ability of sesame oil extracts and sesame isolated lignans to scavenge the 2, 2′-azinobis-(3-ethylbenzothiazoline-6-sulphonic acid) diammonium salt radical cation ($ABTS^{\bullet+}$) and observed that sesamol showed the strongest radical scavenging activity, followed by the sesaminol-enriched sesame extract, the sesame oil extract and sesamin. Therefore, it was assumed that those lignans carrying a phenolic hydroxyl group are more potent radical scavengers than those without it. A synergistic protective effect between sesaminol and γ-tocopherol has been shown to contribute to the stability of sunflower oil (Dachtler *et al.* 2003).

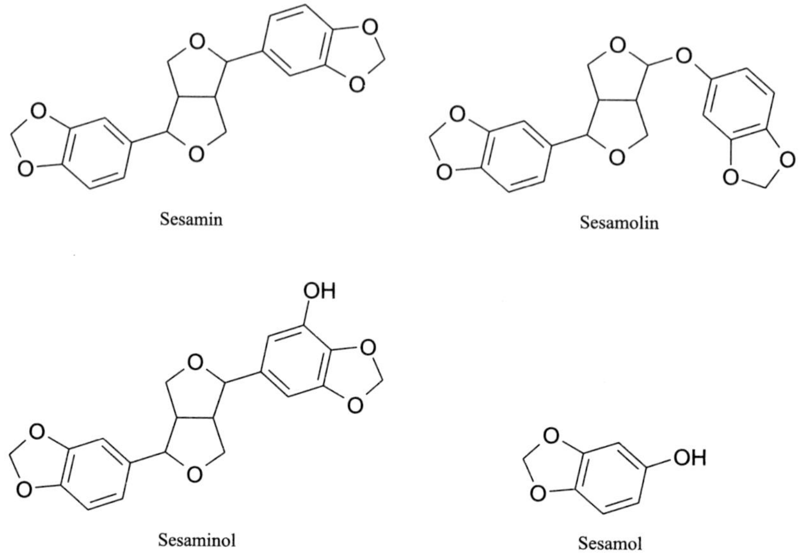

Fig. 4.12 Structures of several sesame oilseed antioxidant compounds.

Sesaminol has been shown to inhibit the lipid peroxidation of human LDL *in vitro*: it scavenged peroxyl radicals either derived from 2, 2′-azobis(2-amidinopropane) dihydrochloride (AAPH) in aqueous condition, or induced by 2, 2′-azobis(22,4-dimethylvaleronitrile) in methanol solution; these results suggested that sesaminol can be highly hydrophobic or lipophilic, and probably acted within the LDL particle (Kang *et al.* 2000). Sesamol was also shown to be a superoxide and hydroxyl radicals scavenger (Kaur and Saini 2000).

Monoterpenes and sesquiterpenes, either oxygenated or not, are also widely distributed in nature. Alone or together with phenolic compounds of low molecular weight, they constitute the great majority of essential oils components. Some studies have been conducted on the antioxidant activity of essential oils (Barata *et al.* 1998, Grassman *et al.* 2000), but it was only recently that studies began to be extended to their individual components. Giuseppe *et al.* (2000) reported the antioxidant activities of about 100 compounds, usually occurring in volatile oils, in two model systems, using α-tocopherol as a reference compound. One of the conclusions drawn by the authors is that sesquiterpenes showed a low, if any, antioxidant effect. Thymol and carvacrol (Fig. 4.13), although considered monoterpenes, share an aromatic structure and therefore showed significant activity. As to what concerns other monoterpenes, the activity was, in general, lower than that exhibited by the phenolic compounds tested. Exceptions were observed for some monoterpenic hydrocarbons

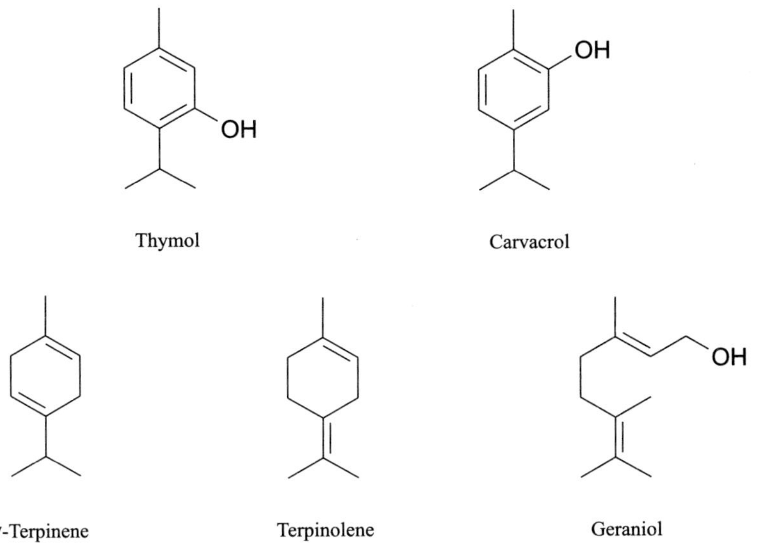

Fig. 4.13 Representative structures of terpenic compounds.

(terpinolene, α-terpinene, γ-terpinene, sabinene) and for allylic alcohols such as nerol and geraniol. According to the authors' suggestion, these compounds could act as chain breaking antioxidants by giving hydrogen from allylic regions of their molecules.

The radical scavenging properties of citrus essential oils and their components were also investigated using DPPH, with Trolox as the reference antioxidant (Choi *et al.* 2000). For 34 kinds of citrus oils, the results obtained were similar to or stronger than that of Trolox. Compounds with higher scavenging activities were, once more, geraniol, terpinolene, and γ-terpinene.

Other essential oils and their components have been studied using other experimental models (Grassman *et al.* 2001, 2002). From the compounds tested, the ones that showed highest activity were α-pinene, limonene, γ-terpinene, and terpinolene. As can be concluded, from among the three mentioned works, γ-terpinene revealed itself to be an effective antioxidant; this fact corroborates the previously mentioned hypothesis advanced by Giuseppe and Baratta (2000).

Some diterpenes, mainly found in spices such as sage and rosemary, have also been revealed to be efficient antioxidants. It is a group of compounds that are synthesized via the mevalonate pathway, but that have acquired aromatic properties in one of their rings during their genesis, usually known as **phenolic diterpenes**. This group includes compounds like carnosic acid, carnosol and rosmanol (Fig. 4.14). Their antioxidant

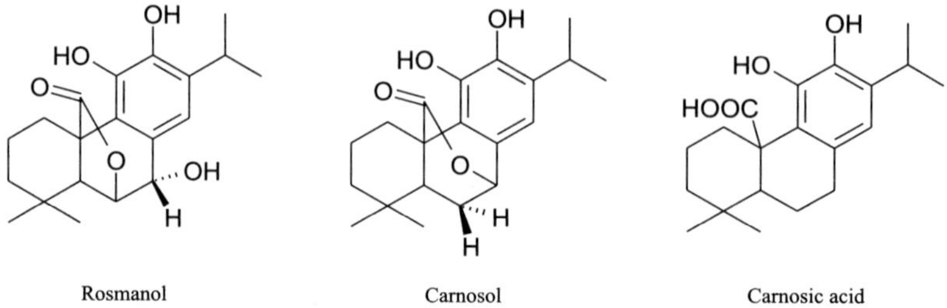

Fig. 4.14 Representative structures of phenolic diterpenes.

properties are due to the presence of phenolic groups and they act as chain breaking antioxidants, just as other phenolic compounds do (Cuvelier *et al.* 1994, Cuppett and Hall 1998).

Organosulphur Compounds

Allium species, such as onion and garlic, are characterized by the presence of organosulphur compounds. The main constituents of this group in fresh bulbs are cysteine derivatives, as those represented in Fig. 4.15, which are non-volatile compounds. When the bulbs are crushed, cooked or subjected to extraction, most of these compounds are subjected to enzymatic and thermal degradation giving rise to volatile substances, mainly di- and trisulfides and cyclic dithiins, responsible for their characteristic odour (Higuchi *et al.* 2003, Banerjee *et al.* 2003). In garlic, the best studied of the two mentioned species, the main compound is alliin, that is transformed into alliicin, the main component of alcoholic extracts,

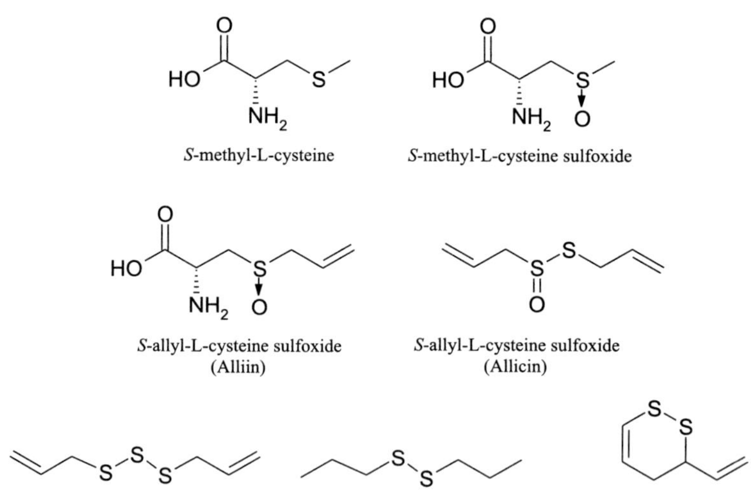

Fig. 4.15 Representative structures of sulphur-containing compounds of *Allium* spp.

which in turn can be transformed into other derivatives like 3,4-dihydro-3-vilyl-1, 2-dithiin; for more complete information on the complex chemistry of organosulphur compounds of allium species and their enzymatic derivatives, see Banerjee *et al.* (2003) and Higuchi *et al.* (2003).

Several studies have been developed with extracts and pure compounds, and an effective antioxidant activity has been demonstrated. The mechanisms through which this ability is exerted are diverse. According to the findings of Yeh and Liu (2001), diallyl sulfide, diallyl disulfide, S-ethyl cysteine and N-acetyl cysteine exert their antioxidant activity by modulating the activity of antioxidant enzymes such as 3-hydroxy-3-methylglutaryl-CoA reductase, glutathione-S-transferase, and catalase.

Yin *et al.* (2002) have studied the non-enzymatic antioxidant capabilities of these same four compounds in a liposome system. They evaluated the lipophilicity, reducing power, and iron chelating capability of these organosulphur compounds and their antioxidant interactions with α-tocoferol. The conclusion was that the non enzymatic activities of those four compounds were mainly attributed to their reducing power and to interactions with biomembranes and/or other antioxidant agents.

The antioxidative activity of S-alk(en)yl-L-cysteines and their sulfoxides, volatile alk(en)yl disulfides and trisulfides and vinildithiins against lipid hydroperoxide (LOOH) formation in human low-density lipoproteins (LDL) was evaluated and structure-antioxidative activity relationships were established (Higuchi *et al.* 2003). The authors demonstrated that organic sulfides and sulfoxides were able to scavenge hydroxyl and peroxide radicals. It was also concluded that sulphur and alk(en)yl substituents (methyl, propyl, and allyl) in S-alk(en)yl-L-cisteine and its sulfoxide are very significant in terms of the antioxidative activity for human LDL. In the case of trisulfides, it seems that an allyl double bond and the length of the S-substituted alk(en)yl group are important for an higher antioxidant activity. Furthermore, the antioxidant activity is influenced by the number of sulphur atoms in molecules, since trisulfides present higher activity than disulfides. The sulphur-containing compound that exhibited the highest antioxidative activity was 3, 4-dihydro-3-vinyl-1,2-dithiin.

Special attention has been paid to allicin (allyl 2-propenethiosulfinate) (Fig. 4.15), the principal bioactive compound present in aqueous garlic extract. In *in vitro* studies (Rabinkov *et al.* 1998) allicin exhibited radical trapping ability but *in vivo* studies demonstrated that it is rapidly metabolized into different compounds. Allicin reacts with L-cysteine and glutathione to form S-allylmercaptocysteíne and S-allylmercapoglutathione, respectively. S-allylmercaptocysteine suppressed hepatic cytochrome P450 2E1 and S-allyl mercaptoglutathione exhibited high antioxidant properties, also by the scavenging of hydroxyl radicals. Thus, the bioactive effect of allicin inside the body is attributed to its rapid reaction with thiol containing proteins, rather than any antioxidant effect (Banerjee *et al.* 2003).

Iridoids and Secoiridoids

Iridoids are cyclopentan-[c]-pyranic monoterpenes while secoiridoids have the pyran ring open, most of them occurring as glycosides. Only a few members of this group have shown antioxidant activity.

Picroliv is a mixture of two iridoids glycosides, picroside-I and kutkoside (Fig. 4.16) (1:1.5, w/w), purified from the roots and rhizomes of *Picrorhiza kurroa* (Rajeshkumar and Kuttan 2001). It has been shown to possess antioxidant properties which appear to be mediated through inhibition of superoxide dismutase and xanthine oxidase and also by chelating metal ions (Chandler *et al.* 1992).

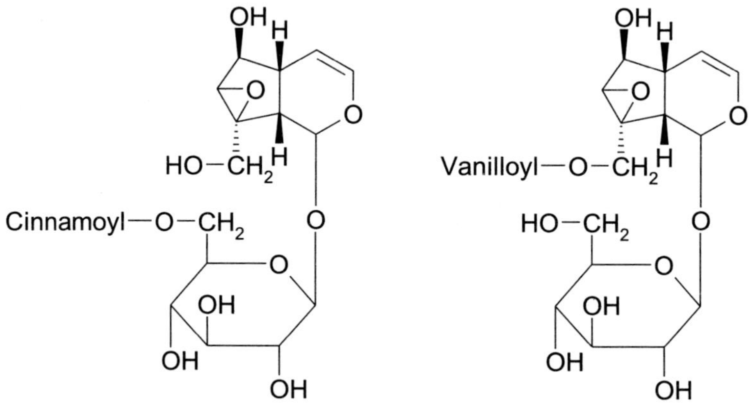

Picroside I Kutkoside

Fig. 4.16 Iridoid constituents of picroliv: picroside I and kutkoside.

Gentiopicroside (Fig. 4.17), a secoiridoid glycoside present in *Centaurium erythraea* has shown a free radical scavenging activity by DPPH assay, although this activity is much lower than that of quercetin (Kumarasamy *et al.* 2003).

The secoiridoid oleuropein (Fig. 4.17) present in the leaves and fruits (and respective oil) of *Olea europaea*, is a heterosidic ester of β-glucosylated elenolic acid and 3,4-dihydroxyphenylethanol (hydroxytyrosol). The antioxidant activity of oleuropein is mainly due to the hydroxytyrosol moiety in its structure, because of their *o*-dihydroxy substitution (Visioli and Galli 1998a). Therefore, oleuropein shows the properties already mentioned for hydroxytyrosol.

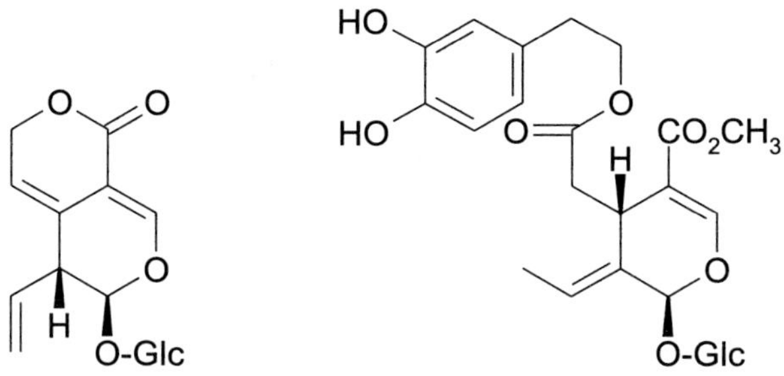

Gentiopicroside Oleuropein

Fig. 4.17 Secoiridoids: gentiopicroside and oleuropein.

Phytosterols

Phytosterols contained in vegetable oils have been documented to possess antioxidant activity by interaction with oil surfaces and inhibiting oxidation. Sterols may oxidize at oil surfaces and inhibit propagation by acting as hydrogen donors. A proportional correlation was found by Ramadan *et al.* (2003) among the radical scavenging activities, toward galvinoxyl and DPPH radicals, of black cumin, coriander and niger seed oils and their sterol levels. The same conclusions were found when the resistance to autoxidation of the seed oils referred to was investigated (Ramadan and Mörsel 2004).

Sitosterol, campesterol, stigmasterol and Δ^5-avenasterol (Fig. 4.18) were the major sterols found in sesame oil unsaponifiable matter. Considerable differences were noticed in the relative levels of these sterols after roasting, with an increase of ethylidene sterols (Δ^5-and Δ^7-avenasterol). These ethylidene side-chain sterols protect oil from oxidation during prolonged heating at high temperatures (Mohamed and Awatif 1998). The antioxidant activity of these compounds has been attributed to the formation of an allylic free radical, followed by isomerization to a relatively stable tertiary free radical. It should be noted that these ethylidene group-containing compounds show little or no antioxidant activity at room temperature, or at even 120°C, which is probably due to the necessity of high partial oxygen pressures for the formation of a peroxyl radical from a stable hindered radical derived from Δ^5-avenasterol (Kochhar 2000).

Phytic Acid

Phytic acid (myo-inositol hexaphosphoric acid) (Fig. 4.19) is an abundant plant constituent, particularly in cereals, nuts, oil seeds, legumes, pollen and spores. Several studies relate its antioxidant activity with the ability

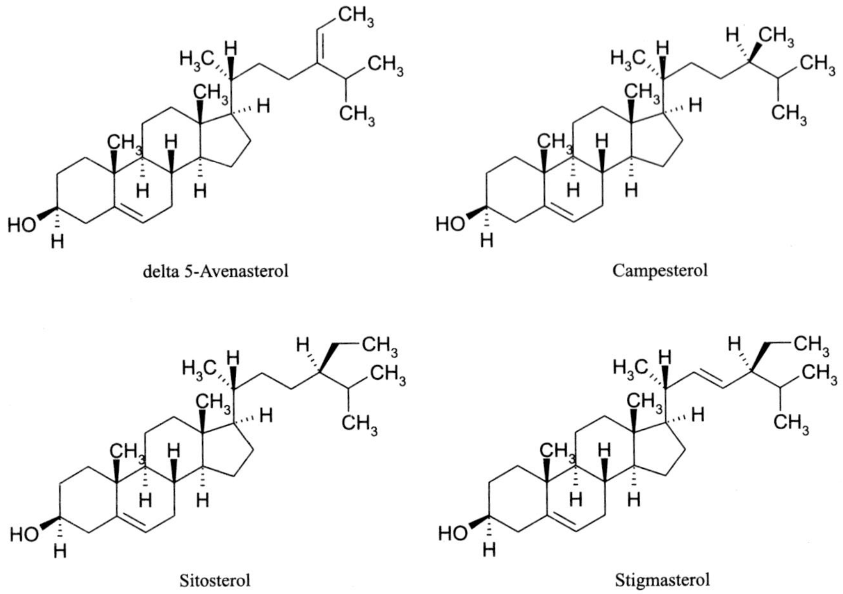

Fig. 4.18 Structures of several seed oil sterols.

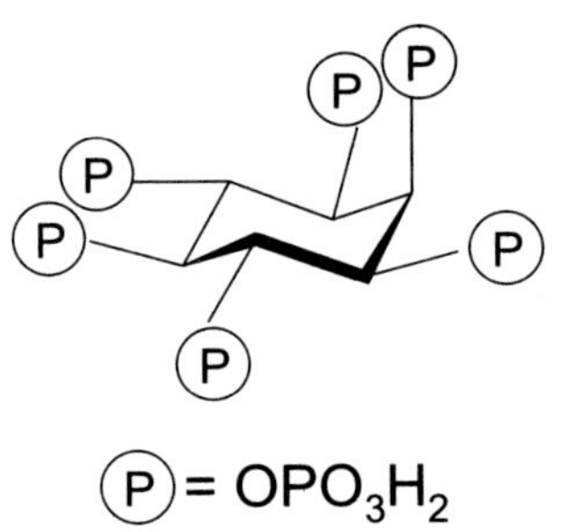

Fig. 4.19 Structure of phytic acid.

to inhibit the generation of reactive oxygen species by chelating metal ions (Midorikawa *et al.* 2001, Obata 2003).

Recently Muraoka and Miura (2004) reported that phytic acid inhibits non-competitively the activity of xanthine oxidase; when comparing it with myo-inositol it was concluded that the phosphate groups have a crucial role in this inhibitory activity. In this study it was demonstrated that phytic acid interferes with the formation of ADP-iron-oxygen complexes which initiate lipid peroxidation.

Betalains

Betalains are among the most important plant pigments, including two classes of compounds: betacyanins, which are red violet, and betaxanthins, which are yellow (Fig. 4.20). Their most common representative is betanin (betanidin 5-*O*-*β*-glucoside) (Fig. 4.20) from the red beet (Stintzing and Carle 2004).

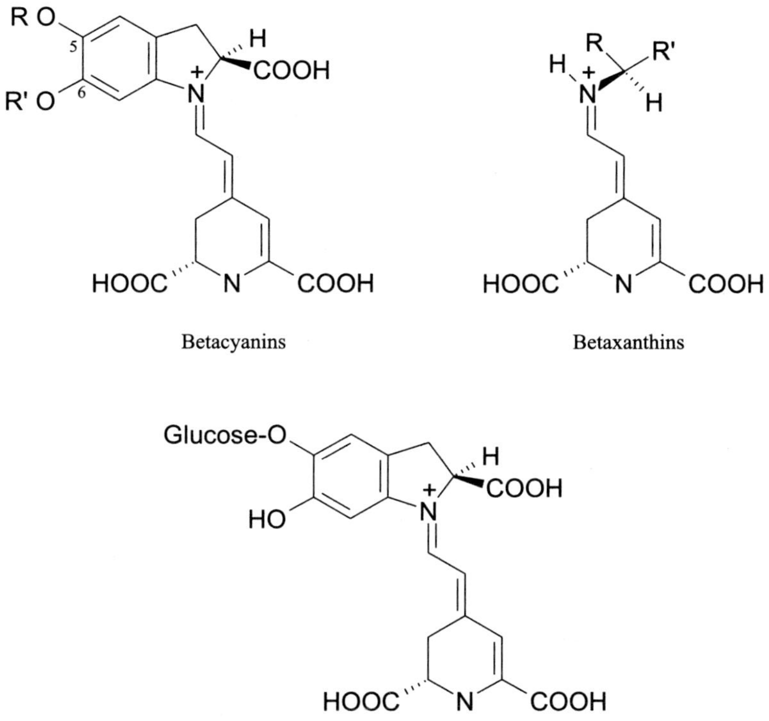

Betanin

Fig. 4.20 Basic structure of betacyanins and betaxanthins, and betanin.

Strong antioxidant effects of red beet betacyanins in model systems of lipid peroxidation catalyzed by cytochrome c, H_2O_2-activated metmyoglobin, iron redox cycle, and lipoxygenase, in microemulsions, membranes, and LDL were observed (Kanner *et al.* 2001). Pedreño and Escribano (2001) showed that betanin had a higher activity at acidic pHs and a lower activity at higher temperatures, against the 2, 2′-azinobis-(3-ethylbenzothiazoline-6-sulphonic acid) diammonium salt radical cation ($ABTS^{\bullet+}$). However, Escribano *et al.* (1998) observed that both betacyanins and betaxanthins showed greater activity against $ABTS^{\bullet+}$ at neutral and basic pHs than at acidic pH conditions. The antiradical activity of betacyanins is greater than that of betaxanthins, which could be explained by the facility to withdraw one electron from their molecules and the stability of the corresponding radicals. The withdrawal of an electron from betacyanins is relatively easy, since this electron belongs to the phenolic oxygen, the betacyanin radicals being stabilized by delocalization of the unpaired electron through the aromatic ring. In the case of betaxanthin, the electron abstracted can only be one from the conjugated π-orbitals and this loss is hindered by the positive charge of the nitrogen atom (Escribano *et al.* 1998). Betalains from prickly pear have also been implicated in the antioxidant activity of fruit extracts observed against $ABTS^{\bullet +}$ (Butera *et al.* 2002).

As to what concerns the structure-activity relationships of betalains, a screening study for the evaluation of free radical scavenging ability of betacyanins and betaxanthins from plants of Amaranthaceae showed that the antioxidant potential was related to structural features of the respective betalains. In betaxanthins, an increasing number of hydroxy and imino residues improved antioxidant capacity. In betacyanins, glycosylation reduced activity while acylation generally raised the antioxidant potential. Furthermore, 5-*O*-glycosylated structures exhibited lower potential than 6-*O*-glycosylated betacyanins (Cai *et al.* 2003).

Alkaloids

A precise definition of **alkaloids** is always complex and difficult. Currently, alkaloids are defined as plant compounds with an alkaline behaviour, with one or more nitrogenous atoms (most often in a heterocycle ring), and usually bearing marked physiological action and some toxicity on man or other animals. Other more or less similar definitions are possible because there is no clear-cut boundary between alkaloids and naturally occurring complex amines. As a consequence, alkaloids are a large and chemically heterogeneous group. It is within this group that man has found the most interesting compounds, as far as the pharmacological effect is concerned and, therefore, they gave rise to numerous drugs now in use. From the definition itself, it can be seen that some toxicity (to a higher or lower degree) is always associated with them.

Nevertheless, some alkaloids are present in our diet like the steroidal alkaloids of Solanaceae (tomatoes, potatoes, egg plant, red and green peppers) and the capsaicinoids from *Capsicum* spp. (red and green peppers). On the other hand, some plants used in popular medicine, like boldo and greater celandine, have alkaloids in their composition. For this, and probably for mere scientific curiosity, many alkaloids have been tested for their antioxidant capacity.

Many alkaloids bear a phenolic nature (Figs. 4.21-4.24); it is not, therefore, surprising that they exhibit radical scavenging and metal chelating activities, when an *ortho*-dihydroxyl group is present. However, given the heterogeneity of alkaloids, several antioxidant mechanisms have been found in this group.

An ability to quench singlet oxygen was found in brucine and strychnine (Fig. 4.21) by Larson and Marley (1984). These authors propose a mechanism for this ability based on the rigid cage-like structure of the molecule. These same alkaloids were studied by Tripathi and Chaurasia (2000) who found them to be able to chelate iron. Other indole alkaloids, such as reserpine and vincamine (Fig. 4.21), with basic tertiary nitrogen atoms and fully conjugated indole nuclei, also effectively quenched the oxygen singlet. Simple monosubstituted indoles, such as 5-hydroxyindole and 5-methoxyindole, are weakly quenchers of singlet oxygen. The higher quenching activity of 5,6-dimethoxyindole, comparing it with that of 5-methoxyindole, demonstrates the importance of electron-donating groups to the singlet oxygen quenching activity (Larson and Marley 1984). Other alkaloids to which a singlet oxygen scavenging ability has been attributed are atropine and eucatropine (Larson and Marley 1984).

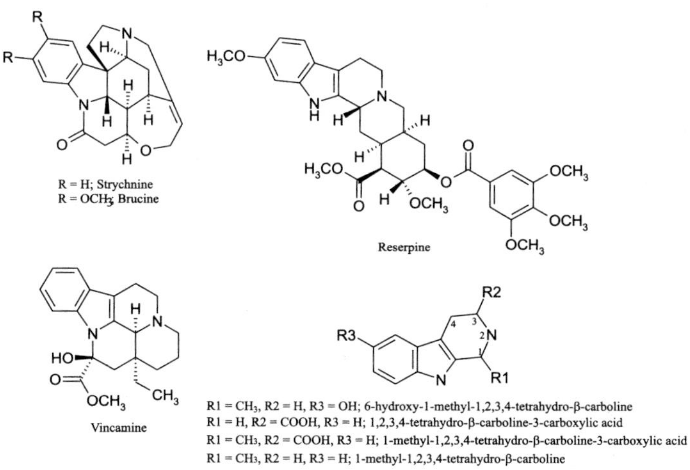

Fig. 4.21 Structures of several indole alkaloids.

Tetraydro-*β*-carbolines (Fig. 4.21) are indole alkaloids that are naturally occurring in some fruit and fruit juices, like tomato, peach, pear and apple. It was demonstrated that some compounds in this group acted as antioxidants and free radical scavengers in the ABTS assay (Herraiz and Galisteo 2003).

The benzylisoquinoline group is the most important group of alkaloids, due to the amount of compounds and the variety of structures and pharmacological actions exhibited by their members. This group integrates the simple benzylisoquinoline, bisbenzylisoquinoline, aporphines, protoberberines and morphinanes.

Boldine (Fig. 4.22) is a very much studied aporphine alkaloid, found in the leaves and bark of boldo (*Peumus boldus* Mol.), a herbal remedy used for the treatment of liver ailments. This compound, to which is mainly attributed the pharmacological action of boldo, has been shown to be an effective hydroxyl radical scavenging agent in both biological and non-biological systems (Cederbaum *et al.* 1992, Ubeda *et al.* 1993, Kringstein and Cederbaum 1995, Jang *et al.* 2000). In Herman and Cassels' (1994) opinion, these findings, associated with the low toxicity of boldine may open a broad range of uses for this compound both in medicine and industry.

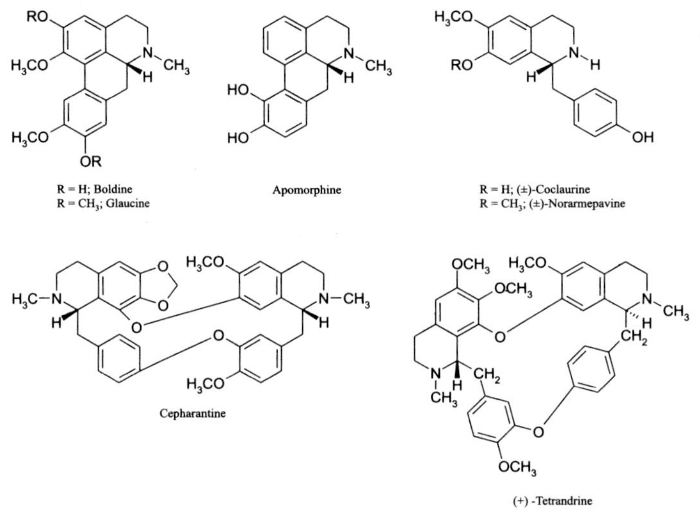

Fig. 4.22 Structures of several benzylisoquinoline alkaloids.

Cassels *et al.* (1995) compared the antioxidant activities of the aporphines boldine, glaucine and apomorphine with the benzyltetrahydroisoquinolines (±)-coclaurine and (±)-norarmepavine (Fig. 4.22) in a brain homogenate autoxidation model. They found that the aporphines (boldine, glaucine, and apomorphine) showed a much more potent antioxidant activity than the other simple benzylisoquinolines (±)-coclaurine and (±)-norarmepavine), suggesting that the antioxidant activity of these compounds is related to the presence of the biphenyl system rather than phenol groups. The antioxidant activity in non-catecholic benzylisoquinoline alkaloids depends upon the presence of a considerably conjugated π electron system as in the aporphine nucleus, and of a hydrogen atom which may easily be removed to afford a free radical, stabilized by spin delocalization over the π system. This labile hydrogen atom may be from a phenol group, although benzylic hydrogen can apparently accomplish the same role, if it is in the neighbourhood of a nitrogen lone electron pair, as happens in aporphines, which could further delocalize the unpaired electron. That seems to be the case of the non-phenolic aporphine glaucine.

The bisbenzylisoquinoline alkaloid cepharanthine (Fig. 4.22) presents a potent antiperoxidative effect, due to its direct radical scavenging activity. This alkaloid scavenged hydroxyl radicals in solutions at neutral pH, but not in acidic solutions (Kogure *et al.* 1999, 2003). Tetrandrine, another bisbenzyltetrahydroisoquinoline alkaloid showed hydroxyl and superoxide radicals scavenging activity and a significant inhibition of lipid peroxidation (Seow *et al.* 1988, Shi *et al.* 1995).

The benzophenantridine alkaloids sanguinarine and chelerythrine (Fig. 4.23) are potent inhibitors of 5- and 12 –lipoxygenase by a non-redox mechanism (Vavreckova *et al.* 1996). Sanguinarin also inhibits the oxidative burst on DMSO-differentiated HL-60 cells, not by a ROS scavenging activity but, most likely, as a result of the inhibition of the NADPH oxidase activity and, in part, by preventing the formation of NADPH oxidase protein complex (Vrba *et al.* 2004).

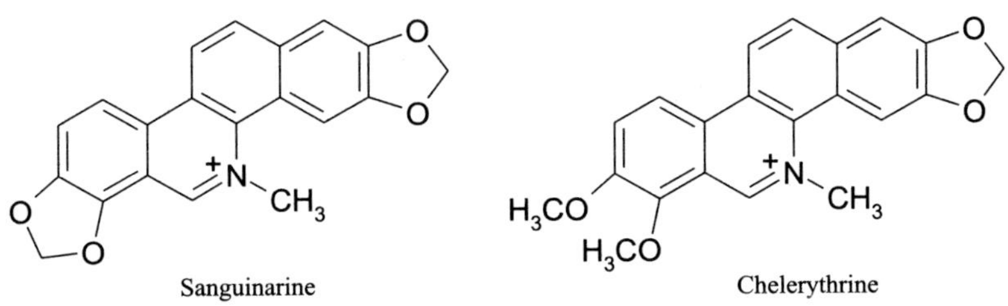

Fig. 4.23 Structures of two benzophenantridine alkaloids.

Capsaicin (Fig. 4.24) and dihydrocapsaicin are the principal compounds responsible for the pungent effects of some peppers (*Capsicum* spp.). These compounds show a considerable antioxidant activity, since they can inhibit iron-mediated lipid peroxidation and copper-dependent oxidation of low-density lipoprotein, due to their capacity to form complexes with reduced metals and to act as hydrogen donors (Murakami *et al.* 2001). Capsaicin has also been shown to prevent the oxidation of oleic acid at cooking temperatures (in this case, it seems that the amide group plays the major role in the antioxidant activity) (Henderson and Henderson 1992), as well as the formation of lipid hydroperoxides from the autoxidation of linoleic acid (it seems that this antioxidant behaviour is due primarily to the phenolic moiety of the molecule) (Henderson and Slickman 1999). In spite of all these reports about the action of capsaicin in the inhibition of various lipid peroxidations, the total mechanism of the antiperoxidative effect of this alkaloid remained unclear until the

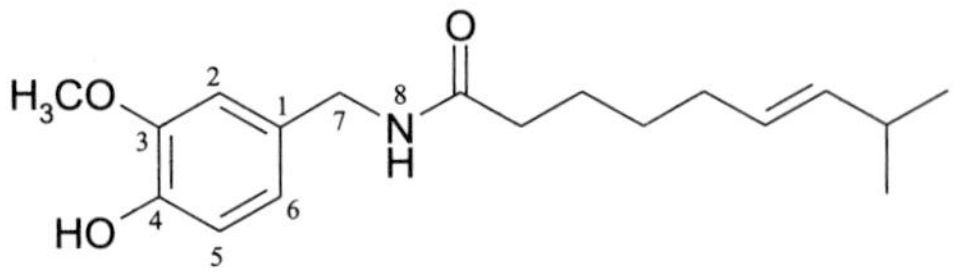

Fig. 4.24 Structure of capsaicin.

study of Kogure *et al.* (2002), concerning the effect of capsaicin on lipid peroxidation in the mitochondrial membrane and against DPPH radicals in solution and in membrane. In this study it was found that the C7-benzyl carbon of capsaicin (and not the phenolic OH-group) was responsible for the scavenging activity, and that the higher reactivity of this benzyl carbon could be due to the resonance effect of the acetamide portion located adjacent to the C7-benzyl carbon.

The use of capsaicinoids as food antioxidants is limited by their strong pungency and nociceptive activity, but, recently, capsiate and dihydrocapsiate, two non-alkaloidic capsinoids present in a non-pungent cultivar of *Capsicum annuum* L. and devoid of pungency, have shown an antioxidant activity comparable to that of their corresponding capsaicinoids (capsaicin and dihydrocapsaicin). This finding can turn them into interesting food antioxidants. These compounds act as chain-braking antioxidants in scavenging lipid peroxyl radicals, which is due to their capacity to donate hydrogen atoms and delocalize the resulting radical sites (Rosa *et al.* 2002). More recently Ochi *et al.* (2003) have obtained two capsaicinoid-type compounds from *Capsicum annuum* L. which, in contrast to capsaicin, are not pungent. These compounds are 6″-dihydro-5′, 5″-dicapsaicin and ω-hydroxycapsaicin and the first showed almost the same antioxidant activity as capsaicin.

SOURCES OF ANTIOXIDANTS FOR THE FOOD INDUSTRY

Several industrial products, from rubber to food, cosmetics and medicines, need antioxidant additives in order to extend their shelf lives.

From the many plant metabolites with antioxidant activity referred to in the previous section, only a few can be used as food antioxidants. In fact, a compound, natural or not, can only be used as a food additive if it fulfils several requirements. The first and most important of them is a lack of toxicity, that is, it must be accepted as "generally recognized as safe" (GRAS). Approval of an additive for food use requires extensive toxicological studies, in order to evaluate its possible mutagenic, teratogenic and carcinogenic effects. An antioxidant is considered safe if its LD_{50} is higher than 1000 mg/kg body weight and should not have any significant effect on the growth of the experimental animals, in long-term studies, at a level of 100 times the level proposed for human consumption. Toxicological assays are also extended to their oxidized forms and to their reaction products with food constituents. Besides this first requisite, it must also exhibit several other properties, such as (i) can not impart foreign colour, odour, or flavour (even after long storage), (ii) to be effective for at least one year, at a temperature of 25-30°C and (iii) should be easy to incorporate in the products to be preserved and effective at low concentrations (0.01% or less) (Rajalakshmi and Narasimhan 1995).

The choice of an adequate antioxidant for a given food product is not an easy task: owing to differences in molecular structure, antioxidants have different solubility and their effectiveness may be modified by different food products and under different processing and handling conditions. A given antioxidant, or mixture of antioxidants, may have its efficacy altered by the presence of pro-oxidants and/or antioxidants already intrinsic to the foodstuff or produced during processing (Rajalakshmi and Narasimhan 1995, Miková 2001). Therefore, assays must be performed for each food product because it is not possible to transpose the results already obtained with another one.

The first goal of industry, as far as antioxidants (or other additives) are concerned, is to use a substance that can bring together all the mentioned characteristics and that is also cheap. It happens that most often synthetic substances are cheaper than natural ones. Besides, synthetic antioxidants are produced as pure substances or extracts of constant composition and properties, which allows for their easy application. On the contrary, natural substances are available from raw materials with variable composition and their purification is an expensive process. In recent years, consumers have been opting for products with "all natural" labels, based on the conviction that natural means safe. Consequently the area of natural antioxidants developed enormously, mainly because of the increasing limitations on the use of synthetic ones.

Codex Alimentarius is a collection of international standards, listing those antioxidants that have been evaluated by the Joint FAO/WHO Expert Committee on Food Additives (JECFA). However, the use of additives in foods is ruled by regulatory laws of each country that may, or not, adopt that entire list.

In the European Union (EU) the experts in the matter are part of the Scientific Committee for Food (SCF). At the moment, the list of authorized antioxidants in the EU include mainly natural compounds. Rigorously speaking, most of them are not strictly natural but nature identical. These compounds are supplied in a relatively pure state, are cheap, and easy to handle. Some of them are afterwards subjected to chemical transformations in order to increase the solubility in several media. Really natural compounds are also used but mainly as standard extracts, with a pre-defined range of specified antioxidants.

A question that deserves to be mentioned at this point is related to the racemates. Most natural compounds exhibit stereogenic centres and exist as only one isomer. In some circumstances the substitution of the natural isomer by a racemic mixture may entail a lower activity, since usually only one isomer can link to the biological substrate. For example, naturally occurring α-tocopherol is an optically active compound, since it exhibits three quiral centres, and is now referred to as 2*R*,4'*R*,8'*R*-α-tocopherol (formerly *d*-α-tocopherol), whereas synthetic tocopherol is a mixture of stereoisomers and is referred as *all-rac*-α-tocopherol (formerly *dl*-α-tocopherol). The natural isomer is selectively recognised by a specific transfer protein, which makes it the most potent vitamin E homologue. Although it is assumed that all members of the racemic mixture maintain their antioxidant activity, this is one further reason that leads the public to prefer natural compounds.

In the following lines there are some brief references to the sources of natural antioxidants permitted by the laws of the EU (Directive 95/2/CE from 20th February amended by Directive 2003/114/CE, Official Journal of the European Communities, 29th January 2004, page 58).

L-Ascorbic acid (E300) used as an antioxidant is not exactly a natural compound because in fact industrially it is most often synthesized from glucose. It may be esterified on an industrial scale to improve the solubility in fats and oils. Ascorbyl palmitate or stearate are currently used and are listed as food antioxidants. Isoascorbic acid (the *D* epimer of ascorbic acid, also called erythorbic acid) is practically deprived of vitamin activity but it is still active as a synergist. It has not been described from plants, but it can be obtained from sucrose or produced by several *Penicillium* species. Citric acid (E330) is produced by microbial synthesis on a large scale and is applied in pure crystalline form. It is also used as isopropyl citrate. Tartaric acid (E334) is isolated from grapes or its residues after wine production and is also supplied in a crystalline form. Either of these acids can be used as their sodium, potassium, or calcium salts. Mixtures of citric or ascorbic acids, or their esters, with phenolic antioxidants and monoacylglycerols, glycols, or any other dissolving agents are commercially available.

Other plant chelating agents can also be used. The most widely known natural metal chelator is phytic acid, which occurs mainly in plant materials. It was considered to be a negative factor from the standpoint of nutrition as it reduces the availability of calcium and iron. On the other hand, the chelating activity is sometimes appreciated as it reduces the pro-oxidant activity of iron and other heavy metal ions. Phytic acid (E391) is reported in the *Codex Alimentarius* but was not adopted by the EU.

Lecithin (E322), a mixture of the diglycerides of stearic, palmitic, and oleic acids, linked to the choline ester of phosphoric acid, is also endowed with good antioxidant activity, acting as a synergist. The product of the commerce is predominantly soybean lecithin, obtained as a by-product in the manufacture of soybean oil.

The list of additives in the area of tocopherols include α-(E307), γ-(E308), and δ-tocopherol (E309) and a mixed tocopherols concentrate (E 306). Vitamin E can be isolated from natural sources (plants, vegetables, and even meat) or can be prepared in the laboratory. Esterified forms of tocopherols, such as acetate, succinate, and nicotinate, are also made in the laboratory and commercialized. However, tocopherols from natural sources are still used. The most commonly used sources of natural mixtures of tocopherols are the deodorizer sludges, which are distillates obtained during the deodorization of vegetable oils. These distillates are complex mixtures and contain sterols, sterol esters, triacylglicerols, tocopherols and tocotrienols; the concentration of tocopherols in these mixtures is usually 8-9% of unsaponifiable matter. Tocopherols present in these mixtures can afterwards be separated from the other components and purified. These tocopherol concentrates recommended as antioxidants (E306) are mixtures with relatively high contents of γ- and δ-tocopherol (obtained from soybean oil), but α-tocopherol is also present. The total tocopherol concentration usually ranges between 30% and 80%, while the remaining is constituted by triacylglycerols (Pokorny and Korczak 2001). Mixtures of tocopherols with other antioxidants, with which they can act in a synergistic way, are also available.

Although carotenoids are good antioxidants in certain conditions, their effectiveness in materials exposed to light is low because they are susceptible to auto-oxidation and quickly destroyed during the free radical oxidation process. Their effectiveness in unsaturated lipids exposed to light is only achieved if other antioxidants are present; otherwise carotenoids may exert a pro-oxidative effect. (Yanishlieva-Maslarova 2001). That is probably the reason why carotenoids, although non-toxic compounds, are not listed as food antioxidants. However, carotenoids are listed as food colourings and, if other antioxidants are added, as is usually the case, they can also function as antioxidants. β-Carotene is available as an almost pure compound (not less than 96 %) and can be used in several products, namely in cosmetics. Canthaxanthine (E161g), luteine (E161b), beta-apo-carotenal (E160e), and licopene (E160d) are carotenoids extracted from several natural sources. They are not pure compounds: carotenoids are mixed with other colouring matters, such as oils, fats, and aromatic compounds, obtained by co-extraction of the natural source and they are usually available dissolved in edible oils or fats. Vegetable extracts can also be utilized. *Bixa orellana* seeds are used to obtain an extract (E160b) rich in bixin and norbixin. An alkaline or an oil extract of *B. orellana* can also be used, as well as a mixture of bixin and norbixin obtained by extraction with organic solvents.

An extract of *Capsicum annuum* is also permitted as food colouring (E160c). It is obtained by extraction from fruits (with or without seeds) and is rich in capsanthin and capsorubin. Some algae, when developed in controlled conditions, are able to accumulate reasonable amounts of carotenoids and are used with that aim. *Dunaliella salina* and *Haematococcus pluvialis* are used to produce β-carotene and astaxanthin, respectively.

As it happens with carotenoids, chlorophylls are not listed as antioxidants but as food colourings. Since they can act as antioxidants, a slight reference is also made here. They can be extracted from several plant materials by means of a restricted number of organic solvents. In the course of the extraction, magnesium can be partly or totally removed. The obtained extract also contains carotenoids, fats, and oils, but at least 10% must be chlorophylls. The chlorophyllins are prepared by saponification and purification of the aforementioned extract, and sodium or potassium salts are obtained in a degree of purity of not less than 95%. Copper complexes of chlorophylls and chlorophyllins can also be used.

Amino acids are common food components and they have been found to be efficient antioxidants. The addition of pure amino acids to foodstuffs would be a good choice but, as they are too expensive, they are usually substituted for by the cheaper protein hydrolysates. Besides amino acids, these hydrolysates contain peptides and stable antioxidants from the starting material (like phenolics), which enlarges their sphere of antioxidant action. According to European law, amino acids (with the exception of glutamic acid, glycine, cysteine, and cystine) and protein hydrolysates are not considered food additives and, therefore, they do not have to be mentioned on the labels. They may be prepared by enzymatic or acid-catalysed hydrolysis of cheap proteins, such as oilseed extracted meal (most often soybean meal), wheat gluten, and many other food materials and wastes. These protein hydrolysates are used as food seasoning ingredients, but they are also suitable as an additive to dried beef or chicken soup, stew, ground meat, fish, or other foods where flavour of protein hydrolysates is not an objection (Pokorny and Korczak 2001). In general, these hydrolysates are used together with other antioxidants with which they can act as synergists: on chelating the metals from the medium they extend the efficiency of chain breaking antioxidants.

So far, the only phenolics listed as food antioxidants are the derivatives of gallic acid (propyl, octyl, dodecyl, and ethyl galhates, E310-E313) and guaiac (E314), which is not a pure compound but a resin obtained from *Guaicum officinalis*. However, one must have in mind that many spices, or their extracts, are used as food seasoning products and are rich in phenolic compounds that impart to these extracts their antioxidant activities.

Spices, herbs, and their extracts are traditional food ingredients and, when added to foods, they do not need to be declared. Their use is only limited by the aroma and flavour given to the foodstuff and by the tradition of a given population. Since the prehistoric era they have been used to give colour, aroma, and flavour and also to extend the shelf life of a given prepared food. Today we know that this lengthening is due to an antimicrobial and/or to an antioxidation action. Their composition is now relatively well studied and their antioxidant activity is mainly due to phenolic compounds, although other classes of antioxidants may also be present. Many investigations have been carried out on spices and their components (Yanishlieva-Maslarova and Heinonen 2001) and excellent results were obtained especially with plants from the Labiatae, like sage, rosemary, thyme, and oregano. They are usually more active in mixtures once it is possible to profit from their synergism. For instance, the stabilization of mashed potatoes was achieved with more success with a mixture of rosemary, thyme, and marjoram than with α-tocopherol or with tert-butylated hydroquinone (TBHQ) (Baardseth 1989).

Because the isolation of pure natural compounds is expensive, future trends in natural antioxidants development may be in the utilization of extracts of several natural sources (Gazzani *et al.* 1998, Kähkönen *et al.* 1999). Many agricultural and industrial residues may be explored to prepare safe and cheap extracts rich in antioxidants. Several potential sources are already known for their composition, safety and efficacy. Potato peel waste, rape of olive, grape seeds, and grape pomace peels have been studied for their potential as cheap sources and reviewed by Moure *et al.* (2001). In the same review we can find information on apple and pomace, citrus seeds and peels, carrot pulp wastes, old tea leaves, cocoa by-products, and soybean molasses; all these residual wastes are rich in polypenolic compounds. Wastes from brewing seeds, grains, and/or germs are other possibilities that can be explored (Moure *et al.* 2001).

Natural sources of hydroxytyrosol are the olive mill waste waters, by-products of olive oil production, whose purified extracts contain a high proportion of this compound (Visioli *et al.* 1999). Also with the purpose of obtaining hydroxytyrosol, Briante *et al.* (2002) reported a method to produce high amounts of this compound from oleuropein that is present in extracts of *Olea europaea* leaves, using the partially purified recombinant β-glycosidase from *Sulfobolus solfataricus*, expressed in *Escherichia coli*, after it has been immobilised on a chitosan support.

Product developers and industrialists are aware of the limitations of these sources. All these raw materials have slightly different compositions, depending on numerous factors, such as variety, maturation stage, geographical origin, climatic conditions, harvesting data, storage period, and environmental and technological factors. In addition to this group of factors strictly related to the material itself, it is necessary to consider the fact that the nature of the solvent and the extractive methods used also play an important role in the composition of the final extract. After the extract is ready and incorporated into the final product, care must still be taken with light, temperature, and concentration, which are fundamental factors affecting storage stability.

Despite these limitations, much work has already been done in the area of plant extracts that can be an important stimulus for future research. If public demands continue along the track of natural products, extracts from plant materials may be the future of antioxidants.

REFERENCES

Aleynik, S.I., M.A. Leo, M.K. Aleynik, and C.S. Lieber. 1999. Alcohol-induced pancreatic oxidative stress: protection by phospholipid repletion. *Free Radic. Biol. Med.* 26: 609-619.

Aleynik, S.I., M.A. Leo, X. Ma, M.K. Aleynik, and C.S. Lieber. 1997. Polyenylphosphaticylcholine prevents carbon tetrachloride-induced lipid peroxidation while it attenuates liver fibrosis. *J. Hepatol.* 27: 554-561.

Ali, S., and D.A. Mann. 2004. Signal transduction via the NF-kappaB pathway: a targeted treatment modality for infection, inflammation and repair. *Cell. Biochem. Funct.* 22: 67-79.

Antolovich, M., P.D. Prenzler, E. Patsalides, S. McDonald, and K. Robards. 2002. Methods for testing antioxidant activity. *Analyst* 127: 183-98.

Arrigo, A.P. 1999. Gene expression and the thiol redox state. *Free Radic. Biol. Med.* 27: 936-44.

Arnao, M.B., A. Cano, and M. Acosta. 2001. The hydrophilic and lipophilic contribution to total antioxidant activity. *Food Chem.* 73: 239-244.

Arnao, M.B., A. Cano, and M. Acosta. 1999. Methods to measure the antioxidant activity in plant material. A comparative discussion. *Free Radic. Res.* 31: S89-S96.

Arnao, M.B. 2000. Some methodological problems in the determination of antioxidant activity using chromogen radicals: A practical case. *Trends Food Sci. Technol.* 11: 419-421.

Arora, A., R.K. Sairam, and G.C. Srivastava. 2002. Oxidative stress and antioxidative system in plants. *Curr. Sci.* 82: 1227-1238.

Arts, M.J.T.J., G.R.M.M. Haenen, L.C. Wilms, S.A.J.N. Beetstra, C.G.M. Heijnen, H.-P. Voss, and A. Bast. 2002. Interactions between flavonoids and proteins: effect on the total antioxidant capacity. *J. Agric. Food Chem.* 50: 1184-1187.

Aruoma, O.I. 1996. Characterization of drugs as antioxidant prophylactics. *Free Radic. Biol. Med.* 20: 675-705.

Aruoma, O.I. 1999. Free radicals, antioxidants and international nutrition. *Asia Pacific J. Clin. Nutr.* 8: 53-63.

Aruoma, O.I. 2003a. Methodological considerations for characterizing potential antioxidant actions of bioactive components in plant foods. *Mutat. Res.* 523-524: 9-20.

Aruoma, O.I. 2003b. Characterization of drugs as antioxidant prophylactics. *Free Radic. Biol. Med.* 20: 675-705.

Aruoma, O.I., J.P.E. Spencer, D. Warren, P. Jenner, J. Butler, and B. Halliwell. 1997. Characterization of food antioxidants, illustrated using commerical garlic and ginger preparations. *Food Chem.* 60: 149-156.

Aruoma, O.I., M. Deiana, A. Jenner, B. Halliwell, K. Karpakash, S. Banni, F.P. Corongui, M. A. Dessi, and R. Aeschbach. 1998. Effect of hydroxytyrosol found in extra virgin olive oil on oxidative DNA damage and on low-density lipoprotein oxidation. *J. Agric. Food Chem.* 46: 5181-5187.

Azzi A., K.J.A. Davies, and F. Kelly. 2004. Free radical biology-Terminology and critical thinking. *FEBS Lett.* 558: 3-6.

Baardseth, P. 1989. Effect of selected antioxidants on the stability of dehydrated mashed potatoes. *Food Addit. Contam.* 6: 201-207.

Babior, M.B. 2000. Phagocytes and oxidative stress. *Am. J. Med.* 109: 33-44.

Banerjee, S.K., P.K. Mukherjee, and S.K. Maulik. 2003. Garlic as an antioxidant: the good, the bad and the ugly. *Phytoter. Res.* 17: 97-106.

Barata, M.T., H.J. Dorman, S.G. Deans, A.C. Figueiredo, J.G. Barroso, and G. Ruberto. 1998. Antimicrobial and antioxidant properties od some commercial essential oils. *Flavour Frag. J.* 13: 235-244.

Bast, A., G.R.M.M. Haenen, and C.J.A. Doelman. 1991. Oxidants and antioxidants: state of the art. *Am. J. Med.* 91: 2S-13S.

Beckman, K.B., and B.N. Ames. 1998. The free radical theory of aging. *Physiol. Rev.* 78: 547-581.

Belitz, D., and W. Grosch. 1999. *Food Chemistry*, 2nd ed., Springer, New York.

Beyer, R.E. 1994. The role of ascorbate in antioxidant protection of biomembranes: Interaction with vitamin E and coenzyme Q. *J. Bioenerg. Biomembr.* 26: 349-358.

Böhm, F., R. Edge, D.J. McGarvey, and T.G. Truscott. 1998a. Beta-carotene with vitamins E and C offers synergistic cell protection against NO_x. *FEBS Lett.* 436: 387-389.

Böhm, F., R. Edge, L. Lange, and T.G. Truscott. 1998b. Enhanced protection of human cells against ultraviolet light by antioxidant combinations involving dietary carotenoids. *J. Photochem. Photobiol. B: Biol.* 44: 211-215.

Bors, W., and C. Michel. 1999. Antioxidant capacity of flavanols and gallate esters: pulse radiolysis studies. *Free Radic. Biol. Med.* 27: 1413-1426.

Briante, R., F. La Cara, F. Febbraio, M. Patumi, and R. Nucci. 2002. Bioactive derivatives from oleuropein by a biotransformation on *Olea europea* leaf extracts. *J. Biotechnnol.* 93: 109-119.

Brigelius-Flohé, R.F., J. Kelly, J.T., Salonen, J. Neuzil, J.-M. Zingg, and A. Azzi. 2002. The european perspective on vitamin E: current knowledge and future concepts. *Am. J. Clin. Nut.* 76: 703-716.

Britton, G. 1991. Carotenoids. *Met. Plant Biochem.* 7: 473-518.

Britton, G. 1976. Biosynthesis of Carotenoids. Pages 262-327 in *Chemistry and Biochemistry of Plant Pigments*, Vol. 1, T.W. Goodwin, ed., Academic Press, London.

Britton, G. 1995. Structure and properties of carotenoids in relation to function. *FASEB J.* 9: 1551-1558.

Brown, J.E., H. Khodr, R.C. Hider, and C.A. Rice-Evans. 1998. Structural dependence of flavonoid interactions with Cu^{2+} ions: implications for their antioxidant properties. *Biochem. J.* 330: 1173-1178.

Buettner, G.R., and B.A. Jurkiewicz. 1996. Catalytic metals, ascorbate and free radicals: combinations to avoid. *Radiat. Res.* 145: 532-541.

Burda, S., and W. Oleszek. 2001. Antioxidant and antiradical activities of flavonoids. *J. Agric. Food Chem.* 49: 2774-2779.

Burton, G.W., and K.U. Ingold. 1984. *β*-Carotene: an unusual type of lipid antioxidant. *Science* 224: 569-73.

Butera, D., L. Tesoriere, F. Di Gaudio, A. Bongiorno, M. Allegra, A.M. Pintaudi, R. Koiien, and M.A. Livrea. 2002. Antioxidant activities of Sicilian prickly pear (*Opuntia ficus indica*) fruit extracts and reducing properties of its betalains: betanin and indicaxanthin. *J. Agric. Food Chem.* 50: 6895-6901.

Cai, Y., M. Sun, and H. Corke. 2003. Antioxidant activity of betalains from plants of the Amaranthaceae. *J. Agric. Food Chem.* 51: 2288-2291.

Cai, Y., Q. Luo, M. Sun, and H. Corke. 2004. Antioxidant activity and phenolic composition of 112 traditional Chinese medicinal plants associated with anticancer. *Life Sci.* 74: 2157-2184.

Campbell, J.D., M. Cole, B. Bunditrutavorn, and A.T. Vell. 1999. Ascorbic acid is a potent inhibitor of various forms of T cell apoptosis. *Cell Immunol.* 194: 1-5.

Cantuti-Castelvetri, I., B. Shukitt-Hale, and J.A. Joseph. 2000. Neurobehavioral aspects of antioxidants in aging. *Int. J. Dev. Neurosci.* 18: 367-381.

Caporaso, N. 2003. The molecular epidemiology of oxidative damage to DNA and cancer. *J. Natl. Cancer Inst.* 95: 1263-1265.

Carr, A.C., and B. Frei. 1999. Does vitamin C act as pro-oxidant under physiological conditions? *FASEB J.* 13: 1007-1024.

Carvalho, F., J.A. Duarte, M.J. Neuparth, H. Carmo, E. Fernandes, F. Remiao, and M.L. Bastos. 2001. Hydrogen peroxide production in mouse tissues after acute d-amphetamine administration. Influence of monoamine oxidase inhibition. *Arch. Toxicol.* 75: 465-469.

Cassels, B.K., A. Asencio, P. Conget, H. Sepeisky, L.A. Videla, and E.A. Lissi. 1995. Sctructure-antioxidant activity relationships in benzylisoquinoline alkaloids. *Pharmacol. Res.* 31: 103-107.

Cederbaum, A.I., E. Kukielka, and H. Speisky. 1992. Inhibition of rat liver microsomal lipid peroxidation by boldine. *Biochem. Pharmacol.* 44: 1765-1772.

Chandler, R., N.K. Kapoor, and B.N. Dhawan. 1992. Picroliv, picroside-I and kutkoside from *Picrorhiza kurroa* are scavengers of superoxide anions. *Biochem. Pharmacol.* 44: 180-183.

Chan, A.C. 1993. Partners in defense, vitamin E and vitamin C. *Can. J. Physiol. Pharmacol.* 71: 725-731.

Chang, L.-W., W.-J. Yen, S.C. Huang, and P.D. Duh. 2002. Antioxidant activity of sesame coat. *Food Chem.* 78: 347-354.

Chen, H.M., K. Muramoto, F. Yamauchi, K. Fujimoto, and K. Nokihara. 1998. Antioxidative properties of histidine-containg peptides designed from peptide fragments found in the digests of a soybean protein. *J. Agric. Food Chem.* 46: 49-53.

Chen, K., J. Suh, A.C. Carr, J.D. Marrow, J. Zeind, and B. Frei. 2000. Vitamin C suppresses lipid damage in vivo even in the presence of iron. *Am. J. Physiol. Endocrinol. Metab.* 279: E1406-1412.

Cho, K.J., S.H. Han, B.Y. Kim, S.G. Hwang, K.K. Park, and A.S. Chung. 2000. Chlorophyllin suppression of lipopolysaccharide-induced nitric oxide production in RAW 264.7 cells. *Toxicol. Appl. Pharmacol.* 166: 120-127.

Choi, H.R., J.S. Choi, Y.N. Han, S.J. Bae, and H.Y. Chung. 2002. Peroxynitrite scavenging activity of herb extracts. *Phytother. Res.* 16: 364-367.

Choi, H-S., H.S. Song, H. Ukeda, and M. Sawamura. 2000. Radical scavenging activities of citrus essential oils and their components: detection using 1,1-diphenylpicrylhydrazyl. *J. Agric Food. Chem.* 48: 4156-4161.

Conn, P.F., W. Schalch, and T.G. Truscott. 1991. The singlet oxygen and carotenoid interaction. *J. Photochem. Photobiol. B*, 11: 41-47.

Cooke, M.S., M.D. Evans, M. Dizdaroglu, and J. Lunec. 2003. Oxidative DNA damage: mechanisms, mutation, and disease. *FASEB J.* 17: 1195-1214.

Cos, P., L. Ying, M. Calomme, J.P. Hu, K. Cimanga, B.V. Poel, L. Pieters, A.J. Vlietinck, and D.V. Berghe. 1998. Structure-activity relationship and classification of flavonoids as inhibitors of xanthine oxidase and superoxide scavengers. *J. Nat. Prod.* 61: 71-76.

Croft, K.D. 1998. The chemistry and biological effects of flavonoids and phenolic acids. *Ann. N. Y. Acad. Sci.* 854: 435-442.

Cuppett, S.L., and C.A Hall. 1998. Antioxidant activity of the Labiatae. *Adv. Food Nutr. Res.* 42: 245-271.

Cuvelier, M-E., C. Berset, and H. Richard. 1994. Antioxidant constituents in sage (*Salvia oficinalis*). *J. Agric Food Chem.* 42: 665-669.

Dachtler, M., F.H.M. van de Put, F.V. Stijn, C.M. Beindorff, and J. Fritsche. 2003. On-line LC-NMR-MS characterization of sesame oil extracts and assessment of their antioxidant activity. *Eur. J. Lipid Sci. Technol.* 105: 488-496.

Decker, E.A. 1997. Phenolics: prooxidants or antioxidants? *Nutr. Rev.* 55: 396-398.

Deiana, M., O.I. Aruoma, M.L. Bianchi, J.P. Spencer, H. Kaur, B. Halliwell, R. Aeschbach, S. Banni, M.A. Dessi, and F.P. Corongui. 1999. Inhibition of peroxynitrile dependent DNA base modification and tyrosine nitration by the extra virgin olive oil-derived antioxidant hydroxytyrosol. *Free Radic. Biol. Med.* 26: 762-769.

Di Mascio, P., S. Kaiser, and H. Sies. 1989. Lycopene as the most efficient biological carotenoid singlet oxygen quencher. *Arch. Biochem. Biophys.* 274: 532-538.

Di Mascio, P., T.P.A. Devasagayam, S. Kaiser, and H. Sies. 1990. Carotenoids, tocopherols, and thiols as biological singlet molecular oxygen quenchers. *Biochem. Soc. Trans.* 18: 1054-1056.

Dodge, J.A. 1985. The nutritional state and nutrition. *Acta Paediatr. Scand. Suppl.* 317: 31-37.

Egorov, S.Y., E.G. Kurella, A.A. Boldrynev, and A.A. Kasnovski. 1992. The quenching of singlet molecular oxygen by carnosine and anserine in aqueous solution. *Bioorg. Khim.* 18: 142-144.

Egorov, S.Y., E.G. Kurella, A.A. Boldyrev, and A.A. Krasnovsky. 1997. Quenching of singlet oxygen by carnosine and related antioxidants. Monotoring 1270 nm phosphorescence in aqueous media. *Biochem. Mol. Biol. Int.* 41: 687-694.

Embola, C.W., O.S. Sohn, E.S. Fiala, and J.H. Weisburger. 2002. Induction of UDP-glucuronosyltransferase 1 (UDP-GT1) gene complex by green tea in male F344 rats. *Food Chem. Toxicol.* 40: 841-844.

Endo, Y., R. Usuki, and T. Keneda. 1984. Prooxidant activities of chlorophylls and their decomposition products on the photoxidation of methyl linolenate. *J. Am. Oil Chem. Soc.* 61: 781-784.

Escribano, J., M.A. Pedreño, F. García-Carmona, and R. Muñoz. 1998. Characterization of the antiradical activity of betalains from *Beta vulgaris* L. roots. *Phytochem Anal.* 9: 124-127.

Everet, S.A, M.F. Dennis, K.B. Patel, S. Maddix, S.C. Kundu, and R.L. Wilson. 1996. Scavenging of nitrogen dioxide, thyil and sulphonyl free radicals by the nutritional antioxidant β-carotene. *J. Biol. Chem.* 271: 3988-94.

Fang, J.-G., M. Lu, Z.-H. Chen, H.-H. Zhu, Y. Li, L. Yang, L.-M. Wu, and Z.-L. Liu. 2002. Antioxidant effect of resveratrol and its analogues against the free-radical-induced peroxidation of linoleic acid in micelles. *Chem. Eur. J.* 8: 4191-4198.

Feki, M., M. Souissi, and A. Mebazaa. 2001. Vitamin E: structure, metabolism and functions. *Ann. Med. Intern.* 152: 384-391.

Feng, Q., T. Kumagai, Y. Torii, Y. Nakamura, T. Osawa, and K. Uchida. 2001. Anticarcinogenic antioxidants as inhibitors against intracellular oxidative stress. *Free Radic. Res.* 35: 779-788.

Fernandes, E.R., F.D. Carvalho, F.G. Remião, M.L. Bastos, M.M. Pinto, and O.R. Gottlieb. 1995. Hepatoprotective activity of xanthones and xanthonolignoids against *tert*-butyl-hydroperoxide-induced toxicity in isolated rat hepatocytes-Comparison with silybin. *Pharm. Res.* 12: 1756-1760.

Fernandes, E., S.A. Toste, J.L.F.C. Lima, and S. Reis. 2003. The metabolism of sulindac enhances its scavenging activity against reactive oxygen and nitrogen species. *Free Radic. Biol. Med.* 35: 1008-1017.

Floyd, R.A., J.J. Watson, and P.K. Wong. 1984. Sensitive assay of hydroxyl free radical formation utilizing high pressure liquid chromatography with electrochemical detection of phenol and salicylate hydroxylation products. *J. Biochem. Biophys. Methods* 10: 221-35.

Fornai, F., M. Saviozzi, S. Piaggi, M. Gesi, G.U. Corsini, G. Malvaldi, and A.F. Casini. 1999. Localization of a glutathione-dependent dehydroascorbate reductase within the central nervous system of the rat. *Neuroscience* 94: 937-948.

Foti, M., M. Piatelli, M.T. Baratta, and G. Ruberto. 1996. Flavonoids, coumarins and cinnamic acids as antioxidants in a micellar system. Structure-activity relationship. *J. Agric. Food Chem.* 44: 497-501.

Frei, B. 1997. Vitamin C as an antiatherogen: mechanism of action. Pages 163-182 in *Vitamin C in Health and Disease*, L. Packer, and J. Fuchs, eds., Marcel Dekker, Inc., New York.

Fridovich, I. 1997. Superoxide anion radical ($O_2^{\cdot-}$), superoxide dismutase, and related matters. *J. Biol. Chem.* 272: 18515-18517.

Fryer, M.J. 1992. The antioxidant effects of thylakoid vitamin E (α-tocopherol). *Plant Cell Environ.* 15: 381-392.

Fukumoto, L.R., and G. Mazza. 2000. Assessing antioxidant and prooxidant activities of phenolic compounds. *J. Agric. Food Chem.* 48: 3597-3604.

Fukuzawa, K., Y. Inokami, A. Tokumura, J. Terao, and A. Suzuki. 1998a. Singlet oxygen scavenging by alpha-tocopherol and beta-carotene: kinetic studies in phospholipid membranes and ethanol solution. *Biofactors* 7: 31-40.

Fukuzawa, K., Y. Inokami, A. Tokumura, J. Terao, and A. Suzuki. 1998b. Rate constants for quenching singlet oxygen and for inhibiting lipid peroxidation of carotenoids and alpha-tocopherol in liposomes. *Lipids* 33: 751-756.

Gazzani, G., A. Papetti, G. Massolini, and M. Daglia. 1998. Anti- and prooxidant activity of water soluble components of some common diet vegetables and the effect of thermal treatment. *J. Agric. Food Chem.* 46: 4118-4122.

Gertz, C., S. Klostermann, and S.P. Kochhar. 2000. Testing and comparing oxidative stability of vegetable oils and fats at frying temperature. *Eur. J. Lipid Sci. Technol.* 102: 543-551.

Giuseppe, R., and M.T. Baratta. 2000. Antioxidant activity of selected essential oil components in two lipid model systems. *Food Chem.* 69: 167-174.

Goto, S., K. Kogure, K. Abe, Y. Kimata, K. Kitahama, E. Yamashita, and H. Terada. 2001. Efficient radical trapping at the surface and inside the phospholipid membrane is responsible for high potent antiperoxidative activity of the carotenoid asthaxanthin. *Biochim. Biophys. Acta* 1512: 251-258.

Grassman, J., D. Schneider, D. Weiser, and E.F. Elstner. 2001. Antioxidative effects of lemon oil and its components on copper induced oxidation of low density lipoprotein. *Drug Res.* 51: 799-805.

Grassman, J., H. Susanne, and F.E. Erich. 2002. Plant's defence and its benefits for animals and medicine: role of phenolics and terpenoids in avoiding oxygen stress. *Plant Physiol. Biochem.* 40: 471-478.

Grassman, J., S. Hippeli, K. Dormisch, U. Rohnert, N. Beuscher, and E.F. Elstner. 2000. Antioxidant properties of essential oils: possible explanation for their anti-inflammatory effects. *Drug Res.* 50: 135-139.

Halliwell, B. 1995. Antioxidant characterization. Methodology and mechanism. *Biochem. Pharmacol.* 49: 1341-1348.

Halliwell, B. 1996. Vitamin C: antioxidant or pro-oxidant in vivo? *Free Radic. Res.* 25: 439-454.

Halliwell, B., and J.M.C. Gutteridge. 1999. *Free Radicals in Biology and Medicine,* Oxford University Press, Oxford.

Hässig, A., W.X. Liang, H. Schwabl, and K. Stampfli. 1999. Flavonoids and tannins: plant-based antioxidants with vitamin character. *Med. Hypotheses* 52: 479-481.

Heber, D., and Q.Y. Lu. 2002. Overview of mechanism of action of lycopene. *Exp. Biol. Med. (Maywood)* 227: 920-923.

Heim, K.E., A.R. Tagliaferro, and D.J. Bobilya. 2002. Flavonoid antioxidants: chemistry, metabolism and structure-activity relationships. *J. Nutr. Biochem.* 13: 572-584.

Henderson, D.E., and A.M. Slickman. 1999. Quantitative HPLC determination of the antioxidant activity of capsaicin on the formation of lipid hydroperoxides of linoleic acid: a comparative study against BHT and melatonin. *J. Agric. Food Chem.* 47: 2563-2570.

Henderson, D.E., and S.K. Henderson. 1992. Thermal decomposition of capsaicin.1. Interactions with oleic acid at high temperatures. *J. Agric. Food Chem.* 40: 2263-2268.

Herman, S., and B.K. Cassels. 1994. Boldo and boldine: an emerging case of natural drug development. *Pharmacol. Res.* 29: 1-12.

Herraiz, T., and J. Galisteo. 2003. Tetrahydro-β-carboline alkaloids occur in fruits and fruit juices. Activity as antioxidants and radical scavengers. *J. Agric. Food. Chem.* 51: 7156-7161.

Higuchi, O., K. Tateshita, and H. Nishimura. 2003. Antioxidative activity of sulphur-containing compounds in *Allium* species for human low-density lipoprotein (LDL) oxidation *in vitro*. *J. Agric. Food Chem.* 51: 7208-7214.

Hopia, A., and M. Heinonen. 1999. Antioxidant activity of flavonol aglycones and their glycosides in methyl linoleate. *JAOCS* 76: 139-144.

Hoult, J.R.S., and M. Payá. 1996. Pharmacological and biochemical actions of simple coumarins: natural products with therapeutic potential. *Gen. Pharmacol.* 27: 713-722.

Huang, S.-W., E.N. Frankel, and J.B. German. 1995. Effects if individual tocopherols and tocopherol mixtures on the oxidative stability of corn oil triglycerides. *J. Agric. Food Chem.* 43: 2345-2350.

Huang, S.-W., E.N. Frankel, and J.B. German. 1994. Antioxidant activity of α-and γ-tocopherols in bulk oils and in oil-in-water emulsions. *J. Agric. Food Chem.* 42: 2108-2114.

Irache, J.M., F.A. Vega, and I. Ezpeleta. 1992. Antioxidants in some pharmaceuticals, cosmetics and food from the European market. *Pharm. Acta Helv.* 67: 152-155.

Jadhav, S.J., S.S. Nimbalkar, A.D. Kulkarni, and D.L. Madhavi. 1995. Lipid oxidation in biological and food systems. Pages 5-63 in *Food Antioxidants, Technological, Toxicological and Health Perspectives*, D.L. Madhavi, S.S. Deshpande, and D.K. Salunkhe, eds., Marcel Dekker, Inc, New York.

Jang, Y.Y., J.H. Song, Y.K. Shin, E.S. Han, and C.S. Lee. 2000. Protective effect of Boldine on oxidative mitochondrial damage in streptozotocin-induced diabetic rats. *Pharmacol. Res.* 42: 361-371.

Jorgensen, K., and L.H. Skibsted. 1993. Carotenoids scavenging of radicals. Effect of carotenoid structure and oxygen partial pressure on antioxidant activity. *Z. Lebensm. Unters Forsch.* 196: 423-9.

Jung, M.Y., S.H. Yoon, H.O. Lee, and D.B. Min. 1998. Singlet oxygen and ascorbic acid effects on dimethyldissulfide and off-flavour in skim milk exposed to light. *J. Food Sci.* 63: 408-12.

Jung, Y.J., E. Lee, and D.B. Min. 1991. α-, β-, γ-Tocopherol effect on chlorophyll photosensithized oxidation of soybean oil. *J. Food Sci.* 45: 183-187.

Kähkönen, M.P., and M. Heinonen. 2003. Antioxidant activity of anthocyanins and their aglycons. *J. Agric. Food Chem.* 51: 628-633.

Kähkönen, M.P., A.I. Hopia, H.J. Vuoela, J.-P. Rauha, K. Pihlaja, T.S. Kujala, and M. Heinonen. 1999. Antioxidant activity of plant extracts containing phenolic compounds. *J. Agric. Food Chem.* 47: 3954-3962.

Kaneko, T., N. Baba, and M. Matsuo. 2003. Protection of coumarins against linoleic acid hydroperoxide-induced cytotoxicity. *Chem. Biol. Interact.* 142: 239-254.

Kang, M.-H., M. Naito, K. Sakai, K. Uchida, and T. Osawa. 2000. Mode of action of sesame lignans in protecting low-density lipoprotein against oxidative damage in vitro. *Life Sci.* 66: 161-171.

Kanner, J., S. Harel, and R. Granit. 2001. Betalains-a new class of dietary cationized antioxidants. *J. Agric. Food Chem.* 49: 5178-5185.

Kaur, I.P., and A. Saini. 2000. Sesamol exhibits antimutagenic activity against oxygen species mediated mutagenicity. *Mutation Res.* 470: 71-76.

Kitts, D.D., Y.V. Yuan, A.N. Wijewickreme, and L.U. Thompson. 1999. Antioxidant activity of the flaxseed lignan secoisolariciresinol diglycoside and its mammalian lignan metabolties enterodiol and enterolactone. *Mol. Cel. Biochem.* 202: 91-100.

Kochhar, S.P. 2000. Stabilisation of frying oils with natural antioxidative components. *Eur. J. Lipid Sci. Technol.* 102: 552-559.

Kogure, K., K. Tsuchiya, K. Abe, M. Akasu, T. Tamaki, K. Fukuzawa, and H. Terada. 2003. Direct radical scavenging by the bisbenzylisoquinoline alkaloid cepharanthine. *Biochim. Biophys. Acta* 1622: 1-5.

Kogure, K., S. Goto, K. Abe, C. Ohiwa, M. Akasu, and H. Terada. 1999. Potent antiperoxidation activity of the bisbenzylisoquinoline alkaloid cepharantine: the amine moiety is responsible for its pH-dependent radical scavenge activity. *Biochim. Biophys. Acta* 1426: 133-142.

Kogure, K., S. Goto, M. Nishimura, M. Yasumoto, A. Kazutoyo, C. Ohiwa, H. Sassa, T. Kusumi, and H. Terada. 2002. Mechanism of potent antiperoxidative effect of capsaicin. *Biochim. Biophys. Acta* 1573: 84-92.

Kohen, R., Y. Yamamoto, K.C. Cundy, and B.N. Ames. 1988. Antioxidant activity of carnosine, homocarnosine and anserine present in muscle and brain. *Proc. Natl. Acad. Sci. USA* 85: 3175-3179.

Kohen, R., and A. Nyska. 2002. Oxidation of biological systems: Oxidative stress phenomena, antioxidants, redox reactions, and methods for their quantification. *Toxicol Pathol.* 30: 620-50.

Kringstein, P., and A.I. Cederbaum. 1995. Boldine prevents human liver microsomal lipid peroxidation and inactivation of cytochrome P450 2EI. *Free Radic. Biol. Med.* 18: 559-563.

Kumarasamy, Y., L. Nahar, and S.D. Sarker. 2003. Bioactivity of gentiopicroside from the aerial parts of *Centaurium erythraea*. *Fitoterapia* 74: 151-154.

Lampi, A.M., L. Kataja, A. Kamal-Eldin, and V. Piironen. 1999. Antioxidant activities of α- and γ-tocopherols in the oxidation of rapessed oil triacylglycerols, *JAOCS* 76: 749-755.

Landrum, J.T., and R.A. Bone. 2001. Lutein, zeaxanthin and the macular pigment. *Arch. Biochem. Biophys.* 385: 28-40.

Laranjinha, J.A.N., L.M. Almeida, and V.M.C. Madeira. 1994. Reactivity of dietary phenolic acids with peroxyl radicals: antioxidant activity upon low density lipoprotein peroxidation. *Biochem. Pharmacol.* 48: 487-494.

Larson, R.A. 1988. The antioxidants of higher plants. *Phytochemistry* 27: 969-978.

Larson, R.A., and K.A. Marley. 1984. Quenching of singlet oxygen by alkaloids and related nitrogen heterocycles. *Phytochemistry* 10: 2351-2354.

Laughton, M.J., P.J. Evans, M.A. Moroney, J.R.S. Hoult, and B. Halliwell. 1991. Inhibition of mammalian 5-lipoxygenase and cyclo-oxygenase by flavonoids and phenolic dietary additives—relationship to antioxidant activity and to iron ion-reducing ability. *Biochem. Pharmacol.* 42: 1673-1681.

Lee K.H., M.Y. Jung, and S.Y. Kim. 1998. Effects of ascorbic acid on the light-induced riboflavin degradation and colour changes in milks. *J. Agric. Food Chem.* 46: 407-410.

Liao, K.-L., and M.-C. Yin. 2000. Individual and combined antioxidant effects of seven phenolic agents in human erythrocyte membrane ghosts and phosphatidylcholine liposome systems: importance of the partition coefficient. *J. Agric. Food Chem.* 48: 2266-2270.

Lima, C., F. Carvalho, E. Fernandes, M.L. Bastos, P.C. Santos-Gomes, M. Fernandes-Ferreira, and C. Pereira-Wilson. 2004. Evaluation of toxic/protective effects of the essential oil of *Salvia officinalis* on freshly isolated rat hepatocytes. *Toxicol. In Vitro* 18: 457-465.

Liu, S., G.S. Timmins, H. Shi, C.M. Gasparovic, and K.J. Liu. 2004. Application of *in vivo* EPR in brain research: monitoring tissue oxygenation, blood flow, and oxidative stress. *NMR Biomed.* 17: 327-334.

Mäkinen, M., A. Kamal-Eldin, A.M. Lampi, and A. Hopia. 2000. Effects of α- and γ- tocopherols on formation of hydroperoxides and two decomposition products from methyl linolenate. *JAOCS* 77: 801-806.

Mallikarjuna, G., S. Dhanalakshmi, R.P. Singh, C. Agarwal, and R. Agarwal. 2004. Silibinin protects against photocarcinogenesis via modulation of cell cycle regulators, mitogen-activated protein kinases, and akt signaling. *Cancer Res.* 64: 6349-6356.

Martin, A., and B. Frei. 1997. Both intracellular and extracellular vitamin C inhibit atherogenic modifications of LDL by human vascular endothelial cells. *Atheroscler. Thromb. Vasc. Biol.* 17: 1583-1590.

Martínez-Tomé, M., F.G. Carmona, and M.A. Murci. 2001. Comparison of the antioxidant and pro-oxidant activities of broccoli amino acids with those of common food additives. *J. Sci. Food Agric.* 81: 1019-1026.

Mathews-Roth, M.M. 1993. Carotenoids in erythropoietic protoporphyria and other photosensity diseases. *Ann. N. Y. Acad. Sci.* 691: 127-138.

Matsuda, H., T. Morikawa, I. Toguchida, J.-Y. Park, S. Harima, and M. Yoshikawa. 2001. Antioxidant constituents from rhubarb: structural requirements of stilbenes for the activity and structures of two new anthraquinone glucosides. *Bioorg. Med. Chem.* 9: 41-50.

Midorikawa, K., M. Murata, S. Oikawa, Y. Hiraku, and S. Kawanishi. 2001. Protective effect of phytic acid on oxidative DNA damage with reference to cancer chemoprevention. *Biochem. Biophys. Res. Commun.* 288: 552-557.

Miková, K. 2001. The regulation of antioxidants in food. Pages 267-284 in *Antioxidants in Food*, J. Pokorny, N. Yanisshlieva, and M. Gordon, eds., Woodhead Publishing Ltd, CRC Press, Boca Raton, Florida.

Miller, N.J., J. Sampson, L.P. Candeias, P.M. Bramley, and C.A. Rice-Evans. 1996. Antioxidant activities of carotenes and xanthophylls. *FEBS Letters* 384: 240-242.

Moure, A., J.M. Cruz, D. Franco, J.M. Dominguez, J. Sineiro, H. Domínguez, M.J. Nuñez, and J.C. Parajó. 2001. Natural antioxidants from residual sources. *Food Chem.* 72: 145-171.

Mohamed, H.M.A., and I.I. Awatif. 1998. The use of sesame oil unsaponifiable matter as a natural antioxidant. *Food Chem.* 62: 269-276.

Mortensen, A., L.H. Skibsted, J. Sampson, C. Rice-Evans, and S.A. Everett. 1997. Comparative mechanisms and rates of free radical scavenging by carotenoid antioxidants. *FEBS Lett.* 418: 91-97.

Moure, A., J.M. Cruz, D. Franco, J.M. Dominguez, J. Sineiro, H. Dominguez, M.J. Nuñez, and J. Carlos Parajó. 2001. Natural antioxidants from residual sources. *Food Chem.* 72: 145-171.

Murakami, K., M. Ito, H.H. Htay, R. Tsubouchi, and M. Yoshibo. 2001. Antioxidant effect of capsaicinoids on the metal-catalyzes lipid peroxidation. *Biomed. Res.* 22: 15-17.

Muraoka, S., and T. Miura. 2004. Inhibition of xanthine oxidase by phytic acid and its antioxidative action. *Life Sci.* 74: 1691-1700.

Myhrstad, M.C.W., H. Carlsen, O. Nordström, R. Blomhoff, and J.Ø. Moskaug. 2002. Flavonoids increase the intracellular glutathione level by transactivation of the γ-glutamylcysteine synthetase catalytical subunit promoter. *Free Radic. Biol. Med.* 32: 386-393.

Naidu, K.A. 2003. Vitamin C in human health ad disease is still a mystery? An overview. *Nutr. J.* 2: 7.

Natella, F., M. Nardini, M. Di Felice, and C. Scaccini. 1999. Benzoic and cinnamic acid derivatives as antioxidants: structure-activity relation. *J. Agric. Food Chem.* 47: 1453-1459.

Navder, K.P., and C.S. Lieber. 2002. Dilinoleoylphosphatidylcholine is responsible for the beneficial effects of polyenylphosphatidylcholine on ethanol-induced mitochondrial injury in rats. *Biochem. Biophys. Res. Commun.* 291: 1109-1112.

Navder, K.P., E. Baraona, and C.S. Lieber. 2000. Dilinoleoylphosphatidylcholine protects human low density lipoproteins against oxidation. *Atherosclerosis* 152: 89-95.

Niemeyer, H.B., and M. Metzler. 2003. Differences in the antioxidant activity of plant and mammalian lignans. *J. Food Eng.* 56: 255-256.

Niki, E. 1987a. Lipid antioxidants: how they may act in biological systems. *Br. J. Cancer Suppl.* 8: 153-7.

Niki, E. 1987b. Interaction of ascorbate and alpha-tocopherol. *Ann. N. Y. Acad. Sci.* 498: 186-199.

Niki, E., N. Noguchi, H. Tsuchihashi, and N. Gotoh. 1995. Interactions among vitamin C, vitamin E and β-carotene. *Am. J. Clin. Nut.* 62: 1322S-6S.

Niki, E., and N. Noguchi. 2000. Evaluation of antioxidant capacity. What capacity is being measured by which method? *IUBMB Life* 50: 323-329.

Obata, T. 2003. Phytic acid suppresses 1-methyl-4-phenylpiridinium ion-induced hydroxyl radical generation in rat striatum. *Brain Res.* 978: 241-244.

Ochi, T., Y. Takaischi, K. Kogure, and I. Yamauti. 2003. Antioxidant activity of a new capsaicin derivative from *Capsicum annuum*. *J. Nat. Prod.* 66: 1094-1096.

Ong, T.M., W.Z. Whong, J. Stewart, and H.E. Brockman. 1986. Chlorophyllin: a potent antimutagenic against environmental and dietary complex mixtures. *Mutat. Res.* 173: 111-115.

Oomah, D., and G. Mazza. 1999. Health benefits of phytochemicals from selected Canadian crops. *Trends Food Sci. Tech.* 10: 193-198.

Ottley, C. 2000. Vitamin and mineral supplements: who needs them? *Nursing Standard* 14: 42-45.

Packer, L. 1992. Interactions among antioxidants in health and disease: vitamin E and its redox cycle. *Proc. Soc. Exp. Biol. Med.* 200: 271-276.

Park K.K., J.H. Park, Y.J. Jung, and W.Y. Chung. 2003. Inhibitory effects of chlorophyllin, hemin and tetrakis(4-benzoic acid)porphyrin on oxidative DNA damage and mouse skin inflammation induced by 12-O-tetradenaoylphorbol-13-acetate as a possible anti-tumor promoting mechanism. *Mutat. Res.* 542: 89-97.

Parr, A.J., and G.P. Bolwell. 2000. Phenols in the plant and in man. The potential for possible nutritional enhancement of the diet by modifying the phenols content or profile. *J. Sci. Food Agric.* 80: 985-1012.

Payá, M., B. Halliwell, and J.R.S. Hoult. 1992. Interactions of a series of coumarins with reactive oxygen species. Scavenging of superoxide, hypochlorous acid and hydroxyl radicals. *Biochem. Pharmacol.* 44: 205-214.

Pedreño, M.A., and J. Escribano. 2001. Correlation between antiradical activity and stability of betanine from *Beta vulgaris* L. roots under different pH, temperature and light conditions. *J. Sci. Food Agric.* 81: 627-631.

Pietta, P.-G. 2000. Flavonoids as antioxidants. *J. Nat. Prod.* 63: 1035-1042.

Pires das Neves, R.N., F. Carvalho, M. Carvalho, E. Fernandes, E. Soares, M.L. Bastos, M.L. Pereira. 2004. Protective activity of hesperidin and lipoic acid against sodium arsenite acute toxicity in mice. *Toxicol. Pathol.* 32: 527-535.

Pokorny, J., and J. Korczak. 2001. Preparation of natural antioxidants. Pages 311-330 in *Antioxidants in Food*, J. Pokorny, N. Yanisshlieva., and M. Gordon, eds., Woodhead Publishing Ltd, CRC Press, Boca Raton, Florida.

Pokorny, J., and S. Schmidt. 2001. Natural antioxidant functionality during food processing. Pages 331-354 in *Antioxidants in Food*, J. Pokorny, N. Yanisshlieva., and M. Gordon, eds., Woodhead Publishing Ltd, CRC Press, Boca Raton, Florida.

Poulsen, H.E., B.R. Jensen, A. Weimann, S.A. Jensen, M. Sorensen, and S. Loft. 2000. Antioxidants, DNA damage and gene expression. *Free Radic. Res.* 33: S33-39.

Prasad, K. 1997. Hydroxyl radical-scavenging property of secoisolariciresinol diglucoside (SDG) isolated from flax-seed. *Mol. Cel. Biochem.* 168: 117-123.

Proteggente, A.R, A. Rehman, B. Halliwell, and C.A. Rice-Evans. 2000. Potential problems of ascorbic acid and iron supplementation: pro-oxidant effect in vivo? *Biochem. Biophys. Res. Commun.* 277: 535-540.

Puerta, R., M.E.M. Dominguez, V. Ruíz-Gutierrez, J.A. Flavill, and J.R.S. Hoult. 2001. Effects of virgin olive oil phenolics on scavenging of reactive nitrogen species and upon nitregic neurotransmission. *Life Sci.* 69: 1213-1222.

Rabinkov, A., T. Miron, L. Konstantinovski, M. Wilchek, D. Mirelman, and L. Weiner. 1998. The mode of action of allicin: trapping of radicals and interaction with thiol containg proteins. *Biochim. Biophys. Acta* 1379: 233-244.

Rajalakshmi, D., and S. Narasimhan. 1995. Food Antioxidants: Sources and Methods of Evaluation. Pages 65-157 in *Food Antioxidants, Technological, Toxicological and Health Perspectives*, D.L. Madhavi, S.S. Deshpande, and D.K. Salunkhe, eds., Marcel Dekker, Inc, New York.

Rajeshkumar, N.V., and R. Kuttan. 2001. Protective Effect of Picroliv, the active constituent of *Picrorhiza kurroa*, against chemical carcinogenesis in mice. *Teratog. Carcinog. Mutag.* 21: 303-313.

Ramadan, M.F., and J.-T. Mörsel. 2004. Oxidative stability of black cumin (*Nigella sativa* L.), coriander (*Coriandrum sativum* L.) and niger (*Guizotia abyssinica* Cass.) crude seed oils upon stripping. *Eur. J. Lipid Sci. Technol.* 106: 35-43.

Ramadan, M.F., L.W. Kroh, and J.-T. Mörsel. 2003. Radical scavenging activity of black cumin (*Nigella sativa* L.), coriander (*Coriandrum sativum* L.) and niger (*Guizotia abyssinica* Cass.) crude seed oils and oil fractions. *J. Agric. Food Chem.* 51: 6961-6969.

Rice-Evans, C.A., N.J. Miller, and G. Paganga. 1996. Structure-antioxidant activity relationships of flavonoids and phenolic acids. *Free Radic. Biol. Med.* 20: 933-956.

Rice-Evans, C.A., N.J. Miller, P.G. Bolwell, P.M. Bramley, and J.B. Pridham. 1995. The relative antioxidant activities of plant-derived polyphenolic flavonoids. *Free Radic. Res.* 22: 375-383.

Rosa, A., M. Deiana, V. Casu, S. Paccagnini, G. Appendino, M. Ballero, and M.A Dessi. 2002. Antioxidant activity of capsinoids. *J. Agric. Food Chem.* 50: 7396-7401.

Ryu, K., N. Ide, H. Matsuura, and Y. Itakura. 2001. N*a*-(1-deoxy-D-fructos-1-yl)-L-arginine, an antioxidant compound identified in aged garlic extract. *J. Nutr.* 131: 972S-976S.

Sánchez-Moreno, C. 2002. Review: Methods Used to Evaluate the Free Radical Scavenging Activity in Foods and Biological Systems. *Food Sci. Tech. Int.* 8: 121-137.

Schafer, F.Q., H.P. Wang, E.E. Kelley, K.L. Cueno, S.M. Martin, and G.R. Buettner. 2002. Comparing β-carotene, vitamin E and nitric oxide as membrane antioxidants. *Biol. Chem.* 382: 671-81.

Schumann, J., J. Prockl, A.K. Kiemer, A.M. Vollmar, R. Bang, and G. Tiegs. 2003. Silibinin protects mice from T cell-dependent liver injury. *J. Hepatol.* 39: 333-340.

Scott, G. 1988. Potential toxicological problems associated with antioxidants in plastics and rubber consumables. *Free Radic Res. Commun.* 5: 141-147.

Searle, A.J., and R.L. Wilson. 1983. Stimulation of microsomal lipid peroxidation by iron and cysteine. Characterization and the role of free radicals. *Biochem. J.* 212: 549-554.

Seow, W.K., A. Ferrante, S.Y. Li, and Y.H. Thong. 1988. Antiphagocytic and antioxidant properties of plant alkaloid tetrandrine. *Int. Arch. Appl. Immunol.* 85: 404-409.

Shi, X., Y. Mao, U. Saffiotti, L. Wang, Y. Rojanasakul, S.S. Leonard, and V. Vallyathan. 1995. Antioxidant activity of tetrandrine and its inhibition of quartz-induced lipid peroxidation. *J. Toxic. Environ. Health.* 46: 233-48.

Shimpuku, H., Y. Tachi, M. Shinohara, and K. Ohura. 2000. Effect of vitamin E on the degradaton of hydron peroxide in cultured human umbilical vein endothelial cells. *Life Sci.* 68: 353-359.

Sies, H. 1993. Strategies of antioxidant defence. *Eur. J. Biochem.* 215: 213-219.

Stadtman, E.R., and R.L. Levine. 2003. Free radical-mediated oxidation of free amino acids and amino acid residues in proteins. *Amino Acids* 25: 207-218.

Stahl, W., and H. Sies. 2003. Antioxidant activity of carotenoids. *Mol. Aspects Med.* 24: 345-51.

Stahl, W., A Junghans, B. de Boer, E. Driomina, K. Briviba, and H. Sies. 1998. Carotenoid mixtures protect multilamellar liposomes against oxidative damage: synergistic effects of lycopene and lutein. *FEBS Lett.* 427: 305-308.

Stintzing, F.C., and R. Carle. 2004. Functional properties of anthocyanins and betalains in plants, food, and in human nutrition. *Trends Food Sci. Technol.* 15: 19-38.

Suzuki, Y.J., H.J. Forman, and A. Sevanian. 1997. Oxidants as stimulators of signal transduction. *Free Radic. Biol. Med.* 22: 269-285.

Tapiero, H., K.D. Tew, G. Nguyen Ba, and G. Mathé. 2002. Polyphenols: do they play a role in the prevention of human pathologies? *Biomed. Pharmacother.* 56: 200-207.

Terao, J. 1989. Antioxidant activity of β-carotene-related carotenoids in solution. *Lipids* 24: 659-61.

Terwel, L., and J.C. van der Hoeven. 1985. Antimutagenic activity of some naturally occurring compounds towards cigarette-smoke condensate and benzo[a]pyrene in the Salmonella/microsome assay. *Mutat. Res.* 152: 1-4.

Tripathi, Y.B., and S. Chaurasia. 2000. Interaction of *Strychnos nux-vomica* products and iron: with reference to lipid peroxidation. *Phytomedicine* 7: 523-528.

Tsaliki, E., V. Lagouri, and G. Doxastakis. 1999. Evaluation of the antioxidant activity of lupin seed flour and derivatives (*Lupinus albus* ssp. *Graecus*). *Food Chem.* 65: 71-75.

Tuck, K.L., and P.J. Hayball. 2002. Major phenolic compounds in olive oil: metabolism and health effects. *J. Nutr. Biochem.* 13: 636-644.

Ubeda, A., C. Montesinos, M. Paya, C. Cio, and M.J. Alcaraz. 1993. Antioxidant action of benzylisoquinoline alkaloids. *Free Radic. Res. Commun.* 18: 167-175.

Usuki, R., Y. Endo, and T. Keneda. 1984. Prooxidant activities of chlorophylls and pheophytins on the photoxidation of edible oils. *Agric. Biol. Chem.* 48: 991-994.

Valentão, P., E. Fernandes, F. Carvalho, P.B. Andrade, R.M. Seabra, and M.L. Bastos. 2001. Antioxidant activity of *Centaurium erythraea* infusion evidenced by its superoxide radical scavenging and xanthine oxidase inhibitory activity. *J. Agric. Food Chem.* 49: 3476-3479.

Valentão, P., E. Fernandes, F. Carvalho, P.B. Andrade, R.M. Seabra, and M.L. Bastos. 2002a. Antioxidative properties of cardoon (*Cynara cardunculus* L.) infusion against superoxide radical, hydroxyl radical and hypochlorous acid. *J. Agric. Food Chem.* 50: 4989-4993.

Valentão, P., E. Fernandes, F. Carvalho, P.B. Andrade, R.M. Seabra, and M.L. Bastos. 2002b. Antioxidant activity of *Hypericum androsaemum* infusion: scavenging activity against superoxide radical, hydroxyl radical and hypochlorous acid. *Biol. Pharm. Bull.* 25: 1320-1323.

Valentão, P., E. Fernandes, F. Carvalho, P.B. Andrade, R.M. Seabra, and M.L. Bastos. 2002c. Studies on the antioxidant activity of *Lippia citriodora* infusion: scavenging effect on superoxide radical, hydroxyl radical and hypochlorous acid. *Biol. Pharm. Bull.* 25: 1324-1327.

Valentão, P., E. Fernandes, F. Carvalho, P.B. Andrade, R.M. Seabra, and M.L. Bastos. 2003. Hydroxyl radical and hypochlorous acid scavenging activity of small centaury (*Centaurium erythraea*) infusion. A comparative study with green tea (*Camellia sinensis*). *Phytomedicine* 10: 517-522.

Valentão, P., M. Carvalho, E. Fernandes, F. Carvalho, P.B. Andrade, R.M. Seabra, and M.L. Bastos. 2004a. Protective activity of *Hypericum androsaemum* infusion against *tert*-butyl-hydroperoxide-induced oxidative damage in isolated rat hepatocytes. *J. Ethnopharmacol.* 92: 79-84.

Valentão, P., M. Carvalho, F. Carvalho, E. Fernandes, M.L. Pereira, P.B. Andrade, R.M. Seabra, and M.L. Bastos. 2004b. *Hypericum androsaemum* infusion increases *tert*-butyl-hydroperoxide-induced mice hepatotoxicity *in vivo*. *J. Ethnopharmacol.* 94: 345-351.

Valgimigli, L., G.F. Pedulli, and M. Paolini. 2001. Measurement of oxidative stress by EPR radical-probe technique. *Free Radic. Biol. Med.* 31: 708-716.

van Dam, S. 2002. Oxidative stress and diabetic neuropathy: Pathophysiological mechanisms and treatment perspectives. *Diabetes Metab. Res. Rev.* 18: 176-184.

Vavreckova, C., I. Gawlik, and K. Muller. 1996. Benzophenanthridine alkaloids of *Chelidonium majus*; I. Inhibition of 5- and 12-lipoxygenase by a non-redox mechanism. *Planta Med.* 62: 397-401.

Velioglu, Y.S., G. Mazza, L. Gao, and B.D. Oomah. 1998. Antioxidant activity and total phenolics in selected fruits, vegetables and grain products. *J. Agric. Food Chem.* 46: 4113-4117.

Vendemiale, G., I. Grattagliano, and E. Altomare. 1999. An update on the role of free radicals and antioxidant defence in human disease. *Int. J. Clin. Lab. Res.* 29: 49-55.

Visioli, F., and C. Galli. 1998a. Olive oil phenols and their potential effects on human health. *J. Agric. Food Chem.* 46: 4292-4296.

Visioli, F., and C. Galli. 1998b. The effect of minor constituents of olive oil in cardiovascular disease: new findings. *Nutr. Rev.* 56: 142-147.

Visioli, F., A. Romani, N. Mulinacci, S. Zarini, D. Conte, F.F. Vincieri, and C. Galli. 1999. Antioxidant and other biological activities of olive mill waste waters. *J. Agric. Food Chem.* 47: 3397-3401.

Visioli, F., C. Galli, G. Galli, and D. Caruso. 2002. Biological activities and metabolic fate of olive oil phenols. *Eur. J. Lipid Sci. Technol.* 104: 677-684.

Visioli, F., F.F. Vincieri, and C. Galli. 1995. "Waste waters" from olive oil production are rich in natural antioxidants. *Experientia* 51: 32-34.

Visioli, F., G. Bellomo, and C. Galli. 1998. Free radical-scavenging properties of olive oil polyphenols. *Biochem. Biophys. Res. Commun.* 247: 60-64.

Vrba, J., J. Hrbác, J. Ulrichová, and M. Modrianský. 2004. Sanguinarine is a potent inhibitor of oxidative burst in FMSO-differentiated HL-60 cells by a non-redox mechanism. *Chem. Biol. Interact.* 147: 35-47.

Waine, C. 2004. Vitamin and mineral supplements. *Pharm. J.* 267: 352-354.

Wang, X., and P.J. Quinn. 1999. Vitamin E and its function in membranes. *Prog. Lipid. Res.* 38: 309-336.

Wheeler, D.S., and W.J. Wheeler. 2004. The medicinal chemistry of tea. *Drug Develop. Res.* 61: 45-65.

Wu, H.-C., C.-H. Shiau, H.-M. Chen, and T.-K. Chiou. 2003. Antioxidant activities of carnosine, anserine, some free amino acids and their combination. *J. Food Drug Anal.* 2: 148-153.

Yang, C.S., J.M. Landau, M.-T. Huang, and H.L. Newmark. 2001. Inhibition of carcinogenesis by dietary polyphenolic compounds. *Annu. Rev. Nutr.* 21: 381-406.

Yanishlieva, N.V., and E.M. Marinova. 2001. Stabilisation of edible oils with natural antioxidants. *Eur. J. Lipid Sci. Technol.* 103: 752-767.

Yanishlieva-Maslarova, N.V. 2001. Inhibiting oxidation. Pages 22-70 in *Antioxidants in Food,* J. Pokorny, N. Yanisshlieva, and M. Gordon, eds., Woodhead Publishing Ltd, CRC Press, New York.

Yanishlieva-Maslarova, N.V., and I.M. Heinonen. 2001. Sources of natural antioxidants: vegetables, fruits, herbs, spices and teas. Pages 210-263 in *Antioxidants in Food,* J. Pokorny, N. Yanisshlieva, and M. Gordon, eds., Woodhead Publishing Ltd, CRC Press, Boca Raton, Florida.

Yeh, Y.Y., and L. Liu. 2001. Cholesterol-lowing effect of garlic extracts and organosulphur compounds: human and animals studies. *J. Nutr.* 131: 989S-993S.

Yin, M.C., S.W. Hwang, and K.C. Chan. 2002. Nonenzymatic antioxidant activity of four organosulphur compounds derived from garlic. *J. Agric. Food Chem.* 50: 6143-6147.

Yokozawa, T., C.P. Chen, E. Dong, T. Tanaka, G.-I. Nonaka, and I. Nishioka. 1998. Study on the inhibitory effect of tannins and flavonoids against the 1,1-diphenyl-2-picrylhydrazyl radical. *Biochem. Pharmacol.* 56: 213-222.

Yong, S.H., and M. Karel. 1978. Reactions between peroxidizing lipids and histidyl residue analogues: enhancement of lipid oxidation and browning by 4-methylimidazole. *Lipids* 13: 1-5.

Zhang, H.-Y., and L.-F. Wang. 2004. Theoretical elucidation of structure-activity relationship for coumarins to scavenge peroxyl radical. *J. Mol. Struct.* (*Theochem*) 673: 199-202.

Zheng, W., and S.Y. Wang. 2001. Antioxidant activity and phenolic compounds in selected herbs. *J. Agric. Food Chem.* 49: 5165-5170.

5

Antimicrobial Activity of Terpenes Isolated from African Plants

VERONIQUE SEIDEL AND ROGER WAIGH

Department of Pharmaceutical Sciences, University of Strathclyde, 27 Taylor Street, Glasgow G4 0NR, United Kingdom

INTRODUCTION

In contrast to their value in providing a range of pharmaceuticals for other purposes, higher plants have so far failed to provide much in the way of antimicrobials. There are valid reasons for continuing the search: plants normally protect themselves against bacterial, fungal and probably viral attack. The range of secondary metabolites that they produce to protect themselves is huge, exceeding the scope of human imagination, and presents a far richer source of active compounds than, for example, 'combinatorial chemistry'. Finally, higher plants are used very widely in developing countries for the treatment of microbial infections.

The potential value of studying antimicrobial natural products is twofold. Obviously, new drugs may be found, either to be used themselves or as leads. Second, the true value of ethnic remedies may be tested; those that are not useful may be disregarded and those that do work could be developed for people who are unable to afford medicines from elsewhere.

In the specific context of antimicrobials, with the exception of antivirals, screening methods are simple, compared to those for many classes of medicine. Provided that facilities are available for the culture of microbial species, the process of testing extracts or isolated compounds need involve very small quantities of test material and is economic in its use of nutrient media. The widespread availability of titre plates with at least 96 wells, with the use of an indicator that changes colour when metabolised by viable organisms, means that minimum inhibitory concentrations (MICs) can be determined very rapidly (Eloff 1998, Drummond and Waigh 2000). The plates can be washed and sterilised for re-use. It is a surprising feature of antimicrobial tests, worldwide, that many are carried out using a method that would have been familiar to Alexander Fleming in the 1920's. Such methods are slow, inaccurate, use relatively large quantities of valuable material and do not provide the MIC directly.

There have been some very clever developments of methodology for bioassay-guided screening, for example using spray reagents for direct visualisation of biological activity on TLC plates. The major caveat in adopting such an approach is the waste of plant material that is entailed. If material is collected, often from inaccessible places, identified and catalogued, there is a persuasive argument that maximum value should be extracted from it. This means isolating as many of the secondary metabolites as possible, for evaluation in as many screens as possible. Using modern chromatographic techniques, notably preparative HPLC, it is possible to isolate microgram amounts of a large number of compounds from almost any plant. Using microtitre methods, these can be tested individually at appropriate concentrations for any chosen kind of activity. In the present context, this could mean testing against a range of bacteria and fungi, with the option of going back to the compound for evaluation against further microbial species as the need arises.

MONOTERPENES

Antibacterials

Most of the reports that we have found relate to the evaluation of mixtures of monoterpenes in 'essential oils'. It is therefore not possible to associate the activity with any specific component(s), which are numerous. In some cases oils contained both mono-and sesqui-terpenes, in which case they are included under the monoterpenes.

In an assessment of the activity of the bark exudate of *Commiphora tenuis*, activity was found down to an oil concentration of 0.5% against methicillin resistant *Staphylococcus aureus* and at higher concentrations against *Enterococcus faecalis* and *Proteus mirabilis* (Asres *et al*. 1998). It should be noted that a concentration of 0.5% equates to 5,000 μg/mL, which is not very active by the standards of antibacterials used clinically. The major constituents of the oil were α-pinene (60.8%), α–thujene (8.9%), β-pinene (8.8%), sabinene (6.3%), limonene (5.5%), 3-carene (3.7%) and β-myrcene (1.8%). All the others were present at less than 1%, the majority as traces. It is likely that the activity resides in the major constituents, but there is the fascinating possibility that one of the lesser compounds is very active. The *C. tenuis* bark exudate is used to treat animal wounds, particularly on camels.

The essential oils isolated from the needles of the Moroccan pine species, *Pinus halepensis, P. canariensis, P. pinaster, P. pinea* and *P. brucia*, were investigated for their chemical composition (Hmamouchi *et al.* 2001). The oil of *P. halepensis* was found to be rich in the monoterpenes α-pinene (23.3%) and myrcene (16.3%). The former was also found in *P. pinea* (37%) while the oil of *P. brucia* contained mostly β-pinene (33.6%). The main components of *P. canariensis* and *P. pinaster* were the sesquiterpenes germacrene D (62.5%) and β-caryophyllene (22.2%), respectively. The essential oils, α-pinene, myrcene and β-caryophyllene were tested for their antimicrobial activity against 21 strains of bacteria using the agar dilution method. Only *P. pinaster* and *P. pinea* oils showed activity against all organisms. None of the pure terpenes was effective, apart from β-caryophyllene which showed very weak activity at 500 µg/mL against the Gram positive *Sarcina lutea*. This suggests that other minor compounds in the oils are important in contributing to the antimicrobial activity.

A study of the essential oils from 15 Congolese plants, against 12 species of bacteria isolated from human specimens, showed a wide range of activity from a total of 43 terpenes (Cimanga *et al.* 2002). These were not all represented in all the plant species and the proportions varied considerably. In 6 of 10 species of *Eucalyptus* the major component was 1,8-cineole, which the authors considered responsible for most of the activity. In the oil from *Aframomum stipulatum* the major component was thymol, in the oil from *Cympobogon citratus* it was geranial, from *Monodora myristica* it was α-phellandrene and from *Ocimum americanum* and *Ocimum gratissimum* it was again thymol. Unfortunately the authors used zones of inhibition around impregnated paper discs on agar plates and thus could not arrive at MICs. The zones of inhibition were large enough to encourage the authors to state that their results 'can justify and partly support the use of these medicinal plants as traditional remedies for the treatment of some infections'.

The essential oils from *Lippia chevalieri* and *L. multiflora*, grown in Burkina Fasso, were tested against nine species of bacteria: *Bacillus cereus, E. faecalis, Escherichia coli, Listeria innocua, P. mirabilis, Salmonella enterica, Shigella dysenterica, Staphylococcus aureus* and *Staphylococcus camorum* (Bassole *et al.* 2003). Of a total of 25 volatile constituents, *p*-cymene and thymol were major, along with 2-phenethyl propionate. A serial dilution method using microtitre plates allowed the determination of MICs, which were given as percentages. The greatest potency was seen against *Salmonella enterica* and *Staphylococcus aureus*, both at 0.06%, or 600 µg/mL.

Overall, the results tended to justify the traditional use of the leaves as a tea to treat boils, diarrhoea and as a mouth disinfectant.

From Cameroon, a study of the essential oils of *Ocimum gratissimum* leaves and *Zanthoxylum xanthoxyloides* fruits using a disc diffusion method against seven species of bacteria showed zones of inhibition in every case (Ngassoum *et al.* 2003). The major components were expected to be thymol for *O. gratissimum* and α-pinene from *Z. zanthoxyloides*, which correlates with the above studies and similar comments apply.

Unusually in this context, a study from Morocco on the essential oil of *Mentha suavolens* included an evaluation of some of the major constituents as individual compounds; against 14 species of bacteria and fungi the MICs for the oil were in the range 0.34-2.77 ppm, less than 3 μg/mL. Different populations of *M. suavolens* produce oils with different amounts of the major constituents, pulegone, piperitone oxide and piperitenone oxide, with consequent differences in potency against the test organisms, although all three compounds were active when tested individually (Oumzil *et al.* 2002).

The aerial parts of *Myrothamnus flabellifolius* are used for colds and other respiratory diseases. In South Africa, the smoke from the burnt leaves is inhaled to treat chest pains and the leaves are chewed for mouth infections. Eighty-five compounds were identified in the leaf essential oil (Viljoen *et al.* 2002). The major constituents were characterised as pinocarvone (11%), *trans*-pinocarveol (19%), limonene (6%), *trans-p*-menth-1-(7)-8-diene-2-ol (7%), and *cis-p*-menth-1-(7)-8-diene-2-ol (6%). Antibacterial activity was observed against *Pseudomonas aeruginosa, E. coli, Staphylococcus aureus, Enterococcus faecalis, P. vulgaris, Serratia odorifera* and *Salmonella typhimurium* in a disc diffusion assay. The authors conclude that this could explain the traditional use of *M. flabellifolius* in the treatment of infections; see below for an account of the antifungal activity. It should be noted that the cidal concentrations, at 0.0625-0.25% (up to 2500 μg/mL), are very high by normal antibiotic standards.

The fruits of four Cameroonian plants, used in traditional medicine for a variety of ailments, were investigated for their essential oils (Tatsadjieu *et al.* 2003). The main constituents of the oil of *Xylopia aethiopica* were identified as β-pinene (18.3%), terpinen-4-ol (8.9%), sabinene (7.2%), α-phellandrene (7.1%), α-terpineol (4.1%), *trans*-α-ocimene (3.2%) and *trans*-β-ocimene (3.1%). The fruits of *Monodora myristica* yielded mainly α-phellandrene (29.2%), *p*-cymene (11.2%), α-pinene (9.7%), *cis*-sabinol (6.9%) and limonene (6.8%). *Zanthoxylum leprieurii* contained *trans*-β-ocimene (42.9%), α-terpinolene (9.3%), 3-δ-carene (6.5%), limonene (5.5%), myrcene (4.7%), α-pinene (3.9%) and *p*-cymene (3.3%). The oil from *Zanthoxylum xanthoxyloides* was rich in α-pinene (38.2%), *trans*-β-ocimene (5.4%), citronellol (3.3.%), sabinene (3.1%), myrcene (3.1%), limonene

(3%), citronellyl acetate (3%) and α-phellandrene (2.6%). The oils of *X. aethiopica, Z. xanthoxyloides* and *M. myristica* showed activity against several organisms in a disk diffusion assay but determination of MICs revealed only very weak activity for *X. aethiopica* and *Z. xanthoxyloides*, at or above 2,500 μg/mL, figures that were not compared to any positive control. See below for an account of the antifungal activity.

The leaves of *Lippia multiflora* (commonly known throughout west tropical Africa as 'bush tea') are widely used in folk medicine, particularly for common colds and chest complaints. Hexane extracts from the leaves (collected in Nigeria) were subjected to a bioassay-guided fractionation using a range of bacteria (*Pseudomonas aeruginosa, Staphylococcus aureus, E. coli, Salmonella typhi, Bacillus subtilis*) and the yeast *Candida albicans* in an agar well diffusion assay (Kunle *et al*. 2003). The monoterpene carvacrol **1**, isolated from the most active extract and a well known antimicrobial, was subsequently tested against the same organisms using a disk diffusion method, with positive results against all except *P. aeruginosa*. However, this could not be compared to the positive control since the latter was tested using a different methodology.

The oil from *Osmitopsis asteriscoides* showed weak activity against nine bacterial species (Viljoen *et al*. 2003). The antifungal activity is described below.

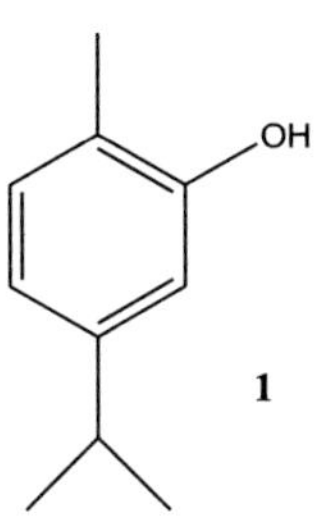

1

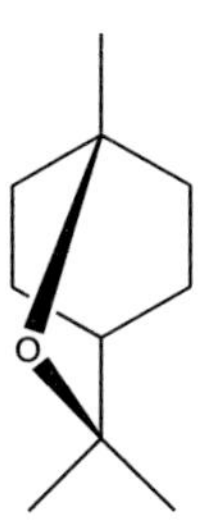

2

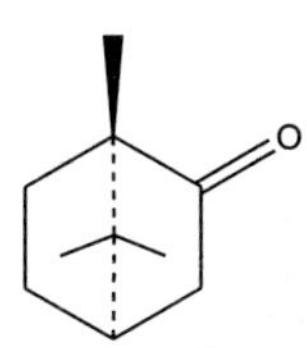

3

Antifungals

The leaves of *Hoslundia opposita*, a small shrub growing in Zimbabwe, were investigated for their essential oil (Gundidza *et al*. 1992). The main constituents were identified as α-terpinene (4%), terpinen-4-ol (19%), α-terpineol (23%), terpinolene (28%) and β-humulene (17%). Antifungal activity of the oil was observed against *Aspergillus niger*. The testing would need to be repeated to determine MICs.

The essential oils from 13 Cameroonian plants were investigated for their antifungal activity (Amvam Zollo *et al*. 1998). The amounts of oil (minimal inhibitory quantities, MIQs) required to inhibit fungal growth were recorded according to the microatmosphere method. This used a disk impregnated with essential oil and placed on the lid of an inverted culture plate. Among the plants tested, *Ocimum basilicum* (whole plant),

O. gratissimum (leaves) and *Thymus vulgaris* (whole plant) showed the strongest antifungal activity. The main constituents present in the oil of *O. gratissimum* were thymol (46.2%), γ-terpinene (20%), *p*-cymene (7%). *Thymus vulgaris* was rich in thymol (27.2%), *p*-cymene (23.6%), γ-terpinene (22.7%) and carvacrol (3.3%) while *O. basilicum* contained linalol (50.8%), eugenol (13.5%) and 1,8-cineole (3.1%). A subsequent broth microdilution assay was performed to determine MICs which were in the range 78-10,000 ppm (78-10,000 μg/mL) against *Candida albicans, Cryptococcus neoformans, Aspergillus flavus, Aureobasidium pullulans, Microsporum gypseum* and *Trichophyton rubrum*. However, this could not be compared to any positive controls since the latter were tested using a different methodology.

Various extracts of the Moroccan plant *Cistus ladaniferus* (rockrose) were investigated for their antifungal activity against *Aspergillus niger, Botrytis cinerea, Mucor racemosus* and *Verticillium albo-atrum* (Mrabet *et al.* 1999). Following agar dilution assays, the greatest activity was observed for the essential oil of rockrose with MICs in the range of 4,000-10,000 μg/mL against all fungi. Chemical analysis of the oil revealed that it was rich in the sesquiterpene viridiflorol (20%) and the monoterpene bornyl acetate (16%).

In South Africa, the smoke from the burnt leaves of *Myrothamnus flabellifolius* is inhaled to treat chest pains and the leaves are chewed for mouth infections (see above for antibacterial activity). Antifungal activity was observed against *Candida albicans, Cryptococcus neoformans* and *Aspergillus niger* in a disc diffusion assay (Viljoen *et al.* 2002). The oil was fungicidal against *C. albicans*. The authors conclude that this could explain the traditional use of *M. flabellifolius* in the treatment of fungal infections. The same caveat applies as to the antibacterial activity, in that the potency of the oil is far lower than would be acceptable for antibiotics in regular clinical use.

The oil of *X. aethiopica, Z. xanthoxyloides* and *M. myristica* showed antifungal activity against *Aspergillus flavus*, in a disk diffusion assay. Determination of MICs revealed weak activity for *X. aethiopica* and *Z. xanthoxyloides* (at 4,000 and 3,000 μg/mL, respectively), figures that were not compared to any positive control (Tatsadjieu *et al.* 2003).

Osmitopsis asteriscoides is used in South Africa (South Western Cape region) for chest complaints, inflammation, cuts and swellings. The antimicrobial activity of the leaf essential oil was tested in a disc diffusion assay (Viljoen *et al.* 2003). It displayed weak antifungal activity against *C. albicans* and *Cryptococcus neoformans*; lower than the positive control. When tested against *C. albicans* and *C. neoformans* in a broth microdilution assay, the oil displayed MICs of 16,000 and 8000 μg/mL, respectively. Time-kill studies using *C. albicans* revealed a rapid fungicidal activity at concentrations ranging from 0.5-2% (v/v). The two main constituents of the oil were identified as the monoterpenes, 1,8-cineole **2** and (-)-camphor **3**. Both exhibited fungicidal activity separately and higher efficacy when combined as a 5:1 ratio (similar to the ratio found in the oil). The authors conclude that the high concentration of **2** and **3** in the oil and the observed synergy provide a rationale for the traditional uses of *O. asteriscoides*. This is only true if the oil is used at the high concentrations tested.

SESQUITERPENES

Antibacterials

Of five sesquiterpenes isolated from *Brachylaena hutchinsii* and tested against *Streptococcus mutans* and *Brevibacterium ammoniagenes*, only **4** showed significant potency, with MIC 25 μg/mL (Vieira *et al*. 1991).

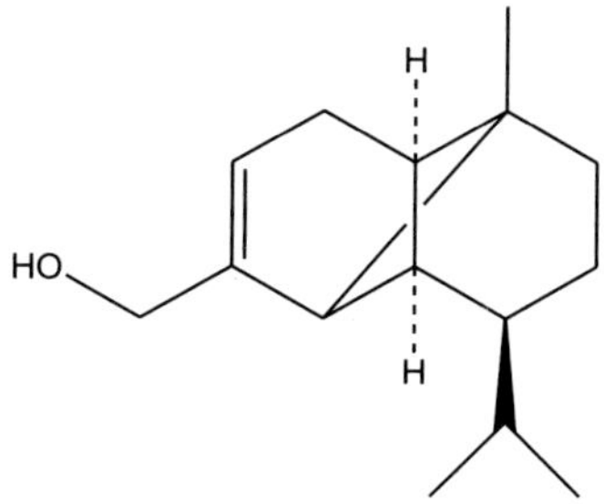

4

The possible use of *Vernonia amygdalina* as a medicinal agent by wild chimpanzees led to the evaluation of 10 of its constituent sesquiterpenes against *Bacillus subtilis* and *Micrococcus luteus* (Jisaka *et al*. 1993). Although no MICs were measured, a paper disc/agar method led to the conclusion that compounds **5**, **6**, **7**, **8**, **9** and **10** were the most active. The presence of α,β-unsaturated carbonyl groups is likely to lead to formation of covalent adducts with thiol groups on proteins, an effect which may be expected to cause non-selective toxicity.

The cyclofarnesone **11** from *Premna oligotricha* showed modest activity (MIC 50-100 μg/mL) against *Bacillus pumilus, B. subtilis, Staphylococcus aureus* and *Streptococcus faecalis*. It was inactive against *E. coli* and *Pseudomonas aeruginosa* (Habtemariam *et al*. 1993).

The reputation of decoctions of the bark of *Warburgia salutaris* as a treatment for respiratory complaints and for skin ailments led to the evaluation of crude extracts for antibacterial activity and then to a bioassay-guided fractionation, looking for the active principles (Rabe and van Stade 2000). The bioautographic method employed *Staphylococcus aureus*

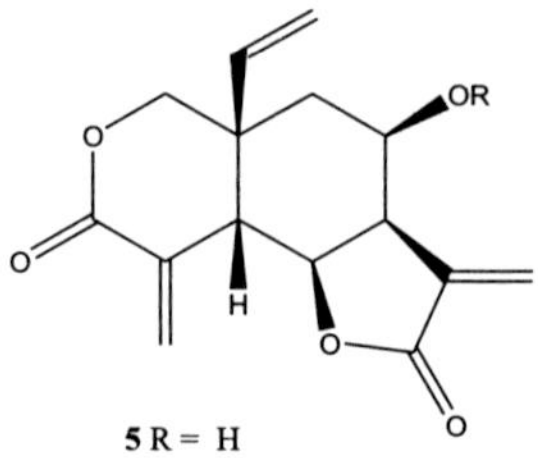

5 R = H
6 R = COC(=CH_2)CH_2OH

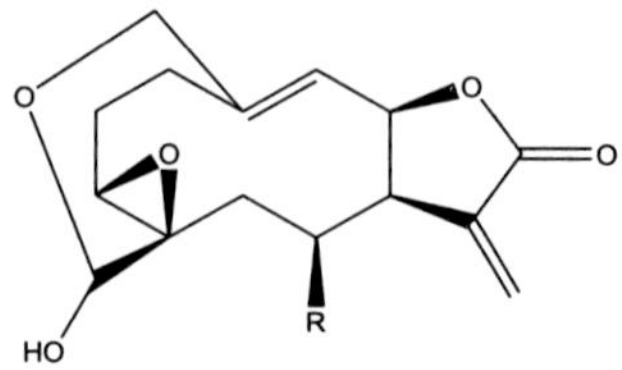

7 R = OCOC(=CH_2)CH_3
8 R = OCOC(=CH_2)CH_2OH

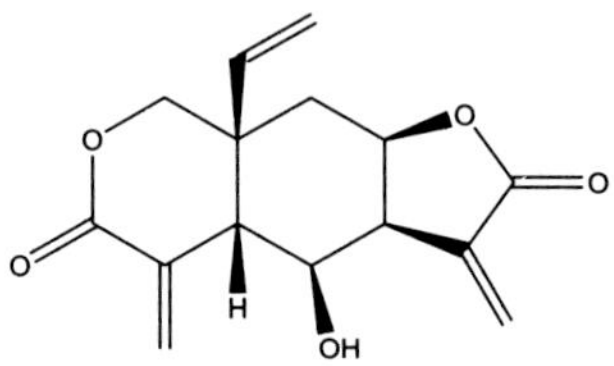

9

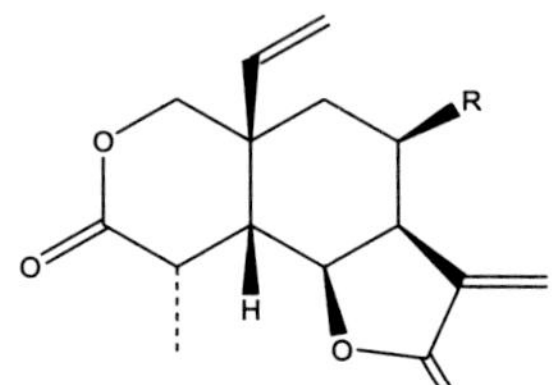

10 R = OCOC(=CH_2)CH_2OH

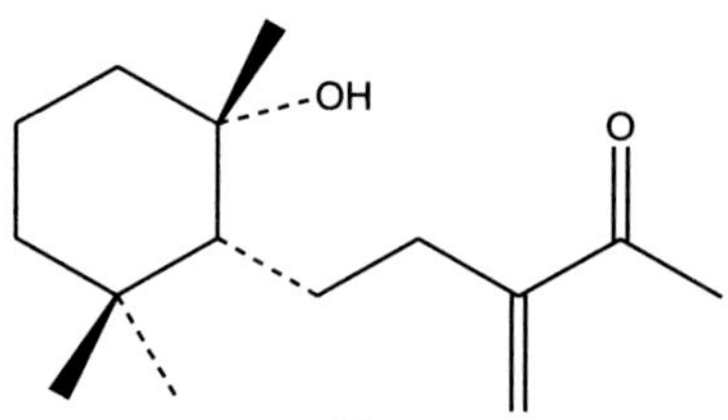

11

as the test organism on TLC plates and this was followed by microtitre determination of MICs for muzigadial **12** against *S. aureus, Bacillus subtilis, Staphylococcus epidermidis, Micrococcus luteus, E. coli* and *Klebsiella pneumoniae*. The most sensitive organisms were *S. aureus* and *B. subtilis,* with MICs of 12.5 μg/mL. For comparison, the MIC for these two organisms against neomycin was 0.2-0.4 μg/mL. It is of course possible that the use of a different detection organism might have shown the presence of other active compounds; this is an inherent problem with bioassay-guided methods.

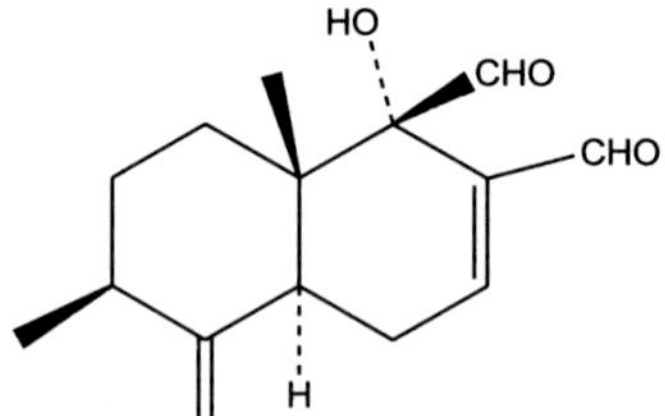

12

A study was carried out to investigate the chemical composition and antimicrobial activity of the essential oils obtained from the leaves and fruits of *Amoora rohituka* growing in Egypt (Aboutabi *et al.* 2000). Chromatographic analysis revealed that both oils were rich in sesquiterpenes. *β*-Caryophyllene (36.2%) and *β*-eudesmol (11.4%) were identified as the major compounds in the leaf oil; *γ*-cadinene (11.7%) in the fruit oil. Antimicrobial screening was performed by the disc diffusion method against a good selection of pathogenic bacterial and fungal strains. Both oils were 'very active' against *E. coli, Micrococcus flavoroscus, Listeria monocytogenes* and *C. albicans*. The leaf oil had a better activity against *Staphylococcus epidermidis* and *Streptococcus faecalis.* The authors state that in some cases the oils exhibited a higher antibacterial activity than rifamycin SV (positive control). Such a comparison is not valid if rifamycin SV is not the antibiotic of choice for some test organisms.

A similar approach led the same group to the isolation of three sesquiterpenes from the leaves of *Vernonia colorata,* again following reported uses in traditional medicine (Rabe *et al.* 2002). Compounds **13**, **14** and **15** all showed moderate activity against *B. subtilis,* with **13** and **15** showing modest activity against *S. aureus;* none of the MICs was less than 100 μg/mL. Given that all three compounds possess three very reactive functional groups, the low level of activity is surprising and may indicate that higher potency could be found against other organisms.

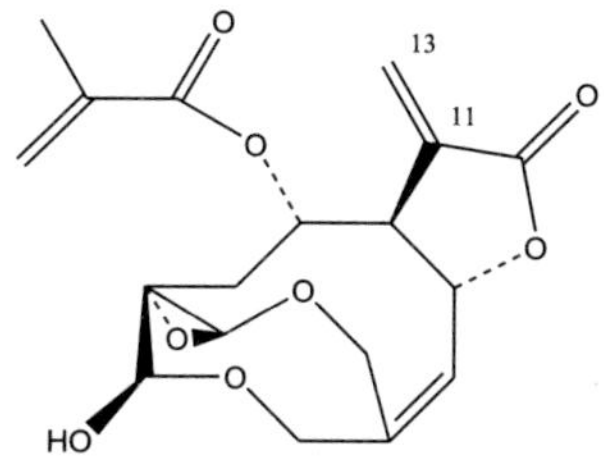

13
14 is 11,13-dihydro

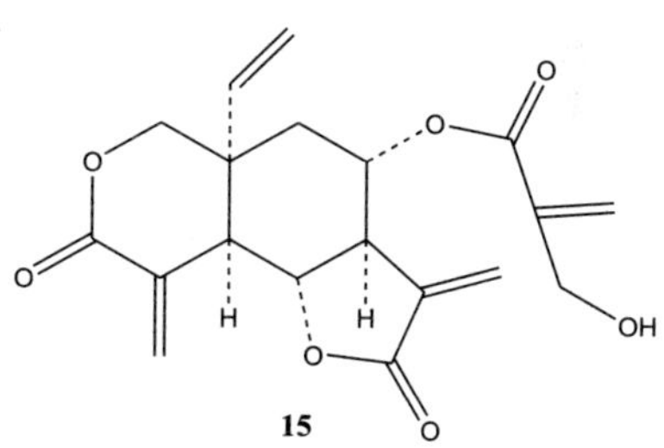

15

Similarly low potencies were reported for the essential oils of *Helichrysum kraussii* and *H. rugulosum,* growing in South Africa, although the sesquiterpene *β*-caryophyllene oxide, a major constituent of the oil of *H. rugulosum,* was found to have an MIC of 73 μg/mL against *Staphylococcus aureus* (Bougatsos *et al.* 2003).

Antifungals

Selected East African plants were screened for antibacterial and antifungal activities (Kubo and Taniguchi 1988). The extracts from the stem bark of two trees, *Warburgia ugandensis* and *W. stuhlmannii* (Canellaceae), were investigated for their active constituents. Both *Warburgia* species are used in East Africa for various purposes, including the treatment of diarrhoea and malaria. A series of sesquiterpene dialdehydes, polygodial **16**, warburganal **17** and muzigadial **12**, were isolated. Strong antifungal activity was observed for **12, 16** and **17**, particularly against *Saccharomyces cerevisiae, Candida utilis* and *Sclerotina libertiana* with MICs in the range 0.78-3.13 μg/mL. Compound **16** was 2-8 times more active than **17** and **12** against *Saccharomyces cerevisiae, Candida utilis, Sclerotina libertiana, Schizosaccharomyces pombe, Hansenula anomala, Mucor mucdeo, Rhizopus chinensis, Aspergillus niger* and *Penicillium crustosum*. When tested against *S. cerevisiae*, **16** displayed a fungicidal rather than fungistatic effect and induced lesions on the fungal membrane. Additionally, a higher activity was observed when **16** was combined with actinomycin D. The authors suggested that **16** could increase fungal membrane permeability.

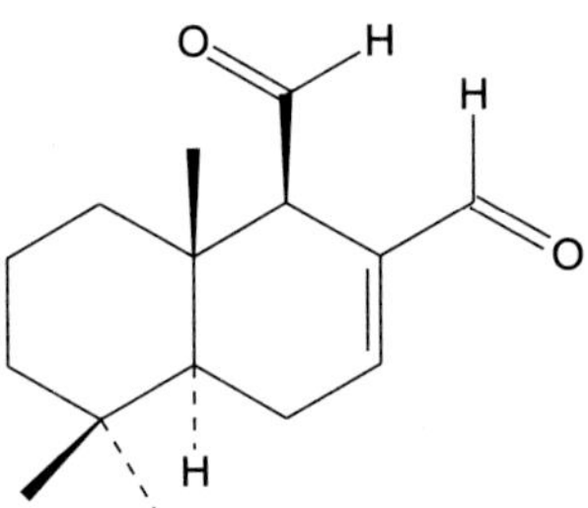

16

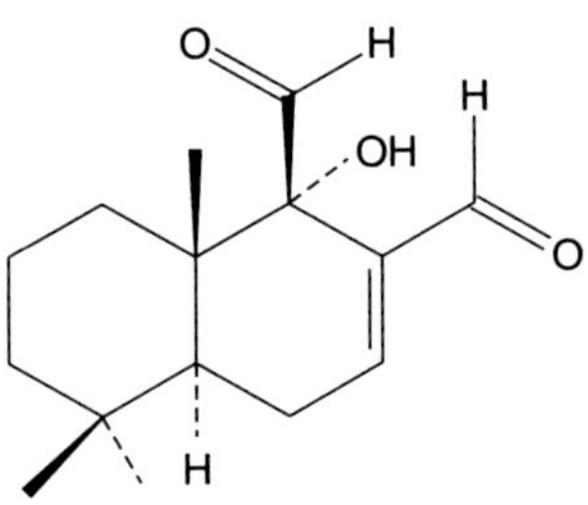

17

The susceptibilities of a range of fungal species to polygodial **16** were investigated using a broth dilution method (Lee *et al.* 1999, Kubo *et al.* 2001). Strong fungicidal activity was observed against *Trichophyton mentagrophytes, T. rubrum, Penicillium marneffei, Saccharomyces cerevisiae, Cryptococcus neoformans, Candida utilis, C. krusei* and *C. albicans*. Activity against *C. albicans* was increased under acidic conditions with MIC and MFC (minimal fungicidal concentration) 0.39 μg/mL at pH 3.0 compared to 1.56 and 3.13 μg/mL at pH 7.0, respectively. Time-kill studies on *C. albicans* revealed a strong and rapid fungicidal activity in the presence of **16**. Only a weak fungicidal effect was observed against *Candida lipolytica, C. tropicalis, Aspergillus fumigatus, A. flavus* and *Penicillium chrysogenum*. The authors suggest that **16** represents a promising antifungal lead compound due to its fungicidal effect, its potentially novel mode of action and its ability to kill *C. albicans* especially in environments which are naturally acidic in the human body (e.g., vagina, skin, mouth). Studies on the mechanism of action suggest that it could act as a surfactant, disrupting the structure of the fungal membrane and possibly denaturing H^+-ATPase proteins.

DITERPENES

Antibacterials

Xylopia aethiopica (Annonaceae) is a tree widely used in West African traditional medicine. In Ghana, the fruits and seeds are employed to treat bronchitis and dysentery and are also added as a spice to many food preparations. In a 1977 study, five compounds were isolated from the fresh ripe fruits (Boakye-Yiadom *et al.* 1977). One of them was characterised as xylopic acid **18**, which showed activity against *B. subtilis*, *S. aureus* and *P. aeruginosa* in an agar diffusion assay. Xylopic acid also displayed some activity against *Candida albicans*. This appears to provide a rationale for the traditional use of *X. aethiopica* fruits as part of suppositories and douches to alleviate vaginal candidiasis.

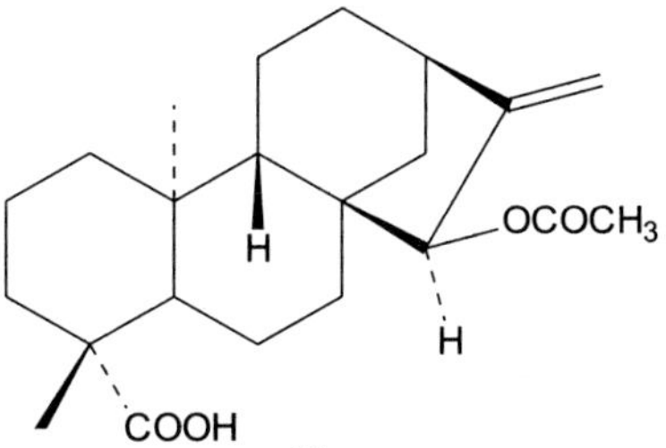

18

Tetradenia riparia, a labiate, is widely cultivated in Rwanda and used to treat almost every kind of human malady, from malaria to angina. It is also used to help preserve foodstuffs in traditional silos, which may be an indicator of antifungal activity. The isolation of 8(14),15-sandaracopimaradiene-7α,18-diol **19** from the chloroform extract of the leaves permitted the determination of MICs against 14 bacteria and 11 fungi, with values in the range 6-12 μg/mL for *Shigella dysenteriae*, *B. subtilis*, *Mycobacterium smegmatis*, *S. aureus*, *S. epidermidis*, *Streptococcus pyogenes* and *Candida albicans* (Van Puyvelde *et al.* 1986).

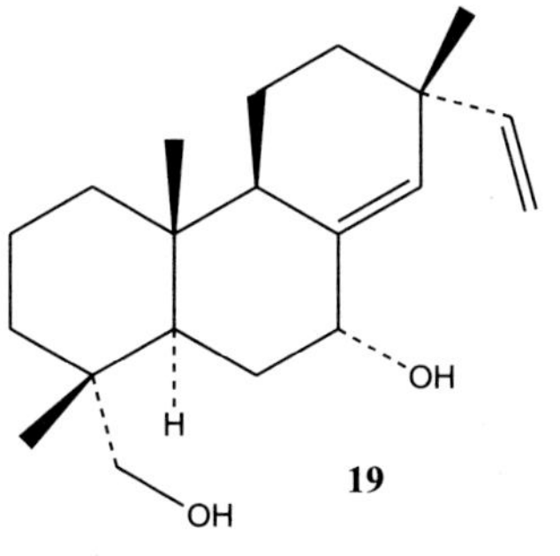

19

A screening programme of 15 Rwandese medicinal plants used by traditional healers to treat pulmonary diseases gave positive results against *M. tuberculosis* for three plants, *Bidens pilosa, Pantas longiflora* and *Tetradenia riparia*. In this study no fractionation was carried out, although the active principle of *T. riparia* was assumed to be **19**, despite the relatively low activity of **19** which was also tested and gave MIC 100 µg/mL (Van Puyvelde *et al.* 1994).

The leaves of *Premna schimperi* are used to treat superficial infections in Ethiopia. A bioassay-guided fractionation using *S. aureus* and *B. subtilis* resulted in the isolation of a new diterpene, (5R,8R,9S,10R)-12-oxo-ent-3,13(16)-clerodien-15-oic acid **20** with MICs of 20 and 25 µg/mL respectively against the two organisms (Habtemariam *et al.* 1990). It is possible that the organisms causing the infections in Ethiopia might respond differently to **20** and also that the use of different organisms might yield different active compounds.

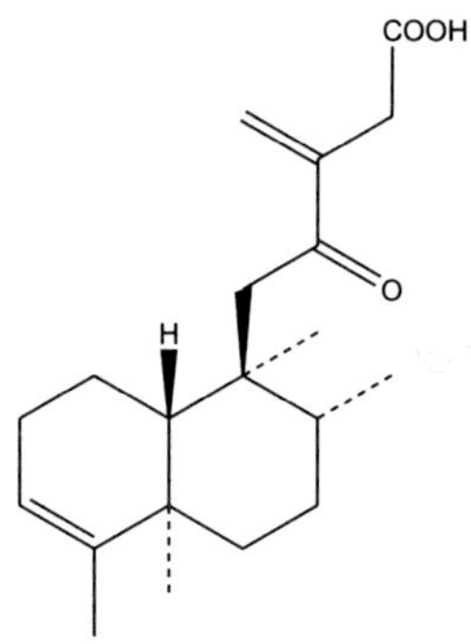

20

The same group investigated a closely related species, *Premna oligotricha*, which is used in Ethiopia both as chewing sticks and to 'sterilise' milk containers. They found two antibacterial agents, **21** and **22**, the latter being more active against *B. subtilis, B. pumilus, S. aureus* and *Streptococcus faecalis*, with MICs in the range 3.75-30 µg/mL (Habtemariam *et al.* 1992).

The seeds of *Aframomum aulacocarpos*, growing in Cameroon, were investigated in a search for sources of the diterpene dialdehyde aframodial (Ayafor *et al.* 1994). Three diterpenes 12α, 17-epoxy-3β-hydroxy-8(9), 13-labdadien-16,15-olide (aulacocarpinolide) **23**, methyl 8β,17:14ξ, 15-diepoxy-3β-hydroxy-12*E*-labden-16-oate (aulacocarpin A) **24** and methyl 8β, 17:14ξ, 15-diepoxy-3β,6b-dihydroxy-12*E*-labden-16-oate (aulacocarpin B) **25**, along with aframodial, were isolated. A serial dilution assay was carried out, giving MIC 25 µg/mL against *B. subtilis* for all three compounds. Antifungal activity was observed for com-

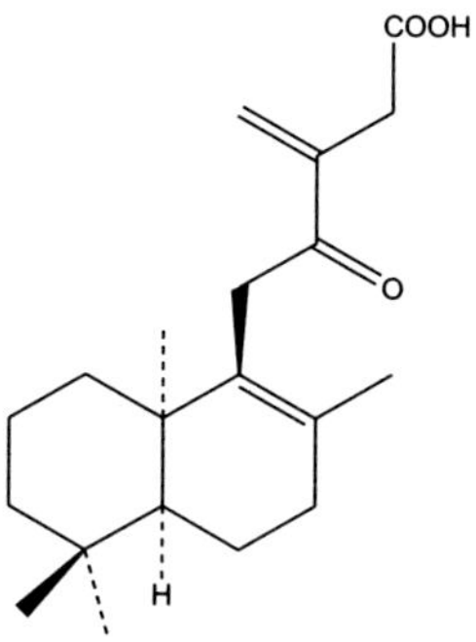

21

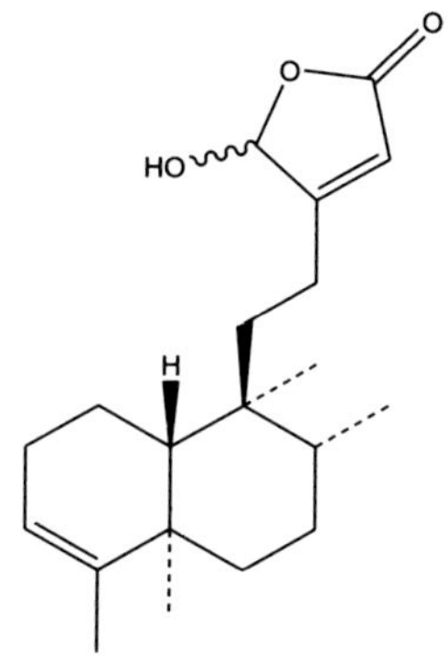

22

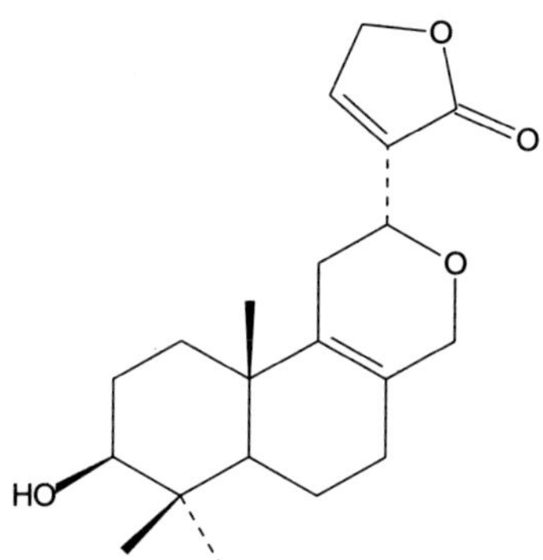

23

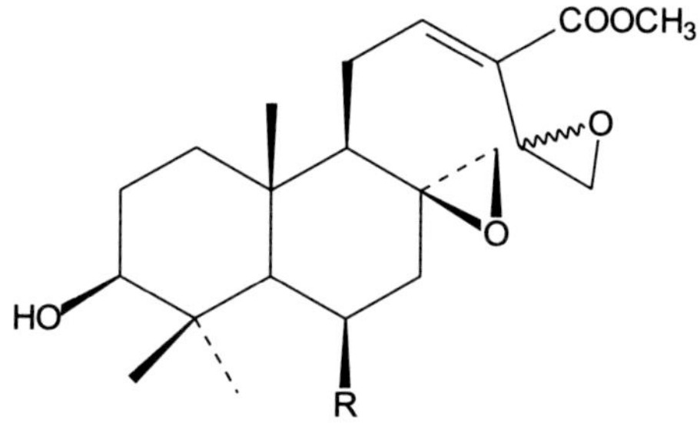

24 R=H
25 R=OH

pounds **23-25** against *Mucor miehei* (MICs 50 μg/mL) in an agar diffusion assay. No positive controls were mentioned in the study.

The diterpene lanigerol **26**, from *Salvia lanigera* growing in Egypt, has been reported to be weakly active against *B. subtilis*, *S. aureus* and *Micrococcus luteus*, but inactive against Gram-negative bacteria and yeasts (El-Lakany *et al.* 1995).

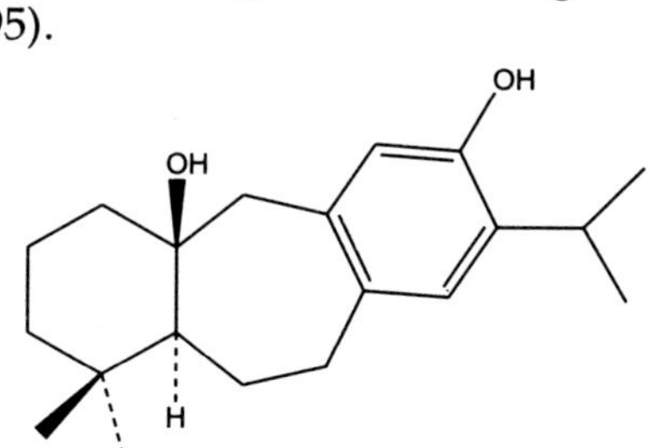

26

Good activity against *M. tuberculosis* was found in a range of phytols, initiated by the isolation of *(E)-phytol* **27** as the active constituent of *Leucas volkensii*, a herb used in Kenya to treat various pulmonary disorders (Rajab *et al.* 1998). For comparison, *(Z)-phytol* **28** was isolated from a commercially available mixture and **27** was converted to derivatives **29-32**. Compounds **27, 28** and **29** were most active, with MIC 2 μg/mL. The authors point out that this level of potency is comparable to that of ethambutol, which is clinically useful.

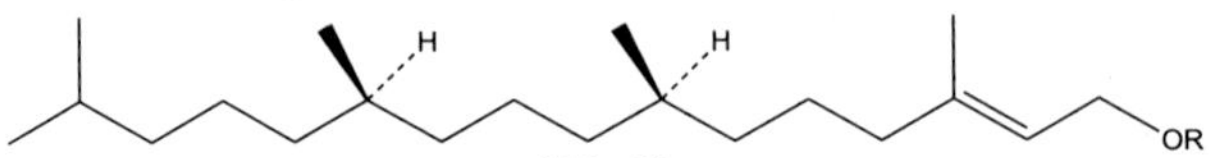

27 R = H
32 R = Ac

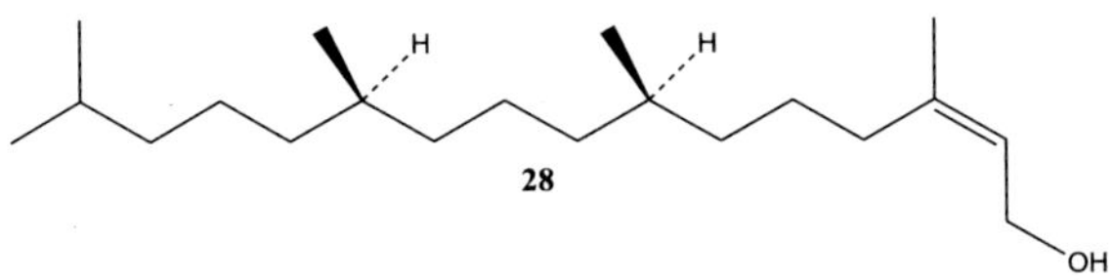

28

29

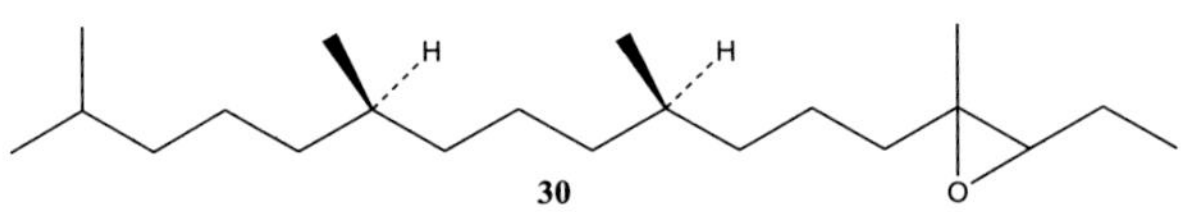

30

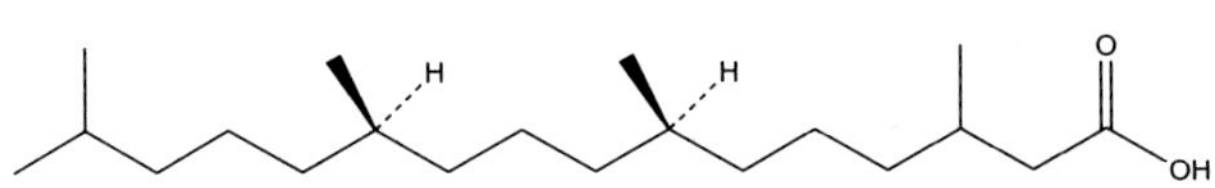

31

Antifungals

The root bark of *Bobgunnia madagascariensis* (syn. *Swartzia madagascariensis*), growing in Zimbabwe, was investigated as part of a search for new antifungal compounds (Schaller *et al.* 2000). The diterpene **33** was characterised as an active component using TLC bioautography with *Cladosporium cucumenirum* and *Candida albicans* to guide the isolation. Subsequent screening for antifungal activity in a broth microdilution assay revealed that **33** was very active against several *Candida* species with MIC (0.19 μg/mL) lower than some standard antifungal drugs.

Further investigation of the root bark of *B. madagascariensis*, using bioautography with *Cladosporium cucumenirum* and *Candida albicans*, led to the isolation of a mixture of two other diterpenes, **34** and **35.** A minimum amount of '0.5 μg' of the 1/1 mixture was required to inhibit fungal growth. A subsequent broth microdilution assay against *C. albicans* revealed good antifungal activity for the mixture with IC_{50} 7 μg/mL while **33** had IC_{50} 0.2 μg/mL (Schaller *et al.* 2001).

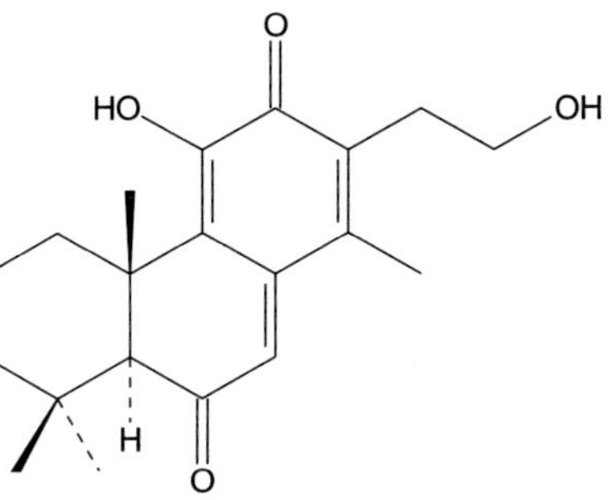

33

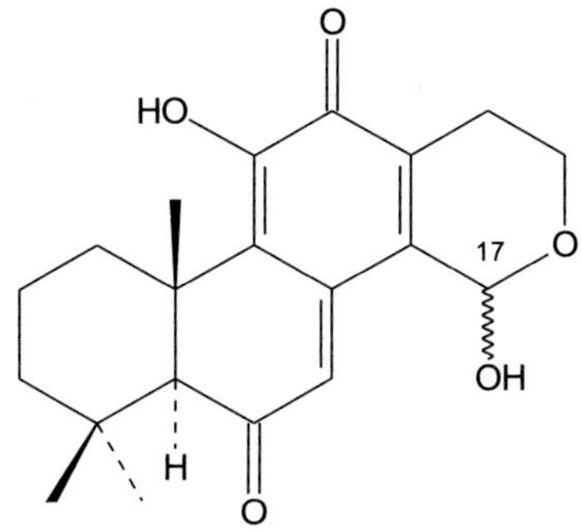

34 and **35** (Diastereoisomers at position 17)

Two diterpenes, fusicoserpenol A **36** and dolabeserpenoic acid A **37**, were isolated from the leaves of *Hypoestes serpens*, growing in Madagascar, using a bioautographic TLC assay with *Cladosporium cucumenirum* and *Candida albicans*. Inhibition of fungal growth was observed with minimum amounts of '2 and 5 µg' for **36** and **37**, respectively (Rasoamiaranjanahary *et al.* 2003a). This was comparable to results obtained for the positive control. However, the testing would need to be repeated to express values as MICs and allow comparison with other data.

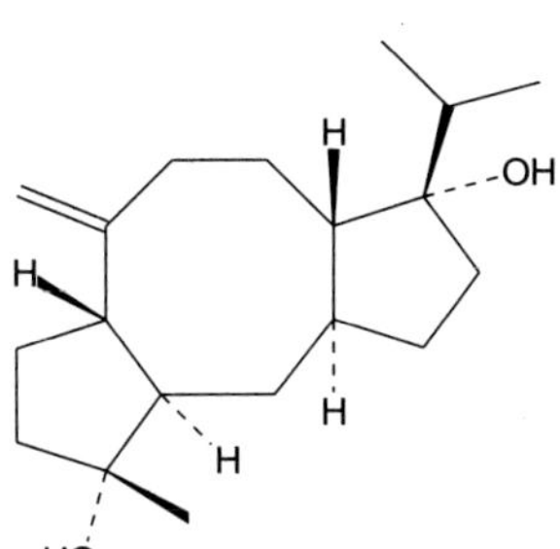

36

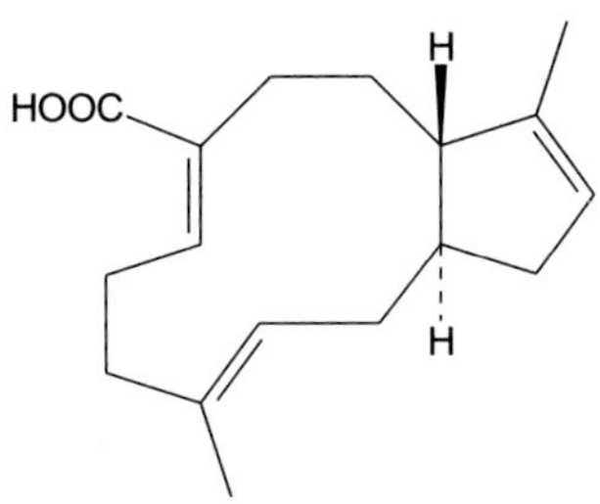

37

Further investigation of *Hypoestes serpens* leaves in the search for antifungal compounds using the aforementioned assay led to the isolation of five other diterpenes, 7β-hydroxyisopimara-8, 15-dien-14-one **38**, 14α-hydroxyisopimara-7, 15-dien-1-one **39**, 1β,14α-dihydroxyisopimara-7,15-diene **40**, 7β-hydroxyisopimara-8(14),15-dien-1-one **41** and 7β-acetoxyisopimara-8(14),15-dien-1-one **42**. Compounds **39** and **41** were the most active (minimum amounts required to inhibit fungal growth '0.5 µg') and superior to the positive control (Rasoamiaranjanahary *et al.* 2003b). The activity of **38** was comparable to the positive control while other compounds were weakly active. The authors conclude that the

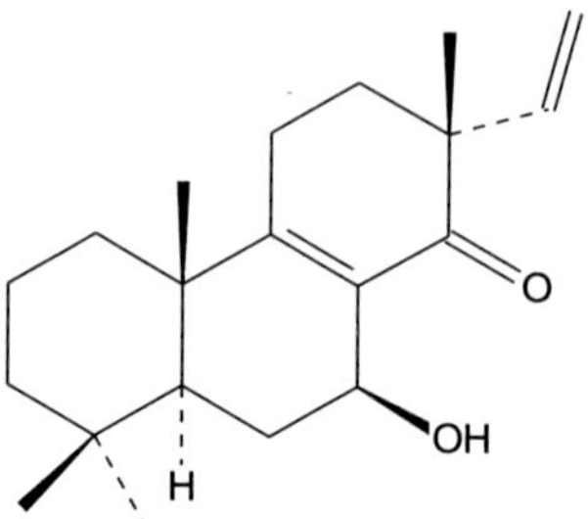

38

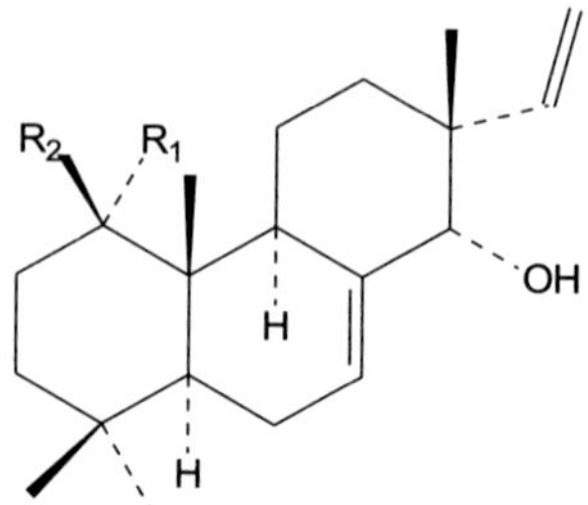

39 R_1=R_2=O
40 R_1=H, R_2=OH

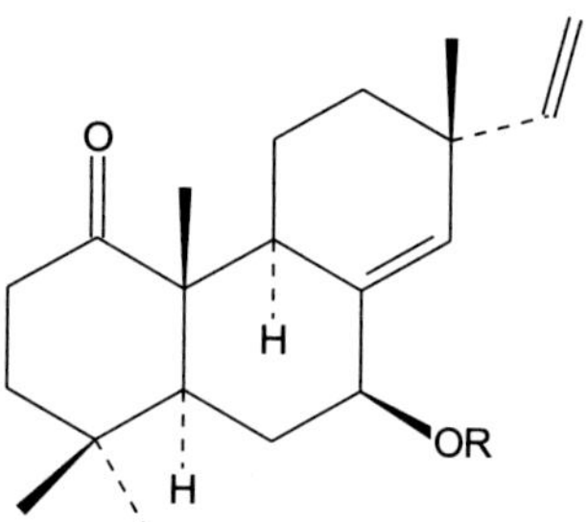

41 R=H
42 R=$COCH_3$

antifungal effect observed against *C. albicans* supports the traditional use of *Hypoestes serpens* in the treatment of vaginitis. However, the same comment as for **36** and **37** applies to the interpretation of the results.

More than 75 plants selected on the basis of their ethnobotanical use in Cameroon were screened for their antifungal activity (Salah *et al.* 2003). Two species, *Macaranga monandra* and *Pterygota kamerunensis,* displayed activity worthy of further investigation. Bioassay-guided fractionation of *M. monandra* led to the isolation of two *neo*-clerodane type diterpenes, kolavenic acid **43** and 2-oxo-kolavenic acid **44**.

Bioautography on TLC plates sprayed with spore suspensions of *Colletotrichum acucatum, C. fragariae* or *C. gloeosporioides* was used to direct the isolation. Compound **43** inhibited the growth of *C. fragariae* (at 100 and 50 µg/mL) and *C. gloeosporioides* (100 µg/mL). Growth inhibition for **44** could not be clearly measured. A microtitre broth dilution assay was further used to evaluate antifungal activity against *C. acucatum, C. fragariae, C. gloeosporioides, Botrytis cinerea, Fusarium oxysporum, Phomopsis viticola and P. obscurans*. Only moderate inhibition of the growth of *B. cinerea* and *Phomopsis* species (≤ 50% of the activity observed for standard fungicides) was detected for **43** and **44** at the concentrations tested (0.3, 3 and 30 µM).

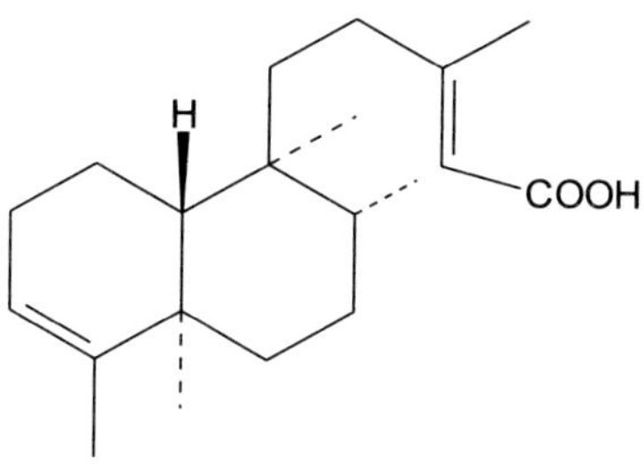

43

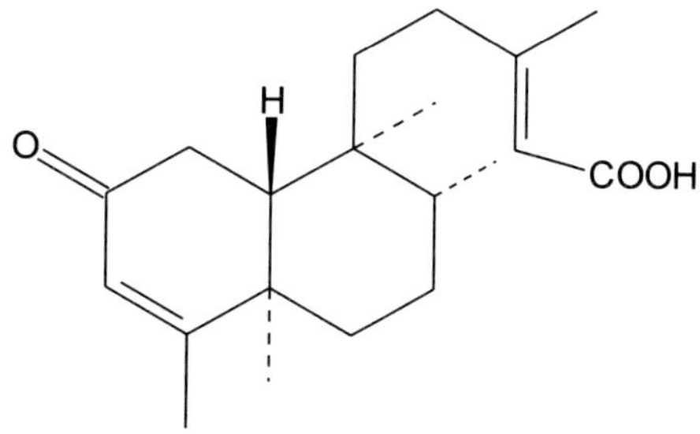

44

Antivirals

Several diterpenoids were isolated from the aerial parts of *Euphorbia paralias*, a common species of *Euphorbia* growing in Egypt (Abdelgaleil *et al.* 2001). Antiviral activity against the human immunodeficiency virus HIV-1 replication was determined by the inhibition of virus-induced cytopathicity on MT-4 cells using a colorimetric assay. The diterpene 5, 8, 14-triacetoxy-3-benzoyloxy-15-hydroxy-9-oxo-paraliane **45** showed moderate antiviral activity (EC_{50} 14 µg/mL). No comparison was made with a positive control.

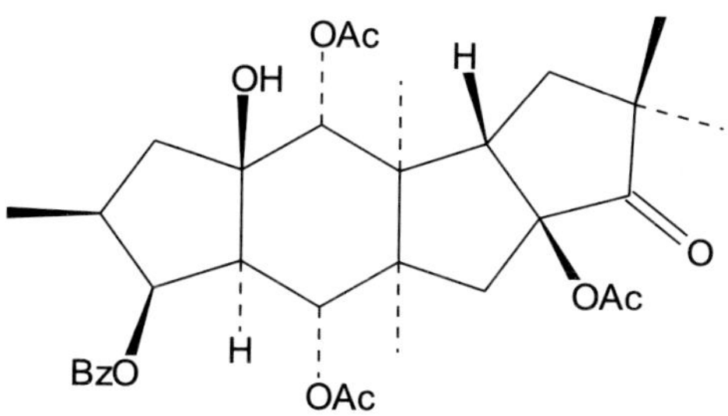

45

As part of a study on diterpenoid-containing plants, several diterpenes, including 7β-hydroxy fagonene **46,** were isolated from the aerial parts of *Fagonia boveana*, collected in Egypt (Gedara *et al.* 2003). Antiviral activity of the isolated compounds was evaluated by determining IC_{50} values (50% protection from the cytopathic effect of *Herpes simplex* type 1 virus on infected VERO cells). Compound **46** showed antiviral activity with IC_{50} 10 µg/mL (31.05 nM). The authors conclude that due to its low cytotoxicity, **46** represents a good potential antiviral agent. No positive control was used in the study.

B. subtilis and *Pseudomonas fluorescens*, with slightly less activity against the fungus *Cladosporium cucumerinum*. Claimed activities, in the range '2.5-20 μg', would need to be re-measured in a microtitre assay to be comparable with other data. Compound **53** did not show any antifungal activity.

A study of the antimicrobial constituents of two species of Combretaceae from Zimbabwe, *Combretum imberbe* and *Terminalia stuhlmannii*, yielded five pentacyclic triterpenes **58-62** which were tested against five microorganisms (Katerere *et al.* 2003). Antibacterial activity was generally modest compared to the control antibiotic streptomycin, except for imberbic acid **58** which had MIC 1.56 μg/mL against *Mycobacterium fortuitum* and 3.13 μg/mL against *Staph. aureus*, for which streptomycin had values of 1.56 and 0.78 μg/mL respectively. The measured level of potency is sufficient to justify further tests against *Mycobacterium tuberculosis*, in particular. It may be noted that imberbic acid is the major component of the leaves of *C. imberbe*, a large tree growing widely in sandy soils in southern Africa; it should be possible to obtain relatively large quantities of the compound if required, without destroying the trees. The antifungal activity of the compounds isolated in this study is described below.

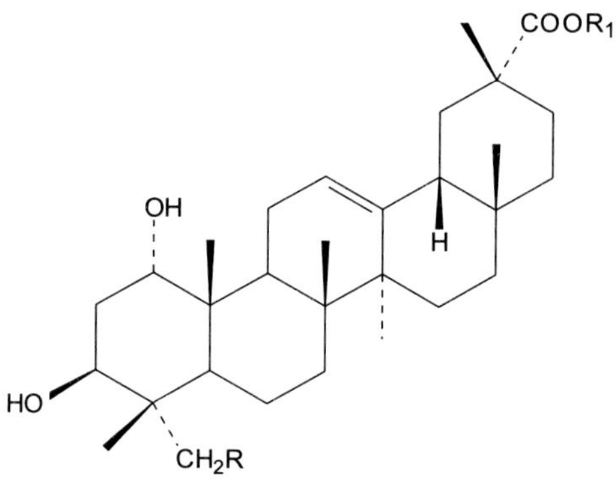

58 R=H,R_1=H
59 R=4-Ac-O-Rhamnose, R_1=H
60 R=3,4-Ac-O-Rhamnose, R_1=H
61 R=3,4-Ac-O-Rhamnose, R_1=Rhamnose
62 R=4-Ac-O-Rhamnose, R_1= Rhamnose

The fruits of *Embelia schimperi* are used by the Maasai of East Africa as an antibacterial and anthelminthic. The stem bark was extracted, giving the triterpenoids **63-67**, which were subjected to a very crude agar plate assay using 4 bacterial species. There was no indication of activity against *E. coli*, *Pseudomonas purida* or *B. subtilis* but some indication of activity against a *Rhodococcus* species (Machocho *et al.* 2003).

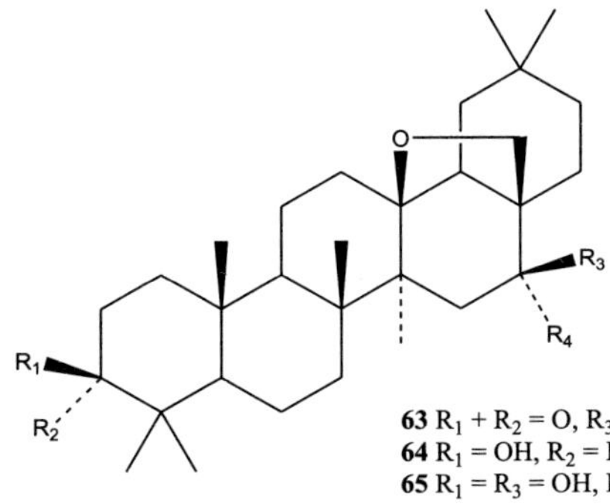

63 $R_1 + R_2 = O$, $R_3 + R_4 = O$
64 $R_1 = OH$, $R_2 = H$, $R_3 + R_4 = O$
65 $R_1 = R_3 = OH$, $R_1 = R_4 = H$

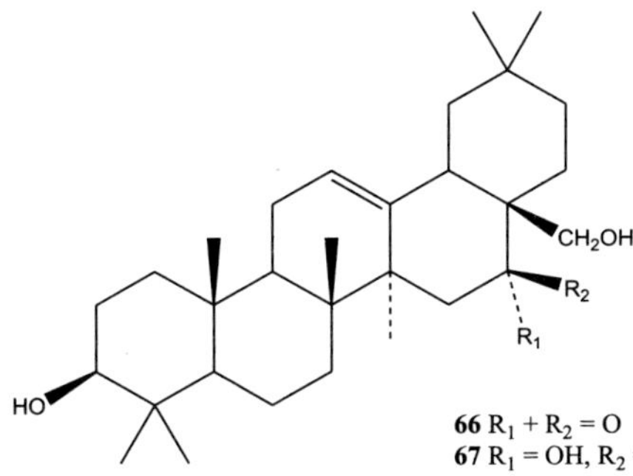

66 $R_1 + R_2 = O$
67 $R_1 = OH$, $R_2 = H$

Antifungals

Antifungal activity against the plant pathogenic fungus *Cladosporium cucumerinum* was observed for sakurasosaponin **68,** a triterpene saponin isolated from the leaves of *Rapanea melanophloeos* growing in Zimbabwe (Ohtani *et al*. 1993). Activity was determined using a bioautography TLC assay. It was found that **68** was antifungal at '1 μg'. This study would need to be repeated in order to express values as MICs and allow comparison with other data.

Maesa lanceolata is a shrub traditionally used in Rwanda to treat a range of diseases, including some of infectious origin (e.g., conjunctivitis, pertussis, impetigo, dysentery, gastro-enteritis, viral hepatitis, poliomyelitis, measles, chickenpox). It is also used to prevent cholera in East Africa. On screening Rwandan plants for antimicrobial and antiviral properties, the investigation of a leaf extract of *M. lanceolata* led to the isolation of two mixtures of triterpenic saponins, identified as maesasaponin mixture A and B in a 1/5 ratio, respectively (Sindambiwe *et al*. 1998, 1999). No antifungal activity was detected for maesasaponin mixture A when tested in an agar dilution assay against *Candida albicans, Aspergillus fumigatus, A. niger, Epidermophyton floccosum, Microsporum canis, M. langeroni, Microides interdigitalis* and *Trichophyton rubrum.* A moderate fungistatic activity against *M. langeroni* (MIC 250 μg/mL), *C. albicans*

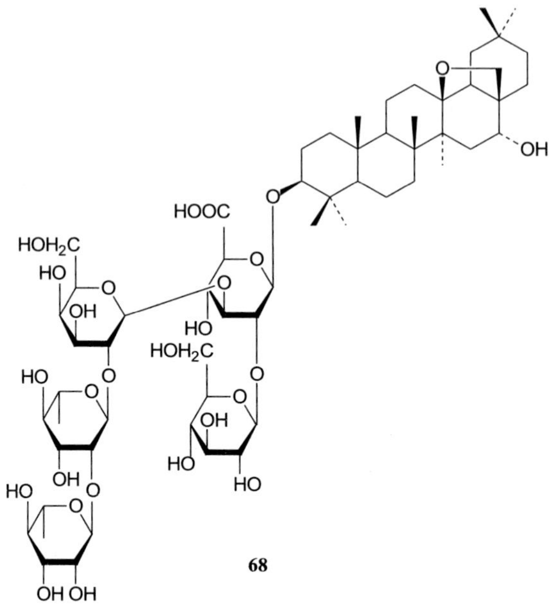

68

and *M. canis* (100 μg/mL), and *E. floccosum, M. interdigitalis, T rubrum* (MIC 50 μg/mL) was observed for maesasaponin mixture B.

Five pentacyclic triterpenes were isolated from two species of Combretaceae collected in Zimbabwe (Katerere *et al*. 2003). Imberbic acid **58**, 1*α*,3*β*-hydroxyimberbic acid-23-*O*-*α*-L-4-acetyl-rhamnopyranoside **59**, 1*α*,3*β*-hydroxyimberbic acid-23-*O*-*α*-L-3,4-diacetyl-rhamnopyranoside **60** and 1*α*,3*β*-hydroxyimberbic acid-23-*O*-*α*-[L-3,4-diacetyl-rhamnopyranosyl]-29-*O*-*α*-rhamnopyranoside **61** were isolated from the leaves of *Combretum imberbe.* Compound **59** and 1*α*,3*β*-hydroxyimberbic acid-23-*O*-*α*-[L-4-acetyl-rhamnopyranosyl]-29-*O*-*α*-rhamnopyranoside **62** were isolated from *Terminalia stuhlmannii* stem bark. Individual compounds were tested against *Candida albicans* in a microtitre dilution assay. Antifungal activity, comparable to the positive control, was observed for compound **62** with MIC 12.5 μg/mL. Compounds **58** and **60** showed weaker activity with MICs of 100 and 50 μg/mL, respectively. Compounds **59** and **61** had MICs > 100 μg/mL.

Antivirals

Arjunolic acid **69** was isolated from the rhizomes of *Cochlospermum tinctorium* growing in Guinea. Compound **69** and three methylated and/or acetylated derivatives **70-72** were screened *in vitro* on Epstein-Barr virus early antigen (EBV-EA) activation in Raji cells induced by 12-*O*-tetradecanoylphorbol-13-acetate (TPA). All compounds showed in-

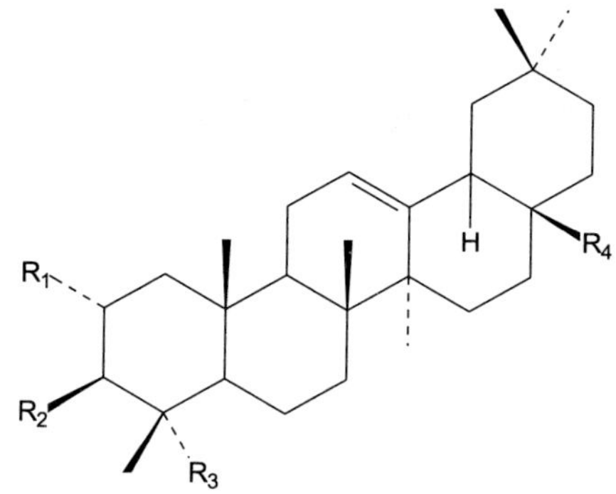

69 $R_1=R_2=R_3=OH$, $R_4=COOH$
70 $R_1=R_2=R_3=OH$, $R_4=COOMe$
71 $R_1=R_2=R_3=OAc$, $R_4=COOH$
72 $R_1=R_2=R_3=OAc$, $R_4=COOMe$

hibitory effects on EBV-EA activation. Compounds **71** and **72**, in particular, could be promising antitumor promoters (Diallo *et al*. 1989).

Previous investigation of the aerial parts of the Egyptian species *Astragalus spinosus* had revealed the presence of the triterpene glycoside astragaloside II **73** (Abdallah *et al*. 1993). In this study, **73** showed antiviral activity as determined by the inhibition of cytopathicity on T_4 lymphocytes infected by the HIV virus ($EC_{50}=2.5\ 10^{-5}$ M). No positive control was used in the study.

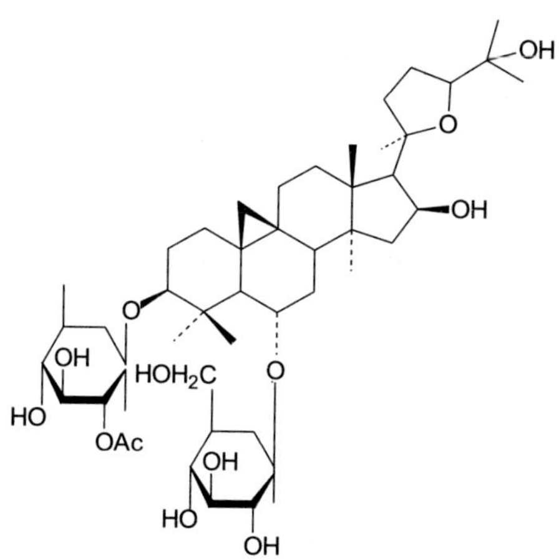

73

The pentacyclic triterpenes 1β-hydroxymaprounic acid 3-*p*-hydroxybenzoate **74** and 2α-hydroxymaprounic acid 2,3-*bis*-*p*-hydroxybenzoate **75** (later renamed 1β-hydroxyaleuritolic acid 3-*p*-

hydroxybenzoate and 2α-hydroxyaleuritolic acid 2, 3-*bis-p*-hydroxybenzoate, respectively) were isolated from the roots of *Maprounea africana* collected in Tanzania (Pengsuparp *et al.* 1994, 1995, Chaudhuri *et al.* 1995). Compounds **74** and **75**, together with their respective hydrolysis products **76** and **77**, showed potent inhibition of HIV-1 reverse transcriptase (RT) with IC_{50} in the range of 3-5 μM. All compounds were more active against HIV-1 than HIV-2 RT. Further studies on **74** revealed that the inhibition involved non-specific binding to the RT enzyme at non-substrate binding sites.

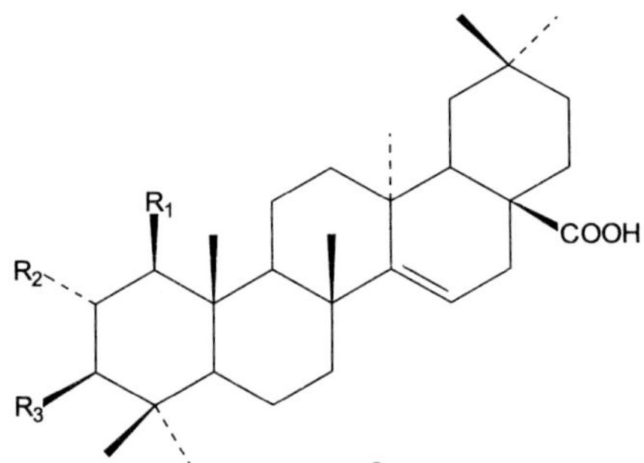

74 R_1=OH, R_2=H, R_3=

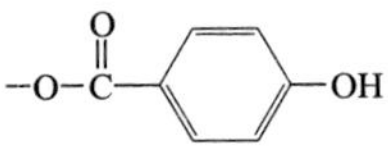

75 R_1=H, R_2=R_3=

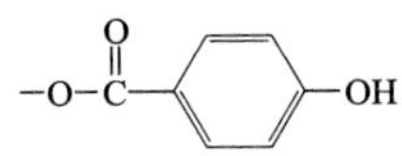

76 R_1=R_3=OH, R_2=H

77 R_1=H, R_2=R_3=OH

On screening Rwandan plants for antiviral properties, the leaf extract of *Maesa lanceolata* had revealed a virucidal effect against a range of enveloped viruses such as *Herpes simplex* type 1 (HSV-1), measles, Semliki forest and vesicular stomatitis viruses (Sindambiwe *et al.* 1998). Bioassay-guided fractionation (using HSV-1 as the target model) led to the isolation of two saponin mixtures, maesasaponin mixture A and B (obtained in a 1/5 ratio). A moderate extracellular virucidal activity against HSV-1 and -2, vesicular stomatitis and measles was observed for the maesasaponin mixture B (compounds **78-83**). The choice of a different model virus to guide the isolation might have resulted in the identification of other active compounds.

Maesasaponin mixture A showed virucidal activity against *Herpes simplex* type 1 (HSV-1), type 2 (HSV-2) and vesicular stomatitis viruses. The composition of the mixture was not detailed (Sindambiwe *et al.* 1999).

Further chromatographic analysis of the maesasaponin B mixture from *Maesa lanceolata* led to the characterisation of 10 saponins, maesasaponin I **84**, maesasaponin II **85**, maesasaponin III **86**,

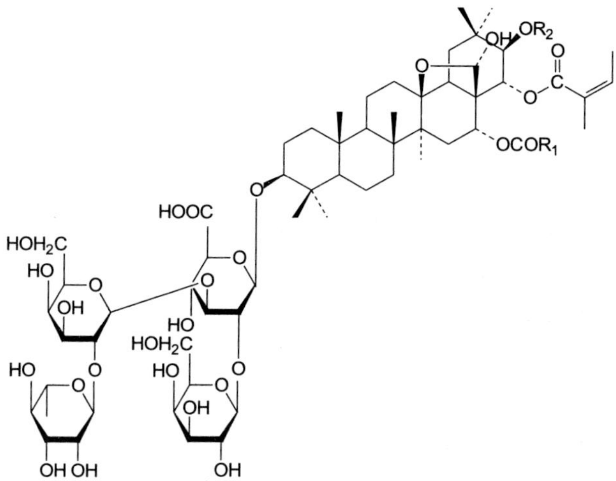

78 $R_1=C_2H_5, R_2=H$
79 $R_1=C_3H_7, R_2=H$
80 $R_1=C_4H_7, R_2=H$
81 $R_1=C_4H_9, R_2=H$
82 $R_1=C_2H_5, R_2=COCH_3$
83 $R_1=C_3H_7, R_2=COCH_3$

maesasaponin IV_2 **87**, maesasaponin IV_3 **88**, maesasaponin V_2 **89**, maesasaponin V_3 **90**, maesasaponin VI_2 **91**, maesasaponin VI_3 **92**, maesasaponin VII_1 **93**. Extracellular virucidal screening was determined against HSV-1 (Apers *et al.* 1999, 2001). Results were expressed as the reduction factor of the viral titer (RF) (i.e, ratio of viral titers in the virus control and in the presence of the maximal non-toxic dose of compound). Compounds **84, 88** and **91** showed moderate activity with RF $\geq 10^3$ at 50 µg/mL. Compound **90** showed strong activity with RF $\geq 10^3$ at 10 µg/mL. Other saponins were only weakly active (RF $\leq 10^2$ at 50 µg/mL). Structure-activity relationship studies suggest that an enhanced virucidal effect is observed for 21, 22-diacylated compounds. Protection against the cytopathic effect of HIV type 1 and -2 viruses on MT-4 cells was determined using a microtray colorimetric assay. No activity was observed either for the mixture or for the isolated compounds due to high cytotoxicity.

Quassia africana (Simaroubaceae) is traditionally used in Western Africa in the treatment of dysentery, diarrhoea, fevers and malaria. Investigation of the root bark of *Q. africana* (collected in the Democratic Republic of Congo) led to the isolation of the quassinoid simalikalactone D **94**. This compound was tested for antiviral activity against a selection of viruses including *Herpes simplex* type 1 (HSV-1), vesicular stomatitis (VSV), Coxsackie B2 (COX B2), measles, poliomyelitis 1 and Semliki

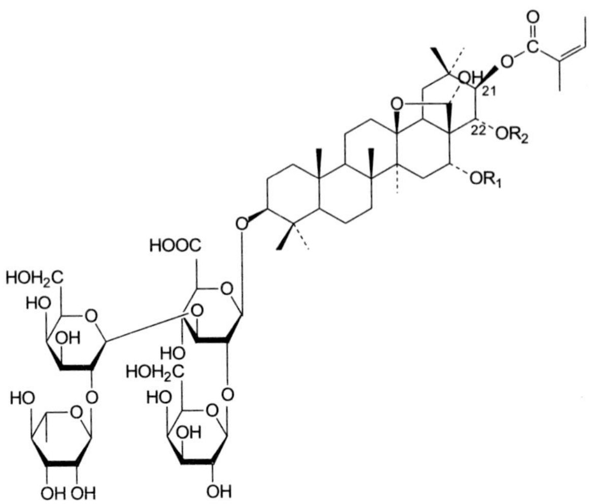

84 R_1=H, R_2=H
85 R_1=$COCH_3$, R_2=H
86 R_1=H, R_2=$COCH_3$
87 R_1=R_2= $COCH_3$
88 R_1=H, R_2=$COCH_2CH_3$
89 R_1=$COCH_3$, R_2=$COCH_2CH_3$
90 R_1=H, R_2=$COCH_2CH_2CH_3$

91 R_1=H, R_2= 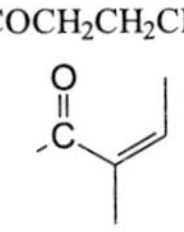

92 R_1=$COCH_3$, R_2= $COCH_2CH_2CH_3$

93 R_1=$COCH_3$, R_2= 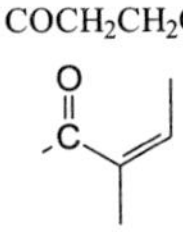

forest viruses (Apers *et al.* 2002). Cytotoxicity assays were run in parallel in order to express antiviral activity as viral titer reduction factors (RF) (ratio of the viral titer reduction in the absence and presence of the maximal non-toxic dose). Compound **94** displayed antiviral activity against HSV-1 and VSV (RF 10^2 at 0.2 μg/mL) and against Poliomyelitis 1 (RF 10^4 at 2 μg/mL) and Semliki forest viruses (RF 10^2 at 2 μg/mL). It did not, however, exhibit any extracellular virucidal activity against Poliomyelitis, Semliki Forest and VSV viruses. It did not exhibit any anti-HIV effect.

Following previous reports of the anti-HIV activity of fruit extracts, a bioassay-guided fractionation of *Tieghemella heckelii* (a tree growing in the Ivory Coast) was carried out (Gosse *et al.* 2002). Two saponins, arganine C **95** and tieghemelin **96** were identified as the constituents responsible for the activity using a cell-cell fusion assay which meas-

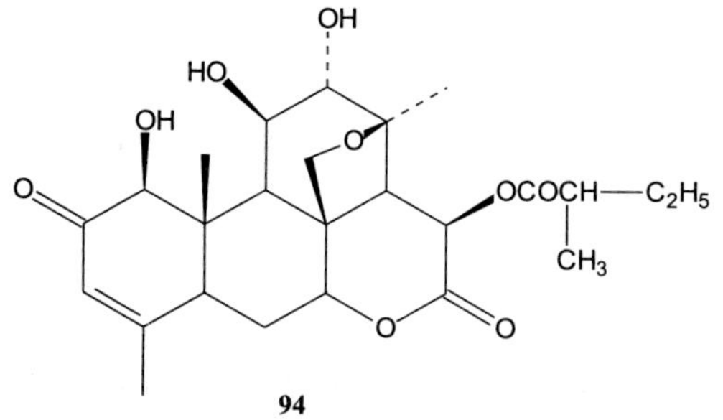

94

ured HIV-1 entry into HeLa-CD4 cells. Compound **95** showed 100% inhibition of syncytium formation at 20 μM and was twice as active as **96** at the same concentration. Removal of the glycosidic moieties resulted in a loss of activity. The authors conclude that due to low cytotoxicity, **95** is a promising antiviral lead compound.

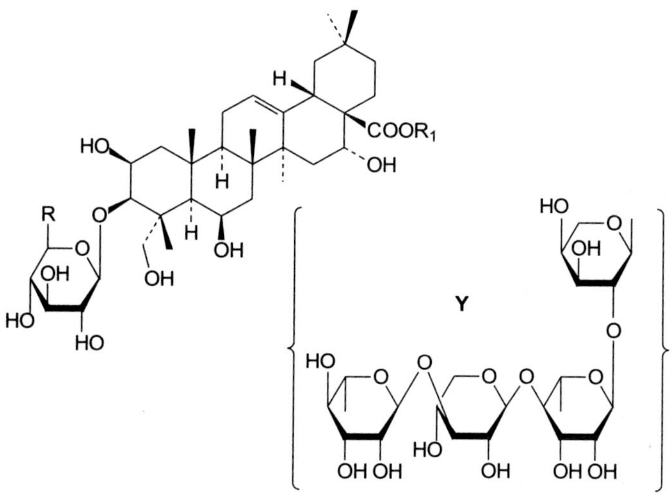

95 R=CH_2OH , R_1=Y
96 R=CH_2OH , R_1=Y

CONCLUSIONS

First, a plea: even in the 21st Century, researchers are using screening methods that make their results impossible to compare with either earlier or contemporary literature. In view of recent developments in screening methodology (Eloff 1998, Drummond and Waigh 2000), any laboratory that is serious about antibacterial or antifungal testing should measure MICs in growth medium and not rely on disk diffusion or cup plates. The test should be conducted on the organism(s) of interest and not on a catch-all species which happens to be convenient. Finally, it is necessary to use a positive control, in this case a known antibiotic with good activity, tested concomitantly with the unknowns.

This survey shows that there are a number of terpene-derived secondary metabolites from African higher plants that would justify further study. The past failure to exploit this resource can be attributed to a number of factors, few of which are associated with scientific issues; two concern the potential profitability to a producer. The costs of development of a new pharmaceutical are enormous, but publication of the discovery of a new active compound in the open literature makes patent protection difficult or impossible. Academic researchers have a difficult choice, to publish their best results or to take a chance that a drug company will be sufficiently interested to attempt to develop their compound to marketability, perhaps failing at a late stage after several years. The failure to exploit the triterpene acids described above may be compared to the successful development and marketing of fusidic acid, isolated from a fungal broth and patented in accord with normal practice for the discovery of antibiotics.

The other profitability factor is associated with the geographical occurrence of the disease that may be treated. Tuberculosis, for example, is primarily a disease of developing countries in which the people cannot afford to pay for the development of medicines according to protocols established in North American and European practice. Perhaps the best way forward is for developing countries themselves to carry out investigative development of plants that have been shown to be active in screening programs. Extracts and single compounds could be tested in human disease, under medical supervision. The result would probably be effective medicines at very low cost, avoiding the litigation-conscious restraints of the developed world. It is scarcely logical to refuse to test medicines because they may cause toxic side-effects, when the sufferers are going to die without treatment. It is possible that the discovery of successful treatments for diseases that are widespread in developing countries might feed back into western/northern medicine, as e.g., MRSA and TB increase their impact on healthcare systems.

REFERENCES

Abdallah, R.M., N.M. Ghazi, N.A. El-Sebakhy, A. Pirillo, and L. Verotta. 1993. Astragalosides from Egyptian *Astragalus spinosus* Vahl. *Pharmazie* 48: 452-454.

Abdelgaleil, S.A.M., S.M.I. Kassem, M. Doe, M. Baba, and M. Nakatani. 2001. Diterpenoids from *Euphorbia paralias*. *Phytochemistry* 58: 1135-1139.

Aboutabi, E.A., F.S. El-Sakhawy, M.M. Fathy, and R.M.A. Megid. 2000. Composition and antimicrobial activity of the leaf and fruit oils from *Amoora rohituka* Wight. et Arn. *J. Essent. Oil Res.* 12: 635-638.

Amvam Zollo, P.H., L. Biyiti, F. Tchoumbougnang, C. Menut, G. Lamaty, and P. Bouchet. 1998. Aromatic plants of tropical Central Africa. Part XXXII. Chemical composition and antifungal activity of thirteen essential oils from aromatic plants of Cameroon. *Flavour Frag. J.* 13: 107-114.

Apers, S., T.E. De Bruyne, M. Clayes, A.J. Vlietinck, and L. Pieters. 1999. New acylated triterpenoid saponins from *Maesa lanceolata*. *Phytochemistry* 52: 1121-1131.

Apers, S., S. Baronikova, J.B. Sindambiwe, M. Witvrouw, E. De Clercq, D. Vanden Berghe L., E. Van Marck, A. Vlietinck, and L. Pieters. 2001. Antiviral, haemolytic and molluscicidal activities of triterpenoid saponins from *Maesa lanceolata*: establishment of structure-activity relationships. *Planta Med.* 67: 528-532.

Apers, S., K. Cimanga, D. Vanden Berghe, E. Van Meenen, A. Otshudi Longanga, A. Foriers, A. Vlietinck, and L. Pieters. 2002. Antiviral activity of Simalilalactone D, a quassinoid from *Quassia africana*. *Planta Med.* 68: 20-24.

Asres, K., A. Tei, G. Moges, F. Sporer, and M. Wink. 1998. Terpenoid composition of the wound-induced bark exudate of *Commiphora tenuis* from Ethiopia. *Planta Med.* 64: 473-475.

Ayafor, J.F., M.H.K. Tchuendem, B. Nyasse, F. Tillequin, and H. Anke. 1994. Novel bioactive diterpenoids from *Aframomum aulacocarpos*. *J. Nat. Prod.* 57: 917-923.

Bassole, I.H.N., A.S. Ouattara, R. Nebie, C.A.T. Ouattara, Z.I. Kabore, and S.A. Traore. 2003. Chemical composition and antibacterial activities of the essential oils of *Lippia chevalieri* and *Lippia multiflora* from Burkina Fasso. *Phytochemistry* 62: 209-212.

Boakye-Yiadom, K., N.I.Y. Fiagbe, and J.S.K. Ayim. 1977. Antimicrobial properties of some West African medicinal plants IV. Antimicrobial activity of xylopic acid and other constituents of the fruits of *Xylopia aethiopica* (Annonaceae). *Lloydia* 40: 543-545.

Bougatsos, C., J.J.M. Meyer, P. Magiatis, C. Vagias, and I. B.Chinou. 2003. Composition and antimicrobial activity of the essential oils of *Helichrysum kraussii* and *H. rugulosum* from South Africa. *Flavour Fragr. J.* 18: 48-51.

Cantrell, C.L., M.S. Rajab, S.G. Franzblau, and N.H. Fischer. 1999. Antimycobacterial triterpenes from *Melia volkensii*. *J. Nat. Prod.* 62: 546-548.

Chaudhuri, S.K., F. Fullas, D.M. Brown, M.C. Wani, M.E. Wall, L. Cai, W. Mar, S.K. Lee, Y. Luo, K. Zaw, H.H.S. Fong, J.M. Pezzuto, and A.D. Kinghorn. 1995. Isolation and structural elucidation of pentacyclic triterpenoids from *Maprounea africana*. *J. Nat. Prod.* 58: 1-9.

Cimanga, K., K. Kambu, L. Tona, S. Apers, T. De Bruyne, N. Hermans, J. Totte, L. Pieters, and A.J. Vlietinck. 2002. Correlation between chemical composition and antibacterial activity of essential oils of some aromatic medicinal plants growing in the Democratic Republic of Congo. *J. Ethnopharmacol.* 79: 213-220.

Conrad, J., B. Vogler, I. Klaiber, G. Roos, U. Walter, and W. Kraus. 1998. Two triterpene esters from *Terminalia macroptera* bark. *Phytochemistry* 48: 647-650.

Diallo, B., M. Vanhaelen, R. Vanhaelen-Fastre, T. Konoshima, M. Kozuka, and H. Tokuda. 1989. Studies on inhibitors of skin-tumor promotion. Inhibitory effects of triterpenes from *Cochlospermum tinctorium* on Epstein-Barr virus activation. *J. Nat. Prod.* 52: 879-881.

Drummond, A.J., and R.D. Waigh. 2000. The development of microbiological methods for phytochemical screening. *Recent. Res. Devel. Phytochem.* 4: 143-152.

El-Lakany, A.M., M.S. Abdel-Kader, N.N. Sabri, and F.R. Stermitz. 1995. Lanigerol: a new antimicrobial icetaxane diterpene from *Salvia lanigera*. *Planta Med.* 61: 559-560.

Eloff, J.N. 1998. A sensitive and quick microplate method to determine the minimal inhibitory concentration of plant extracts for bacteria. *Planta Med.* 64: 711-713.

Gedara, S.R., O.B. Abdel-Halim, S.H. El-Sharkawy, O.M. Salama, T.W. Shier, and A. F. Halim. 2003. New erythroxane-type diterpenoids from *Fagonia boveana* (Hadidi) Hadidi & Graf. *Z. Naturforsch.* 58: 23-32.

Gosse, B., J. Gnabre, R.B. Bates, C.W. Dicus, P. Nakkiew, and R.C.C. Huang. 2002. Antiviral saponins from *Tieghemella heckelii*. *J. Nat. Prod.* 65: 1942-1944.

Gundidza, G.M., S.G., Deans, K.P. Svoboda, and S. Mavi. 1992. Antimicrobial activity of essential oil from *Hoslundia opposita*. *Centr. Afr. J. Med.* 38: 290-293.

Habtemariam, S., A.I. Gray, G.W. Halbert, and P.G. Waterman. 1990. A novel antibacterial diterpene from *Premna schimperi*. *Planta Med.* 56: 187-189.

Habtemariam, S., A.I. Gray, and P.G. Waterman. 1992. Antibacterial diterpenes from the aerial parts of *Premna oligotricha*. *Planta Med.* 58: 109-110.

Habtemariam, S., A.I. Gray, and P.G. Waterman. 1993. A new antibacterial sesquiterpene from *Premna oligotricha*. *J. Nat. Prod.* 56: 140-143.

Hmamouchi, M., J. Hamamouchi, M. Zouhdi, and J.M. Bessiere. 2001. Chemical and antimicrobial properties of essential oils of five Moroccan Pinaceae. *J. Essent. Oil Res.* 13: 298-302.

Hostettman-Kaldas, M., and K. Nakanishi. 1979. Moronic acid, a simple triterpenoid keto acid with antimicrobial activity isolated from *Ozoroa mucronata*. *Planta Med.* 37: 358-360.

Jisaka, M., H. Ohigashi, K. Takegawa, M.A. Huffman, and K. Koshimizu. 1993. Antitumoral and antimicrobial activities of bitter sesquiterpene lactones of *Vernonia amygdalina*, a possible medicinal plant used by wild chimpanzees. *Biosci. Biotech. Biochem.* 57: 833-834.

Katerere, D.R., A.I. Gray, R.J. Nash, and R.D. Waigh. 2003. Antimicrobial activity of pentacyclic triterpenes isolated from African Combretaceae. *Phytochemistry* 63: 81-88.

Kubo, I., and M. Taniguchi. 1988. Polygodial, an antifungal potentiator. *J. Nat. Prod.* 51: 22-29.

Kubo, I., K. Fugita, and S.H. Lee. 2001. Antifungal mechanism of polygodial. *J. Agric. Food. Chem.* 49: 1607-1611.

Kunle, O., J. Okogun, E. Egamana, E. Emojevwe, and M. Shok. 2003. Antimicrobial activity of various extracts and carvacrol from *Lippia multiflora* leaf extract. *Phytomedicine* 10: 59-61.

Lee, S.H., J.R. Lee, C.S. Lunde, and I. Kubo. 1999. *In vitro* antifungal susceptibilities of *Candida albicans* and fungal pathogens to polygodial, a sesquiterpene dialdehyde. *Planta Med.* 65: 204-208.

Machocho, A.K., P.C. Kiprono, S. Grinberg, and S. Bittner. 2003. Pentacyclic triterpenoids from *Embelia schimperi*. *Phytochemistry* 62: 573-577.

Mrabet, N., H. Lahlou, and B. Benjilali. 1999. Effet de quelques extraits du ciste ladanifere du Maroc (*Cistus ladaniferus* L.) sur la croissance de quatre champignons. *Cryptogamie Mycol.* 20: 23-33.

Ngassoum, M.B., J.J. Essia-Ngang, L.N. Tatsadjieu, L. Jirovetz, G. Buchbauer, and O. Adjoudji. 2003. Antimicrobial study of the essential oils of *Ocimum gratissimum* leaves and *Zanthoxylum xanthoxyloides* fruits from Cameroon. *Fitoterapia* 4: 284-287.

Ohtani, K, S. Mavi, and K. Hostettman. 1993. Molluscicidal and antifungal triterpenoid saponins from *Rapanea melanophloeos* leaves. *Phytochemistry* 33: 83-86.

Oumzil, H., S. Ghoulami, M. Rhajaoui, A. Ilidrissi, S. Fkih-Tetouani, M. Faid, and A. Benjouad. 2002. Antibacterial and antifungal activity of essential oils of *Mentha suavolens*. *Phytother. Res.* 16: 727-731.

Pengsuparp, T., L. Cai, H.H.S. Fong, A.D. Kinghorn, J.M. Pezzuto, M.C. Wani, and M.E. Wall. 1994. Pentacyclic triterpenes derived from *Maprounea africana* are potent inhibitors of HIV-1 reverse transcriptase. *J. Nat. Prod.* 57: 415-418.

Pengsuparp, T., L. Cai, H. Constant, H.H.S. Fong, L-Z. Lin, A.D. Kinghorn, J.M. Pezzuto, G.A. Cordell, K., Ingolfsdottir, H. Wagner, and S.H. Hughes. 1995. Mechanistic evaluation of new plant-derived compounds that inhibit HIV-1 reverse transcriptase. *J. Nat. Prod.* 58: 1024-1031.

Rabe, T., and J. van Stade. 2000. Isolation of an antibacterial sesquiterpenoid from *Warburgia salutaris*. *J. Ethnopharmacol.* 73: 171-174.

Rabe, T., D. Mulholland, and J. van Staden. 2002. Isolation and identification of antibacterial compounds from *Vernonia colorata* leaves. *J. Ethnopharmacol.* 80: 91-94.

Rajab, S.M., C.L. Cantrell, S.G. Franzblau, and N.H. Fischer. 1998. Antimycobacterial activity of (E)-phytol and derivatives: a preliminary structure-activity study. *Planta Med.* 64: 2-4.

Rasoamiaranjanahary, L., A. Marston, D. Guilet, K. Schenk, F. Randimbivololona, and K. Hostettmann. 2003a. Antifungal diterpenes from *Hypoestes serpens* (Acanthaceae). *Phytochemistry* 62: 333-337.

Rasoamiaranjanahary, L., D. Guilet, A. Marston, F. Randimbivololona, and K. Hostettmann. 2003b. Antifungal isopimaranes from *Hypoestes serpens*. *Phytochemistry* 64: 543-548.

Salah, M.A., E. Bedir, N.J. Toyang, I.A. Khan, M.D. Harries, and D.E. Wedge. 2003. Antifungal clerodane diterpenes from *Macaranga monandra* (L) Muell. et Arg. (Euphorbiaceae). *J. Agric. Food Chem.* 51: 7607-7610.

Schaller, F., L. Rahalison, N. Islam, O. Potterat, K. Hostettmann, H. Stoeckli-Evans, and S. Mavi. 2000. A new potent antifungal 'quinone-methide' diterpene with a cassane skeleton from *Bobgunnia madagascariensis*. *Helv. Chim. Acta* 83: 407-413.

Schaller, F., J.L. Wolfender, K. Hostettmann, and S. Mavi. 2001. New antifungal 'quinone methide' diterpenes from *Bobgunnia madagascariensis* and study of their interconversion by LC/NMR. *Helv. Chim. Acta* 84: 222-229.

Sindambiwe, J.B., M. Calomme, S. Geerts, L., Pieters, A.J. Vlietinck, and D.A. Vanden Berghe. 1998. Evaluation of biological activities of triterpenoid saponins from *Maesa lanceolata*. *J. Nat. Prod.* 61: 585-590.

Sindambiwe, J.B., M. Calomme, P. Cos, J. Totte, L. Pieters, A. Vlietinck, and D. Vanden Berghe. 1999. Screening of seven selected Rwandan medicinal plants for antimicrobial and antiviral activities. *J. Ethnopharmacol.* 65: 71-77.

Tatsadjieu, L.N., J.J. Essia Ngang, M.B. Ngassoum, and F.-X. Etoa. 2003. Antibacterial and antifungal activity of *Xylopia aethiopica, Monodora myristica, Zanthoxylum xanthoxyloides* and *Zanthoxylum leprieurii* from Cameroon. *Fitoterapia* 74: 469-472.

Van Puyvelde, L., S. Nyrankuliza, R. Panebianco, Y. Boily, I. Geizer, B. Sebikali, N. De Kimpe, and N. Schamp. 1986. Active principles of *Tetradenia riparia*. 1. Antimicrobial activity of 8(14),15-sandaracopimaradiene-7α,18-diol. *J. Ethnopharmacol.* 17: 269-275.

Van Puyvelde, L., J.D. Ntawukiliyayo, and F. Portaels. 1994. *In vitro* inhibition of mycobacteria by Rwandese medicinal plants. *Phytother. Res.* 8: 65-69.

Vieira, P.C., M. Himejima, and I. Kubo. 1991. Sesquiterpenoids from *Brachylaena hutchinsii*. *J. Nat. Prod.* 54: 416-420.

Viljoen, A.M., M.E. Klepser, E.J. Ernst, D. Keele, E. Roling, S. van Vuuren, B. Demirci, K.H. C. Baser, and B-E van Wyk. 2002. The composition and antimicrobial activity of the essential oil of the resurrection plant *Myrothamnus flabellifolius*. *S. Afr. J. Bot.* 68: 100-105.

Viljoen, A., S. van Vuuren, S., E. Ernst, M. Klepser, B. Demirci, H. Baser, and B.-E. van Wyk. 2003. *Osmitopsis asteriscoides* (Asteraceae)-the antimicrobial activity and essential oil composition of a Cape-Dutch remedy. *J. Ethnopharmacol.* 88: 137-143.

6

Antimalarial and Antifungal Alkaloids from Plants

D. CHUCK DUNBAR, XING-CONG LI AND ILIAS MUHAMMAD

National Center for Natural Products Research, Research Institute of Pharmaceutical Sciences, School of Pharmacy, University of Mississippi, University, Mississippi 38677, U.S.A.

INTRODUCTION

Humans have used plant materials for medicines since before recorded time. These remedies consisted of whole plants or rather crude extracts until quite recently. In the last one hundred years or so humans began to purify natural compounds and to produce compounds synthetically for use as drugs. Despite all of the modern advances in synthetic techniques, knowledge of receptors, molecular modeling and database searching we are still dependent upon natural compounds as the basis for new drug discoveries. More than half of the small molecule "New Chemical Entities" (NCE) reported in the period of 1981-2002 were natural compounds, derivatives of natural compounds or synthetic compound whose structure was based upon a natural compound. Interestingly, during this same time period pharmaceutical companies deemphasized natural product screening in favor of combinatorial libraries and the number of NCE hit a twenty year low and is still declining (Newman *et al*. 2003).

Alkaloids have played an important role in the history of natural compound derived medicines. Reserpine, the first antihypertensive drug, was discovered from *Rauwolfia serpentine*, and morphine, the basis of narcotic analgesics comes from the opium poppy, *Papaver somniferum*. Farnsworth (1984) has reported 46 plant- derived drugs that are widely used in Western medicine and nearly half of these are alkaloids. With this rich history of providing mankind with drugs or drug leads, there is little doubt plants will continue to be an excellent source for future medicines. In this review we will first discuss antimalarial alkaloids and subsequently antifungal alkaloids recently isolated from plant sources.

ANTIMALARIAL ALKLOIDS

Malaria

Malaria is a blood disease caused by a single-celled parasite *Plasmodium* spp. and is spread through the bite of the female *Anopheles* mosquito. Malaria threatens more than forty percent of the world's population, causes 300 million acute illnesses and at least one million deaths annually according to the World Health Organization. This is not a new disease however, and the collection of plant material for the treatment of malaria has been practiced since ancient times. In China *Artemisia annua* infusions have been used for over 2000 years. Recently, the active constituent, artemisinin was isolated from *A. annua* (Klayman 1985). The South American natives used the bark of *Cinchona* trees for treatment of the malaria and the bark was being shipped back to Europe for this purpose by the 1600's. Quinine **(1)** was isolated as the active constituent of *Cinchona* bark in 1820. The actual cause of malaria, however, was unknown until the discovery of the parasite (*Plasmodium* sp.) by Alphonose Laveran in 1889. In 1897, Sir Ronald Ross demonstrated that the mosquito was the vector for the parasite (Shuler 1985). The Nobel Prize in Physiology or Medicine was awarded to Ross in 1902 and to Laveran in 1907.

Malaria in humans is now known to be caused by four species of Protozoa from the genus *Plasmodium*, which belongs to the class Sporozoa. Plasmodia have an asexual stage during their life cycle in humans known as schizogony and a sexual stage, sporogony, in the female *Anopheles* mosquito (Shuler 1985). Table 6.1 contains the four *Plasmodium* species that infect humans and the percent of reported cases caused by each organism (Goldsmith and Heyneman 1989). The 5% unknown etiology occurs because most malarial infections occur in countries with poor medical facilities and therefore the species is not always determined.

Table 6.1 *Plasmodium* species that cause malaria in humans

Organism	*Incidence of Malaria*
P. vivax	59%
P. ovale	2%
P. falciparum	29%
P. malariae	5%
unknown etiology	5%

Malaria caused by *P. falciparum* is the most dangerous because of its ability to spread through the other organs of the body. In the brain it causes cerebral malaria which leads to neurological damage or death. The two of the most important drug classes used to treat cerebral malaria today are based on ancient remedies discovered from plants. Artemisinins (ART) are endoperoxide terpenoids from which the drugs artemether, arteether and artesunate were developed. Quinine is a quinoline alkaloid from which the important antimalarial drugs chloroquine (CQ), amodiaquine, mefloquine and halofantrine were derived. Currently parenteral treatment with quinine **(1)** or artemether is the only course of treatment for acute cerebral malaria (Mturi *et al.* 2003).

Most of the major drugs used to combat malarial infections have been foiled by the organism in some part of the world. This fact opens the possibility that a "super resistant" malaria strain will develop that is untreatable with current drugs. Development of new drugs has to occur faster than resistance to prevent such a catastrophe. Since plants have provided mankind with leads in the past, it is logical to look to plants for additional antimalarial leads. This part of the work will focus on recent antimalarial alkaloids reported from plants.

Antimalarial Assays

The methods used for *in vitro* antimalarial assays vary and care must be taken when comparing activity from different laboratories. Measurements used for IC_{50} values can be derived from many different methods such as direct microscopic counting of the parasite, colorimetric reading as in the lactate dehydrogenase assay (Makler *et al.* 1993), radiometric reading like the tritium hypoxanthin incorporation assay or flow cytometry (Van der Heyde *et al.* 1995). The *Plasmodium* strain used in the bioassay also affects the activity of drugs. *In vivo* testing using a mouse model (*Plasmodium berghei*) is widely used but again the protocols will differ and direct comparison is dubious. Comparisons should be drawn from the relative activity of the new compound to that of the control drug used in the bioassay.

Antimalarial Alkaloids from Plants

Alkaloids are the most important plant based secondary metabolites for the discovery and development of antimalarial drugs. All the antimalarial drugs currently in therapeutic use were discovered from plant sources, and later modified to different analogs for practical use as needed. Several major reviews on antimalarial natural products have been published so far dealing with different classes of secondary metabolites, including alkaloids (Fournet and Munoz 2002, Kayser *et al.* 2002, Schwikkard and

(*Contd.*)

Ancistrotectoriline A (**40**)	*Ancistrocladus tanzaniensis*	0.5 μg/mL (K1) 14 μg/mL (3D7)	Bringmann *et al*. 2004
chloroquine	Control	0.055 μg/mL (K1) 0.01 μg/mL (3D7)	Bringmann *et al*. 2004
Habropetaline A (**41**)	*Triphyophyllum peltatum*	0.005 μg/mL (K1) 0.0023 μg/mL (NF54)	Bringmann *et al*. 2003a
Ancistrogriffines A (**42**)	*Ancistrocladus griffithii*	0.074 μg/mL (K1)	Bringmann *et al*. 2003a
Ancistrogriffines C (**43**)	*Ancistrocladus griffithii*	0.419 μg/mL (K1)	Bringmann *et al*. 2003a
Chloroquine		0.065 μg/mL (K1)	Bringmann *et al*. 2003a
Ancistrogriffithine A (**44**)	*Ancistrocladus griffithii*	0.035 μg/mL (K1)	Bringmann *et al*. 2002c
Ancistrocongoline A (**45**)	*Ancistrocladus congolensis*	0.214 μg/mL (K1)	Bringmann *et al*. 2002a
Ancistrocongoline B (**46**)	*Ancistrocladus congolensis*	0.158 μg/mL (K1)	Bringmann *et al*. 2002a
Ancistrocongoline C (**47**)	*Ancistrocladus congolensis*	3.002 μg/mL (K1)	Bringmann *et al*. 2002a
Ancistrocongoline D (**48**)	*Ancistrocladus congolensis*	1.983 μg/mL (K1)	Bringmann *et al*. 2002a
korupensamine A (**49**)	*Ancistrocladus congolensis*	0.164 μg/mL (K1)	Bringmann *et al*. 2002a
artemisinine	Control	0.001 μg/mL (K1)	Bringmann *et al*. 2002a
Dioncophylline E (**50**)	*Dioncophyllum thollonii*	0.022 μg/mL (NF54) 0.021 μg/mL (K1)	Bringmann *et al*. 2002b
artemisinin	Control	0.0028 μg/mL (NF54) 0.0011 μg/mL (K1)	Bringmann *et al*. 2002b
Isostrychnopentamine (**52**)	*Strychnos usambarensis*	0.120 μM (FCA 20) 0.104 μM (FCB1-R) 0.152 μM (W2)	Frederich *et al*. 2004b
Strychnogucine B (**54**)	*Strychnos icaja*	0.617 μM (FCA 20) 0.085 μM (W2)	Frederich *et al*. 2001a
Sungucine (**55**)	*Strychnos icaja*	7.8 μM (FCA 20) 10.1 μM (W2)	Frederich *et al*. 2001a Frederich *et al*. 2000
Strychnopentamine (**56**)	*Strychnos usambarensis*	0.117 μM (FCA 20) not tested (FCB1-R) 0.145 μM (W2)	Frederich *et al*. 2004a,b
Chrysopentamine (**57**)	*Strychnos usambarensis*	0.579 μM (FCA 20) 0.550 μM (FCB1-R) 0.507 μM (W2)	Frederich *et al*. 2004a,b
Quinine (**1**)	Control	0.269 μM (FCA 20) 0.200 μM (FCB1-R) 0.413 μM (W2)	Frederich *et al*. 2004a,b

(*Contd.*)

(*Contd.*)

Chloroquine	Control	0.011 μM (FCA 20) 0.032 μM (FCB1-R) 0.284 μM (W2)	Frederich *et al.* 2004a,b
Strychnohexamine (**60**)	*Strychnos icaja*	1 μM (FCA 20)	Frederich *et al.* 2004a,b
Chloroquine	Control	0.02 μM (FCA 20)	Frederich *et al.* 2004a,b
Quinine (**1**)	Control	0.269 μM (FCA 20)	Frederich *et al.* 2004a,b
Strychnogucine C (**61**)	*Strychnos icaja*	16.05 μM (FCA 20)	Philippe *et al.* 2003
Strychnogucine A (**63**)	*Strychnos icaja*	2.310 μM (FCA 20) μM (W2)	Philippe *et al.* 2003
Isosungucine (**64**)	*Strychnos icaja*	1.315 μM (FCA 20) 0.265 μM (W2)	Philippe *et al.* 2003
18-Hydroxyisosungucine (**65**)	*Strychnos icaja*	0.847 μM (FCA 20) 0.140 μM (W2)	Philippe *et al.* 2003
2-Methoxy-13-methylrutaecarpine (**70**)	*Araliopsis tabouensis*	1.8 μg/mL (D6) > 4.7 μg/mL (W2)	Herrera *et al.* 2001b
5,8,13,14-Tetrahydro-2-methoxy-14-methyl-5-oxo-7H -indolo[2′,3′:3,4] pyrido[2,1-b]quinazolin -6-ium chloride (**70a**)	*Araliopsis tabouensis*	3.3 μg/mL (D6) > 4.7 μg/mL (W2)	Herrera *et al.* 2001b
Pleiocarpine (**71**)	*Pleiocarpa mutica*	>200 μM (K1) Active *in vivo*	Addae-Kyereme *et al.* 2001
Pleiocarpamine (**71a**)	*Pleiocarpa mutica*	17.6 μM (K1)	Addae-Kyereme *et al.* 2001
Pleiomutinine (**72**)	*Pleiocarpa mutica*	5.2 μM (K1)	Addae-Kyereme *et al.* 2001
Flavopereirine (**73**)	*Geissospermum sericeum*	11.53 μM (K1) 1.83 μM (T9-96)	Steele *et al.* 2002
Chloroquine	Control	0.003 μg/mL (D6) 0.038 μg/mL (W2) 0.52 μM (K1)	Steele *et al.* 2002
Decursivine (**74**)	*Rhaphidophora decursiva*	3.9 μg/mL (D6) 4.4 μg/mL (W2)	Zhang *et al.* 2002
Canthin-6-one (**75**)	*Eurycoma longifolia*	2.2 μg/mL (W2)	Kuo et al. 2003
1,3,5-Trihydroxy-2,8-bis (3-methylbut-2-enyl)-10-methyl-9-acridone (**80**)	*Swinglea glutinosa*	5.3 μM (Nigerian) 7.9 μM (FcM29)	Weniger *et al.* 2001
5-Hydroxynoracronycine (**81**)	*Swinglea glutinosa*	11.1 μM (Nigerian) 10.2 μM (FcM29)	Weniger *et al.* 2001
1,3,5-Trihydroxy-4-methoxy-2-(3-methylbut -2-enyl)-10-methyl-9-acridone (**82**)	*Swinglea glutinosa*	2.5 μM (Nigerian) 2.0 μM (FcM29)	Weniger *et al.* 2001

(*Contd.*)

(Contd.)

1,3,5-Trihydroxy-4-methoxy-10-methylacridone (**83**)	*Swinglea glutinosa*	33 μM (Nigerian) 27.1 μM (FcM29)	Weniger *et al.* 2001
Chloroquine	Control	0.07 μM (Nigerian) 0.45 μM (FcM29)	Weniger *et al.* 2001
Normelicopicine (**84**)	*Teclea trichocarpa*	8.25 μM (HB3) 14.7 μM (K1) *In vivo* 32.1%	Muriithi *et al.* 2002
Arborinine (**85**)	*Teclea trichocarpa*	3.85 μM (HB3) 9.34 μM (K1)	Muriithi *et al.* 2002
Skimmianine (**86**)	*Teclea trichocarpa*	47.5 μM (HB3) 59.0 μM (K1)	Muriithi *et al.* 2002
Melicopicine (**87**)	*Teclea trichocarpa*	>100 μM (HB3) >100 μM (K1)	Muriithi *et al.* 2002
Tecleanthine (**88**)	*Teclea trichocarpa*	23.2 μM (HB3) 53 μM (K1)	Muriithi *et al.* 2002
6-Methoxytecleanthine (**89**)	*Teclea trichocarpa*	32.3 μM (HB3) 56.9 μM (K1)	Muriithi *et al.* 2002
Chloroquine	Control	0.028 μM (HB3) 0.63 μM (K1)	Muriithi *et al.* 2002
(+)-Haemanthamine (**90**)	*Cyrtanthus elatus*	0.67 μg/mL (NF54) Chl 0.004 Art 0.002	Herrera *et al.* 2001a
(+)-Haemanthamine (**90**)	*Pancratium maritimum, Leucojum aestivum, Narcissus tazetta*	0.703 μg/mL (T9.96) 0.433 μg/mL (K1)	Sener *et al.* 2003
(+)-6α-Hydroxyhaemanthamine (**91**)	*Pancratium maritimum, Leucojum aestivum, Narcissus tazetta*	0.348 μg/mL (T9.96) 0.352 μg/mL (K1)	Sener *et al.* 2003
(+)-3-Epihydroxybulbispermine (**92**)	*Pancratium maritimum, Leucojum aestivum, Narcissus tazetta*	1.139 μg/mL (T9.96) 0.553 μg/mL (K1)	Sener *et al.* 2003
(–)-lycorine (**93**)	*Pancratium maritimum, Leucojum aestivum, Narcissus tazetta*	1.026 μg/mL (T9.96) 0.379 μg/mL (K1)	Sener *et al.* 2003
(–)-crinine (**94**)	*Pancratium maritimum, Leucojum aestivum, Narcissus tazetta*	2.11 μg/mL (T9.96) 1.65 μg/mL (K1)	Sener *et al.* 2003
(-)-galanthamine (**95**)	*Pancratium maritimum, Leucojum aestivum, Narcissus tazetta*	4.38 μg/mL (T9.96) 15.93 μg/mL (K1)	Sener *et al.* 2003
(+)-tazettine (**96**)	*Pancratium maritimum, Leucojum aestivum, Narcissus tazetta*	5.42 μg/mL (T9.96) 5.08 μg/mL (K1)	Sener *et al.* 2003

(Contd.)

(*Contd.*)

Mefloquin	Control	3.94 μg/mL (K1)	Sener *et al.* 2003
Chloroquin	Control	6.06 μg/mL (K1)	Sener *et al.* 2003
Galanthine (**97**)	*Zephyranthes citrina*	0.20 μg/mL Chl 0.04	Herrera *et al.* 2001a
Haemanthimine (**98**)	*Cyrtanthus elatus*	0.70 μg/mL (NF54)	Herrera *et al.* 2001a
Budmunchiamine L4 (**99**)	*Albizia adinocephala*	14 μM (plasmepsin II enzyme inhibition)	Ovenden *et al.* 2002
Budmunchiamine L5 (**100**)	*Albizia adinocephala*	15 μM (plasmepsin II enzyme inhibition)	Ovenden *et al.* 2002
Harman (**101**)	*Guiera senegalensis*	2.2 μg/mL (D6) 1.3 μg/mL (W2)	Ancolio *et al.* 2002
Tetrahydroharman (**102**)	*Guiera senegalensis*	3.9 μg/mL (D6) 1.4 μg/mL (W2)	Ancolio *et al.* 2002

Quinoline alkaloids: *quinolone, tetrahydroquinoline, benzoquinolizidine*

The quinoline class of alkaloids contributes to the major portion of antimalarial drugs for therapeutic use. Though the crude extract was used for hundreds of years, quinine **(1)** was not isolated until 1820 by Joseph Pelletier and Joseph Caventou, and the structure determination took another 100 years. The first written efficacy of *Cinchona* bark was recorded by Father Calancha in the *Chronicle of St. Augustine* in 1633. By 1884, 25 additional quinine alkaloids had been isolated and an additional 6 between 1884 and 1941. There are eight major *Cinchona* alkaloids which occur as four pairs of enantiomers, the (–)-isomers [quinine **(1)** series] have the 2*S*,3*R* configurations, and the (+)- isomers [quinidine **(1a)** series] have the 2*R*,3*S* configuration.

The blood schizontocidal activity, using drug-specific *in vitro* inhibition of schizont maturation assay, of the four main *Cinchona* alkaloids against *P. falciparum* was compared in 46 fresh parasite isolates collected in northwestern Thailand, an area where quinine **(1)** alone is no longer able to eliminate malarial infections. Quinidine **(1a)** showed the highest blood schizontocidal activity, followed by cinchonine **(2)**, cinchonidine **(2a)** and finally quinine (i.e. mean EC_{50} values 80, 104, 144, and 225 nM, respectively). The EC_{99} values for quinidine (861 nM) was within the therapeutic concentration range and those for quinine and cinchonidine (8040 and 6531 nM) were beyond it (Knauer *et al.* 2003).

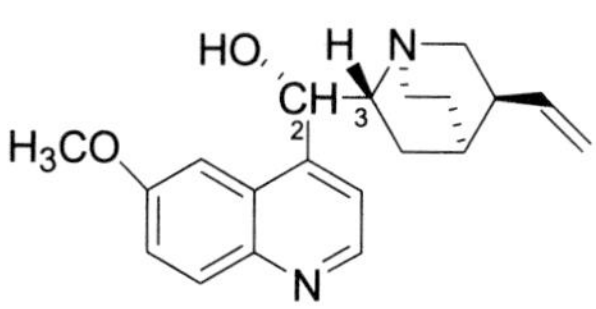

(-)-Quinine (2S,3R) (**1**)
(+)-Quinidine (2R,3S) (**1a**)

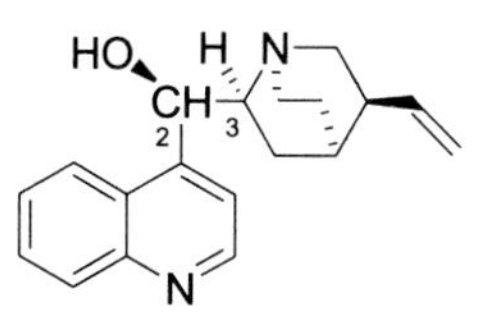

Cinchonine (2R,3S) (**2**)
Cinchonidine (2S,3R) (**2a**)

malarial activity and were found to be inactive (Vennerstrom and Klayman 1988). Interestingly, **19** showed activity *in vivo* against *P. chabaudi*, in contrast to its reported lack of activity against *P. berghei* in mice (McCall *et al.* 1994). Berberine **(19)** isolated from *Arcangelisia flava* inhibited telomerase activity in a dose-dependent manner over the range of 30-300 μM, indicating that *P. falciparum* telomerase might be a potential target for future malaria chemotherapy (Sriwilaijareon *et al.* 2002). Berberine was also identified as the active constituent of Vietnamese medicinal plant *Coscinium fenestratum*, and showed an IC_{50} value of 0.5 μg/mL against *P. falciparum* FCR3 strain (Tran *et al.* 2003). It is intriguing to note that **19** is the major constituent of the widely used dietary supplement containing Goldenseal (*Hydrastis canadensis*) (Abourashed and Khan 2001).

In a clinical study involving **19** (Sheng *et al.* 1997) 215 patients with CQ-resistant malaria were randomized into three groups. The first (berberine) group of 82 patients was given pyrimethamine+**19**, the second (tetracycline) group of 64 patients was given pyrimethamine+tetracycline and the third (cotrimoxazole) group consisting 69 patients was given pyrimethamine+cotrimoxazole. In the berberine group, the clearance rate of asexual parasitoid was 74.4% after treatment (*vs.* 67.2% and 47.8% for tetracycline and cotrimoxazole groups). Berberine **(19)** showed more effectiveness in clearing the parasite than either tetracycline or cotrimoxazole and the combination of pyrimethamine+berberine gave the best results for CQ resistant malaria.

A series of 12 aporphinoids were tested alone or in combination with CQ, against *P. falciparum in vitro*. N-methyl oliveridine was most potent against the FcM29 strain with IC_{50} value of 0.17 μg/mL. Nuciferine, pachyconphine and cassythicine potentiated the effect of CQ against CQ-resistant D6 strain. An additive effect was displayed by other aporphinoids (Fournet and Munoz 2002). Crostparine **(20)**(Fig. 6.1), isolated from the annonaceous plant *Uvaria klaineana* Engels and Diels, showed antiplasmodial effects on CQ-sensitive and resistant (K1 and FcB1) Thai strains of *P. falciparum* with IC_{50} values of 3.2, 2.1 and 3.4 μg/mL, respectively (Akendengue *et al.* 2002, 2003). Duguevalline, a novel aporphinoid alkaloid from *Duguetia vallicola* (Annonaceae), and four known alkaloids, namely cleistopholine **(21)**, O-methylmoschatoline, (-)-oliveroline **(22)** and (-)-oliveridine, were isolated. Among them, **21** and **22** exhibited antiplasmodial activity with IC_{50} values of 17.8 and 14.9 μg/mL against *P. falciparum* (Perez *et al.* 2004). Twenty-six pure plant-derived compounds including alkaloids, quinones, flavonoids, and coumarins, obtained from five species of Central American plants associated with traditional uses, were tested *in vitro* for antiprotozoal and cytotoxic activities (Camacho *et al.* 2004). Benzo[g]isoquinoline-5,10-dione **(23)**, isolated from *Cephaelis* *camponutans* (syn. *Psychotria camponutans*) by Solis *et al.* (1995), had an IC_{50} value of 4.02 μM against *P. falciparum* K1 strain as reported by Camacho *et al.* (2004).

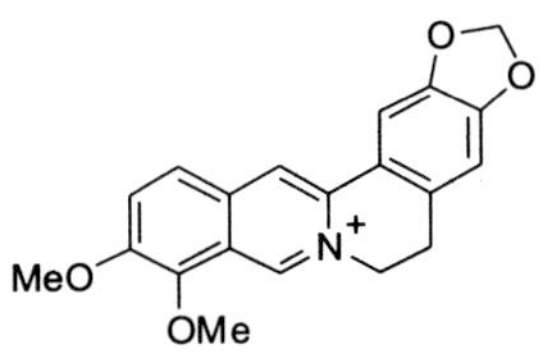
Berberine (**19**)

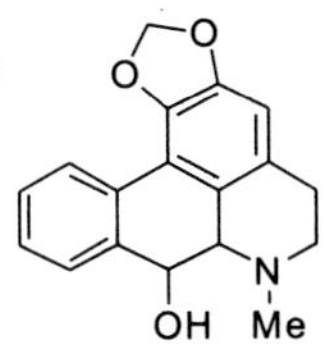
Cleistopholine (**21**)

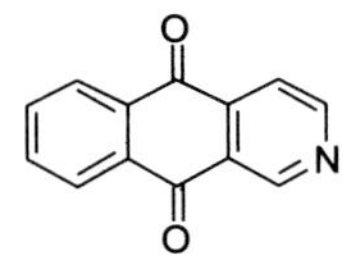
Oliveroline (**22**)

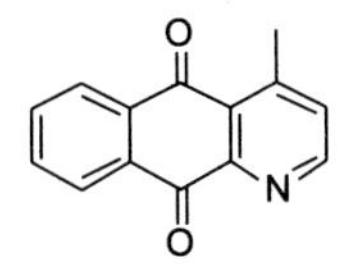
Benzo[g]isoquinoline-5,10-dione (**23**)

Bioassay-guided isolation of Peruvian *Duguetia hadrantha* (Annonaceae) yielded two 4,5-dioxo-1-azaaporphinoids, hadranthine A **(24)** and hadranthine B, and the known imbiline-1 **(25)**, sampangine **(26)** and 3-methoxysampangine **(27)**. Compounds **24**, **26** and **27** demonstrated potent *in vitro* antimalarial activity against *P. falciparum* D6 and W2 clone with IC_{50} values between 0.068 and 0.420 μg/mL, while **25** was less active. Sampangine also inhibited cell aggregation with a MIC value of <0.15 μg/mL, while 3-methoxysampangine was only weakly active (Muhammad *et al.* 2001). In addition, two new antimalarial bis-dehydroaporphines, bis-7,7′-dehydro-8,8′-dimethoxyannonaine **(28)** and bis-7,7′-dehydro-10,10′-dimethoxyannonaine **(29)**, were isolated recently from the root of another annonaceous plant *Polyalthia debilis* (Kanokmedhakul *et al.* 2003). **28** and **29** demonstrated moderate activity against multi-drug resistant *P. falciparum* K1 with IC_{50} of 5.4 and 4.1 μg/mL, respectively.

Bisbenzylisoquinolines (Bis-BIQ)

Bis-BIQ's are present in members of the Annonaceae, Berberidaceae, Lauraceae, Menispermeceae, Monimiaceae and Ranunculaceae plant families, and investigated widely for the discovery of antimalarial alkaloids. Twenty four bis-BIQ alkaloids were screened for antiplasmodial, antiamoebic and cytotoxic activities using *in vitro* microtests (Marshall *et al.* 1994). Eight of them had antiplasmodial activity with IC_{50} of <1 μM against a multidrug resistant strain of *P. falciparum* (*vs.* IC_{50} of CQ was 0.2 μM). None of the 24 bis-BIQ's exhibited significant cytotoxicity against the KB cell line, the most toxic being berbamine, with an IC_{50} of 17.8 μM (*vs.* IC_{50} of podophyllotoxin was 0.008 μM). Angerhofer *et al.* (1999) reported 53 bis-BIQ's for antiplasmodial activity against CQ-sensitive and -resistant clones of *P. falciparum*. More than half of the compounds isolated from *Cyclea barbata, C. atjehensis, Stephania pierrei, S. erecta, Pachygone dasycarpa, Hernandia peltata, Curare candicans, Albertisia*

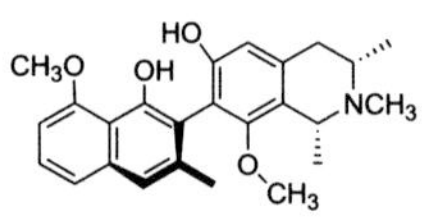

Ancistrotanzanine C (**33**)

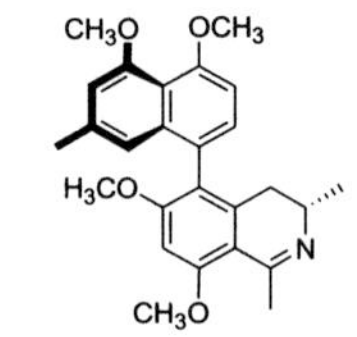

O-Methylancistrocladinine (**34**)

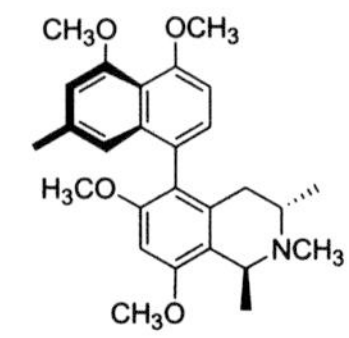

O,N-Dimethylancistrocladine (**35**)

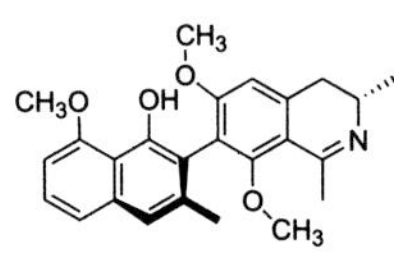

Ancistrotectorine (**36**)

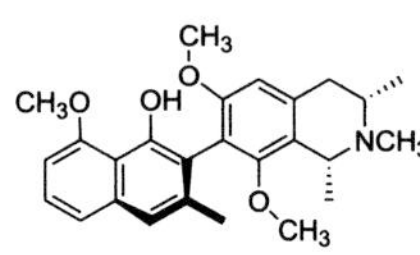

Ancistrocladidine (**37**)

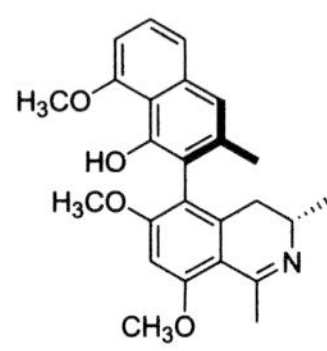

Ancistrotanzanines A (**38**)

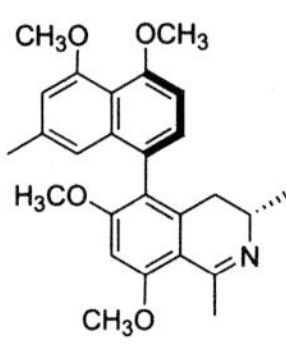

Ancistrotanzanines B (**39**)

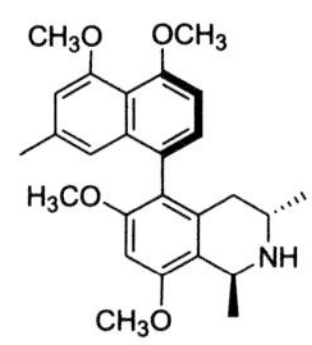

Ancistrotectoriline A (**40**)

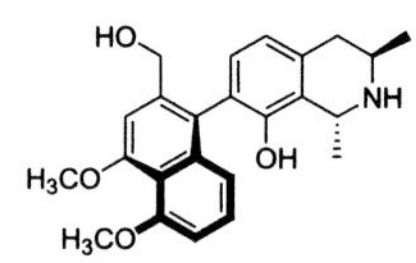

Habropetaline A (**41**)

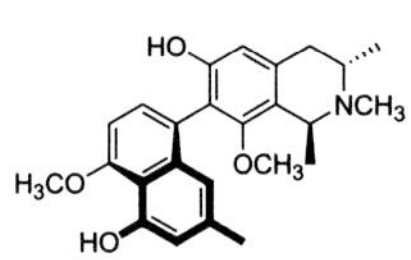

Ancistrogriffines A (**42**)

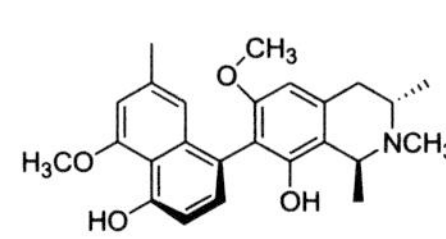

Ancistrogriffines C (**43**)

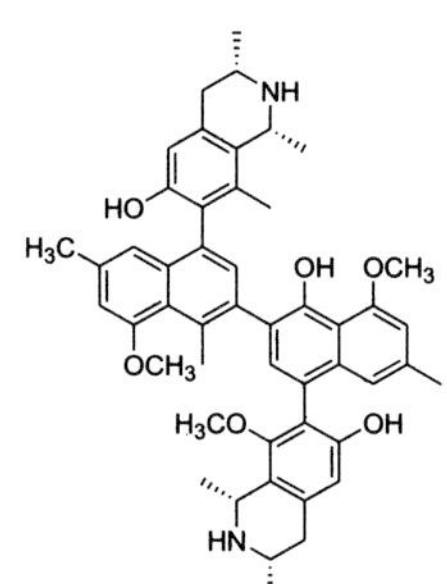

Ancistrogriffithine A (**44**)

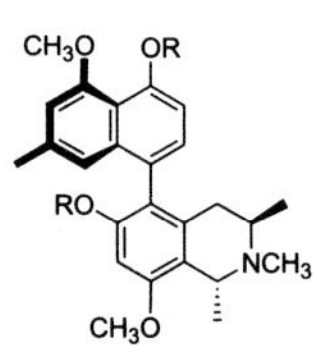

Ancistrocongolines A R=H (**45**)
Ancistrocongolines B R=CH3 (**46**)

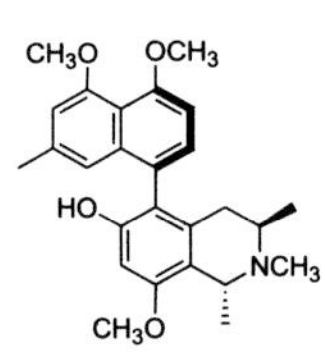

Ancistrocongolines C (**47**)

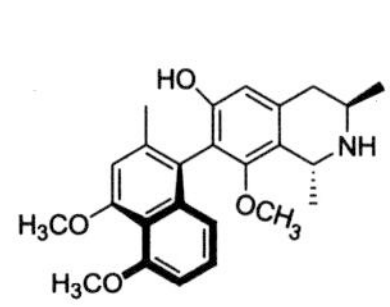

Ancistrocongolines D (**48**)

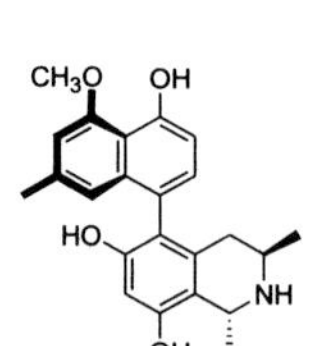

Korupensamine A (**49**)

Fig. 6.2 Naphthylisoquinoline alkaloids from plants.

is inactive against other protozoal pathogens *T. brucei* rhodesience, *T. cruzi* and *L. donovani*.

Three new monomeric NIQ's, ancistrogriffines A **(42)**, B and C **(43)**, and the first dimer of a 7,8′-coupled NIQ, ancistrogriffithine A **(44)**, have been detected by online screening of *A. griffithii*. Ancistrogriffithine A **(44)**, **42** and **43**, and the known ancistrocladine and hamatine were structurally assigned using online methods without previous isolation. Both **44** and **42** are active against *P. falciparum* (IC_{50} 0.035 and 0.074 μg/mL against K1 strain, respectively). Furthermore, the latter compound showed good activity against *L. donovani* (Bringmann *et al.* 2002c). Four new NIQ's, ancistrocongolines A-D **(45-48)** were isolated from *A. congolensis*, along with the known compound korupensamine A **(49)**. Their biological activities against the tropical diseases malaria, Leishmaniasis, Chagas disease, and African sleeping sickness were evaluated (Bringmann *et al.* 2002a). Among these, **49** showed good activity against *P. falciparum* K1 strain (IC_{50} 0.164 μg/mL).

Dioncophylline E **(50)**, isolated from *D. thollonii*, displayed an unusual 7,3′-linkage between the naphthalene and the isoquinoline portions and was the first alkaloid that occurs as a mixture of two configurationally semi-stable atropo-diastereomers. Potent antimalarial activity was exhibited by **50** against both NF54 and K1 strains of *P. falciparum* (IC_{50} 0.022 and 0.021 μg/mL, respectively), while it had weak antitrypanosomal activity against *T. cruzi* and moderate activity against *T. brucei* rhodesiense. Furthermore, four additional NIQ's, previously known from the related plant species *T. peltatum*, have been identified in *D. thollonii* (Bringmann *et al.* 2002b). The total synthesis of dioncophylline B **(51)** was reported by Bringmann *et al.* (2001).

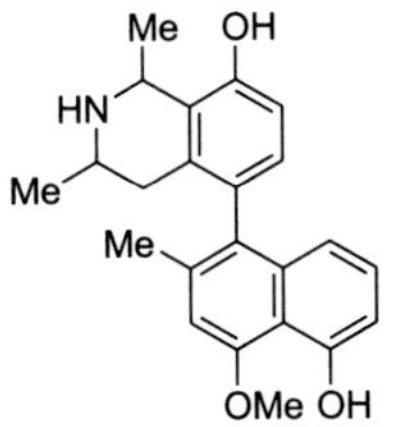

Dioncophylline C **(32)**

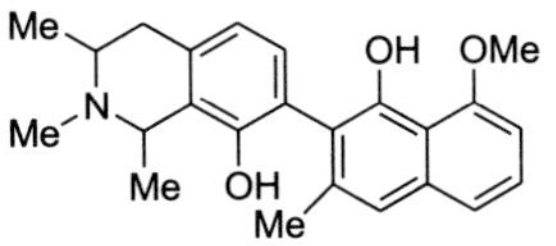

Dioncophylline E **(50)**

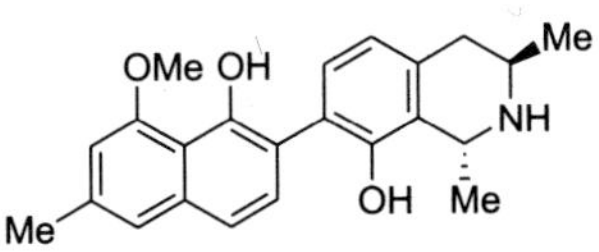

Dioncophylline B **(51)**

Three new NIQ's, ancistrolikokines A-C, as well as the known korupensamine A **(49)** have been isolated and structurally assigned from *A. likoko*. Their 5,8′-coupling hints at a close biogenetic relationship of *A. likoko* to other Central African *Ancistrocladus* species. These compounds showed strong to moderate antimalarial activities when evaluated *in vitro* against the NF54 and K1 strains of *P. falciparum* (Bringmann *et al.* 2000).

Indole alkaloids: *indolomonoterpene* (IMT), *bisindole* (*Strychnos*), *indoloquinolines, indoloquinazolidines, β-carbolines, other indoles*

The antiplasomodial and cytotoxic activities of IMT alkaloids isolated from African *Strychnos* species (Loganiaceae) were recently reviewed by Frederich *et al.* (2002, 2003). Several indole alkaloids were tested against three *P. falciparum* strains (CQ -resistant W2 and FCB-1, and CQ -sensitive FCA). A few of the compounds demonstrated activities under the micromolar range: isostrychnopentamine **(52)** (Fig 6.3) from *S. usambarensis*, has an IC_{50} near 0.1 μM against all strains, ochrolifuanine A **(53)**, from *S. potatorum*, has an IC_{50} between 0.1 and 0.5 μM, and strychnogucine B **(54)**, from *S. icaja*, has an IC_{50} between 0.1 and 0.6 μM. Concerning SAR, some level of lipophilia and at least two basic amino centers as well as certain steric cluttering (for bis- or tris- indolic compounds) are necessary for antiplasmodial activity. The possible strychnine-like toxicity of sungucine **(55)** derivatives was investigated because such activity could be a limiting factor for their potential use as antimalarial drugs. Fortunately, these bis-indoles were inactive on glycine

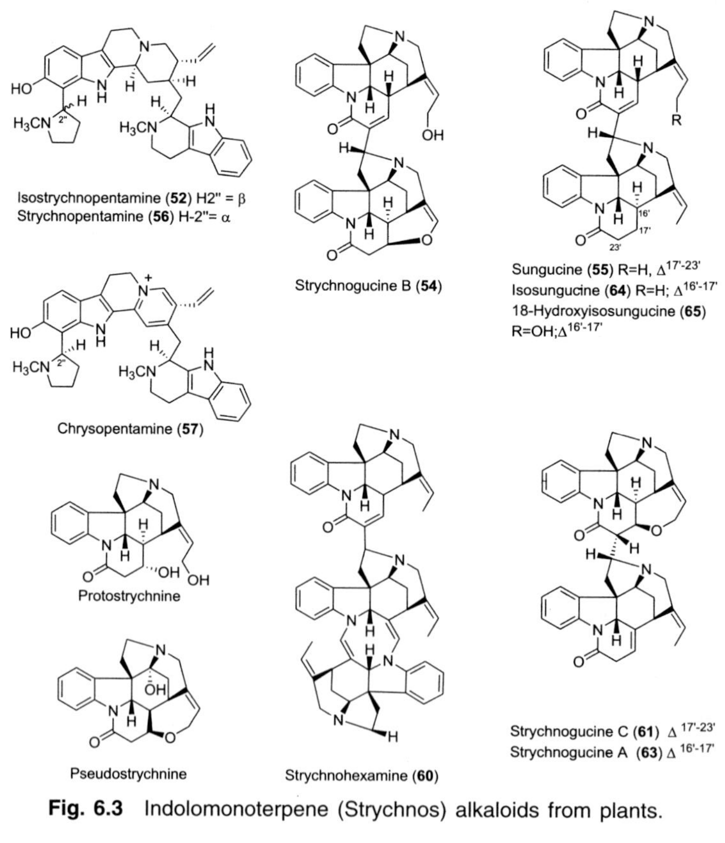

Fig. 6.3 Indolomonoterpene (Strychnos) alkaloids from plants.

receptors (Frederich *et al.* 2003). The stage-dependent susceptibility of *Plasmodium* to a short exposure of **52** was investigated, where the ring stage was shown to be the most sensitive one, but all stages were affected by treatment. By fluorescence microscopy, it was shown that **52** did not accumulate inside the food vacuole of the parasite. The ED_{50} value of **52** was found to be 30 mg/kg/day by the IP route (after 4 days treatment) against *P. berghei* NK173 and *P. vinckei* petteri murine strains (Frederich *et al.* 2004a, b).

A new derivative of strychnopentamine **(56)**, named chrysopentamine **(57)**, was isolated from the leaves of *S. usambarensis*. It presented a OH-substitution at C-14 and an aromatization of the ring D of **56** leading to anhydronium base properties and exhibited strong antiplasmodial properties (IC_{50} >1 μM) (Frederich *et al.* 2004a). Alkaloids derived from strychnobrasiline **(58)** were prepared for therapeutic uses as antitumor and antimalarial agents. Thus, strychnobrasiline derivative **59** was prepared *via* a multistep synthetic sequence began with a concomitant ring opening/lactol formation reaction of **58** with LiAlH4 in THF. The alkaloid **59** in combination with CQ was tested *in vitro* for antimalarial activity against *P. falciparum* (Trigalo *et al.* 2003).

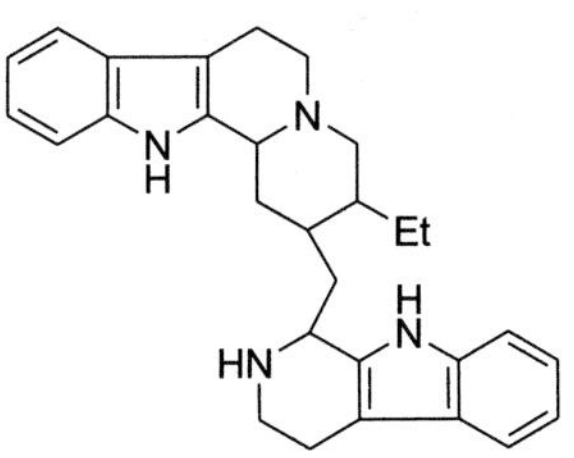

Ochrolifuanine A (**53**)

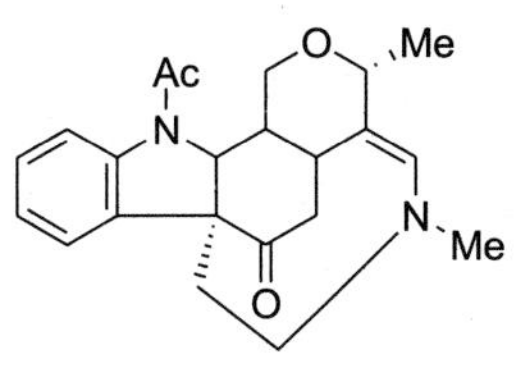

Strychnobrasiline (**58**)

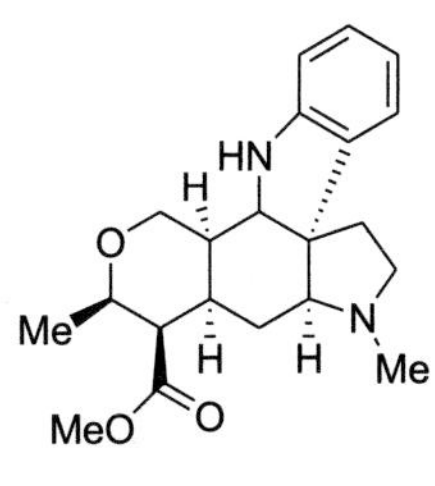

59

Philippe *et al.* (2003) have isolated the first naturally occurring trimeric IMT alkaloid strychnohexamine **(60)** and the other IMT's protostrychnine, genostrychnine and pseudostrychnine, as well as a new bisindolic alkaloid, named strychnogucine C **(61)** from *S. icaja*. The trimeric alkaloid **60** demonstrated an antiplasmodial IC_{50} against the *P. falciparum* FCA line of 1 μM.

The *in vitro* antiplasmodial activities of **69** alkaloids from various *Strychnos* species were evaluated against CQ-resistant and -sensitive lines of *P. falciparum* (Frederich *et al.* 2002). The most active alkaloids were also tested for cytotoxicity against HCT-116 colon cancer cells to determine their antiplasmodial selectivity. Alkaloids presenting four types of bisindole skeleton exhibited potent and selective activities against *Plasmodium*. They were sungucine-type (IC_{50} values 80 nM-10 μM), longicaudatine-type (IC_{50} values 0.5-10 μM), matopensine-type (IC_{50} values 150 nM - 10 μM), and usambarine-type alkaloids. Within the last structural type, dihydrousambarensine exhibited a 30-fold higher activity against the CQ-resistant strain (IC_{50} value at 32 nM) than against the CQ-sensitive strain.

Eight naturally occurring monoindoles were evaluated *in vitro* for their ability to inhibit *P. falciparum* growth and, in drug combination, to reverse the resistance of a CQ-resistant strain of *P. falciparum*. None of these indoles has significant intrinsic antiplasmodial activity (IC_{50} >10 μM or 5 μg/mL). However, three alkaloids [icajine **(62)**, isoretuline and strychnobrasiline **(58)**] did reverse CQ resistance at concentration between 2.5 and 25 μg/mL (IF of 12.82 for isoretuline on W2 strain). The Interaction Factor (IF) equals 2, <2, or >2 for additive, antagonistic or synergistic effects of alkaloids on CQ inhibition, respectively. Icajine and isoretuline were also assessed *in vitro* for their mefloquine-potentiating activity on a mefloquine-resistant strain of *P. falciparum*. Only icajine **(62)** proved to be synergistic with mefloquine (IF = 15.38) (Frederich *et al.* 2001b).

A reinvestigation of *S. icaja* roots has yielded two tertiary quasi-symmetrical bisindoles strychnogucines A **(63)** and B **(54)**. Compound **54** was highly active *in vitro* and **63** moderately active against four strains of *P. falciparum*. Strychnogucine B **(54)** was more active against a CQ-resistant strain than against a CQ-sensitive one (IC_{50} value 80 nM against W2 strain). In addition, **54** showed a selective antiplasmodial activity with 25-180 times greater activity toward *P. falciparum*, relative to cultured human cancer cells (KB) or human fibroblasts (WI38) (Frederich *et al.* 2001a). An earlier reinvestigation of *S. icaja* Baillon by the same group (Frederich *et al.* 2000) reported vomicine, isostrychnine and of three new sungucine analogs, isosungucine **(64)**, 18-hydroxysungucine and 18-hydroxyisosungucine **(65)**. This last compound showed a selective antiplasmodial activity, with > 100-fold greater toxicity towards *P. falciparum*, relative to KB, HeLa or WI38 lines (Frederich *et al.* 2000).

Malagashanine (**66**) was isolated from the Malagasy *Strychnos* species traditionally used as CQ adjuvants in the treatment of chronic malaria. It demonstrated weak *in vitro* antiplasmodial activity (IC_{50} = 146.50±02 μM), but displayed marked *in vitro* CQ-potentiating action against CQ-resistant strains FcM29, W2 and FcB1 of *P. falciparum*, but had no action against the susceptible strain 3D7 and two sensitive isolates. The selective enhancing action of **66** on quinolines (chloroquine, quinine and mefloquine), aminoacridines (quinacrine and pyronaridine) and halofantrine was observed, all of which are believed to exert their effect by binding with hematin and specifically act with CQ on the old trophozoite stage of the *P. falciparum* life cycle (Rafatro *et al.* 2000a). In a related study, two new metabolites of **66** were isolated from rat urine and human liver microsomes (Rafatro *et al.* 2000b).

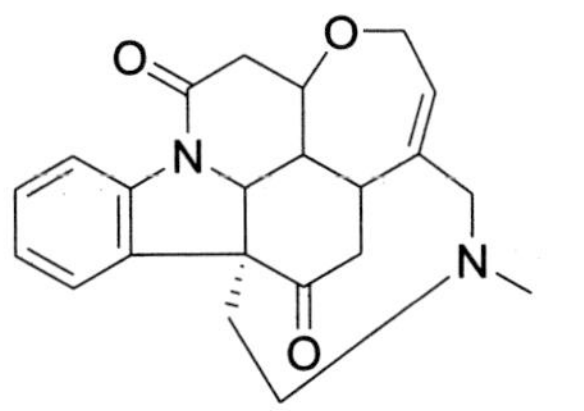

Icajine (**62**)

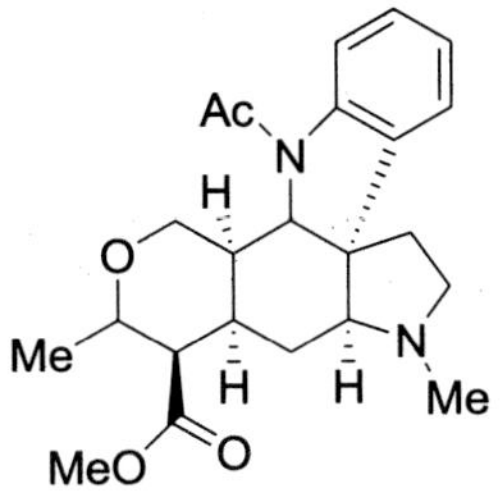

Malagashanine (**66**)

Indoloquinolines (IQ) and Other Indoles

The roots of *Cryptolepis sanguinolenta* (Periplocaceae) have been investigated for their chemical compounds since 1931. The *in vitro* and *in vivo* antiplasmodial activity of cryptolepine (**67**) and neocryptolepine (**68**) and related IQ alkaloids from *C. sanguinolenta* were studied by various research groups (Cimanga *et al.* 1997, Paulo *et al.* 2000). Three different extracts and four alkaloids from the root bark have been assessed *in vitro* against *P. falciparum* K-1, D-6 and W-2 strains. **67** and its hydrochloride, 11-hydroxycryptolepine, and **68** showed strong activity against *P. falciparum* CQ-resistant and sensitive strains (IC_{50} value between 0.027 and 0.041 μg/mL.). *In vivo* tests on infected mice showed that **67**-hydrochloride exhibited a significant chemosuppressive effect against *P. berghei yoelii* and *P. berghei*, while **67** had the same effect against *P. berghei yoelii* only. These compounds were regarded as potential lead structures for the development of antimalarials (Cimanga *et al.* 1997).

Cryptolepine (**67**) is a DNA intercalator with cytotoxic properties. The antiplasmodial mechanism of **67** is likely to be due, at least in part, to a CQ-like action that does not depend on intercalation into DNA. A number of cryptolepine analogs displayed potent activities against both CQ-sensitive and -resistant strains of *P. falciparum*, and also have in common with CQ the inhibition of *β*-hematin formation in cell-free systems. Several compounds also displayed activity against *P. berghei* in mice, the most potent being 2,7-dibromocryptolepine, which suppressed parasitemia by 89% at a dose of 12.5 mg/kg/day/ip (Wright *et al.* 2001). Cryptolepinoic acid, its methyl ester, as well as **67**, hydroxycryptolepine and quindoline were isolated from *C. sanguinolenta* which is used in traditional medicine in Guinea-Bissau (Paulo *et al.* 2000). Aqueous and EtOH extracts of the leaves and roots and 7 alkaloids isolated from those extracts were tested *in vitro* against *P. falciparum* K1 and T996 (CQ-sensitive clone). **67** was again the most active alkaloid tested against *P. falciparum* [IC_{50} values

0.23 μM (K1) and 0.059 μM (T996) *vs.* IC_{50} valutes of 0.26 μM (K1) and 0.019 μM (T996) for CQ], followed by the indolobenzazepine alkaloid cryptoheptine (**69**) with an IC_{50} of 0.8 μM (K1) and 1.2 μM (T996). Cryptolepinoic acid showed no activity while its ethyl ester derivative was active against *P. falciparum* K1 (IC_{50}=3.7 μM). All the IQ's showed cross-resistance with CQ but not the indolobenzazepine alkaloid **69**.

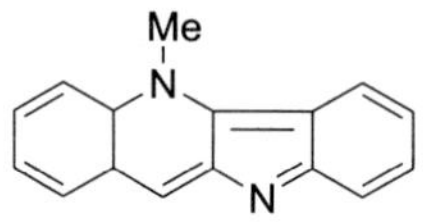

Cryptolepine (**67**)

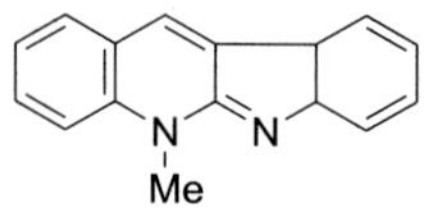

Neocryptolepine(**68**)

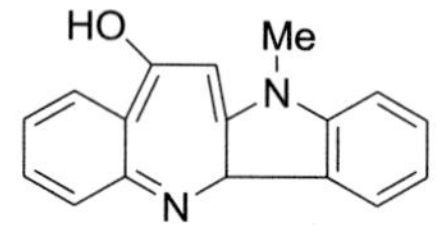

Cryptoheptine (**69**)

Synthesis, cytotoxicity and antimalarial activity of new neocryptolepine (**68**) derivatives have been recently reported (Jonckers *et al.* 2002). According to this study some derivatives of **68** showed higher activity than **67** and its bromo analog, 2-bromocryptolepine, which also showed more pronounced cytotoxicity due to higher affinity for DNA. The 2-bromoneocryptolepine is the most promising lead compound as it showed the most selective activity at 4 μM against CQ-resistant *P. falciparum* in the absence of cytotoxicity (>32 μM).

Indoloquinazolidin and carbolines

Three new indoloquinazolidine-type alkaloids, 8,13-dihydro-2-methoxyindolo[2′,3′:3,4]pyrido[2,1-b]quinazolin-5(7H)-one, 2-methoxy-13-methyl-rutaecarpine (**70**) and 5,8,13,14-tetrahydro-2-methoxy-14-methyl-5-oxo-7H-indolo[2′,3′:3,4]pyrido[2,1-b]quinazolin-6-ium chloride (**70a**), (Fig. 6.4) were isolated from *Araliopsis tabouensis* (Rutaceae), together with three other alkaloids including **11** and **12**. The IC_{50} value of compound **70** and **70a** were between 1.8 and >4.7 μg/mL against *P. falciparum* D6 and W2 clones (Christopher *et al.* 2003). Extracts from the Annonceous plants *Pleiocarpa mutica*, *Cleistopholis patens*, and *Uvaria chamae* were found to have significant antiplasmodial activity (Addae-Kyereme *et al.* 2001). Five alkaloids, pleiocarpine (**71**), kopsinine, pleiocarpamine, eburnamine, and pleiomutinine (**72**) were isolated from the roots of *P. mutica*. **72** was most active against *P. falciparum* with IC_{50} value of 5 μM, while **71** was inactive against malaria parasites *in vitro*, but demonstrated moderate activity against *P. berghei* in mice (25 mg/kg/daily for 4 days reduced parasitemia by 28.5%). The indole alkaloid geissoschizoline and two new derivatives, geissoschizoline N4-oxide and 1,2-dehydrogeissoschizoline, were obtained from the bark of *Geissospermum sericeum* together with the

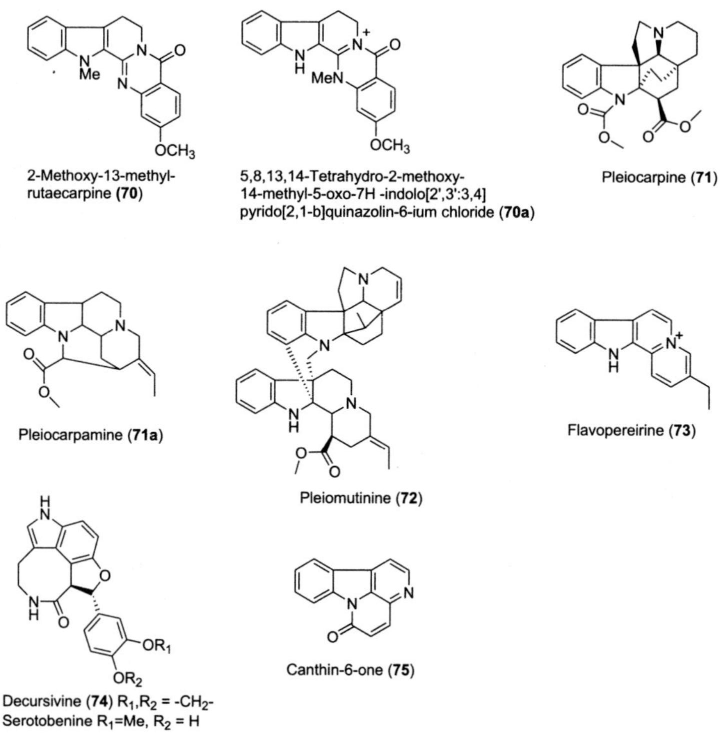

Fig. 6.4 Indole derived alkaloids from plants.

β-carboline alkaloid flavopereirine (**73**). The *in vitro* IC_{50} value of **73** was found to be 11.53 and 1.83 μM against CQ-resistant (K1) and -sensitive (T9-96) *P. falciparum*, respectively. Their cytotoxicity was determined in a human (KB) cell line (Steele *et al.* 2002).

Antimalarial bioassay-directed fractionation led to the isolation of an active indole alkaloid decursivine (**74**) from the leaves and stems of *Rhaphidophora decursiva* Schott (Araceae). Its antimalarial activity was observed with IC_{50} values of 3.93 and 4.41 μg/mL against the D6 and W2 clones of *P. falciparum*, respectively (Zhang *et al.* 2002). Three new and 19 known *β*-carboline alkaloids were isolated from the roots of *Eurycoma longifolia*. Canthin-6-one (**75**) and 9-hydroxycanthin-6-one demonstrated weak antimalarial activity against the *P. falciparum* W2 clone with IC_{50} values 2.2 and 2.3 μg/mL, respectively, and also showed cytotoxicity against human lung (A-549) and breast (MCF-7) cancer cell lines (Kuo *et al.* 2003).

Quinazoline alkaloids: *febrifugine, isofebrifugine*

Since 1942 six alkaloids have been isolated from the Chinese medicinal plant Chang Shan (*Dichroa febrifuga*), of which three of them, namely *α*-, *β*- and *γ*-dichroine, are isomeric quinazoline derivatives (Fu and Jung

1948). These highly toxic molecules continue to attract considerable attention due to their potent antimalarial activity, intricate structural problems and relative instability. The two alkaloids, febrifugine and isofebrifugine, isolated by Koepfli *et al.* (1947, 1950) from the same source are identical with β- and α-dichroine, respectively. The total synthesis of febrifugine and isofebrifugine was first reported in 1952 (Baker *et al.* 1952), but it was until 1999 that the correct stereochemistry of febrifugine (**76**) and isofebrifugine (**77**) (Fig. 6.5) was established by asymmetric total synthesis (Kobayashi *et al.* 1999, Katoh *at al.* 2004). The elegant work of Kobayashi *et al.* (1999) necessitated that the old steriochemical assignments of these alkaloids (**78** and **79**) had to be abandoned, and both the *trans* relative and (2R, 3S)-absolute stereochemistry of (+)-febrifugine (**76**) has recently been proven beyond doubt (Kobayashi *et al.* 1999). The antimalarial activity of all enantiomers against *P. falciparum* was determined (Table 6.3) and it is very fascinating to note that febrifugine (**76**) and isofebrifugine (**77**) were found to be >21-26 -fold more potent than their corresponding enantiomers **78** and **79**, respectively. Therefore, they are perhaps the most interesting quinazoline alkaloids of all, but unfortunately toxicity in humans, such as the strong emetic effect, has precluded their clinical use against malaria. A number of interesting analogs have been prepared (Kikuchi *et al.* 2002) and the SAR demonstrated that the 4-quinazolinone ring is essential for the activity of febrifugine.

Table 6.3 Antimalarial activity of **76** and **79**, and their enantiomers **77** and **78** (Kobayashi *et al.* 1999)

Compound	EC_{50} *(P. falciparum)*	EC_{50}*(FM3A cell)*	*Selectivity*
76	$(7.6 \pm 0.4) \times 10^{-11}$	$(2.1 \pm 0.1) \times 10^{-7}$	2763
78	$(2.0 \pm 0.3) \times 10^{-7}$	$(2.1 \pm 0.1) \times 10^{-5}$	105
77	$(2.9 \pm 0.3) \times 10^{-10}$	$(7.3 \pm 1.4) \times 10^{-7}$	2517
79	$(1.6 \pm 0.2) \times 10^{-7}$	$(1.9 \pm 0.1) \times 10^{-5}$	119

Recently, alkaloids **76** and **77** have been confirmed as constituents of *Hydrangea chinensis* (Saxifragaceae) (Khalil *et al.* 2003) and *H. macrophylla* Seringe var. *Otaksa* (Ishih *et al.* 2003). The mixture of **76** and **77** demonstrated quite different responses and survival rates in mice infected with *P. yoelii* XL17, *P. berghei* MK65 and *P. chabaudi* AS, the animals generally succumbing to *P. berghei* infection as parasite levels increased after an initial check in growth. The hot water extract of *H. macrophylla* Seringe var. *Otaksa* leaves have demonstrated the survival of mice infected with *P. falciparum* (Ishih *et al.* 2003a). The combination effect of CQ with a mixture of **76** and **77** were evaluated against CQ resistant *P. berghei* NK65 in ICR mice. Malaria parasite given in the mice treated with CQ and alkaloids (**76**+**77**) completely decreased on day 6 and diminished thereafter (Ishih *et al.* 2003).

Acridone Alkaloids

A new prenylated acridone alkaloid, 1,3,5-trihydroxy-2,8-bis(3-methylbut-2-enyl)-10-methyl-9-acridone (**80**) (Fig. 6.5), was isolated from the stem bark of *Swinglea glutinosa* (Rutaceae) along with three known acridones, 5-hydroxynoracronycine (**81**), 1,3,5-trihydroxy-4-methoxy-2-(3-methylbut-2-enyl)-10-methyl-9-acridone (**82**), and 1,3,5-trihydroxy-4-methoxy-10-methylacridone (**83**). Alkaloids **80** and **82** demonstrated IC_{50} values between 2.0 and 7.9 μM against CQ-sensitive and -resistant *P. falciparum* strains (Weniger *et al.* 2001). Seven alkaloids have been isolated from *Teclea trichocarpa* (Rutaceae) including four, normelicopicine (**84**),

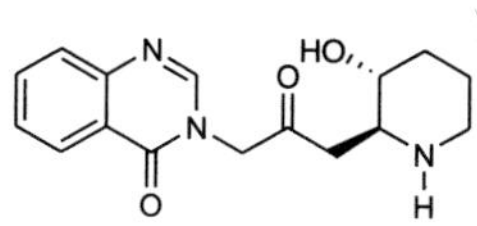

Febrifugine (**76**)

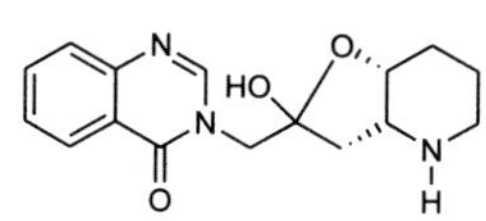

Isofebrifugine (**77**)

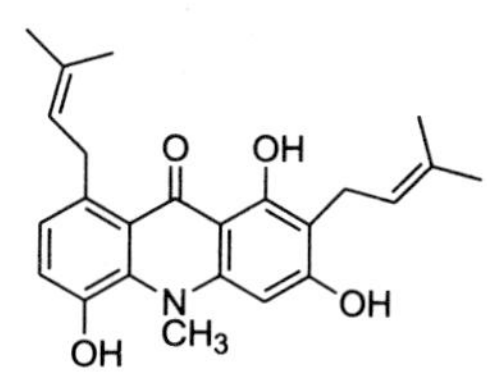

1,3,5-Trihydroxy-2,8-bis (3-methylbut-2-enyl)-10-methyl-9-acridone (**80**)

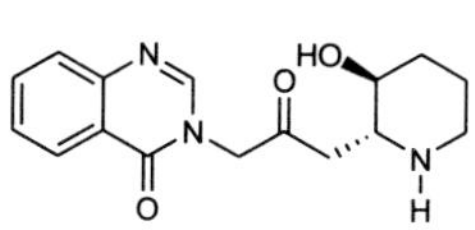

78

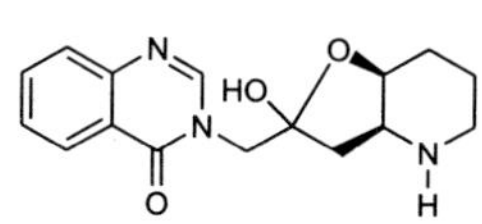

79

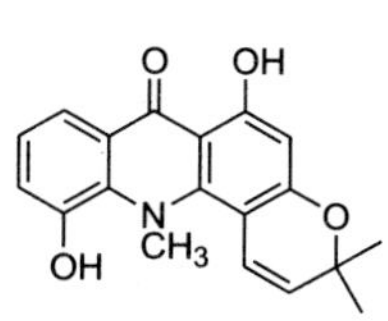

5-Hydroxynoracronycine (**81**)

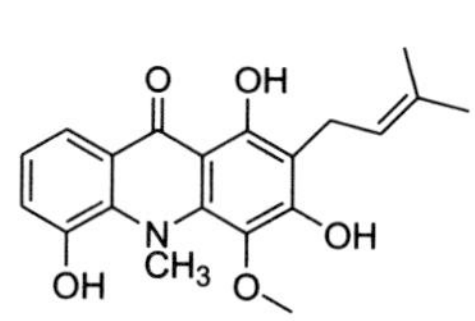

1,3,5-Trihydroxy-4-methoxy-2-(3-methylbut-2-enyl)-10-methyl-9-acridone (**82**)

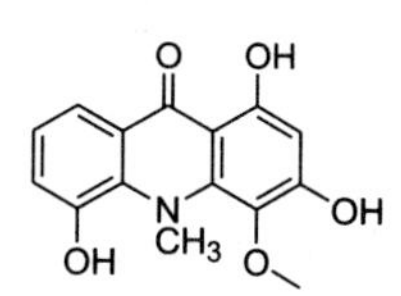

1,3,5-Trihydroxy-4-methoxy-10-methylacridone (**83**)

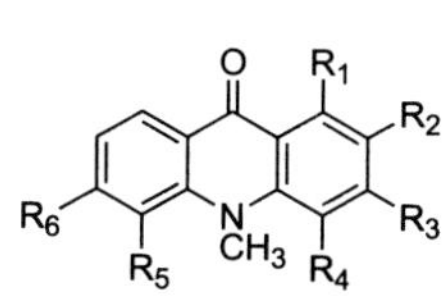

Normelicopicine (**84**) R_1=OH; R_2,R_3,R_4=OCH_3;R_5,R_6=H
Arborinine (**85**) R_1=OH;R_2,R_3=OCH_3;R_4,R_5,R_6=H
Melicopicine (**87**) R_1,R_2,R_3,R_4=OCH_3;R_5,R_6=H
Tecleanthine (**88**) R_1=OCH_3, R_2,R_3= -OCH_2O-;R_4,R_6=H; R_5=OCH_3
6-Methoxytecleanthine (**89**) R_1=OCH_3; R_2,R_3= -OCH_2O--; R_4=H;R_5,R_6=OCH_3

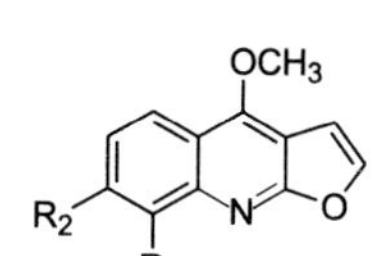

Skimmianine (**86**) R_1,R_2=OCH_3
Dictamnine R_1,R_2=H

Fig. 6.5 Quinazoline and acridone alkaloids from plants.

inhibition of total protein synthesis (Ancolio *et al.* 2002). The potent *in vitro* synergistic antimalarial interaction between **101** or **102**, and extracts from alkaloidal plants used for treating malaria was reported (Azas *et al.* 2002). Three combinations demonstrate a strong, synergistic, inhibitory effect on *in vitro* plasmodial development and are devoid of cytotoxicity towards human cells, validating their use in association in traditional medicine.

The Role of Alkaloids in Combination Therapy

The increase in resistance of *P. falciparum* to standard treatment and discovery of multidrug resistant strains has created a challenge to antimalarial chemotherapy. Alkaloids are valuable constituents of the combination therapy currently under development in several countries, especially China and Viet Nam. There is a need for antimalarial combination treatment containing artemisinin derivatives in order to preserve their significant therapeutic advantage and delay the emergence of resistance to artemisinins (WHO report, 2001). Currently in China treatment that uses the alkaloid derived drugs piperaquine, pyronaridine, amodiaquine and mefloquine in the combination therapy with artemisinin derivatives are in different phases of clinical trials. However, some combinations contain partner drugs with proven effectiveness, the long-term objective of delaying resistance is to deploy combinations where neither partner drug is available as a monotherapy.

The search for alkaloids known as resistant modifiers or chemosensitizers to overcome drug resistance in *Plasmodium* species has become a new strategy in discovery programs. Alkaloids were found to be the most interesting source of resistant modifiers. Two *Strychnose* alkaloids strychnobrasiline (**58**) and malagashanine (**67**) demonstrated CQ potentiating activity (Rasoanaivo *et al.* 1994). The monoindole alkaloid icajine (**63**), at a concentration non-toxic to human cells, reverses resistance to mefloquine in the F32 mefloquine-resistant strain of *P. falciparum in vitro*, and chloroquine in the W2 and PFB chloroquine-resistant strains. On the other hand, its analog isoretuline exhibited a more peculiar profile, it is at the same time antagonistic (low concentrations) and synergistic (high concentrations) when used in combination (Frederich *et al.* 2001b). Very few compounds, such as penfluridol, have demonstrated the ability to reverse resistance to both chloroquine and mefloquine.

Antifungal alkaloids

A book chapter entitled "Antifungal alkaloids" by Clark and Hufford (1992) in The Alkaloid series has reviewed antifungal alkaloids from higher plants, marine and microorganisms. In a later review

"Antimicrobially active alkaloids" by Verpoorte (1998), about 300 alkaloids from higher plants with antibacterial and antifungal activity have been described. As a continuation of reviewing antifungal alkaloids from plants, we have attempted to encompass the literature that has not been covered by the reviews mentioned above and some recent articles (Kuc 1992, Wink 1998, Cowan 1999). It must be pointed out that the results of *in vitro* antifungal susceptibility testing, e.g., minimum inhibitory concentration (MIC) values, are markedly influenced by the methods used and hence antifungal potencies in different papers are difficult to compare (Clark and Hufford 1992, Okunade *et al.* 1994). In addition, some reports do not even include a positive control for comparison of antifungal activity. Compounds that are claimed to be antifungal actually may be only marginally active. Therefore, in this chapter we mostly deal with the alkaloids from plants with significant antifungal activity when compared with control drugs.

Antifungal Alkaloids Against Human Related Pathogens

Although about 300,000 fungal species exist in nature, only 200 or so species may be well adapted to a parasitic mode of existence, and a few of them infect man (Richardson 1991). These pathogens can be classified into truly pathogenic fungi, opportunistic fungi, and dematophyte fungi. Interestingly, most individuals affected by the truly pathogenic fungi have mild signs or subclinical manifestations. In contrast, opportunistic fungal infections almost always produce significant diseases and are therefore the major interest in current antifungal drug discovery.

Opportunistic pathogens, such as *Candida* species, *Cryptococcus neoformans* and *Aspergillus fumigatus*, cause systemic fungal infections in immunocompromised hosts (AIDS, cancer, and organ transplant patients). Over the past few decades, the incidence of these kinds of infections has increased rapidly as the population of immunocompromised patients expands and as the patterns of mycoses evolve (White *et al.* 1998, Wong-Beringer and Kriengkauykiat 2003). Moreover, the mortality of opportunistic infections usually is very high. For example, 38-50% mortality was reported in disseminated candidiasis (Kontoyiannis *et al.* 2003).

Our current antifungal armamentarium includes five major drug classes: polyenes (interacting with fungal ergosterol, thereby disrupting the cytoplasmic membrane), azoles (inhibiting 14α-sterol demethylase in ergosterol biosynthesis pathway), allylamines (inhibiting squalene epoxidase in ergosterol synthesis pathway), echinocandins (inhibiting 1,3-glucan synthase which is required for fungal cell wall biosynthesis) flucytosine (incorporated into RNA, and thus inhibiting DNA synthesis) (Georgopapadakou and Walsh 1996, White *et al.* 1998, Kontoyiannis *et al.*

2003). Representatives from the five classes are amphotericin B, fluconazole, terbinafine, caspofungin, and flucytosine, respectively. For example, amphotericin B remains, after 40 years of use, the first choice in many cases for the immediate control of life-threatening disseminated mycoses due to its strong, broad-spectrum antifungal potency, although the newly introduced caspofungin and voriconazole (approved in 2002) have shown comparable therapeutic effects with lower toxicity (Wong-Beringer and Kriengkauykiat 2003, Keating and Figgitt 2003). Fluconazole (Diflucan, Pfizer) has dominated the market for systemic antifungal drugs in recent years (Brookman *et al.* 2003). However, the existing antifungal drugs each suffer one or more major shortcomings (e.g., significant dose-limiting toxicities for amphotericin B, fungistatic mechanism of action and development of resistance for the azole drugs, limitation of its use to dermatophytoses for terbinafine, and lack of activity in cryptococcosis for caspofungin). Therefore, the need for continued efforts for the discovery of novel antifungals, preferably with novel mechanisms of action, is clear, particularly in light of the emerging resistance and increasing occurrence of inherently resistant species as human pathogens (White *et al.* 1998).

The steroidal alkaloid solacongestidine (**103**) isolated from the solanaceous plant *Solanum congestiflorum* (Sato *et al.* 1969) has demonstrated significant *in vitro* antifungal activity against several important human pathogens including *C. albicans* and *Cr. neoformans* and shows *in vivo* efficacy in a mouse mode infected with *C. albicans* (Kusano *et al.* 1987a). Solacongestidine at 40 μM exhibited 59% inhibition of cholesterol synthesis from [24,25-^{3}H]-24,25-dihydrolanosterol (18 μM). Some inhibitory effects on cholesterol synthesis were also found in the analogs solafloridine, solasodine, tomatillidine and tomatidine (Kusano *et al.* 1987b). These results indicate that the antifungal mechanism of solacongestidine may be associated with the biosynthetic pathways of fungal sterol synthesis.

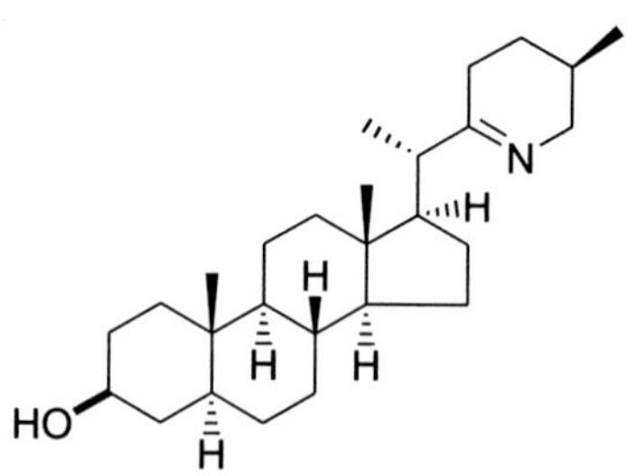

Solacongestidine (**103**)

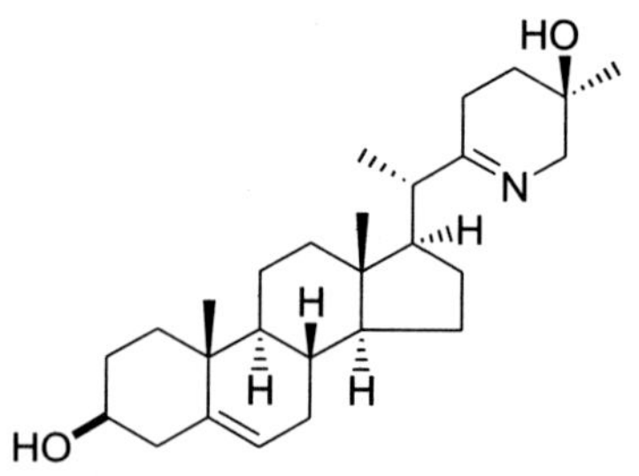

25β-Hydroxyverazine (**104**)

25β-Hydroxyverazine (**104**) isolated from Suriname plant *Solanum surinamense* (Solanaceae) is a new addition to antifungal steroidal alkaloids (Abdel-Kader *et al.* 1998). It showed good *in vitro* activity against *C. albicans*, with MIC value only slightly less than those of the clinically used antifungal drugs amphotericin B and ketoconazole. These authors further synthesized four simple imines (**107a-107d**) (Scheme 1) for structure-activity relationship studies. The synthesis was carried out via cyclization of ketoamides **106**, which were prepared by addition to a Grignard reagent to the lactam 1-(*tert*-butoxycarbonyl)-1-azacyclohexan-2-one **105**. These compounds were completely inactive, indicating that the streroidal skeleton is necessary for antifungal activity. It is noted that **103** and **104** are the most active naturally occurring antifungal steroidal alkaloids reported so far against several fungal pathogens including *C. albicans*.

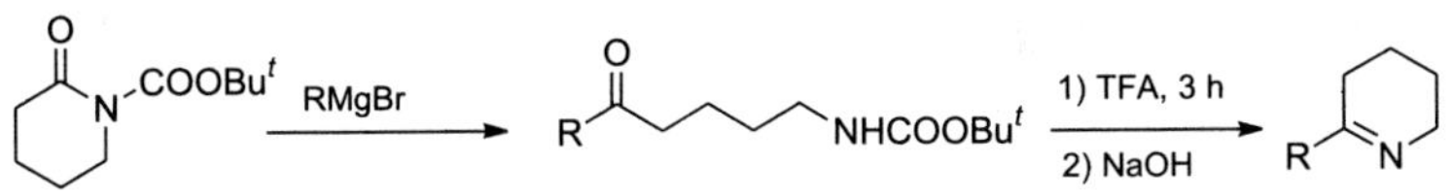

105

106a R = n-C_9H_{19}
106b R = n-C_3H_7
106c R = C_6H_5
106d R = p-ClC_6H_4

107a R = n-C_9H_{19}
107b R = n-C_3H_7
107c R = C_6H_5
107d R = p-ClC_6H_4

Scheme 1 Synthesis of cyclic imines **107a-107d**.

Inspired by the antifungal activity of steroidal alkaloids, Chen *et al.* (1998) prepared seven analogs in a multiple step sequence using the commercially available steroid, 3β-acetoxy-5,16-pregnadiene-20-one (**108**) as starting material. Preliminary *in vitro* antifungal testing against *C. albicans*, *Cr. neoformans* and *Aspergillus fumigatus* showed that four compounds (**109-112**) were equivalent or more effective than the control drug fluconazole. The selectivity of **112** may be doubtful due to the hydrazine functional group in the molecule. The synthesis of **109-111** is illustrated in Scheme 2 which involves reduction of the double bond at C-16 on the pregnadiene skeleton of **118** and sequential modification of the C-17 side chain via intermediates **113-116**.

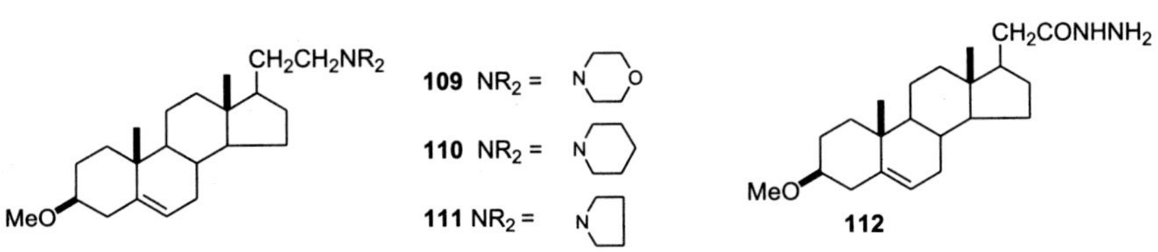

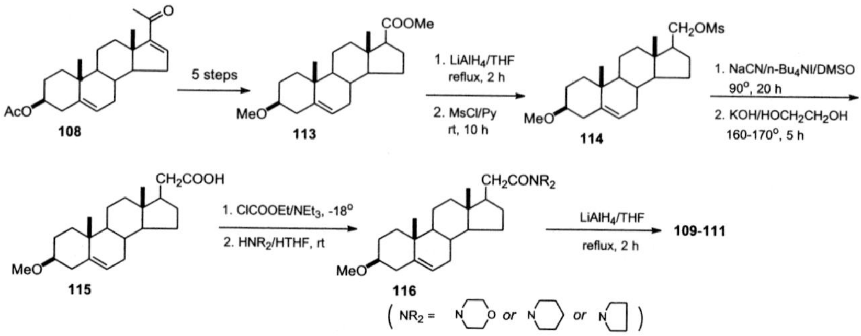

Scheme 2 Synthesis of steroidal analogs **109-111**.

A novel antifungal indolizinium alkaloid, 6,8-didec-(1Z)-enyl-5,7-dimethyl-2,3-dihydro-1*H*-indolizinum (**117**), as the trifluoacetic acid salt was isolated from the wood of *Aniba panurensis* (Lauraceae) (Klausmeyer *et al.* 2004). This lead resulted from a high-throughput screening of 140,000 crude plant, marine and fungal extracts from the Developmental Therapeutic Program (DTP) Natural Products Library in the National Institutes of Health. Compound **117** was a potent inhibitor against azole-resistant strain of *C. albicans*, with MIC_{50} of ~0.25 µg/mL, but with limited activity against bacteria. Compound **117** also strongly inhibited 20 recent antibiotic resistant isolates, including *C. albicans*, *C. lusitaniae*, *C. glabrata*, *C. krusei*, *C. parapsilosis*, and *Cr. neoformans*, with MICs in the range of 0.25-4 µg/mL. Its potency against *C. krusei* equaled that of amphotericin B and exceeded the potency of fluocytosine and fluconazole by factors of 16 and 32, respectively. However, compound **117** possessed potent hemolytic activity in the same molarity range as antifungal activity. It was also non-selectively cytotoxic to the DTP 60 human tumor cell lines, with an LC_{50} of 10^{-5} M. These properties preclude further development of this compound as antifungal agents.

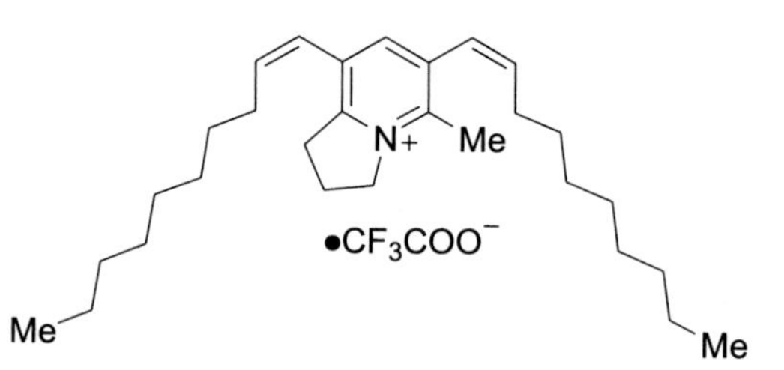

6,8-Didec-(1*Z*)-enyl-5,7-dimethyl-2,3-dihydro-1*H*-indolizinu (**117**)

A few species of the tribe Psychotrieae (Rubiaceae) contain a series of structurally related alkaloids consisting of oligomers of two to eight pyrrolidinoindoline units, two of which are coupled through the alicyclic carbons C-3a and C-3a′, while the remaining units are coupled to the dimer through one alicyclic carbon (C-3a) and one aromatic carbon (C-7). Nine such alkaloids, holdgkinsine A (**118**), quadrigemine C (**119**), quadrigemine H (**120**), psychotridine C (**121**), isopsychotridine E (**122**), vatine (**123**), vatine A (**124**), vatamine (**125**), and vatamidine (**126**) isolated from the aerial parts and stem bark of *Calycodendron milnei* were evaluated for their antifungal activity using a tube dilution method by comparison with those of positive controls amphotericin B and streptomycin (Saad *et al.* 1995). All the compounds showed, to various degrees, activity against several fungal pathogens and yeasts (Table 6.4). Compound **119** possessed strong activity against most of the tested fungi. In addition, antibacterial, antiviral, and cytotoxic activities of these compounds were also evaluated. It should be mentioned that compound **119** also had a strong cytotoxic activity against African green monkey kidney Vero cells.

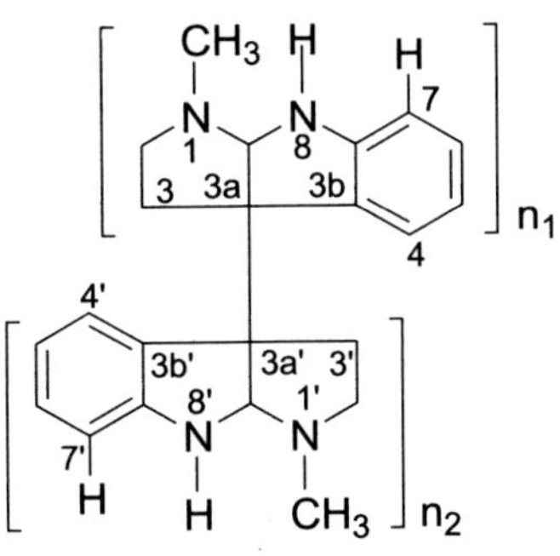

	n_1	n_2
Holdgkinsine A (**118**)	1	2
Quadrigemine C (**119**)	2	2
Quadrigemine H (**120**)	1	3
Psychotridine C (**121**)	2	3
Isopsychotridine E (**122**)	1	4
Vatine (**123**)	2	4
Vatine A (**124**)	2	4
Vatamine (**125**)	2	5
Vatamidine (**126**)	2	6

Table 6.4 Antifungal activity of pyrrolidinoline alkaloids **118-126** (MIC, μg/mL).

	Aspergillus flavus	*Aspergillus fonsecaeus*	*Penicillium stipitatum*	*Trichophyton concentricum*	*Candida rugosa*	*Candida lipolytica*	*Rhodotorula rubra*	*Saccharomyces cerevisiae*
118	25	25	50	10	100	50	5	50
119	10	10	10	5	25	10	5	25
120	25	25	25	50	25	10	10	25
121	50	100	50	50	25	50	50	75
122	50	25	100	75	50	25	ND	75
123	25	25	50	75	10	25	10	50
124	50	25	25	75	50	25	75	75
125	10	25	25	75	75	50	100	100
126	75	50	100	50	25	25	50	25
Amphotericin B	5	4	5	3	5	4	4	5
Streptomycin	6	5	4	5	4	5	3	5

Three protoberberine alkaloids, berberine (**19**), palmatrine (**127**) and jatrorrhizine (**128**) isolated from *Mahonia aquifolium* (Berberidaceae) were screened for their inhibitory activity as iodide salt forms against a variety of dermatophytes and *C. albicans* and *C. tropicalis* using the *in vitro* agar dilution plate method with a 6-day incubation time (Volleková *et al.* 2003). Jatrorrhizine was found to be the most effective against all fungal species tested (MIC ranges from 62.5 to 125 μg/mL), while the positive control fluconazole gave a MIC range of 12.5 to >100 μg/mL. It was suggested that **128** may serve as a lead for further studies to develop new antifungal agents.

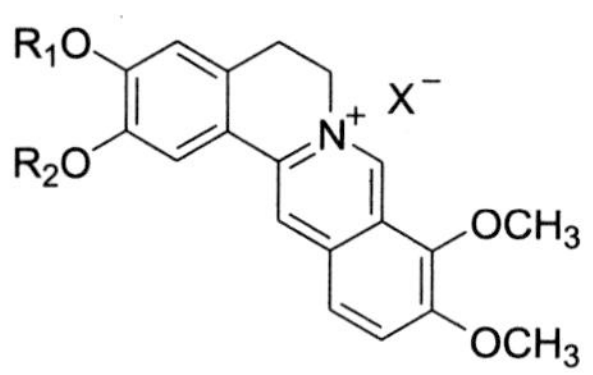

	R_1	R_2
Palmatrine (**127**)	H	CH_3
Jatrorrhizine (**128**)	CH_3	CH_3

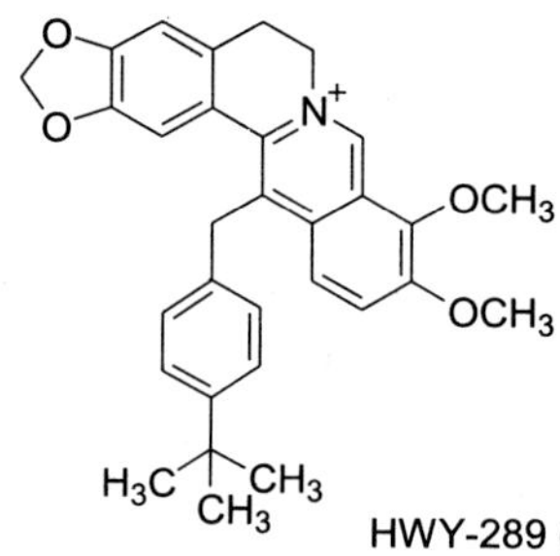

HWY-289 (**129**)

Park *et al.* (1999) have also reported the antifungal activity of **19** and **127** using a microdilution method, both showing moderate or marginal activity against *C. albicans*, *C. parapsilosis*, *C. glabrata*, *C. guilliermondii* and *C. krusei* when compared to miconazole and amphotericin B. They have demonstrated that their antifungal activity against *C. albicans* are associated with the inhibition of sterol 24-methyl transferase and chitin synthase, two key enzymes in the pathways of ergosterol and chitin biosynthesis, respectively. Berberine (**19**) chloride was also reported in an earlier report (Okunade *et al.* 1994) to possess good *in vitro* activity against *C. albicans* and *Cr. neoformans* with MICs ranging from 0.39 to 1.56 μg/mL. However, it was inactive in a mouse model of disseminated cryptococcosis.

It is worthwhile to point out that a semi-synthetic protoberberine derivative, HWY-289 (**129**), has been selected as potential antifungal lead from 515 synthetic and semi-synthetic analogs (Park *et al.* 2001). This compound exhibited potent anti-*Candida* activity with a MIC of 1.56 μg/mL for both *C. albicans* and *C. krusei* and a MIC of 6.25 μg/mL for *C. guilliermondii*. These *in vitro* potencies are comparable to those of amphotericin B with MICs in the range of 0.39-0.78 μg/mL for the three *Candida* strains. Mechanism of action studies indicated that **129** targets multiple sites of *C. albicans*, including sterol biosynthesis pathway involving an indirect effect on sterol 24-methyl transferase and cell wall biosynthesis by inhibing chitin synthase isoenzymes CaCHS1 and

CACHS2. The lack of toxicity in rats warrants further *in vitro* and *in vivo* investigation of **129** as potential antifungal agents.

A TLC bioauthography-guided fractionation of an alkaloidal extract of the stem barks of *Zanthoxylum chiloperone* var. *angustifolium* (Rutaceae) resulted in the isolation of two indolonaphthyridinone alkaloids, canthin-6-one (**75**) and 5-methoxycanthin-6-one (**130**) (Thouvenel *et al.* 2003). Compound **75** was also previously reported to have antifungal activity (Mitscher *et al.* 1972). The MIC values of the two compounds were determined in 24-well microplates using dextrose yeast extract dilution method adding the red phenol as reagent. Compound **75** exhibited a broad spectrum of activities against *A. fumigatus, A. niger, A. terreus, C. albicans, C. tropicalis, C. glabrata, Cr. neoformans, Geotrichum candidum, Saccharomyces cerevisiae, Trichosporon beigelii, T. cutaneum* and *Trichophyton mentagrophytes* var. *interdigitale* with MICs values between 5.3 to 46 μM. Compound **130** was only active against *T. mentagrophytes* var. *interdigitale* with a MIC value of 12.3 μM. Ketoconazole was used as a positive control with MICs values of 5.3 to 46 μM.

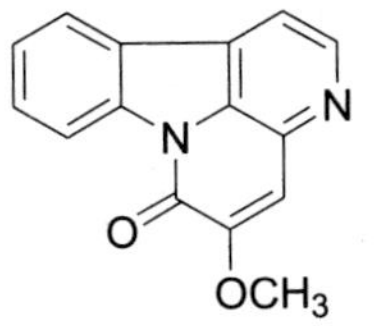

5-Methoxycanthin-6-one (**130**)

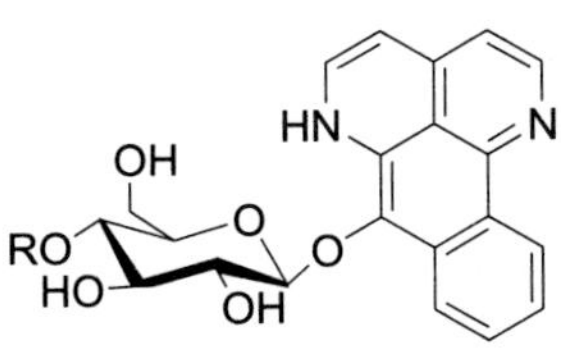

131 R = H
132 R = CH_3

The antifungal activity and preparation methods of the copyrine alkaloid sampagine (**26**) and its analogs of this type have been extensively reviewed (Clark and Hufford 1992). New progress in this regard is the production of two sampagine glycosides (**131** and **132**) by microbial transformation of **26** using a standard two-stage fermentation technique (Orabi *et al.* 1999). Compounds **131** and **132** demonstrated significant *in vitro* activity against *Cr. neoformans*, but were inactive against *C. albicans*. The MIC of **131** against *Cr. neoformans* is 0.2 μg/mL and of **132** is 0.78 μg/mL. By comparison, the MIC of **26** ranges from 3.12 μg/mL for *C. albicans* to *Cr. neoformans*. However, **132** is inactive *in vivo* in a mouse model of cryptococcosis.

Recently, two new sampagine analogs, hadranthine A (**24**) and hadranthine B (**25**), have been isolated from the stem bark of *Duguetia hadrantha* (Annonaceae), along with the known imbiline-1 (**26**). Using a modified antifungal protocol of the National Committee for Clinical Laboratory Standards method (1997), compounds **24** and **26** showed marginal activity against *C. albicans* with IC_{50} values of 4.5 and 9.0 μg/mL, respectively, while amphotericin B gave an IC_{50} of 0.011 μg/mL (Muhammad *et al.* 2001).

Antifungal Alkaloids Against Agriculture Related Pathogens

Agricultural fungal pathogens are a major cause for the pre-harvest losses of crop production particularly in developing countries. Synthetic fungicides have been used successfully for the control of various fungal diseases of crop plants for a number of years. This, however, has resulted in human hazards, resistance in pathogens and environmental pollution. Increasing concern over these negative effects has highlighted the need for new agricultural fungicides or alternative methods for the control of fungal diseases (Maurya *et al.* 2001, Lee *et al.* 2001). Some of the approaches that are currently being pursued include biological control, genetic engineering for evolving resistant varieties and use of induced resistance by biotic and abiotic means (Lyon *et al.* 1995). Research efforts on plant derived fungicides for agriculture has now been intensified as it is generally believed that the secondary metabolism of plants has evolved to protect them from attack by microbial pathogens (Benner, 1993). This is evident by the large number of anti-infective drugs that are derived from plants (Newman *et al.* 2003).

Benzophenanthridine alkaloids, corynoline (**133**) and acetylcorynoline (**134**), isolated from the aerial parts of *Corydalis incise* (Papaveraceae) are active against the agricultural pathogen *Cladosporium herbarum* (Ma *et al.* 1999). The assay method used is a TLC bioautography and the activity is defined by minimum amount of compound (MAC) required to inhibit the growth of *C. herbarum* on TLC plates. Both **133** and **134** gave a MAC value of 3 μg, while the MAC of the positive control propiconazole was 0.1 μg. In another report by same authors (Ma *et al.* 2000), papavorientine (**135**), dihydrosanguinarine (**136**), dihydrochelerythrine (**137**), chelidonine (**138**) isolated from the fresh roots of *Papaver orientale* (Papaveraceae) exhibited activity against *C. herbarum* with MACs of 3, 4, 6, and 10 μg, respectively.

Furoquinoline alkaloids, dictamnine (**139**) and haplopine (**140**), isolated from the root bark of *Dictamnus dasycarpus* (Rutaceae) were evaluated for thier antifungal activity against the plant pathogen *Cladosporium cucumerinum* using a TLC bioautography method described above (Zhao *et al.* 1998). The MACs of 139 and 140 were 0.6 and 10 μg, respectively, while the MAC of the positive control propiconazole was 0.1 μg.

Bioassay guided fractionation using an agar diffusion method of the ethanol extract of the roots and rhizomes of *Veratrum taliense* (Liliaceae) afforded eight antifungal steroidal alkaloids (Zhou *et al.* 2003). Their *in vitro* antifungal activity were compared with that of the positive control triadimefon which produced MICs of 120 and 80 μg/mL, respectively, against two plant pathogens *Phytophthora capisis* and *Rhizoctonia cerealis*. The novel secocevine type alkaloids, neoceratalines A (**141**) and B (**142**) showed moderate activity against the two pathogens with MICs around

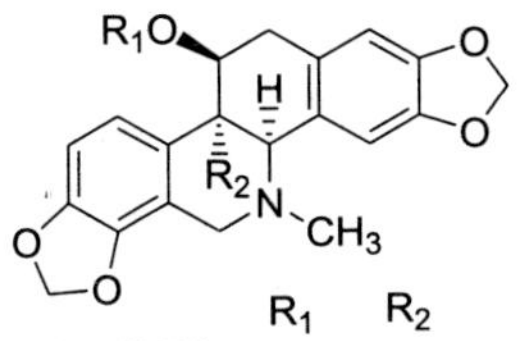

	R_1	R_2
Corynoline (**133**)	CH_3	H
Aetylcorynoline (**134**)	CH_3	OAc
Chelidonine (**138**)	H	H

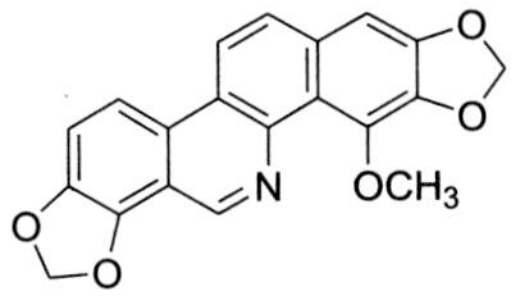

Papavorientine (**135**)

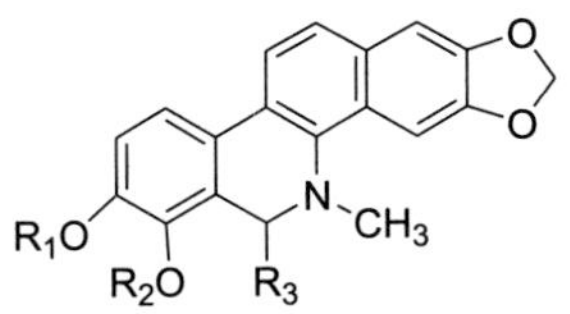

	R_1	R_2	R_3
Dihydrosanguinarine (**136**)	-CH_2-		OCH_3
Dihydrochelerythrine (**137**)	CH_3	CH_3	H

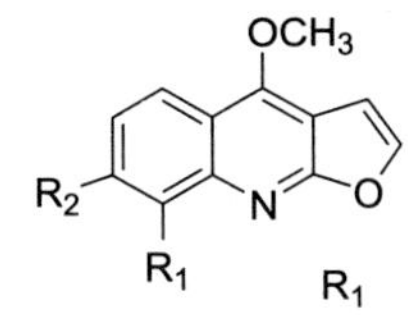

	R_1	R_2
Dictamnine (**139**)	H	H
Haplopine (**140**)	OH	OCH_3

200 μg/mL. Two jerveratrum-type, jervine (**143**) and jervine 3-*O*-*β*-D-glucopyranoside (**144**), and four verazine type, stenophylline B (**145**), stenophylline B 3-*O*-*β*-D-glucopyranoside (**146**), veramiline 3-*O*-*β*-D-glucopyranoside (**147**), veramitaline (**148**), exhibited strong activity against *P. capisis* with MICs of 80, 120, 120, 160, 80 and 160 μg/mL, respectively. Compounds **145-147** inhibited *R. cerealis* with MICs of 160, 120 and 120, respectively. Preliminary structure-activity relationship was also briefly discussed in this report.

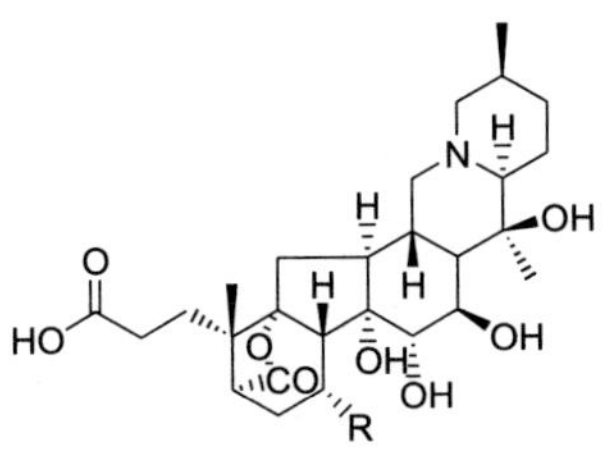

Neoceratatine A (**141**) R = H
Neoceratatine B (**142**) R = OH

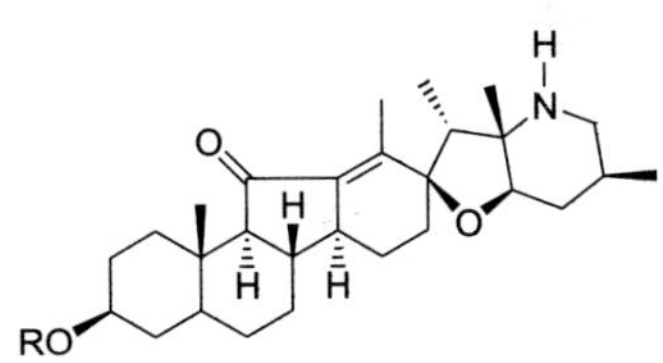

Jervine (**143**) R = H
Jervine 3-*O*-β-D-glucopyranoside (**144**) R = Glc

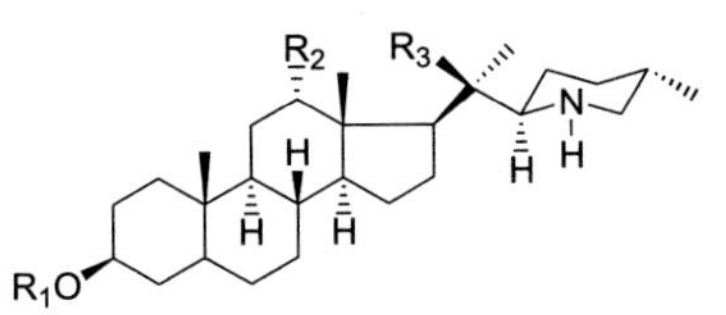

Stenophylline B (**145**)	$R_1 = R_2 = H, R_3 = OH$
Stenophylline B 3-*O*-β-D-glucopyranoside (**146**)	$R_1 = Glc, R_2 = H, R_3 = OH$
Veramiline 3-*O*-β-D-glucopyranoside (**147**)	$R_1 = Glc, R_2 = R_3 = H$
Veramitaline (**148**)	$R_1 = R_3 = H, R_2 = OH$

Two piperidine alkaloids, pipernonaline (**149**) (Lee *et al.* 2001) and piperocatadecalidine (**150**), (Park *et al.* 2003) were isolated from long pepper, *Piper longum* (Piperaceae) and evaluated for antifungal activity against six phytopathogenic fungi, *Pyricularia grisea, Rhizoctonia solani, Botrytis cineria, Phytophthora infestans, Puccinia recondita* and *Erysiphe graminis* using an *in vivo* whole plant method and compared with the positive controls chlorothalonil, dichlofluanid and mancozeb. Compound **149** selectively inhibited *P. recondita* with 91% and 80% control values at the concentrations of 0.5 and 0.25 mg/mL, respectively, while **150** showed activity against *P. recondita* with 100, 87.6, and 56.3% inhibitions of fungal growth on the wheat leaf surface at the concentrations of 1.0, 0.5, and 0.25 mg/mL, respectively. By comparison, potent fungicidal activity (> 90% control value) was observed for mancozeb against *P. recondita* at 50 μg/mL.

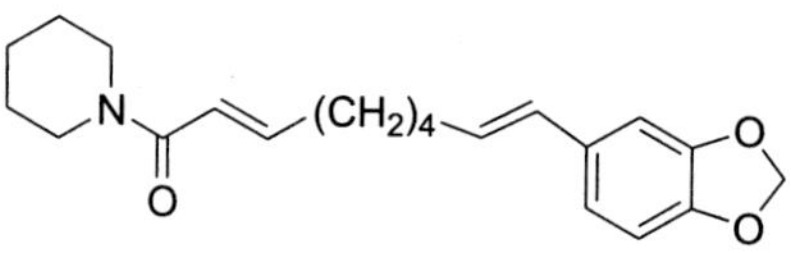

Pipernonaline (**149**)

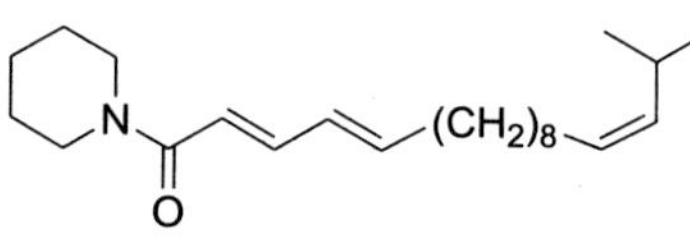

Piperocatadecalidine (**150**)

The quaternary indoloquinolizine alkaloid, Δ^3-alstovenine (**151**), isolated from *Alstonia venenata* inhibited the spore germination of *Cercospora* sp., *Helminthosporium maydis, H. sativum, H. oryzae, H. frumentacei, Erysiphe polygoni, E. pisi, Albugo candida, Alternaria brassicicola, Puccinia ruelliae, Curvularia lunata* and *C. verruculosa* at concentrations of 0.25 to 1.0 mg/mL. *Fusarium udum* and several *Alternaria* species were found to be resistant to this alkaloid (Singh *et al.* 1999). Analog venenatine (**152**) isolated from the same plant in an aqueous acetic acid solution inhibited spore germination of all the 10 tested fungi at the concentrations of 0.25 to 2 mg/mL (Singh *et al.* 2000). *F. udum, A. brassicicola, Ustilago cynodontis* and *Aspergillus flavus* were sensitive to this compound, exhibiting germination levels below 10% at 200 ppm. The spore germination and colony development of the parasitic fungus *E. pisi*, which causes powdery mildew in pea (*Pisum sativum*), on excised pea leaves was also significantly affected.

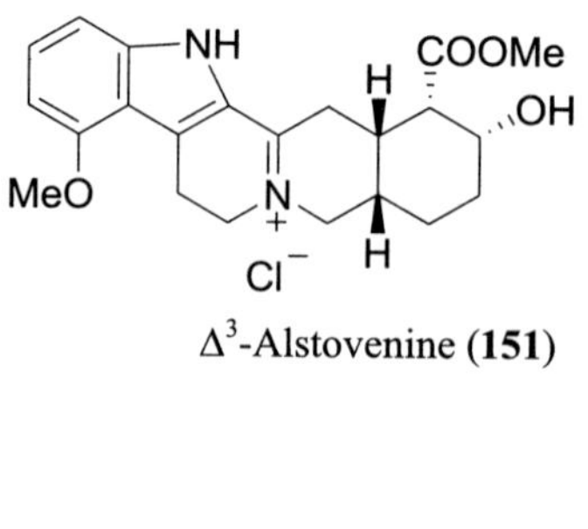

Δ^3-Alstovenine (**151**)

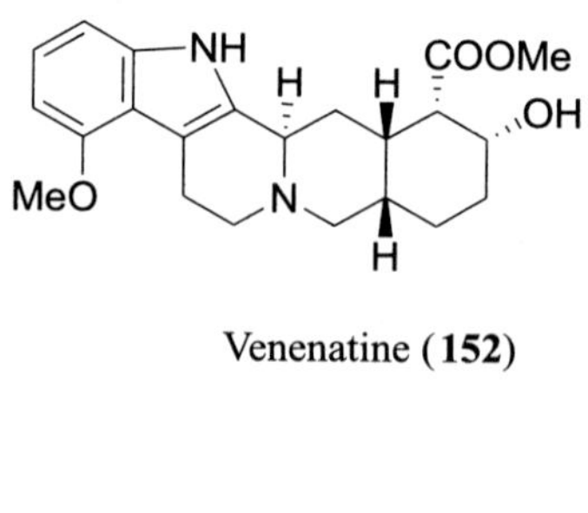

Venenatine (**152**)

The dibenzoquinolizine alkaloid, tetrahydropalmatine (**153**), isolated from *Corydalis charophylla* (Fumariaceae) was assayed against spore germination of some saprophytic and pathogenic fungi including *Alternaria solani, A. brassicicola, A. brassicae, A. alternata, Erysiphe pisi, Curvularia lunata, C. pallescens, C. maculans, Curvularia* sp., *Colletorichum* sp., *C. musae, Helminthosporium echinoclova, H. pennisetti, H. spiciferum,* and *Heterosporium* sp. (Maurya *et al.* 2001). It inhibited spore germination of all the fungi tested. *Colletotrichum* spp. *C. lunata, H. spiciferum* and *Heterosporium* sp. were most sensitive as complete inhibition of spore germination was observed at 200 ppm. Analog (–)-corypalmine (**154**) isolated from the same plant also showed good activity against the above pathogens. Complete inhibition on spore germination of *Heterosporium* sp. and *Ustilago cynodontis* were observed at 200 ppm (Maurya *et al.* 2002).

N-Methylhydrasteine hydroxylactum (**155**) and 1-methoxyberberine chloride (**156**) isolated from *Corydalis longipes* (Papaveraceae) showed high efficacy individually (in concentrations of 50-150 ppm) and also in a 1:1 mixture against spore germination of *Alternaria alternate, A. brassicae, Curvularia maculans, Curcularia* sp., *Colletotrichum gloeosporioides, Colletotrichum* sp., *Helminthosporium spiciferum, H. pennisetti, Helminthosporium* sp. and *Ustilago cynodontis* (Singh *et al.* 2003). It was noted that the inhibitory doses of the two compounds are relatively low when compared to **151, 153, 154,** and berberine iodide (Sarma *et al.* 1999). However, it should be mentioned that no positive controls were included in these studies.

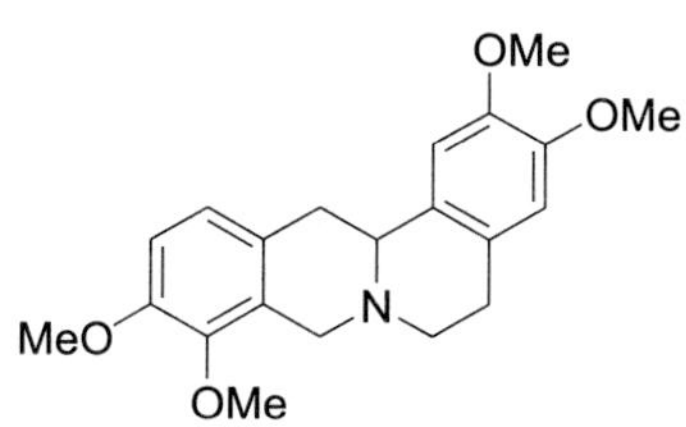

Tetrahydropalmatine (**153**)

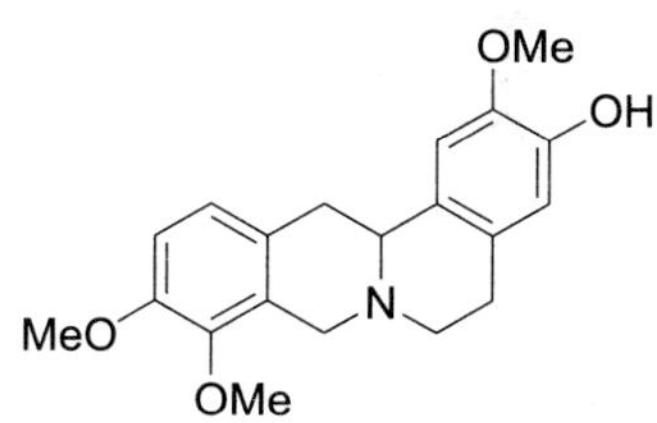

(-)-Corypalmine (**154**)

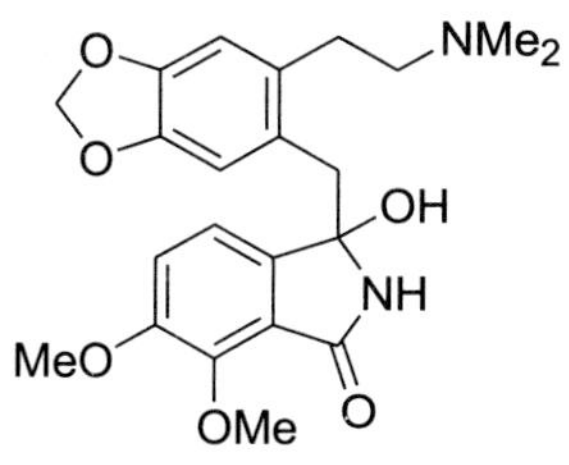

N-Methylhydrasteine hydroxylactum (**155**)

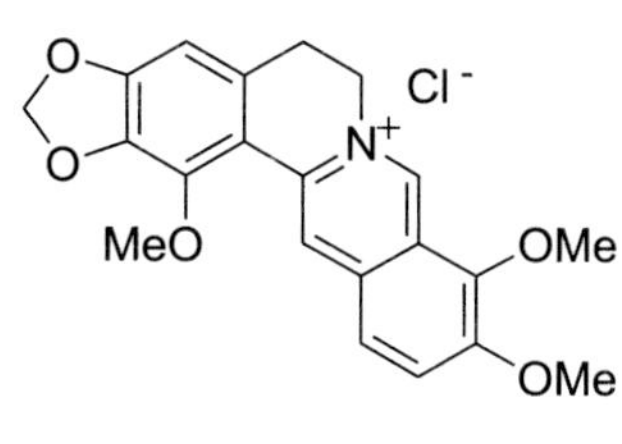

1-Methoxyberberine chloride (**156**)

SUMMARY

Alkaloids are one of the most fascinating classes of natural products and have played a vital role in drug discovery as prototype drugs, leads, or biological probes (Clark 1996, Newman *et al.* 2000, 2003). Historically plants have provided the best leads for antimalarial drugs. The two mainstays for treating cerebral malaria, quinine and artemisinin, would likely never have been created by combinatorial chemistry or rational drug design because of the intricate structural and stereochemical problems. Thus, for the foreseeable future, screening of natural products will likely provide the richest and most diverse source of lead antimalarial compounds. New screening techniques such as looking for synergistic activity and target-based screens should decrease the time required for the discovery of active compounds. The sequencing of the genome of *P. falciparum*, as well as its mosquito vector *A. gambiae* will provide new target-based bioassays for identifying novel antimalarials.

Finding new active compounds is only the beginning, following up on promising compounds is also very important. Most malaria research is conducted in academic settings where a comprehensive drug development program is not in place and further development work is dependent upon acquisition of financial support through grants, which requires an investigator taking the initiative to secure such funding. Two funding initiatives have already increased the amount of research into malaria drugs. The Roll Back Malaria (RBM) global partnership was launched in 1998 by WHO, UNICEF, UNDP and the World Bank to provide a coordinated international approach to fighting malaria. RBM's stated goal is to halve the burden of malaria by 2010. The Medicines for Malaria Venture (MMV) is a nonprofit organization created to discover, develop and deliver new antimalarial drugs through effective public-private partnerships. These sources of funding are making possible the rather costly development of promising leads produced in screening efforts.

On the other hand, it can be seen that significant discoveries have not been made regarding the identification of antifungal alkaloids from plants, particularly in the pharmaceutical aspect of developing drug candidates for the treatment of fungal infections. Over the past few years, new biological technologies such as genomic sequencing of fungal pathogens have facilitated the discoveries of new molecular antifungal targets. It is anticipated that such developments coupled with classical cell-based assays will ultimately expedite the discovery of new antifungal leads from the huge reservoir of alkaloids existing in nature.

ACKNOWLEDGEMENT

This work was supported in part by the NIH Grant No. AI 27094 and the USDA Agricultural Research Service Specific Cooperative Agreement No. 58-6408-2-0009.

REFERENCES

Abdel-Kader, M.S., B.D. Bahler, S. Malone, M.C.M. Werkhoven, F.V. Troon, D.J.H. Wisse, I. Bursuker, K.M. Eddermann, S.W. Mamber, and D.G.I. Kingston. 1998. DNA-Damaging steroidal alkaloids from *Eclipta alba* from the Suriname rainforest. *J. Nat. Prod.* 61: 1202-1208. A correction for the plant name has been made from *Eclipta alba* to *Solanum surinamense* in *J. Nat. Prod.* 63: 1184-1184, 2000.

Abourashed, E.A., and I.A. Khan. 2001. High performance liquid chromatography determination of hydrastine and berberine in dietary supplements containing Goldenseal. *J. Pharm. Sci.* 90: 817-822.

Addae-Kyereme, J., S.L. Croft, H. Kendrick, and C.W. Wright. 2001. Antiplasmodial activities of some Ghanaian plants traditionally used for fever/malaria treatment and of some alkaloids isolated from *Pleiocarpa mutica; in vivo* antimalarial activity of pleiocarpine. *J. Ethnopharmacol.* 76: 99-103.

Akendengue, B., E. Ngou-Milama, F. Roblot, A. Laurens, R. Hocquemiller, P. Grellier, and F. Frappier. 2002. Antiplasmodial activity of *Uvaria klaineana*. *Planta Med.* 68: 167-169.

Akendengue, B., E. Ngou-Milama, H. Bourobou-Bourobou, J. Essouma, F. Roblot, C. Gleye, A. Laurens, R. Hocquemiller, P. Loiseau, and C. Bories. 2003. Acaricidal activity of *Uvaria versicolor* and *Uvaria klaineana* (Annonaceae). *Phytother. Res.* 17: 364-367.

Ancolio, C., N. Azas, V. Mahiou, E. Ollivier, C.D. Giorgio, A. Keita, P. Timon-David, and G. Balansard. 2002. Antimalarial activity of extracts and alkaloids isolated from six plants used in traditional medicine in Mali and Sao Tome. *Phytother. Res.*16: 646-649.

Angerhofer, C.K., H. Guinaudeau, V. Wongpanich, J.M. Pezzuto, and G.A. Cordell. 1999. Antiplasmodial and cytotoxic activity of natural bisbenzylisoquinoline alkaloids. *J. Nat. Prod.* 62: 59-66.

Ayafor, J.F., B.L. Sondengam, B.T. Ngadjui, and B. Tchaleu. 1982. Quinoline and indolopyridoquinazoline alkaloids from *Vepris louisii*. *Phytochemistry* 21: 2733-2736.

Azas, N., N. Laurencin, F. Delmas, C. Di Giorgio, M. Gasquet, M. Laget, and P. Timon-David. 2002. Synergistic *in vitro* antimalarial activity of plant extracts used as traditional herbal remedies in Mali. *Parasitol. Res.* 88: 165-71.

Baker, B.R., R.E. Schaub, F.J. McEvoy, and J.H. Williams. 1952. An Antimalarial alkaloid from *hydrangea*. XII. Synthesis of 3-[b-keto-g-(3-hydroxy-2-piperidyl)propyl]-4-quinazolone, the alkaloid. *J. Org. Chem.* 17: 132-40.

Benner, J.P. 1993. Pesticidal compounds from higher plant. *Pestic. Sci.* 39: 95-102.

Bringmann, G., M. Wenzel, M. Rubenacker, M. Schaffer, M. Ruckert, and L.A. Assi. 1998. Acetogenic isoquinoline alkaloids. Part 108. Dioncophylline D and 8-O-methyldioncophylline D, 7,8′-coupled naphthylisoquinoline alkaloids from *Triphyophyllum peltatum*. *Phytochemistry* 49: 1151-1155.

Bringmann, G., C. Guenther, W. Saeb, J. Mies, A. Wickramasinghe, V. Mudogo, and R. Brun. 2000. Ancistrolikokines A-C: new 5,8′-coupled naphthylisoquinoline alkaloids from *Ancistrocladus likoko*. *J. Nat. Prod.* 65: 1333-1337.

Bringmann, G., C. Gunther, E.-M. Peters, and K. Peters. 2001. Acetogenic isoquinoline alkaloids. Part 147. First total synthesis of the 7,6′-coupled antifungal naphthylisoquinoline alkaloid dioncophylline B. *Tetrahedron* 57: 1253-1259.

Bringmann, G., K. Messer, R. Brun, and V. Mudogo. 2002a. Ancistrocongolines A-D, new naphthylisoquinoline alkaloids from *Ancistrocladus congolensis*. *J. Nat Prod.* 65: 1096-1101.

Bringmann, G., K. Messer, K. Wolf, J. Muhlbacher, M. Grune, R. Brun, and A.M. Louis. 2002b. Dioncophylline E from *Dioncophyllum thollonii*, the first 7,3′-coupled dioncophyllaceous naphthylisoquinoline alkaloid. *Phytochemistry* 60: 389-397.

Bringmann, G., M. Wohlfarth, H. Rischer, J. Schlauer, and R. Brun. 2002c. Extract screening by HPLC coupled to MS-MS, NMR, and CD: a dimeric and three monomeric naphthylisoquinoline alkaloids from *Ancistrocladus griffithii*. *Phytochemistry* 61: 195-204.

Bringmann, G., M. Dreyer, J.H. Faber, P.W. Dalsgaard, D. Strk, J.W. Jaroszewski, H. Ndangalasi, F. Mbago, R. Brun, M. Reichert, K. Maksimenka, and S.B. Christensen. 2003. Ancistrotanzanine A, the first 5,3′-coupled naphthyl-isoquinoline alkaloid, and two further, 5,8′-linked related compounds from the newly described species *Ancistrocladus tanzaniensis*. *J. Nat. Prod.* 66: 1159-1165.

Bringmann, G., K. Messer, B. Schwobel, R. Brun, and A.S Ake Assi. 2003a. Habropetaline A, an antimalarial naphthylisoquinoline alkaloid from *Triphyophyllum peltatum*. *Phytochemistry* 62: 345-349.

Bringmann, G., M. Dreyer, J.H. Faber, P.W. Dalsgaard, D. Strk, J.W. Jaroszewski, H. Ndangalasi, F. Mbago, R. Brun, and S.B. Christensen. 2004. Acetogenic isoquinoline alkaloids. Part 158. Ancistrotanzanine C and related 5,1′- and 7,3′-coupled naphthylisoquinoline alkaloids from *Ancistrocladus tanzaniensis*. *J. Nat. Prod.* 67: 743-748.

Brookman, J., M. Bromley, and W. Gissane. 2003. A rational approach to the design of systemic antifungal drugs. *Innovations Pharm. Tech.* 3: 30-38.

Camacho, M. del R., J.D. Phillipson, S.L. Croft, V. Yardley, and P.N. Solis. 2004. In vitro antiprotozoal and cytotoxic activities of some alkaloids, quinones, flavonoids, and coumarins. *Planta Med.* 70: 70-72.

Chen, Y.H., Q.T. Zhou, and D.L. Bai. 1998. Synthesis of several analogues of antifungal steroid alkaloids. *Acta Pharm. Sinica* 33: 436-441.

Christopher, E., E. Bedir, C. Dunbar, I.A. Khan, C.O. Okunji, B.M. Schuster, and M.M. Iwu. 2003. Indoloquinazoline alkaloids from *Araliopsis tabouensis*. *Helv. Chim. Acta* 86: 2914-2918.

Cimanga, K., T. De Bruyne, L. Pieters, A.J. Vlietinck, and C.A. Turger. 1997. *In vitro* and *in vivo* antiplasmodial activity of cryptolepine and related alkaloids from *Cryptolepis sanguinolenta*. *J. Nat. Prod.* 60: 688-691.

Clark, A.M., and C.D. Hufford. 1992. Antifungal alkaloids. Pages 117-150 in *The Alkaloids*, Vol. 42, Chapter 2, G.A. Cordell, ed., Academic Press, Orlando, Florida.

Clark, A.M. 1996. Natural products as a resource for new drugs. *Pharm. Res.* 13: 1133-1141.

Cordell, G.A., C.K. Angerhofer, and J.M. Pezzuto. 1994. Recent studies on cytotoxic, anti-HIV and antimalarial agents from plants. *Pure Appl. Chem.* 66: 2283-2286.

Cowan, M.M. 1999. Plant products as antimicrobial agents. *Clin. Microbiol. Rev.* 12: 564-582.

Farnsworth, N.R. 1984. The role of medicinal plants in drug development. Natural products and drug development, Alfred Benzon Symposium 20. Eds. P. Krogsgaard-Larson, S. Brogger Christensen, H. Munksgaard, Copenhagen.

Fournet, A., and V. Munoz. 2002. Natural products as trypanocidal, antileishmanial, and antimalarial drugs. *Curr. Top. Med. Chem.* 2: 1215-1237.

Francois, G., G. Timperman, W. Eling, L.A. Assi, J. Holenz, and G. Bringmann. 1997. Naphthylisoquinoline alkaloids against malaria:evaluation of the curativepotentials of dioncophylline C and dioncopeltine A against *Plasmodium berghei in vivo*. *Antimicrob. Agents Chemother.* 41: 2533-2539.

Frederich, M., M.-C. De Pauw, G. Llabres, M. Tits, M.-P. Hayette, V. Brandt, J. Penelle, P. De Mol, and L. Angenot. 2000. New antimalarial and cytotoxic sungucine derivatives from *Strychnos icaja* roots. *Planta Med.* 66: 262-269.

Frederich, M., M.-C. De Pauw, C. Prosperi, M. Tits, V. Brandt, J. Penelle, M.-P. Hayette, P. DeMol, and L. Angenot. 2001a. Strychnogucines A and B, two new antiplasmodial 2001b bisindole alkaloids from *Strychnos icaja*. *J. Nat. Prod.* 64: 12-16.

Frederich, M., M.-P. Hayette, M. Tits, P. De Mol, and L. Angenot. 2001b. Reversal of chloroquine and mefloquine resistance in *Plasmodium falciparum* by the two monoindole alkaloids, icajine and isoretuline. *Planta Med.* 67: 523-527.

Frederich, M., M.-J. Jacquier, P. Thepenier, P. De Mol, M. Tits, G. Philippe, C. Delaude, L. Angenot, and M. Zeches-Hanrot. 2002. Antiplasmodial activity of alkaloids from various *Strychnos* species. *J. Nat. Prod.* 65: 1381-1386.

Frederich, M., M. Tits, and L. Angenot. 2003. Indole alkaloids from *Strychnos* species and their antiplasmodial and cytotoxic activities. *Chem. Nat. Comp.* (*Khimiya Prirodnykh Soedinenii*) 39: 513-519.

Frederich, M., A. Cristino, Y.H. Choi, R. Verpoorte, M. Tits, L. Angenot, E. Prost, J.-M. Nuzillard, and M. Zeches-Hanrot. 2004a. Chrysopentamine, an antiplasmodial anhydronium base from *Strychnos usambarensis* leaves. *Planta Med.* 70: 72-76.

Frederich, M., M. Tits, E. Goffin, G. Philippe, P. Grellier, P. De Mol, M.-P. Hayette, and L. Angenot. 2004b. *In vitro* and *in vivo* antimalarial properties of isostrychnopentamine, an indolomonoterpenic alkaloid from *Strychnos usambarensis*. *Planta Med.* 70: 520-525.

Fu, F.Y., and C.S. Jang. 1948. Chemotherapeutic studies on ch'ang shan, *Dichroa febrifuga*. III. Potent antimalarial alkaloids from ch'ang shan. *Sci. Technol. China* 1: 56-61.

Funayama, S., K. Murata, and T. Noshita. 2001. Quinoline alkaloids of *Orixa japonica*. *Heterocycles* 54: 1139-1148.

Gantier, J.C., A. Fournet, M.H. Munos, and R. Hocquemiller. 1996. The effect of some 2-substituted quinolines isolated from *Galipea longiflora* on *Plasmodium vinckei petteri* infected mice. *Planta Med.* 62: 285-286.

Georgopapadakou, N.H., and T.J. Walsh. 1996. Antifungal agents: chemotherapeutic targets and immunologic strategies. *Antimicrob. Agents Chemother.* 40: 279-291.

Goldsmith, R., and D. Heyneman. 1989. Tropical medicine and parasitology. Appleton and Lange, Norwalk, Connecticut.

Haruki, K., P.G. Bray, M. Ono, and S.A. Ward. 2000. Potent enhancement of the sensitivity of *Plasmodium falciparum* to chloroquine by the bisbenzylisoquinoline alkaloid cepharanthin. *Antimicrob. Agents Chemother.* 44: 2706-2708.

Herrera, M.R., A.K. Machocho, R. Brun, F. Viladomat, C. Codina, and J. Bastida. 2001a. Crinane and lycorane type alkaloids from *Zephyranthes citrina*. *Planta Med.* 67: 191-193.

Herrera, M.R., A.K. Machocho, J.J. Nair, W.E. Campbell, R. Brun, F. Viladomat, C. Codina, and J. Bastida. 2001b. Alkaloids from *Cyrtanthu elatus*. *Fitoterapia* 72: 444-448.

Ishih, A., T. Miyase, K. Ohori, and M. Terada. 2003. Different responses of three rodent plasmodia species, *Plasmodium yoelii* 17XL, *P. berghei* NK65 and *P. chabaudi* AS on treatment with febrifugine and isofebrifugine mixture from *Hydrangea macrophylla* var. Otaksa leaf in ICR mice. *Phytother. Res.* 17: 650-656.

Ishih, A., T. Miyase, and M. Terada. 2003a. Comparison of antimalarial activity of the alkaloidal fraction of *Hydrangea macrophylla* var. otaksa leaves with the hot-water extract in ICR mice infected with *Plasmodium yoelii* 17 XL. *Phytother. Res.* 17: 633-639.

Iwasa, K., Y. Nishiyama, M. Ichimaru, M. Moriyasu, H.-S. Kim, Y. Wataya, T. Yamori, T. Takashi, and D.-U. Lee. 1999. Structure-activity relationships of quaternary protoberberine alkaloids having an antimalarial activity. *Eur. J. Med. Chem.* 34: 1077-1083.

Jacquemond-Collet, I., F. Benoit-Vical, X. Mustofa, A. Valentin, E. Stanislas, M. Mallie, and I. Fouraste. 2002. Antiplasmodial and cytotoxic activity of galipinine and other tetrahydroquinolines from *Galipea officinalis*. *Planta Med.* 68: 68-69.

Jonckers, T.H.M., S. Van Miert, K. Cimanga, C. Bailly, P. Colson, M.-C. DePauw-Gillet, H. van den Heuvel, M. Claeys, F. Lemiere, E.L. Esmans, J. Rozenski, L. Quirijnen, L. Maes, R. Dommisse, G.L.F. Lemiere, A. Vlietinck, and L. Pieters. 2002. Synthesis, cytotoxicity, and antiplasmodial and antitrypanosomal activity of new neocryptolepine derivatives. *J. Med. Chem.* 45: 3497-3508.

Kanokmedhakul, S., K. Kanokmedhakul, D. Yodbuddee, and N. Phonkerd. 2003. New antimalarial bis-dehydroaporphine alkaloids from *Polyalthia debilis*. *J. Nat. Prod.* 66: 616-619.

Katoh, M., R. Matsune, H. Nagase, and T. Honda. 2004. Stereocontrolled synthesis of a potent antimalarial alkaloid, (+)-febrifugine. *Tetrahedron Lett.* 45: 6221-6223.

Kayser, O., A.F. Kiderlen, and S.L. Croft. 2002. Natural products as potential antiparasitic drugs. *Studies in Natural Products Chemistry 26* (*Bioactive Natural Products*), Part G, 779-848.

Keating, G.M., and D.P. Figgitt. 2003. Caspofungin: a review of its use in esophageal candidiasis, invasive candidiasis and invasive aspergillosis. *Drugs* 63: 2235-2263.

Khalil, A.T., F.-R. Chang, Y.-H. Lee, C.-Y. Chen, C.-C. Liaw, P. Ramesh, S.-S.F. Yuan, and Y.-C. Wu. 2003. Chemical constituents from the *Hydrangea chinensis*. *Arch. Pharm. Res.* 26: 15-20.

Kikuchi, H., H. Tasaka, S. Hirai, Y. Takaya, Y. Iwabuchi, H. Ooi, S. Hatakeyama, H.-S. Kim, Y. Wataya, and Y. Oshima. 2002. Potent antimalarial febrifugine analogues against the *Plasmodium* malaria parasite. *J. Med. Chem.* 45: 2563-2570.

Klausmeyer, P., G.N. Chmurny, T.G. McCloud, K.D. Tucker, and R.H. Shoemaker. 2004. A novel antimicrobial indolizinium alkaloid from *Aniba panurensis*. *J. Nat. Prod.* 67: 1732-1735.

Klayman, D.L 1985. Qinghaosu (Artemisinin): An antimalaial drug from China. *Science* 228: 1049.

Knauer, A., J. Sirichaisinthop, F.F. Reinthaler, G. Wiedermann, G. Wernsdorfer, and W.H. Wernsdorfer. 2003. *In vitro* response of *Plasmodium falciparum* to the main alkaloids of Cinchona in northwestern Thailand. *Wien. Klin. Wochenschr.* 115 (Suppl. 3), 39-44.

Koepfli, J.B., J.A. Brockman Jr., and J. Moffat. 1950. Structure of febrifugine and isofebrifugine. *J Am. Chem. Soc.* 72: 3323.

Kobayashi, S., M. Ueno, R. Suzuki, H. Ishitani, H.-S. Kim, and Y. Wataya. 1999. Catalytic asymmetric synthesis of antimalarial alkaloids febrifugine and isofebrifugine and their biological activity. *J. Org. Chem.* 64: 6833-6841.

Koepfli, J.B., J.F. Mead, and J.A. Brockman, Jr. 1947. Alkaloid with high antimalarial activity from *Dichroa febrifuga*. *J. Am. Chem. Soc.* 69: 1837.

Kontoyiannis, D.P., E. Mantadakis, and G. Samonis. 2003. Systemic mycoses in the immunocompromised host: an update in antifungal therapy. *J. Hosp. Infect.* 53: 243-258.

Kuc, J. 1992. Antifungal compounds from plants. Pages 159-184 in *Phytochemical Resources for Medicine and Agriculture*, H.N. Nigg, and D. Seigler, eds., Plenum Press, New York.

Kuo, P.-C., L.-S. Shi, A.G. Damu, C.-R. Su, C.-H Huang, C.-H. Ke, J.-B. Wu, A.-J. Lin, K.F. Bastow, K.-H. Lee, and T.-S. Wu. 2003. Cytotoxic and antimalarial β-carboline alkaloids from the roots of *Eurycoma longifolia*. *J. Nat. Prod.* 66: 1324-1327.

Kusano, G., A. Takahashi, K. Sugiyama, and S. Nozoe. 1987a. Antifungal properties of solanum alkaloids. *Chem. Pharm. Bull.* 35: 4862-4867.

Kusano, G., A. Takahashi, K. Sugiyama, S. Nozoe, Y. Sonoda, and Y. Sato. 1987b. Solanum alkaloids as inhibitors of enzymic conversion of dihydrolanosterol into cholesterol. *Chem. Pharm. Bull.* 35: 4321-4325.

Lee, S.E., B.S. Park, M.K. Kim, W.S. Choi, H.T. Kim, K.Y. Cho, S.G. Lee, and H.S. Lee. 2001. Fungicidal activity of pipernonaline, a piperidine alkaloid derived from long pepper, *Piper longum* L., against phytopathogenic fungi. *Crop Prot*. 20: 523-528.

Lyon, G.P., T. Reglinski, and A.C. Newton. 1995. Novel disease control compounds: the potential to immunize plants against infections. *Plant Pathol*. 44: 407-427.

Ma, W.G., Y. Fukushi, and S. Tahara. 1999. Fungitoxic alkaloids from Hokkaido Corydalis species. *Fitoterapia* 70: 258-265.

Ma, W.G., Y. Fukushi, S. Tahara, and T. Osawa. 2000. Fungitoxic alkaloids from Hokkaido Papaveraceae. *Fitoterapia* 71: 527-534.

Makler, M.T., J.M. Ries, J.A. Williams, J.E. Bancroft, R.C. Piper, B.L. Gibbins, and D.J. Hinrichs. 1993. Parasite lactate dehydrogenase as an assay for *Plasmodium falciparum* drug sensitivity. *Am. J. Trop. Med. Hyg*. 48: 739-741.

Mambu, L., M.-T. Martin, D. Razafimahefa, D. Ramanitrahasimbola, P. Rasoanaivo, and F. Frappier. 2000. Spectral characterization and antiplasmodial activity of bisbenzylisoquinolines from *Isolona ghesquiereina*. *Planta Med*. 66: 537-540.

Marshall, S.J., P.F. Russell, C.W. Wright, M.M. Anderson, J.D. Phillipson, G.C. Kirby, D.C. Warhurst, and P.L. Schiff, Jr. 1994. *In vitro* antiplasmodial, antiamoebic, and cytotoxic activities of a series of bisbenzylisoquinoline alkaloids. *Antimicrob. Agents Chemother*. 38: 96-103.

Maurya, S., J.S. Srivastava, R.N. Jha, V.B. Panday, and U.P. Singh. 2001. Effect of tetrahydropalmatine, an alkaloid on spore germination of some fungi. *Mycobiology* 29: 142-144.

Maurya, S., J.S. Srivastava, R.N. Jha, V.B. Panday, and U.P. Singh. 2002. Efficacy of alkaloid (-)-corypalmine against spore germination of some fungi. *Folia Microbiol*. 47: 287-290.

McCall, D.L.C., J. Alexander, J. Barber, R.G. Jaouhari, A. Satoskar, and R.D. Waigh. 1994. The first protoberberine alkaloid analog with *in vivo* antimalarial activity. *Bioorg. Med. Chem. Lett*. 4: 1663-1666.

Mitscher, L.A., H.D. Showalter, M.T. Shipchandler, R.P. Leu, and J.L. Beal. 1972. Antimicrobial agents from higher plants. IV. *Zanthoxylum elephantiasis*. Isolation and identification of canthin-6-one. *Lloydia* 35: 177-180.

Mturi, N., C.O. Musumba, B.M. Wamola, B.R. Ogutu, and R.J.C. Newton. 2003. Cerebral malaria: Optimizing management. *CNS Drugs* 17: 153-165.

Muhammad, I., D.C. Dunbar, S. Takamatsu, L.A. Walker, and A.M. Clark. 2001. Antimalarial, cytotoxic , and antifungal alkaloids from *Duguetia hadrantha*. *J. Nat. Prod*. 64: 559-562.

Muhammad, I., C.D. Dunbar, S.I. Khan, B.L. Tekwani, E. Bedir, S. Takamatsu, D. Ferreira, and L.A. Walker. 2003. Antiparasitic Alkaloids from *Psychotria klugii*. *J. Nat. Prod*. 66: 962-967.

Murata, K., K. Seya, I. Miki, H. Junke, S. Motomura, and Y. Oshima. 1997. Pharmacological properties of pteleprenine, a quinoline alkaloid from *Orixa japonica* (Rutaceae), in guinea pig ileum and canine left atrium. *Japn. J. Pharm*. 110 (Suppl 1): 148-152.

Muriithi, M.W., W.-R, Abraham, J. Addae-Kyereme, I. Scowen, S.L. Croft, P.M. Gitu, H. Kendrick, E.N.M. Njagi, and C.W. Wright. 2002. Isolation and *in vitro* antiplasmodial activities of alkaloids from *Teclea trichocarpa*: *In vivo* antimalarial activity and X-ray crystal structure of normelicopicine. *J. Nat. Prod*. 65: 956-959.

National Committee on Clinical Laboratory Standards. 1997. Methods for Dilution. Antimicrobial Susceptibility Test for Bacteria That Grow Aerobically. Approved Standard M7-A. 4th ed., NCCLS, Wayne, PA.

Newman, D.J., G.M. Cragg, and K.M. Snader. 2000. The influence of natural products upon drug discovery. *Nat. Prod. Rep*.17: 215-234.

Newman, D.J., G.M. Cragg, and K.M. Snader. 2003. Natural products as sources of new drugs over the period 1981-2002. *J. Nat. Prod*. 66: 1022-1037.

Okunade, A.I., C.D. Hufford, M.D. Richardson, J.R. Peterson, and A.M. Clark. 1994. Antimicrobial properties of alkaloids from *Xanthorhiza simplicissima*. *J. Pharm. Sci*. 83: 404-406.

Orabi, K.Y., E. Li, A.M. Clark, and C.D. Hufford. 1999. Microbial transformation of sampangine. *J. Nat. Prod*. 62: 988-992.

Ovenden, S.P.B., S. Cao, C. Leong, H. Flotow, M.P. Gupta, A.D. Buss, and M.S. Butler. 2002. Spermine alkaloids from *Albizia adinocephala* with activity against *Plasmodium falciparum* plasmepsin II. *Phytochemistry* 60: 175-177.

Park, K.S., K.C. Kang, J.H. Kim, D.J. Adams, T.N. Johng, and Y.K. Paik. 1999. Differential inhibitory effects of protoberberines on sterol and chitin biosyntheses in *Candida albicans*. *J. Antimicrob. Chemother*. 43: 667-674.

Park, K.S., K.C. Kang, K.Y. Kim, P.Y. Jeong, J.H. Kim, D.J. Adams, J.H. Kim, and Y.K. Paik. 2001. *J. Antimicrob. Chemother*. 47: 513-519.

Park, B.-S., D.-Y. Kim, P.J. Rosenthal, S.-C. Huh, B.J. Lee, E.-J. Park, S.-M. Kim, J.-E. Kim, M.-H. Kim, T.-L. Huh, Y.-J. Choi, K.-H. Suh, W.-S. Choi, and S.-E. Lee. 2002. Synthesis and evaluation of new antimalarial analogues of quinoline alkaloids derived from *Cinchona ledgeriana* moens ex trimen. *Bioorg. Med. Chem. Lett*. 12: 1351-1355.

Park, B.S., W.S. Choi, and S.E. Lee. 2003. Antifungal activity of piperoctadecalidine, a piperidine alkaloid derived from long pepper, *Piper longum* L., against phytopathogenic fungi. *Agr. Chem. Biotech*. 46: 73-75.

Paulo, A., E.T. Gomes, J. Steele, D.C. Warhurst, and P.J. Houghton. 2000. Antiplasmodial activity of *Cryptolepis sanguinolenta* alkaloids from leavesand roots. *Planta Med*. 66: 30-34.

Perez, E., J. Saez, S. Blair, X. Franck, and B. Figadere. 2004. Isoquinoline alkaloids from *Duguetia vallicola* stem bark with antiplasmodial activity. *Lett.Org. Chem*. 1: 102-104.

Philippe, G., P. De Mol, M. Zeches-Hanrot, J.-M. Nuzillard, M.-H. Tits, L. Angenot, and M. Frederich. 2003. Indolomonoterpenic alkaloids from *Strychnos icaja* roots. *Phytochemistry* 62: 623-629.

Rafatro, H., D. Ramanitrahasimbola, P. Rasoanaivo, S. Ratsimamanga-Urverg, A. Rakoto-Ratsimamanga, and F. Frappier. 2000a. Reversal activity of the naturally occurring chemosensitizer malagashanine in *Plasmodium* malaria. *Biochem. Pharm*. 59: 1053-1061.

Rafatro, H., R.K. Verbeeck, T.Y.R. Gougnard, P.J.M. De Jonghe, P. Rasoanaivo, A. Laurent, and G. Lhoest. 2000b. Isolation from rat urine and human liver microsomes and identification by electrospray and nanospray tandem mass spectrometry of new malagashanine metabolites. *J. Mass Spec*. 35: 1112-1120.

Rasoanaivo, P., S. Ratsimamanga-Urverg, R. Milijaona, H. Rafatro, A. Rakoto- Ratsimamanga, C. Galeffi, and M. Nicoletti. 1994. *In vitro* and *in vivo* chloroquine-potentiating action of Strychnos myrtoides alkaloids against chloroquine-resistant strains of *Plasmodium malariae*. *Planta Med*. 60: 13-16.

Richardson, M.D. 1991. Opportunistic and pathogenic fungi. *J. Antimicrob. Chemother*. 28, Suppl. A, 1-11.

Saad, H.E., E.S. Saleh, and W.T. Shier. 1995. Biological activities of pyrrolidinoindoline alkaloids from *Calycodendron milnei*. *Planta Med*. 61: 313-316.

Sarma, B.K., V.B. Pandey, G.D. Mishra, and U.P. Singh. 1999. Antifungal activity of berberine iodide, a constituent of *Fumaria indica*. *Folia Microbiol*. 44: 164-166.

Sato, Y., Y. Sato, H. Kaneko, E. Bianchi, and H. Kataoka. 1969. Alkaloids from *Solanum congestiflorum*. *J. Org. Chem*. 34: 1577-1582.

Saxena, S., N. Pant, D.C. Jain, and R.S. Bhakuni. 2003. Antimalarial agents from plant sources. *Curr. Sci*. 85: 1314-1329.

Schwikkard, S., and F.R. van Heerden. 2002. Antimalarial activity of plant metabolites. *Nat. Prod. Rep*. 19: 675-692.

Sener, B., I. Orhan, and J. Satayavivad. 2003. Antimalarial activity screening of some alkaloids and the plant extracts from Amaryllidaceae. *Phytother. Res.* 17: 1220-1223.

Sharma, P., and J.D. Sharma. 2001. A review of plant species assessed *in vitro* for antiamoebic or both antiamoebic and antiplasmodial properties. *Phytother. Res.* 15: 1-17.

Sheng, W.D., M.S. Jiddawi, X.Q. Hong, and S.M. Abdulla. 1997. Treatment of chloroquine-resistant malaria using pyrimethamine in combination with berberine, tetracycline or cotrimoxazole. *East Afr. Med. J.* 74: 283-284.

Shuler, A.V. 1985. Malaria: Ancient scourge of humankind. in Malaria: Meeting the Global Challenge. Oelgeschlager, Gunn & Hain Inc. Boston, Massachusetts.

Singh, N.V., S. Azmi, S. Maurya, U.P. Singh, R.N. Jha, and V.B. Pandey. 2003. Two plant alkaloids isolated from *Corydalis longipes* as potential antifungal agents. *Folia Microbiol.* 48: 605-609.

Singh, S.K., B.K. Sarma, J.S. Srivastava, U.P. Singh, and A.B. Ray. 1999. Antifungal activity of Δ^3-alstovenine, a plant alkaloid isolated from *Alstonia venenata. Folia Microbiol.* 44: 510-512.

Singh, U.P., B.K. Sarma, P.K. Mishra, and A.B. Ray. 2000. Antifungal activity of venenatine, an indole alkaloid isolated from *Alstonia venenata. Folia Microbiol.* 45: 173-176.

Solis P.N., C. Lang'at, M.P. Gupta, G.C. Kirby, D.C. Warhusrst, and J.D. Phillipson. 1995. Bioactive compounds from *Psychotria camponutans. Planta Med.* 61: 62-5.

Sriwilaijareon, N., S. Petmitr, A. Mutirangura, M. Ponglikitmongkol, and P. Wilairat. 2002. Stage specificity of *Plasmodium falciparum* telomerase and its inhibition by berberine. *Parasitol. Int.* 51: 99-103.

Steele, J.C.P., N.C. Veitch, G.C. Kite, M.S.J. Simmonds, and D.C. Warhurst. 2002. Indole and *β*-carboline alkaloids from *Geissospermum sericeum. J. Nat. Prod.* 65: 85-88.

Thouvenel, C., J.C. Gantier, P. Duret, C. Fourneau, R. Hocquemiller, M.E. Ferreira, A.R. de Arias, and A. Fournet. 2003. Antifungal compounds from *Zanthoxylum chiloperone* var. *angustifolium. Phytother. Res.* 17: 678-680.

Tran, Q.L., Y. Tezuka, J.-Y. Ueda, N.T. Nguyen, Y. Maruyama, K. Begum, H.-S. Kim, Y. Wataya, and Q.K. Tran. 2003. *In vitro* antiplasmodial activity of antimalarial medicinal plants used in Vietnamese traditional medicine. *J. Ethnopharm.* 86: 249-52.

Trigalo, A.F., F. Frappier, P. Rasoanaivo, and S. Ratsimamanga. 2003. Preparation of strychnobrasiline derived alkaloids for use in pharmaceutical compositions as antitumor and antimalarial agents. Fr. Demande 35 pp. FR 2839513 A1 20031114 Patent written in French. Application: FR 2002-5740 20020507.

Tshibangu, J.N., A.D. Wright, and G.M. Konig. 2003. HPLC isolation of the anti-plasmodially active bisbenzylisoquinone alkaloids present in roots of *Cissampelos mucronata. Phytochem. Anal.* 14: 13-22.

Van der Heyde, H.C., M.M. Elloso, W.J. Vande, K. Schell, and W. Weidanz, 1995. Use of hydroethidine and flow cytometry to assess the effects of leucocytes on the malarial parasite. *Clin. Diagn. La. Immunol.* 2: 417-425.

Vennerstrom, J.L., and D.L. Klayman. 1988. Protoberberine alkaloids as antimalarials. *J. Med. Chem.* 31: 1084-1087.

Verpoorte, R. 1998. Antimicrobially active alkaloids. Pages 397-433 in *Alkaloids: Biochemistry, Ecology, and Medicinal Applications,* M.F. Roborts, and M. Wink, eds., Plenum Press, New York.

Volleková, A., D. Košt'álová, V. Kettmann, and J. Tóth. 2003. Antifungal activity of *Mahonia aquifolium* extract and its major protoberberine alkaloids. *Phytother. Res.* 17: 834-837.

Weniger, B., B.-H. Um, A. Valentin, A. Estrada, A. Lobstein, R. Anton, M. Maille, and M. Sauvain. 2001. Bioactive acridone alkaloids from *Swinglea glutinosa. J. Nat. Prod.* 64: 1221-1223.

White, T.C., R.A. Bowden, and K.A. Marr. 1998. Clinical, cellular, and molecular factors that contribute to antifungal drug resistance. *Clin. Microbiol. Rev.* 11: 382-402.

Wink, M. 1998. Modes of Action of Alkaloids. Pages 301-326 in *Alkaloids: Biochemistry, Ecology, and Medicinal Applications*, M.F. Roborts, and M. Wink, eds., Plenum Press, New York.

Wong-Beringer, A., and J. Kriengkauykiat. 2003. Systemic antifungal therapy: new options, new challenges. *Pharmacotherapy* 23: 1441-1462.

World Health Organization, Regional office for the Western Pacific. Report:Meeting on Antimalarial Drug Development, Shanghai, China, 16-17 November, 2001.

Wright, C.W., J. Addae-Kyereme, A.G. Breen, J.E. Brown, M.F. Cox, S.L. Croft, Y. Goekcek, H. Kendrick, R.M. Phillips, and P.L. Pollet. 2001. Synthesis and evaluation of cryptolepine analogues for their potential as new antimalarial agents. *J. Med. Chem.* 44: 3187-3194.

Wright, C.W., S.J. Marshall, P.F. Russell, M.M. Anderson, J.D. Phillipson, G.C. Kirby, D.C Warhurst, and P.L. Schiff, Jr. 2000. *In vitro* antiplasmodial, antiamoebic, and cytotoxic activities of some monomeric isoquinoline alkaloids. *J. Nat. Prod.* 63: 1638-1640.

Zhang, H., S.X. Qiu, P. Tamez, G.T. Tan, Z. Aydogmus, N. Van Hung, N.M. Cuong, C. Angerhofer, D.D. Soejarto, J.M. Pezzuto, and H.H.S. Fong. 2002. Antimalarial agents from plants II. Decursivine, a new antimalarial indole alkaloid from *Rhaphidophora decursiva*. *Pharm. Biol.* 40: 221-224.

Zhao, W.M., J.L. Wolfender, K. Hostettmann, R. Xu, and G. Qin. 1998. Antifungal alkaloids and limonoid derivatives from *Dictamnus dasycarpus*. *Phytochemistry* 47: 7-11.

Zhou, C.X., J.Y. Liu, W.C. Ye, C.H. Liu, and R.X. Tan. 2003. Neoverataline A and B, two antifungal alkaloids with a novel carbon skeleton from *Veratrum taliense*. *Tetrahedron* 59: 5743-5747.

7

Anticancer Compounds from Higher Plants

HIDEJI ITOKAWA,[a] KOICHI TAKEYA[b] AND KUO-HSIUNG LEE[a]

[a]Natural Products Laboratory, School of Pharmacy, University of North Carolina, Chapel Hill, North Carolina 27599-7360, USA

[b]Tokyo University of Pharmacy and Life Science, 1432-1 Horinouchi, Hachioji, Tokyo, 192-0392 Japan

INTRODUCTION

Numerous anticancer agents have been isolated from natural sources, especially from microorganisms and plants. No single compound type is useful for all cancer therapy. However, various types of substances, e.g., alkaloids, lignans, terpenes, and steroids (Itokawa 1988), are effective against different cancers and tumors.

Higher plants have yielded many effective, clinically used anticancer agents, including *Vinca* alkaloids, *Taxus* diterpenes, *Camptotheca* alkaloids, and *Podophyllum* lignans. Accordingly, many prior reviews have discussed anticancer or antitumor compounds from higher plants (Lee 1993, Itokawa *et al.* 1999, 2000). Recent research on the development of antitumor agents from Chinese herbal medicines has also been reported (Tang *et al.* 2003 a,b). Current reviews have discussed antitumor agents from natural products (Lee 2004, Cragg and Newman 2004).

Suffness and Douros (1982) suggested the following definitions to avoid confusion in terminology. **CYTOTOXICITY** is used when compounds or extracts show activity against tumor cell lines. **ANTITUMOR** or **ANTINEOPLASTIC** indicates that the materials are effective *in vivo* in experimental systems. **ANTICANCER** refers to compounds that are active clinically against human cancers.

The search for potential anticancer agents from natural sources has been guided mainly by bioassays. Thus, development of novel, clinically useful anticancer agents is highly dependent on the bioassay screening systems as well as the sample sources. Initially, L1210, P388 and KB cells were used in preliminary screening for antineoplastic activity. To increase accuracy, antineoplastic activity has more recently been evaluated by using a panel of human cancer cell lines, i.e., an *in vitro* Phase II study as used by the National Cancer Institute (NCI), Washington, DC.

In this chapter, we have divided our discussion of anticancer and antineoplastic compounds by plant species. A brief background and the most current research discoveries are included. The seminal discoveries of taxol and camptothecin by Wall and Wani (1996) are discussed in order to demonstrate how these two compounds have influenced the further development of naturally occurring, natural product-derived, and synthetic entities as potential drugs that interact either with tubulin or topoisomerase I (Wall and Wani 1996, Oberlies and Kroll 2004). Accordingly, tubulin interactive agents and topoisomerase I inhibitors that are in preclinical and clinical development have been reported (Cragg and Newman 2004).

Anticancer Groups

Vinca spp.

Vinca alkaloids are famous worldwide as anticancer agents. Originally, these compounds were isolated from *Vinca rosea* (= *Catharanthus roseus*) (Apocynaceae). In Madagascar, they are used in folk medicine as hypotensors, astringents, emetics, and agents to inhibit milk secretion. Moreover, native peoples in West Indian Islands use *Vinca* spp. to depress blood sugar. When an extract of this plant was given non-orally, leucopenia and indirect inhibition of cellular nuclear division were observed, leading to the discovery of the anticancer activity.

Among the more than 60 kinds of alkaloids isolated from *Vinca* spp., vinblastine (**1**) and vincristine (**2**) are the most active compounds. The former is effective against Hodgkin's lymphoma and the latter against acute childhood leukemia. Among numerous analogues designed to have activity against other tumor types or fewer side effects, navelbine (vinorelbine) (**3**) was developed by Burroughs Wellcome and is used against non-small cell lung and advanced breast cancers (Potier 1989, Jenks and Smigel 1995). This compound is a synthetic analogue of **1**, but has an eight-membered, rather than a nine-membered, C ring and a dehydrated D ring (Potier 1989).

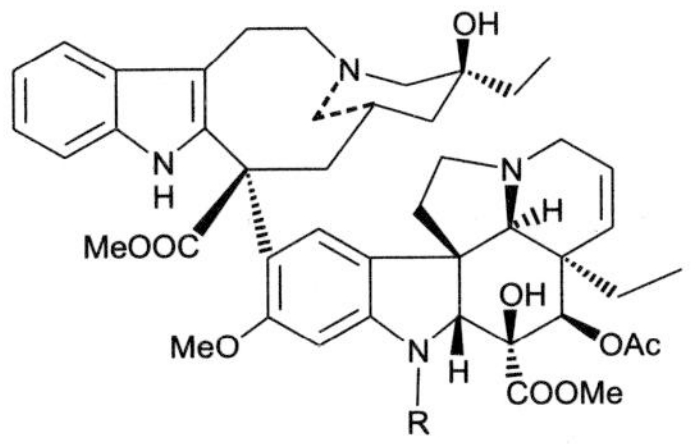

1 Vinblastine (Velban) R = CH_3
2 Vincristine (Oncovin) R = CHO

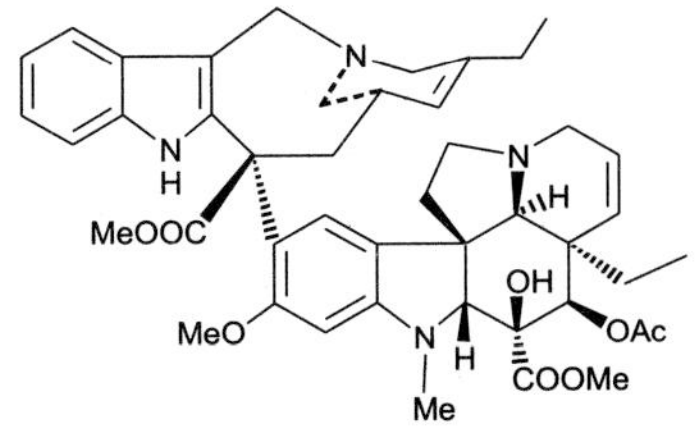

3 Vinorelbine (Navelbine)

Fig. 7.1 Main vinca alkaloids used clinically.

Taxus *spp.*

Recently, *Taxus* alkaloids were reviewed in the book *"Taxus, genus Taxus"*, edited by Itokawa and Lee (2003).

Taxol (paclitaxel) (**4**), a unique compound with a taxane ring, was isolated from the bark of *Taxus brevifolia* (Taxaceae) by Wall and Wani, the foremost researchers in the discovery of the anticancer potential of taxol (Wani *et al.* 1971, Wall and Wani 1996). It showed substantial anticancer activity in patients with solid tumors that were refractory to standard chemotherapy. At first, a limited natural supply of taxol severely restricted the full exploration of its antineoplastic potential. Various approaches to increase the supply were explored: surveying various *Taxus* species for optimal taxol content, improving extraction methods, developing alternative renewable resources, and improving semi-synthesis from baccatin III. As a result, many related compounds were discovered or developed as antineoplastic agents.

Paclitaxel (**4**) has a unique mode of anticancer action (it promotes the assembly of microtubules and consequently inhibits mitosis) and is active against breast, brain, tongue, endometrial, and ovarian cancers (Cragg and Suffness 1988, Kingston 1991). Clinically used docetaxel (taxotere) (**5**) is synthesized from the more readily available 10-deacetylbaccatin III (**6**), which is found in the needles of the European yew tree *Taxus baccata* L. (Lavelle *et al.* 1993).

As mentioned in the book *"Taxus genus Taxus"* edited by Itokawa and Lee (2003), many natural taxoids have been isolated. Dantaxusins C (**7**) and D (**8**) were isolated from *Taxus yunnanensis* (Shinozaki *et al.* 2002). Three C-14 oxygenated taxanes, hongdoushans A-C (**9-11**), were also found in the wood of *T. yunnanensis* (Banskota *et al.* 2002). Taxumairol Q (**12**) and 13-*O*-acetyl wallifoliol (**13**) were isolated from the leaves and twigs of *T. sumatrana* (Shen *et al.* 2002). Seven minor taxane derivatives were isolated from the needles of *T. canadensis*. Four (**14-17**) of these natural taxanes are *O*-glycosylated (**17** being the only reported taxane with a glucose on ring C), and two (**18** and **19**) have a dimethylamino C-5 side chain. An unusual double bond at C-5(6) has been identified in taxane **20** (Shi *et al.* 2003a). Ten taxanes (**21-30**) were isolated from the methanol extract of the needles of the Canadian yew, *Taxus canadensis* (Shi *et al.* 2003b). Three new taxoids (**31-33**) with oxetane rings were isolated from *Taxus chinensis* (Wang *et al.* 2004).

Recently, the design, synthesis, and bioactivities of conjugates of taxol and 3,17β-estradiol have been reported (Liu *et al.* 2004). In addition, structure-activity relationships of ring C-secotaxoids were recently published (Appendino *et al.* 2004).

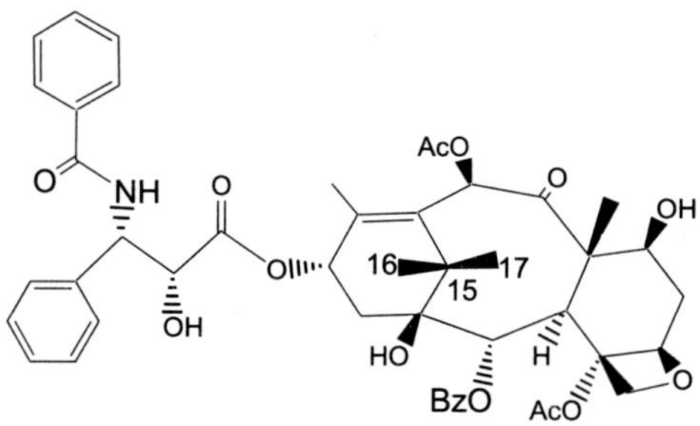

Taxol (Paclitaxel) (**4**)

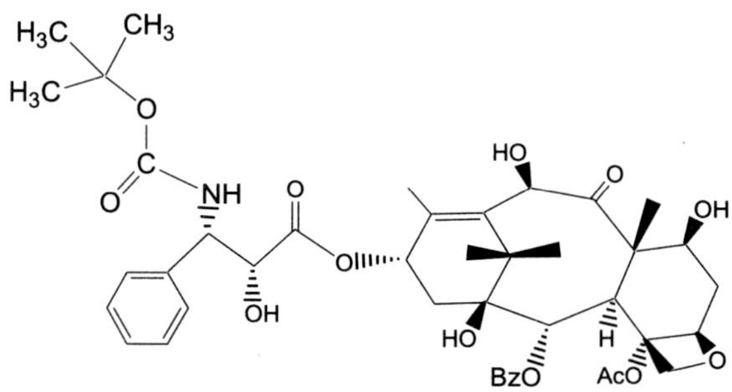

Taxotere (docetaxel) (**5**)

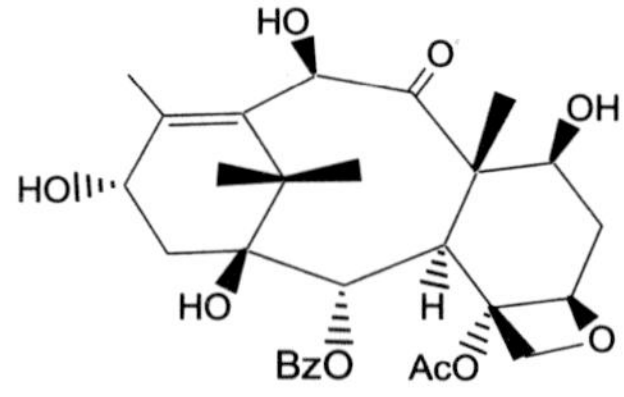

10-Deacetylbaccatin III (**6**)

Fig. 7.2 Main taxoids.

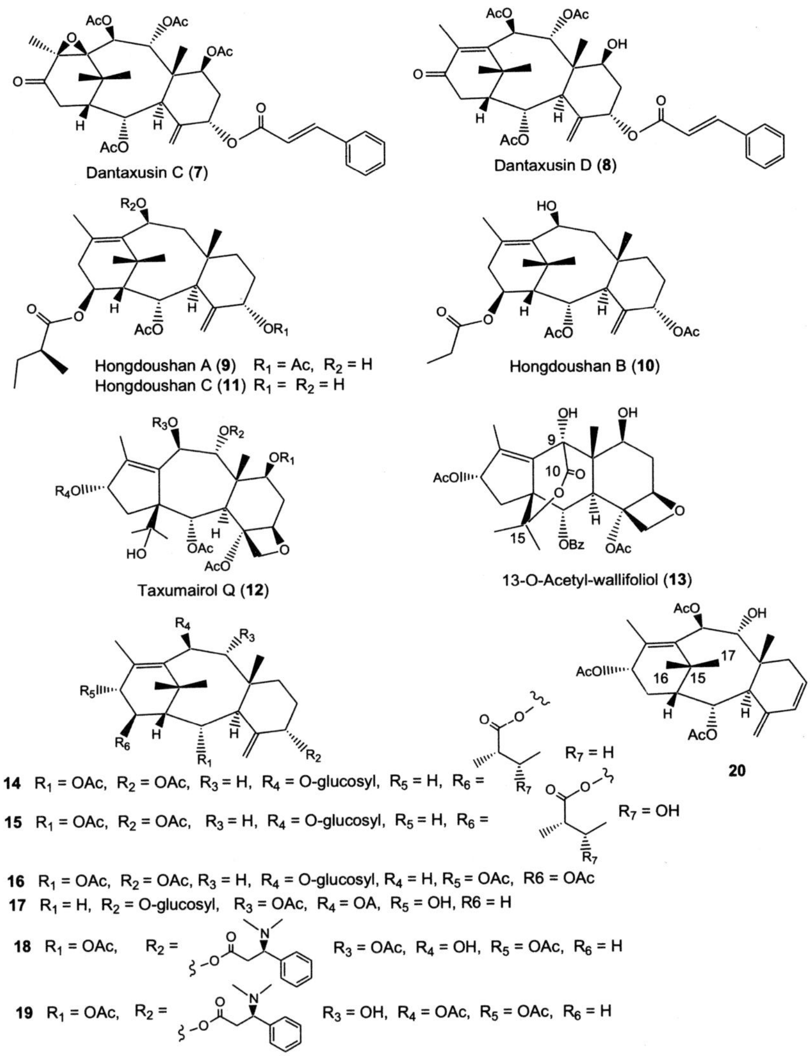

Fig. 7.3 Newly isolated taxoids after 2002.

Camptotheca spp.

Thomas *et al.* (2004) recently published a useful review of camptothecin. Camptothecin (CPT) (**34**) is a potent antitumor agent isolated from *Camptotheca acuminata* (Nyssaceae). Its structure was determined to be a pentacyclic alkaloid and was first reported in 1966 (Wall *et al.* 1966, Wall 1993). The discovery that the primary cellular target of CPT is type I DNA topoisomerase created renewed interest in the drug (Hsiang *et al.* 1985, 1989, Covey *et al.* 1989).

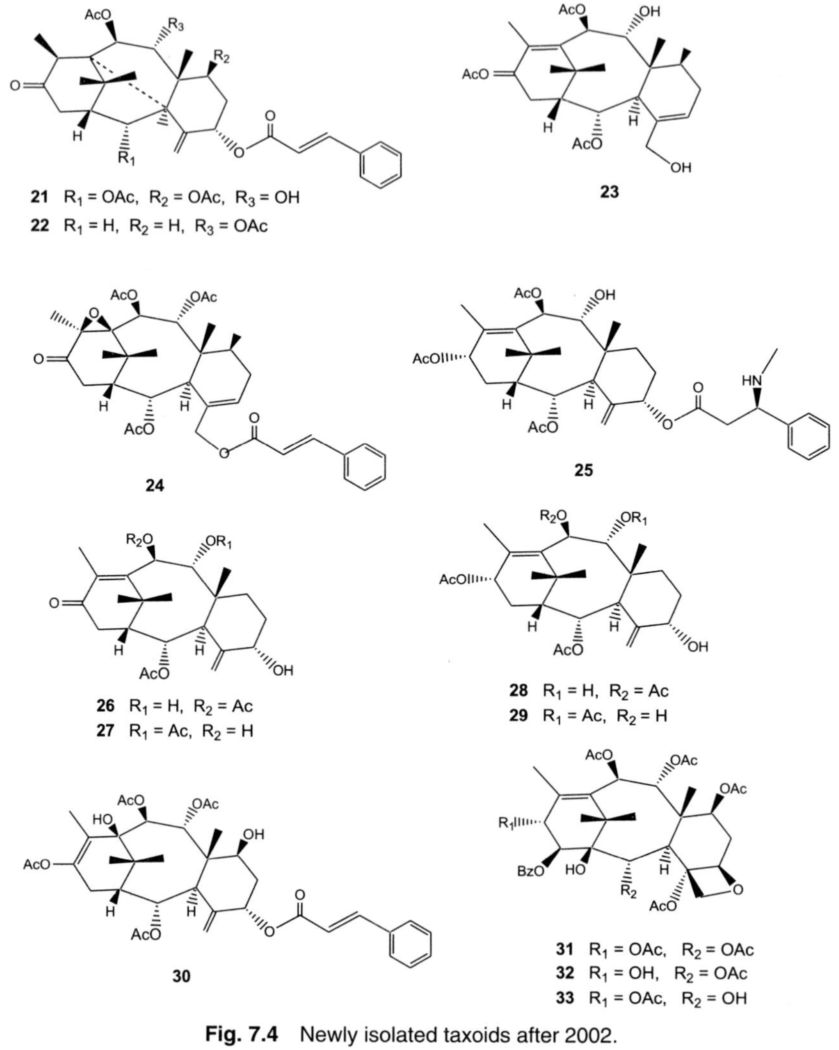

Fig. 7.4 Newly isolated taxoids after 2002.

Advances in the medicinal chemistry of CPT resulted in the semi-synthetic, more water-soluble analogues topotecan (Hycamtin, **35**) and irinotecan (Camptosar, **36**), which are used clinically for the treatment of colon and ovarian cancers, respectively (Giovanella *et al.* 1989, Saltz *et al.* 2000, Ozols 2000, Vanhoefer *et al.* 2001). Additional CPT analogues are under investigation, and are also of interest in combination regimens as radiation sensitizers (Lerchen 1999, Caffo 2001, Kim and Lee 2002, Dallavalle *et al.* 2004).

The clinical application of CPTs has been discussed in several excellent reviews (Takimoto *et al.* 1998, O'Leary and Muggia 1998, Ulukan and Swaan 2002, Garcia-Carbonero and Supko 2002).

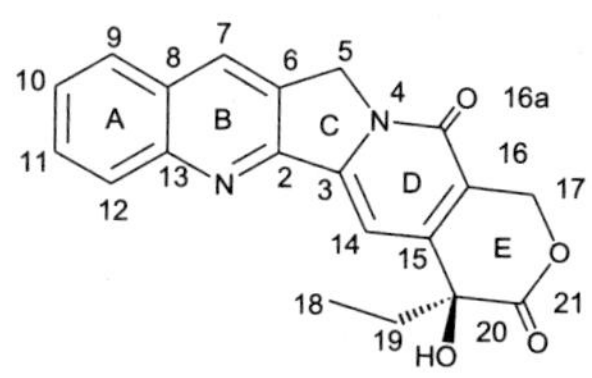

34

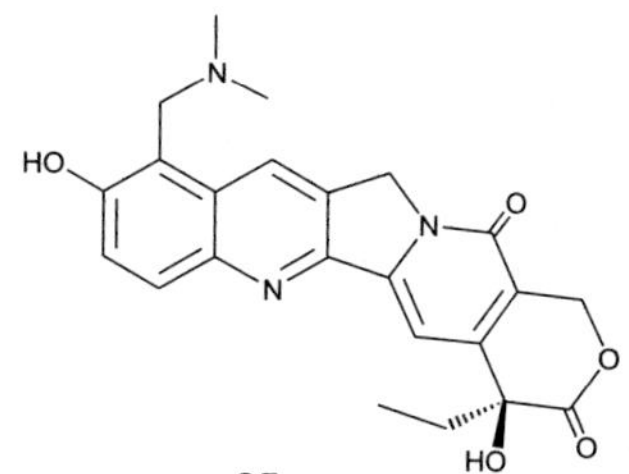

35

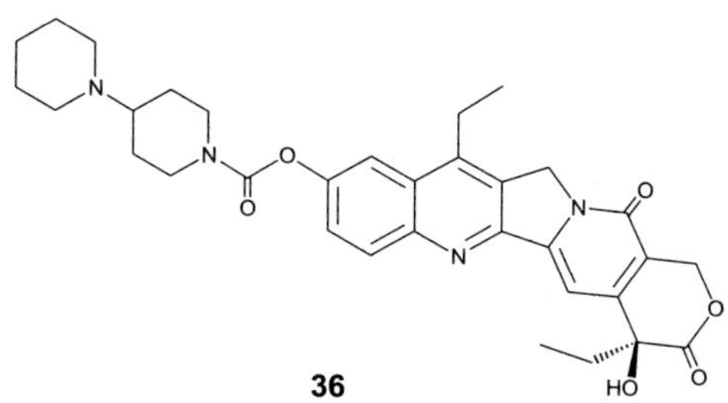

36

Fig. 7.5 The sturctures of camptothecines.

Podophyllum spp.

Lee and Xiao (2004) recently reviewed podophyllotoxins and related analogs. The genus *Podophyllum* (Berberidaceae), including the American *P. peltatum* L. and Indian or Tibetan *P. emodi* Wall (syn. *P. hexandrum* Royle), has been used for centuries for its medicinal properties. The indigenous populations of North America and the Himalayas have long valued *Podophyllum* plants for their cathartic and cholagogue properties. Podophyllin, the alcoholic extract of *Podophyllum* rhizomes, was listed in the first American Pharmacopoeia (1820), and was later introduced into European Pharmacopoeias. However, it was removed from the U.S. Pharmacopoeia in 1942 due to its undesirable toxicity (Imbert 1998).

Podophyllotoxin (**37**) is a representative lignan isolated in 1880 from rhizomes of *Podophyllum peltatum*. It is an aryltetralinlactone cyclolignan, and its skeleton is a flat, rigid, five-ring system. Podophyllotoxins are widespread in the plant kingdom and not limited to the *Berberidaceae* family. Key chemical structures include podophyllotoxin (**37**), deoxypodophyllotoxine (**38**), 4′-demethylpodophyllotoxin (**39**), 4′-demethylepipodophyllotoxin (**40**), α-peltatin (**41**), β-peltatin (**42**), and their corresponding glycosides.

Because podophyllotoxin was found to inhibit cell division, possible antineoplastic activity was examined. The biological assessment of **37** was followed by discovery of its mode of action and was followed by chemical modification of this natural prototype and successful development of the clinically useful anticancer drugs etoposide (**43**) and teniposide (**44**).

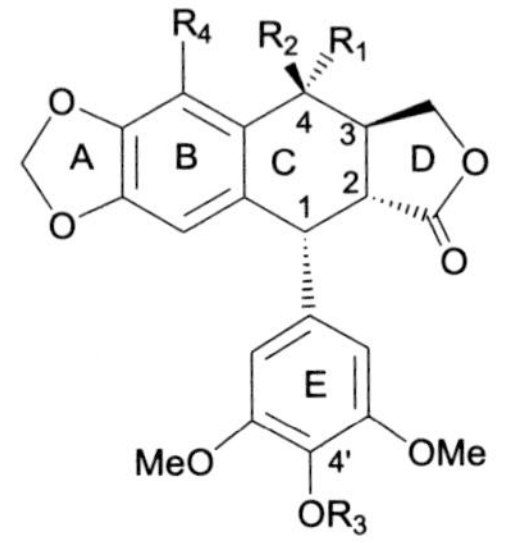

	R1	R2	R3	R4
Podophyllotoxin (**37**)	OH	H	Me	H
Deoxypodophyllotoxin (**38**)	H	H	Me	H
4'-Demethylpodophyllotoxin (**39**)	OH	H	H	H
4'-Demethylepipodophyllotoxin (**40**)	H	OH	H	H
α-Peltatin (**41**)	H	H	H	OH
β-Peltatin (**42**)	H	H	Me	OH

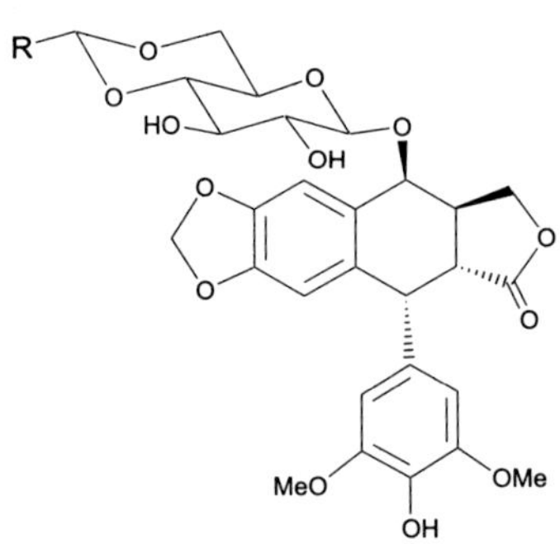

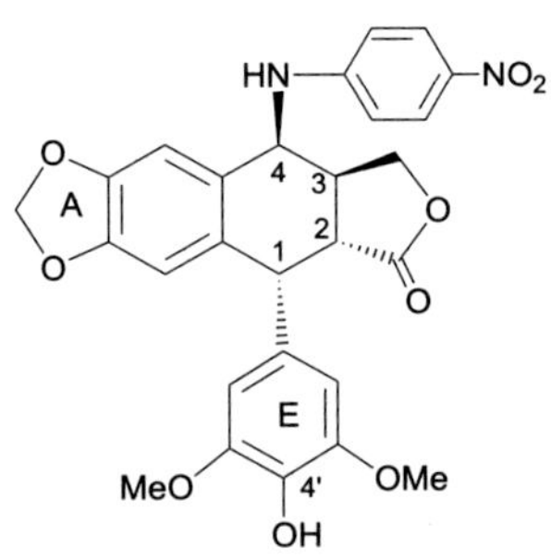

43 R = Me

44 R =

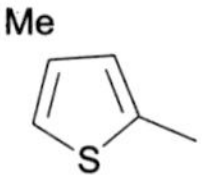

GL-331 (**45**)

Fig. 7.6 Podophyllotoxins.

However, because etoposide still has limitations, extensive structure-activity relationship, enzyme interaction, and computational studies have been performed to generate new compounds using drug improvement principles. The rational design of improved etoposide analogs has made use of new computational strategies (Cho *et al.* 1996, Xiao *et al.* 2002). In particular, several series of 4-alkylamino and 4-arylamino epipodophyllotoxin analogues were synthesized from the natural product podophyllotoxin (Lee *et al.* 1989).

Compared with etoposide, several synthetic compounds have shown a similar or increased percent inhibition of DNA topo II activity, percent protein-linked DNA breakage (Chang *et al.* 1991), and particularly, increased cytotoxicity in etoposide-resistant cell lines. GL-331 (**45**) (Wang *et al.* 1990), which contains a *p*-nitroanilino moiety at the 4β position of etoposide, has emerged from the preclinical development as a clinical trials candidate.

Like etoposide, GL-331 (**45**) inhibits DNA topoisomerase II, causing DNA double strand breakage and G2-phase arrest. Formulated **45** has good stability and biocompatibility and similar pharmacokinetic profiles to those of etoposide (Wang *et al.* 1990). After being patented by Genelabs Technologies, Inc., GL-331 entered and completed phase I clinical trials as an anticancer drug at the M.D. Anderson Cancer Center. It showed good antitumor efficacy (Liu *et al.* 1993) against four tumor types (non-small and small cell lung, colon, and head/neck cancers). Side effects were minimal with the major toxicity being cytopenia. The maximum tolerated dose (MTD) was declared 300 mg/m^2, while that for etoposide was 140 mg/m^2. Other advantages of GL-331 (**45**) over etoposide include (a) greater *in vitro* and *in vivo* activity, (b) a shorter synthesis and, thus, easier manufacture, and (c) evidence of activity in refractory tumors because it overcomes drug-resistance in many cancer cell lines (KB/VP-16, KB/VCR, P388/ADR, MCF-7/ADR, L1210/ADR, HL60/ADR, and HL60/VCR) (Liu *et al.* 1993).

In summary, **45** is an exciting chemotherapeutic candidate, which illustrates successful preclinical drug development from plant natural products.

Cephalotaxus spp.

Cephalotaxus alkaloids and homoharringtonine were recently reviewed (Itokawa *et al.* 2004). The genus *Cephalotaxus* contains coniferous evergreen trees and shrubs that are indigenous to Asia. Historically, the bark has been used in China as traditional medicine for a variety of indications. Through screening assays and preliminary clinical trials, Chinese investigators discovered that the total alkaloid fraction from *Cephalotaxus fortunei* Hook. F (Cephalotaxaceae) possessed definite antitumor activity (Huang *et al.* 1983). Homoharringtonine (**46**, HHT) (cephalotaxine 4-methyl-2-hydroxy-4-methylpentyl butanedioate), one of the most active *Cephalotaxus* alkaloids, was obtained by alcoholic extraction from the Chinese evergreen tree *C. harringtonia* K. Koch var. *harringtonia* (Grem *et* *al.* 1988). Other *Cephalotaxus* alkaloids have been isolated from various *Cephalotaxus* spp. (Paudler *et al.* 1963, Paudler and McKay 1973, Spencer *et al.* 1976).

The parent compound cephalotaxine (**47**) and its esters represent a group of alkaloids with a unique ring system. Although cephalotaxine is devoid of antitumor activity, ester alkaloids derived from cephalotaxine are of particular anticancer interest. The most active compounds in this series are homoharringtonine (**46**) and harringtonine (**48**), isolated by Powell *et al.* (1970). In addition to **46** and **48**, two other active ester alkaloids isoharringtonine (**49**) and deoxyharringtonine (**50**) occur in *C. harringtonia* (Powell *et al.* 1972). The Japanese species *C. harringtonia* (Knight) Koch. f. *drupacea* (Sieb. & Zucc.) Kitamura (= *C. harringtonia* var. *drupacea* (Sieb. & Zucc.) = *C. drupacea* Sieb. & Zucc.) contains drupacine (**51**) (Powell *et al.* 1974), in addition to the characteristic *Cephalotaxus* alkaloids **46–50**. In the last several years, more than 20 *Cephalotaxus* alkaloids have been isolated from various *Cephalotaxus* spp. (Takano *et al.* 1996a-e, 1997).

Homoharringtonine (**46**) has been investigated in Phase II clinical trials as an anticancer agent at the National Cancer Institute (NCI), USA. However, its side effects still remain an issue. Thus, the authors have continued to study new components from this plant species and to develop new analogs on the basis of the structure-activity relationships of these alkaloids.

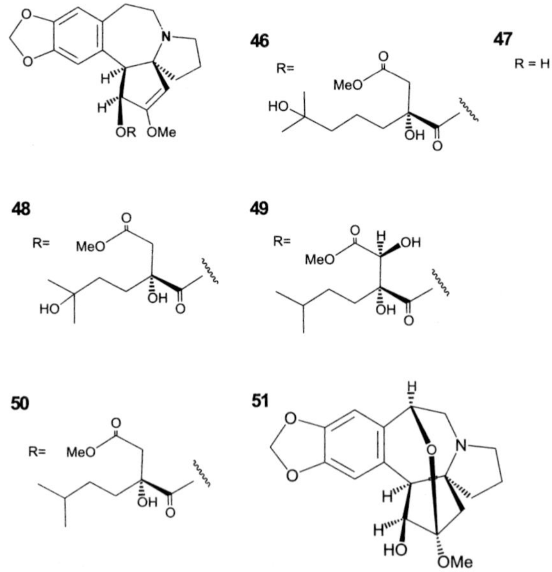

Fig. 7.7 Cephalotaxus alkaloids.

OTHER ANTINEOPLASTIC (OR ANTITUMOR) GROUPS (CANDIDATES AS ANTICANCER DRUGS)

Colchicum spp.

The alkaloid colchicine (**52**) was isolated from *Colchicum autumnale* L. (Liliaceae) and has long been and is still used medicinally, currently to treat gout and familial Mediterranean fever (**53**). Colchicine and thiocolchicine (SCH_3 rather than OCH_3 at C-10) are mitotic inhibitors, which inhibit polymerization of tubulin with IC_{50} values of 1.5 and 0.65 μM, respectively (Brossi 1984). Although they show antileukemic activity, they are too toxic to use as anticancer agents. Thus, the Natural Product Laboratory (NPL) at the University of North Carolina (UNC) has synthesized many new analogues that might overcome this deficiency. The C-7 acetoamide group on the B ring has been a target of modification and has been replaced by ketone (**54**, thiocolchicine), hydroxyl (**55**), and ester (**56, 57**) groups (Shi *et al.* 1997). Thiocolchicone and the (–)-aS, 7S optically pure enantiomers [the C-7 alcohol, (–)-**55**, and its acetate, (–)-**56**, and isonicotinoate, (–)-**57**, esters] were equally or more active (TTP IC_{50} values ranging from 0.56 to 0.76 μM) than thiocolchicine. In a C-ring modification, deaminodeoxy colchinol-7-one thiomethy ether or allo-ketone (**58**) was synthesized by reacting thiocolchicone (**53**) with aniline, causing ring contraction and forming a six membered C ring rather than the seven membered ring found in the natural product. This analogue was equipotent with its seven-membered ring parent compound. Three related ring contracted colchicinoids (**59-61**), with OCH_3 rather than SCH_3 substitution on the C ring, showed excellent activities in drug sensitive and resistant KB cell lines (Shi *et al.* 1998). Investigation of demethylated A-ring colchicines analogues led to a very interesting result. Full tubulin binding affinity and inhibition of tublin polymerization require all three phenolic groups in the A ring to be methylated. However, in a topo II inhibition assay, 1,2,3-demethylthiocolchicine (**62**) was more potent than etoposide (Guan *et al.* 1998). Thus, tridemethylated colchicines and thiocolchicines are a new class of DNA topo II inhibitory agents. Two compounds (**63, 64**) with an acetoamide group at C-7 and amino or amide groups at C-10 showed topo II inhibitory and cytotoxic activity with ED_{50} values of 0.36 and 0.48 μg/mL against the MCF-7 breast cancer cell line and 0.72 and 0.98 μg/ml against the CAKI-1 kidney cancer cell line (Bastow *et al.* 1993).

Maytenus spp.

Commonly known as "Cangorosa", *Maytenus ilicifolia* Mart. ex Reiss. (Celastraceae) is used in South America for its analgesic, antipyretic,

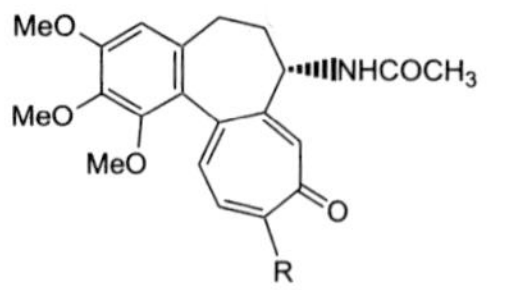

52 R= OMe
53 R= SCH_3

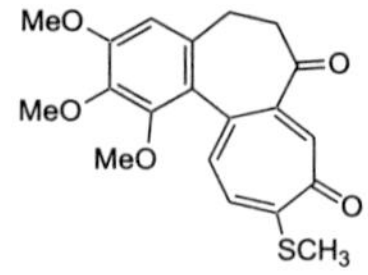

54

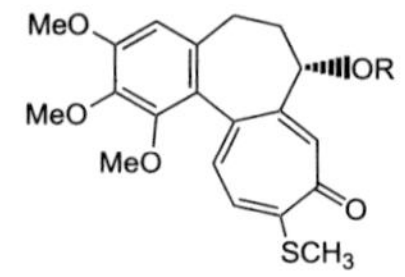

55 R= H
56 R=$COCH_3$
57 R= —CO– 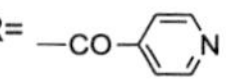

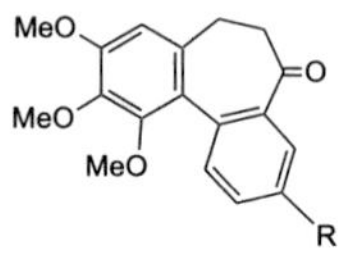

58 R= SCH_3
59 R= OMe

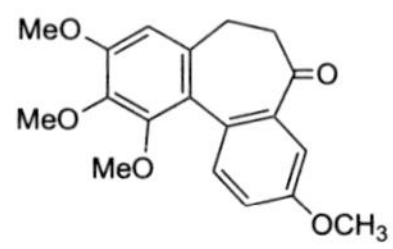

60

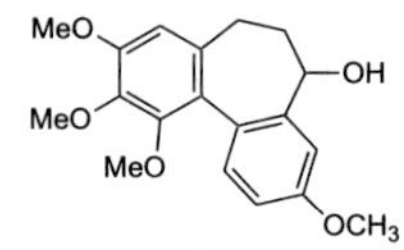

61

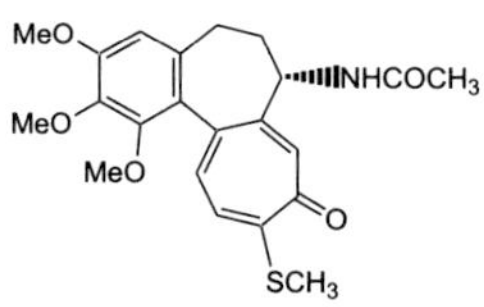

62

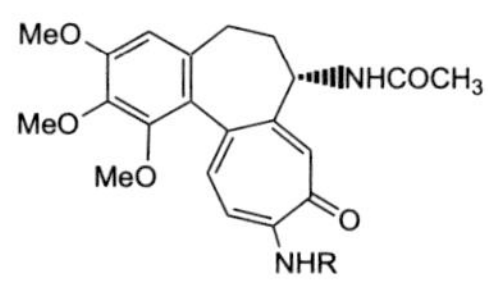

63 R= CH_3
64 R= COC_6H_5Cl

Fig. 7.8 Colchicine and related compounds.

antiseptic, and anticancer effects, and, particularly in Paraguay, for birth control. Maitenin and pristimerin have been identified as toxic principles.

In 1990-91, Itokawa *et al.* conducted the following fractionation and isolation of cytotoxic principles of *M. ilicifolia*. The methanolic extract of the plant was partitioned between chloroform and water, and then *n*-butanol and water. Purification of each extract was guided by cytotoxic assays and led to the isolation of friedelane- and pristimerin-type triterpenes. From the chloroform soluble extract, D:A-friedoolean-24-al-3-en-3-ol-2-on-29-oic acid (**65**), D:A-friedoolean-1-en-29-ol-3-one (**66**), maytenoic acid (**67**), D:B-friedoolean-5-en-3β,29-diol (**68**), D:A-friedoolean-29-ol-3-one (**69**), pristimerin (**70**) and salaspermic acid (**71**), and the triterpene dimers cangorosin A (**72**), isocangorosin A (**73**), 6′,7′-dihydroisocangorosin A (**74**) and cangorosin B (**75**) were isolated. The *n*-butanol soluble extract yielded isopristimerin III (**76**) and isotingenone III (**77**).

The isolation of **65** from *M. ilicifolia* is noteworthy from a biogenetic viewpoint. In **65**, the carbon at position-24 is oxidized to a formyl moiety; thus, suggesting a possible biogenetic route from friedelin-to pristimerin-type triterpenes, and furthermore to triterpene dimers by the oxidative elimination of the 24-Me. In addition, the cytotoxic activities of isolated

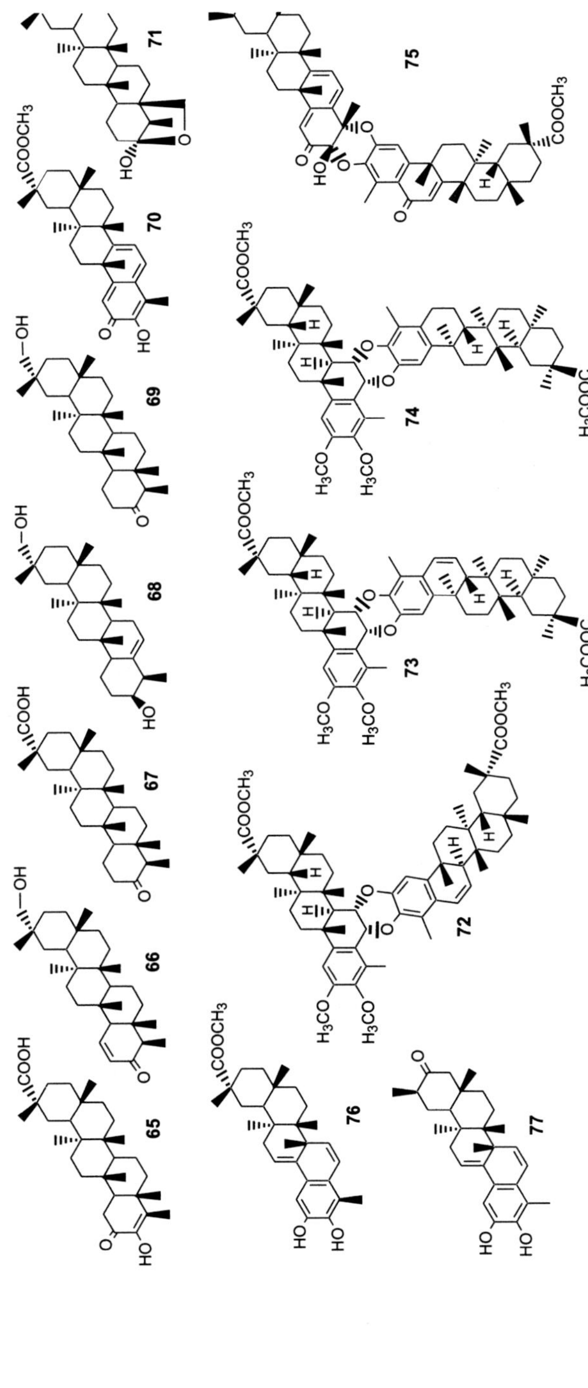

Fig. 7.9 The structures of triterpenoids from *M. ilicifolia*.

compounds against V-79, KB and P388 cells were examined. The results are summarized in Table 1. Pristimerin (**70**) exhibited the strongest activity against all three cell lines.

Recently, new triterpenes were isolated from *M. chuchuhuasca* (Shirota *et al.* 2004a) along with sesquiterpenes from the same plant (Shirota *et al.* 2004b). New compounds were also reported from *M. diversifolia* (Nakao, *et. al.* 2004) and *M. ilicifolia* (Ohsaki *et al.* 2004). Antitumor promoting sesquiterpenes were isolated from *M. cuzcoina* (Gonzalez *et al.* 2000).

Table 7.1 Cytotoxic activity of triterpenes from *Maytenus ilicifolia.*

	IC_{50} values (μg/mL)		
Compound	*V-79 cells*	*KB cells*	*P388 cells*
D:A Friedoolean-1-en-29-ol-3-one (**66**)	>100	>100	>100
D:A Friedooleanan-3-on-29-oic acid (**67**)	38	12	23
D:A Friedoolean-5-en-3,29-diol (**68**)	>100	110	95
D:A Friedooleanan-29-ol-3-one (**69**)	110	100	98
Pristimerin (**70**)	0.13	0.23	0.052
Isopristimerin III (**76**)	9.4	1.7	2.0
Isotingenone III (**77**)	1.4	1.1	0.18

Rubia spp.

Reubiaceous plants occur commonly *Rubia akane* in Japan, *R. cordifolia* in China, and *R. tinctorum* in Europe. The former two show antineoplastic activity, but the latter does not (Itokawa 1988, Itokawa *et al.* 2000).

RAs (named after *Rubia akane*) are bicyclic hexapeptides (**78–85**) as shown in Fig. 10. RA-VII (**84**) was the most potent component among 16 RAs isolated from natural sources. Cell growth and inhibitory effects were examined against KB cells, P388 lymphocytic leukemia cells, and MM2 mammary carcinoma cells. Since RA-V (**82**) possesses a reactive phenolic hydroxyl group on the Tyr-6 residue, its derivation has extensively been studied. A number of alkyl and acyl groups were introduced into this position. Interestingly, analogs possessing longer alkyl or acyl groups retained potent antitumor activity, with the n-hexyl derivative showing the strongest antitumor activity in an *in vivo* assay. These results suggested that modification at this position may lead to more biologically promising analogs compared with peptide RA-VII (**84**). Subsequently, the synthetic TIA-356 (**86**) was developed as a more potent compound.

Moreover, three epimers (at Ala-1, Ala-2, and Ala-4) of RA-VII (**84**) were synthesized and tested against various tumors, but were less potent than the natural compound (Hitotsuyanagi *et al.* 2001, 2003).

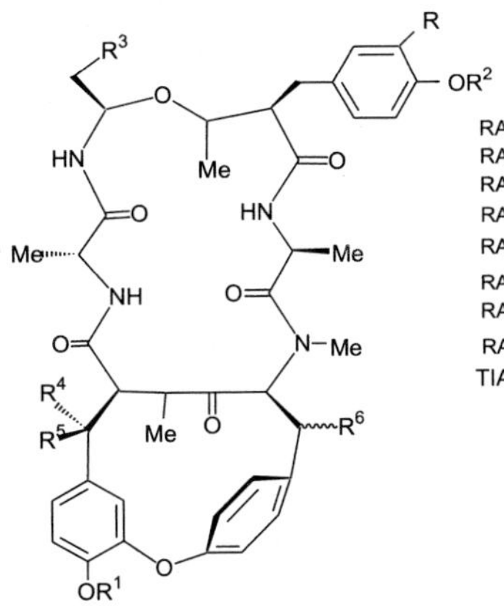

	R^1	R^2	R^3	R^4	R^5	R^6	R^7
RA-I (**78**)	H	Me	OH	H	H	H	H
RA-II (**79**)	Me	H	H	H	H	H	H
RA-III (**80**)	Me	Me	OH	H	H	H	H
RA-IV (**81**)	Me	Me	H	OH	H	H	H
RA-V (**82**)	H	Me	H	H	H	H	H
RA-VI (**83**)	Me	Me	OH	H	H	H	H
RA-VII (**84**)	Me	Me	H	H	H	H	H
RA-VIII(**85**)	Me	Me	Me,OH	H	H	H	H
TIA-356 (**86**)	N(i-Pr)$_2$	Me	H	H	H	H	H

Fig. 7.10 RAs from *Rubia cordifolia.*

Withania spp.

In several Asian countries, the leaves of *Withania somnifera* (Solanaceae) are used in the treatment of tumors and inflammation.

Twelve withanolides isolated from *W. somnifera* (Jayaprakasam *et al.* 2003) were tested for *in vitro* human tumor cell antiproliferative activity against lung, colon, CNS, and breast tumor cell lines as determined by the MTT assay (Tian *et al.* 2001). In general, the withanolides were most inhibitory to lung cells and least inhibitory to colon cells, and showed similar trends in antiproliferative activity against all four tumor cell lines. The compounds represented three structural groups, withaferin A-, and viscosalactone B-, and physaglin D- type. Potency followed the rank order given, and withaferin A (**87**) and its acetyl derivative were the most potent compounds. Furthermore, the double bond in the A ring contributed significantly to the antiproliferative activity as seen by comparison with the corresponding saturated dihydrowithaferin A (**88**). The double bond and the C-27 hydroxyl in the lactone ring of withanolides are also important to the activity, as 27-desoxy-24,25-dihydrowithaferin A (**89**) and compound **90**, which is glucosylated at C-27, were much less active than withaferin A (**87**).

Recently, a *W. somnifera* extract and its bioactive constituent withanolide D were found to significantly inhibit metastasis of B16F-10 melanoma cells in the lung using, C57BL/6 mice (Leyon and Kuttan 2004).

Four new withanolides were recently isolated; two from *W. coasgulans* (Atta-ur-Rahman *et al.* 2003) and two from *W. somnifera* (Choudhary *et al.* 2004).

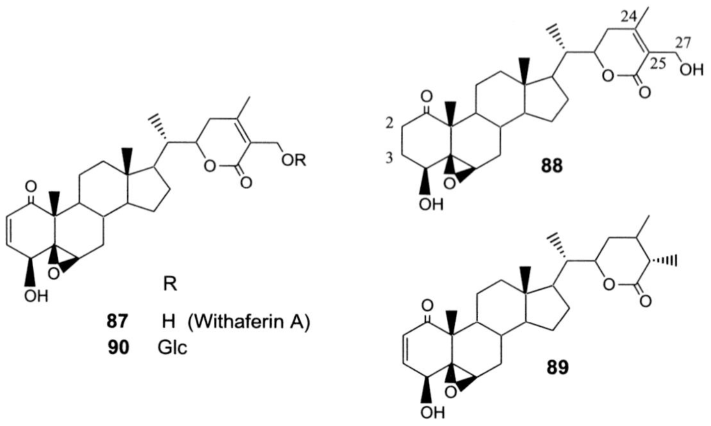

Fig. 7.11 Withanolides.

***Stephania* spp.**

Tetrandrine (**91**) is a bis-benzylisoquinoline alkaloid isolated from the roots of *Stephania tetrandra* (Menispermaceae). It induces apoptosis in human T-cell lines, lung carcinoma cells, and hepatoblastoma cells. In a mechanistic study (Oh and Lee 2003), induction of apoptosis was determined by morphological analysis and a terminal deoxynucleotidyl transferase-mediated dUTP nick-endlabeling assay. In HepG2 cells treated with **91**, p53 was upregulated and Bc-XL was downregulated, Bid and Bax were cleaved, cytochrome c was released, and caspases 9, 3 and then 8 were activated (the activation of caspase 8 after earlier activation of caspases 9 and 3 might act as an amplification loop for activation of upstream signals such as Bid cleavage or cytochrome c release). The actions induced by tetrandrine as well as cell death were blocked by treatment with either a broad-spectrum caspase inhibitor or a caspase 8-specific inhibitor. These data indicate that the mitochondrial pathway is primarily involved in tetrandrine-induced apoptosis and that tetrandrine (**91**) may be a likely therapy for hepatocellular carcinoma.

Other mechanistic studies have demonstrated that tetrandrine-induced apoptosis of U937 cells is also associated with activation of PKC-δ (Jang *et al.* 2004). In cells pretreated with a specific PKC-δ inhibitor before drug treatment, cell survival was promoted and both drug-induced apoptosis and caspase-3 activation were blocked. Mutual regulation between capsases and PKC-δ was supported by the finding that tetrandrine induced caspase-dependent generation of a catalytically active fragment of PKC-δ, which in turn likely was involved in the activation of caspases.

Cephäranthine (**92**) is a biscoculaurine-type alkaloid isolated from *Stephania cepharantha*. Cepharanthine significantly enhanced the *in vitro* thermosensitivity of FSa-11 cells with heating at 44 °C and increased cepharanthine concentration. When single cells were observed with time-lapse microscopy, cepharanthine treatment was found to accelerate both apoptosis (specifically, apoptotic budding) and necrosis (shown by staining with propidium iodide). Cepharanthine (**92**) was suggested to be a potential thermosensitizer to increase apoptosis of tumor cells (Wang *et al.* 2004).

The ability of cepharanthine (**92**) to potentiate doxorubicine chemotherapy was also evaluated (Nakajima *et al.* 2004). Using drug sensitivity testing with the MTT assay, co-administration of **92** with doxorubicin significantly enhanced cytotoxic effects, but only in doxorubicin-resistant, not doxorubicin-sensitive, heptocellular carcinoma cells. Cepharanthine (**92**) inhibited doxorubicine efflux and, thus, significantly increased the intracellular drug concentration. Thus, chemotherapy of drug-resistant hepatocellular carcinoma could benefit from the co-administration of **92** with doxorubicin.

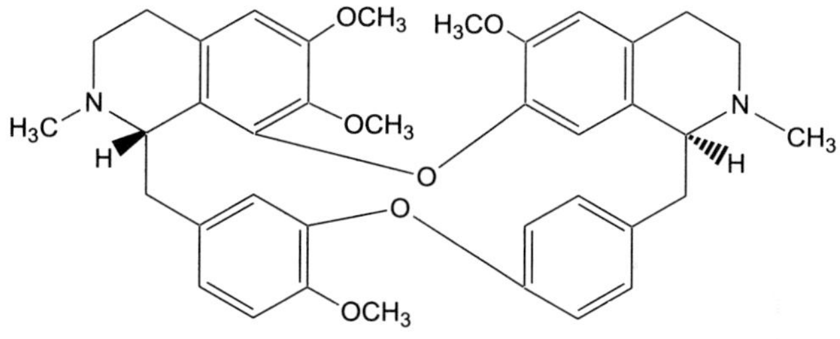

Tetrandrine (**91**)

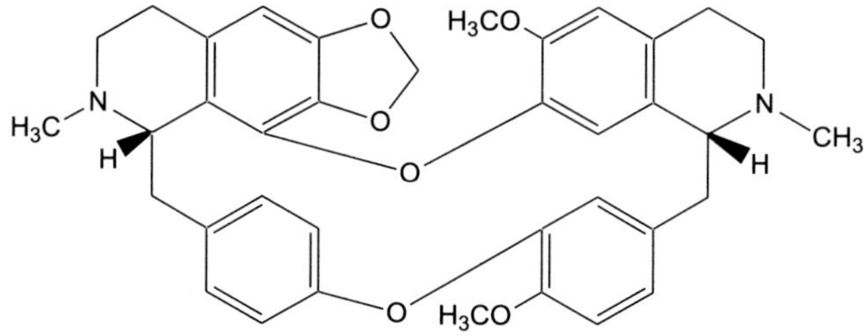

Cepharanthine (**92**)

Fig. 7.12 Biscoculaurine alkaloids.

Curcuma spp.

Curcuma species contain the yellow phenolic diarylheptanoid curcumin (**93**), which is the major pigment in turmeric. Curcumin and its analogues show various biological activities, including cytotoxicity (Aggarwal *et al.* 2003). Several synthetic curcumin analogues, including **94** and **95**, showed potent antiandrogenic activities against two human prostate cancer cell lines, PC-3 and DU-145, and were superior to hydroxyflutamide, which is the antiandrogen of choice for the treatment of prostate cancer (Ohtsu *et al.* 2002). In addition, the sesquiterpene curcumol (**96**), obtained from *Curcuma aromatica*, was effective against cancer of the uterine cervix (Lee 1993).

The root of *Curcuma xanthorrhiza* (called Temu Lawak in Indonesia) has been utilized as a tonic in southeastern Asia and as a choleretic drug in Europe. An *n*-hexane extract of this plant was found to be active against sarcoma 180A in mice and was further fractionationed by repeated column chromatography to give several antitumor bisabolane sesquiterpenes, α-curcumene (**97**), ar-turmerone (**98**) and xanthorrhizol

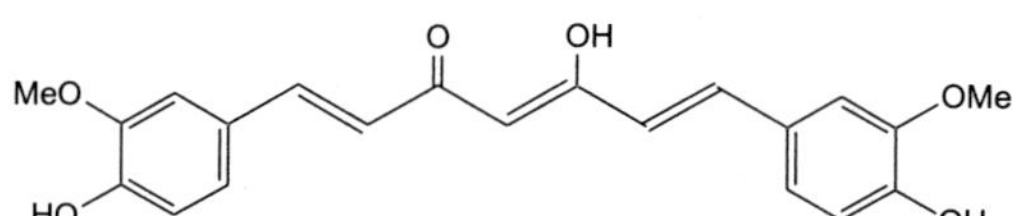

93

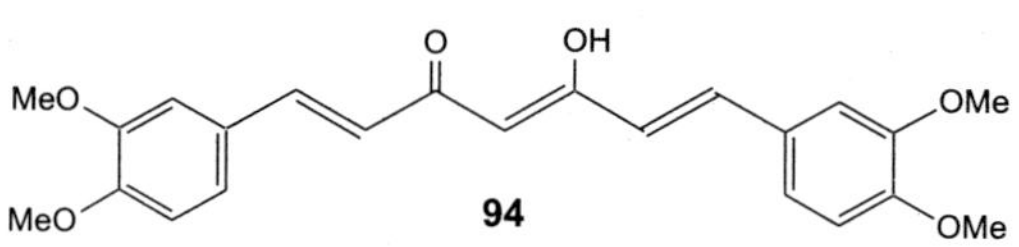

94

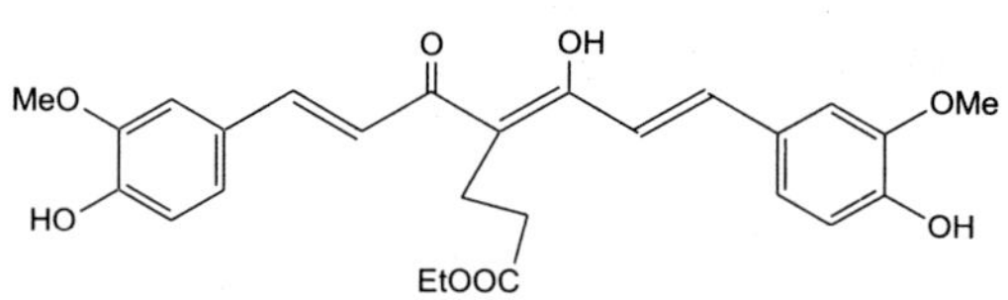

95

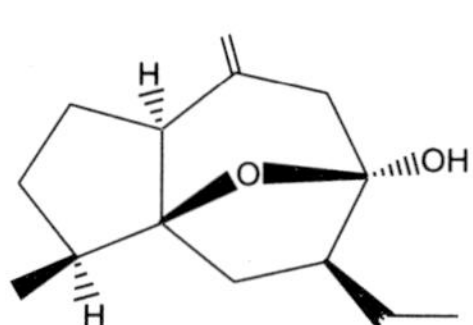

Curcumol (**96**)

Fig. 7.13 Diarylheptanoids and curccumol.

(**99**) as well as minor related compounds. Their structures and antitumor data against sarcoma 180A in mice are shown in Table 7.2. α-Curcumene exhibited a dose-dependent effect against this cancer type, and ar-turmerone and xanthorrhizol were less active. On the other hand, α-curcumene showed no significant activity against P388 lymphocytic leukemia in mice at doses of 50 to 200 mg/kg.

Table 7.2 Antitumor activity of 97-99 against Sarcoma 180A in mice.

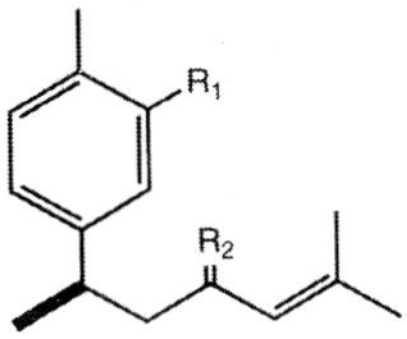

97: R_1 = H, R_2 = H_2
98: R_1 = H, R_2 = O
99: R_1 = OH, R_2 = H_2

Compound	*Dose (mg/kg)*	*BWC*	*PCV/TV*	*GR(%)*	*Assessment*
97	10	+4.2	0.31	97	–
	20	+ 2.1	0.16	31	++
	50	+ 2.4	0.00	0	+++
98	25	+ 3.1	0.39	77	–
	50	+ 2.7	0.12	22	++
99	50	+ 3.6	0.13	18	++

Brucea spp.

The quassinoid bruceantin (**100**) was first isolated from the tree *Brucea antidysenterica* (Simaroubaceae), which is used in Ethiopia to treat cancer. Bruceantin showed activity against B16 melanoma, colon 38, and L1210 and P388 leukemia in mice, prompting Phase I and II clinical trials. However, clinical development was halted after no objective tumor regressions were observed in these trials. Recently, bruceantin has been reinvestigated in various leukemia, lymphoma, and myeloma cell lines, where both induction of cell differentiation and down-regulation of c-MYC were observed. Bruceantin-treatment of HL-60 and RPMI 8226 cell lines also induced apoptosis, which was associated with the caspase and mitochondrial pathways. In an *in vivo* study using RPMI 8226 human-SCID xenografts, bruceantin induced regression in early as well as advanced tumors, without overt toxicity. Apoptosis was significantly promoted in tumors derived from animals treated with bruceantin. The results described above indicate that bruceantin merits investigation for clinical efficacy against hematological malignancies (Cuendet and Pezzuto 2004).

A new quassinoid, yadanziolide S (**101**), and 10 known compounds were isolated from the seeds of the traditional Chinese medicinal herb, *B. javanica* (Su *et al.* 2002). Yadanziolide S (**101**) is the first quassinoid isolated from this plant that does not have a methyleneoxy bridge between C-8 and C-13. All compounds were evaluated for their ability to induce human promyelocytic leukemia (HL-60) cell differentiation, and to inhibit cyclooxygenase-1 (COX-1), -2(COX-2), and 7,12-dimethylbenz[a]anthracene (DMBA)-induced lesions in a mouse mammary model.

Other quassinoids, including brusatol and the bruceosides, from *B. javanica* have been extensively studied for antitumor (Fukamiya *et al.* 1992, Ohnishi *et al.* 1995, Lee 2004 and references cited therein) and cancer chemopreventive (Murakami *et al.* 2004) effects.

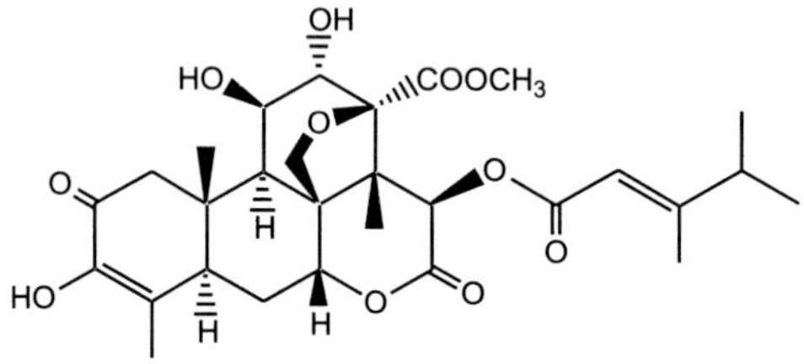

Bruceantin (**100**)

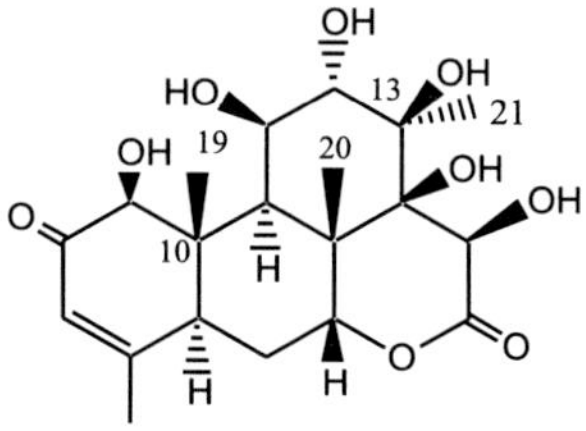

Yadanziolide S (**101**)

Fig. 7.14 Bruceantin and yadanziolide S.

Cocculus spp.

Cocculus trilobus DC. (Menispermaceae), which grows in the mountains of East Asia, has been used in folk medicine as a diuretic, analgesic, and anti-inflammatory crude drug.

Sinococuline was isolated as a new antitumor principle from the stems and rhizomes of *C. tribolus* by Itokawa *et al.* (1987). The methanolic extract of the plant was dissolved in water and then partitioned successively with *n*-hexane and ethyl acetate. The antitumor activity against sarcoma 180 ascites in mice was concentrated in the aqueous-layer residue, which was subjected to Amberlite XAD-2 column chromatography, eluting with a water-methanol gradient. The fraction eluted with water exhibited significant antitumor activity. Preparative TLC of this active fraction on silica gel using a chloroform:diethylamine (3:2) solvent system gave the antitumor morphinane alkaloid, named sinococuline (**102**). The tentative structure was established by various spectroscopic methods and the C_9 configuration was assumed to be *S* from the viewpoint of chemotaxonomy; however, it was confirmed by measuring the CD spectrum (positive maximum at 238 nm). Further, in order to determine the absolute configuration using the exciton chirality rule, sinococuline (**102**) was converted to the 6,7-dibenzoate derivative as shown in Fig. 15. Its CD spectrum showed a negative Cotton effect at 252 nm and the coupling constant between H_6 and H_7 was 3.5Hz. These data were consistent with both structures **104a** (6*S*, 7*S*) and **104b** (6*R*, 7*R*) in Figure 15.

However, the NOE between H_6 and H_{15} supported only the structure **104a**. Sinococuline (**102**) also has antitumor activity against P388 leukemia *in vivo* (Itokawa 1987).

Sinococuline caused a dose-dependent decrease in cell viability and uptake of tritiated-thymidine, -leucine, and -uridine in tumor cells after a 24h exposure. Also, the production of tumor necrosis factor and reactive nitrogen intermediates was reduced and consequently macrophage function was altered. Sinococuline induced apoptosis in human leukemia HL60 cells, but not mouse fibroblast L929 cells. Thus, sinocuoculin is likely a general cytotoxic agent rather than a cell-specific agent (Liu *et al.* 1996).

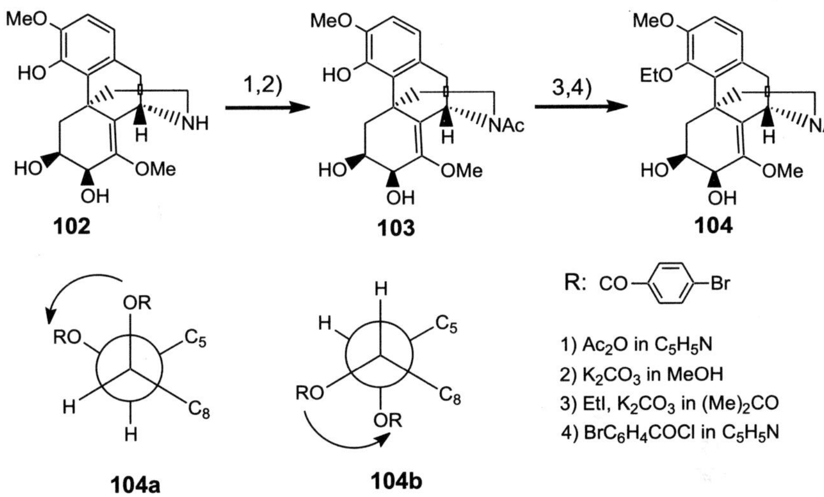

Fig. 7.15 Sinococuline and related compounds.

Rabdosia **spp.**

Several *Rabdosia* species (Labiatae), which are native to China, are used in folk medicine as antitumor or anti-inflammatory agents. *R. rubescens* (Hemsl.) contains diterpene compounds, such as oridonin (**105**) and ponicidin (**106**), which are derived from kaurene. The diterpenes were found to be the cytotoxic principles in *Rabdosia* species (Yuan *et al.* 1997).

Jungermannia **spp.**

Ent-11α-hydroxy-16-kauren-15-one (**107**), isolated from the liverwort, *J. truncata*, showed cytotoxic effects in the HL-60 leukemia cell line. Cells treated with **107** and other related compounds with an enone group at C-15/C-16 showed DNA fragmentation, which is typical of apoptotic induction, and proteolysis of poly(ADP-ribose)polymerase(PARP), which is characteristic of caspase activation. However, when Z-Asp-CH_2-DCB, a broad-spectrum inhibitor of caspases, was also added to drug-treated cells, DNA fragmentation was not observed and the cytotoxic effects were lower. Based on these results, **107** and its related compounds induce apoptosis by activation of caspases. Thus, the cytotoxicity of **107** may be linked, at least in part, to induction of apoptosis (Nagashima *et al.* 2003).

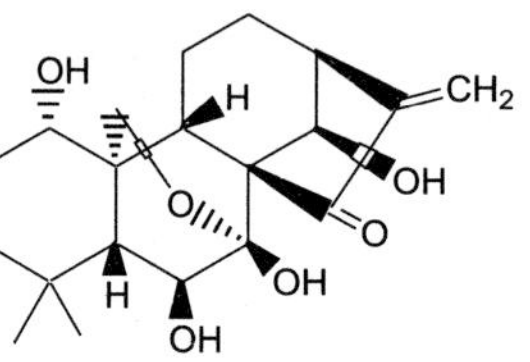

Oridonin (**105**)

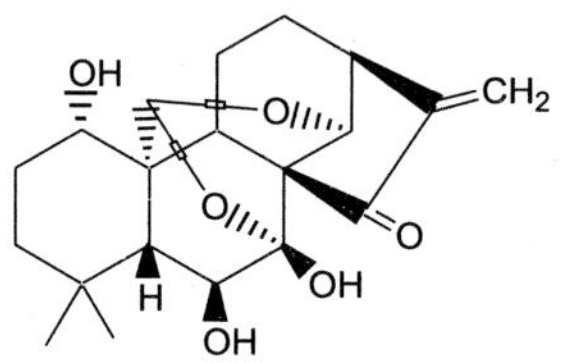

Ponicidin (**106**)

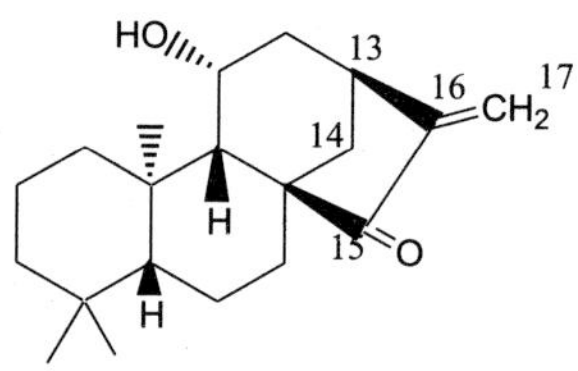

107

Fig. 7.16 Kaurene-type compounds.

Angelica spp.

Angelica keiskei (Umbelliferae) is used as a traditional health food, and the roots and leaves are said to have preventive effects against coronary heart disease, hypertension, and cancer. The root contains chalchone derivatives, such as xanthoangerol and 4-hydroxyderricin (**108**), as primary components. Kimura *et al.* (2004) investigated the cytotoxic effects of the ethanol extract and isolated the active chemical principles. 4-Hydroxyderricin inhibited tumor growth and metastasis to the lung or liver in mice with subcutaneous or intrasplenic Lewis lung carcinoma implants. At 100 μM, it (**108**) also inhibited DNA synthesis in these tumor cells but not in human umbilical vein endothelial cells (HUVECs). When 4-hydroxyderricine (**108**) was administered orally, spleen and thymus weights of tumor-bearing mice did not significantly differ from untreated animals. In the spleen of tumor-removed mice, the numbers of lymphocytes, $CD4^+$, $CD8^+$ and natural killer (NK)-T cells were reduced on treatment with doxorubicin but **108** inhibited the reduction. In summary, the antitumor and antimetastatic effects of **108** may be associated with the immune system and inhibition of angiogenesis (Kimura *et al.* 2004).

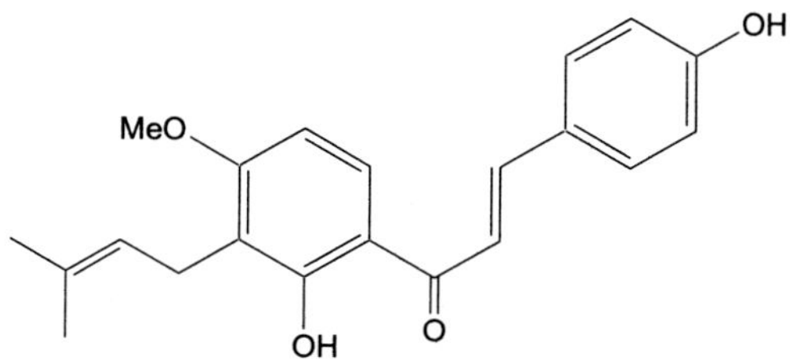

108

Fig. 7.17 4-Hydroxyderricin.

CONCLUSION

Folk or clinical experiences with herbs or herbal prescriptions can suggest the starting point for a potential new drug. After verification of the activity by modern pharmacological assays, isolation of a bioactive natural compound(s), using bioactivity-directed fractionation and isolation (BDFI), is the step. This new compound can then be chemically modified to increase its activity, decrease toxicity, or improve other pharmacological profiles, based on SAR, mechanism of action, computational, and other studies. The most promising drug development targets are selected based on preclinical screening [for instance, in the National Cancer Institute (NCI) *in vitro* human cell line panels and selected *in vivo* xenograft systems] and production, formulation, efficacy, and toxicological studies prior to clinical trials. With these approaches, numerous drugs that are natural products have been discovered.

As discussed in this review, many anticancer agents have already been isolated from higher plants and many candidates, *i.e.*, antitumor and cytotoxic compounds, are undergoing evaluation as potential anticancer agents. The ultimate goal is to identify compounds having strong activity against cancer.

ACKNOWLEDGEMENT

The authors appreciate Dr. Susan L. Natschke for her valuable suggestion. Thanks are also due to partial support from NIH grant CA17625 (K.H. Lee).

REFERENCES

Aggarwal, B.B., A. Kumar, and A.C. Bharti. 2003. Anticancer potential of curcumin: preclinical and clinical studies. *Anticancer Res.* 23: 363–398.

Appendino, G., P. Bettoni, A. Noncovich, O. Sterner, G. Fontana, E. Bombardelli, P. Pera, and R.J. Bernack. 2004. Structure-activity relationship of ring C-secotaxoids. 1. Acylative modifications. *J. Nat. Prod.* 67: 184–188.

Atta-ur-Rahman, Dur-e-Shahwar, A. Naz, and M.I. Choudhary. 2003. Withanolides from *Withania coagulans*. *Phytochemistry* 63: 387–390.

Banskota, A.H., T. Usia, Y. Tezuka, K. Kouda, N.T. Nguen, and S. Kadota. 2002. Three new C-14 oxygenated taxanes from the wood of *Taxus yunnanensis*. *J. Nat. Prod.* 65: 1700–1702.

Bastow, K.F., H. Tatematsu, I.D. Bori, Y. Fukushima, L. Sun, G. Goz, and K.H. Lee. 1993. Induction of reversible protein-linked DNA breaks in human osteogenic sarcoma cells by novel cytocidal colchicine derivatives which inhibit DNA topoisomerase II *in vitro*: absence of cross-resistance in a colchicine-resistant sub-clone. *Bioorg. Med. Chem. Lett.* 3: 1045–1050.

Brossi, A. 1984. *The Alkaloids*, Academic Press, New York.

Caffo, O. 2001. Radiosensitization with chemotherapeutic agents. *Lung Cancer* 34: S81–90.

Chang, J.Y., F.S. Han, S.Y. Liu, H.K. Wang, K.H. Lee, and Y.C. Cheng. 1991. Effect of 4-β-arylamino derivatives of 4'-O-demethylpodophyllotoxin on human DNA topoisomerase II, tubulin polymerization. KB cells, and their resistant variants. *Cancer Res.* 51: 1755–1759.

Choudhary, M.I., S. Yousuf, S.A. Nawaz, and Atta-ur-Rahman. 2004. Cholinesterase inhibiting withanolides from *Withania somnifera*. *Chem. Pharm. Bull.* 52: 1358–1361.

Cho, S.J., A. Tropsha, M. Suffness, Y.C. Cheng, and K.H. Lee. 1996. Three dimensional quantitative structure-activity relationship study of 4'-O-demethylepipodophyllotoxin analogs using the modified CoMFA/q2-GRS approach. *J. Med. Chem.* 39: 1383–1385.

Covey, J.M., C. Jaxel, K.W. Kohn, and Y. Pommier. 1989. Protein-linked DNA strand breaks induced in mammalian cells by camptothecin, an inhibitor of topoisomerase I. *Canc. Res.* 49: 5016–5022.

Cragg, G.M., and M. Suffness. 1988. Metabolism of plant-derived anticancer agents. *Pharmcol. Ther.* 37: 425–461.

Cragg, G.M., and D.J. Newman. 2004 A tale of two tumor targets: topoisomerase I and tubulin. The Wall and Wani contribution to cancer chemotherapy. *J. Nat. Prod.* 67: 233–244.

Cuendet, M., and J.M. Pezzuto. 2004 Antitumor activity of bruceantin: An old drug with new promise. *J. Nat. Prod.* 67: 269–272.

Dallavalle, S., L. Merlini, G. Morini, L. Muso, S. Penco, G.L. Beretta, S. Tinelli, and F. Zunino. 2004. Synthesis and cytotoxic activity of substituted 7-aryliminomethyl derivatives of camptothecin. *Eur. J. Med. Chem.* 39: 507–513.

Fukamiya, N., M. Okano, M. Miyamoto, K. Tagahara, and K.H. Lee. 1992. Antitumor agents, 127. Bruceoside C, a new cytotoxic quassinoid glucoside, and related compounds from *Brucea javanica*. *J. Nat. Prod.* 55: 468–475.

Garcia-Carbonero, R., and J.G. Supko. 2002. Current perspectives on the clinical experience, pharmacology, and continued development of the camptothecins, *Clin. Cancer Res.* 8: 641–661.

Giovanella, B., J.S. Sthelin, M.E. Wall, M.C. Wani, A.W. Nicholas, L.F. Liu, R. Silber, and M. Potmesil. 1989. DNA-topoisomerase I-targeted chemotherapy of human colon cells in xenografts. *Science* 246: 1046–1048.

Gonzalez, A.G., B.M. Tincusi, I.L. Bazzocchi, I.H. Tokuda, H. Nishino, Y. Konoshima, I.A. Jimenez, and A.G. Ravelo. 2000. Anti-tumor promoting effects of sesquiterpenes from *Maytenus cuzcoina* (Celastraceae). *Bioorg. Med. Chem.* 8: 1773–1778.

Grem, J.L., B.D. Cheson, S.A. King, B. Leyland-Jones, and M. Suffness. 1988. Cephalotaxine esters: antileukemic advance or therapeutic failure. *J. Natl. Cancer Inst.* 80: 1095–1103.

Guan, J., X.K. Zhu, Y. Tachibana, K.F. Bastow, A. Brossi, E. Hamel, and K.H. Lee. 1998. Antitumor agents. Part 185. Synthesis and biological evaluation of tridemethylthiocolchicine analogues as novel topoisomerase II inhibitors. *J. Med. Chem.* 41: 1956–1961.

Hitotsuyanagi, Y., Y. Matsumoto, S. Sasaki, K. Yamaguchi, H. Itokawa, and K. Takeya. 2001. A new approach to the epimeric analogue of cyclic peptides: epimerization via oxazoles of RA-VII, an anti-tumor bicyclic hexapeptide from *Rubia* plants. *Tetrahedron Lett.* 42: 1535–1537.

Hitotsuyanagi, Y., S. Sasaki, Y. Matsumoto, K. Yamaguchi, H. Itokawa, and K. Takeya. 2003. Synthesis of [L-Ala-1]RA-VII, [D-Ala-2]RA-VII, and [A-Ala-4]RA-VII by epimerization of RA-VII, an antitumor bicyclic hexapeptide from *Rubia* plants through oxazoles. *J. Amer. Chem. Soc.* 125: 7284–7290.

Hsiang, Y.H., R. Hertzberg, S. Hecht, and L.F. Liu. 1985. Camptothecin induces protein-linked DNA breaks via mammalian DNA topoisomerase I. *J. Biol. Chem.* 260: 14873–14878.

Hsiung, Y.H., L.F. Liu, Wall, M.E. Wall, M.C. Wani, A.W. Nicholas, G. Manikumar, S. Kirschenbaum, R. Silber, and M. Potmesil. 1989. DNA topoisomerase I-mediated DNA cleavage and cytotoxicity of camptothecin analogues. *Cancer Res.* 49: 4385–4389.

Huang, C.C., C.S. Han, X.F. Yue, C.M. Shen, S.W. Wang, F.G. Wu, and B. Xu. 1983. Cytotoxity and sister chromatid exchanges induced in vitro by six anticancer drugs developed in the People's Republic of China, *J. Natl. Cancer Inst.* 71: 841–847.

Imbert, T.F. 1998. Discovery of podophyllotoxins. *Biochimie* 80: 207–222.

Itokawa, H., S. Tsuruoka, K.Takeya, N. Mori, T. Sonobe, S. Kosemura, and T. Hamanaka. 1987. An antitumor morphinane alkaloid, sinococuline, from *Cocculus trilobus*. *Chem. Pharm. Bull.* 35: 1660–1662.

Itokawa, H. 1988. Research on antineoplastic drugs from natural sources especially from higher plants. *Yakugaku Zasshi* 108: 824–841.

Itokawa, H., O. Shirota, H. Morita, K. Takeya, N. Tomioka, and Y. Iitaka. 1990. Triterpene dimers from *Maytenus ilicifolia*. *Tetrahedron Lett.* 31: 6881–6882.

Itokawa, H., O. Shirota, H. Ikuta, H. Morita, K. Takeya, and Y. Iitaka. 1991. Triterpenes from rhizomes of *Maytenus ilicifolia*. *Phytochemistry* 30: 3713–3716.

Itokawa, H., K. Takeya, Y. Hitotsuyanagi, and H. Morita. 1999. Antitumor compounds isolated from higher plants. *Yakugaku Zasshi* 119: 529–583.

Itokawa, H., K. Takeya, Y. Hitotsuyanagi, and H. Morita. 2000. Antitumor compounds isolated from higher plants. *Stud. Nat. Prod. Chem.* 24: 269–350.

Itokawa, H. and K.H. Lee. 2003. *Taxus: genus Taxus*, Taylor & Francis, London.

Itokawa, H., X. Wang, and K.H. Lee. 2004. Homoharringtonine and related compounds, In *Antitumor Agents from Natural Sources*, D. Kingston, G. Cragg, and D. Newman, eds., CRC Press, Boca Raton, Florida, in press.

Jang, B.C., K.J. Lim, J.H. Paik, J.W. Cho, W.K. Baek, M.H. Suh, J.B. Park, T.K. Kwon, J.W. Park, S.P. Kim, D.H. Shin, D.K. Song, J.H. Bae, K.C. Mun, and S.I. Suh. 2004. Tetrandrine-induced apoptosisis mediated by activation of caspases and PKC-δ in U937. *Biochem. Pharmacol.* 67: 1819–1829.

Jayaprakasam, B., Y. Zhang, N.P. Seeram, and M.G. Nair. 2003. Growth inhibition of human tumor cell lines by withanolides from Withania somnifera leaves. *Life Sci.* 74: 125–132.

Jenks, S., and K. Smigel. 1995. Updates: cancer drug approved; new leukemia treatment. *J. Nat. Cancer Inst.* 87: 167–170.

Kim, D.K., and N. Lee. 2002. Recent advances in topoisomerase I-targeting agents, camptothecin analogues. *Mini-Rev. Med. Chem.* 2: 611–619.

Kimura, Y., M. Taniguchi, and K. Baba. 2004. Antitumor and antimetastatic activities of 4-hydroxyderricin isolated from *Angelica keiskei* roots. *Planta Med.* 70: 211-219.

Kingston, D.G. 1991. The chemistry of taxol. *Pharmacol. Ther.* 52: 1–34.

Lavelle, F., F. Gueritte-Vogellein, and D. Guenard. 1993. Taxotere: from yew's needles to clinical practice. *Bull. Cancer* 80: 326–338.

Lee, K.H., Y. Imakura, M. Haruna, S.A. Beers, L.S. Thurston, H.J. Dai, C.H. Chen, S.Y. Liu, and Y.C. Chan. 1989. New cytotoxic 4-alkylamino analogues of 4'-demethyl-epipodophyllotoxin as inhibitors of human DNA topoisomerase II. *J. Nat. Prod.* 52: 606–613.

Lee, K.H. 1993. Antineoplastic agents and their analogues from Chinese traditional medicine. Pages 170-190 in *Human Medicinal Agents from Plants*, A.D. Kinghorn, and M. Balandrin, eds., Amer. Chem. Soc. Symposium Series 534.

Lee, K.H. 2004. Current developments in the discovery and design of new drug candidates from plant natural product lead. *J. Nat. Prod.* 67: 273–283.

Lee, K.H., and Z. Xiao. 2004 The podophyllotoxins and analogs. In *Antitumor Agents from Natural Sources*, D. Kingston, G. Cragg, and D. Newman, eds., CRC Press, Boca Raton, Florida, in press.

Lerchen, H.G. 1999. Camptothecin antitumor agents.*Drugs* 2: 896–906.

Leyon, P.V., and G. Kuttan. 2004. Effect of *Withania somnifera* on B16F-10 melanoma induced metastasis in mice. *Phytother. Res.* 18: 118–122.

Liu, C., J.S. Strobl, S. Bane, J.K. Schilling, M. McCracken, S.K. Chatterjee, R. Rahim-Bata, and D.G.I. Kingston. 2004. Design, synthesis, and bioactivities of steroid-linked taxol analogues as potential targeted drugs for prostate and breast cancer. *J. Nat. Prod.* 67: 152–159.

Liu, S.Y., R. Soikes, J. Chen, T. Lee, G. Taylor, K.M. Hwang, and Y.C. Cheng. 1993. 84th AACR Annual Conference, Orlando, Florida.

Liu, W.K., X.K., Wang, and C.T. Che. 1996. Cytotoxic effects of sinococuline. *Cancer Lett.* 99: 217–224.

Murakami, C., N. Fukamiya, S. Tamura, M. Okano, K.F. Bastow, H. Tokuda, T. Mukainaka, H. Nishino, and K.H. Lee. 2004. Multidrug-resistant cancer cell susceptibility to cytotoxic quassinoids, and cancer chemopreventive effects of quassinoids and canthin alkaloids. *Bioorg. Med. Chem.* 12: 4963–4968.

Nagashima, F., M. Kondoh, M. Kawase, S. Shimizu, H. Osada, M. Fujii, Y. Watanabe, M. Sato, and Y. Asakawa. 2003. Apoptosis-inducing properties of ent-kaurene-type diterpenoids from the liverwort *Jungermannia truncate*. *Planta Med.* 69: 377–379.

Nakajima, A., Y. Yamamoto, K. Taura, K. Hata, M. Fukumoto, H. Uchinami, K. Yonezawa, and Y. Yamaoka. 2004. Beneficial effect of cepharanthine on overcoming drug-resistance of hepatocellular carcinoma. *Intern. J. Oncology*. 24: 635–645.

Nakao, H., K. Senokuchi, C. Umebayashi, K. Kanemaru, T. Masuda, Y. Oyama, and S. Yonemori. 2004. Cytotoxic activity of maytanprine isolated from *Maytenus diversifolia* in human leukemia K562 cells. *Biol. Pharm. Bull.* 27: 1236–1240.

Oberlies, N.H., and D.J. Kroll. 2004 Camptothecin and taxol: Historic achievements in natural products research. *J. Nat. Prod.* 67: 129–135.

Oh, S.H., and B.H. Lee. 2003. Induction of apoptosis in human hepatoblastoma cells by tetrandrine via caspase-dependent Bid cleavage and cytochrome c release. *Biochem. Pharmacol.* 66: 725–731.

Ohnishi, S., N. Fukamiya, M. Okano, K. Tagahara, and K.H. Lee. 1995. Bruceosides D, E, and F, three new cytotoxic quassinoid glucosides from *Brucea javanica*. *J. Nat. Prod.* 58: 1032–1038.

Ohsaki, A., Y. Imai, M. Naruse, S. Ayabe, K. Komiyama, and J. Takashima. 2004. Four new triterpenoids from *Maytenus ilicifolia*. *J. Nat. Prod.* 67: 469–471.

Ohtsu, H., Z. Xiao, J. Ishida, M. Nagai, H.K. Wang, H. Itokawa, C.Y. Su, C. Shih, Y. Lee, M.Y. Tsai, C. Chang, and K.H. Lee. 2002. Antitumor agents 217. Curcumin analogues as novel androgen receptor antagonists with potential as anti-prostate cancer agents. *J. Med. Chem.* 45: 5037–5042.

O'Leary, J., and F.M. Muggia. 1998. Camptothecins: a review of their development and schedules of administration. *Eur. J. Cancer*. 34: 1500–1508.

Ozols, R.F. 2000. Optimum chemotherapy for ovarian cancer. *Int. J. Gynecol. Cancer* 10: 33-37.

Paudler, W.W., G.I. Kerley, and J. McKay. 1963. The alkaloids of *Cephalotaxus drupacea* and *Cephalotaxus fortunei*. *J. Org. Chem.* 28: 2194–2197.

Paudler, W.W., and J. McKay. 1973. The structures of some of the minor alkaloids of *Cephalotaxus fortunei*. *J. Org. Chem.* 38: 2110–2112.

Potier, P. 1989. The synthesis of navelbine prototype of a new series of vinblastine derivatives. *Sem. Oncol.* 16: 2–4.

Powell, R.G., R.V. Madrigal, C.R. Smith, and K.L. Mikolajczak. 1974. Alkaloids of *Cephalotaxus harringtonia* var. drupacea. 11-Hydroxycephalotaxine and drupacine. *J. Org. Chem.* 39: 676–680.

Powell, R.G., D. Weisleder, and C.R. Smith. 1972. Antitumor alkaloids for *Cephalotaxus harringtonia*: structure and activity. *J. Pharm. Sci.* 61: 1227–1230.

Powell, R.G., D. Weisleder, C.R. Smith, and W.K. Rohwedder. 1970. Structures of harringtonine, isoharringtonine, and homoharringtonine. *Tetrahedron Lett.* 11: 815–818.

Saltz, L.B., J.V. Cox, C. Blanke, L.S. Rosen, L. Fehrenbacher, M.J. Moore, J.A. Maroum, S.P. Ackland, P.K. Pitotta, G.L. Elfring, and L.L. Miller. 2000. Irinotecan plus fluorouracil and leucovorin for metastatic colorectal cancer. Irinotecan study group. *N. Engl. J. Med.* 343: 905–914.

Shen, Y.C., S.S. Wang, Y.L. Pan, K.L. Lo, R. Chakraborty, C.T. Chien, Y.H. Kuo, and Y.C. Lin. 2002. New taxane diterpenoids from leaves and twigs *of Taxus sumatrana*. *J. Nat. Prod.* 65: 1848–1852.

Shi, Q.W., F. Sauriol, O. Mamer, and L.O. Zamir. 2003a. New minor taxane derivatives from the needles of *Taxus Canadensis*. *J. Nat. Prod.* 66: 1480–1485.

Shi, Q.W., F. Sauriol, O. Mamer, and L.O. Zamir. 2003b. New taxanes from the needles of *Taxus Canadensis*. *J. Nat. Prod.* 66: 470–476.

Shi, Q., P. Verdier-Pinard, A. Brossi, E. Hamel, A.T. McPhail, and K.H. Lee. 1997. Antitumor agents 172. Synthesis and biological evaluation of novel deacetamidothiocolchicin-7-ols and ester analogs as antitubulin agents. *J. Med. Chem.* 40: 961–966.

Shi, Q., K. Chen, A. Brossi, P. Verdier-Pinard, E. Hamel, A.T. McPhail, and K.H. Lee. 1998. Antitumor agents 184. Syntheses and antitubulin activity of compounds derived from reaction of thiocolchicone with amines: lactams, alcohols, and esters analogs of allothiocolchicinoids. *Helv. Chim. Acta* 81: 1023–1037.

Shigemori, H., and J. Kobayashi. 2004 Biological activity and chemistry of taxoids from the Japanese yew, *Taxus cuspidate*. *J. Nat. Prod.* 67: 245–256.

Shinozaki, Y., N. Fukamia, M. Fukushima, M. Okano, T. Nehira, K. Tagahara, S.X. Zhang, D.C. Zhang, and K.H. Lee. 2002. Dantaxusins C and D, Two novel taxoids from *Taxus yunnanensis*. *J. Nat. Prod.* 65: 371–374.

Shirota, O., S. Sekita, M. Satake, H. Morita, K. Takeya, and H. Itokawa. 2004a. Nine regioisomeric and stereoisomeric triterpene dimers from *Maytenus chuchuhuasca*. *Chem. Pharm. Bull.* 52: 739–746.

Shirota, O., S. Sekita, M. Satake, H. Morita, K. Takeya, and H. Itokawa. 2004b. Two new sesquiterpene pyridine alkaloids from *Maytenus chuchuhuasca*. *Heterocycle* 63: 1891–1896.

Spencer, G.F., R.D. Plattner, and R.G. Powell. 1976. Quantitative gas chromatography and gas chromatography-mass spectrometry of *Cephalotaxus* alkaloids. *J. Chromatogr.* 120: 335–341.

Su, B.N., L.C. Chang, E.J. Park, M. Cuendet, B.D. Santarsiero, A.D. Mesecar, R.G. Mehta, H.H.S. Fong, J.M. Pezzuto, and A.D. Kinghorn. 2002. Bioactive constituents of the seeds of *Brucea javanica*. *Planta Med.* 68: 730–733.

Suffness, M., and J. Douros. 1982. Current status of the NCI plant and animal product program. *J. Nat. Prod.* 45: 1–14.

Takano, I., I. Yasuda, M. Nishijima, Y. Hitotsuyanagi, K. Takeya, and H. Itokawa. 1996a. Drupangtonine, a novel antileukemic alkaloid from *Cepalotaxus harringtonia var. drupacea*. *Bioorg. Med. Chem. Lett.* 6: 1689–1690.

Takano, I., I. Yasuda, M. Nishijima, Y. Hitotsuyanagi, K. Takeya, and H. Itokawa. 1996b. Alkaloids from *Cephalotaxus harringtonia*. *Phytochemistry* 43: 299–303.

Takano, I., I. Yasuda, M. Nishijima, Y. Hitotsuyanagi, K. Takeya, and H. Itokawa. 1996c. New *Cephalotaxus* alkaloids from *Cephalotaxus harringtonia* var. *drupacea.J. Nat. Prod.* 59: 965–967.

Takano, I., I. Yasuda, M. Nishijima, Y. Hitotsuyanagi, K. Takeya, and H. Itokawa. 1996d. Cephalotaxidine, a novel dimeric alkaloid from *Cepharotaxus harringtonia* var. *drupacea*. *Tetrahedron Lett.* 37: 7053–7054.

Takano, I., I. Yasuda, M. Nishijima, Y. Hitotsuyanagi, K. Takeya, and H. Itokawa. 1996e. New oxygenated *Cephalotaxus* alkaloids from *Cephalotaxus harringtonia* var. *drupacea*. *J. Nat. Prod.* 59: 1192–1195.

Takano, I., I. Yasuda, M. Nishijima, Y. Hitotsuyanagi, K. Takeya, and H. Itokawa. 1997a. Ester-type *Cephalotaxus* alkaloids from *Cephalotaxus harringtonia* var. *drupacea*. *Phytochemistry* 44: 735–738.

Takano, I., I. Yasuda, M. Nishijima, Y. Hitotsuyanagi, K. Takeya, and H. Itokawa. 1997b. Modification of the skeleton of homoharringtonine through unusual rearrangements. *J. Org. Chem.* 62: 8251–8254.

Takimoto, C.H., J. Wright, and S.G. Arbuck. 1998. Clinical applications of the camptothecins. *Biochim. Biophys. Acta* 1400: 107–119.

Tang, W., I. Hemm, and B. Bertram. 2003a. Recent development of antitumor agents from Chinese herbal medicines; Part I. Low molecular compounds. *Planta Med.* 69: 97-108.

Tang, W., I. Hemm, and B. Bertram. 2003b. Recent development of antitumor agents from Chinese herbal medicines; Part II. Low molecular compounds. *Planta Med.* 69: 193–201.

Thomas, C. J., N.J. Rahier, and S.M. Hecht. 2004. Camptothecin: current perspectives. *Bioorg. Med. Chem.* 12: 1585–1604.

Tian, Q., E.G. Miller, H. Ahmed, L. Tang, and B.S. Patil. 2001. Differential inhibition of human cancer cell proliferation by citrus limonoids. *Nutr. Cancer* 40: 180-184.

Ulukan, H., and P.W. Swaan. 2002. Camptothecins: a review of their therapeutic potential. *Drugs* 62: 2039–2057.

Vanhoefer, U., A. Harstrick, W. Achterrath, S. Cao, S. Seeber, and Y.M. Rustum. 2001. Irinotecan in the treatment of colorectal cancer: clinical overview. *J. Clin. Oncol.* 19: 1501–1518.

Wall, M.E. 1993. In *Chronicles of Drug Discovery*, ed. D. Lednicer,. Am. Chem. Soc.: Washington D.C., Vol. 3, pp. 327–348.

Wall, M.E., and M.C. Wani. 1996. Camptothecin and taxol: from discovery to clinic, *J. Ethnopharmacol.* 51: 239–253.

Wall, M.E., M.C. Wani, C.E. Cook, K.H. Palmer, A.T. McPhail, and G.A. Sim. 1966. Plant antitumor agents. 1. The isolation and structure of camptothecin, a novel alkaloidal leukemia and tumor inhibitor from *Camptotheca acuminate*. *J. Am. Chem. Soc.* 88: 3888–3890.

Wani, M.C., H.I. Tayler, M.E. Wall, P. Coggon, and A.T. McPhail. 1971. Plant antitumor agents. VI. The isolation and structure of taxol, a novel antileukemia and antitumor agent from *Taxus brevifolia*. *J. Am. Chem. Soc.* 93: 2325–2327.

Wang, F.S., L.Y. Zhao, G. Xu, Q.S. Zhao, and H.D. Sun. 2004. Three new oxetane-ring-containing taxoids from *Taxus chinensis*. *J. Nat. Prod.* 67: 905–907.

Wang, Y., M. Kuroda, X.S. Gao, S. Akaki, J. Asaumi, Y. Okumura, K. Shibuya, S. Kawasaki, I. Joja, H. Kato, K. Himei, S. Dendo, S. Kanazawa, and Y. Hiraki. 2004. Cepharanthine enhances in vitro and in vivo thermosensitivity of a mouse fibrosarcoma. *Int. J. Mol. Med.* 13: 405–411.

Wang, Z.W., Y.H. Kuo, D. Schnur, J.P. Bowen, S.Y. Liu, F.S. Han, J.Y. Chan, Y.C. Chen, and K.H. Lee. 1990. New 4 β-arylamino derivatives of 4′-*O*-demethylepipodophyllotoxin and related compounds as potent inhibitors of human DNA topoisomerase II. *J. Med. Chem.* 33: 2660–2666.

Xiao, Z., Xiao, Y.D., J. Feng, A. Golbraikh, A. Tropsha, and K.H. Lee. 2002. Modeling of epipodophyllotoxin derivatives using variable selection k nearest neighbor QSAR method. *J. Med. Chem.* 45: 2294–2309.

Yuan, K., R. Hu, and M. Yin. 1997. New method for preparing oridonin by column chromatography. *China J. Chin. Mater. Med.* 22: 478–480.

8

Biologically Active Natural Products from Marine Fungi

MICHIO NAMIKOSHI

Department of Ocean Sciences, Tokyo University of Marine Science and Technology, Konan, Minato-ku, Tokyo 108-8477, Japan

INTRODUCTION

Fungi isolated from marine environments are prolific resources of biologically active secondary metabolites. The importance and attractiveness of marine fungi as a source of new biologically active compounds are recognized from the numbers of reviews describing marine fungal metabolites (Bugni and Ireland 2004, Proksch *et al.* 2002, Jensen and Fenical 2002, 2000, Verbist *et al.* 2000, Biabani and Laatsch 1998, Bernan *et al.* 1997, Pietra 1997, Davidson 1995, Liberra and Lindequist 1995, Kobayashi and Ishibashi 1993). The driving force of the chemical study on marine fungi is derived from the successful history of terrestrial fungal products as new medicines and their leads. Traditional studies on terrestrial fungi resulted in the high rate of rediscovery of known compounds. The rate of new compounds versus known compounds of the metabolites from marine fungi is much higher than that of terrestrial species, although significant numbers of marine isolates are taxonomically similar or identical to terrestrial fungi. The ability to produce new secondary metabolites by marine fungi may be acquired by the tolerance or adaptation to marine environments such as salinity and high pressure. Since marine fungi are reproducible by the similar technology of fermentation to that of terrestrial fungi, the supply problems, limiting the application of natural products obtained from marine invertebrates, will not be a problem for commercial application of marine fungal metabolites.

Marine fungi are grouped by an ecological definition but are not of a taxonomic group. The ecological definition of marine fungi was proposed by Kohlmeyer and Kohlmeyer (1979) and accepted by taxonomists (Kohlmeyer and Volkmann-Kohlmeyer 2003): "*Obligate marine fungi* are those that grow and sporulate exclusively in a marine or estuarine habitat; *facultative marine fungi* are those from freshwater or terrestrial milieus able to grow (and possibly also to sporulate) in the marine environment". Although this definition of marine mycology can be useful also for searching for novel biologically active metabolites, researchers working on the chemical study of fungal metabolites tend to isolate the fast-growing species such as *Aspergillus* and *Penicillium*. It is not clear if these fungi are active in the marine environment or if dormant spores or hyphae are being isolated. These fungi are referred to as marine-derived strains. Although it would be interesting to determine if these fungi are adapted to the marine environment from which they were isolated, most importantly they are prolific producers of new secondary metabolites. Therefore, this article discusses new secondary metabolites produced by fungi isolated from marine environments by seawater based media (a few compounds that have been known as synthetic products and were isolated for the first time as natural products are also included in this article).

Cephalosporin C (Fig. 8.1) might be the first compound isolated from the culture of a marine-derived fungus, in 1945. However, the fungus *Acremonium chrrysogenum* was isolated from a sewage outlet off the Sardinia coast and, therefore, considered to be a terrestrial strain. In 1977, Okutani reported a known diketopiperazine antibiotic, gliotoxin (Fig. 8.1), from *Aspergillus* sp. MO–10 isolated from sediment from the Seto Inland Sea in Japan (Okutani 1977). The first metabolite of an obligate marine fungus was siccayne (Fig. 8.1) isolated from the marine basidiomycete *Halocyphina villosa* 573 (Kupka *et al*. 1981). This compound was first isolated from the terrestrial fungus *Helminthosporium siccans* (Ishibashi *et al*. 1968). A few reports on isolation of known compounds were published before the first new metabolite was reported in 1986. Although chemical study of marine fungi was rather rare before the early 1990s, rapid progress in the isolation of new secondary metabolites has been observed since 1994. Vigorous studies provided 437 new secondary metabolites from 149 strains reported in 182 papers up to June, 2004 (Table 8.1 and Organism Index).

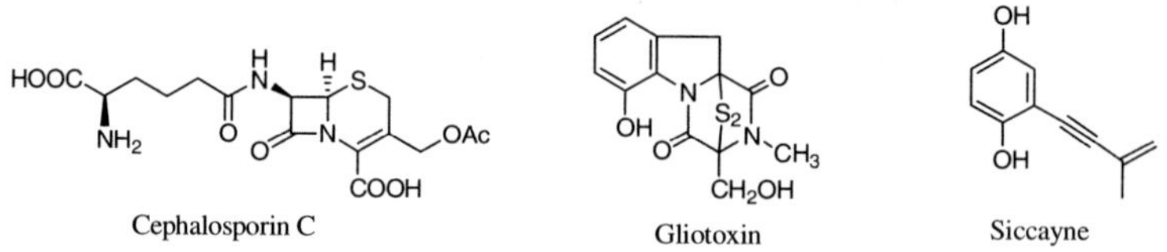

Fig. 8.1 Structures of cephalosporin C, gliotoxin, and siccayne.

Table 8.1 Numbers of reports and new compounds published in scientific journals up to June 2004.

Year	*86*	*87*	*88*	*89*	*90*	*91*	*92*	*93*	*94*	*95*	*96*	*97*	*98*	*99*	*00*	*01*	*02*	*03*	*04 (June)*	*Total*
Reports	1	1	1	4	1	4	2	2	6	7	12	7	20	14	17	18	29	23	13	182
New Compounds	1	1	1	12	4	6	5	3	19	18	24	20	44	31	48	34	79	54	33	437

The new secondary metabolites are divided into four groups in this article. The first group contains the compounds, which showed cytotoxicity to cultured cancer cells and antitumor activity. The compounds in the second group had antimicrobial activities, such as antibacterial, antifungal, and antiviral activities. The third group consists of compounds with other biological activities, which include inhibitory activity against proteins in the cell cycle. The last group contains the compounds which did not show biological activity in the bioassays examined or were not tested.

New Metabolites with Antitumor and Cytotoxic Activity (Compounds 1–145)

Cytotoxicity against cultured cancer cells has been the most popular screen for biological activities of marine natural products. Numerous culture broths of marine fungi have also been tested for cytotoxicity, and, so far, the 145 compounds listed in Table 8.3 were reported to show cytotoxic activity. The structures of these compounds are shown in Figure 8.2.

Five cytotoxic compounds, spiroxins A–E (**1–5**), were purified from the culture extract of an unidentified fungus LL-37H248 isolated from an orange soft coral collected at Vancouver Island in Canada (McDonald *et al.* 1999, absolute stereochemistry: Wang *et al.* 2001). Spiroxins A–E (**1–5**) exhibited cytotoxicity and antibacterial activity against Gram-positive bacteria and were active in a mouse xenograft model. Spiroxin A (**1**) revealed cytotoxicity against a panel of 25 diverse cell lines (mean IC_{50} of 0.09 μg/mL) and *in vivo* activity against an ovarian carcinoma in nude mice (59% growth inhibition after 21 days). The mode of action of **1–5** was explained to be due in part to their effect on DNA cleavage.

Leptosphaeria sp. OUPS-N80 (formerly OUPS-4) isolated from the brown alga *Sargassum tortile* is a prolific fungus and 24 new compounds, leptosins A–S (**6–29**), have been obtained from this strain (Takahashi *et al.* 1994a,b, 1995a,b, Yamada *et al.* 2002a, 2004). The cytotoxicities of leptosins against P388 are listed in Table 8.2. Leptosins A–C (**6–8**), G (**12**), G_1 (**13**), G_2 (**14**), H (**15**), K (**18**), K_1 (**19**), and K_2 (**20**) had strong activity with ED_{50} values in the ng order. Leptosin M (**21**) exhibited

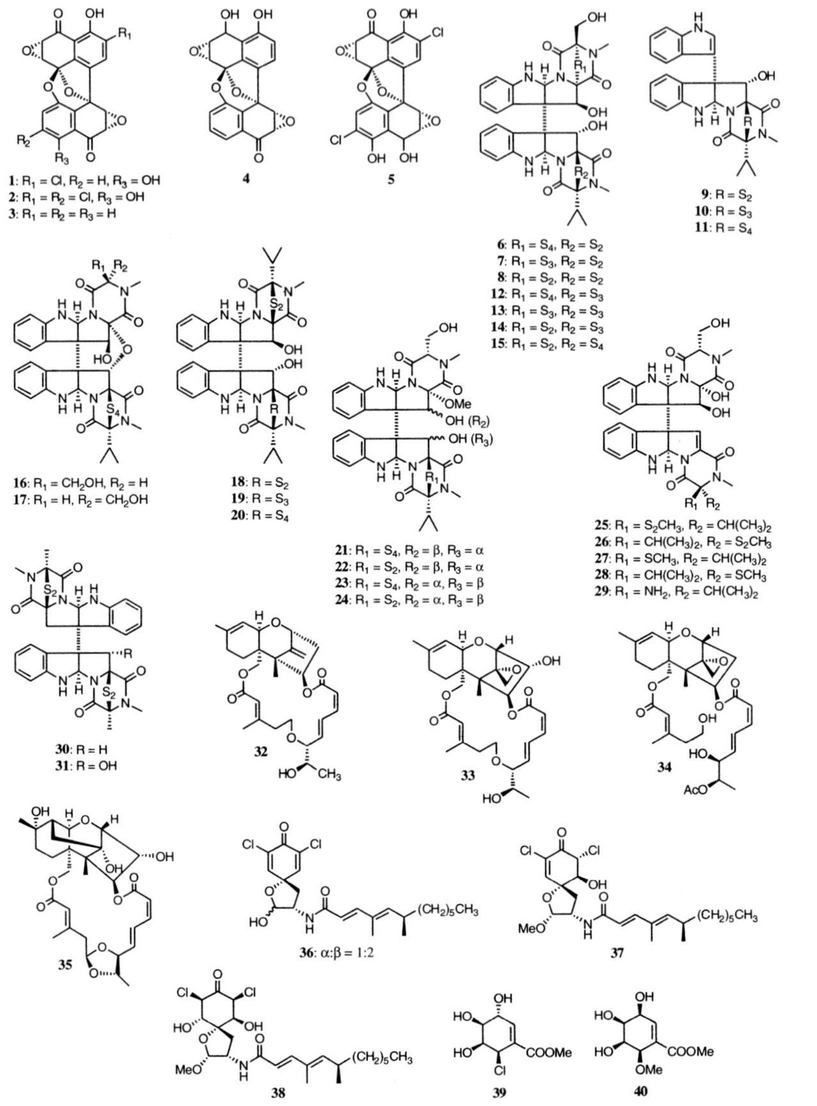

Fig. 8.2.1 Structures of new marine fungal metabolites with cytotoxic/antitumor activity.

appreciable cytotoxicity against the 39 human cancer cell lines, and evaluation of the pattern of differential cytotoxicity using the COMPARE program suggested the possibly different mode of action from that shown by other anticancer drugs. Moreover, **21** inhibited two protein kinases PTK and CaMKII by 40–70% at 10 μg/mL selectively from PKA, PKC, and EGFR and also demonstrated specific inhibition of topoisomerase II (IC_{50} = 59.1 μM) without affecting I (>300 μM). Leptosins O (**25**) and P (**26**), possessing the methyldithio group, showed moderate cytotoxic ac-

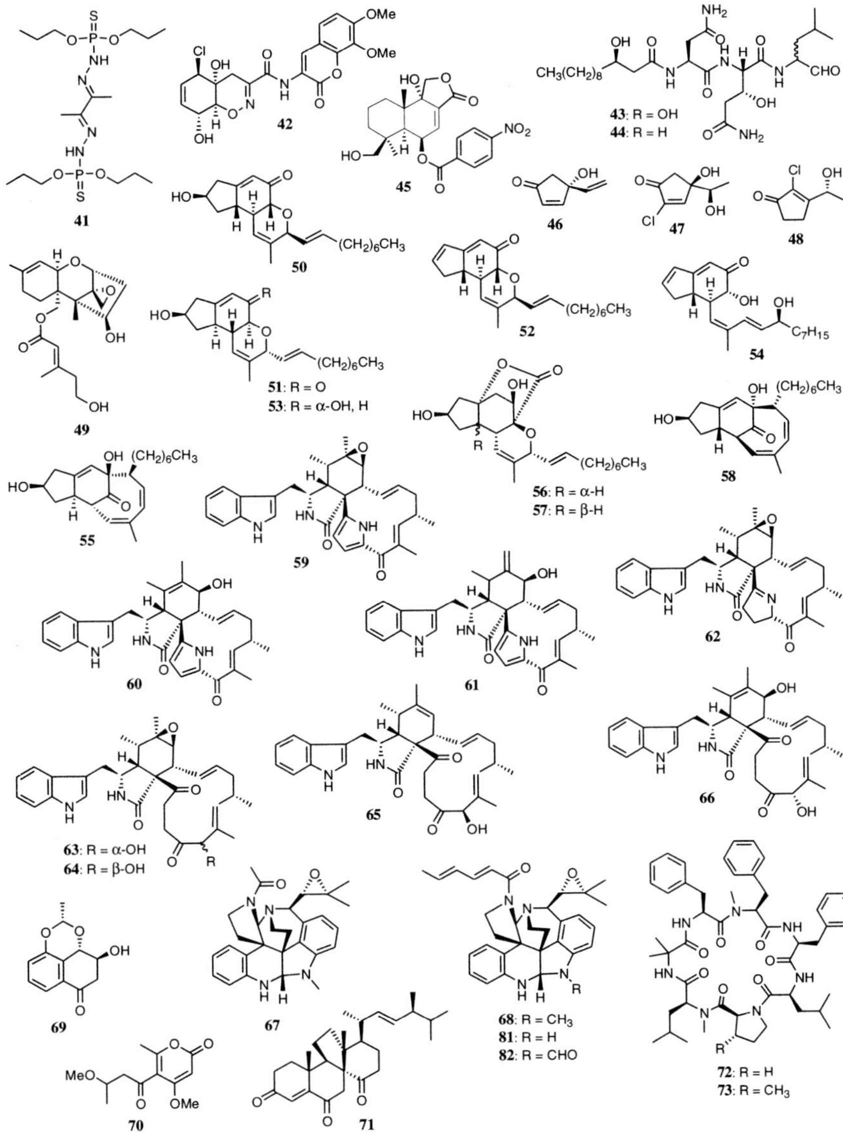

Fig. 8.2.2

tivities against the 39 human cancer cell lines. Two polythio bridges in both units seem to be very important to their toxicity since A–C, G, G_1, G_2, H, K, K_1, and K_2 were one to three orders more potent than other leptosins. Leptosins M (**21**), M_1 (**22**), N (**23**), and N_1 (**24**) lacked a second polythio bridge and were less active. The ether linkage contained in I (**16**) and J (**17**) reduced the activity. Interestingly, the structurally most different compounds (D–F) retained their potent cytotoxicity.

Table 8.2 Cytotoxicity of compounds **6–29**, **36–40**, **46–48**, and **50–68** against P388 murine leukemia cells.

Compound		ED_{50}	*Compound*		ED_{50}	*Compound*		ED_{50}
Name	*No.*	*(μg/mL)*	*Name*	*No.*	*(μg/mL)*	*Name*	*No.*	*(μg/mL)*
Leptosin A	**6**	0.00185	Leptosin N	**23**	0.18	Penostatin C	**52**	1.1
Leptosin B	**7**	0.0024	Leptosin N_1	**24**	0.19	Penostatin D	**53**	11.5
Leptosin C	**8**	0.00175	Leptosin O	**25**	1.1	Penostatin E	**54**	0.9
Leptosin D	**9**	0.086	Leptosin P	**26**	0.1	Penostatin F	**55**	1.4
Leptosin E	**10**	0.046	Leptosin Q	**27**	14.8	Penostatin G	**56**	0.5
Leptosin F	**11**	0.056	Leptosin R	**28**	15.2	Penostatin H	**57**	0.8
Leptosin G	**12**	0.0046	Leptosin S	**29**	10.1	Penostatin I	**58**	1.2
Leptosin G_1	**13**	0.0043	Gymnastatin A	**36**	0.018	Penochalasin A	**59**	0.4
Leptosin G_2	**14**	0.0044	Gymnastatin B	**37**	0.108	Penochalasin B	**60**	0.3
Leptosin H	**15**	0.003	Gymnastatin C	**38**	0.106	Penochalasin C	**61**	0.5
Leptosin I	**16**	1.13	Pericosine A	**39**	0.12	Penochalasin D	**62**	3.2
Leptosin J	**17**	1.25	Pericosine B	**40**	4.0	Penochalasin E	**63**	2.1
Leptosin K	**18**	0.0038	Trichodenone A	**46**	0.21	Penochalasin F	**64**	1.8
Leptosin K_1	**19**	0.0022	Trichodenone B	**47**	1.21	Penochalasin G	**65**	1.9
Leptosin K_2	**20**	0.0021	Trichodenone C	**48**	1.45	Penochalasin H	**66**	2.8
Leptosin M	**21**	1.05	Penostatin A	**50**	0.8	Communesin A	**67**	3.5
Leptosin M_1	**22**	1.40	Penostatin B	**51**	1.2	Communesin B	**68**	0.45

ylene position (C-12, 13) in **32**) showed an 80-fold greater cytotoxicity than **32**.

Three cytotoxic trichothecenes, 3-hydroxyroridin E (**33**), 13′-acetyltrichoverrin B (**34**), and miophytocen C (**35**), were purified from *Myrothecium verrucaria* 973023 (Amagata *et al.* 2003a). This fungus was isolated from the sponge *Spongia* sp. from Hawaii. The cytotoxic properties of compounds **33–35** were assessed against murine and human tumor cell lines using the disc diffusion soft agar colony formation assay, and **33–35** showed significant inhibition zones against all the cell lines. Compounds **33–35** were isolated together with nine known trichothecenes, roridins A, L, and M, verrucarins A and M, trichoverrins A and B, isororidin A, and epiroridin E; and the structure-activity relationships of these 12 compounds were also discussed.

Gymnascella dankaliensis OUPS-N134 isolated from the sponge *Halichondria japonica* collected at Osaka Bay, Japan, afforded eight new cytotoxic compounds, gymnastatins A–C (**36–38**), D (**110**), and E (**111**), dankasterone (**71**), and gymnasterones A (**108**) and B (**109**) (Numata *et al.* 1997a, Amagata *et al.* 1998c,d, 1999). Gymnastatins A–C (**36–38**) exhibited potent cytotoxicity against P388 (Table 8.2). A total synthesis of gymanastatin A (**36**) has been achieved (Gurjar and Bhaket 2000). Dankasterone (**71**) and gymnasterone B (**109**) showed moderate cytotoxicity against P388 (ED_{50} = 2.2 and 1.6 μg/mL, respectively). Gymnasterone A (**108**) and gymnastatins D (**110**) and E (**111**) were less cytotoxic to P388 with ED_{50} values of 10.1, 10.5, and 10.8 μg/mL, respectively.

Table 8.3 New marine fungal metabolites with cytotoxic/antitumor activity.

Compound		*Fungus*	*Reference*
No.	*Name*		
1	Spiroxin A	Unidentified fungus LL-37H248	McDonald *et al.* 1999
2	Spiroxin B		
3	Spiroxin C		
4	Spiroxin D		
5	Spiroxin E		
6	Leptosin A	*Leptosphaeria* sp. OUPS-N80	Takahashi *et al.* 1994a
7	Leptosin B		
8	Leptosin C		
9	Leptosin D		
10	Leptosin E		
11	Leptosin F		
12	Leptosin G		Takahashi *et al.* 1995a
13	Leptosin G_1		
14	Leptosin G_2		
15	Leptosin H		
16	Leptosin I		Takahashi *et al.* 1994b
17	Leptosin J		
18	Leptosin K		Takahashi *et al.* 1995b
19	Leptosin K_1		
20	Leptosin K_2		
21	Leptosin M		Yamada *et al.* 2002a
22	Leptosin M_1		
23	Leptosin N		
24	Leptosin N_1		
25	Leptosin O		Yamada *et al.* 2004
26	Leptosin P		
27	Leptosin Q		
28	Leptosin R		
29	Leptosin S		
30	11,11′-Dideoxyverti-cillin A	*Penicillium* sp. CNC-350	Son *et al.* 1999
31	11′-Deoxyverticillin A		
32	12,13-Deoxyroridin E	*Myrothecium roridum* TUF 98F42	Namikoshi *et al.* 2001
33	3-Hydroxyroridin E	*Myrothecium verrucaria* 973023	Amagata *et al.* 2003a
34	13′-Acetyltrichove-rrin B		
35	Miophytocen C		
36	Gymnastatin A	*Gymnascella dankaliensis* OUPS-N134	Numata *et al.* 1997a
37	Gymnastatin B		
38	Gymnastatin C		
39	Pericosine A	*Periconia byssoides* OUPS-N133	Numata *et al.* 1997b
40	Pericosine B		
41	n/a[1]	*Lignincola laevis*	Abraham *et al.* 1994
42	Trichodermamide B	*Trichoderma virens* CNK266, CNL910	Garo *et al.* 2003
43	*Fellutamide A*	*Penicillium fellutanum*	Shigemori *et al.* 1991
44	Fellutamide B		

(*Contd.*)

(*Contd.*)

45	Insulicolide A	*Aspergillus insulicola*	Rahbæk *et al.* 1997
46	Trichodenone A	*Trichoderma harzianum* OUPS-N115	Amagata *et al.* 1998a
47	Trichodenone B		
48	Trichodenone C		
49	Verrol 4-acetate	*Acremonium neocaledoniae* LCP 98 4255	Laurent *et al.* 2000
50	Penostatin A	*Penicillium* sp. OUPS-N79	Takahashi *et al.* 1996
51	Penostatin B		
52	Penostatin C		
53	Penostatin D		
54	Penostatin E		
55	Penostatin F	*Penicillium* sp. OUPS-N79	Iwamoto *et al.* 1998
56	Penostatin G		
57	Penostatin H		
58	Penostatin I		
59	Penochalasin A		Numata *et al.* 1996
60	Penochalasin B		
61	Penochalasin C		
62	Penochalasin D		Iwamoto *et al.* 2001
63	Penochalasin E		
64	Penochalasin F		
65	Penochalasin G		
66	Penochalasin H		
67	Communesin A		Numata *et al.* 1993
68	Communesin B		
69	Humicolone	*Humicola grisea* 880M/272	Laurent *et al.* 2002
70	Pyrenocine E	*Penicillium waksmanii* OUPS-N127	Amagata *et al.* 1998b
71	Dankasterone	*Gymnascella dankaliensis* OUPS-N134	Amagata *et al.* 1999
72	Scytalidamide A	*Scytalidium* sp. CNC-310	Tan *et al.* 2003
73	Scytalidamide B		
74	Evariquinone	*Emericella variecolor* E-00-6/3	Bringmann *et al.* 2003
75	7-Deacetoxyy-anuthone A	*Penicillium* sp. MFA 577	Li *et al.* 2003a
76	Neomangicol A	*Fusarium heterosporum* CNC-477	Renner *et al.* 1998
77	Neomangicol B		
78	Sansalvamide A	*Fusarium* sp. CNL 292	Belofsky *et al.* 1999
79	*N*-Methylsansal-vamide	*Fusarium* sp. CNL-619	Cueto *et al.* 2000
80	Acetophthalidin	*Penicillium* sp. BM923	Cui *et al.* 1996a
81	Communesin C	*Penicillium* sp.	Jadulco *et al.* 2004
82	Communesin D		
83	(+)-Formylanserinone B	*Penicillium corylophilum* a004181, b004181	Gautschi *et al.* 2004
84	(–)-Epoxyserinone A		
85	(+)-Epoxyserinone B		
86	Varitriol	*Emericella variecolor* M75-2	Malmstrøm *et al.* 2002a
87	Tryprostatin A	*Aspergillus fumigatus* BM939	Cui *et al.* 1995
88	Tryprostatin B		
89	Spirotryprostatin A		Cui *et al.* 1996b

(*Contd.*)

(Contd.)

90	Spirotryprostatin B		
91	Cyclotryprostatin A		Cui *et al.* 1997
92	Cyclotryprostatin B		
93	Cyclotryprostatin C		
94	Cyclotryprostatin D		
95	Macrosphelide E	*Periconia byssoides* OUPS-N133	Numata *et al.* 1997b
96	Macrosphelide F		
97	Macrosphelide G		
98	Macrosphelide H		
99	Macrosphelide I		Yamada *et al.* 2001
100	Macrosphelide L		Yamada *et al.* 2002b
101	Fumiquinazoline A	*Aspergillus fumigatus*	Numata *et al.* 1992
102	Fumiquinazoline B		
103	Fumiquinazoline C		
104	Fumiquinazoline D		Takahashi *et al.* 1995c
105	Fumiquinazoline E		
106	Fumiquinazoline F		
107	Fumiquinazoline G		
108	Gymnasterone A	*Gymnascella dankaliensis* OUPS-N134	Amagata *et al.* 1998c
109	Gymnasterone B		
110	Gymnastatin D	*Gymnascella dankaliensis* OUPS-N134	Amagata *et al.* 1998d
111	Gymnastatin E		
112	Aspergillamide A	*Aspergillus* sp. CNC-120	Toske *et al.* 1998
113	Aspergillamide B		
114	Mangicol A	*Fusarium heterosporum* CNC-477	Renner *et al.* 2000
115	Mangicol B		
116	Mangicol C		
117	Mangicol D		
118	Mangicol E		
119	Mangicol F		
120	Mangicol G		
121	Brocaenol A	*Penicillium brocae* F97S76	Bugni *et al.* 2003
122	Brocaenol B		
123	Fusaperazine A	*Fusarium chlamydosporum* OUPS-N124	Usami *et al.* 2002
124	Virescenoside M	*Acremonium striatisporum* KMM 4401	Afiyatullov *et al.* 2000
125	Virescenoside N		
126	Virescenoside O		Afiyatullov *et al.* 2002
127	Virescenoside P		
128	Virescenoside Q		
129	Aspergillicin A	*Aspergillus carneus* MST-MF156	Capon *et al.* 2003
130	Aspergillicin B		
131	Aspergillicin C		
132	Aspergillicin D		
133	Aspergillicin E		
134	Asperazine	*Aspergillus niger*	Varoglu *et al.* 1997
135	Virescenoside R	*Acremonium striatisporum* KMM 4401	Afiyatullov *et al.* 2004
136	Virescenoside S		

(Contd.)

(*Contd.*)

137 Virescenoside T		
138 Virescenoside U		
139 Kasarin	*Hyphomycetes* sp.	Suenaga *et al.* 2000
140 Harzialactone B	*Trichoderma harzianum* OUPS-N115	Amagata *et al.* 1998a
141 6-Epi-ophiobolin G	*Emericella variecolor* GF10	Wei *et al.* 2004
142 6-Epi-ophiobolin N		
143 Aspergilloxide	*Aspergillus* sp. CNM-713	Cueto *et al.* 2002
144 Herbarin A	*Cladosporium herbarum*	Jadulco *et al.* 2002
145 Herbarin B		

[1]n/a: not available

Eight new cytotoxic compounds, pericosines A (**39**) and B (**40**) and macrosphelides E–I (**95–99**) and L (**100**), and a non-toxic linear analog, *seco*-macrosphelide E (**390**), were produced by *Periconia byssoides* OUPS-N133 isolated from the gastrointestinal tract of the sea hare *Aplysia kurodai* collected at Osaka Bay (Numata *et al.* 1997b, Yamada *et al.* 2001, 2002b, Nakamura *et al.* 2002a). Pericosine A (**39**) exhibited significant cytotoxicity against P388 (ED_{50} = 0.12 μg/mL) and also showed potent *in vivo* antitumor activity, while B (**40**) was less active (ED_{50} = 4.0 μg/mL) (Numata *et al.* 1997b). Macrosphelides E–H (**95–98**) and L (**100**) inhibited the adhesion of HL-60 cells to human umbilical vein endothelial cells (HUVEC) with IC_{50} values of 19.5, 27.2, 22.5, 8.6, and 5.6 μM, respectively. Macrosphelide I (**99**) did not inhibit the adhesion of HL-60 cells to HUVEC and showed weak cytotoxicity against P388 (ED_{50} = 20 μg/mL). A linear compound, *seco*-macrosphelide E (**390**), has the same sequence as E (**95**) and was not cytotoxic. Therefore, the cyclic structure is essential for the toxicity of macrosphelides. Total syntheses of E (Ono *et al.* 2000), F (Nakamura *et al.* 2002b), G (Nakamura *et al.* 2002b, Kobayashi and Wang 2002), and H (Kobayashi and Wang 2002) have been reported. Macrosphelides A–D, J, and K were isolated from the terrestrial fungus *Microsphaeropsis* sp. FO-5050 by Omura's group as cell-adhesion inhibitors (Hayashi *et al.* 1995, Takamatsu *et al.* 1996, 1997, Sunazuka *et al.* 1997, Fukami *et al.* 1999).

An obligate marine fungus *Lignincola laevis* isolated from a marsh grass produced a cytotoxic compound **41** (Abraham *et al.* 1994). Compound **41** has a unique structure (Fig. 8.2) and exhibited strong cytotoxicity against L1210 cells at the 0.25 μg/mL level.

Trichodermamide B (**42**) exhibited significant cytotoxicity against HCT-116 human colon carcinoma with an IC_{50} value of 0.32 μg/mL and also showed moderate antimicrobial activities (see the next section) (Garo *et al.* 2003). Compound **42** was obtained from two strains of *Trichoderma virens* (CNK266 and CNL910) together with non-toxic trichodermamide A (**402**). Two strains, CNK266 and CNL910, were isolated from the green alga *Halimeda* sp. and the ascidian *Didemnum molle*, respectively, in Papua, New Guinea. Trichodermamide A (**402**) has an OH group instead of the Cl group in B (**42**), and, therefore, the Cl group is necessary for the cytotoxicity of **42**.

Two linear peptides, fellutamides A (**43**) and B (**44**), were obtained from *Penicillium fellutanum* isolated from the gastrointestinal tract of the Japanese fish *Apogon endekataenia* (Shigemori *et al*. 1991). Compound **43** exhibited cytotoxicity against KB, P388, and L1210 cells (IC_{50} = 0.5, 0.2, and 0.8 μg/mL, respectively) and was a potent enhancer of nerve growth factor (NGF) synthesis and secretion *in vitro*. A (**43**) stimulated the production of NGF in mouse L-M, rat brain, and rat glial cells by a different mode of action from epinephrine, a known NGF inducer (Yamaguchi *et al*. 1993). Fellutamide B (**44**) showed similar cytotoxicity to A (**43**) against KB, P388, and L1210 cells (IC_{50} = 0.7, 0.1, and 0.7 μg/mL, respectively).

Insulicolide A (**45**) was first purified from the culture of *Aspergillus insulicola* isolated from various algae from the Bahamas (Rahbæk *et al*. 1997) and later from *Aspergillus versicolor* CNC 327 isolated from the green alga *Penicillus capitatus* also from the Bahamas together with three non-cytotoxic compounds **301–303** (Belofsky *et al*. 1998). Compound **45** exhibited a mean IC_{50} of 1.1 μg/mL in the NCI's 60-cell line panel and good cytotoxicity against the colon cancer cell lines HCC-2998 and HCT-116, CNB cancer cell line SNB-75, and the breast cancer cell line BT-549 (IC_{50} = 0.53, 0.44, 0.44, and 0.27 μg/mL, respectively) (Belofsky *et al*. 1998).

Trichoderma harzianum OUPS-N115 isolated from the sponge *Halichondria okadai* at Tanabe Bay, Japan, gave four cytotoxic compounds, trichodenones A–C (**46–48**) and harzialactone B (**140**), and a non-toxic harzialactone A (**400**) (Amagata *et al*. 1998a). Trichodenone A (**46**) was obtained as a scalemic mixture (*R*-enantiomer was predominant). The cytotoxicities of trichodenones A–C (**46–48**) are listed in Table 8.2. Harzialactone B (**140**) had very weak cytotoxicity against P388 (ED_{50} = 60 μg/mL). Total syntheses of trichodenones A–C have been reported (Usami and Numata 1999, Usami *et al*. 2000).

A trichothecene derivative, verrol 4-acetate (**49**), was obtained from the lignicolous fungus *Acremonium neocaledoniae* LCP 98 4255 isolated from driftwood in New Caledonia (Laurent *et al*. 2000). Compound **49** demonstrated cytotoxicity against KB cells (IC_{50} = 0.4 μg/mL), but was, however, much less active than macrocyclic trichothecenes. A macrocyclic ring seems to be important to strengthen the toxicity of trichothecenes.

Penicillium sp. OUPS-N79 (formerly OUPS-79) isolated from the green alga *Enteromorpha intestinalis* at Tanabe Bay is also a prolific fungus, and 19 new compounds, penostatins A–I (**50–58**), penochalasins A–H (**59–66**), and communesins A (**67**) and B (**68**) have been reported (Numata *et al.* 1993, 1996, Takahashi *et al.* 1996, Iwamoto *et al.* 1998, 1999, 2001). The cytotoxicities of compounds **50–68** against P388 cells are listed in Table 8.2. Communesins C (**81**) and D (**82**) were produced together with B (**68**) by *Penicillium* sp. isolated from the Mediterranean sponge *Axinella verrucosa* and showed a moderate antiproliferative activity against several human leukemia cell lines (ED_{50} = 8.2–13.1 and 9.0–16.2 μg/mL, respectively) (Jadulco *et al.* 2004). Communesins possess a novel carbon skeleton, and only one other compound with a similar skeleton, perophoramidine, has been isolated, from the ascidian *Perophora nameii* (Verbitski *et al.* 2002).

Humicolone (**69**) exhibited moderate cytotoxicity against KB cells with IC_{50} values between 1 and 5 μg/mL. This compound was produced by *Humicola grisea* 880M/272 isolated from the core of driftwood at New Caledonia (Laurent *et al.* 2002).

Pyrenocines D (**389**) and E (**70**) were obtained from *Penicillium waksmanii* OUPS-N127 isolated from the brown alga *Sargassum ringgoldianum* (Amagata *et al.* 1998b). Pyrenocine E (**70**) showed moderate cytotoxicity against P388 (ED_{50} = 1.30 μg/mL), while D (**389**) was not toxic. The different side chains remarkably affected the cytotoxicity of these pyrenocines.

Scytalidamide A (**72**) was cytotoxic against HCT-116 cells (IC_{50} = 2.7 μM), and B (**73**) showed moderate activity (IC_{50} = 11.0 μM) (Tan *et al.* 2003). Interestingly, methylation on the proline unit decreased the activity. Scytalidamides A (**72**) and B (**73**) were produced by *Scytalidium* sp. CNC-310 isolated from the surface of the green alga *Halimeda* sp. collected in the Bahamas.

Moderately cytotoxic evariquinone (**74**) was obtained together with non-toxic isoemericellin (**340**) from *Emericella variecolor* E-00-6/3 isolated from the interior of the Mediterranean sponge *Haliclona valliculata* (Bringmann *et al.* 2003). Evariquinone (**74**) had antiproliferative activity toward KB and NCI-H460 cells (60% and 69% inhibition, respectively, at 3.16 μg/mL). Interestingly, the *O*-demethyl derivative of **74** (7-hydroxyemodin) was inactive.

7-Deacetoxyyanuthone A (**75**) exhibited moderate cytotoxicity against five human solid tumor cell lines (ED_{50} = 3.86–10.07 μg/mL) and also showed moderate antibacterial activity (see the next section) (Li *et al.* 2003a). Compound **75** was obtained together with non-toxic farnesylhydroquinone (**256**) and 2,3-dihydro-7-deacetoxyyanuthone A (**384**) from *Penicillium* sp. MFA 577 isolated from the surface of the polymeric cord collected at Bijin Island, Korea. Compound **384** has diol groups instead of the epoxy group in **75**, and, therefore, the epoxide moiety is very important to the cytotoxicity.

Fusarium heterosporum CNC-477 was isolated from driftwood in the Bahamas and produced sesquiterpenes, neomangicols A (**76**), B (**77**), and C (**344**), and the sesterterpenes, mangicols A–G (**114–120**) (Renner *et al.* 1998, 2000). Neomangicol A (**76**) exhibited moderate cytotoxicity against a variety of cancer cell lines with a mean IC_{50} value of 10 μM, especially against human breast (MCF-7) and colon (CACO-2) carcinoma cell lines (IC_{50} = 4.9 and 5.7 μM, respectively). Neomangicol B (**77**) showed modest cytotoxicity against a variety of cancer cell lines (mean IC_{50} = 27 μM) and also showed weak antibacterial activity. Neomangicol C (**344**) was not active. Mangicols A–G (**114–120**) exhibited modest cytotoxicity against human tumor cell lines (GI_{50} = 24.5, 20.4, 17, 24.5, 17.8, 36.3, and 25.1 μM, respectively) (Renner *et al.* 2000). Mangicols A (**114**) and C (**116**) also showed anti-inflammatory activity.

Sansalvamide A (**78**) was gotten from *Fusarium* sp. CNL 292 isolated from the surface of the seagrass *Halodule wrightii* from the Bahamas (Belofsky *et al.* 1999). Compound **78** showed cytotoxicity with a mean IC_{50} value of 27.4 μg/mL in the NCI's 60-cell line panel, especially selectively against colon cancer (COLO 205) and melanoma (SK-MEL-2) cell lines (IC_{50} = 3.5 and 5.9 μg/mL, respectively) and also had antiviral activity (see the next section) (Hwang *et al.* 1999). A total synthesis of **78** has been accomplished (Leo and Silverman 2000). *Fusarium* sp. CNL-619 isolated from the green alga *Avrainvillea* sp. from the US Virgin Islands produced an *N*-methyl derivative of **78**, *N*-methylsansalvamide (**79**), which exhibited moderate cytotoxicity (GI_{50} = 8.3 μM) in the NCI's human tumor cell line screen (Cueto *et al.* 2000).

Acetophthalidin (**80**) was produced by *Penicillium* sp. BM923 isolated from sediment from Shizuoka, Japan (Cui *et al.* 1996a). Compound **80** was readily degraded into an inactive compound, 3,4,6-trihydroxymelleine, during the separation procedure, and the natural compound **80** was recovered by boiling 3,4,6-trihydroxymelleine in an acidic condition. Acetophthalidin (**80**) completely arrested tsFT210 cells in the G2/M phase at a concentration of 6.25 μg/mL. A total synthesis has been reported (Nomoto and Mori 1997).

(+)-Formylanserinone B (**83**), (–)-epoxyserinone A (**84**), (+)-epoxyserinone B (**85**), deoxyanserinone B (**378**), and hydroxymethylanserinone B (**379**) were obtained from *Penicillium corylophilum* a004181 and b004181 (mixture of two strains) isolated from a South Pacific deep water (–1335 m) sediment together with two known compounds, anserinones A and B (Gautschi *et al.* 2004). Interestingly, anserinones A and B were first isolated from the coprophilous fungus *Podospora anserina* as antifungal and antibacterial components (Wang, H.-J. *et al.* 1997). (+)-Formylanserinone B (**83**) significantly inhibited the growth of murine cancer cell lines selectively over murine solid tumor cells. Cytotoxicity of **83** against MDA-MB-435 cells was 2.90 μg/mL (IC_{50}). Compounds **84** and **85** were less active than **83**, and compounds **378** and **379** did not show cytotoxicity.

The Caribbean sponge-derived *Emericella variecolor* M75-2 produced varitriol (**86**), varixanthone (**162**), dihydroterrein (**338**), and varioxirane (**339**) (Malmstrøm *et al.* 2002a). Varitriol (**86**) displayed more than 100-fold greater potency against selected renal, CNS, and breast cancer cell lines in the NCI's 60-cell panel than did the others. Varixanthone (**162**) showed antibacterial activity (see the next section). Since **338** and **339** were isolated after peracetylation, bioassay was not performed.

A series of diketopiperzines, tryprostatins A (**87**) and B (**88**), spirotryprostatins A (**89**) and B (**90**), and cyclotryprostatins A–D (**91–94**), were obtained from *Aspergillus fumigatus* BM939 isolated from a sediment (–760 m) from Shizuoka (Cui *et al.* 1995, 1996b, 1997, Usui *et al.* 1998). Compounds **87–94** inhibited the cell cycle progression of tsFT210 cells at the G2/M phase at 50, 12.5, 197.5, 14.0, 5.6, 19.5, 23.4, and 25.3 μg/mL, respectively. These compounds are novel inhibitors of microtubule assembly through disruption of the microtubule spindles although the activity is rather weak. The effects of tryprostatins and cyclotryprostatins on microtubule assembly *in vitro* and *in situ* were reported (Kondoh *et al.* 1998). Total syntheses of tryprostatins A (**87**) (Gan and Cook 1997, Gan *et* al. 1997, Zhao *et* al. 1998) and B (**88**) (Depew *et al.* 1996, Zhao *et al.* 1998, Schkeryantz *et al.* 1999) have been achieved.

Fumiquinazolines A–G (**101–107**) were obtained from *Aspergillus fumigatus* isolated from the gastrointestinal tract of the marine fish *Pseudolabrus japonicus* collected at Tanabe Bay, Japan, and exhibited modest cytotoxicity against P388 with ED_{50} values of 6.1, 16.0, 52.0, 13.5, 13.8, 14.6, and 17.7 μg/mL (Numata *et al.* 1992, Takahashi *et al.* 1995c). Total syntheses of A (**101**) and B (**102**) (Snider and Zeng 2000) and F (**106**) and G (**107**) (Wang and Ganesan 2000) have been reported.

Aspergillamides A (**112**) and B (**113**) exhibited modest cytotoxicity against HCT-116 cells (IC_{50} = 16 μg/mL) (Toske *et al.* 1998). Aspergillamide A (**112**) was a 1:1 mixture of *cis* and *trans* rotational isomers about the amide bond (*). These compounds were produced by *Aspergillus* sp. CNC-120 isolated from the sediment of a saline lake in the Bahamas.

Penicillium brocae F97S76 isolated from a tissue homogenate of the sponge *Zyzzya* sp. from Fiji gave brocaenols A (**121**), B (**122**), and C (**375**) (Bugni *et al.* 2003). Brocaenols A (**121**) and B (**122**) showed weak cytotoxicity against HCT-116 cells (IC_{50} = 20 and 50 μg/mL), while C (**375**) was not active at 50 μg/mL.

Fusaperazines A (**123**) and B (**343**) were obtained from *Fusarium chlamydosporum* OUPS-N124 isolated from the red alga *Carpopeltis affinis*

(Usami *et al.* 2002). Fusaperazine A (**123**) demonstrated modest cytotoxicity against P388 (ED_{50} = 22.8 μg/mL), but B (**343**) was inactive. Therefore, the phenolic OH group will be important for the cytotoxicity.

Acremonium striatisporum KMM 4401 isolated from superficial mycobiota of the holothurian *Eupentacta fraudatrix* collected in the Sea of Japan, Russian Federation, gave nine compounds, virescenosides M–Q and R–U (**124–128** and **135–138**) (Afiyatullov *et al.* 2000, 2002, 2004). Virescenosides M (**124**) and N (**125**) exhibited cytotoxicity against Ehrlich carcinoma cells (IC_{50} = 10–100 μM) and also showed cytotoxic effect to developing eggs of the sea urchin *Strongylocentrotus intermedius* (MIC_{50} = 2.7–20 μM) (Afiyatullov *et al.* 2000). Virescenosides O–Q (**126–128**) were cytotoxic against Ehrlich carcinoma cells (IC_{50} = 20–100 μM), and P (**127**) had cytotoxic effect to developing eggs of *S. intermedius* (MIC_{50} = 5.0 μM) (Afiyatullov *et al.* 2002). Virescenosides R–U (**135–138**) displayed cytotoxicity against Ehrlich carcinoma cells (IC_{50} = 25–60 μM) (Afiyatullov *et al.* 2004).

Aspergillicins A–E (**129–133**) were produced by *Aspergillus carneus* MST-MF156 isolated from an estuarine sediment in Tasmania, Australia, and exhibited modest cytotoxicity (LD_{99} = 25-50 μg/mL) (Capon *et al.* 2003).

Asperazine (**134**) was obtained from *Aspergillus niger* isolated from the sponge *Hyrtios proteus* collected in Florida and showed leukemia selective cytotoxicity in the Corbett-Valeriote soft agar disc diffusion assay (50 μg/disc) (Varoglu *et al.* 1997).

Hyphomycetes sp. isolated from the zoanthid *Zoanthus* sp. at Amami Island, Japan, afforded kasarin (**139**), which had weak cytotoxicity against P388 (IC_{50} = 34 μg/mL) and antibacterial activity against *Rhodospirillium salexigens* (12 mm zone at 700 μg/disc) (Suenaga *et al.* 2000).

Two new sesterterpenes, 6-epi-ophiobolins G (**141**) and N (**142**) were purified from the extract of *Emericella variecolor* GF10 isolated from marine sediment (–70 m) at Mie, Japan, together with six known ophiobolins (Wei *et al.* 2004). Two new compounds **141** and **142** were cytotoxic against the neuroblastoma cell line, Neuro 2A. This fungus gave different ratios of components in the different culture media.

Aspergillus sp. CNM-713 gave aspergilloxide (**143**), which exhibited little cytotoxicity against HCT-116 (no quantitative data) (Cueto *et al.* 2002).

Herbarins A (**144**) and B (**145**) showed brine shrimp (*Artemia salina*) mortality (75 and 65% at 50 μg, respectively) (Jadulco *et al.* 2002). These compounds were obtained from *Cladosporium herbarum* isolated from the Mediterranean sponge *Aplysina aerophoba*.

New Metabolites with Antimicrobial Activity (Compounds 1–5, 42, 75, 77, 78, and 146–219)

Antibiotics had a successful era of curing infectious diseases, but the development of antibiotic-resistance in bacteria eventually curtailed their effectiveness. Therefore, efficient and continuous search for new antibiotics is necessary to overcome the acquired resistance. Especially, the development of effective antibacterial agents to fight against methicillin-resistant *Staphylococcus aureus* (MRSA), vancomycin-resistant *Enterococcus faecium* (VREF), and multidrug-resistant *Mycobacterium* are urgently required. Antifungal drugs are also needed to use for treatment of immunodeficient patients, such as with cancers and AIDS. A number of resistant fungi against synthetic antifungal agents have been found. Moreover, malarial and viral infections are also important targets for drug discovery. New marine fungal metabolites possessing antimicrobial activities are listed in Table 8.4 and the structures of these compounds are shown in Figure 8.3.

Cytotoxic spiroxins A–E (**1–5**) were reported to show antibacterial activity against Gram-positive bacteria (McDonald *et al.* 1999). Trichodermamide B (**42**) showed moderate antimicrobial activities against amphotericin-resistant *Candida albicans*, MRSA, and VREF with MIC values of ca. 15 μg/mL besides the cytotoxicity (Garo *et al.* 2003). 7-Deacetoxyyanuthone A (**75**) had moderate cytotoxicity and also antibacterial activity against MRSA and multidrug-resistant SA (MIC = 50 μg/mL) (Li *et al.* 2003a). Neomangicol B (**77**) showed weak antibacterial activity against *B. subtilis* (10 mm inhibition zone at 50 μg/disc) (Renner *et al.* 1998). Sansalvamide A (**78**) selectively inhibited the topoisomerase of the poxvirus *Molluscum contagiosum* (MCV) (Belofsky *et al.* 1999, Hwang *et al.* 1999).

Pestalone (**146**) was obtained from *Pestalotia* sp. CNL-365 isolated from the surface of the brown alga *Rosenvingea* sp. from the Bahamas and exhibited potent antibacterial activity against MRSA and VREF with MIC of 37 and 78 ng/mL, respectively (Cueto *et al.* 2001). Compound **146** also showed moderate *in vitro* cytotoxicity (GI_{50} = 6.0 μM) in the NCI's human tumor cell line screen. Interestingly, this antibiotic was produced only when this strain was cultured in the presence of a marine bacterium. This fungus or the marine bacterium alone did not produce pestalone (**146**), and the production of this compound in very low yield was induced by the addition of ethanol (1% v/v) in the culture medium after 24 h of fermentation.

An ascomycete *Kirschsteiniothelia maritima* M9 isolated from submerged wood in the Baltic Sea, Germany, produced ascochital (**147**), which inhibited the growth of *B. subtilis* with an MIC value of 0.5 μg/mL (Kusnick *et al.* 2002).

Table 8.4 New marine fungal metabolites with antimicrobial activity.

Compound		*Fungus*	*Activity*[1]	*Reference*
No.	*Name*			
1	Spiroxin A	unidentified fungus LL-37H248	B (G+)	McDonald *et al.* 1999
2	Spiroxin B		B (G+)	
3	Spiroxin C		B (G+)	
4	Spiroxin D		B (G+)	
5	Spiroxin E		B (G+)	
42	Trichodermamide B	*Trichoderma virens* CNK266, CNL910	B (G+), Y	Garo *et al.* 2003
75	7-Deacetoxyyanuthone A	*Penicillium* sp. MFA 577	B (G+)	Li *et al.* 2003a
77	Neomangicol B	*Fusarium heterosporum* CNC-477	B (G+)	Renner *et al.* 1998
78	Sansalvamide A	*Fusarium* sp. CNL 292	V	Belofsky *et al.* 1999
146	Pestalone	*Pestalotia* sp. CNL-365	B (G+)	Cueto *et al.* 2001
147	Ascochital	*Kirschsteiniothelia maritima* No. M9	B (G+)	Kusnick *et al.* 2002
148	Exophilin A	*Exophiala pisciphila* NI10102	B (G+)	Doshida *et al.* 1996
149	Speradine A	*Aspergillus tamarii* M143	B (G+)	Tsuda *et al.* 2003a
150	Auranticin A	*Preussia aurantiaca* ATCC 14745	B (G+)	Poch and Gloer 1991
151	Auranticin B		B (G+)	
152	Acetyl Sumiki's acid	*Cladosporium herbarum*	B (G+)	Jadulco *et al.* 2001
153	Seragakinone A	unidentified fungus K063	B (G+)	Shigemori *et al.* 1999
154	Unguisin A	*Emericella unguis* M87-2, M90B-10	B (G+)	Malmstrøm 1999
155	Unguisin B		B (G+)	
156	Guisinol		B (G+)	Nielsen *et al.* 1999
157	Secocurvularin	unidentified fungus 951014	B (G+)	Abrell *et al.* 1996a
158	Hirsutanol A	unidentified fungus 95-1005C	B (G+)	Wang *et al.* 1998
159	*Ent*-gloeosteretriol		B (G+)	
160	Isocyclocitrinol A	*Penicillium citrinum* 991084	B (G+)	Amagata *et al.* 2003b
161	22-Acetylisocyclocitrinol A		B (G+)	
162	Varixanthone	*Emericella variecolor* M75-2	B (G+, –)	Malmstrøm *et al.* 2002a
163	Aspergillitine	*Aspergillus versicolor* HBI-2	B (G+, –), Y	Lin *et al.* 2003
164	8-Chloro-9-hydroxy-8,9-deoxyasperlactone	*Aspergillus ostianus* TUF 01F313	B (G+, –)	Namikoshi *et al.* 2003
165	9-Chloro-8-hydroxy-8,9-deoxyasperlactone		B (G+, –)	
166	9-Chloro-8-hydroxy-8,9-deoxyaspyrone		B (G+, –)	

(*Contd.*)

(*Contd.*)

167	Lunatin	*Curvularia lunata*	B (G+, –)	Jadulco *et al.* 2002
168	Corollosporine	*Corollospora maritima* 1069	B (G+, –)	Liberra *et al.* 1998
169	n/a^2	*Halorosellinia oceanica* BCC 5149	M	Chinworrungsee *et al.* 2002
170	Modiolide A	*Paraphaeosphaeria* sp. N-119	M, F	Tsuda *et al.* 2003b
171	Modiolide B		M, F	
172	Stachybotrin A	*Stachybotrys* sp. CS-710-1	B (G+), F	Xu *et al.* 1992
173	Stachybotrin B		B (G+), F	
174	Zopfiellamide A	*Zopfiella latipes* CBS 611.97	B (G+, –), Y	Daferner *et al.* 2002
175	Zopfiellamide B		B (G+, –), Y	
176	Yanuthone A	*Aspergillus niger* F97S11	B (G+)	Bugni *et al.* 2000
177	Yanuthone B		B (G+), Y	
178	Yanuthone C		B (G+, –)	
179	Yanuthone D		B (G+, –), Y	
180	Yanuthone E		B (G+, –)	
181	1-Hydroxyyanuthone A		B (G+, –), Y	
182	1-Hydroxyyanuthone C		B (G+, –)	
183	22-Deacetylyanuth-one A		B (G+, –)	
184	Cladospolide D	*Cladosporium* sp. FT-0012	F	Zhang *et al.* 2001
185	(6*R*,11*S*,12*S*,14*R*)-9,10-Dihydrocolleto-diol	*Varicosporina ramulosa* 195-31	F	Höller *et al.* 1999a
186	(6*R*,11*R*,12*R*,14*R*)-9,10-Dihydrocolleto-diol		F	
187	(3*S*)-3,5-Dihydroxy-phenyl-butan-2-one	*Coniothyrium* sp. 193H77	F	Höller *et al.* 1999b
188	Microsphaeropsisin	*Microsphaeropsis* sp. H5-50	F	Höller *et al.* 1999b
189	Phomopsidin	*Phomopsis* sp. TUF 95F47	F	Namikoshi *et al.* 1997
190	15G256γ	*Hypoxylon oceanicum* LL-15G256	F	Abbanat *et al.* 1998
191	15G256δ		F	Schlingmann *et al.* 1998
192	15G256ε		F	
193	15G256ι		F	Schlingmann *et al.* 2002
194	15G256ω		F	
195	M-3	unidentified fungus M-3	F	Byun *et al.* 2003
196	YM-202204	*Phoma* sp. Q60596	F, Y	Nagai *et al.* 2002
197	Dihydronaphthalen-1(4*H*)-one	*Keissleriella* sp. YS4108	F, Y	Liu *et al.* 2002
198	Keisslone		F, Y	Liu, C.H. *et al.* 2003
199	Hypoxysordarin	*Hypoxylon croceum* M97-25	F, Y	Daferner *et al.* 1999
200	Xestodecalactone B	*Penicillium* cf. *montanense* HBI-3/D	Y	Edrada *et al.* 2002

(*Contd.*)

(*Contd.*)

201	Fumiquinazoline H	*Acremonium* sp. CNC 890	Y	Belofsky *et al.* 2000
202	Fumiquinazoline I		Y	
203	Mactanamide	*Aspergillus* sp. CNC-688	Y	Lorenz *et al.* 1998
204	Ascosalipyrrolidinone A	*Ascochyta salicorniae*	P, H	Osterhage *et al.* 2000
205	Drechslerine E	*Drechslera dematioidea* MOR5 test 2-1	P	Osterhage *et al.* 2002a
206	Drechslerine G		P	
207	Aigialomycin D	*Aigialus parvus* BCC 5311	P	Isaka *et al.* 2002
208	Halorosellinic acid	*Halorosellinia oceanica* BCC 5149	P	Chinworrungsee *et al.* 2001
209	Solanapyrone E	unidentified fungus CNC-159	A	Jenkins *et al.* 1998a
210	Solanapyrone F		A	
211	Exumolide A	*Scytalidium* sp. CNC-389	A	Jenkins *et al.* 1998b
212	Exumolide B		A	
213	Halymecin A	*Fusarium* sp. FE-71-1	A, B (G+)	Chen *et al.* 1996
214	Bipolal	*Bipolaris* sp. F5206	A, B (G+)	Watanaba *et al.* 1995
215	Halovir A	*Scytalidium* sp. CNL240	V	Rowley *et al.* 2003
216	Halovir B		V	
217	Halovir C		V	
218	Halovir D		V	
219	Halovir E		V	

[1]Biological Activity: A, *antialgal;* B, *antibacterial* (G+, Gram-positive; G–, Gram-negative); F, *antifungal;* H, *antihemoflagellate;* M, *antimycobacterial;* P, *antiplasmodial* (*antimalarial*); V, *antiviral;* Y, *antiyeast.*
[2]n/a: not available.

A linear triester, exophilin A (**148**), was purified from mycelia of *Exophiala pisciphila* NI10102 isolated from the sponge *Mycale adhaerens* (Doshida *et al.* 1996). Exophilin A (**148**) showed antibacterial activity against *Enterococcus faecalis, E. faecium, S. aureus,* and MRSA (MIC = 25, 12.5, 50, and 50 μg/mL, respectively). The structure of **148** resembled those of an antialgal halymecin A (**213**) and its analogs halymecins B–E (**270–273**).

Speradine A (**149**) was yielded from *Aspergillus tamarii* M143 isolated from a driftwood in Okinawa, Japan, and was antibacterial against *M. luteus* (MIC = 16.7 μg/mL) (Tsuda *et al.* 2003a). Speradine A (**149**) also inhibited Ca^{2+}-ATPase (IC_{50} = 8 μM) and histone deacetylase (IC_{50} = 100 μg/mL).

Auranticins A (**150**) and B (**151**) were produced by *Preussia aurantiaca* ATCC 14745 originally isolated from mangrove sediment (Poch and Gloer 1991). The fungus was obtained from ATCC and the culture filtrate showed antibacterial activity. Auranticin A (**150**) was much more potent than B (**151**) in antimicrobial assays and accounted for most of the antibacterial activity against *B. subtilis* (5 μg/disc) and *S. aureus* (50 μg/disc) of this fungus.

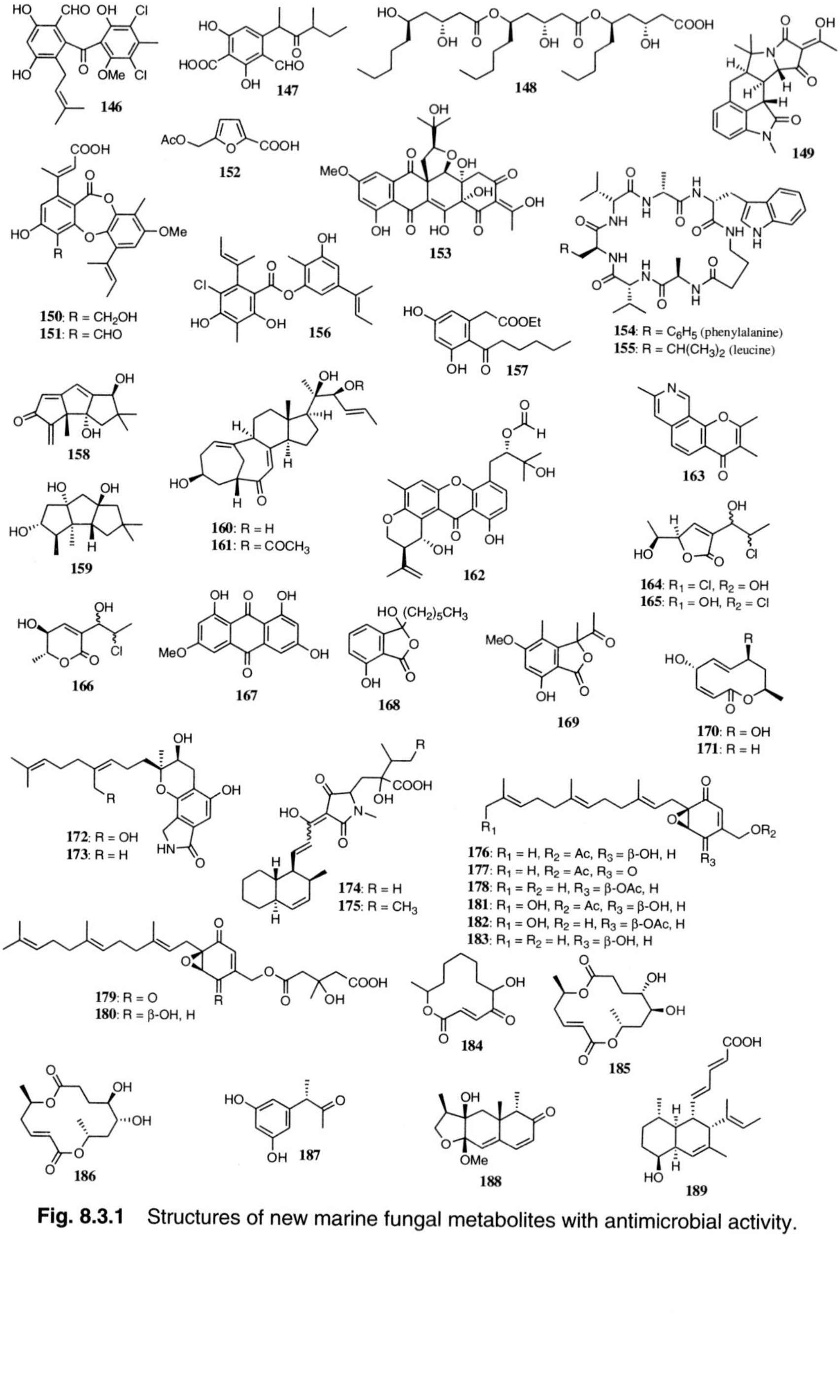

Fig. 8.3.1 Structures of new marine fungal metabolites with antimicrobial activity.

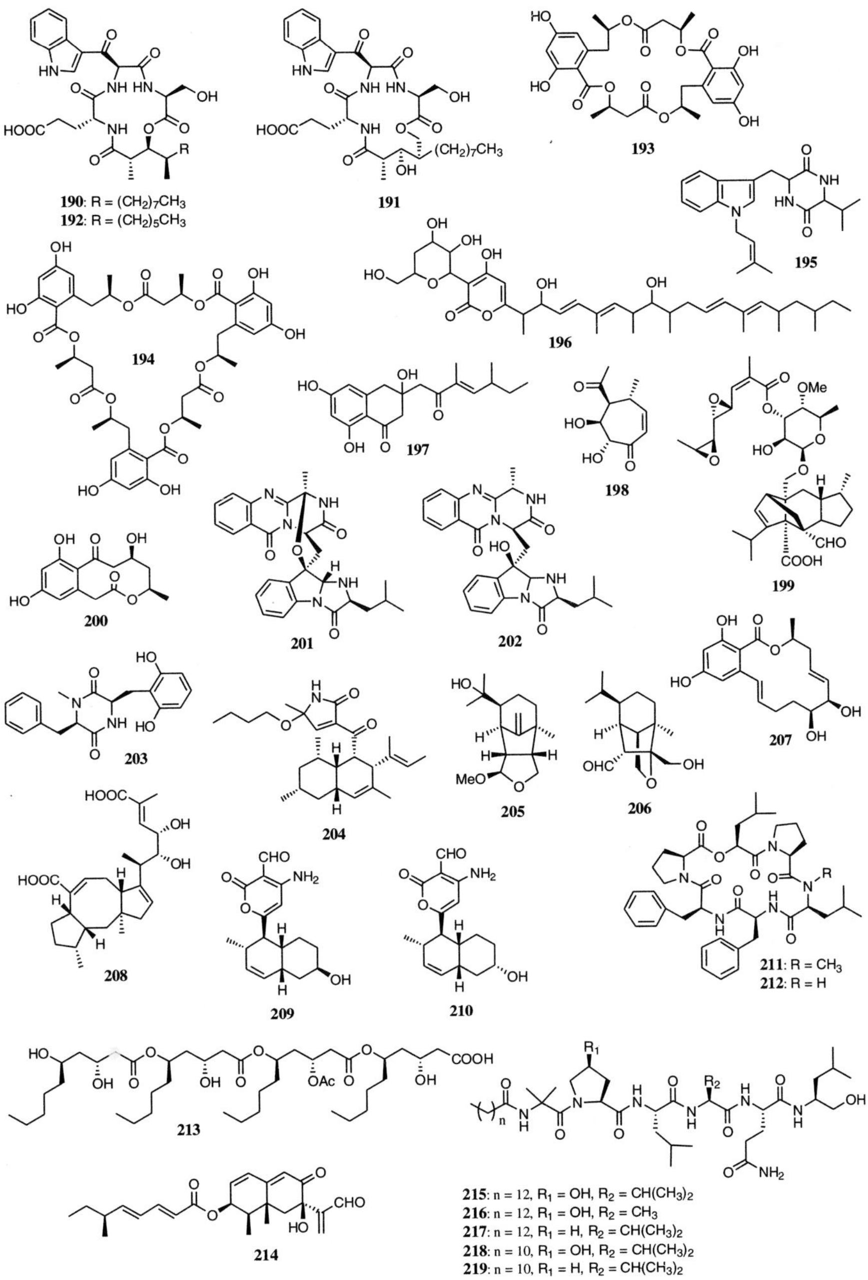
190: R = $(CH_2)_7CH_3$
192: R = $(CH_2)_5CH_3$
191
193
194
195
196
197
198
199
200
201
202
203
204
205
206
207
208
209
210
211: R = CH_3
212: R = H
213
214
215: n = 12, R_1 = OH, R_2 = $CH(CH_3)_2$
216: n = 12, R_1 = OH, R_2 = CH_3
217: n = 12, R_1 = H, R_2 = $CH(CH_3)_2$
218: n = 10, R_1 = OH, R_2 = $CH(CH_3)_2$
219: n = 10, R_1 = H, R_2 = $CH(CH_3)_2$

Fig. 8.3.2

Acetyl Sumiki's acid (**152**) was obtained from *Cladosporium herbarum* isolated from the sponge *Callyspongia aerizusa* collected in Indonesia (Jadulco *et al.* 2001). Compound **152** inhibited the growth of *B. subtilis* and *S. aureus* (7 mm inhibition zones at 5 μg/disc). This fungus also produced non-active compounds, pandangolides 3 (**309**) and 4 (**310**) (Jadulco *et* al. 2001) and herbaric acid (**311**) (Jadulco *et al.* 2002).

Seragakinone A (**153**) was obtained from an unidentified fungus K063 isolated from the red alga *Ceratodictyon spongiosum* from Okinawa (Shigemori *et al.* 1999). This fungus also produced two new linear peptides, dictyonamides A (**238**) and B (**435**) (Komatsu *et al.* 2001). Compound **153** exhibited mild antibacterial activity against *S. aureus, M. luteus, Corynebacterium xerosis,* and *B. subtilis* (MIC = 10, 20, 20, and 41 μg/mL, respectively).

Two cyclic heptapeptides, unguisins A (**154**) and B (**155**), were gotten from two strains of *Emericella unguis* (M87-2 and M90B-10) (Malmstrøm 1999). These strains also produced a chlorinated depside, guisinol (**156**) (Nielsen *et al.* 1999). Strain M87-2 was isolated from a solution of the medusa *Stomolopus meliagris* mixed with water and M90B-10 from the soft part of an unidentified mollusc from Paria Bay in Venezuela. *Emericella unguis* M90A-2 isolated from scrapings of the shell of an unidentified mollusc at the same site gave two new cyclic heptapeptides, unguisins C (**336**) and D (**337**), together with unguisins A and B (Malmstrøm *et al.* 2002b). Feeding experiments of L-Phe and L-Leu stimulated the production of unguisins A and B, respectively, while the addition of L-Ala and L-Ser did not change the pattern of unguisin production by strain M90A-2 (Malmstrøm *et al.* 2002b). Unguisin D (**337**) was formed when L-Leu was added to the culture medium. Unguisins A (**154**) and B (**155**) were reported to show moderate antibacterial activity against *S. aureus* (no quantitative data) (Malmstrøm 1999), but no biological activity of unguisins C (**336**) and D (**337**) was reported (Malmstrøm *et al.* 2002b). Guisinol (**156**) exhibited moderate antibacterial activity against *S. aureus* (5 mg/mL DMSO, 15 μL added to a 4 mm well).

Secocurvularin (**157**) was obtained from an unidentified fungus 951014 isolated from the Indonesian sponge *Spirastrella vagabunda* (Abrell *et al.* 1996a). Compound **157** showed mild antibiotic activity against *B. subtilis* at 200 μg/disc (20% inhibition compared to 100% by tetracycline at 30 μg/disc).

Antibacterial hirsutanol A (**158**) and *ent*-gloeosteretriol (**159**) were produced together with two non-active hirsutanols, B (**428**) and C (**429**), by an unidentified fungus 95-1005C isolated from the sponge *Haliclona* sp. collected at North Sulawesi, Indonesia (Wang *et al.* 1998). Compounds **158** and **159** were reported to show mild antibacterial activity against *B. subtilis* (no quantitative data).

Two novel steroids, isocyclocitrinol A (**160**) and 22 acetylisocyclocitrinol A (**161**), were isolated from sponge-derived *Penicillium citrinum* 991084 (Amagata *et al.* 2003b). These compounds have a unique carbon skeleton, a bicyclo[4.4.1] A/B ring system, which was most likely biosynthesized through a 1,2-migration of a C-5–C-10 bond to construct the C-5–C-18 bond. Compounds **160** and **161** exhibited weak antibacterial activity against *Staphylococcus epidermidis* (MIC = 100 μg/mL each) and *Enterococcus durans* (MIC = 100 μg/mL each). The fungal strain 991084 was isolated from the sponge *Axinella* sp. from Papua New Guinea.

Varixanthone (**162**) exhibited antibacterial activity against *Escherichia coli, Proteus* sp., *B. subtilis* and *S. aureus* (MIC = 12.5 μg/mL), with lower potency to *E. faecalis* (MIC = 50 μg/mL) (Malmstrøm *et al.* 2002a). This compound was obtained from the Caribbean sponge-derived *Emericella variecolor* M75-2 together with varitriol (**86**), dihydroterrein (**338**), and varioxirane (**339**).

Aspergillus versicolor HBI-2 isolated from the sponge *Xestospongia exigua* collected at Bali, Indonesia, yielded an antibacterial metabolite, aspergillitine (**163**), and non-active compounds, aspergiones A–F (**295–300**) (Lin *et al.* 2003). Aspergillitine (**163**) had moderate antibacterial activity against *B. subtilis* (7 mm inhibition zone at 5 μg). Any biological activity of aspergiones A–F was not reported. It is quite confusing that the same authors reported the isolation and different structures of aspergiones A–F in 2001 (Lin, W. *et al.* 2001b). However, the authors did not cite the 2001 paper nor mention the revision of structures in the 2003 paper. These compounds must be the same, and, probably, A (**295**), B (**296**), C (**297**), D (**298**), E (**299**), and F (**300**) reported in 2003 correspond to C, D, A, B, E, and F in the 2001 paper, respectively. Therefore, aspergiones A–F reported in 2001 were omitted.

Three chlorinated compounds **164–166** were obtained from *Aspergillus ostianus* TUF 01F313 isolated from an unidentified sponge at Pohnpei (Namikoshi *et al.* 2003). 8-Chloro-9-hydroxy-8,9-deoxyasperlactone (**164**) exhibited antibacterial activity against *S. aureus* (10.2 mm inhibition zone at 25 μg/disc), *E. coli* (11.6 mm at 50 μg/disc), and the marine bacterium *Ruegeria atlantica* (12.7 mm at 5 μg/disc). 9-Chloro-8-hydroxy-8,9-deoxyasperlactone (**165**) inhibited the growth of *S. aureus* (9.9 mm at 50 μg/disc) and *R. atlantica* (10.1 mm at 25 μg/disc), and 9-chloro-8-hydroxy-8,9-deoxyaspyrone (**166**) was active against *S. aureus* (9.7 mm at 50 μg/disc) and *R. atlantica* (10.5 mm at 25 μg/disc).

Lunatin (**167**) was obtained from *Curvularia lunata* isolated from the Indonesian sponge *Niphates olemda* and exhibited antibacterial activity against *B. subtilis, S. aureus,* and *E. coli* (7.5–9.0 mm inhibition zones at 5μg/disc) (Jadulco *et al.* 2002).

An obligate marine ascomycete, *Corollospora maritima* 1069, isolated from driftwood in Germany produced corollosporine (**168**), which was moderately antibacterial against *S. aureus*, *B. subtilis*, and *E. coli* (MIC = 10, 15, and 50 μg/mL, respectively) (Liberra *et al.* 1998). A chemical synthesis of **168** has been reported (Ohzeki and Mori 2001).

An antimycobacterial compound **169** and non-active compounds, **348**, 17-dehydroxyhalorosellinic acid (**349**), and halorosellins A (**350**) and B (**351**), were isolated from an obligate marine fungus, *Halorosellinia oceanica* BCC 5149 (Chinworrungsee *et al.* 2002). This fungus also produced an antimalarial halorosellinic acid (**208**) and a non-active compound **347** (Chinworrungsee *et al.* 2001). Compound (**169**) exhibited mild antimycobacterial activity against *Mycobacterium tuberculosis* (MIC = 200 μg/mL). Halorosellinic acid (**208**) showed moderate antimalarial activity against *Plasmodium falciparum* (IC_{50} = 13 μg/mL). Interestingly, 17-dehydroxyhalorosellinic acid (**349**) did not show antiplasmodial activity, and, therefore, the 1,2-diol moiety must be necessary for the activity of **208**.

Modiolides A (**170**) and B (**171**) and modiolin (**372**) were produced by *Paraphaeosphaeria* sp. N-119 isolated from the marine horse mussel, *Modiolus auriculatus*, from Okinawa (Tsuda *et al.* 2003b). Modiolides A (**170**) and B (**171**) exhibited antibacterial activity against *M. luteus* (MIC = 16.7 μg/mL) and antifungal activity against *Neurospora crassa* (MIC = 33.3 μg/mL).

Stachybotrins A (**172**) and B (**173**) were obtained from *Stachybotrys* sp. CS-710-1 isolated from a twig submerged in the brackish water of Whitewater Bay, Florida (Xu *et al.* 1992). Compounds **172** and **173** showed antibacterial activity against *B. subtilis* (8 and 10 mm zone at 10 μg/disc, respectively) and inhibited the radial growth of the filamentous fungi *Ascobolus furfuraceus* and *Sordaria fimicola* by 50% at 20 and 10 μg/disc, respectively.

A facultative marine ascomycete, *Zopfiella latipes* CBS 611.97, isolated from sediment of the Indian Ocean at New Delhi, India, produced two antibiotics, zopfiellamides A (**174**) and B (**175**) (Daferner *et al.* 2002). Zopfiellamide A (**174**) inhibited the growth of several Gram-positive and Gram-negative bacteria and yeasts with MIC values of 2–10 μg/mL; and B (**175**) was approximately 5 times less active than A. It is interesting that an extra methyl group at the end of the structure reduced the activity.

Aspergillus niger F97S11 isolated from the orange tunicate *Aplidium* sp. from Fiji yielded eight new antibacterial metabolites, yanuthones A–E (**176–180**), 1-hydroxyyanuthone A (**181**), 1-hydroxyyanuthone C (**182**), and 22-deacetylyanuthone A (**183**) (Bugni *et al.* 2000). These compounds exhibited antimicrobial activity (62–250 μg/disc) against *S. aureus*, MRSA,

VREF, *C. albicans*, or *E. coli* (IMP-mutation). Yanuthone D (**179**) was the most active of these compounds, showing increased activity against MRSA (15, 17, 15, 12, and 9 mm inhibition zones at 62 μg/disc against the above microorganisms, respectively).

An antifungal metabolite, cladospolide D (**184**), was obtained from *Cladosporium* sp. FT-0012 isolated from an unidentified sponge in Pohnpei (Zhang *et al.* 2001). Cladospolide D (**184**) inhibited the growth of *Mucor racemosus* (IC_{50} = 0.15 μg/mL) and *Pyricularia oryzae* (IC_{50} = 29 μg/mL).

(6*R*,11*S*,12*S*,14*R*)-9,10-Dihydrocolletodiol (**185**) and (6*R*,11*R*,12*R*,14*R*)-9,10-dihydrocolletodiol (**186**) were the metabolites of *Varicosporina ramulosa* 195-31 isolated from the brown alga *Cytoseira* sp. from Spain (Höller *et al.* 1999a). Compounds **185** and **186** showed weak antifungal activity against *Eurotium repens* (2 and 1 mm inhibition zone at 50 μg/disc, respectively).

(3*S*)-(3,5-Dihydroxyphenyl)butan-2-one (**187**) and 2-(1(*E*)-propenyl)-octa-4(*E*),6(*Z*)-diene-1,2-diol (**312**) were gotten from *Coniothyrium* sp. 193H77 isolated from the Caribbean sponge *Ectyoplasia ferox* (Höller *et al.* 1999b). Compound **187** had antifungal activity at the 50 μg/disc level against *Ustilago violacea* and *Mycotypha microspora*.

Microsphaeropsisin (**188**) was obtained from *Microsphaeropsis* sp. H5-50 isolated from the sponge *Myxilla incrustans* from Germany and showed antifungal activity at the 50 μg/disc level against *Eurotium repens* and *U. violacea* (Höller *et al.* 1999b).

Phomopsis sp. TUF 95F47 isolated from a fallen mangrove branch at Pohnpei gave phomopsidin (**189**) (Namikoshi *et al.* 1997, 2000b, Kobayashi *et al.* 2003). Phomopsidin (**189**) is an antimicrotubule component. It inhibited microtubule assembly with an IC_{50} value of 5.7 μM. This compound exhibited potent antifungal activity against phytopathogenic fungi and weak cytotoxicity. A total synthesis of **189** has been reported (Suzuki *et al.* 2004).

A mangrove fungus, *Hypoxylon oceanicum* LL-15G256, isolated from *Kandelia candel* wood from Shenzen, China, yielded 10 new compounds, 15G256γ (**190**), 15G256δ (**191**), 15G256ε (**192**), 15G256ι (**193**), 15G256ω (**194**), 15G256α-2 (**357**), 15G256β-2 (**358**), 15G256ν (**359**), 15G256o (**360**), and 15G256π (**361**) (Abbanat *et al.* 1998, Schlingmann *et al.* 1998, 2002). Compounds **190–194** exhibited antifungal activity against the phytopathogenic fungus *N. crassa* through inhibition of cell wall biosynthesis. The biological properties of 15G256γ (**190**) have been thoroughly investigated (Albaugh *et al.* 1998), and 15G256γ (**190**) was suggested to act on a novel cell wall target since *in vitro N. crassa* glucan and chitin synthase activities were only inhibited at high concentrations. 15G256γ (**190**) also inhibited the growth of pathogenic dermatophytic fungi and yeast with MIC of 2–16 μg/mL. Linear polyesters 15G256α-2 (**357**), 15G256β-2 (**358**), 15G256ν (**359**), 15G256o (**360**), and 15G256π (**361**) did not show antifungal activity. Therefore, the macrocyclic polyester ring is essential for the activity of 15G256ι (**193**) and 15G256ω (**194**).

An antifungal compound M-3 (**195**) was produced by an unidentified fungus M-3 isolated from the laver *Porphyra yezoensis* in Chiba, Japan, and exhibited the potent activity against *P. oryzae* (MIC = 0.36 μM) (Byun *et al.* 2003).

Phoma sp. Q60596 isolated from the Okinawan sponge *Halichondria japonica* produced YM-202204 (**196**) and a known analog (Nagai *et al.* 2002). Compound **196** demonstrated potent antifungal activities against *C. albicans* (IC_{80} = 6.25 μg/mL), *Cryptococcus neoformans* (IC_{80} = 1.56 μg/mL), *A. fumigatus* (IC_{80} = 12.5 μg/mL), and *Saccharomyces cerevisiae* (IC_{80} = 1.56 μg/mL) and also inhibited glycosyl-phosphatidyl-inositol (GPI)-anchoring in yeast cells.

Two antifungal metabolites, dihydronaphthalen-1(4*H*)-one (**197**) and keisslone (**198**) were obtained from *Keissleriella* sp. YS4108 isolated from a Yellow Sea sediment (–50 m) at Sheyang Port, China (Liu *et al.* 2002, 2003). Compounds **197** and **198** displayed antifungal activity against *C. albicans* (MIC = 40 and 50 μg/mL, respectively), *Tricophyton rubrum* (20 and 70 μg/mL, respectively), and *A. niger* (80 and 40 μg/mL, respectively).

A facultative marine fungus, *Hypoxylon croceum* M97-25, isolated from driftwood in a mangrove estuary at Everglades, Florida, yielded an antifungal hypoxysordarin (**199**) and a non-active hypoxylactone (**355**) (Daferner *et al.* 1999). Hypoxysordarin (**199**) exhibited the activity against several yeast and filamentous fungal species (MIC = 0.5–20 μg/mL) through a specific inhibition of protein synthesis.

Xestodecalactones A (**376**), B (**200**), and C (**377**) were the metabolites of *Penicillium* cf. *montanense* HBI-3/D isolated from the sponge *Xestospongia exigua* collected at Bali (Edrada *et al.* 2002). Xestodecalactone B (**200**) was active against *C. albicans* showing inhibition zones of 25, 12, and 7 mm at 100, 50, and 20 μM, respectively, while A (**376**) and C (**377**) did not show any activity. Therefore, the presence and stereochemistry of one OH group is critical for the antifungal activity.

Fumiquinazolines H (**201**) and I (**202**) were obtained together with oxepinamides A (**261**), B (**268**), and C (**269**) from *Acremonium* sp. CNC 890 isolated from the surface of the tunicate *Ecteinascidia turbinata* from the Bahamas (Belofsky *et al.* 2000). Compounds **201** and **202** had very weak antifungal activity against *C. albicans* at 500 μg/mL in a broth microdilution assay. A total synthesis of fumiquinazoline I (**202**) has been accomplished (Snider and Zeng 2000).

Mactanamide (**203**) was gotten from *Aspergillus* sp. CNC-688 isolated from the surface of a decomposing sample of the brown alga

Sargassum sp. from the Philippines (Lorenz *et al*. 1998). Compound **203** showed very weak fungistatic activity against *C. albicans*.

An endophytic obligate marine fungus, *Ascochyta salicorniae*, isolated from the green alga *Ulva* sp. from the North Sea produced ascosalipyrrolidinones A (**204**) and B (**286**) and ascosalipyrone (**285**) (Osterhage *et al*. 2000). Ascosalipyrrolidinone A (**204**) exhibited antimalarial activity against two strains of *P. falciparum* (IC_{50} = 378 and 736 ng/mL), antihemoflagellate activity against two species of *Trypanosoma* (MIC = 1.1 and 30 μg/mL), and moderate cytotoxicity (2.2–3.7 μg/mL). Compound **204** also inhibited tyrosine kinase ($p56^{lck}$) activity (30% inhibition at 40 μg/mL). Ascosalipyrrolidinone B (**286**) did not show any biological activity. Therefore, the carbon chain at the ether moiety is critical for the activity of **204**.

Ten new compounds, drechslerines A–D (**327–330**), E (**205**), F (**331**), and G (**206**), 9-hydroxyhelminthosporol (**326**), isosativenetriol (**332**), and sativene epoxide (**333**), were obtained from the algicolous fungus *Drechslera dematioidea* MOR5 test 2-1 isolated from the inner tissue of the Mediterranean red alga *Liagora viscida* (Osterhage *et al*. 2002a). Two compounds, drechslerines E (**205**) and G (**206**), showed antimalarial activity among these 10 metabolites. Drechslerines E (**205**) and G (**206**) inhibited the growth of two strains of *P. falciparum* with IC_{50} at 5.1 μg/mL or lower concentrations.

A lignicolous mangrove ascomycete, *Aigialus parvus* BCC 5311, gave seven new metabolites, aigialomycins A–C (**274–276**), D (**207**), and E (**277**), aigialone (**278**), and aigialospirol (**279**) (Isaka *et al*. 2002, Vongvilai *et al*. 2004). Among these compounds, only aigialomycin D (**207**) exhibited moderate antimalarial activity against *P. falciparum* (IC_{50} = 6.6 μg/mL). Aigialomycin D (**207**) also showed weak cytotoxicity against two cancer cell lines KB and BC-1 and African green monkey kidney fibroblast (Vero) cells (IC_{50} = 3.0, 18, and 1.8 μg/mL, respectively). Therefore, antimalarial activity may be related to the cytotoxicity.

Solanapyrones E (**209**), F (**210**), and G (**430**) were gotten from an unidentified fungus CNC-159 isolated from the surface of the green alga *Halimeda monile* from the Bahamas (Jenkins *et al*. 1998a). Compounds **209** and **210** showed weak toxicity against the marine unicellular green alga *Dunaliella* sp. (15 and 25% inhibition at 200 μg/mL, respectively). These antialgal metabolites may be produced to play an ecological role in the association between marine fungi and algae. A biological activity of G (**430**) was not tested due to insufficient amounts obtained.

Exumolides A (**211**) and B (**212**) inhibited the growth of the unicellular green alga *Dunaliella* sp. in a liquid medium cell growth assay (27 and 33% inhibition at 20 μg/mL, respectively) (Jenkins *et al*. 1998b). These compounds were obtained from *Scytalidium* sp. CNC-389 isolated from a decaying plant material from the Bahamas.

Fusarium sp. FE-71-1 isolated from the alga *Halumenia dilatata* from Palau gave halymecins A (**213**), B (**270**), and C (**271**); and *Acremonium* sp. FK-N30 isolated from an unidentified marine alga at Kanagawa, Japan, yielded halymecins D (**272**) and E (**273**) (Chen *et al.* 1996). Halymecin A (**213**) showed growth inhibition against the diatom *Skeletonema costatum* (MIC = 4 μg/mL), antibacterial activity against *E. faces*, *Klebsiella pneumoniae*, and *Proteus vulgaris* (MIC = 10 μg/mL), and cytotoxicity against HeLa S3 and BALB3T3 cells (IC_{50} = 6 and 7 μg/mL, respectively). No biological activity of the other compounds was reported. The structures of the halymecins are similar to that of exophilin A (**148**).

Bipolal (**214**) was obtained from *Bipolaris* sp. F5206 isolated from a fallen leaf sample from Japan and showed antimicrobial activity against the marine Gram-positive bacterium *Trichophyton mentagrophytes* (MIC = 1.58 μg/mL) and the diatom *Nitzschia closterium* (12% inhibition at 1.6 x 10^{-2} μmol/mL) (Watanabe *et al.* 1995). The structure of bipolal (**214**) is similar to those of the dendryphiellins.

Five anti-HIV compounds, halovirs A–E (**215–219**), were gotten from *Scytalidium* sp. CNL240 isolated from the Caribbean seagrass *Halodule wrightii* (Rowley *et al.* 2003). Halovirs A–E (**215–219**) are potent inhibitors of HSV-1 and HSV-2. The biological experiments revealed that these compounds directly inactivate the viruses. This mechanism of action suggests that these compounds could be used to prevent HSV transmission.

New Metabolites with Miscellaneous Biological Activity (Compounds 114, 116, and 220–261)

The culture broths of marine fungi have also been utilized for other important screening bioassays. These bioassays mostly use proteins as target molecules for specific biological activities. New marine fungal metabolites with interesting biological activities are listed in Table 8.5 and the structures in Figure 8.4.

Mangicols A (**114**) and C (**116**) showed considerable antiinflammatory activity in the phorbol myristate acetate (PMA)-induced mouse ear edema assay (81 and 57% inhibition of edema at 50 μg/ear, respectively) (Renner *et al.* 2000).

A series of antiinflammatory compounds, phomactins A–G (**220–228**), were obtained from *Phoma* sp. SANK 11486 isolated from the shell of the crab *Chinoecetes opilio* from Japan (Sugano *et al.* 1991, 1994, 1995, 1996). Phomactins A (**220**), B (**221**), B1 (**222**), B2 (**223**), C (**224**), D (**225**), E (**226**), F (**227**), and G (**228**) inhibited PAF-induced platelet aggregation (IC_{50} = 10, 17.0, 9.8, 1.6, 6.4, 0.80, 2.3, 3.9, and 3.2 μM, respectively) and binding of PAF to its receptor (IC_{50} = 2.3, >47.9, 20.0, >22.1, 63.0, 0.12,

Table 8.5 New marine fungal metabolites with miscellaneous biological activity.

Compound No.	Compound Name	*Fungus*	*Activity*[1]	*Reference*
114	Mangicol A	*Fusarium heterosporum* CNC-477	anti-inflammatory	Renner *et al*. 2000
116	Mangicol C			
220	Phomactin A	*Phoma* sp. SANK 11486	PAF-antagonist	Sugano *et al*. 1991
221	Phomactin B			Sugano *et al*. 1994
222	Phomactin B1			
223	Phomactin B2			
224	Phomactin C			
225	Phomactin D			
226	Phomactin E			Sugano *et al*. 1995
227	Phomactin F			
228	Phomactin G			
229	Nafuredin	*Aspergillus niger* FT-0554	anthelmintic	Omura *et al*. 2001
230	Roselipin 1A	*Gliocladium roseum* KF-1040	DGAT-inhibitor	Omura *et al*. 1999
231	Roselipin 1B			
232	Roselipin 2A			
233	Roselipin 2B			
234	10-Hydroxy-18-methoxybetaenone	*Microsphaeropsis* sp. KMPB W-22	tyrosine kinase inhibitor	Brauers *et al*. 2000
235	n/a[2]			
236	n/a			
237	n/a			
238	Dictyonamide A	unidentified fungus K063	CDK4-inhibitor	Komatsu *et al*. 2001
239	Ulocladol	*Ulocladium botrytis* 193A4	$p56^{lck}$-inhibitor	Höller *et al*. 1999a
240	Pulchellalactam	*Corollospora pulchella* ATCC 62554	CD45-inhibitor	Alvi *et al*. 1998
241	Sculezonone A	*Penicillium* sp. K036	DNA polymerase inhibitor	Komatsu *et al*. 2000
242	Sculezonone B			
243	Xyloketal A	*Xylaria* sp. 2508	ACE-inhibitor	Lin, Y. *et al*. 2001a

(*Contd.*)

(*Contd.*)

244	Chlorogentisylquinone	*Phoma* sp. FOM-8108	SMase-inhibitor	Uchida *et al*. 2001
245	Cathestatin C	*Microascus longirostris* SF-73	protease-inhibitor	Yu *et al*. 1996a
246	Phenochalasin A	*Phomopsis* sp. FT-0211	inhibitor of lipid	Tomoda *et al*. 1999a
247	Phenochalasin B		droplet formation	
248	Obionin A	*Leptosphaeria obiones* SAP 14	CNS activity	Poch and Gloer 1989a
249	n/a	*Acremonium* cf. *roseogriseum* M1-11-1	antioxidative	Abdel-Lateff *et al*. 2002
250	n/a			
251	Anomalin A	*Wardomyces anomalus* OS4T3-2-1	antioxidative	Abdel-Lateff *et al*. 2003a
252	Parasitenone	*Aspergillus parasiticus* MFA 153	antioxidative	Son *et al*. 2002a
253	Epicoccone	*Epicoccum* sp.	antioxidative	Abdel-Lateff *et al*. 2003b
254	Golmaenone	*Aspergillus* sp. MFA 212	antioxidative	Li *et al*. 2004a
255	Dihydroxyisoechinulin A		antioxidative	Li *et al*. 2004b
256	Farnesylhydroquinone	*Penicillium* sp. MFA 577	antioxidative	Son *et al*. 2002b
257	Terreusinone	*Aspergillus terreus* MFA 460	UV-A absorption	Lee *et al*. 2003
258	Epolactaene	*Penicillium* sp. BM 1689-P	neuritogenic	Kakeya *et al*. 1995
259	Aspermytin A	*Aspergillus* sp. MF297	neurite growth	Tsukamoto *et al*. 2004
260	Paecilospirone	*Paecilomyces* sp. TUF 94F79	antimicrotubule	Namikoshi *et al*. 2000a
261	Oxepinamide A	*Acremonium* sp. CNC 890	anti-inflammatory	Belofsky *et al*. 2000

[1]Biological activity: PAF, platelet aggregation factor; DGAT, diacylglycerol transferase; CDK4, cyclin-dependent kinase 4; $p56^{lck}$, tyrosine kinase; CD45, protein tyrosine phosphatase; ACE, acetylcholine esterase; SMase, sphingomyelinase; CNS, central nervous system.
[2]n/a: not available.

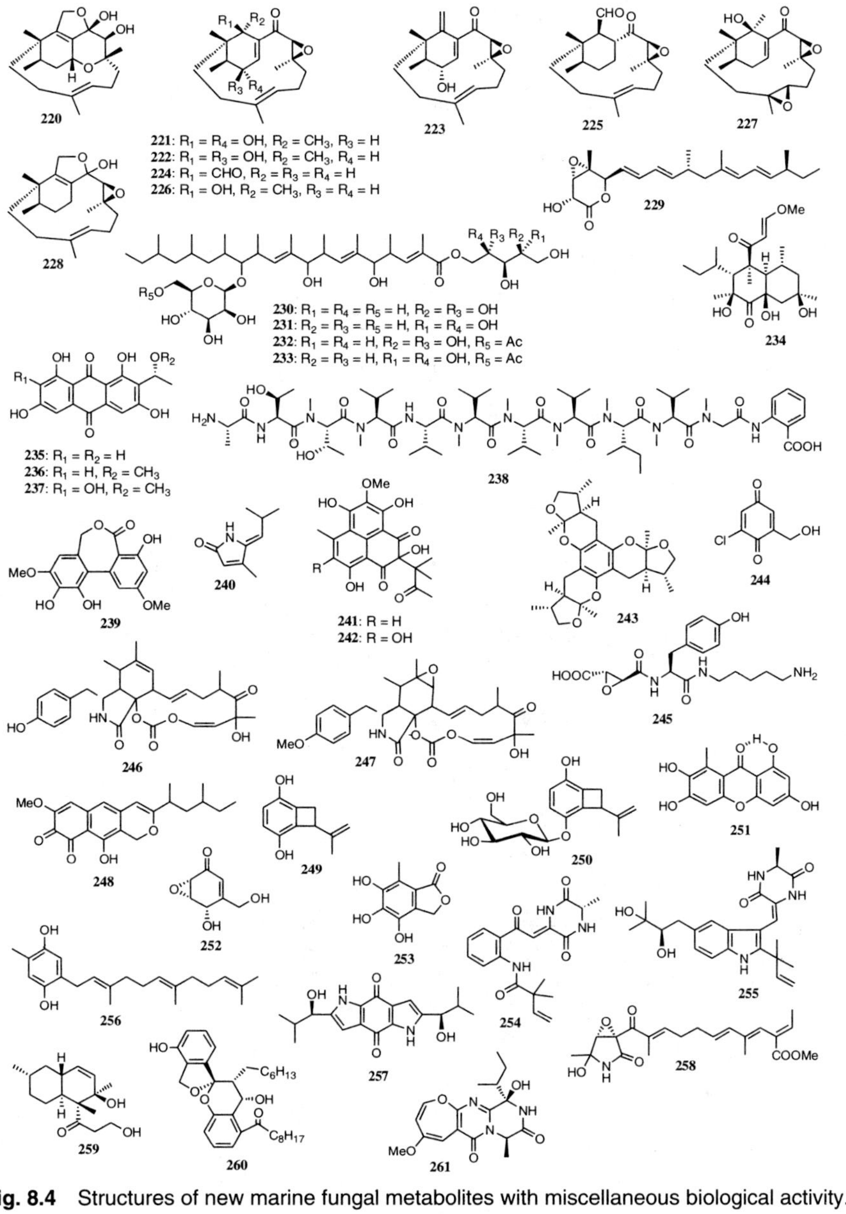

Fig. 8.4 Structures of new marine fungal metabolites with miscellaneous biological activity.

5.19, 35.9, and 0.38 μM, respectively). Phomactin D (**225**) was most potent among these phomactins. Additional structure-activity relationships were studied utilizing natural and synthetic derivatives (Sugano *et al.* 1996). Phomactin C (**224**) was also isolated from terrestrial *Phoma* sp. ATCC 74077 (Chu *et al.* 1992, 1993). Total syntheses of A (**220**) (Diaper *et* *al*. 2003), D (**225**) (Miyaoka *et al*. 1996), and G (**228**) (Goldring and Pattenden 2004) have been achieved.

Nafuredin (**229**) was isolated from Palauan sponge-derived *Aspergillus niger* FT-0554 as a highly selective inhibitor of NADH-fumarate reductase (NFRD) of the pig roundworm *Ascaris suum* (Omura *et al*. 2001, Ui *et al*. 2001, Takano *et al*. 2001a). This compound showed antihelminthic activity against *Haemonchus contortus* in *in vivo* studies. The yield of nafuredin (**229**) increased with increase of the seawater concentration, and maximal production was observed with higher concentrations of seawater (75–100%). A total synthesis has been accomplished (Takano *et al*. 2001b).

Gliocladium roseum KF-1040 isolated from an unidentified red alga from Yap, Micronesia, produced four related compounds, roselipins (Omura *et al*. 1999, Tomoda *et al*. 1999c, Tabata *et al*. 1999). Roselipins 1A (**230**), 1B (**231**), 2A (**232**), and 2B (**233**) inhibited diacylglycerol acyl transferase (DGAT) with the IC_{50} values of 17, 15, 22, and 18 μM, respectively.

Four tyrosine kinase inhibitors, 10-hydroxy-18-methoxybetaenone (**234**) and compounds **235**, **236**, and **237**, were obtained together with three non-active compounds, 10-hydroxy-18-[*N*-(2-naphthyl)-*N*-phenylamino]betaenone (**368**) and microsphaerones A (**369**) and B (**370**), from *Microsphaeropsis* sp. KMPB W-22 isolated from the Mediterranean sponge *Aplysina aerophoba* (Brauers *et al*. 2000, Wang *et al*. 2002). Compounds **234**, **235**, **236**, and **237** inhibited the activity of tyrosine kinases PKC-ε (IC_{50} = 36.0, 18.5, 27.0, and 54.0 μM, respectively), CDK4 (IC_{50} = 11.5, 43.5, 22.5, and 37.5 μM, respectively), and EGF-R (IC_{50} = 10.5, 37.5, 27.5, and 41.0 μM, respectively).

Two linear peptides, dictyonamides A (**238**) and B (**435**), were purified from mycelia of an unidentified fungus K063 isolated from the Okinawan red alga *Ceratodictyon spongiosum* (Komatsu *et al*. 2001). This fungus also yielded an antibacterial seragakinone A (**153**) (Shigemori *et al*. 1999). Dictyonamide A (**238**) showed inhibitory activity on CDK 4 with an IC_{50} of 16.5 μg/mL, while B (**435**), a glucoside derivative of A, was not active. Therefore, a hydrophilic glycosidation reduced the activity.

Ulocladium botrytis 193A4 isolated from the Caribbean sponge *Callyspongia vaginalis* gave ulocladol (**239**) (Höller *et al*. 1999a). Ulocladol (**239**) inhibited tyrosine kinase ($p56^{lck}$), reducing the enzyme activity to 7% at 20 μg/mL in an ELISA-based assay.

Pulchellalactam (**240**) was obtained from *Corollospora pulchella* ATCC 62554 isolated from driftwood and exhibited dose-dependent inhibition of CD45 protein tyrosine phosphatase (IC_{50} = 124 μg/mL) (Alvi *et al*. 1998). Syntheses of **240** and related compounds have been reported (Bessho *et al*. 2004).

Two phenalenone derivatives, sculezonones A (**241**) and B (**242**), were obtained from *Penicillium* sp. K036 isolated from the bivalve *Mytilus coruscus* from Okinawa (Komatsu *et al.* 2000). Sculezonone A (**241**) inhibited bovine DNA polymerases α, β, γ, and ε with IC_{50} values of 20, 17, 25, and 105 μM, respectively, and B (**242**) with 12, 90, 18.5, and 50 μM, respectively (Perpelescu *et al.* 2002).

A mangrove fungus, *Xylaria* sp. 2508, isolated from seeds of an angiosperm tree of the South China Sea, Hong Kong, yielded an enzyme inhibitor, xyloketal A (**243**), and six non-active compounds, xyloketals B–E (**408–411**), (*p*-hydroxycinnamoyl)but-2,3-dienyl ether (**406**) and xyloallenolide A (**407**) (Lin, Y. *et al.* 2001a). Xyloketal A (**243**) inhibited acetylcholine esterase activity at 1.5 μM. Xyloketals B–E (**408–411**) have fewer ether rings and did not inhibit this esterase activity.

Chlorogentisylquinone (**244**) was produced by *Phoma* sp. FOM-8108 isolated from sea sand from Kanagawa, Japan (Uchida *et al.* 2001, Yamaguchi *et al.* 2002). Compound **244** inhibited sphingomyelinase activity of rat brain membranes (IC_{50} = 1.2 μM). Fermentation studies revealed that the production of chlorogentisylquinone (**244**) required seawater in the culture medium. The yield of **244** increased with increase of the seawater concentration, and full production was observed with higher concentrations of seawater (75–100%). It was suggested that **244** is produced to extrude cellular chloride ions at higher salt concentrations.

Cathestatin C (**245**) was gotten from *Microascus longirostris* SF-73 isolated from an unidentified sponge from New Zealand and exhibited potent and irreversible *in vitro* inhibition against papain and cathepsins B and L with IC_{50} values of 2.5, 5.0, and 15.0 nM, respectively (Yu *et al.* 1996a).

Phenochalasins A (**246**) and B (**247**) from *Phomopsis* sp. FT-0211 isolated from an unidentified marine sponge collected at Pohnpei inhibit lipid droplet formation in mouse peritoneal macrophages (Tomoda *et al.* 1999a,b, Namatame *et al.* 2000). Phenochalasin B (**247**) inhibited the synthesis of cholesteryl ester (IC_{50} = 0.2 μM) and triacylglycerol (IC_{50} = 0.38 μM) in the cytosolic lipid droplet. Phenochalasin A (**246**) also inhibited the synthesis of cholesteryl ester (IC_{50} = 0.61 μM) but not triacylglycerol.

Obionin A (**248**) from an ascomycete *Leptosphaeria obiones* SAP 14 isolated from the coastal marsh grass *Spartina alterniflora* in Georgia, USA, was found during a screen of central nervous system activity (Poch and Gloer 1989a). Compound **248** inhibited binding of the dopamine D-1 selective ligand to bovine corpus striatum membrane (IC_{50} = 2.5 μg/mL).

An algicolous fungus, *Acremonium* cf. *roseogriseum* M1-11-1, isolated from tissues of the Mediterranean brown alga *Cladostephus spongius* yielded eight new compounds, **249**, **250**, and **262–267** (Abdel-Lateff *et al.* 2002). Compound **249** exhibited significant radical scavenging effect (85.5% inhibition at 25.0 μg/mL) in the *α,α*-diphenyl-*β*-picrylhydrazyl radical scavenging assay (DPPH assay), and also inhibited peroxidation of linolenic acid (35.5% at 37.0 μg/mL) in the thiobarbituric acid reactive substances method (TBARS assay). Compound **250** showed a modest radical scavenging effect (17.5% inhibition at 25.0 μg/mL) in the DPPH assay. A hydrophilic glucoside reduced the activity. Compounds **262–267** were not active.

Anomalins A (**251**) and B (**405**) are metabolites of *Wardomyces anomalus* OS4T3-2-1 isolated from the green alga *Enteromorpha* sp. collected in the Baltic Sea, Germany (Abdel-Lateff *et al.* 2003a). Anomalin A (**251**) exhibited significant antioxidative property in the DPPH (94.7% radical scavenging effect at 25.0 μg/mL) and TBARS (17.0% at 7.4 μg/mL) assays, and also showed 100% inhibition of $p56^{lck}$ tyrosine kinase activity at 200 μg/mL. Anomalin B (**405**) was not active. Interestingly, the substitution of one OH group at the C-4 position significantly reduced the antioxidant activity.

Parasitenone (**252**) was obtained from *Aspergillus parasiticus* MFA 153 isolated from the surface of the red alga *Carpopeltis cornea* in Korea and exhibited mild radical scavenging activity (IC_{50} = 57.0 μM) in the DPPH assay (Son *et al.* 2002a).

An ascomycete *Epicoccum* sp. isolated from the brown alga *Fucus vesiculosus* collected at the German coast of the North Sea yielded an antioxidant epicoccone (**253**) (Abdel-Lateff *et al.* 2003b). Epicoccone (**253**) exhibited 95% radical scavenging effect at 25.0 μg/mL in the DPPH assay and also inhibited the peroxidation of linolenic acid (62.1% inhibition at 37.0 μg/mL) in the TBARS assay.

Two diketopiperazine derivatives, golmaenone (**254**) and dihydroxyisoechinulin A (**255**), were gotten from *Aspergillus* sp. MFA 212 isolated from the red alga *Lomentaria catenata* from Ulsan City, Korea (Li *et al.* 2004a,b). Compound **254** exhibited significant radical scavenging activity (IC_{50} = 20 μM) in the DPPH assay, which was about the same as the activity of vitamin C (IC_{50} = 22.7 μM), and also showed UV-A protecting activity (ED_{50} = 90 μM), which was more potent than the sunscreen agent oxybenzone (ED_{50} = 350 μM) (Li *et al.* 2004a). Compound **255** was also active in the DPPH assay (IC_{50} = 20 μM) and protected UV-A (ED_{50} = 130 μM) (Li *et al.* 2004b).

An antioxidant compound, farnesylhydroquinone (**256**), was obtained together with a cytotoxic 7-deacetoxyyanuthone A (**75**) and a non-active 2,3-dihydro-7-deacetoxyyanuthone A (**384**) from *Penicillium* sp. MFA 577 isolated from the surface of the polymeric cord at Bijin Island, Korea (Son *et al.* 2002b). Although farnesylhydroquinone (**256**) has been transformed from its benzoquinone derivative by reduction (Bohlman *et al.* 1975), this was the first report of this compound being a natural product. The benzoquinone derivative was also isolated from the same strain. Compound **256** exhibited potent radical scavenging activity (IC_{50} = 12.5 μM) in the DPPH assay.

An algicolous fungus, *Aspergillus terreus* MFA 460, isolated from the surface of the red alga *Halymenia acuminata* at Bijin Island yielded terreusinone (**257**) (Lee *et al.* 2003). Compound **257** showed UV-A absorbing activity with an ED_{50} value of 70 μg/mL.

Epolactaene (**258**) was gotten from *Penicillium* sp. BM 1689-P isolated from sediment from Uchiura Bay, Japan, and exhibited neuritogenic activity in human neuroblastoma SH-SY5Y cells at 2.5–10 μg/mL in a dose-dependent manner (Kakeya *et al.* 1995, 1997). *In vitro* biochemical actions of synthetic epolactaene (**258**) and shorter alkyl chain derivatives were examined (Mizushima *et al.* 2000). These compounds selectively inhibited the activities of mammalian DNA polymerases α and β and human DNA topoisomerase II (IC_{50} = 25, 94, and 10 μM, respectively). The longer alkyl chain in epolactaene (**258**) was suggested to have an important role in the activity. The inhibitory activity against these enzymes is presumably related to their neuritogenic activity in SH-SY5Y cells.

A heptaketide, aspermytin A (**259**), was gotten from *Aspergillus* sp. MF297 isolated from the mussel *Mytilus edulis* collected at Toyama Bay, Japan (Tsukamoto *et al.* 2004). Aspermytin A (**259**) induced neurite outgrowth in rat pheochromocytoma (PC-12) cells at 50 μM.

Paecilomyces sp. TUF 94F79 isolated from seawater (–10 m) in Yap, Micronesia, gave paecilospirone (**260**), which had a weak inhibitory activity on microtubule assembly (20% inhibition at 50 μM) (Namikoshi *et al.* 2000a,b).

Oxepinamide A (**261**) was obtained from *Acremonium* sp. CNC 890 isolated from the surface of the tunicate *Ecteinascidia turbinata* from the Bahamas together with two antifungal metabolites, fumiquinazolines H (**201**) and I (**202**), and two non-active compounds, oxepinamides B (**268**) and C (**269**) (Belofsky *et al.* 2000). Oxepinamide A (**261**) exhibited good anti-inflammatory activity in a topical resiniferatoxin-induced mouse ear edema assay (82% inhibition of edema at 50 μg/ear). Two structurally related compounds, oxepinamides B (**268**) and C (**269**), did not show biological activity. Therefore, the free OH group and its stereochemistry are very important for the activity of **261**.

New Metabolites with no Biological Activity (Compounds 262–437)

The new marine fungal metabolites listed in Table 8.6 have been reported to have no biological activity in the bioassays utilized or no bioassays were employed. The structures of these compounds are shown in Figure 8.5. Some compounds, which were isolated together with biologically active compounds, are discussed in the above three sections.

Since only limited bioassays were examined, biological activities of these compounds have not been effectively screened. These are interesting compounds to test in the search for new drug candidates. Total syntheses of dendryphiellin C (**317**) (Akao *et al.* 1999), leptosphaerin (**363**) (Pallenberg and White 1986, Rollin 1987, White *et al.* 1989), harzialactone A (**400**) (Mereyala *et al.* 2000), epoxysorbicillinol (**401**) (Wood *et al.* 2001), and xyloketal D (**410**) (Krohn and Riaz 2004) have been reported.

Kojic acid dimethyl ether (**280**) (Li *et al.* 2003b) was isolated together with the known inhibitor kojic acid (IC_{50} = 12 μM) during screening for tyrosinase inhibitors. Compound **280** was isolated for the first time as a natural product, although it had been synthesized from kojic acid (Poonia and Yadav 1978).

^{1}H and ^{13}C NMR data for chlorocarolides A (**287**) and B (**288**) (Abrell *et al.* 1996b) were identical to those for 9-chloro-8-hydroxy-8,9-deoxyasperlactone (**165**) and 8-chloro-9-hydroxy-8,9-deoxyasperlactone (**164**), respectively (Namikoshi *et al.* 2003).

Compounds **301–303** were obtained together with insulicolide (**45**), which was the responsible component for the majority of antitumor activity of the crude extract (Belofsky *et al.* 1998).

Since bipolal (**214**) (Watanabe *et al.* 1995), an aldehyde analog of dendryphiellin E (**319**), showed antibacterial and antidiatom activities as mentioned above and a structurally similar compound KM-01 has been isolated as a brassinolide-inhibitor from the terrestrial fungus *Drechslera avenae* (Kim *et al.* 1994), dendryphiellins (**313, 316–324**) (Guerriero *et al.* 1988, 1989, 1990) may have similar biological activities.

Leptosphaerin (**363**) (Schiehser *et al.* 1986, Pallenberg and White 1986, Rollin 1987, White *et al.* 1989) was the first compound reported as a new metabolite obtained from a marine fungus. Isolation and structure elucidation (although it was the wrong structure) of leptosphaerin were presented by J. Clardy in the NATO Conference on Marine Natural Products held October 11–17, 1976, at Jersey, U.K. (Structural studies on marine natural products).

Remisporine A (**398**) is the first example of a cyclopentadiene derivative as a natural product (Kong and Carter 2003). This compound is unstable and spontaneously dimerizes through Diels-Alder reaction

Table 8.6 New marine fungal metabolites for which no biological activity has so far been detected or reported.

Compound		*Fungus*	*Reference*
No.	*Name*		
262	n/a[1]	*Acremonium* cf. *roseogriseum* M1-11-1	Abdel-Lateff *et al*. 2002
263	n/a		
264	n/a		
265	n/a		
266	n/a		
267	n/a		
268	Oxepinamide B	*Acremonium* sp. CNC 890	Belofsky *et al*. 2000
269	Oxepinamide C		
270	Halymecin B		
271	Halymecin C	*Fusarium* sp. FE-71-1	Chen *et al*. 1996
272	Halymecin D		
273	Halymecin E	*Acremonium* sp. FK-N30	
274	Aigialomycin A		
275	Aigialomycin B	*Aigialus parvus* BCC 5311	Isaka *et al*. 2002
276	Aigialomycin C		
277	Aigialomycin E		
278	Aigialone		Vongvilai *et al*. 2004
279	Aigialospirol		
280	Kojic acid dimethyl ether	*Altenaria* sp. MFA 898	Li *et al*. 2003b
281	(–)-Hexylitaconic acid	*Apiospora montagnei*	Klemke *et al*. 2004
282	Apiosporic acid		
283	9-Hydroxyhexylitaconic acid 1-methyl ester		
284	Myrocin A		
285	Ascosalipyrone	*Ascochyta salicorniae*	Osterhage *et al*. 2000
286	Ascosalipyrrolidinone B		
287	Chlorocarolide A	*Aspergillus* cf. *ochraceus* 941026	Abrell *et al*. 1996b
288	Chlorocarolide B		

(*Contd.*)

(*Contd.*)

289	Asperic acid	*Aspergillus niger* 94-1212	Varoglu and Crews 2000
290	Circumdatin G	*Aspergillus ochraceus*	Dai *et al.* 2001
291	12-Acetylaspergillol	*Aspergillus versicolor*	Lin *et al.* 2001a
292	Aspergillodiol		
293	Aspergillol		
294	Aspergillone		
295	Aspergione A	*Aspergillus versicolor* HBI-2	Lin *et al.* 2003
296	Aspergione B		
297	Aspergione C		
298	Aspergione D		
299	Aspergione E		
300	Aspergione F		
301	14-Hydroxy-6β-*p*-nitrobenzoylcinnamolide	*Aspergillus versicolor* CNC 327	Belofsky *et al.* 1998
302	7α,14-Dihydroxy-6β-*p*-nitrobenzoylconfertifolin		
303	6β,7α-Dihydroxy-14-*p*-nitrobenzoylconfertifolin		
304	(+)-2,4-Dimethyl-4,5-dihydro-furan-3-carboxaldehyde	*Asteromyces cruciatus* H5-81	Höller *et al.* 1999a
305	n/a	*Aureobasidium pullulans* (yeast)	Shigemori *et al.* 1998
306	n/a		
307	Orcinotriol		
308	Scorpinone	*Bispora*-like fungus NN 6654	Miljikovic *et al.* 2001
309	Pandangolide 3	*Cladosporium herbarum*	Jadulco *et al.* 2001
310	Pandangolide 4		
311	Herbaric Acid		Jadulco *et al.* 2002
312	2-(1(*E*)-propenyl)-octa-4(*E*),6(*Z*)-diene-1,2-diol	*Coniothyrium* sp. 193H77	Höller *et al.* 1999b
313	Dendryphiellin A	*Dendryphiella salina*	Guerriero *et al.* 1988
314	Dendryphiellic acid A		Guerriero *et al.* 1989
315	Dendryphiellic acid B		
316	Dendryphiellin B		
317	Dendryphiellin C		

(*Contd.*)

(*Contd.*)

318	Dendryphiellin D		
319	Dendryphiellin E		
320	Dendryphiellin F		
321	Dendryphiellin G		
322	Dendryphiellin A1		Guerriero *et al*. 1990
323	Dendryphiellin E1		
324	Dendryphiellin E2		
325	Glycer-1-yl dendryphiellate A		
326	9-Hydroxyhelminthosporol	*Drechslera dematioidea* MOR5 test 2-1	Osterhage *et al*. 2002a
327	Drechslerine A		
328	Drechslerine B		
329	Drechslerine C		
330	Drechslerine D		
331	Drechslerine F		
332	Isosativenetriol		
333	Sativene epoxide		
334	Butoxyl-spiciferin	*Drechslera hawaiiensis* P22-96	Edrada *et al*. 2000
335	Spiciferol A		
336	Unguisin C	*Emericella unguis* M90A-2	Malmstrøm *et al*. 2002b
337	Unguisin D		
338	Dihydroterrein	*Emericella variecolor* M75-2	Malmstrøm *et al*. 2002a
339	Varioxirane		
340	Isoemericellin	*Emericella versicolor* E-00-6/3	Bringmann *et al*. 2003
341	Epicoccamide	*Epicoccum purpurascens* N7-9	Wright *et al*. 2003
342	Eutypoid A	*Eutypa* sp. 424	Lin *et al*. 2002a
343	Fusaperazine B	*Fusarium chlamydosporum* OUPS-N124	Usami *et al*. 2002
344	Neomangicol C	*Fusarium heterosporum* CNC-477	Renner *et al*. 1998
345	JM 47	*Fusarium* sp. MOBCOF-1	Jiang *et al*. 2002
346	Neurosporaxanthin glucoside	*Fusarium* sp. T-1	Sakaki *et al*. 2002
347	4-Methyl-2-hexylidene-3-methylsuccinate	*Halorosellinia oceanica* BCC 5149	Chinworrungsee *et al*. 2001

(*Contd.*)

(*Contd.*)

348	n/a		Chinworrungsee *et al*. 2002
349	17-Dehydroxyhalorosellinic acid		
350	Halorosellin A		
351	Halorosellin B		
352	Helicascolide A	*Helicascus kanaloanus* ATCC 18591	Poch and Gloer 1989b
353	Helicascolide B		
354	Hortein	*Hortaea werneckii*	Brauers *et al*. 2001
355	Hypoxylactone	*Hypoxylon croceum* M97-25	Daferner *et al*. 1999
356	8-Hydroxy-9-oxo-7(11)-eremophilien-12,8-olide	*Hypoxylon oceanicum* 326	Li *et al*. 2001
357	15G256α-2	*Hypoxylon oceanicum* LL-15G256	Schlingmann *et al*. 2002
358	15G256β-2		
359	15G256ν		
360	15G256o		
361	15G256π		
362	Isoculmorin	*Kallichroma tethys* 6175	Alam *et al*. 1996
363	Leptosphaerin	*Leptosphaeria oraemaris*	Schiehser *et al*. 1986
364	Obioninene		Miller and Savard 1989
365	Leptosphaerolide		Guerriero *et al*. 1991
366	n/a	*Microsphaeropsis olivacea* SF-10	Yu *et al*. 1996b
367	n/a		
368	10-Hydroxy-18-[*N*-(2-naphthyl)-*N*-phenylamino]betaenone	*Microsphaeropsis* sp. KMPB W-22	Brauers *et al*. 2000
369	Microsphaerone A		Wang *et al*. 2002
370	Microsphaerone B		
371	Deoxynortrichoharzin	*Paecilomyces* cf. *javanica* 961331	Rahbæk *et al*. 1998
372	Modiolin	*Paraphaeosphaeria* sp. N-119	Tsuda *et al*. 2003b
373	Petrosifungin A	*Penicillium brevicompactum*	Bringmann *et al*. 2004
374	Petrosifungin B		
375	Brocaenol C	*Penicillium brocae* F97S76	Bugni *et al*. 2003
376	Xestodecalactone A	*Penicillium* cf. *montanense* HBI-3/D	Edrada *et al*. 2002

(*Contd.*)

(*Contd.*)

377	Xestodecalactone C		
378	Deoxyanserinone B	*Penicillium corylophilum* a004181, b004181	Gautschi *et al.* 2004
379	Hydroxymethylanserinone B		
380	386-A	*Penicillium* sp. 386	Shao *et al.* 1999
381	Penicillazine		Lin *et al.* 2000
382	CNC-350.1	*Penicillium* sp. CNC-350	Son *et al.* 1999
383	Coruscol A	*Penicillium* sp. K029	Kagata *et al.* 2000
384	2,3-Dihydro-7-deacetoxy-yanuthone A	*Penicillium* sp. MFA 577	Li *et al.* 2003a
385	NI15501A	*Penicillium* sp. NI15501	Onuki *et al.* 1998
386	3,7-Dimethyl-1,8-dihydroxy-6-methoxyisochroman	*Penicillium steckii* M23B-7 (IBT 20952)	Malmstrøm *et al.* 2000
387	Tanzawaic acid E		
388	Tanzawaic acid F		
389	Pyrenocine D	*Penicillium waksmanii* OUPS-N127	Amagata *et al.* 1998b
390	*Seco*-macrosphelide E	*Periconia byssoides* OUPS-N133	Nakamura *et al.* 2002a
391	Phomoxide	*Phoma* sp. CNC-651	Liu, Z. *et al.* 2003
392	Phomoxin		
393	5-Hydroxyramulosin	*Phoma tropica* AZ14-4	Osterhage *et al.* 2002b
394	Pitholide A	*Pithomyces* sp. 969053	Wang, G.-Y.-S. *et al.* 1997
395	Pitholide B		
396	Pitholide C		
397	Pitholide D		
398	Remisporine A	*Remispora maritima* 22G295	Kong and Carter 2003
399	Trichoharzin	*Trichoderma harzianum*	Kobayashi *et al.* 1993
400	Harzialactone A	*Trichoderma harzianum* OUPS-N115	Amagata *et al.* 1998a
401	Epoxysorbicillinol	*Trichoderma longibrachiatum* 961015	Sperry *et al.* 1998
402	Trichodermamide A	*Trichoderma virens* CNK266, CNL910	Garo *et al.* 2003
403	Enalin A	*Verruculina enalia* No. 2606	Lin *et al.* 2002b
404	Enalin B		
405	Anomalin B	*Wardomyces anomalus* OS4T3-2-1	Abdel-Lateff *et al.* 2003a

(*Contd.*)

(*Contd.*)

406	n/a	*Xylaria* sp. 2508	Lin, Y. *et al*. 2001b
407	Xyloallenolide A		
408	Xyloketal B		Lin, Y. *et al*. 2001a
409	Xyloketal C		
410	Xyloketal D		
411	Xyloketal E		
412	Zopfinol	*Zopfiella marina* SANK 21274	Kondo *et al*. 1987
413	Methyl 2,4-dibromo-5-oxo-2-decenoate	unidentified fungus	Li, X. *et al*. 2004
414	Methyl 2,4-dibromo-5-oxo-3-decenoate		
415	1893A	unidentified fungus 1893	Chen *et al*. 2003
416	1893B		
417	n/a	unidentified fungus 2516	Yin *et al*. 2002
418	n/a		
419	n/a		
420	No. 2524-1	unidentified fungus 2524	Li, H. *et al*. 2004
421	No. 2524-2		
422	Avicennin A	unidentified fungus 2533	Wan *et al*. 2001
423	Chloriolin A	unidentified fungus 92902	Cheng *et al*. 1994
424	Chloriolin B		
425	Chloriolin C		
426	Demethyl nectriapyrone A	unidentified fungus 92948	Abrell *et al*. 1994
427	Nectriapyrone B		
428	Hirsutanol B	unidentified fungus 95-1005C	Wang *et al*. 1998
429	Hirsutanol C		
430	Solanapyrone G	unidentified fungus CNC-159	Jenkins *et al*. 1998a
431	*Iso*-cladospolide B	unidentified fungus I96S215	Smith *et al*. 2000
432	Pandangolide 1		
433	Pandangolide 2		
434	*Seco*-patulolide		
435	Dictyonamide B	unidentified fungus K063	Komatsu *et al*. 2001
436	Ergosteryl myristate	unidentified fungus MF001	Lee *et al*. 1999
437	N_b-Acetyltryptamine	unidentified fungus MFA 936	Li *et al*. 2003

[1]n/a: not available

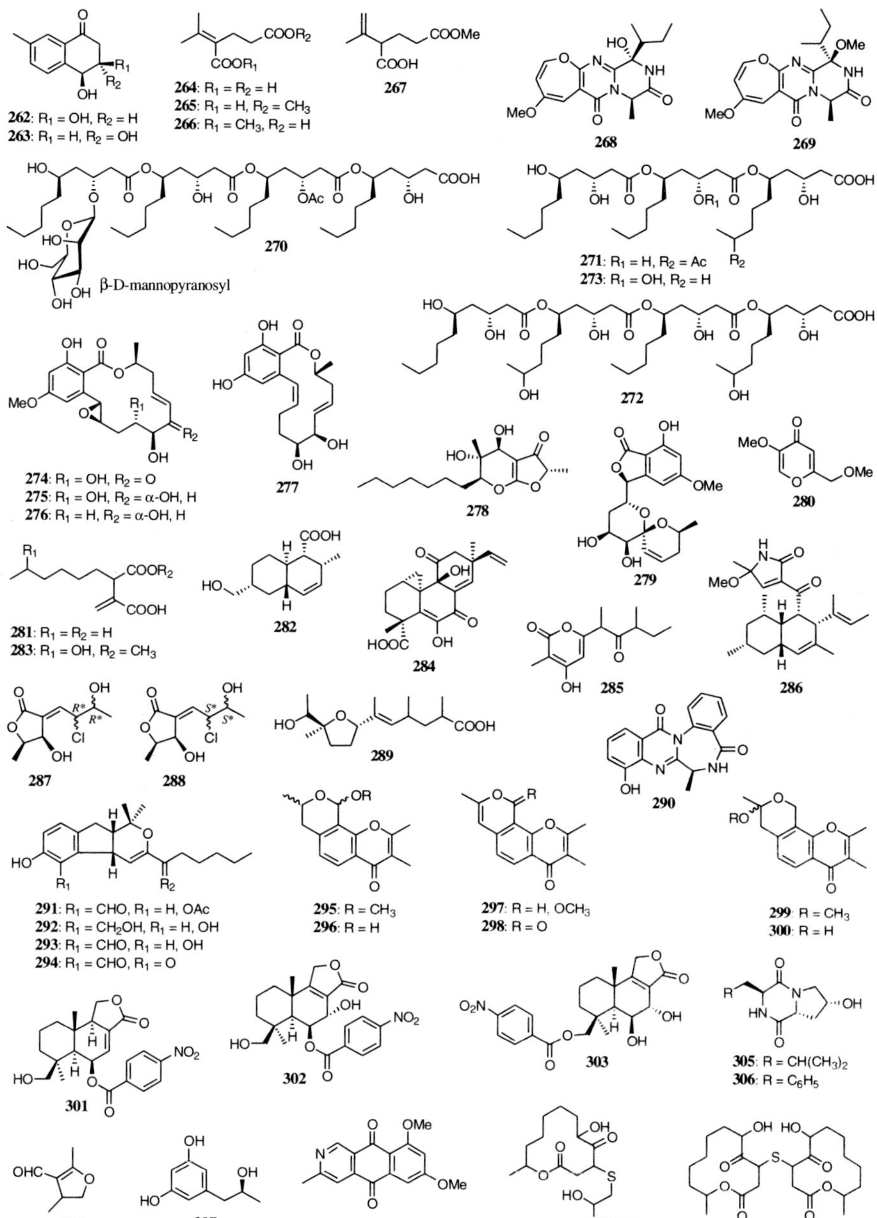

Fig. 8.5.1 Structures of new marine fungal metabolites that no biological activity has so far been detected or reported.

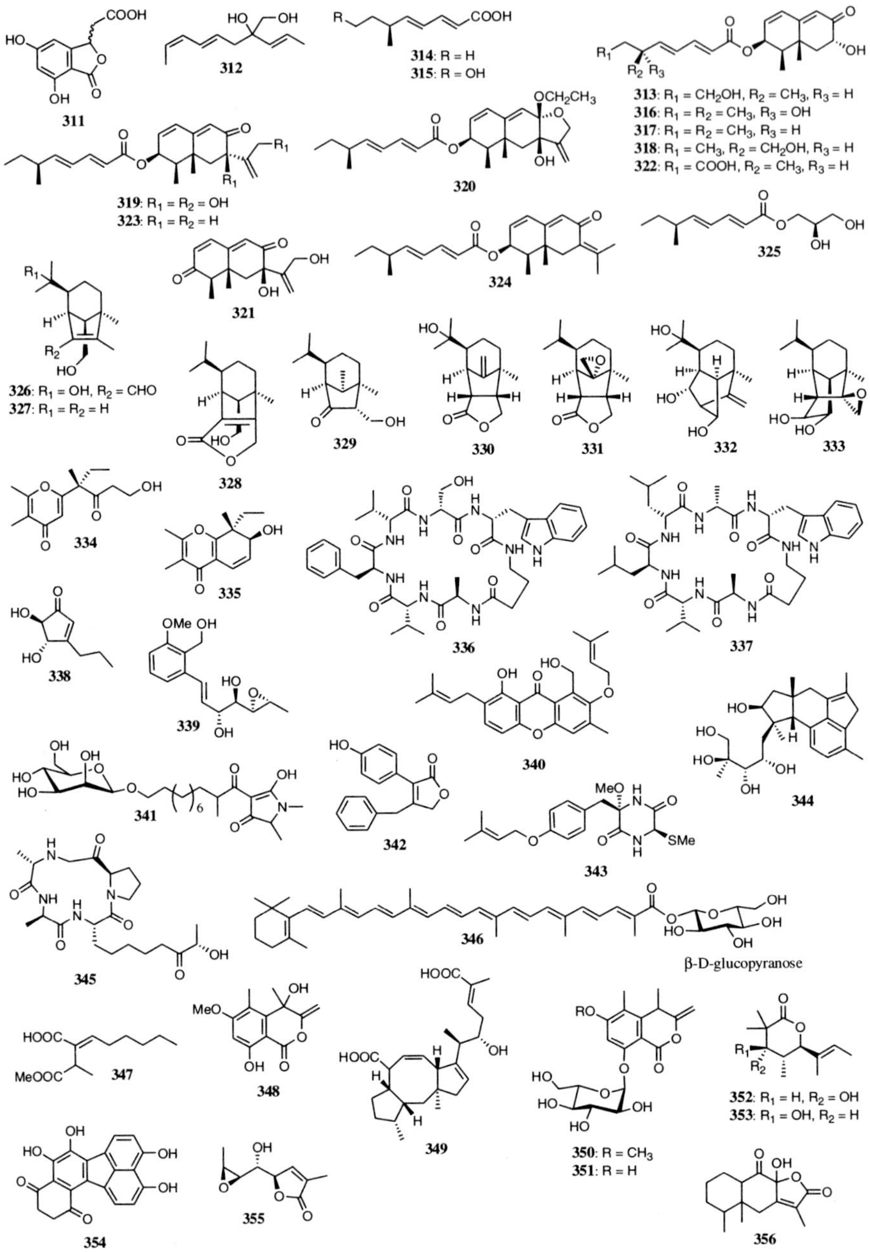
311
312
314: R = H
315: R = OH
313: R1 = CH2OH, R2 = CH3, R3 = H
316: R1 = R2 = CH3, R3 = OH
317: R1 = R2 = CH3, R3 = H
318: R1 = CH3, R2 = CH2OH, R3 = H
322: R1 = COOH, R2 = CH3, R3 = H
319: R1 = R2 = OH
323: R1 = R2 = H
320
321
324
325
326: R1 = OH, R2 = CHO
327: R1 = R2 = H
328
329
330
331
332
333
334
335
336
337
338
339
340
341
342
343
344
345
346
β-D-glucopyranose
347
348
349
350: R = CH3
351: R = H
352: R1 = H, R2 = OH
353: R1 = OH, R2 = H
354
355
356

Fig. 8.5.2

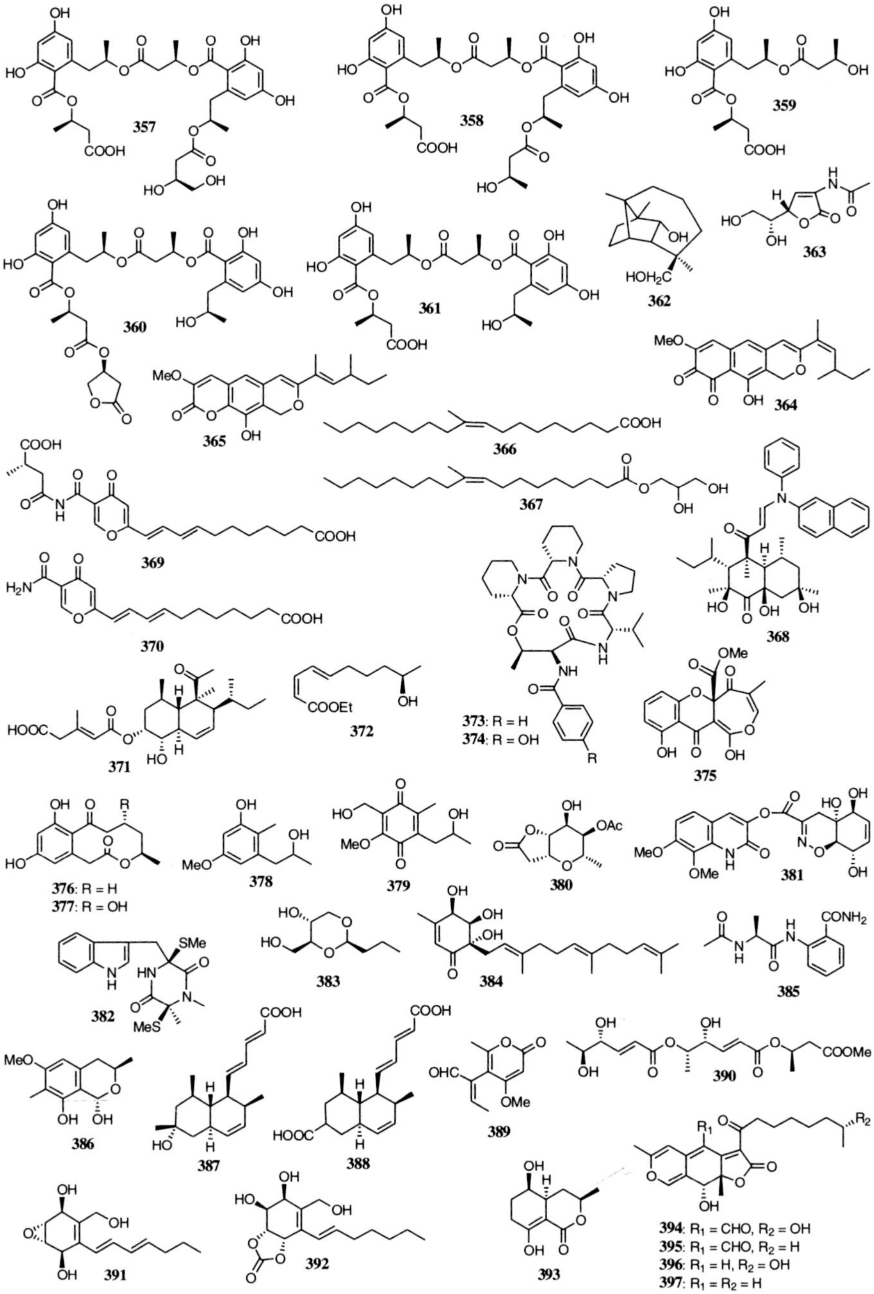
357
358
359
360
361
362
363
364
365
366
367
368
369
370
371
372
373: R = H
374: R = OH
375
376: R = H
377: R = OH
378
379
380
381
382
383
384
385
386
387
388
389
390
391
392
393
394: R1 = CHO, R2 = OH
395: R1 = CHO, R2 = H
396: R1 = H, R2 = OH
397: R1 = R2 = H

Fig. 8.5.3

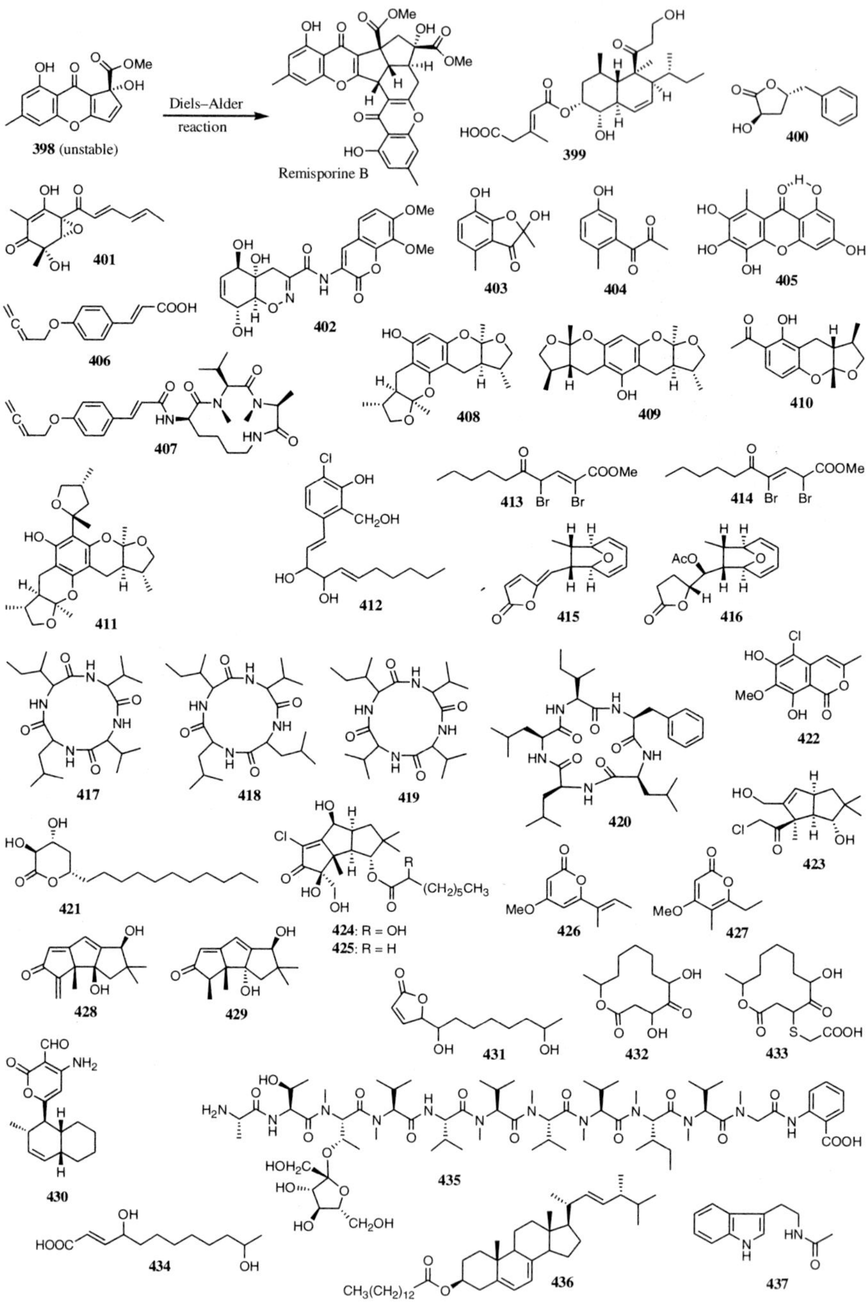
398 (unstable)
Diels–Alder
reaction
Remisporine B
399
400
401
402
403
404
405
406
407
408
409
410
411
412
413
414
415
416
417
418
419
420
421
422
423
424: R = OH
425: R = H
426
427
428
429
430
431
432
433
434
435
436
437

Fig. 8.5.4

to give the stereospecific product B during the isolation process. Although N_b-acetyltryptamine (**437**) has been obtained as the biotransformed product (Yang and Cordell 1997), this was the first report of the isolation as a natural product (Li *et al.* 2003).

CONCLUSIONS

Marine fungi (and marine-derived fungi) have clearly demonstrated their importance as potent sources of new biologically active secondary metabolites. However, new cultivation methods for isolating diverse species of marine fungi are required. Marine fungi living in extreme environments, such as deep sea (high pressure) and high salinity, and symbiotic/commensal relationships with other marine organisms will be interesting targets for future study. Adaptation to the specific conditions will change the metabolic pathways of fungi and produce new secondary metabolites. The ecological roles of marine fungi in marine and estuarine habitats will also be interesting study subjects. The chemical ecological studies on marine fungi will provide novel secondary metabolites, which can be used for screening for new drugs. Therefore, close cooperation between marine mycologists and natural product chemists is necessary to enlarge the scope of chemical study on marine fungi.

It is known that fungi give different ratios of secondary metabolites and even produce different compounds in different culture media. A few reports examined the response of secondary metabolisms to seawater concentrations (Bugni and Ireland 2004, Yamaguchi *et al.* 2002, Masuma *et al.* 2001, Miller and Savard 1989). The results showed that secondary metabolite production is sensitive to salt concentration. Fermentation experiments of nafredin (**229**) and chlorogentisylquinone (**244**) showed similar results, as mentioned above. Therefore, cultures in media of different seawater concentrations and in different nutrients are worth examining for production of biologically active metabolites.

It is highly to be expected that chemical studies on marine fungi will increase dramatically and that new potent biologically active metabolites will be discovered and utilized as new medicinal agents.

REFERENCES

Abbanat, D., M. Leighton, W. Maiese, E.B. Jones, C. Pearce, and M. Greenstein. 1998. Cell wall active antifungal compounds produced by the marine fungus *Hypoxylon oceanicum* LL-15G256. I. Taxonomy and fermentation. *J. Antibiot.* 51: 296–302.

Abdel-Lateff, A., C. Klemke, G.M. König, and A.D. Wright. 2003a. Two new xanthone derivatives from the algicolous marine fungus *Wardomyces anomalus*. *J. Nat. Prod.* 66: 706–708.

Abdel-Lateff, A., K.M. Fisch, A.D. Wright, and G.M. König. 2003b. A new antioxidant isobenzofuranone derivative from the algicolous marine fungus *Epicoccum* sp. *Planta Med.* 69: 831–834.

Abdel-Lateff, A., G.M. König, K.M. Fisch, U. Höller, P.G. Jones, and A.D. Wright. 2002. New antioxidant hydroquinone derivatives from the algicolous marine fungus *Acremonium* sp. *J. Nat. Prod.* 65: 1605–1611.

Abraham, S.P., T.D. Hoang, M. Alam, and E.B.G. Jones. 1994. Chemistry of the cytotoxic principles of the marine fungus *Lignincola laevis*. *Pure Appl. Chem.* 66: 2391–2394.

Abrell, L.M., B. Borgeson, and P. Crews. 1996a. A new polyketide, secocurvularin, from the salt water culture of a sponge derived fungus. *Tetrahedron Lett.* 37: 8983–8984.

Abrell, L.M., B. Borgeson, and P. Crews. 1996b. Chloro polyketides from the cultured fungus (*Aspergillus*) separated from a marine sponge. *Tetrahedron Lett.* 37: 2331–2334.

Abrell, L.M., X.-C. Cheng, and P. Crews. 1994. New nectriapyrones by salt water culture of a fungus separated from an indo-pacific sponge. *Tetrahedron Lett.* 35: 9159–9160.

Afiyatullov, S.S., A.I. Kalinocsky, T.A. Kuznetsova, M.V. Pivkin, N.G. Prokofeva, P.S. Dmitrenok, and G.B. Elyakov. 2004. New glycosides of the fungus *Acremonium striatisporum* isolated from a sea cucumber. *J. Nat. Prod.* 67: 1047–1051.

Afiyatullov, S.S., A.I. Kalinocsky, T.A. Kuznetsova, V.V. Isakov, M.V. Pivkin, P.S. Dmitrenok, and G.B. Elyakov. 2002. New diterpene glycosides of the fungus *Acremonium striatisporum* isolated from a sea cucumber. *J. Nat. Prod.* 65: 641–644.

Afiyatullov, S.S., T.A. Kuznetsova, V.V. Isakov, M.V. Pivkin, N.G. Prokof¢eva, and G.B. Elyakov. 2000. New diterpenic altrosides of the fungus *Acremonium striatisporum* isolated from a sea cucumber. *J. Nat. Prod.* 63: 848–850.

Akao, H., H. Kiyota, T. Nakajima, and T. Kitahara. 1999. Synthesis of dendryphiellin C, a trinor-sesquiterpene from a marine source. *Tetrahedron* 55: 7757–7770.

Alam, M., E.B.G. Jones, M.B. Hossain, and D. van der Helm. 1996. Isolation and structure of isoculmorin from the marine fungus *Kallichroma tethys*. *J. Nat. Prod.* 59: 454–456.

Albaugh, D., G. Albert, P. Bradford, V. Cotter, J. Froyd, J. Gaughran, D.R. Kirsch, M. Lai, A. Rehnig, E. Sieverding, S. Silverman. 1998. Cell wall active antifungal compounds produced by the marine fungus *Hypoxylon oceanicum* LL-15G256. III. Biological properties of 15G256γ. *J. Antibiot.* 51: 317–322.

Alvi, K.A., A. Casey, and B.G. Nair. 1998. Pulchellalactam: a CD45 protein tyrosine phosphatase inhibitor from the marine fungus *Corollospora pulchella*. *J. Antibiot.* 51: 515–517.

Amagata, T., C. Rath, J.F. Rigot, N. Tarlov, K. Tenney, F.A. Valeriote, and P. Crews. 2003a. Structures and cytotoxic properties of trichoverroids and their macrolide analogues produced by saltwater culture of *Myrothecium verrucaria*. *J. Med. Chem.* 46: 4342–4350.

Amagata, T., A. Amagata, K. Tenney, F.A. Valeriote, E. Lobkovsky, J. Clardy, and P. Crews. 2003b. Unusual C25 steroids produced by a sponge-derived *Penicillium citrinum*. *Organic Lett.* 5: 4393–4396.

Amagata, T., M. Doi, M. Tohgo, K. Minoura, and A. Numata. 1999. Dankasterone, a new class of cytotoxic steroid produced by a *Gymnascella* species from a marine sponge. *Chem. Commun.* 1321–1322.

Amagata, T., Y. Usami, K. Minoura, T. Ito, and A. Numata. 1998a. Cytotoxic substances produced by a fungal strain from a sponge: physico-chemical properties and structures. *J. Antibiot.* 51: 33–40.

Amagata, T., K. Minoura, and A. Numata. 1998b. Cytotoxic metabolites produced by a fungal strain from a *Sargassum* alga. *J. Antibiot.* 51: 432–434.

Amagata, T., K. Minoura, and A. Numata. 1998c. Gymnasterones, novel cytotoxic metabolites produced by a fungal strain from a sponge. *Tetrahedron Lett.* 39: 3773–3774.

Amagata, T., M. Doi, T. Ohta, K. Minoura, and A. Numata. 1998d. Absolute stereostructures of novel cytotoxic metabolites, gymnastatins A-E, from a *Gymnascella* species separated from a *Halichondria* sponge. *J. Chem. Soc., Perkin Trans. 1* 3585–3599.

Belofsky, G.N., M. Anguera, P.R. Jensen, W. Fenical, and M. Kock. 2000. Oxepinamides A-C and fumiquinazolines H—I: bioactive metabolites from a marine isolate of a fungus of the genus *Acremonium*. *Chem. Eur. J.* 6: 1355–1360.

Belofsky, G.N., P.R. Jensen, and W. Fenical. 1999. Sansalvamide: a new cytotoxic cyclic depsipeptide produced by a marine fungus of the genus *Fusarium*. *Tetrahedron Lett.* 40: 2913–2916.

Belofsky, G.N., P.R. Jensen, M.K. Renner, and W. Fenical. 1998. New cytotoxic sesquiterpenoid nitrobenzoyl esters from a marine isolate of the fungus *Aspergillus versicolor*. *Tetrahedron* 54: 1715–1724.

Bernan, V.S., M. Greenstein, and W.M. Maiese. 1997. Marine microorganisms as a source of new natural products. *Adv. Appl. Microbiol.* 43: 57–90.

Bessho, J., Y. Shimotsu, S. Mizumoto, N. Mase, H. Yoda, K. Takabe. 2004. Convenient synthesis of pulchellalactam, a CD45 protein tyrosine phosphatase inhibitor from the marine fungus *Corollospora pulchella*, and its related compounds. *Heterocycles* 63: 1013–1016.

Biabani, M.A.F., and H. Laatsch. 1998. Advances in chemical studies on low-molecular-weight metabolites of marine fungi. *J. Prakt. Chem.* 340: 589–607.

Bohlman, F., C. Zdero, and A. Suwita. 1975. A new sesquiterpene quinone from *Seseli* species. *Chem. Ber.* 108: 2818–2821.

Brauers, G., R. Ebel, R. Edrada, V. Wray, A. Berg, U. Gräfe, and P. Proksch. 2001. Hortein, a new natural product from the fungus *Hortaea werneckii* associated with the sponge *Aplysina aerophoba*. *J. Nat. Prod.* 64: 651–652.

Brauers, G., R.A. Edrada, R. Ebel, P. Proksch, V. Wray, A. Berg, U. Gräfe, C. Schächtele, F. Totzke, G. Finkenzeller, D. Marme, J. Kraus, M. Münchbach, M. Michel, G. Bringmann, and K. Schaumann. 2000. Anthraquinones and betaenone derivatives from the sponge-associated fungus *Microsphaeropsis* species: novel inhibitors of protein kinases. *J. Nat. Prod.* 63: 739–745.

Bringmann, G., G. Lang, S. Steffens, and K. Schaumann. 2004. Petrosifungins A and B, novel cyclodepsipeptides from a sponge-derived strain of *Penicillium brevicompactum*. *J. Nat. Prod.* 67: 311–315.

Bringmann, G., G. Lang, S. Steffens, E. Gunther, K. Schaumann. 2003. Evariquinone, isoemericellin, and stromemycin from a sponge derived strain of the fungus *Emericella variecolor*. *Phytochemistry* 63: 437–443.

Bugni, T.S., and C.M. Ireland. 2004. Marine-derived fungi: a chemically and biologically diverse group of microorganisms. *Nat. Prod. Rep.* 21: 143–163.

Bugni, T.S., V.S. Bernan, M. Greenstein, J.E. Janso, W.M. Maiese, C.L. Mayne, and C.M. Ireland. 2003. Brocaenols A-C: novel polyketides from a marine derived *Penicillium brocae*. *J. Org. Chem.* 68: 2014–2017.

Bugni, T.S., D. Abbanat, V.S. Bernan, W.M. Maiese, M. Greenstein, R.M. van Wagoner, and C.M. Ireland. 2000. Yanuthones: novel metabolites from a marine isolate of *Aspergillus niger*. *J. Org. Chem.* 65: 7195–7200.

Byun, H.-G., H. Zhang, M. Mochizuki, K. Adachi, Y. Shizuri, W.-J. Lee, and S.-K. Kim. 2003. Novel antifungal diketopiperazine from marine fungus. *J. Antibiot.* 56: 102–106.

Capon, R.J., C. Skene, M. Stewart, J. Ford, R.A.J. O'Hair, L. Williams, E. Lacey, J.H. Gill, K. Heiland, and T. Friedel. 2003. Aspergillicins A-E: five novel depsipeptides from the marine-derived fungus *Aspergillus carneus*. *Org. Biomol. Chem.* 1: 1856–1862.

Chen, C., N. Imamura, M. Nishijima, K. Adachi, M. Sakai, and H. Sano. 1996. Halymecins, new antimicroalgal substances produced by fungi isolated from marine algae. *J. Antibiot.* 49: 998–1005.

Chen, G., Y. Lin, L. Wen, L.L.P. Vrijmoed, and E.B.G. Jones. 2003. Two new metabolites of a marine endophytic fungus (no. 1893) from an estuarine mangrove on the South China Sea coast. *Tetrahedron* 59: 4907–4909.

Cheng, X.-C., M. Varoglu, L. Abrell, P. Crews, E. Lobkovsky, and J. Clardy. 1994. Chloriolins A–C, chlorinated sesquiterpenes produced by fungal cultures separated from a *Jaspis* marine sponge. *J. Org. Chem.* 59: 6344–6348 [Erratum: 1995. 60: 1478].

Chinworrungsee, M., P. Kittakoop, M. Isaka, R. Chanphen, M. Tanticharoen, and Y. Thebtaranonth. 2002. Halorosellins A and B, unique isocoumarin glucosides from the marine fungus *Halorosellinia oceanica*. *J. Chem. Soc., Perkin Trans. 1* 2473–2476.

Chinworrungsee, M., P. Kittakoop, M. Isaka, A. Rungrod, M. Tanticharoen, and Y. Thebtaranonth. 2001. Antimalarial halorosellinic acid from the marine fungus *Halorosellinia oceanica*. *Bioorg. Med. Chem. Lett.* 11: 1965–1969.

Chu, M., I. Truumees, I. Gunnarsson, W.R. Bishop, W. Kreutner, A.C. Horan, M.G. Patel, V.P. Gullo, and M.S. Puar. 1993. A novel class of platlet activating factor antagonists from *Phoma* sp. *J. Antibiot.* 46: 554–563.

Chu, M., M.G. Patel, V.P. Gullo, I. Truumees, M.S. Puar, and A.T. McPhail. 1992. Sch 47918: A novel PAF antagonist from the fungus *Phoma* sp. *J. Org. Chem.* 57: 5817–5818.

Cueto, M., P.R. Jensen, and W. Fenical. 2002. Aspergilloxide, a novel sesterterpene epoxide from a marine-derived fungus of the genus *Aspergillus*. *Org. Lett.* 4: 1583–1585.

Cueto, M., P.R. Jensen, C. Kauffman, W. Fenical, E. Lobkovsky, and J. Clardy. 2001. Pestalone, a new antibiotic produced by a marine fungus in response to bacterial challenge. *J. Nat. Prod.* 64: 1444–1446.

Cueto, M., P.R. Jensen, and W. Fenical. 2000. *N*-Methylsansalvamide, a cytotoxic cyclic depsipeptide from a marine fungus of the genus *Fusarium*. *Phytochemistry* 55: 223–226.

Cui, C.-B., H. Kakeya, and H. Osada. 1997. Novel mammalian cell cycle inhibitors, cyclotryprostatins A~D, produced by *Aspergillus fumigatus*, which inhibit mammalian cell cycle at G2/M phase. *Tetrahedron* 53: 59–72.

Cui, C.-B., M. Ubukata, H. Kakeya, R. Onose, G. Okada, I. Takahashi, K. Isono, and H. Osada. 1996a. Acetophthalidin, a novel inhibitor of mammalian cell cycle, produced by a fungus isolated from sea sediment. *J. Antibiot.* 49: 216–219.

Cui, C.-B., H. Kakeya, and H. Osada. 1996b. Novel mammalian cell cycle inhibitors, spirotryprostatins A and B, produced by *Aspergillus fumigatus*, which inhibit mammalian cell cycle at G2/M phase. *Tetrahedron* 52: 12651–12666.

Cui, C.-B., H. Kakeya, G. Okada, R. Onose, M. Ubukata, I. Takahashi, K. Isono, and H. Osada. 1995. Tryprostatins A and B, novel mammalian cell inhibitors produced by *Aspergillus fumigatus*. *J. Antibiot.* 48: 1382–1384.

Daferner, M., T. Anke, and O. Sterner. 2002. Zopfiellamides A and B, antimicrobial pyrrolidinone derivatives from the marine fungus *Zopfiella latipes*. *Tetrahedron* 58: 7781–7784.

Daferner, M., S. Mensch, T. Anke, and O. Sterner. 1999. Hypoxysordarin, a new sordarin derivative from *Hypoxylon croceum*. *Z. Naturforsch.* 54c: 474–480.

Dai, J.-R., B.K. Carte, P.J. Sidebottom, A.L.S. Yew, S.-B. Ng, Y. Huang, and M.S. Butler. 2001. Circumdatin G, a new alkaloid from the fungus *Aspergillus ochraceus*. *J. Nat. Prod.* 64: 125–126.

Davidson, B.S. 1995. New dimensions in natural products research: cultured marine microorganisms. *Curr. Opin. Biotechnol.* 6: 284–291.

Depew, K.M., S.J. Danishefsky, N. Rosen, and L. Sepp-Lorenzino. 1996. Total synthesis of tryprostatin B: generation of a nucleophilic prenylating species from a prenylstannane. *J. Am. Chem. Soc.* 118: 12463–12464.

Diaper, C.M., W.P.D. Goldring, and G. Pattenden. 2003. A total synthesis of (±)-phomactin A. *Org. Biomol. Chem.* 1: 3949–3956.

Doshida, J., H. Hasegawa, H. Onuki, and N. Shimidzu. 1996. Exophilin A, a new antibiotic from a marine microorganism *Exophiala pisciphila*. *J. Antibiot.* 49: 1105–1109.

Edrada, R.A., M. Heubes, G. Brauers, V. Wray, A. Berg, U. Gräfe, M. Wohlfarth, J. Mühlbacher, K. Schaumann, Sudarsono, G. Bringmann, and P. Proksch. 2002. Online analysis of xestodecalactones A–C, novel bioactive metabolites from the fungus *Penicillium* cf. *montanense* and their subsequent isolation from the sponge *Xestospongia exigua*. *J. Nat. Prod.* 65: 1598–1604.

Edrada, R.A., V. Wray, A. Berg, U. Grafe, Sudarsono, G. Brauers, and P. Proksch. 2000. Novel spiciferone derivatives from the fungus *Drechslera hawaiiensis* isolated from the marine sponge *Callyspongia aerizusa*. *Z. Naturforsch.* 55c: 218–221.

Fukami, A., Y. Taniguchi, T. Nakamura, M.-C. Rho, K. Kawaguchi, M. Hayashi, K. Komiyama, and S. Omura. 1999. New members of the macrosphelides from *Microsphaeropsis* sp. FO-5050 IV. *J. Antibiot.* 52: 501–504.

Gan, T., and J.M. Cook. 1997. Enantiospecific total synthesis of tryprostatin A. *Tetrahedron Lett.* 38: 1301–1304.

Gan, T., R. Liu, P. Yu, S. Zhao, and J.M. Cook. 1997. Enantiospecific synthesis of optically active 6-methoxytryptophan derivatives and total synthesis of tryprostatin A. *J. Org. Chem.* 62: 9298–9304.

Garo, E., C.M. Starks, P.R. Jensen, W. Fenical, E. Lobkovsky, and J. Clardy. 2003. Trichodermamides A and B, cytotoxic modified dipeptides from the marine-derived fungus *Trichoderma virens*. *J. Nat. Prod.* 66: 423–426.

Gautschi, J.T., T. Amagata, A. Amagata, F.A. Valeriote, S.L. Mooberry, and P. Crews. 2004. Expanding the strategies in natural product studies of marine-derived fungi: a chemical investigation of *Penicillium* obtained from deep water sediment. *J. Nat. Prod.* 67: 362–367.

Goldring, W.P.D., and G. Pattenden. 2004. Total synthesis of (±)-phomactin G, a platelet activating factor antagonist from the marine fungus *Phoma* sp. *Org. Biomol. Chem.* 2: 466–473.

Guerriero, A., M. D'Ambrosio, V. Cuomo, and F. Pietra. 1991. A novel degraded polyketidic lactone, leptosphaerolide, and its likely diketone precursor, leptosphaerodione. Isolation from cultures of the marine ascomycete *Leptosphaeria oraemaris* (Linder). *Helv. Chim. Acta* 74: 1445–1450.

Guerriero, A., V. Cuomo, F. Vanzanella, and F. Pietra. 1990. A novel glycerol ester (glyceryl dendryphiellate A), a trinor-eremophilane (dendryphiellin A1), and eremophielanes (dendryphiellin E1 and E2) from the marine deuteromycete *Dendryphiella salina* (Sutherland) Pugh et Nicot. *Helv. Chim. Acta* 73: 2090–2096.

Guerriero, A., M. D'Ambrosio, V. Cuomo, F. Vanzanella, and F. Pietra. 1989. Novel trinor-eremophilanes (dendryphiellin B, C, and D), eremophilanes (dendryphiellin E, F, and G), and branched C_9-carboxylic acid (dendryphiellic Acid A and B) from the marine deuteromycete *Dendryphiella salina* (Sutherland) Pugh et Nicot. *Helv. Chim. Acta* 72: 438–446.

Guerriero, A., M. D'Ambrosio, V. Cuomo, F. Vanzanella, and F. Pietra. 1988. Dendryphiellin A, the first fungal trinor-eremophilane. Isolation from the marine deuteromycete *Dendryphiella salina* (Sutherland) Pugh et Nicot. *Helv. Chim. Acta* 71: 57–61.

Gurjar, M.K., and P. Bhaket. 2000. Total synthesis of a novel cytotoxic metabolite gymnastatin A. *Heterocycles* 53: 143–149.

Hayashi, M., Y.-P. Kim, H. Hiraoka, M. Natori, S. Takamatsu, T. Kawakubo, R. Masuma, K. Komiyama, and S. Omura. 1995. Macrosphelide, a novel inhibitor of cell-cell adhesion molecule. I. Taxonomy, fermentation, isolation and biological activities. *J. Antibiot.* 48: 1435–1439.

Hwang, Y., D. Rowley, D. Rhodes, J. Gertch, W. Fenical, and F. Bushman. 1999. Mechanism of inhibition of a poxvirus topoisomerase by the marine natural product sansalvamide A. *Mol. Pharmacol.* 55: 1049–1053.

Höller, U., G.M. König, and A.D. Wright. 1999a. A new tyrosine kinase inhibitor from a marine isolate of *Ulocladium botrytis* and new metabolites from the marine fungi *Asteromyces cruciatus* and *Varicosporina ramulosa*. *Eur. J. Org. Chem.* 2949–2955.

Höller, U., G.M. König, and A.D. Wright. 1999b. Three new metabolites from marine-derived fungi of the genera *Coniothyrium* and *Microsphaeropsis*. *J. Nat. Prod.* 62: 114–118.

Isaka, M., C. Suyarnsestakorn, M. Tanticharoen, P. Kongsaeree, and Y. Thebtaranonth. 2002. Aigialomycins A-E, new resorcylic macrolides from the marine mangrove fungus *Aigialus parvus*. *J. Org. Chem.* 67: 1561–1566.

Ishibashi, K., K. Nose, T. Shindo, M. Arai, and H. Mishima. 1968. Siccayne: a novel acetylenic metabolite of *Helminthosporium siccans*. *Ann. Rep. Sankyo Res. Lab.* 20: 76–79.

Iwamoto, C., T. Yamada, Y. Ito, K. Minoura, and A. Numata. 2001. Cytotoxic cytochalasans from a *Penicillium* species separated from a marine alga, *Tetrahedron* 57: 2997–3004.

Iwamoto, C., K. Minoura, T. Oka, T. Ohta, S. Hagashita, and A. Numata. 1999. Absolute stereostructures of novel cytotoxic metabolites, penostatins A-E, from a *Penicillium* species separated from an *Enteromorpha* alga. *Tetrahedron* 55: 14353–14368.

Iwamoto, C., K. Minoura, S. Hagishita, K. Nomoto, and A. Numata. 1998. Penostatins F-I, novel cytotoxic metabolites from a *Penicillium* species separated from an *Enteromorpha* marine alga. *J. Chem. Soc., Perkin Trans.* 1: 449–456.

Jadulco, R., R.A. Edrada, R. Ebel, A. Berg, K. Schaumann, V. Wray, K. Steube, and P. Proksch. 2004. New communesin derivatives from the fungus *Penicillium* sp. derived from the Mediterranean sponge *Axinella verrucosa*. *J. Nat. Prod.* 67: 78–81.

Jadulco, R., G. Brauers, R.A. Edrada, R. Ebel, V. Wray, Sudarsono, and P. Proksch. 2002. New metabolites from sponge-derived fungi *Curvularia lunata* and *Cladosporium herbarum*. *J. Nat. Prod.* 65: 730–733.

Jadulco, R., P. Proksch, V. Wray, Sudarsono, A. Berg, and U. Gräfe. 2001. New macrolides and furan carboxylic acid derivative from the sponge-derived fungus *Cladosporium herbarum*. *J. Nat. Prod.* 64: 527–530.

Jenkins, K.M., S.G. Toske, P.R. Jensen, and W. Fenical. 1998a. Solanapyrones E–G, antialgal metabolites produced by a marine fungus. *Phytochemistry* 49: 2299–2304.

Jenkins, K.M., M.K. Renner, P.R. Jensen, and W. Fenical. 1998b. Exumolides A and B: antimicroalgal cyclic depsipeptides produced by a marine fungus of the genus *Scytalidium*. *Tetrahedron Lett.* 39: 2463–2466.

Jensen, P.R., and W. Fenical. 2002. Secondary metabolites from marine fungi. Pages 293–315 in *Fungi in Marine Environments*, K.D. Hyde, ed., Fungal Diversity Press, Hong Kong.

Jensen, P.R., and W. Fenical. 2000. Marine microorganisms and drug discovery: current status and future potential. Pages 6–29 in *Drugs from the Sea*, N. Fusetani, ed., Karger, Basel.

Jiang, Z., M.O. Barret, K.G. Boyd, D.R. Adams, A.S. Boyd, and J.G. Burgess. 2002. JM47, a cyclic tetrapeptide HC-toxin analogue from a marine *Fusarium* species. *Phytochemistry* 60: 33–38.

Kagata, T., H. Shigemori, Y. Mikami, and J. Kobayashi. 2000. Coruscol A, a new metabolite from the marine-derived fungus *Penicillium* Species. *J. Nat. Prod.* 63: 886–887.

Kakeya, H., C. Onozawa, M. Sato, K. Arai, and H. Osada. 1997. Neuritogenic effect of epolactaene derivatives on human neuroblastoma cells which lack high-affinity nerve growth factor receptors. *J. Med. Chem.* 40: 391–394.

Kakeya, H., I. Takahashi, G. Okada, K. Isono, and H. Osada. 1995. Epolactaene, a novel neuritogenic compound in human neuroblastoma cells, produced by a marine fungus. *J. Antibiot.* 48: 733–735.

Kim, S.-K., K. Mizuno, M. Hatori, and S. Marumo. 1994. A brassinolide-inhibitor KM-01, its isolation and structure elucidation from a fungus *Drechslera avenae*. *Tetrahedron Lett.* 35: 1731–1734.

Klemke, C., S. Kehraus, A.D. Wright, and G.M. König. 2004. New secondary metabolites from the marine endophytic fungus *Apiospora montagnei*. *J. Nat. Prod.* 67: 1058–1063.

Kobayashi, H., S. Meguro, T. Yoshimoto, and M. Namikoshi. 2003. Absolute structure, biosynthesis, and anti-microtubule activity of phomopsidin, isolated from a marine-derived fungus *Phomopsis* sp. *Tetrahedron* 59: 455–459.

Kobayashi, J., and M. Ishibashi. 1993. Bioactive metabolites of symbiotic marine microorganisms. *Chem. Rev.* 93: 1753–1769.

Kobayashi, M., H. Uehara, K. Matsunami, S. Aoki, and I. Kitagawa. 1993. Trichoharzin, a new polyketide produced by the imperfect fungus *Trichoderma harzianum* separated from the marine sponge *Micale cecilia*. *Tetrahedron Lett.* 34: 7925–7928.

Kobayashi, Y., and Y.G. Wang. 2002. Synthesis of macrosphelides H and G. *Tetrahedron Lett.* 43: 4381–4384.

Kohlmeyer, J., and E. Kohlmeyer. 1979. *Marine Mycology: The Higher Fungi*. Academic Press, New York.

Kohlmeyer, J., and B. Volkmann-Kohlmeyer. 2003. Fungi from coral reefs: a commentary. *Mycol. Res.* 107: 386–387.

Komatsu, K., H. Shigemori, and J. Kobayashi. 2001. Dictyonamides A and B, new peptides from marine-derived fungus. *J. Org. Chem.* 66: 6189–6192.

Komatsu, K., H. Shigemori, Y. Mikami, and J. Kobayashi. 2000. Sculezonones A and B, two metabolites possessing a phenalenone skeleton from a marine-derived fungus *Penicillium* species. *J. Nat. Prod.* 63: 408–409.

Kondo, M., T. Takayama, K. Furuya, M. Okudaira, T. Hayashi, and M. Kinoshita. 1987. A nuclear magnetic resonance study of zophinol isolated from *Zopfiella marina*. *Sankyo Kenkyusho Nempo* (*Ann. Rep. Sankyo Res. Lab.*) 39: 45–53.

Kondoh, M., T. Usui, T. Mayumi, and H. Osada. 1998. Effects of tryprostatin derivatives on microtubule assembly *in vitro* and *in situ*. *J. Antibiot.* 51: 801–804.

Kong, F., and G.T. Carter. 2003. Remisporine B, a novel dimeric chromenone derived from spontaneous Diels-Alder reaction of remisporine A. *Tetrahedron Lett.* 44: 3119–3122.

Krohn, K., and M. Riaz. 2004. Total synthesis of (+)-xyloketal D, a secondary metabolite from the mangrove fungus *Xylaria* sp. *Tetrahedron Lett.* 45: 293–294.

Kupka, J., T. Anke, W. Steglich, and L. Zechlin. 1981. Antibiotics from basidiomycetes. XI. The biological activity of siccayne, isolated from the marine fungus *Halocyphina villosa* J. & E. Kohlmeyer. *J. Antibiot.* 34: 298–304.

Kusnick, C., R. Jansen, K. Liberra, and U. Lindequist. 2002. Ascochital, a new metabolite from the marine ascomycete *Kirschsteiniothelia maritima*. *Pharmazie* 57: 510–512.

Laurent, D., G. Guella, I. Mancini, M.-F. Roquebert, F. Farinole, and F. Pietra. 2002. A new cytotoxic tetralone derivative from *Humicola grisea*, a filamentous fungus from wood in the Southeastern lagoon of New Caledonia. *Tetrahedron* 58: 9163–9167.

Laurent, D., G. Guella, M.-F. Roquebert, F. Farinole, I. Mancini, and F. Pietra. 2000. Cytotoxins, mycotoxins and drugs from a new deuteromycete, *Acremonium neocaledoniae*, from the Southwestern lagoon of New Caledonia. *Planta Med.* 66: 63–66.

Lee, D.I., J.S. Choi, M.-R. Yang, W.K. Lee, D.-S. Kim, H.D. Choi, and B.W. Son. 1999. Ergosteryl myristate, a new ergosterol derivative from unidentified marine algicolous fungus. *Nat. Prod. Sci.* 5: 93–96.

Lee, S.M., X.F. Li, H. Jiang, J.G. Cheng, S. Seong, H.D. Choi, and B.W. Son. 2003. Terreusinone, a novel UV-A protecting dipyrroloquinone from the marine algicolous fungus *Aspergillus terreus*. *Tetrahedron Lett.* 44: 7707–7710.

Lee, Y., and R. B. Silverman. 2000. Rapid, high-yield, solid-phase synthesis of the antitumor antibiotic sansalvamide A using a side-chain-tethered phenylalanine building block. *Org. Lett.* 2: 3743–3746.

Li, H., Y.-C. Lin, J.-H. Yao, L.L.P. Vrijmoed, and E.B.G. Jones. 2004. Two new metabolites from the mangrove endophytic fungus no. 2524. *J. Asian Nat. Prod. Res.* 6: 185–191.

Li, H., Y. Lin, L. Wang, S. Zhou, and L.L.P. Vrijmoed. 2001. Metabolites of marine fungus *Hypoxylon oceanicum* from the South China Sea. *Acta Scientiarum Naturalium Universitatis Sunyatsuni* 40: 70.

Li, X., S.-K. Kim, J.S. Kang, H.D. Choi, and B.W. Son. 2004. Polyketide and sesquiterpenediol metabolites from a marine-derived fungus. *Bull. Korean Chem. Soc.* 25: 607–608.

Li, X., H.D. Choi, J.S. Kang, C.-O. Lee, and B.W. Son. 2003a. New polyoxygenated farnesylcyclohexenones, deacetoxyyanuthone A and its hydro derivative from the marine-derived fungus *Penicillium* sp. *J. Nat. Prod.* 66: 1499–1500.

Li, X., J.H. Jeong, K.T. Lee, J.R. Rho, H.D. Choi, J.S. Kang, and B.W. Son. 2003b. γ-Pyrone derivatives, kojic acid methyl ethers from a marine-derived fungus *Altenaria* sp. *Arch. Pharm. Res.* 26: 532–534.

Li, Y., X. Li, S.-K. Kim, J.S. Kang, H.D. Choi, J.R. Rho, and B.W. Son. 2004a. Golmaenone, a new diketopiperazine alkaloid from the marine-derived fungus *Aspergillus* sp. *Chem. Pharm. Bull.* 52: 375–376.

Li, Y., X. Li, J.S. Kang, H.D. Choi, and B.W. Son. 2004b. New radical scavenging and ultraviolet-A protecting prenylated dioxopiperazine alkaloid related to isoechinulin A from a marine isolate of the fungus *Aspergillus*. *J. Antibiot.* 57: 337–340.

Li, Y., X.F. Li, D.S. Kim, H.D. Choi, and B.W. Son. 2003. Indolyl alkaloid derivatives, N_b-acetyltryptamine and oxaline from a marine-derived fungus. *Arch. Pharm. Res.* 26: 21–23.

Liberra, K., R. Jansen, and U. Lindequist. 1998. Corollosporine, a new phthalide derivative from the marine fungus *Corollospora maritima* Wetderm. 1069. *Pharmazie* 53: 578–581.

Liberra, K., and U. Lindequist. 1995. Marine fungi – a prolific resource of biologically active natural products? *Pharmazie* 50: 583–588.

Lin, W., G. Brauers, R. Ebel, V. Wray, A. Berg, Sudarsono, and P. Proksch. 2003. Novel chromone derivatives from the fungus *Aspergillus versicolor* isolated from the marine sponge *Xestospongia exigua*. *J. Nat. Prod.* 66: 57–61.

Lin, W., J. Li, H.Z. Fu, and P. Proksch. 2001a. Four novel hydropyranoindeno-derivatives from marine fungus *Aspergillus versicolor*. *Chin. Chem. Lett.* 12: 435–438.

Lin, W., H.Z. Fu, J. Li, and P. Proksch. 2001b. Novel chromone derivatives from marine fungus *Aspergillus versicolor* isolated from the sponge *Xestospongia exigua*. *Chin. Chem. Lett.* 12: 235–238.

Lin, Y., H. Li, G. Jian, S. Zhou, L.L.P. Vrijmoed, and E.B.G. Jones. 2002a. A novel γ-lactone, eutypoid-A and other metabolites from marine fungus *Eutypa* sp. (#424) from the South China Sea. *Ind. J. Chem.* 41B: 1542–1544.

Lin, Y., X. Wu, Z. Deng, J. Wang, S. Zhou, L.L.P. Vrijmoed, and E.B.G. Jones. 2002b. The metabolites of the mangrove fungus *Verruculina enalia* no. 2606 from a salt lake in the Bahamas, *Phytochemistry* 59: 469–471.

Lin, Y., X. Wu, S. Feng, G. Jiang, J. Luo, S. Zhou, L.L.P. Vrijmoed, E.B.G. Jones, K. Krohn, K. Steingrover, and F. Zsila. 2001a. Five unique compounds: xyloketals from mangrove fungus *Xylaria* sp. from the South China Sea coast. *J. Org. Chem.* 66: 6252–6356.

Lin, Y., X. Wu, S. Feng, G. Jiang, S. Zhou, L.L.P. Vrijmoed, and E.B.G. Jones. 2001b. A novel *N*-cinnamoylcyclopeptide containing an allenic ether from the fungus *Xylaria* sp. (strain # 2508) from the South China Sea. *Tetrahedron Lett.* 42: 449–451.

Lin, Y., Z. Shao, G. Jiang, S. Zhou, J. Cai, L.L.P. Vrijmoed, and E.B.G. Jones. 2000. Penicillazine, a unique quinolone derivative with 4*H*-5,6-dihydro-1,2-oxazine ring system from the marine fungus *Penicillium* sp. (strain #386) from the South China Sea. *Tetrahedron* 56: 9607–9609.

Liu, C.H., J.Y. Liu, L.L. Huang, W.X. Zou, and R.X. Tan. 2003. Absolute configuration of keisslone, a new antimicrobial metabolite from *Keissleriella* sp. YS4108, a marine filamentous fungus. *Planta Med.* 69: 481–483.

Liu, C.H., J.C. Meng, W.X. Zou, L.L. Huang, H.Q. Tang, and R.X. Tan. 2002. Antifungal metabolite with a new carbon skeleton from *Keissleriella* sp. YS4108, a marine filamentous fungus. *Planta Med.* 68: 363–365.

Liu, Z., P.R. Jensen, and W. Fenical. 2003. A cyclic carbonate and related polyketides from a marine-derived fungus of the genus *Phoma*. *Phytochemistry* 64: 571–574.

Lorenz, P., P.R. Jensen, and W. Fenical. 1998. Mactanamide, a new fungistatic diketopiperazine produced by a marine *Aspergillus* sp. *Nat. Prod. Lett.* 12: 55–60.

Malmstrøm, J. 1999. Unguisins A and B: new cyclic peptides from the marine-derived fungus *Emericella unguis*. *J. Nat. Prod.* 62: 787–789.

Malmstrøm, J., C. Christophersen, A.F. Barrero, J.E. Oltra, J. Justicia, and A. Rosales. 2002a. Bioactive metabolites from a marine-derived strain of the fungus *Emericella variecolor*. *J. Nat. Prod.* 65: 364–367.

Malmstrøm, J., A. Ryager, U. Antoni, and P.H. Nielsen. 2002b. Unguisin C, a GABA-containing cyclic peptide from the fungus *Emericella unguis*. *Phytochemistry* 60: 869–872.

Malmstrøm, J., C. Christophersen, and J.C. Frisvad. 2000. Secondary metabolites characteristic of *Penicillium citrinum, Penicillium steckii* and related species. *Phytochemistry* 54: 301–309.

Masuma, R., Y. Yamaguchi, M. Noumi, S. Omura, and M. Namikoshi. 2001. Effect of sea water concentration on hyphal growth and antimicrobial metabolite production in marine fungi. *Mycoscience* 42: 455–459.

McDonald, L.A., D.R. Abbanat, L.R. Barbieri, V.S. Bernan, C.M. Discafani, M. Greenstein, K. Janota, J.D. Korshalla, P. Lassota, M. Tischler, and G.T. Carter. 1999. Spiroxins, DNA cleaving antitumor antibiotics from a marine-derived fungus. *Tetrahedron Lett.* 40: 2498–2492.

Mereyala, H.B., M. Joe, and R.R. Gadikota. 2000. Synthesis of harzialactone A and its isomers from D-glucose and assignment of absolute stereochemistry. *Tetrahedron: Asymmetry* 11: 4071–4081.

Miljkovic, A., P.G. Mantle, D.J. Williams, and B. Rassing. 2001. Scorpinone: a new natural azaanthraquinone produced by a *Bispora*-like tropical fungus. *J. Nat. Prod.* 64: 1251–1253.

Miller, J.D., and M.E. Savard. 1989. Antibiotic activity of the marine fungus *Leptosphaeria oraemaris*. *Proc. N. S. Inst. Sci.* 39: 51–58.

Miyaoka, H., Y. Saka, S. Miura, and Y. Yamada. 1996. Total synthesis of phomactin D. *Tetrahedron Lett.* 37: 7107–7110.

Mizushima, Y., S. Kobayashi, K. Kuramochi, S. Nagata, F. Sugawara, and K. Sakaguchi. 2000. Epolactaene, a novel neuritogenic compound in human neuroblastoma cells, selectively inhibits the actives of mammalian DNA polymerases and human DNA topoisomerase II. *Biochem. Biophys. Res. Commun.* 273: 784–788.

Nagai, K., K. Kamigiri, H. Matsumoto, Y. Kawano, M. Yamaoka, H. Shimoi, M. Watanabe, and K. Suzuki. 2002. YM-202204, a new antifungal antibiotic produced by marine fungus *Phoma* sp. *J. Antibiot.* 55: 1036–1041.

Nakamura, H., M. Ono, T. Yamada, A. Numata, and H. Akita. 2002a. Determination of the absolute stereostructure of *seco*-macrosphelide E produced by a fungal strain from a sea hare. *Chem. Pharm. Bull.* 50: 303–306.

Nakamura, H., M. Ono, Y. Shida, and H. Akita. 2002b. New total synthesis of (+)-macrosphelides C, F and G. *Tetrahedron: Asymmetry* 13: 705–713.

Namatame, I., H. Tomoda, M. Arai, and S. Omura. 2000. Effect of fungal metabolites cytochalasans on lipid droplet formation in mouse macrophages. *J. Antibiot.* 53: 19–25.

Namikoshi, M., R. Negishi, H. Nagai, A. Dmitrenok, and H. Kobayashi. 2003. Three new chlorine containing antibiotics from a marine-derived fungus *Aspergillus ostianus* collected in Pohnpei. *J. Antibiot.* 56: 755–761.

Namikoshi, M., K. Akano, S. Meguro, I. Kasuga, Y. Mine, T. Takahashi, and H. Kobayashi. 2001. A new macrocyclic trichothecene, 12,13-deoxyroridin E, produced by the marine-derived fungus *Myrothecium roridum* collected in Palau. *J. Nat. Prod.* 64: 396–398.

Namikoshi, M., H. Kobayashi, T. Yoshimoto, and S. Meguro. 2000a. Paecilospirone, a unique spiro[chroman-2,1′(3′*H*)-isobenzofuran] derivative isolated from tropical marine fungus *Paecilomyces* sp. *Chem. Lett.* 308–309.

Namikoshi, M., H. Kobayashi, T. Yoshimoto, S. Meguro, and K. Akano. 2000b. Isolation and characterization of bioactive metabolites from marine-derived filamentous fungi collected from tropical and sub-tropical coral reefs. *Chem. Pharm. Bull.* 48: 1452–1457.

Namikoshi, M., H. Kobayashi, T. Yoshimoto, and T. Hosoya. 1997. Phomopsidin, a new inhibitor of microtubule assembly produced by *Phomopsis* sp. isolated from coral reef in Pohnpei. *J. Antibiot.* 50: 890–892.

Nielsen, J., P.H. Nielsen, and J.C. Frisvad. 1999. Fungal depside, guisinol, from a marine derived strain of *Emericella unguis*. *Phytochemistry* 50: 263–265.

Nomoto, S., and K. Mori. 1997. Synthesis of acetophthalidin, a fungal metabolite which inhibits the progression of the mammalian cell cycle. *Liebigs Ann. Chem.* 721–723.

Numata, A., T. Amagata, K. Minoura, and T. Ito. 1997a. Gymnastatins, Novel cytotoxic metabolites produced by a fungal strain from a sponge. *Tetrahedron Lett.* 38: 5675–5678.

Numata, A., M. Iritani, T. Yamada, K. Minoura, E. Matsumura, T. Yamori, and T. Tsuruo. 1997b. Novel antitumor metabolites produced by a fungal strain from a sea hare. *Tetrahedron Lett.* 38: 8215–8218.

Numata, A., C. Takahashi, Y. Ito, K. Minoura, T. Yamada, C. Matsuda, and K. Nomoto. 1996. Penochalasins, a novel class of cytotoxic cytochalasans from a *Penicillium* species separated from a marine alga: structure determination and solution conformation. *J. Chem. Soc., Perkin Trans.* 1: 239–245.

Numata, A., C. Takahashi, Y. Ito, T. Takada, K. Kawai, Y. Usami, E. Matsumura, M. Imachi, T. Ito, and T. Hasegawa. 1993. Communesins, cytotoxic metabolites of a fungus isolated from a marine alga. *Tetrahedron Lett.* 34: 2355–2358.

Numata, A., C. Takahashi, T. Matsushita, T. Miyamoto, K. Kawai, Y. Usami, E. Matsumura, M. Inoue, H. Ohishi, and T. Shingu. 1992. Fumiquinazolines, novel metabolites of a fungus isolated from a saltfish. *Tetrahedron Lett.* 23: 1621–1624.

Ohzeki, T., and K. Mori. 2001. Synthesis of corollosporine, an antibacterial metabolite of the marine fungus *Corollospora maritima*. *Biosci. Biotechnol. Biochem.* 65: 172–175.

Okutani, K. 1977. Gliotoxin produced by a strain of *Aspergillus* isolated from marine mud. *Bull. Jpn. Soc. Sci. Fish.* 43: 995–1000.

Omura, S., H. Miyadera, H. Ui, K. Shiomi, Y. Yamaguchi, R. Masuma, T. Nagamitsu, D. Takano, T. Sunazuka, A. Harder, H. Kölbl, M. Namikoshi, H. Miyoshi, K. Sakamoto, and K. Kita. 2001. An anthelmintic compound, nafuredin, shows selective inhibition of complex I in helminth mitochondria. *Proc. Natl. Acad. Sci.*, USA 98: 60–62.

Omura, S., H. Tomoda, N. Tabata, Y. Ohyama, T. Abe, and M. Namikoshi. 1999. Roselipins, novel fungal metabolites having a highly methylated fatty acid modified with a mannose and an arabinitol. *J. Antibiot.* 52: 586–589.

Ono, M., H. Nakamura, F. Konno, and H. Akita. 2000. Total syntheses of macrosphelides (+)-A, (–)-A and (+)-E. *Tetrahedron: Asymmetry* 11: 2753–2764.

Onuki, H., H. Miyashige, H. Hasegawa, and S. Yamashita. 1998. NI15501A, a novel anthranilamide derivative from a marine fungus *Penicillium* sp. *J. Antibiot.* 51: 442–444.

Osterhage, C., G.M. König, U. Höller, and A.D. Wright. 2002a. Rare sesquiterpenes from the algicolous fungus *Drechslera dematioidea*. *J. Nat. Prod.* 65: 306–313.

Osterhage, C., G.M. König, P.G. Jones, and A.D. Wright. 2002b. 5-Hydroxyramulosin, a new natural product produced by *Phoma tropica*, a marine-derived fungus isolated from the alga *Fucus spiralis*. *Planta Med.* 68: 1052–1054.

Osterhage, C., R. Kaminsky, G.M. König, and A.D. Wright, 2000. Ascosalipyrrolidinone A, an antimicrobial alkaloid, from the obligate marine fungus *Ascochyta salicorniae*. *J. Org. Chem.* 65: 6412–6417.

Pallenberg, A.J., and J.D. White. 1986. The synthesis and absolute configuration of (+)-leptosphaerin. *Tetrahedron Lett.* 27: 5591–5594.

Perpelescu, M., J. Kobayashi, M. Furuta, Y. Ito, S. Izuta, M. Takemura, M. Suzuki, and S. Yoshida. 2002. Novel phenalenone derivatives from a marine-derived fungus exhibit distinct inhibition spectra against eukaryotic DNA polymerases. *Biochemistry* 41: 7610–7616.

Pietra, F. 1997. Secondary metabolites from marine microorganisms: bacteria, protozoa, algae and fungi. Achievements and prospects. *Nat. Prod. Rep.* 14: 453–464.

Poch, G.K., and J.B. Gloer. 1991. Auranticins A and B: two new depsidiones from a mangrove isolate of the fungus *Preussis aurantiaca*. *J. Nat. Prod.* 54: 213–217.

Poch, G.K., and J.B. Gloer. 1989a. Obionin A: a new polyketide metabolite from the marine fungus *Leptosphaeria obiones*. *Tetrahedron Lett.* 30: 3483–3486.

Poch, G.K., and J.B. Gloer. 1989b. Helicascolides A and B: new lactones from the marine fungus *Helicascus kanaloanus*. *J. Nat. Prod.* 52: 257–260.

Poonia, N.S., and B.P. Yadav. 1978. Coordinative role of alkali cations in organic synthesis. Selective methylations of 5-hydroxy-2-hydroxymethyl-γ-pyrone. *J. Org. Chem.* 43: 2842–2844.

Proksch, P., R.A. Edrada, and R. Ebel. 2002. Drugs from the sea – current status and microbiological implications. *Appl. Microbiol. Biotechnol.* 59: 125–134.

Rahbæk, L., S. Sperry, J.E. Piper, and P. Crews. 1998. Deoxynortrichoharzin, a new polyketide from the saltwater culture of a sponge-derived *Paecilomyces* fungus. *J. Nat. Prod.* 61: 1571–1573.

Rahbæk, L., C. Christophersen, J. Frisvad, H.S. Bengaard, S. Larsen, and B.R. Rassing. 1997. Insulicolide A: a new nitrobenzoyloxy-substituted sesquiterpene from the marine fungus *Aspergillus insulicola*. *J. Nat. Prod.* 60: 811–813.

Renner, M.K., P.R. Jensen, and W. Fenical. 2000. Mangicols: structures and biosynthesis of a new class of sesterterpene polyols from a marine fungus of the genus *Fusarium*. *J. Org. Chem.* 65: 4843–4852.

Renner, M.K., P.R. Jensen, and W. Fenical. 1998. Neomangicols: structures and absolute stereochemistries of unprecedented halogenated sesterterpenes from a marine fungus of the genus *Fusarium*. *J. Org. Chem.* 63: 8346–8354.

Rollin, P. 1987. A Short Biomimetic Synthesis of Leptosphaerin. *Tetrahedron Lett.* 28: 3813–3814.

Rowley, D.C., S. Kelly, C.A. Kauffman, P.R. Jensen, and W. Fenical. 2003. Halovirs A–E, new antiviral agents from a marine-derived fungus of the genus *Scytalidium*. *Bioorg. Med. Chem.* 11: 4263–4274.

Sakaki, H., H. Kaneno, Y. Sumiya, M. Tsushima, W. Miki, N. Kishimoto, T. Fujita, S. Matsumoto, S. Komemushi, and A. Sawabe. 2002. A new carotenoid glycosyl ester isolated from a marine microorganism, *Fusarium* Strain T-1. *J. Nat. Prod.* 65: 1683–1684.

Schiehser, G.A., J.D. White, G.K. Matsumoto, J.O. Pezzanite, and J. Clardy. 1986. The structure of leptosphaerin. *Tetrahedron Lett.* 27: 5587–5590.

Schkeryantz, J.M., J.C.G. Woo, P. Siliphaivanh, K.M. Depew, and S.J. Danishefsky. 1999. Total synthesis of gypsetin, deoxybrevianamide E, brevianamide E, and tryprostatin S: novel constructions of 2,3-disubstituted indoles. *J. Am. Chem. Soc.* 121: 11964–11975.

Schlingmann, G., L. Milne, and G.T. Carter. 2002. Isolation and identification of antifungal polyesters from the marine fungus *Hypoxylon oceanicum* LL-15G256. *Tetrahedron* 58: 6825–6835.

Schlingmann, G., L. Milne, D.R. Williams, and G.T. Carter. 1998. Cell wall active antifungal compounds produced by the marine fungus *Hypoxylon oceanicum* LL-15G256. II. Isolation and structure determination. *J. Antibiot.* 51: 303–316.

Shao, Z., Y. Lin, G. Jiang, S. Zhou, L.L.P. Vrijmoed, and E.B.G. Jones. 1999. The novel compound with the skeleton of furanopyran from marine fungus *Penicillium* sp. from the South China Sea. *Acta Scientiarum Naturalium Universitatis Sunyatsuni* 38: 131–132.

Shigemori, H., K. Komatsu, Y. Mikami, and J. Kobayashi. 1999. Seragakinone A, a new pentacyclic metabolite from a marine-derived fungus. *Tetrahedron* 55: 14925–14930.

Shigemori, H., M. Tenma, K. Shimazaki, and J. Kobayashi. 1998. Three new metabolites from the marine yeast *Aureobasidium pullulans*. *J. Nat. Prod.* 61: 696–698.

Shigemori, H., S. Wakuri, K. Yazawa, T. Nakamura, T. Sasaki, and J. Kobayashi. 1991. Fellutamides A and B, cytotoxic peptides from a marine fish-possessing fungus *Penicillium fellutanum*. *Tetrahedron* 47: 8529–8534.

Smith, C.J., D. Abbanat, V.S. Bernan, W.M. Maiese, M. Greenstein, J. Jompa, A. Tahir, and C.M. Ireland. 2000. Novel polyketide metabolites from a species of marine fungi. *J. Nat. Prod.* 63: 142–145.

Snider, B.B., and H. Zeng. 2000. Total synthesis of (–)-fumiquinazolines A, B, and I. *Org. Lett.* 2: 4103–4106.

Son, B.W., J.S. Choi, J.C. Kim, K.W. Nam, D.-S. Kim, H.Y. Chung, J.S. Kang, and H.D. Choi. 2002a. Parasitenone, a new epoxycyclohexenone related to gabosine from the marine-derived fungus *Aspergillus parasiticus*. *J. Nat. Prod.* 65: 794–795.

Son, B.W., J.C. Kim, H.D. Choi, and J.S. Kang. 2002b. A radical scavenging farnesylhydroquinone from a marine-derived fungus *Penicillium* sp. *Arch. Pharm. Res.* 25: 77–79.

Son, B.W., P.R. Jensen, C.A. Kauffman, and W. Fenical. 1999. New cytotoxic epidithiodioxopiperazines related to verticillin A from a marine isolate of the fungus *Penicillium*. *Nat. Prod. Lett.* 13: 213–222.

Sperry, S., G.J. Samuels, and P. Crews. 1998. Vertinoid polyketides from the saltwater culture of the fungus *Trichoderma longibrachitum* separated from a *Haliclona* marine sponge. *J. Org. Chem.* 63: 10011–10014.

Suenaga, K., S. Aoyama, W. Xi, H. Arimoto, K. Yamaguchi, K. Yamada, T. Tsuji, A. Yamada, and D. Uemura. 2000. Isolation and structure of kasarin, a novel azetinone compound, isolated from a marine microorganism. *Heterocycles* 52: 1033–1036.

Sugano, M., A. Sato, K. Saito, S. Takaishi, Y. Matsushita, and Y. Iijima. 1996. Structure–activity relationships of phomactin derivatives as platelet activating factor antagonists. *J. Med. Chem.* 39: 5281–5284.

Sugano, M., A. Sato, Y. Iijima, K. Furuya, H. Kuwano, and T. Hata. 1995. Phomactin E, F, and G: new phomactin-group PAF antagonists from a marine fungus *Phoma* sp. *J. Antibiot.* 48: 1188–1190.

Sugano, M., A. Sato, Y. Iijima, K. Furuya, H. Haruyama, K. Yoda, and T. Hata. 1994. Phomactins, novel PAF antagonists from a marine fungus *Phoma* sp. *J. Org. Chem.* 59: 564–569.

Sugano, M., A. Sato, Y. Iijima, T. Oshima, K. Furuya, H. Kuwano, T. Hata, and H. Hanzawa. 1991. Phomactin A: a novel PAF antagonist from a marine fungus *Phoma* sp. *J. Am. Chem. Soc.* 113: 5463–5464.

Sunazuka, T., T. Hirose, Y. Harigaya, S. Takamatsu, M. Hayashi, K. Komiyama, S. Omura, P.A. Sprengeler, and A.B. Smith, III. 1997. Relative and absolute stereochemistries and total synthesis of (+)-macrosphelides A and B, potent, orally bioavailable inhibitor of cell-cell adhesion. *J. Am. Chem. Soc.* 119: 10247–10248.

Suzuki, T., K. Usui, Y. Miyake, M. Namikoshi, and M. Nakada. 2004. First total synthesis of antimitotic compounds, (+)-phomopsidin. *Org. Lett.* 6: 553–556.

Tabata, N., Y. Ohyama, H. Tomoda, T. Abe, M. Namikoshi, and S. Omura. 1999. Structure elucidation of roselipins, inhibitors of diacylglycerol acyltransferase produced by *Gliocladium roseum* KF-1040. *J. Antibiot.* 52: 815–826.

Takahashi, C., A. Numata, T. Yamada, K. Minoura, S. Enomoto, K. Konishi, M. Nakai, C. Matsuda, and K. Nomoto. 1996. Penostatins, novel cytotoxic metabolites from a *Penicillium* species separated from a green alga. *Tetrahedron Lett.* 37: 655–658.

Takahashi, C., Y. Takai, Y. Kimura, A. Numata, N. Shigematsu, and H. Tanaka. 1995a. Cytotoxic metabolites from a fungal adherent of a marine alga. *Phytochemistry* 38: 155–158.

Takahashi, C., K. Minoura, T. Yamada, A. Numata, K. Kushida, T. Shingu, S. Hagishita, H. Nakai, T. Sato, and H. Harada. 1995b. Potent cytotoxic metabolites from a *Leptosphaeria* species. Structure determination and conformational analysis. *Tetrahedron* 51: 3483–3498.

Takahashi, C., T. Matsushita, M. Doi, K. Minoura, T. Shingu, Y. Kumeda, and A. Numata. 1995c. Fumiquinazolines A–G, novel metabolites of a fungus separated from a *Pseudolabrus* marine fish. *J. Chem. Soc., Perkin Trans. 1* 2345–2353.

Takahashi, C., A. Numata, Y. Ito, E. Matsumura, H. Araki, H. Iwaki, and K. Kushida. 1994a. Leptosins, antitumor metabolites of a fungus isolated from a marine alga. *J. Chem. Soc., Perkin Trans. 1* 1859–1864.

Takahashi, C., A. Numata, E. Matsumura, K. Minoura, H. Eto, T. Shingu, T. Ito, T. Hasegawa. 1994b. Leptosins I and J, cytotoxic substances produced by a *Leptosphaeria* sp. Physico-chemical properties and structures. *J. Antibiot.* 47: 1242–1249.

Takamatsu, S., H. Hiraoka, Y.-P. Kim, M. Hayashi, M. Natori, K. Komiyama, and S. Omura. 1997. Macrosphelides C and D, novel inhibitors of cell adhesion. *J. Antibiot.* 50: 878–880.

Takamatsu, S., Y.-P. Kim, M. Hayashi, H. Hiraoka, M. Natori, K. Komiyama, and S. Omura. 1996. Macrosphelide, a novel inhibitor of cell-cell adhesion molecule. II. Physico-chemical properties and structural elucidation. *J. Antibiot.* 49: 95–98.

Takano, D., T. Nagamitsu, H. Ui, K. Shiomi, Y. Yamaguchi, R. Masuma, I. Kuwajima, and S. Omura. 2001a. Absolute configuration of nafuredin, a new specific NADH-fumarate reductase inhibitor. *Tetrahedron Lett.* 42: 3017–3020.

Takano, D., T. Nagamitsu, H. Ui, K. Shiomi, Y. Yamaguchi, R. Masuma, I. Kuwajima, and S. Omura. 2001b. Total synthesis of nafuredin, a selective NADH-fumarate reductase inhibitor. *Org. Lett.* 3: 2289–2291.

Tan, L.T., X.C. Cheng, P.R. Jensen, and W. Fenical. 2003. Scytalidamides A and B, new cytotoxic cyclic heptapeptides from a marine fungus of the genus *Scytalidium*. *J. Org. Chem.* 68: 8767–8773.

Tomoda, H., I. Namatame, S. Si, K. Kawaguchi, R. Masuma, M. Namikoshi, and S. Omura. 1999a. Phenochalasins, inhibitors of lipid droplet formation in mouse macrophages, produced by *Phomopsis* sp. FT-0211. *J. Antibiot.* 52: 851–856.

Tomoda, H., I. Namatame, N. Tabata, K. Kawaguchi, S. Si, and S. Omura. 1999b. Structure elucidation of fungal phenochalasins, novel inhibitors of lipid droplet formation in mouse macrophages. *J. Antibiot.* 52: 857–861.

Tomoda, H., Y. Ohyama, T. Abe, N. Tabata, M. Namikoshi, Y. Yamaguchi, R. Masuma, and S. Omura. 1999c. Roselipins, inhibitors of diacylglycerol acyltransferase, produced by *Gliocladium roseum* KF-1040. *J. Antibiot.* 52: 689–694.

Toske, S.G., P.R. Jensen, C.A. Kauffman, and W. Fenical. 1998. Aspergillamides A and B: modified cytotoxic tripeptides produced by a marine fungus of genus *Aspergillus*. *Tetrahedron* 54: 13459–13466.

Tsuda, M., T. Mugishima, K. Komatsu, T. Sone, M. Tanaka, Y. Mikami, M. Shiro, M. Hirai, Y. Ohizumi, and J. Kobayashi. 2003a. Speradine A, a new pentacyclic oxindole alkaloid from a marine-derived fungus *Aspergillus tamarii*. *Tetrahedron* 59: 3227–3230.

Tsuda, M., T. Mugishima, K. Komatsu, T. Sone, M. Tanaka, Y. Mikami, and J. Kobayashi. 2003b. Modiolides A and B, two new 10-membered macrolides from a marine-derived fungus. *J. Nat. Prod.* 66: 412–415.

Tsukamoto, S., S. Miura, Y. Yamashita, and T. Ohta. 2004. Aspermytin A: a new neurotrophic polyketide isolated from a marine-derived fungus of the genus *Aspergillus*. *Bioorg. Med. Chem. Lett.* 14: 417–420.

Uchida, R., H. Tomoda, M. Arai, and S. Omura. 2001. Chlorogentisylquinone, a new neutral sphingomyelinase inhibitor, produced by a marine fungus. *J. Antibiot.* 54: 882–889.

Ui, H., K. Shiomi, Y. Yamaguchi, R. Masuma, T. Nagamitsu, D. Takano, T. Sunazuka, M. Namikoshi, and S. Omura. 2001. Nafuredin, a novel inhibitor of NADH-fumarate reductase, produced by *Aspergillus niger* FT-0554. *J. Antibiot.* 54: 234–238.

Usami, Y., and A. Numata. 1999. Total synthesis of trichodenones A–C. *Synlett* 723–724.

Usami, Y., S. Aoki, T. Hara, and A. Numata. 2002. New dioxopiperazine metabolites from a *Fusarium* species separated from a marine alga. *J. Antibiot.* 55: 655–659.

Usami, Y., T. Ikura, T. Amagata, and A. Numata. 2000. First total syntheses and configurational assignments of cytotoxic trichodenones A–C. *Tetrahedron: Asymmetry* 11: 3711–3725.

Usui, T., M. Kondoh, C.-B. Cui, T. Mayumi, and H. Osada. 1998. Tryprostatin A, a specific and novel inhibitor of microtubule assembly. *Biochem. J.* 333: 543–548.

Varoglu, M., and P. Crews. 2000. Biosynthetically diverse compounds from a saltwater culture of sponge-derived *Aspergillus niger*. *J. Nat. Prod.* 63: 41–43.

Varoglu, M., T.H. Corbett, F.A. Valeriote, and P. Crews. 1997. Asperazine, a selective cytotoxic alkaloid from a sponge-derived culture of *Aspergillus niger*. *J. Org. Chem.* 62: 7078–7079.

Verbist, J.-F., C. Sallenave, and Y.-F. Pouchus. 2000. Marine fungal substances. Pages 979–1092 in *Studies in Natural Products Chemistry, Vol. 24*, A. Rahman, ed., Elsevier Science, New York.

Verbitski, S.M., C.L. Mayne, R.A. Davis, G.P. Concepcion, and C.M. Ireland. 2002. Isolation, structure determination, and biological activity of a novel alkaloid, perophoramidine, from the Philippine ascidian *Perophora namei*. *J. Org. Chem.* 67: 7124–7126.

Vongvilai, P., M. Isaka, P. Kittakoop, P. Srikitikulchai, P. Kongsaeree, and Y. Thebtaranonth. 2004. Ketene acetal and spiroacetal constituents of the marine fungus *Aigialus parvus* BCC 5311. *J. Nat. Prod.* 67: 457–460.

Wan, J., Y. Lin, X. Wu, S. Zhou, and L.L.P. Vrijmoed. 2001. Avicennin A, a new isocoumarine from mangrove endophytic fungus no. 2533 from the South China Sea. *Acta Scientiarum Naturalium Universitatis Sunyatsuni* 40: 127.

Wang, C.-Y., B.-G. Wang, G. Brauers, H.-S. Guan, P. Proksch, and R. Ebel. 2002. Microsphaerones A and B, two novel γ-pyrone derivatives from the sponge-derived fungus *Microsphaeropsis* sp. *J. Nat. Prod.* 65: 772–775.

Wang, G.-Y.-S., L.M. Abrell, A. Avelar, B.M. Borgeson, and P. Crews. 1998. New hirsutane based sesquiterpenes from salt water cultures of a marine sponge-derived fungus and the terrestrial fungus *Coriolus consors*. *Tetrahedron* 54: 7335–7342.

Wang, G.-Y.-S., B.M. Borgeson, and P. Crews. 1997. Pitholides A-D, polyketides from a marine tunicate-derived culture of *Pithomyces* sp. *Tetrahedron Lett.* 38: 8449–8452.

Wang, H., and A. Ganesan. 2000. Total synthesis of the fumiquinazoline alkaloids: solution-phase studies. *J. Org. Chem.* 65: 1022–1030.

Wang, H.-J., K.B. Gloer, and J.B. Gloer. 1997. Anserinones A and B: new antifungal and antibacterial benzoquinones from the coprophilous fungus *Podospora anserina*. *J. Nat. Prod.* 60: 629–631.

Wang, T., O. Shirota, K. Nakanishi, N. Berova, L.A. McDonald, L.R. Barbieri, and G.T. Carter. 2001. Absolute stereochemistry of the spiroxins. *Can. J. Chem.* 79: 1786–1791.

Watanabe, N., A. Fujita, N. Ban, A. Yagi, H. Etoh, K. Ina, and K. Sakata. 1995. Bipolal: antimicroalgal compound isolated as a candidate for marine antifouling produced by *Bipolaris* sp. F5206. *J. Nat. Prod.* 58: 463–466.

Wei, H., T. Itoh, M. Kinoshita, Y. Nakai, M. Kurotaki, and M. Kobayashi. 2004. Cytotoxic sesterterpenes, 6-epi-ophiobolin g and 6-epi-ophiobolin n, from marine derived fungus *Emericella variecolor* GF10. *Tetrahedron* 60: 6015–6019.

White, J.D., R.A. Badger, H.S. Kezar, III, A.J. Pallenberg, and G.A. Schiehser. 1989. Structure, synthesis and absolute configuration of leptosphaerin, a metabolite of the marine ascomycete *Leptosphaeria oraemaris*. *Tetrahedron* 45: 6631–6644.

Wood, J.L., B.D. Thompson, N. Yusuff, D.A. Pflum, and M.S.P. Mattäus. 2001. Total synthesis of (±)-epoxysorbicillinol. *J. Am. Chem. Soc.* 123: 2097–2098.

Wright, A.D., C. Osterhage, and G.M. König. 2003. Epicoccamide, a novel secondary metabolite from a jellyfish-derived culture of *Epicoccum purpurascens*. *Org. Biomol. Chem.* 1: 507–510.

Xu, X., F. S. de Guzman, J.B. Gloer, and C.A. Shearer. 1992. Stachybotrins A and B: novel bioactive metabolites from a brackish water isolate of the fungus *Stachybotrys* sp. *J. Org. Chem.* 57: 6700–6703.

Yamada, T., C. Iwamoto, N. Yamagaki, T. Yamanouchi, K. Minoura, S. Hagishita, and A. Numata. 2004. Leptosins O–S, cytotoxic metabolites of a strain of *Leptosphaeria* sp. isolated from a marine alga. *Heterocycles* 63: 641–653.

Yamada, T., C. Iwamoto, N. Yamagaki, T. Yamanouchi, K. Minoura, T. Yamori, Y. Uehara, T. Andoh, K. Umemura, and A. Numata. 2002a. Leptosins M–N_1, cytotoxic metabolites from a *Leptosphaeria* species separated from a marine alga. structure determination and biological activities. *Tetrahedron* 58: 479–487.

Yamada, T., M. Iritani, K. Minoura, A. Numata, Y. Kobayashi, and Y.-G. Wang. 2002b. Absolute stereostructures of cell adhesion inhibitors, macrosphelides H and L, from *Periconia byssoides* OUPS-N133. *J. Antibiot.* 55: 147–154.

Yamada, T., M. Iritani, M. Doi, K. Minoura, T. Ito, and A. Numata. 2001. Absolute stereostructures of cell-adhesion inhibitors, macrosphelides C, E–G and I, produced by a *Periconia* species separated from an *Aplysia* sea hare. *J. Chem. Soc., Perkin Trans.* 1: 3046–3053.

Yamaguchi, K., T. Tsuji, S. Wakui, K. Yazawa, K. Kondo, H. Shigemori, and J. Kobayashi. 1993. Stimulation of nerve growth factor synthesis and secretion by fellutamide A *in vitro*. *Biosci. Biotech. Biochem.* 57: 195–199.

Yamaguchi, Y., R. Masuma, R. Uchida, M. Arai, H. Tomoda, and S. Omura. 2002. *Phoma* sp. FOM-8108, a producer of gentisylquinones, isolated from sea sand. *Mycoscience* 43: 127–133.

Yang, S.-W., and G.A. Cordell. 1997. Metabolism studies of indole derivatives using a staurosporine producer, *Streptomyces staurosporeus*. *J. Nat. Prod.* 60: 44–48.

Yin, W., Y. Lin, S. Zhou, and L.L.P. Vrijmoed. 2002. The cyclopeptides from marine fungus 2516 from the South China Sea. *Acta Scientiarum Naturalium Universitatis Sunyatsuni* 41: 56.

Yu, C.M., J.M. Curtis, J.A. Walter, J.L.C. Wright, S.W. Ayer, J. Kaleta, L. Querengesser, Z.R. Fathi-Afshar. 1996a. Potent inhibitors of cysteine proteases from the marine fungus *Microascus longirostris*. *J. Antibiot.* 49: 395–397.

Yu, C.-M., J.M. Curtis, J.L.C. Wright, S.W. Ayer, and Z.R. Fathi-Afshar. 1996b. An unusual fatty acid and its glyceride from the marine fungus *Microsphaeropsis olivacea*. *Can. J. Chem.* 74: 730–735.

Zhang, H., H. Tomoda, N. Tabata, H. Miura, M. Namikoshi, Y. Yamaguchi, R. Masuma, and S. Omura. 2001. Cladospolide D, a new 12-membered macrolide antibiotic produced by *Cladosporium* sp. FT-0012. *J. Antibiot.* 54: 635–641.

Zhao, S., T. Gan, P. Yu, and J.M. Cook. 1998. Total synthesis of tryprostatin A and B as well as their enantiomers. *Tetrahedron Lett.* 39: 7009–7012.

Organism Index

(New natural products isolated from marine fungi up to June, 2004)

Fungus	*Substrate*	*Structure #*
Acremonium cf. *roseogriseum* M1-11-1	brown alga	**249, 250, 262, 263, 264, 265, 266, 267**
Acremonium neocaledoniae LCP 98 4255	driftwood	**49**
Acremonium sp. CNC 890	tunicate	**201, 202, 261, 268, 269**
Acremonium sp. FK-N30	alga	**272, 273**
Acremonium striatisporum KMM 4401	sea cucumber	**124, 125, 126, 127, 128, 135, 136, 137, 138**
Aigialus parvus BCC 5311	mangrove	**207, 274, 275, 276, 277, 278, 279**
Altenaria sp. MFA 898	green alga	**280**
Apiospora montagnei	red alga	**281, 282, 283, 284**
Ascochyta salicorniae	green alga	**204, 285, 286**
Aspergillus carneus MST-MF156	sediment	**129, 130, 131, 132, 133**
Aspergillus cf. *ochraceus* 941026	sponge	**287, 288**
Aspergillus fumigatus	fish	**101, 102, 103, 104, 105, 106, 107**
Aspergillus fumigatus BM939	sediment	**87, 88, 89, 90, 91, 92, 93, 94**
Aspergillus insulicola	algae	**45**
Aspergillus niger	sponge	**134**
Aspergillus niger 94-1212	sponge	**287**
Aspergillus niger F97S11	tunicate	**176, 177, 178, 179, 180, 181, 182, 183**
Aspergillus niger FT-0554	sponge	**229**
Aspergillus ochraceus	sediment	**290**
Aspergillus ostianus TUF 01F313	sponge	**164, 165, 166**
Aspergillus parasiticus MFA 153	red alga	**252**
Aspergillus sp. CNC-120	sediment	**112, 113**
Aspergillus sp. CNC-688	brown alga	**203**
Aspergillus sp. CNM-713	n/a	**143**
Aspergillus sp. MF297	bivalve	**259**
Aspergillus sp. MFA 212	red alga	**254, 255**
Aspergillus tamarii M143	driftwood	**149**
Aspergillus terreus MFA 460	red alga	**257**
Aspergillus versicolor	sponge	**291, 292, 293, 294**
Aspergillus versicolor CNC 327	green alga	**301, 302, 303**
Aspergillus versicolor HBI-2	sponge	**163, 295, 296, 297, 298, 299, 300**
Asteromyces cruciatus H5-81	sponge	**304**
Aureobasidium pullulans (yeast)	sponge	**305, 306, 307**
Bipolaris sp. F5206	leaf	**214**
Bispora-like fungus NN 6654	sediment	**308**
Cladosporium herbarum	sponge	**144, 145, 152, 309, 310, 311**
Cladosporium sp. FT-0012	sponge	**184**
Coniothyrium sp. 193H77	sponge	**187, 312**
Corollospora maritima 1069	driftwood	**168**
Corollospora pulchella ATCC 62554	driftwood	**240**
Curvularia lunata	sponge	**167**
Dendryphiella salina	n/a	**313, 314, 315, 316, 317, 318, 319, 320, 321, 322, 323, 324, 325**

(*Contd.*)

(*Contd.*)

Drechslera dematioidea MOR5 test 2-1	red alga	**205, 206, 326, 327, 328, 329, 330, 331, 332, 333**
Drechslera hawaiiensis P22-96	sponge	**334, 335**
Emericella unguis M87-2, M90B-10	medusa, mollusc	**154, 155, 156**
Emericella unguis M90A-2	mollusc	**336, 337**
Emericella variecolor GF10	sediment	**141, 142**
Emericella variecolor M75-2	sponge	**86, 162, 338, 339**
Emericella variecolor E-00-6/3	sponge	**74, 340**
Epicoccum purpurascens N7-9	jellyfish	**341**
Epicoccum sp.	brown alga	**253**
Eutypa sp. 424	mangrove	**342**
Exophiala pisciphila NI10102	sponge	**148**
Fusarium chlamydosporum OUPS-N124	red alga	**125, 343**
Fusarium heterosporum CNC-477	driftwood	**76, 77, 114, 115, 116, 117, 118, 119, 120, 344**
Fusarium sp. CNL 292	seagrass	**78**
Fusarium sp. CNL-619	green alga	**79**
Fusarium sp. FE-71-1	alga	**213, 270, 271**
Fusarium sp. MOBCOF-1	green alga	**345**
Fusarium sp. T-1	seawater	**346**
Gliocladium roseum KF-1040	red alga	**230, 231, 232, 233**
Gymnascella dankaliensis OUPS-N134	sponge	**36, 37, 38, 71, 108, 109, 110, 111**
Halorosellinia oceanica BCC 5149	n/a	**169, 208, 347, 348, 349, 350, 351**
Helicascus kanaloanus ATCC 18591	mangrove	**352, 353**
Hortaea werneckii	sponge	**354**
Humicola grisea 880M/272	driftwood	**69**
Hyphomycetes sp.	zoanthid	**139**
Hypoxylon croceum M97-25	driftwood	**199, 355**
Hypoxylon oceanicum 326	n/a	**356**
Hypoxylon oceanicum LL-15G256	mangrove	**190, 191, 192, 193, 194, 357, 358, 359, 360, 361**
Kallichroma tethys 6175	n/a	**362**
Keissleriella sp. YS4108	sediment	**197, 198**
Kirschsteiniothelia maritima No. M9	wood	**147**
Leptosphaeria obiones SAP 14	grass	**248**
Leptosphaeria oraemaris	n/a	**363**
Leptosphaeria oraemaris	wood	**364**
Leptosphaeria oraemaris	stem of plant	**365**
Leptosphaeria sp. OUPS-N80	brown alga	**6, 7, 8, 9, 10, 11, 12, 13, 14, 15, 16, 17, 18, 19, 20, 21, 22, 23, 24, 25, 26, 27, 28, 29**
Lignincola laevis	grass	**41**
Microascus longirostris SF-73	sponge	**245**
Microsphaeropsis olivacea SF-10	sponge	**366, 367**
Microsphaeropsis sp. H5-50	sponge	**188**
Microsphaeropsis sp. KMPB W-22	sponge	**234, 235, 236, 237, 368, 369, 370**

(*Contd.*)

(*Contd.*)

Myrothecium roridum TUF 98F42	wood	**32**
Myrothecium verrucaria 973023	sponge	**33, 34, 35**
Paecilomyces cf. *javanica* 961331	sponge	**371**
Paecilomyces sp. TUF 94F79	sea water	**260**
Paraphaeosphaeria sp. N-119	bivalve	**170, 171, 372**
Penicillium brevicompactum	sponge	**373, 374**
Penicillium brocae F97S76	sponge	**121, 122, 375**
Penicillium cf. *montanense* HBI-3/D	sponge	**200, 376, 377**
Penicillium citrinum 991084	sponge	**160, 161**
Penicillium corylophilum a004181 and b004181	sediment	**83, 84, 85, 378, 379**
Penicillium fellutanum	fish	**43, 44**
Penicillium sp.	sponge	**81, 82**
Penicillium sp. 386	sand	**380, 381**
Penicillium sp. BM 1689-P	sediment	**258**
Penicillium sp. BM923	sediment	**80**
Penicillium sp. CNC-350	green alga	**30, 31, 382**
Penicillium sp. K029	bivalve	**383**
Penicillium sp. K036	bivalve	**241, 242**
Penicillium sp. MFA 577	polymeric cord	**75, 256, 384**
Penicillium sp. NI15501	sediment	**385**
Penicillium sp. OUPS-N79	green alga	**50, 51, 52, 53, 54, 55, 56, 57, 58, 59, 60, 61, 62, 63, 64, 65, 66, 67, 68**
Penicillium steckii M23B-7	tunicate	**386, 387, 388**
Penicillium waksmanii OUPS-N127	brown alga	**70, 389**
Periconia byssoides OUPS-N133	sea hare	**39, 40, 95, 96, 97, 98, 99, 100, 390**
Pestalotia sp. CNL-365	brown alga	**146**
Phoma sp. CNC-651	microbial mat	**391, 392**
Phoma sp. FOM-8108	sand	**244**
Phoma sp. Q60596	sponge	**196**
Phoma sp. SANK 11486	shell of crab	**220, 221, 222, 223, 224, 225, 226, 227, 228**
Phoma tropica AZ14-4	brown alga	**393**
Phomopsis sp. FT-0211	sponge	**246, 247**
Phomopsis sp. TUF 95F47	mangrove	**189**
Pithomyces sp. 969053	tunicate	**394, 395, 396, 397**
Preussia aurantiaca ATCC 14745	mangrove sediment	**150, 151**
Remispora maritima 22G295	wood	**398**
Scytalidium sp. CNC-310	green alga	**72, 73**
Scytalidium sp. CNC-389	plant	**211, 212**
Scytalidium sp. CNL240	seagrass	**215, 216, 217, 218, 219**
Stachybotrys sp. CS-710-1	twig	**172, 173**
Trichoderma harzianum	sponge	**399**
Trichoderma harzianum OUPS-N115	sponge	**46, 47, 48, 140, 400**
Trichoderma longibrachiatum 961015	sponge	**401**
Trichoderma virens CNK266, CNL910	green alga, tunicate	**42, 402**
Ulocladium botrytis 193A4	sponge	**239**

(*Contd.*)

(*Contd.*)

Varicosporina ramulosa 195-31	brown alga	**185, 186**
Verruculina enalia No. 2606	mangrove	**403, 404**
Wardomyces anomalus OS4T3-2-1	green alga	**251, 405**
Xylaria sp. 2508	seed of mangrove	**243, 406, 407, 408, 409, 410, 411**
Zopfiella latipes CBS 611.97	sediment	**174, 175**
Zopfiella marina SANK 21274	sediment	**412**
unidentified fungus	red alga	**413, 414**
unidentified fungus 1893	mangrove	**415, 416**
unidentified fungus 2516	sediment	**417, 418, 419**
unidentified fungus 2524	seed of mangrove	**420, 421**
unidentified fungus 2533	mangrove	**422**
unidentified fungus 92902	sponge	**423, 424, 425**
unidentified fungus 92948	sponge	**426, 427**
unidentified fungus 95-1005C	sponge	**158, 159, 428, 429**
unidentified fungus 951014	sponge	**157**
unidentified fungus CNC-159	green alga	**209, 210, 430**
unidentified fungus I96S215	sponge	**431, 432, 433, 434**
unidentified fungus K063	red alga	**153, 238, 435**
unidentified fungus LL-37H248	soft coral	**1, 2, 3, 4, 5**
unidentified fungus MF001	brown alga	**436**
unidentified fungus MFA 936	red alga	**437**
unidentified fungus M-3	red alga	**195**

Formula Index

(New natural products isolated from marine fungi up to June 2004)

Formula	*Structure #*	*Formula*	*Structure #*	*Formula*	*Structure #*
$C_7H_5O_3Cl$	**244**	$C_{12}H_{14}N_2O$	**437**	$C_{16}H_{12}O_7$	**235**
$C_7H_8O_2$	**46**	$C_{12}H_{15}N_3O_3$	**385**	$C_{16}H_{12}O_8$	**375**
$C_7H_8O_4$	**252**	$C_{12}H_{16}O_4$	**386**	$C_{16}H_{13}NO_4$	**308**
$C_7H_9O_2Cl$	**48**	$C_{12}H_{16}O_5$	**70, 379**	$C_{16}H_{14}O_7$	**239**
$C_7H_9O_3Cl$	**47**	$C_{12}H_{18}O_4$	**184**	$C_{16}H_{14}O_8$	**121**
$C_7H_{10}O_2$	**304**	$C_{12}H_{20}O_3$	**352, 353, 372**	$C_{16}H_{16}O_4$	**297**
$C_7H_{10}O_3$	**140**	$C_{12}H_{20}O_4$	**325, 347, 431**	$C_{16}H_{18}O_4$	**295, 299**
$C_8H_8O_5$	**152**	$C_{12}H_{20}O_5$	**283, 432**	$C_{16}H_{19}N_3O_2S_2$	**382**
$C_8H_{10}O_4$	**280**	$C_{12}H_{22}O_4$	**434**	$C_{16}H_{20}O_5$	**416**
$C_8H_{11}NO_5$	**363**	$C_{13}H_{12}O_3$	**406**	$C_{16}H_{22}O_4$	**188**
$C_8H_{11}O_5Cl$	**39**	$C_{13}H_{14}O_5$	**169, 348**	$C_{16}H_{22}O_5$	**157**
$C_8H_{12}O_3$	**338**	$C_{13}H_{16}N_2O_3S_2$	**123**	$C_{16}H_{22}O_8$	**95, 96, 97, 100**
$C_8H_{12}O_4$	**264**	$C_{13}H_{18}O_4$	**285**	$C_{16}H_{24}O_8$	**99**
$C_8H_{16}O_4$	**383**	$C_{13}H_{20}O_3$	**282**	$C_{16}H_{26}O_3$	**205, 259**
$C_9H_8O_5$	**253**	$C_{14}H_{10}O_6$	**251**	$C_{16}H_{26}O_5$	**278**
$C_9H_{12}O_3$	**307, 427**	$C_{14}H_{10}O_7$	**405**	$C_{16}H_{26}O_7S$	**309**
$C_9H_{12}O_4$	**355**	$C_{14}H_{14}O_3$	**415**	$C_{16}H_{28}O_4$	**289**
$C_9H_{13}NO$	**240**	$C_{14}H_{16}O_5$	**376, 401**	$C_{16}H_{30}O_4$	**421**
$C_9H_{13}O_4Cl$	**164, 165, 166,**	$C_{14}H_{16}O_6$	**200, 377**	$C_{16}H_{36}N_4O_4S_2P_2$	**41**
	287, 288	$C_{14}H_{16}N_2O_3$	**305**	$C_{17}H_{13}N_3O_3$	**290**
$C_9H_{13}O_6$	**40**	$C_{14}H_{18}O_3$	**335**	$C_{17}H_{14}O_3$	**342**
$C_9H_{14}O_2$	**314**	$C_{14}H_{18}O_4$	**168**	$C_{17}H_{14}O_7$	**236**
$C_9H_{14}O_3$	**315**	$C_{14}H_{20}O_4$	**334, 391**	$C_{17}H_{14}O_8$	**122, 237**
$C_9H_{14}O_4$	**265, 266, 267**	$C_{14}H_{21}O_3Cl$	**423**	$C_{17}H_{21}N_3O_5$	**261, 268**
$C_{10}H_7O_6$	**311**	$C_{14}H_{22}O_6$	**185, 186**	$C_{17}H_{21}NO_3$	**430**
$C_{10}H_8O_5$	**80**	$C_{14}H_{22}O_6S$	**433**	$C_{17}H_{21}NO_4$	**209, 210**
$C_{10}H_{10}O_3$	**404**	$C_{14}H_{24}O_2$	**327, 329**	$C_{17}H_{21}NO_5$	**370**
$C_{10}H_{10}O_4$	**403**	$C_{15}H_{10}O_6$	**167**	$C_{17}H_{22}O_7$	**250**
$C_{10}H_{10}O_5$	**145**	$C_{15}H_{12}O_6$	**398**	$C_{17}H_{26}O_9$	**390**
$C_{10}H_{12}O_3$	**187, 426**	$C_{15}H_{13}O_4$	**298**	$C_{18}H_{22}O_9$	**351**
$C_{10}H_{14}O_3$	**171**	$C_{15}H_{13}NO_2$	**163**	$C_{18}H_{22}N_2O_4$	**257**
$C_{10}H_{14}O_4$	**170, 198, 393**	$C_{15}H_{16}O_4$	**296, 300**	$C_{18}H_{23}O_6$	**207**
$C_{10}H_{14}O_6$	**380**	$C_{15}H_{18}O_3$	**158, 428**	$C_{18}H_{23}N_3O_5$	**269**
$C_{11}H_9O_5Cl$	**422**	$C_{15}H_{18}O_4$	**321, 410**	$C_{18}H_{24}O_4$	**388**
$C_{11}H_{12}O_2$	**249**	$C_{15}H_{18}O_6$	**147**	$C_{18}H_{24}O_8$	**98**
$C_{11}H_{12}O_3$	**262, 263, 400**	$C_{15}H_{20}O_3$	**429**	$C_{18}H_{24}O_9$	**359**
$C_{11}H_{12}O_4$	**389**	$C_{15}H_{20}O_4$	**356**	$C_{18}H_{24}N_2O_4S$	**343**
$C_{11}H_{14}O_5$	**84, 85**	$C_{15}H_{20}O_5$	**86, 339**	$C_{18}H_{25}O_4Cl$	**412**
$C_{11}H_{16}O_3$	**378**	$C_{15}H_{22}O_3$	**328, 330, 331**	$C_{18}H_{25}N_3O_6$	**245**
$C_{11}H_{16}O_3Br_2$	**413, 414**	$C_{15}H_{22}O_6$	**392**	$C_{18}H_{26}O_3$	**387**
$C_{11}H_{18}O_2$	**312**	$C_{15}H_{23}N_3O_5$	**139**	$C_{19}H_{20}N_2O_4$	**203**
$C_{11}H_{18}O_4$	**281**	$C_{15}H_{24}O_3$	**206, 326, 332,**	$C_{19}H_{21}N_3O_4$	**254**
$C_{11}H_{18}N_2O_3$	**306**		**333**	$C_{19}H_{22}O_8$	**274, 277, 279**
$C_{12}H_{12}O_4$	**69**	$C_{15}H_{26}O_2$	**362**	$C_{19}H_{24}O_5$	**197**
$C_{12}H_{12}O_5$	**144**	$C_{15}H_{26}O_3$	**159**	$C_{19}H_{24}O_7$	**276**
$C_{12}H_{14}O_5$	**83**	$C_{16}H_{12}O_6$	**74**	$C_{19}H_{24}O_8$	**275**

(*Contd.*)

(Contd.)

$C_{19}H_{24}O_9$	**350**	$C_{22}H_{32}O_3$	**50, 51, 54, 55,**	$C_{25}H_{40}O_4$	**118**
$C_{19}H_{36}O_2$	**366, 367**		**58, 75**	$C_{25}H_{40}O_5$	**116**
$C_{20}H_{10}O_7$	**3**	$C_{22}H_{32}O_4$	**183, 229**	$C_{25}H_{42}O_4$	**117**
$C_{20}H_{10}O_8Cl_2$	**5**	$C_{22}H_{34}O_3$	**53**	$C_{25}H_{42}O_5$	**114, 115**
$C_{20}H_{12}O_6$	**354**	$C_{22}H_{34}O_4$	**384**	$C_{26}H_{26}O_{12}$	**153**
$C_{20}H_{12}O_7$	**4**	$C_{22}H_{36}O_6$	**234**	$C_{26}H_{28}O_8$	**162**
$C_{20}H_{19}N_2O_8Cl$	**42**	$C_{22}H_{40}N_4O_4$	**417**	$C_{26}H_{36}O_5$	**320**
$C_{20}H_{20}O_8$	**241**	$C_{23}H_{25}O_5Cl$	**156**	$C_{26}H_{36}O_6$	**411**
$C_{20}H_{20}O_9$	**242**	$C_{23}H_{27}N_3O_5$	**92**	$C_{26}H_{37}NO_6$	**175**
$C_{20}H_{20}N_2O_9$	**381, 402**	$C_{23}H_{30}O_5$	**291**	$C_{26}H_{38}O_8$	**137, 138**
$C_{20}H_{22}O_5$	**365**	$C_{23}H_{31}NO_4$	**173**	$C_{26}H_{40}O_8$	**127, 136**
$C_{20}H_{22}O_6$	**284**	$C_{23}H_{31}NO_4Cl_2$	**36**	$C_{26}H_{40}O_9$	**124**
$C_{20}H_{26}O_5$	**408, 409**	$C_{23}H_{31}NO_5$	**172**	$C_{26}H_{42}O_7$	**128**
$C_{20}H_{28}O_3$	**223, 224**	$C_{23}H_{32}O_7$	**49**	$C_{26}H_{42}O_8$	**126**
$C_{20}H_{30}O_3$	**225, 226, 228**	$C_{23}H_{34}O_5$	**56, 57**	$C_{26}H_{42}O_9$	**125**
$C_{20}H_{30}O_4$	**220, 221, 222,**	$C_{23}H_{34}NO_5Cl$	**110, 111**	$C_{27}H_{27}N_5O_4$	**201, 202**
	227	$C_{23}H_{35}O_6Cl$	**425**	$C_{27}H_{36}O_6$	**243**
$C_{20}H_8O_8Cl_2$	**2**	$C_{23}H_{35}O_7Cl$	**424**	$C_{27}H_{38}O_5$	**161**
$C_{20}H_9O_8Cl$	**1**	$C_{23}H_{42}N_4O_4$	**418**	$C_{27}H_{41}NO_3$	**204**
$C_{21}H_{18}N_4O_2$	**106, 107**	$C_{24}H_{21}N_5O_4$	**103, 104**	$C_{27}H_{49}N_5O_7$	**44**
$C_{21}H_{20}O_6Cl_2$	**146**	$C_{24}H_{22}O_8$	**151**	$C_{27}H_{49}N_5O_8$	**43**
$C_{21}H_{21}N_3O_3$	**90**	$C_{24}H_{23}N_5O_4$	**101, 102**	$C_{28}H_{32}O_{12}$	**193**
$C_{21}H_{21}N_3O_4$	**94**	$C_{24}H_{24}O_8$	**150**	$C_{28}H_{32}N_4O_2$	**67**
$C_{21}H_{22}O_5$	**364**	$C_{24}H_{30}O_5$	**214**	$C_{28}H_{33}NO_7$	**246**
$C_{21}H_{22}N_2O_4$	**149**	$C_{24}H_{31}N_3O_4$	**255**	$C_{28}H_{34}O_{13}$	**361**
$C_{21}H_{23}N_3O_4$	**93**	$C_{24}H_{32}O_3$	**323, 324**	$C_{28}H_{34}N_4O_3$	**112, 113**
$C_{21}H_{24}O_5$	**248**	$C_{24}H_{32}O_5$	**177, 319**	$C_{28}H_{38}O_8$	**179**
$C_{21}H_{25}N_3O_2$	**88**	$C_{24}H_{34}O_5$	**176, 178**	$C_{28}H_{40}O_3$	**109**
$C_{21}H_{26}O_4$	**294**	$C_{24}H_{34}O_6$	**181, 182**	$C_{28}H_{40}O_8$	**180**
$C_{21}H_{26}O_5$	**397**	$C_{24}H_{35}NO_3$	**286**	$C_{28}H_{49}O_5$	**71**
$C_{21}H_{26}O_6$	**322, 396**	$C_{24}H_{35}NO_5Cl_2$	**37**	$C_{29}H_{35}NO_8$	**247**
$C_{21}H_{27}N_3O_2$	**195**	$C_{24}H_{36}O_6$	**371**	$C_{29}H_{38}O_7$	**32**
$C_{21}H_{27}NO_6$	**258**	$C_{24}H_{37}NO_6Cl_2$	**38**	$C_{29}H_{38}O_9$	**33**
$C_{21}H_{28}O_4$	**293, 317**	$C_{24}H_{38}O_8S$	**310**	$C_{29}H_{38}O_{10}$	**35**
$C_{21}H_{28}O_5$	**313, 316, 318**	$C_{25}H_{24}N_4O_3S_2$	**9**	$C_{29}H_{40}N_4O_5$	**407**
$C_{21}H_{30}O_3$	**189**	$C_{25}H_{24}N_4O_3S_3$	**10**	$C_{29}H_{51}NO_9$	**341**
$C_{21}H_{30}O_4$	**292**	$C_{25}H_{24}N_4O_3S_4$	**11**	$C_{30}H_{29}N_6O_4S_4$	**30**
$C_{21}H_{34}N_4O_6$	**345**	$C_{25}H_{25}N_5O_5$	**105**	$C_{30}H_{29}N_6O_5S_4$	**31**
$C_{21}H_{38}N_4O_4$	**419**	$C_{25}H_{28}O_5$	**340**	$C_{30}H_{56}O_{10}$	**148**
$C_{22}H_{25}N_3O_4$	**89**	$C_{25}H_{34}O_2$	**141**	$C_{30}H_{56}O_{11}$	**273**
$C_{22}H_{25}N_3O_5$	**91**	$C_{25}H_{35}NO_6$	**174**	$C_{31}H_{34}N_4O_2$	**81**
$C_{22}H_{25}NO_7$	**301**	$C_{25}H_{36}O_2$	**142**	$C_{31}H_{42}O_{10}$	**34**
$C_{22}H_{25}NO_8$	**45, 302, 303**	$C_{25}H_{36}O_4$	**160**	$C_{31}H_{42}N_4O_9$	**192**
$C_{22}H_{26}O_6$	**395**	$C_{25}H_{36}O_5$	**344, 349**	$C_{32}H_{32}N_6O_7S_4$	**8, 16, 17**
$C_{22}H_{26}O_7$	**394**	$C_{25}H_{36}O_6$	**208**	$C_{32}H_{32}N_6O_7S_5$	**7, 14**
$C_{22}H_{27}N_3O_3$	**87**	$C_{25}H_{37}O_3Cl$	**76**	$C_{32}H_{32}N_6O_7S_6$	**6, 13, 15**
$C_{22}H_{27}NO_8$	**369**	$C_{25}H_{37}O_5Br$	**77**	$C_{32}H_{32}N_6O_7S_7$	**12**
$C_{22}H_{30}O_2$	**52**	$C_{25}H_{38}O_7$	**399**	$C_{32}H_{34}N_4O_3$	**82**
$C_{22}H_{32}O_2$	**256**	$C_{25}H_{40}O_3$	**119, 120, 143**	$C_{32}H_{35}N_3O_3$	**59, 60, 61**

(Contd.)

(Contd.)

Compound Name Index

(New natural products isolated from marine fungi up to June, 2004)

Name	#	*Name*	#	*Name*	#
15G256<a>-2	**357**	Anomalin B	**405**	Aspermytin A	**259**
15G256<b>-2	**358**	Apiosporic acid	**282**	Auranticin A	**150**
15G256<g>	**190**	Ascochital	**147**	Auranticin B	**151**
15G256<d>	**191**	Ascosalipyrone	**285**	Avicennin A	**422**
15G256<e>	**192**	Ascosalipyrrolidinone A	**204**	Bipolal	**214**
15G256<i>	**193**	Ascosalipyrrolidinone B	**286**	Brocaenol A	**121**
15G256<n>	**359**	Asperazine	**134**	Brocaenol B	**122**
15G256<o>	**360**	Aspergillamide A	**112**	Brocaenol C	**375**
15G256<p>	**361**	Aspergillamide B	**113**	Butoxyl-spiciferin	**334**
15G256<w>	**194**	Aspergillicin A	**129**	Cathestatin C	**245**
386-A	**380**	Aspergillicin B	**130**	Chloriolin A	**423**
1893A	**415**	Aspergillicin C	**131**	Chloriolin B	**424**
1893B	**416**	Aspergillicin D	**132**	Chloriolin C	**425**
Acetophthalidin	**80**	Aspergillicin E	**133**	Chlorocarolide A	**287**
12-Acetylaspergillol	**291**	Aspergillitine	**163**	Chlorocarolide B	**288**
22-Acetylisocyclocitrinol A	**161**	Aspergillodiol	**292**	8-Chloro-9-hydroxy-8,9-	**164**
Acetyl Sumiki¢s acid	**152**	Aspergillol	**293**	deoxyasperlactone	
13′-Acetyltrichoverrin B	**34**	Aspergillone	**294**	9-Chloro-8-hydroxy-8,9-	**165**
Aigialomycin A	**274**	Aspergilloxide	**143**	deoxyasperlactone	
Aigialomycin B	**275**	Aspergione A	**295**	9-Chloro-8-hydroxy-8,9-	**166**
Aigialomycin C	**276**	Aspergione B	**296**	deoxyaspyrone	
Aigialomycin D	**207**	Aspergione C	**297**	Chlorogentisylquinone	**244**
Aigialomycin E	**277**	Aspergione D	**298**	Circumdatin G	**290**
Aigialone	**278**	Aspergione E	**299**	Cladospolide D	**184**
Aigialospirol	**279**	Aspergione F	**300**	CNC-350.1	**382**
Anomalin A	**251**	Asperic acid	**289**	Communesin A	**67**

(*Contd.*)

(*Contd.*)

Communesin B	**68**	Dictyonamide A	**238**	Enalin B	**404**
Communesin C	**81**	Dictyonamide B	**435**	*Ent*-gloeosteretriol	**159**
Communesin D	**82**	11,11′-Dideoxyverticillin A	**30**	Epicoccamide	**341**
Corollosporine	**168**	(6*R*,11*S*,12*S*,14*R*)-9,10-	**185**	Epicoccone	**253**
Coruscol A	**383**	Dihydrocolletodiol		6-Epi-ophiobolin G	**141**
Cyclotryprostatin A	**91**	(6*R*,11*R*,12*R*,14*R*)-9,10-	**186**	6-Epi-ophiobolin N	**142**
Cyclotryprostatin B	**92**	Dihydrocolletodiol		Epolactaene	**258**
Cyclotryprostatin C	**93**	2,3-Dihydro-7-deacetoxyyanuthone	**384**	(–)-Epoxyserinone A	**84**
Cyclotryprostatin D	**94**	A		(+)-Epoxyserinone B	**85**
Dankasterone	**71**	Dihydronaphthalen-1(4*H*)-one	**197**	Epoxysorbicillinol	**401**
7-Deacetoxyyanuthone A	**75**	Dihydroterrein	**338**	Ergosteryl myristate	**436**
22-Deacetylyanuthone A	**183**	Dihydroxyisoechinulin A	**255**	Eutypoid A	**342**
17-Dehydroxyhalorosellinic acid	**349**	7α,14-Dihydroxy-6β-*p*-	**302**	Evariquinone	**74**
Demethyl nectriapyrone A	**426**	nitrobenzoylconfertifolin		Exophilin A	**148**
Dendryphiellic acid A	**314**	6β,7α-Dihydroxy-14-*p*-	**303**	Exumolide A	**211**
Dendryphiellic acid B	**315**	nitrobenzoylconfertifolin		Exumolide B	**212**
Dendryphiellin A	**313**	(3*S*)-3,5-Dihydroxyphenyl-butan-2-	**187**	Farnesylhydroquinone	**256**
Dendryphiellin A1	**322**	one		Fellutamide A	**43**
Dendryphiellin B	**316**	(+)-2,4-Dimethyl-4,5-dihydro-furan-	**304**	Fellutamide B	**44**
Dendryphiellin C	**317**	3-carboxaldehyde		(+)-Formylanserinone B	**83**
Dendryphiellin D	**318**	3,7-Dimethyl-1,8-dihydroxy-6-	**386**	Fumiquinazoline A	**101**
Dendryphiellin E	**319**	methoxyisochroman		Fumiquinazoline B	**102**
Dendryphiellin E1	**323**	Drechslerine A	**327**	Fumiquinazoline C	**103**
Dendryphiellin E2	**324**	Drechslerine B	**328**	Fumiquinazoline D	**104**
Dendryphiellin F	**320**	Drechslerine C	**329**	Fumiquinazoline E	**105**
Dendryphiellin G	**321**	Drechslerine D	**330**	Fumiquinazoline F	**106**
Deoxyanserinone B	**378**	Drechslerine E	**205**	Fumiquinazoline G	**107**
Deoxynortrichoharzin	**371**	Drechslerine F	**331**	Fumiquinazoline H	**201**
12,13-Deoxyroridin E	**32**	Drechslerine G	**206**	Fumiquinazoline I	**202**
11′-Deoxyverticillin A	**31**	Enalin A	**403**	Fusaperazine A	**123**

(*Contd.*)

(*Contd.*)

Fusaperazine B	**343**	Herbarin B	**145**	JM 47	**345**
Glycer-1-yl dendryphiellate A	**325**	(–)-Hexylitaconic acid	**281**	Kasarin	**139**
Golmaenone	**254**	Hirsutanol A	**158**	Keisslone	**198**
Guisinol	**156**	Hirsutanol B	**428**	Kojic acid dimethyl ether	**280**
Gymnastatin A	**36**	Hirsutanol C	**429**	Leptosin A	**6**
Gymnastatin B	**37**	Hortein	**354**	Leptosin B	**7**
Gymnastatin C	**38**	Humicolone	**69**	Leptosin C	**8**
Gymnastatin D	**110**	9-Hydroxyhelminthosporol	**326**	Leptosin D	**9**
Gymnastatin E	**111**	9-Hydroxyhexylitaconic acid 1-	**283**	Leptosin E	**10**
Gymnasterone A	**108**	methyl ester		Leptosin F	**11**
Gymnasterone B	**109**	10-Hydroxy-18-methoxybetaenone	**234**	Leptosin G	**12**
Halorosellin A	**350**	Hydroxymethylanserinone B	**379**	Leptosin G_1	**13**
Halorosellin B	**351**	10-Hydroxy-18-[*N*-(2-naphthyl)-*N*-	**368**	Leptosin G_2	**14**
Halorosellinic acid	**208**	phenylamino]betaenone		Leptosin H	**15**
Halovir A	**215**	14-Hydroxy-6*b*-*p*-nitrobenzoylcinna-	**301**	Leptosin I	**16**
Halovir B	**216**	molide		Leptosin J	**17**
Halovir C	**217**	8-Hydroxy-9-oxo-7(11)-eremophilien-	**356**	Leptosin K	**18**
Halovir D	**218**	12,8-olide		Leptosin K_1	**19**
Halovir E	**219**	5-Hydroxyramulosin	**393**	Leptosin K_2	**20**
Halymecin A	**213**	3-Hydroxyroridin E	**33**	Leptosin M	**21**
Halymecin B	**270**	1-Hydroxyyanuthone A	**181**	Leptosin M_1	**22**
Halymecin C	**271**	1-Hydroxyyanuthone C	**182**	Leptosin N	**23**
Halymecin D	**272**	Hypoxylactone	**355**	Leptosin N_1	**24**
Halymecin E	**273**	Hypoxysordarin	**199**	Leptosin O	**25**
Harzialactone A	**400**	Insulicolide A	**45**	Leptosin P	**26**
Harzialactone B	**140**	*Iso*-cladospolide B	**431**	Leptosin Q	**27**
Helicascolide A	**352**	Isoculmorin	**362**	Leptosin R	**28**
Helicascolide B	**353**	Isocyclocitrinol A	**160**	Leptosin S	**29**
Herbaric acid	**311**	Isoemericellin	**340**	Leptosphaerin	**363**
Herbarin A	**144**	Isosativenetriol	**332**	Leptosphaerolide	**365**

(*Contd.*)

9

Antioxidant Metabolites from Marine Derived Fungi

Katja M. Fisch[1], Ahmed Abdel-Lateff[2] and Gabriele M. König[3]

[1]Molecular Biology Laboratory, Department of General Surgery, University Hospital Muenster, Waldeyerstr. 1,48149 Muenster, Germany
[2]Pharmacognosy Department, Faculty of Pharmacy, El-Menia University, El-Menia, Egypt.
[3]Institute for Pharmaceutical Biology, University of Bonn, Nussallee 6, 53115 Bonn, Germany

OXIDATIVE STRESS AND ANTIOXIDANTS

Oxidation is the transfer of electrons from an atom or molecule (oxidizable substrate) to another (electron acceptor or prooxidant) and represents an essential part of aerobic life. Oxygen itself is the most important electron acceptor in the respiratory chain, a biochemical process which produces energy in the form of adenosine triphosphate. Oxidative stress which leads to an excess of oxidation, however, plays an important role in the development of several diseases and ageing (Halliwell and Gutteridge 1999, Diplock *et al.* 1998). Atherosclerosis, for example, involves peroxidation of lipoproteins thus forming so-called foam cells (Bankson *et al.*, 1993), and oxidative damage of DNA can lead to mutations and cancer (Bankson *et al.* 1993, Halliwell 1999, Wiseman and Halliwell 1996). Neurodegeneration as seen in Alzheimer's disease, cardiovascular diseases other than atherosclerosis, and diabetes are also associated with oxidative stress (Dhalla *et al.* 2000, Vendemiale *et al.* 1999). Sies (1991) defined oxidative stress as: "a disturbance in the prooxidant-antioxidant balance in favour of the former, leading to potential damage". The most prominent pro-oxidants in living organisms are represented by reactive oxygen and nitrogen species (ROS/RNS), as shown in Table 9.1.

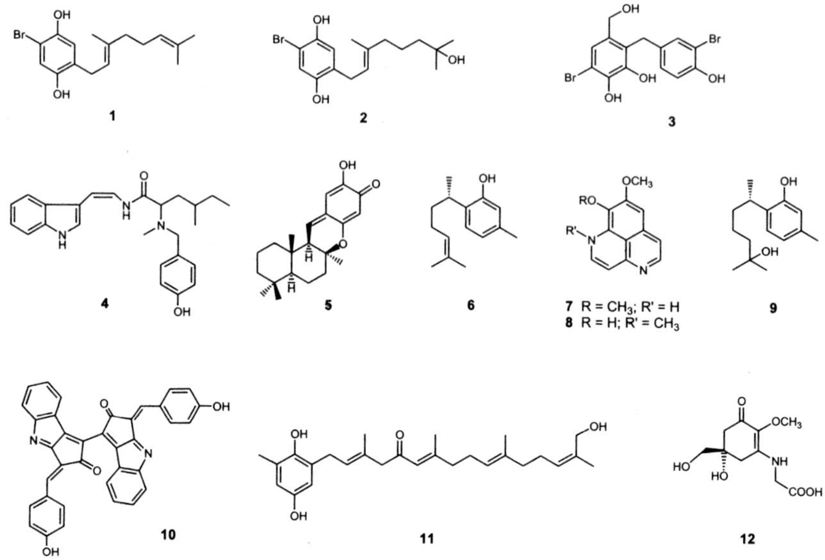

Fig. 9.1 Antioxidants from marine organisms.

investigated non-phenolic marine antioxidants are mycosporine-like amino acids (MAAs), low molecular weight, water soluble compounds, containing a cyclohexenone or cyclohexenimine chromophore with a UV maximum between 310-360nm. They prevent the oxidation of substrates mainly by absorbing UV-B light. MAAs occur in a wide range of marine organisms (Cockel and Knowland 1999), e.g., cyanobacteria, red algae, dinoflagellates, corals, and marine invertebrates such as sea anemones, brine shrimp, and sea urchins. They were also found in terrestrial organisms like fungi or on the surface of apples exposed to solar radiation (reviewed by Bandaranayake 1998). Besides its activity as UV absorbance agent, Dunlap and Yamamoto (1995) found for mycosporine-glycine (**12**) a moderate antioxidative activity. For gadusol and 6-deoxygadusol, MAA related compounds with UV maxima below 300 nm, strong antioxidative activities were reported (Bandaranayake 1998). Further non-phenolic marine structures with antioxidative activities are ovothiols, isolated from eggs of marine invertebrates, reported to be more active than glutathione (Holler and Hopkins 1988, Shapiro 1991) and to possess glutathione peroxidase activity (Turner *et al.* 1988), and poecillanosine, a compound isolated from the sponge *Poecillastra* sp. (Natori *et al.* 1997). Recently, antioxidative properties of the common algal metabolite dimethylsulphoniopropionate (DSMP) and its degradation products were described (Sunda *et al.* 2002).

EVALUATION OF MARINE FUNGAL METABOLITES FOR ANTIOXIDATIVE ACTIVITY

Fungi produce a great variety of secondary metabolites with a predominance of phenolic polyketides. It is thus not surprising that screening efforts aimed at finding antioxidative compounds in fungi resulted in the identification of active metabolites (Table 9.2, Fig. 9.2) (Aoyama *et al.* 1982, Oeda *et al.* 1993, Yen and Lee 1996, Li *et al.* 2001, Keller *et al.* 2002, Murcia *et al.* 2002). Simple phenolic compounds were found, such as 2,3-dihydroxy benzoic acid (**13**) isolated from *Penicillium roquefortii* (Hayashi *et al.* 1995) as well as *USF-406A* (**14**), *USF-406B* (**15**), *USF-406C* (**16**) and *USF-406D* (**17**) isolated from *Mortierella* sp. (Hirota *et al.* 1997). Another study reported the isolation of the three antioxidants, curvulic acid (**18**) from *Penicillium* sp., citrinin (**19**), and protocatechuic acid (**20**), from unidentified fungi (Aoyama *et al.* 1982). An alkaloidal compound namely pyridoxatin (**21**) was obtained from *Acremonium* sp. BX86 (Teshima *et al.* 1991), and investigation of a *Penicillium* sp. led to the isolation of phenopyrrozin (**22**) (Shiomi *et al.* 1995).

Table 9.2 Examples of antioxidants isolated from terrestrial and marine derived (M) fungi.

Fungi species	*Compounds isolated*	*Reference*
Keissleriella sp. (M)	Exopolysaccharide	Sun *et al.* 2004
Epicoccum sp. (M)	Isobenzofuranone	Abdel-Lateff *et al.* 2003a
Wardomyces anomalus (M)	Xanthones	Abdel-Lateff *et al.* 2003b
Acremonium sp.	Pyridoxatin	Teshima *et al.* 1991
Agrocybe aegerita	Fatty acid fraction	Zhang *et al.* 2003
Acremonium sp.(M)	Hydroquinones	Abdel-Lateff *et al.* 2002
Albatrellus ovinus	Neogrifolin derivatives	Nukata *et al.* 2002
Aspergillus candidus	3,3′-Dihydroxyterphenyllin; 3-Hydroxyterphenyllin; Candidusin B	Yen *et al.* 2001
Aspergillus saitoi	Hydroxyflavanones (biotechnological formation frorm flavone glycosides)	Miyake *et al.* 2003
Aspergillus sp. (M)	Golmaneone (Diketopiperazine alkaloid); Neoechinulin A	Li *et al.* 2004
Cercospora nicotianae	Pyridoxine and derivatives	Bilski *et al.* 2000
Ceriporiopsis subvermispora	Alkylitaconates	Watanabe *et al.* 2002
Cladosporium cladosporidae	Melanins	Byshneva and Senchuk 2001
Cortinarius hercynicus	2-(3,4-Dihydroxyphenyl)-glycine	Keller *et al.* 2002
Emericella falconensis	Falconensone A and B	Takashashi *et al.* 2000
Ganoderma lucidum	Amino-polysaccharide fraction	Lee *et al.* 2001
Lyophyllum connatum	Connatine	Keller *et al.* 2002
Mortierella sp.	USF-406A-D	Hirota *et al.* 1997
Penicillium roquefortii	2,3-Dihydroxy benzoic acid	Hayashi *et al.* 1995
Penicillium sp.	Curvulic acid	Aoyama *et al.* 1982

(*Contd.*)

Pestalotiopsis microspora	Isopestacin	Strobel *et al.* 2002
Puccinia distincta	Carotenoids	Davoli and Weber 2002
Several	Melanin pigments	Shcherba *et al.* 2000
Streptomyces sp.	USF-142A, 3-(2-methyl-phenyl)-propionamide-3-one	Hirota *et al.* 1994
Trichoderma sp.	Bisorbicillinol	Abe *et al.* 1998, Abe and Hirota 2002
Unidentified	Citrinin and protocatechuic acid	Aoyama *et al.* 1982

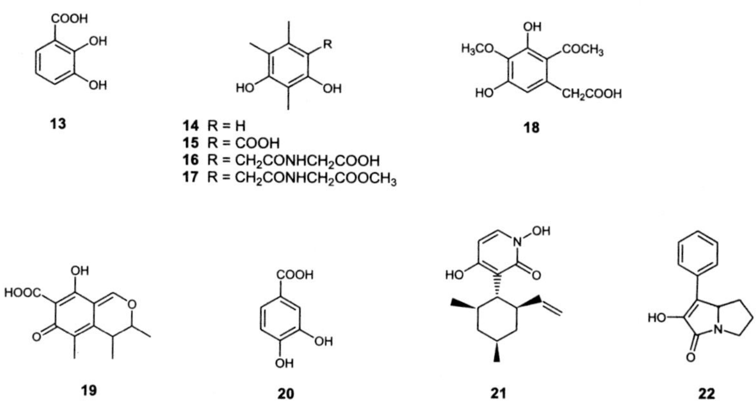

Fig. 9.2 Antioxidants from terrestrial fungi.

The vast majority of reports on antioxidative compounds in fungi relate to terrestrial fungi, including edible fungi and soil-borne fungi. Terrestrial fungi were for decades a promising source of novel and biologically active secondary metabolites. Today, however increasingly known metabolites are obtained from these sources. The discovery of new compounds is thus more likely if the focus is placed on fungi from other ecological surroundings than soil, e.g., salt water. Thus marine fungi emerged as an interesting group of microorganisms for natural products research. As is obvious from Table 9.2, to date only a few marine-derived fungi were evaluated for the production of antioxidative compounds. Our efforts to identify antioxidative marine compounds thus concentrated on marine fungi. All of the investigated fungi were endophytes of marine macroalgae. The isolation of *fungal endophytes* from algae can be performed by placing algal thalli, after surface sterilisation to remove unwanted epibionts, on suitable agar media. Most of the isolated endophytic fungi are facultative and not obligate marine fungi, distinguished by the biosynthesis of structurally diverse natural products (Bugni and Ireland 2004). This is not surprising considering the complex ecological situation of the endophyte. It might excrete compounds that benefit its host, e.g., defense of pathogens, and at the same time has to secure its living space inside the algal tissue.

Using the thiobarbituric acid-reactive substances (*TBARS*) method and the diphenyl-picrylhydrazyl radical (*DPPH*) assay we tested several marine fungal metabolites and extracts derived from the species *Acremonium* sp., *Wardomyces anomalus, Epicoccum* sp., and *Apiospora montagnei* for their antioxidative and *radical scavenging* activities. The *in vitro* observed antioxidative activity may also have ecological significance in the complex interaction between fungi and algal host.

The fungal strain *Acremonium* sp., closely related to *A. roseogriseum* (S. B. Saksena) W. Gams based on morphological characteristics, was isolated from the brown alga *Cladostephus spongius* (Hudson) C. Agardh, collected at the Spanish coast (Moraira, Mediterranean Sea). Isolation and structure elucidation of six pure compounds and two mixtures of closely related natural products (Fig. 9.3) (**23-33**) were described (Abdel-Lateff *et al.* 2002). Compounds **23, 24** and **33** are hydroquinone derivatives, **25** and **26** dihydronaphthalenones, **27-30** acyclic carboxylic acid derivatives, **31** a 2,3-dihydroxybenzofuran, and **32** a chroman derivative.

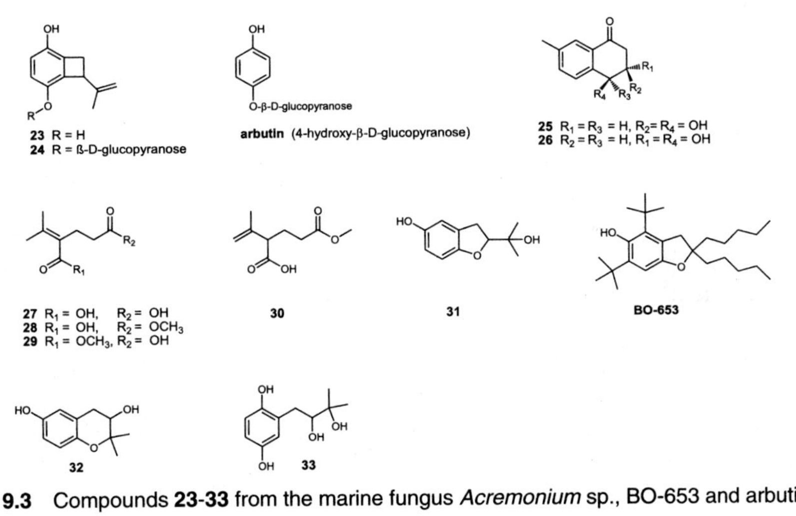

Fig. 9.3 Compounds **23-33** from the marine fungus *Acremonium* sp., BO-653 and arbutin.

Compounds **23, 24,** and **31-33** showed significant DPPH radical scavenging effects (85.5, 17.5, 85.8, 72.9, 90.2%, respectively, at 25.0 μg/mL, Tables 9.3 and 9.4), with **23, 31, 32,** and **33** being also able to inhibit peroxidation of linolenic acid (35.5, 15.9, 9.2, and 16.6%, respectively, at 37.0 μg/mL). Compounds **23, 31, 32** and **33** showed similar radical scavenging activities in the DPPH test. As a probable explanation for the similar radical scavenging activities of **31** and **32** the opening of the chroman ring of **32** and the 2,3-dihydrobenzofuran ring of **31** may be

Table 9.3 Antioxidative effects (TBARS Assay)[a] of fungal compounds.

Item Tested	% *Inhibition*[b]		
	3.7 μg/mL	*7.4 μg/mL*	*37.0 μg/mL*
23	n.t.[c]	13.9	35.5
24	n.t.	–10.6	–10.6
25 and **26**	n.t.	–22.5	–18.5
27	n.t.	–9.0	–18.1
28-30	n.t.	–8.6	–16.3
31	n.t.	–11.8	15.9
32	n.t.	–9.8	9.2
33	n.t.	2.0	16.6
34	n.t.	17.0	37.0
35	n.t.	–7.4	–11.1
36	n.t.	2.6	7.4
37	21.3	31.7	62.1
38	4.6	–0.4	2.3
39 and **40**	9.1	7.4	14.4
41	9.5	10.7	18.4
42	14.1	15.2	30.4
(-)-(3*R*)-Mellein	–3.3 (2.8 μg/mL)	2.2 (5.6 μg/mL)	–1.3 (29.2 μg/mL)
(-)-(3*R*)-8-Methoxymellein	5.5 (3.1 μg/mL)	2.7 (6.2 μg/mL)	1.0 (31.5 μg/mL)
Arbutin	n.t.	–27.6	–25.2
BHT[d]	54.2	63.8	73.9

[a]TBARS = Thiobarbituric acid method.
[b]% Inhibition = 100-(A sample[e] – A sample blank) × 100/(A control - A blank).
[c]n.t. = not tested.
[d]BHT = Butylated hydroxytoluene.
[e]A = Absorbance of test blank and control solutions are measured at both 532 and 600 nm.

suggested. This would lead to a hydroquinone or a semiquinone moiety which can further react to a quinone, as expected also for compounds **23** and **33**. This mechanism is known for α-tocopherol. Whereas their DPPH scavenging capacity was similar, the antioxidant activity for **33** was about half of that of **23** in the TBARS test at a concentration of 37.0 μg/mL. Since both molecules contain a hydroquinone moiety an influence of the substitution pattern can be presumed. Compound **24** and the reference compound arbutin, both possessing only one phenolic hydroxyl group, showed very weak radical scavenging activity in the DPPH test and prooxidant properties in the TBARS assay. Hisatomi *et al.* (2000) also found very low radical scavenging activity for arbutin in the DPPH test. However, the authors also found low antioxidative activity using a TBARS method slightly different from the one applied in the current study. Hisatomi's findings are not consistent with the prooxidant activity seen in our TBARS assay. Mixture **25** and **26**, compound **27** and mixture **28-30** were close to inactive in the DPPH test and prooxidative in the TBARS assay. The compound 2,3-dehydro-5-hydroxy-2,2-dipentyl-

Table 9.4 Radical scavenging effects (DPPH Assay)[a] of fungal compounds.

	% Scavenging[b]				
Item Tested	*5*[c]	*25*	*50*	*100*	*500*
23	26.0	85.5	90.0	90.0	90.1
24	5.6	17.5	27.7	43.0	79.9
25 and **26**	n.t.	n.t.	–1.6	2.8	12.1
27	n.t.	n.t.	–0.5	3.8	15.3
28-30	n.t.	n.t.	3.4	10.3	35.1
31	35.3	85.8	89.1	90.2	90.4
32	25.7	72.9	85.1	89.1	89.9
33	n.t.	90.2	90.7	89.9	82.9
34	n.t.	94.7	94.8	95.2	95.4
35	n.t.	6.2	12.9	25.3	90.6
36	n.t.	30.7	53.4	87.7	97.0
37	46.2	95.2	94.9	94.7	n.t.
38	n.t.	5.9	10.6	22.2	n.t.
39 and **40**	n.t.	1.8	1.7	3.1	n.t.
41	n.t.	5.7	9.6	15.9	n.t.
42	n.t.	22.4	28.4	35.8	n.t.
(–)-(3*R*)-Mellein	–1.5	0.5	–0.9	n.t.	n.t.
(–)-(3*R*)-8-Methoxymellein	–1.5	–0.5	1.5	n.t.	n.t.
Arbutin	1.9	18.3	23.5	52.6	78.6
BHT[e]	6.0	24.2	41.1	63.8	89.8

[a]DPPH = α, α-Diphenyl-β-picrylhydrazyl.
[b]Scavenging % = 100 –(A sample[f] × 100 / A control[f])
[c]Concentrations in µg/mL.
[d]n.t. = not tested.
[e]BHT = Butylated hydroxytoluene.
[f]Absorbance of sample and control measured at 517 nm.

4,6-di-tert-butylbenzofuran (BO-653) with the same skeletal structure as compound **31** was developed by the company Chugai as an antiatherogenic drug and is currently undergoing phase II clinical trials (http://www.clinicaltrials.gov/ct/show/NCT00055510?order=1). The activity of BO-653 is mainly due to its antioxidant properties, which were extensively studied in different test systems (Yoshida *et al.* 2002). The published *in vitro* activities were comparable to α-tocopherol indicating BO-653 to be a powerful antioxidant consistent with its chemical structure. The structural similarity of BO-653 to compond **31** led to the suggestion that the antioxidant activity of both molecules is similar.

The marine-derived fungus, *Wardomyces anomalus* Brooks & Hansford (W) was isolated from the green alga *Enteromorpha* sp., collected around Fehmarn Island in the Baltic Sea. Radical scavenging activity of the EtOAc extract was observed using DPPH as indicator (71.9% at 500 µg/mL). Successive fractionation of the EtOAc extract was done and details of isolation and structure elucidation of the pure compounds were published (Abdel-Lateff *et al.* 2003b). The fractionation yielded four xanthones, including **34** and **35** and one furan derivative (**36**). Compounds **34** and **36** had significant DPPH radical scavenging effects (94.7, 30.7%, respectively, at 25.0 μg/mL). Both compounds also inhibited peroxidation of linolenic acid (17.0, and 2.6%, respectively, at 7.4 μg/mL) (Tables 9.3 and 9.4).

Xanthone derivatives are widespread in nature, commonly occurring in a number of higher plant families (Peres *et al.* 2000, Morel *et al.* 2000). Some fungal species are known as sources of xanthone derivatives, e.g., *Penicillium raistrickii* (Belofsky *et al.* 1998), *Phomopsis* sp. (Isaka *et al.* 2001), *Actinoplanes* sp. (Chu *et al.* 1997), *Ascodesmis sphaerospora* (Hein *et al.* 1998), *Ulocladium botrytis* (Höller *et al.* 1999) and *Humicola* sp. (Tabata *et al.* 1996). Xanthones were shown to have a variety of biological activities, e.g., antimicrobial, antitubercular, antitumoral, antiviral (Peres *et al.* 2000, Morel *et al.* 2000), and antioxidant (Minami *et al.* 1994, 1995, 1996, Gonda *et al.* 2000, Williams *et al.* 1995). Minami and colleagues partly found high activity among the xanthones containing a catechol (1,2-dihydroxybenzyl) moiety. Gonda and colleagues described activity comparable to that of α-tocopherol only for those xanthones, which contained four phenolic hydroxyl groups, whereas compounds containing three phenolic hydroxyl groups were inactive. In our study **34** containing four phenolic hydroxyl groups showed high antioxidative properties in both assay systems, whereas **35** having only three phenolic groups possessed very weak radical scavenging activity in the DPPH test and prooxidant activity in the TBARS assay.

Fungi of the genus *Epicoccum* are commonly associated with beach wrack material. Only a few members of this genus have been studied for their secondary metabolite content (Baute *et al.* 1976, 1978, Frederick *et al.* 1981, Shu *et al.* 1997, Wright *et al.* 2003). The investigated fungus, *Epicoccum* sp., was isolated from the brown alga *Fucus vesiculosus* L., which was collected at the German coast (Tönning, North Sea). Compounds **37**, **38**, **41** and **42** as well as the mixture **39** and **40** (Fig. 9.4) were isolated and their structures published (Abdel-Lateff *et al.* 2003a). All of the compounds are highly oxygenated with **38-40** being furane derivatives and **41** and **42** mellein derivatives.

Compounds **37** and **42** exhibited significant DPPH radical scavenging effects (16.8, 95.2, and 22.4 %, respectively, at 25.0 μg/mL). Additionally, **37**, **41** and **42** were able to inhibit peroxidation of linolenic acid methyl ester (62.1, 18.4, and 30.4%, respectively, at 37.0 μg/mL). In both of the aforementioned assays, **37** possessed the highest activity and still had significant effects at levels between 1-5 μg/mL.

The endophytic fungus *Apiospora montagnei* Saccardo was isolated from the red alga *Polysiphonia violacea* (ROTH.) SPRENG., collected at the German coast (Helgoland, North Sea). This fungus yielded (-)-(3*R*)-

mellein and (-)-(3*R*)-8-methoxymellein (Fig. 9.4), quite common fungal metabolites (Klemke 2004). No activity could be observed for both compounds in both assay systems. As (-)-(3*R*)-mellein contains one phenolic hydroxyl group at least a weak antioxidant activity was expected. An explanation for the absence of any activity of (-)-(3*R*)-mellein could be a hydrogen bond between the hydroxyl group and the proximate carbonyl group, which would make an abstraction of the hydroxyl hydrogen more difficult. Whereas (-)-(3*R*)-mellein and (-)-(3*R*)-8-methoxymellein showed no activity, the related **41** exhibited marked activity in the TBARS assay. The DPPH scavenging activity of **41,** however, was low indicating other than radical scavenging properties to be responsible for its activity in the TBARS assay. There is no good explanation from the structural point of view for the observed big differences in antioxidant activity between these three structures. Further mechanistical studies are required to explain the reason for observed differences. The higher antioxidant activity of **42** is partly explainable by its hydroquinone moiety.

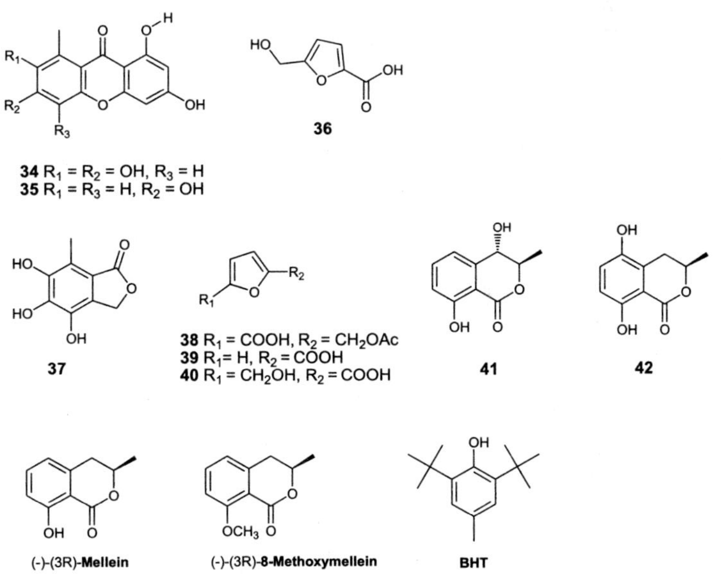

Fig. 9.4 Compounds **34-36** from the marine fungus *Wardomyces anomalus*, **37-42** from *Epicoccum* sp., (-)-(3R)-mellein, (-)-(3R)-8-methoxymellein from the marine fungus *Apiospora montagnei*, and BHT.

CONCLUSIONS

The most potent antioxidant from marine-derived fungi in the current study was compound **37** (Tables 9.3 and 9.4, Fig. 9.4), obtained from the marine endophytic fungus *Epicoccum* sp. This compound is related to the recently described isopestacin and pestacin, isobenzofuran derivatives isolated from *Pestalotiopsis microspora*, an endophytic fungus originating from the rainforests of Papua New Guinea. Isopestacin and pestacin were tested for their antioxidative activity in the total oxyradical scavenging capacity (TOSC) assay and found to be extremely active (Strobel *et al.* 2002). These results are consistent with our findings for compound **37**.

Apart from antioxidative fungal metabolites with new structures and from novel ecological niches, very well known fungus-derived natural products might offer new possibilities. One of the most important groups of fungal metabolites are the statins, e.g., lovastatin and cerivastatin. These polyketides lower low-density lipoprotein (LDL) and total cholesterol levels in blood through inhibition of 3-hydroxy-3-methylglutaryl coenzyme A (HMG CoA) reductase. Recently it was suggested that statins also interfere with the formation of reactive oxygen species. A reduction of ROS results in an increased resistance of LDL to oxidation and thus may prevent the atherogenic action of oxidized LDL (Rosenson 2004). The antioxidative effect of cerivastatin may also be useful for the treatment of Alzheimer´s disease (Siegel-Axel 2003).

These recent results highlight the enormous therapeutic potential of antioxidative fungal metabolites. The marine environment harbours a rich biodiversity of fungal strains that have hardly been investigated. The above discussed results on antioxidative metabolites from marine fungi illustrate that further exploitation of this resource for bioactive natural products is very promising.

ACKNOWLEDGEMENT

Ahmed Abdel-Lateff is greatful for a fellowship of the Egyptian government for performing his PhD studies at the Institute for Pharmaceutical Biology, University of Bonn, Germany.

REFERENCES

Abdel-Lateff, A., K.M. Fisch, A.D. Wright, and G.M. König. 2003a. A new antioxidant isobenzofuranone derivative from the algicolous marine fungus *Epicoccum* sp. *Planta Med.* 69: 831–834.

Abdel-Lateff, A., G.M. König, and A.D. Wright. 2003b. Two new xanthone derivatives from the algicolous marine fungus *Wardomyces anomalus*. *J. Nat. Prod.* 66: 706–708.

Abdel-Lateff, A., G.M. König, K.M. Fisch, U. Höller, P.G. Jones, and A.D. Wright. 2002. New antioxidant hydroquinone derivatives from the algicolous marine fungus *Acremonium* sp. *J. Nat. Prod. 65*: 1605–1611.

Abe., N., and A. Hirota. 2002. Chemical studies of the radical scavenging mechanism of bisorbicillinol using the 1,1-diphenyl-2-picrylhydrazyl radical. *Chem. Comm.* 662–663.

Abe, N., T. Murata, and A. Hirota. 1998. Novel DPPH radical scavengers, bisorbicillinol and demethyltrichodimerol, from a fungus. *Biosci. Biotechnol. Biochem.* 62: 661–666.

Alscher, R., and J.L. Hess, 1993. *Antioxidants in Higher Plants* CRC Press, Boca Raton, Florida.

Aoyama, T., Y. Nakakita, M. Nakagawa, and H. Sakai. 1982. Screening for antioxidants of microbial origin. *Agric. Biol. Chem.* 46: 2369–2371.

Bandaranayake, W.M. 1998. Mycosporines: are they nature's sunscreens? *Nat. Prod. Rep.* 15: 159–172.

Bankson, D.D., M. Kestin, and N. Rifai. 1993. Role of free-radicals in cancer and atherosclerosis. *CLIN LAB MED.* 13: 463–480.

Baute, M.A., G. Deffieux, R. Baute, and A. Neveu. 1978. New antibiotics from fungus *Epicoccum nigrum* 1. Fermentation, isolation and antibacterial properties. *J. Antibiot.* 11: 1099–1101.

Baute, R., G. Deffieux, M.A. Baute, and M.A. Filleau. 1976. New fungal metabolite in 3,6-epidithio-2,5-dioxopiperazine group-epicorazine-A isolated from *epicoccum nigrum* link strain (Adelomycetes). *Tetrahedron Lett.* 44: 3943–3944.

Belofsky, G.N., B.K. Gloer, B.J. Gloer, T.D. Wicklow, and F.P. Dowd. 1998. New p-terphenyl and polyketide metabolites from the sclerotia of *Penicillium raistrickii. J. Nat. Prod.* 61: 1115–1119.

Bilski, P., M.Y. Li, M. Ehrenshaft, M.E. Daub, and C.F. Chignell. 2000. Vitamin B-6 (pyridoxine) and its derivatives are efficient singlet oxygen quenchers and potential fungal antioxidants. *Photochem. Photobiol.* 71: 129–134.

Borowitzka, M.A. 1999. Commercial production of microalgae: ponds, tanks, tubes and fermenters. *J. Biotech.* 70: 313–321.

Bugni, T.S., and C.M. Ireland. 2004. Marine-derived fungi: a chemically and biologically diverse group of microorganisms. *Nat. Prod. Rep.* 21: 143–163.

Byshneva, L.N., and V.V. Senchuk. 2001. Effects of melanins on lipid peroxidation. *Appl. Biochem. Microbiol.* 37: 93–97.

Chu, M., I. Truumees, R. Mierzwa, J. Terracciano, M. Patel, D. Loebenberg, J.J. Kaminski, P. Das, and S.M. Puar. 1997. Sch 54445: A new polycyclic xanthone with highly potent antifungal activity produced by *Actinoplanes* sp. *J. Nat. Prod.* 60: 525–528.

Cockell, C.S., and J. Knowland. 1999. Ultraviolet radiation screening compounds. *Biol. Rev.* 74: 311–345.

Davoli, P., and R.W.S. Weber. 2002. Identification and quantification of carotenoid pigments in aeciospores of the daisy rust fungus, *Puccinia distincta. Phytochemistry*. 60: 309–313.

Dhalla, N.S., R.M. Temsah, and T. Netticadan. 2000. Role of oxidative stress in cardiovascular diseases. *Hypertension* 18: 655–673.

Diplock, A.T., J.-L. Charleux, G. Crozier-Willi, F.J. Kok, C. Rice-Evans, M. Roberfroid, W. Stahl, and J. Viña-Ribes. 1998. Functional food science and defence against reactive oxidative species. *Brit. J. Nutr.* 80: Suppl. 1, S77–S112.

Dunlap, W.C., and Y. Yamamoto. 1995. Small-molecule antioxidants in marine organisms-antioxidant activity of mycosporine-glycine. *Comp. Biochem. Physiol.* 112B: 105–114.

Fisch, K.M., V. Böhm, G.M. König, and A.D. Wright. 2003. Antioxidative meroterpenoids from the brown alga *Cystoseira crinita. J. Nat. Prod.* 66: 968–975.

Frankel, E.N., A.L. Waterhouse, and J.E. Kinsella. 1993. Inhibition of human LDL oxidation by resveratrol. *Lancet* 341: 454–457.

Frederick, B.C., J.P. Szaniszlo, E.P. Vickrey, D.M. Bentley, and W. Shive. 1981. Production and isolation of siderophores from the soil fungus. *Epicoccum purpurascens. Biochemistry*. 20: 2436–2438.

Gonda, R., T. Takeda, and T. Akiyama. 2000. Studies on the constituents of *Anaxagorea luzonensis* A. GRAY. *Chem. Pharm. Bull.* 48: 1219–1222.

Halliwell, B. 1997. Antioxidants in disease, mechanisms and therapy Pages 1-20 in *Advances in Pharmacology*, Vol. 38, H. Sies, ed., Academic Press , San Diego.

Halliwell, B. 1999. Oxygen and nitrogen are pro-carcinogens, Damage to DNA by reactive oxygen, chlorine and nitrogen species: measurement, mechanism and the effects of nutrition. *Mutat. Res*, 443: 37–52.

Halliwell, B., and J.M.C. Gutteridge, 1989. *Free Radicals in Biology and Medicine*, 2nd Ed., Clarendon Press, Oxford.

Halliwell, B., and J.M.C. Gutteridge. 1999. *Free Radicals in Biology and Medicine*, 3rd Ed., Oxford University Press, Oxford.

Hayashi, K., K. Suzuki, M. Kawaguchi, T. Nakajima, T. Suzuki, M. Numata, and R. Nakamura. 1995. Isolation of an antioxidant from *Penicillium roquefortii* IFO-5956. *Biosci. Biotech. Biochem*. 59: 319–320.

Hein, S.M., B.J. Gloer, B. Koster, and D. Malloch. 1998. Arugosin F: A new antifungal metabolite from the coprophilous fungus *Ascodesmis sphaerospora*. *J. Nat. Prod*. 61: 1566–1567.

Hirota, A., Y. Hirata, Y. Morimitsu, and H. Hirota. 1994. New indophenol-reducing 1,3-dicarbonyl compound from streptomyces. *Biosci. Biotech. Biochem*. 58: 792–793.

Hirota, A., Y. Morimitsu, and H. Hojo. 1997. New antioxidative indophenol-reducing phenol compounds isolated from the *Mortierella* sp. fungus. *Biosci. Biotech. Biochem*. 61: 647–650.

Hisatomi, E., M. Matsui, A. Kobayashi, and K. Kubota. 2000. Antioxidative activity in the pericarp and seed of japanese pepper (*Xanthoxylum piperitum* DC). *J. Agric. Food Chem*. 48: 4924–4928.

Holler, T.P., and P.B. Hopkins. 1988. Ovothiols as biological antioxidant. The thiol groups of ovothiol and glutathione are chemically distinct. *J. Am. Chem. Soc*. 110: 4837–4838.

Höller, U., G.M. König, and A.D. Wright. 1999. A new tyrosine kinase inhibitor from a marine isolate of *Ulocladium botrytis* and new metabolites from the marine fungi *Asteromyces cruciatus* and *Varicosporina ramulosa*. *Eur. J. Org. Chem*. 11: 2949–2955.

Isaka, M., A. Jaturapat, K. Rukseree, K. Danwisetkanjana, M. Tanticharoen, and Y. Thebtaranonth. 2001. Phomoxanthones A and B, novel xanthone dimers from the endophytic fungus *Phomopsis* species. *J. Nat. Prod*. 64: 1015–1018.

Keller, C., M. Maillard, J. Keller, and K. Hostettmann. 2002. Screening of european fungi for antibacterial, antifungal, larvicidal, molluscicidal, antioxidant and free-radical scavenging activities and subsequent isolation of bioactive compounds. *Pharm. Biol*. 40: 518–525.

Klemke, C. 2004. Isolation, Cultivation and Biological Screening of Marine Endophytic Fungi and Structure Elucidation of New Fungal Secondary Metabolites, Ph. D Dissertation at the University of Bonn, Institute for Pharmaceutical Biology.

Lee, J.M., H. Kwon, J.W. Jeong, S.Y. Lee, S.J. Baek, and Y.J. Surh. 2001. Inhibition of lipid peroxidation and oxidative DNA damage by *Ganoderma lucidum Phytother. Res*. 15: 245–249.

Li, S.P., P. Li, T.T.X. Dong, and K.W.K. Tsim. 2001. Anti-oxidation activity of different types of natural *Cordyceps sinensis* and cultured *Cordyceps* mycelia. *Phytomedicin*. 8: 207–212.

Li, Y., X. Li, S.-K. Kim, J.S. Kang, H.D. Choi, J.R. Rho, and B.W. Son. 2004. Golmaneone, e new diketopiperazine alkaloid from marine-derived fungus *Aspergillus* sp. *Chem. Pharm. Bull*. 52: 375–376.

Minami, H., M. Kinoshita, Y. Fukuyama, M. Kodama, T. Yoshizawa, M. Sugiura, K. Nakagawa, and H. Tago. 1994. Antioxidant xanthones from *Garcinia subelliptica Phytochemistry* 36: 501–506.

Minami, H., A. Kuwayama, T. Yoshizawa, and Y. Fukuyama. 1996. Novel prenylated xanthones with antioxidant property from the wood of *Garcinia subelliptica*. *Chem. Pharm. Bull*. 44: 2103–2106.

Minami, H., E. Takahashi, Y. Fukuyama, M. Kodama, T. Yoshizawa, and K. Nakagawa. 1995. Novel xanthones with superoxide scavenging activity from *Garcinia subelliptica*. *Chem. Pharm. Bull.* 43: 347–349.

Miyake, Y., K. Minato, S. Fukumoto, K. Yamamoto, T. Oya-Ito, S. Kawakishi, and T. Osawa. 2003. New potent antioxidative hydroxyflavanones produced with *Aspergillus saitoi* from flavanone glycoside in citrus fruit. *Biosci. Biotech. Biochem.* 67: 1443–1450.

Morel, C., D. Séraphin, M.J. Oger, M. Litaudon, T. Sévenet, P. Richomme, and J. Bruneton. 2000. New xanthones from *Calophyllum caledonicum*. *J. Nat. Prod.* 63: 1471–1474.

Murcia, M.A., M. Martinez-Tome, A.M. Jimenez, A.M. Vera, M. Honrubia, and P. Parras. 2002. Antioxidant activity of edible fungi (truffles and mushrooms): Losses during industrial processing. *J. Food Protect.* 65: 1614–1622.

Natori, T., Y. Kataoka, S. Kato, H. Kawai, and N. Fusetani. 1997. Poecillanosine, a new free radical scavenger from the marine sponge *Poecillastra* spec aff *tenuilaminaris*. *Tetrahedron Lett.* 38: 8349–8350.

Nukata, M., T. Hashimoto, I. Yamamoto, N. Iwasaki, M. Tanaka, and Y. Asakawa. 2002. Neogrifolin derivatives possessing anti-oxidative activity from the mushroom *Albatrellus ovinus*. *Phytochemistry* 59: 731–737.

Oeda, Y., Y. Kawai, N. Inoue, H. Shinano, Y. Nagai, Y. Hidaka, and K. Furube. 1993. Antioxidative effect of an isolate from culture filtrate of *Aspergillus niger*. *Biosci. Biotechnol. Biochem.* 57: 1374–1375.

Palozza, P., and N.I. Krinsky. 1992. Antioxidant effects of carotinoids in vivo and in vitro—an overview. *Methods Enzymol.* 213: 403–420.

Palozza, P. 1998. Prooxidant actions of carotenoids in biologic systems. *Nutr. Rev.* 56: 257–265.

Peres, V., T.J. Nagem, and E.F. Oliveira. 2000. Tetraoxygenated naturally occurring xanthones. *Phytochemistry* 55: 683–710.

Pietta, P.G. 2000. Flavonoids as antioxidants. *J. Nat. Prod.* 63: 1035–1042.

Potterat, O. 1997. Antioxidants and free radical scavengers of natural origin. *Curr. Org. Chem.* 1: 415–440.

Renaud, S., and M. De Lorgeril. 1992. Wine, alcohol, platelets, and the french paradox for coronary heart-disease. *Lancet* 339: 1523–1526.

Rosenson, R.S. 2004. Statins in atherosclerosis: Lipid-lowering agents with antioxidant capabilities. *Atherosclerosis* 173: 1–12.

Roy, D., R.A. Floyd, and J.G. Liehr. 1991. Elevated 8-hydroxydeoxyguanosine levels in DNA of diethylstilbestrol-treated syrian-hamsters-covalent DNA damage by free-radicals generated by redox cycling of diethylstilbestrol. *Cancer Res.* 51: 3882–3885.

Shapiro, B.M. 1991. The control of oxidative stress at fertilization. *Science*. 252: 533–536.

Shcherba, V.V., V.G. Babitskaya, V.P. Kurchenko, N.V. Ikonnikova, and T.A. Kukulyanskaya. 2000. Antioxidant properties of fungal melanin pigments. *Appl. Biochem. Microbiol.* 36: 491–495.

Shiomi, K., H. Yang, Q. Xu, N. Arai, M. Namiki, M. Hayashi, J. Inokoshi, H. Takeshima, R. Masuma, *et al.* 1995. Phenopyrrozin, a new radical scavenger produced by *Penicillium* sp. FO-2047. *J. Antibiot.* 48: 1413–1418.

Shu, Y.Z., Q. Ye, H. Li, F.K. Kadow, A.R. Hussain, S. Huang, R.D. Gustavson, E.S. Lowe, L.P. Chang, M.D. Pirnik, and K. Kodukula. 1997. Orevactaene, a novel binding inhibitor of HIV-1 Rev protein to Rev response element (RRE) from *Epicoccum nigrum* WC47880. *Bioorg. Med. Chem. Lett.* 17: 2295–2298.

Siegel-Axel, D.I. 2003. Cerivastatin: a cellular and molecular drug for the future? *Cell. Mol. Life Sci.* 60: 144–164.

Sies, H., ed. 1991. Oxidative Stress: *Oxidants and Antioxidants*, Academic Press, San Diego.

Stanley, L.L., and M.J. Mazier 1999. Potential explanations for the French paradox. *Nutr. Res.* 19: 3–15.

Strobel, G., E. Ford, J. Worapong, J.K. Harper, A.M. Arif, D.M. Grant, P.C.W. Fung, and R.M.W. Chau. 2002. Isopestacin, an isobenzofuranone from *Pestalotiopsis micospora*, possessing antifungal and antioxidant activities. *Phytochemistry* 60: 179–183.

Sun, C., J.W. Wang, L. Fang, X.D. Gao, and R.X. Tan. 2004. Free radical scavenging and antioxidant activities of EPS2, an exopolysaccharide produced by a marine filamentous fungus *Keissleriella* sp. YS 4108. *Life Sci.* 75: 1063–1073.

Sunda, W., D.J. Kieber, R.P. Kiene, and S. Huntsman. 2002. An antioxidant function for DMSP and DMS in marine algae. *Nature* 418: 317–320.

Tabata, N., H. Tomoda, Y. Iwai, and S. Omura. 1996. Xanthoquinodin B3, a new anticoccidial agent produced by *Humicola* sp. FO-888. *J. Antibiot*. 49: 267–271.

Takahashi, N., K. Tamagawa, K.I. Kawai, and T. Fukui. 2000. Antioxidant properties of a new type of polyene, falconensone A and its derivatives. *Biol. Pharm. Bull.* 23: 989–994.

Takamatsu, S., T.W. Hodges, I. Rajbhandari, W.H. Gerwick, M.T. Hamann, and D.G. Nagle. 2003. Marine natural products as novel antioxidant prototypes. *J. Nat. Prod.*, 66: 605–608.

Teshima, Y., K. Shin-Ya, A. Shimazu, K. Furihata, H.S. Chul, K. Furihata, Y. Hayakawa, K. Nagai, and H. Seto. 1991. Isolation and structural elucidation of pyridoxatin, a free radical scavenger of microbial origin. *J. Antibiot.* 44: 685–687.

Turner, E., L. Hager, and B.M. Shapiro. 1988. Ovothiol replaces glutathione peroxidase as an H_2O_2 scavenger in sea urchin eggs. *Science* 242: 939–941.

Vendemiale, G., I. Grattagliano, and E. Altomare. 1999. An update on the role of free radicals and antioxidant defense in human disease. *Int. J. Clin. Lab. Res.* 29: 49–55.

Watanabe, T., H. Teranishi, Y. Honda, and M. Kuwahara. 2002. A selective lignin-degrading fungus, *Ceriporiopsis subvermispora*, produces alkylitaconates that inhibit the production of a cellulolytic active oxygen species, hydroxyl radical in the presence of iron and H_2O_2. *Biochem. Biophys. Res. Comm.* 297: 918–923.

Williams, P., M. Ongsakul, J. Proudfoot, K. Croft, and L. Beilin. 1995. Mangostin inhibits the oxidative modification of human low density lipoprotein. *Free Rad. Res.* 23: 175–184.

Wiseman, H., and B. Halliwell. 1993. Carcinogenic antioxidants-diethylstilbestrol, hexoestrol and 17 alpha-ethynyloestradiol. *FEBS Lett.* 332: 159–163.

Wiseman, H., and B. Halliwell. 1996. Damage to DNA by reactive oxygen and nitrogen species: Role in inflammatory disease and progression to cancer. *Biochem. J.* 313: 17–29.

Wright, A.D., C. Osterhage, and G.M. König. 2003. Epicoccamide, a novel secondary metabolite from a jellyfish-derived culture of *Epicoccum purpurascens*. *Org. Biomol. Chem.* 1: 507–510.

Yen, G.C., Y.C. Chang, F. Sheu, and H.C. Chiang. 2001. Isolation and characterization of antioxidant compounds from *Aspergillus candidus* broth filtrate. *J. Agric. Food Chem.* 49: 1426–1431.

Yen, G.C., and C.A. Lee. 1996. Antioxidant activity of extracts from mold. *J. Food Protect.* 59: 1327–1330.

Yoshida, Y., N. Noguchi, A. Watanabe, Y. Iwaki, M. Yanagisawa, T. Kodama, O. Cynshi, and E. Niki. 2002. Antioxidant action of 2,2,4,6-tetra-substituted 2,3-dihydro-5-hydroxybenzofuran against lipid peroxidation: effects of substituents and side chain. *Free Rad. Res.* 36: 1171–1178.

Zhang, Y., G.L. Mills, and M.G. Nair. 2003. Cyclooxygenase inhibitory and antioxidant compounds from the fruiting body of an edible mushroom, *Agrocybe aegerita*. *Phytomedicine* 10: 386–390.

10

Antiangiogenic Compounds from Marine Invertebrates

TOMOFUMI MIYAMOTO

Graduate School of Pharmaceutical Sciences, Kyushu University, 3-1-1 Maidashi, Higashi-ku, 812-8582 Fukuoka, Japan

INTRODUCTION

In adults, angiogenesis is strictly restricted to processes associated with wound healing and the female reproductive cycle. When a balance between angiogenic and angiostatic factors is disrupted, however, uncontrolled angiogenic factors are released from tumor cells. These factors go on to stimulate endothelial cell (EC) proliferation, and newly formed endothelial cells break down the extracellular matrix, migrate to cancer cells, and eventually begin to form a lumen. Tumor angiogenesis is well known to play a key role in tumor growth and metastasis. The potential regulators of angiogenesis have been identified, i.e., basic fibroblast growth factor (bFGF), vascular endothelial growth factor (VEGF), matrix metalloproteinases (MMPs), methionine aminopeptidase-2 (MetAP-2), thymidine phosphorylase (TP), and plasminogen activator (PA). VEGF and bFGF stimulate endothelial cell proliferation. MMPs digest the extracellular matrix (ECM) and are responsible for the invasion and metastasis. TP is expressed at higher levels in a wide variety of solid tumors than in the adjacent nonneoplastic tissues. PA is promoted by VEGF and activates the invasion of ECs. Angiogenesis is observed in tumor growth and angiogenesis-dependent diseases such as diabetic retinopathy and arthritis. Moreover, it is known that tumor growth is dependent on a blood supply, i.e., oxygen and nutrients. Consequently, angiogenesis inhibitors may potentially be used as new therapeutic agents for cancer treatment (Folkman 1995). Several angiogenesis inhibitors, such as TNP-470 (Ingber *et al.* 1990), marimastat (Gore *et al.* 1996), and SU5416 (Fong *et al.* 1999), have been developed for targeting endothelial cell proliferation, invasion, and VEGFR signaling pathways, respectively.

The process of angiogenesis is summarized in Figure 10.1. Angiogenic factors, such as VEGF and bFGF released from a primary carcinoma, hyperexpress the PA in the ECs. PA activates the plasminogen to the plasmine. The plasmine not only digests ECM consisting of fibronectin and laminin but also activates the collagenase. ECs easily migrate from the wound ECM to interstitial fluid, proliferate, and form the lumen and new blood vessels.

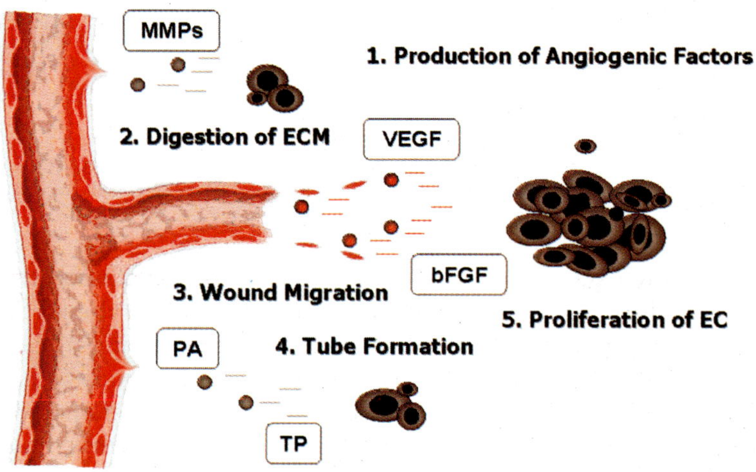

Fig. 10.1 Process of angiogenesis and angiogenic factors.

Recently, new angiogenesis inhibitors have been developed with the above molecular targets. Avastin (bevacizumab, http:// www.gene.com, http://www.avastin.com/), which is the anti-VEGF monoclonal antibody, has been accepted by the U.S. Food and Drug Administration (FDA) to be used in combination with intravenous 5-FU based chemotherapy as a treatment for patients with first-line metastatic cancer of the colon or rectum. Neovastat (AE-941, http://www.aeterna.com/) is a multifunctional antiangiogenic drug, which induces apoptosis against EC, inhibits MMP activities and VEGF signaling pathways isolated from marine cartilage. Neovastat is now the subject of a Phase III trial in non-small cell lung cancer sponsored by the U.S. National Cancer Institute (NCI). The angiogenesis inhibitors being studied are shown in Table 10.1.

Table 10.1 Angiogenesis inhibitors in clinical studies

Code	*Name*	*Mechanism*	*Sponsor*	*Phase*
BB2516	Marimastat	MMP Inhibitor	British-Biotech	III
AG3340	Prinomastat	MMP Inhibitor	Agouron	III
CGS27023A		MMP Inhibitor	Novartis	II
TNP470/AGM1470		MetAP-2 Inhibitor	Takeda/Abbott	II
RhuMadVEGF	Avastin	Anti-VEGF antibody	Genentech	III
Anti-VEGFR1	Angiozyme	Anti-VEGFR1 antibody	Ribozyme	I
	Suramin		Warner-Lambert	II/III
SU5416		Inhibitor of KDR/Flk-1	Sugen	II
AE941	Neovastat		AEtema Lab.	III
	Thalidomide	Inhibitor of TNFa Production	Entremed	II

This chapter tries to give an up-to-date description of angiogenesis inhibitors that have been isolated from marine invertebrates and to point out the fields in which new research might be most rewarding.

POLYSACCHARIDES FROM MARINE ALGAE

Fucoidan

Fucoidan is a sulfated poly-(L-fucopyranose) present in brown marine algae. Chemically oversulphated fucoidans effectively inhibit bFGF-induced migration and tube formation of HUVEC (Soeda *et al.* 1997, 2000). Oversulfated fucoidans increase the bFGF-induced release of PA inhibitor-1 antigen, but not of tissue-type PA antigen in the culture media. PA inhibitor-1 release may play a key role for the prevention of cell migration accompanied by matrix proteolysis.

Laminarin

Laminarin is a β-(1, 3)-glucan present in brown marine algae (*Laminaria*). Laminarin sulfates with different degrees of sulfation are produced by the controlled sulfation of laminarin. Laminarin sulfates inhibit bFGF binding and bFGF-induced proliferation of fetal bovine heart endothelial cells. In addition, laminarin sulfates show the antiangiogenic and anti-tumour activities in the chick chorioallantoic membrane (CAM) assay (Folkman 1975) and murine radiation-induced fibrosarcoma (RIF-1) cells *in vivo* (Hoffman *et al.* 1995, 1996).

Sulfated Galactan

An anticoagulant sulfated galactan present in green marine algae (*Codium cylindricum*) inhibits the tube formation of HUVEC. The sulfated galactan had the same biochemical properties, including chemical composition and anticoagulant activity (Matsubara *et al.* 2003).

MARINE MICROORGANISMS AND PHYTOPLANKTON

Indanone Derivative

An indanone derivative that has been isolated from the tropical marine cyanobacterium *Lyngbya majuscula* inhibits hypoxia-induced activation of the VEGF gene promoter in Hep3B human liver tumor cells at IC_{50} values of 25 μM (Nagle *et al.* 2000). Hypoxic regions are commonly found in solid tumors, and the extent of tumor hypoxia correlates with advanced stages and treatment resistance (Brown 1999).

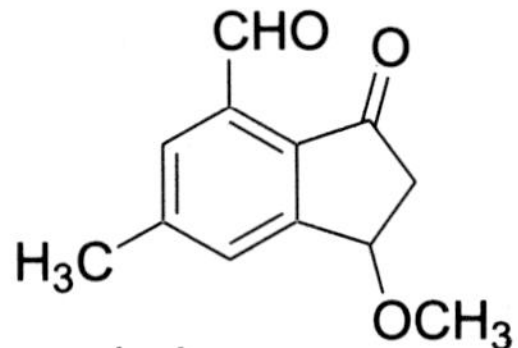

Indanone

Scytonemin

Scytonemin is a yellow-green pigment isolated from the extracellular sheath of cyanobacteria (Proteau *et al.* 1993). Scytonemin inhibits the endothelial cell growth supplement (ECGS)-induced proliferation of HUVEC in a concentration-dependent manner with an IC_{50} value of 5.4 μM. Scytonemin also inhibits recombinant human polo-like kinase 1 phosphorylation of cdc25C, which is an important enzyme for the G_2-M transition and mitotic spindle formation (Stevenson *et al.* 2002).

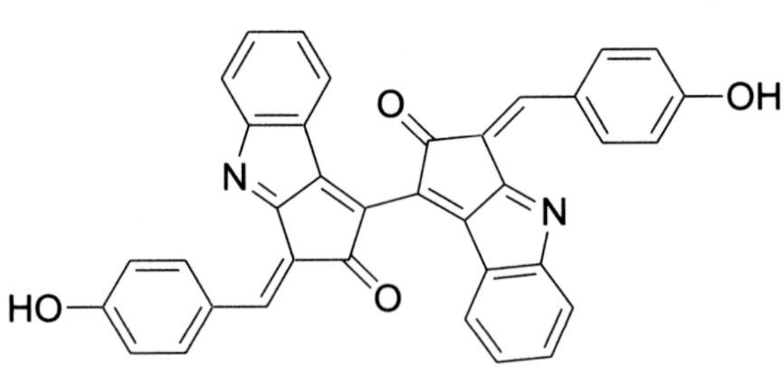

Scytonemin

SPONGES

Motuporamines

Motuporamines are anti-invasion and antiangiogenic alkaloids, have been isolated from the Papua New Guinean marine sponge *Xestospongia exigua* (Kirkpatrick) with the bioassay-guided fractionation. Motuporamine C inhibits cell migration, impairs actin-mediated membrane ruffling, and also reduces *β*1-integrin activation. In addition, motuporamine C shows the antiangiogenic activity in the chick CAM assay at the concentration of 10 μM (Roskelley *et al.* 2001). The structure-activity relationship (SAR) for anti-invasive activity has been evaluated with the natural motuporamines and the synthesizing analogues. As a result, a saturated 15-membered cyclic amine fused to the natural motuporamine diamine side chain is the optimal structure for anti-invasive activity. The SAR data have been combined with the conformational analyses to design a new carbazole analogue that showed good anti-invasion activity (Williams *et al.* 2002).

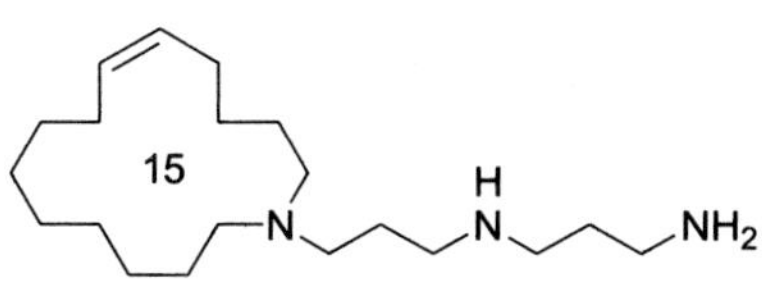

Moutuporamine C

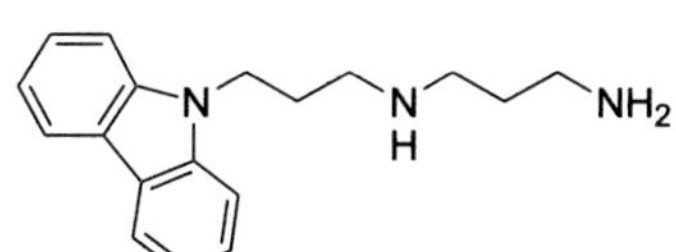

Carbazole analogue

Aeroplysinin-1

Aeroplysinin-1 is an antibacterial bromo compound isolated from the marine sponge *Verongia aerophoba* (Fattorusso *et al.* 1972). Aeroplysinin-1 inhibits the tube formation, the proliferation of EC, the migration and invasion of BAE cell *in vitro*. Aeroplysinin-1 also shows antiangiogenic activity in the chick CAM assay (Rodriguez-Nieto *et al.* 2002).

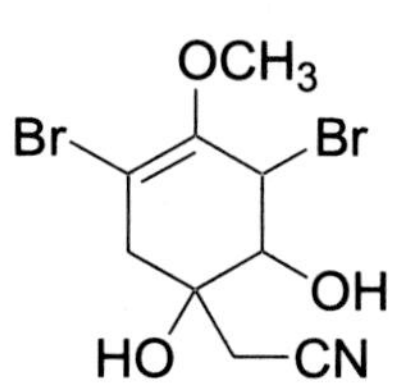

(+)-Aeroplysinin-1

Lembehsterols

Lembehsterols are sulfated sterols, isolated from the marine sponge *Petrosia strongylata*. Lembehsterols A and B show inhibitory activity against TP at the IC_{50} value of 41 and 45 μM, respectively. Most of the known TP inhibitors are nucleoside analogues, and lembehsterols are the first examples having a non-nucleoside skeleton (Aoki *et al.* 2002).

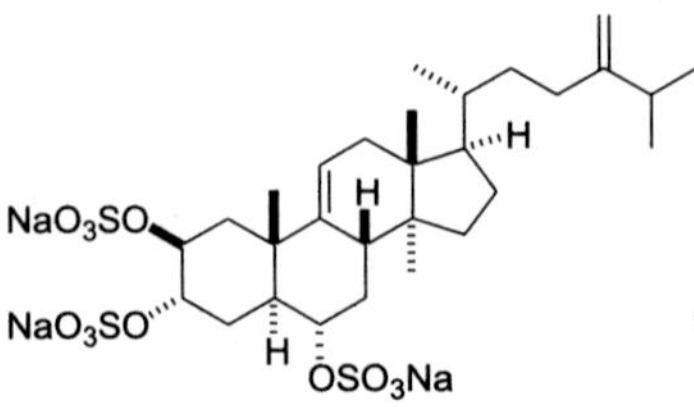

Lembehsterol A

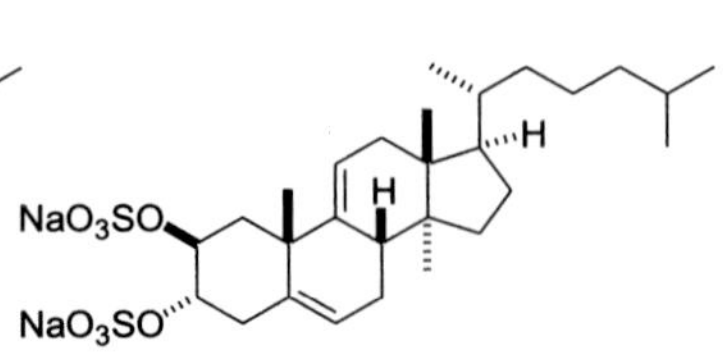

Lembehsterol B

Puupehenones

Isozonal, 8-epipuupehedione, and 8-epi-9, 11-dihydropuupehesione are puupehenone related sesquiterpene, isolated from marine sponges (Fenical *et al.* 1973, Barrero *et al.* 1998, 1999, Hamann *et al.* 1993). These three compounds inhibit EC production of urokinase and invasion, and also show the antiangiogenic activity in the CAM assay at doses equal or lower than 30 nM/egg. 8-Epipuupehedione inhibits EC migration in a dose-dependent manner (Castro *et al.* 2004).

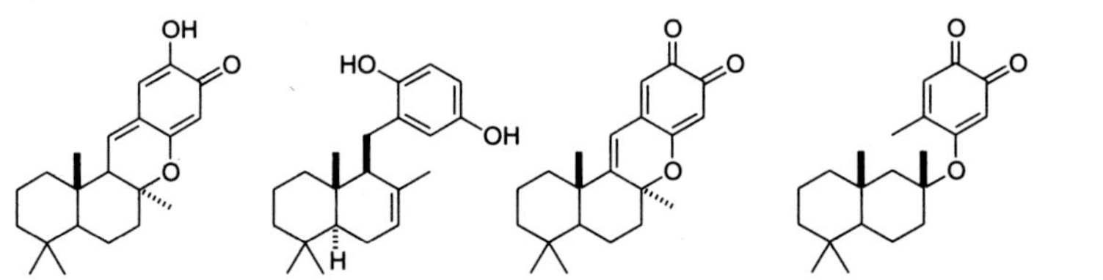

Puupehenone Isozonarol 8-Epipuupehedione 8-Epi-9,11-dihydropuupehedione

Psammaplin A

Psammaplin A is a cytotoxic bromotyrosine derivative isolated from a two-sponge association, *Poecillastra* sp. and *Jaspis* sp. (Jung *et al.* 1995). Psammaplin A was found to inhibit mammalian APN at an IC_{50} value of 18 μM that plays a key role in tumor cell invasion and angiogenesis. Moreover, psammaplin A potently inhibits the proliferation of EC and bFGF-induced invasion and tube formation by EC (Shim *et al.* 2004).

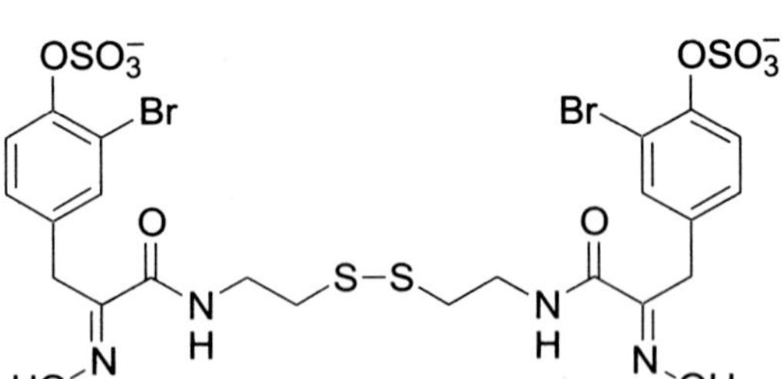

Psammaplin A

1, 3-Dimethylisoguaninium

1, 3-Dimethylisoguaninium is a purine analog isolated from a marine sponge, *Amphimedon paraviridis* (Jeong *et al.* 2003). 1, 3-Dimethylisoguaninium inhibits the bFGF-induced proliferation of BAEC, and reduces tube formation in a time-dependent manner. 1, 3-Dimethylisoguanine has previously reported as being a cytotoxic principle, from the marine sponge *Amphimedon viridis* (Mitchell *et al.* 1997, Chehade *et al.* 1997).

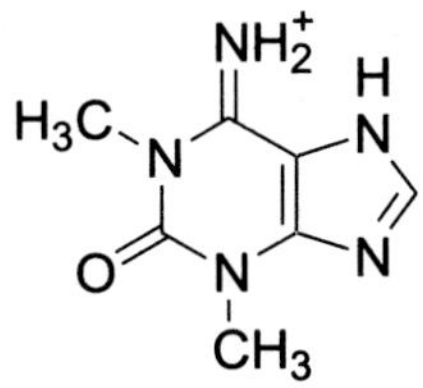

1,3-Dimethylisoguaninium

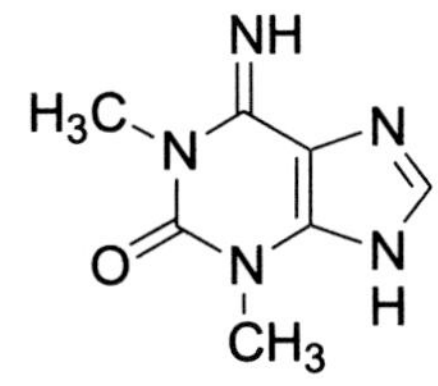

1,3-Dimethylisoguanine

Onnamide A

Onnamaide A is an antiviral metabolite isolated from a marine sponge *Theonella* sp. (Sakemi *et al.* 1988). The bioassay-guided fractionation led to the isolation of an antiangiogenic constituent for onnamaide A from a marine sponge, *Theonella mirabilis* (Miyamoto, unpublished data). Onnamide A inhibits the bFGF-induced proliferation of BAEC at an IC_{50} value of 0.26 nM and also inhibits the wound cell migration and tube formation of BAEC at the dose of 5 nM.

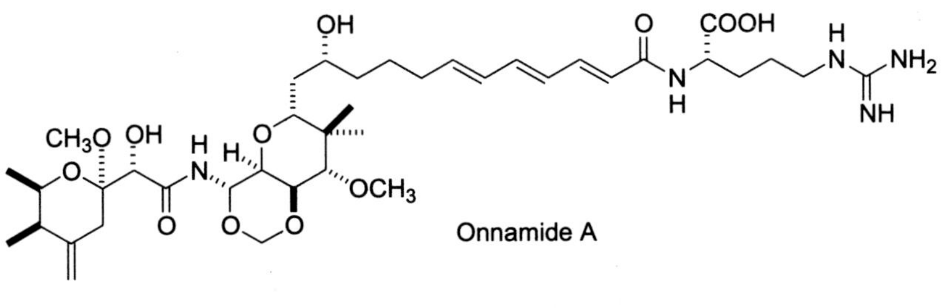

Onnamide A

Haplosamates, Ancornosides, Callysponginol Sulfate, and Ageladine

A series of experiments searching for MMP inhibitors has been reported by Fusetani's group.

Haplosamates A is a phosphorylated sterol sulfate isolated from a marine sponge, *Cribrochalina* sp., with the bioassay-guided isolation. Haplosamates A and related sterol inhibit membrane type 1 MMP (MT1-MMP) with IC_{50} values of 150 and 160 µg/mL (Fujita *et al.* 2001a, b).

Haplosamates A was first reported from Philippine haplosclerid sponges as a steroidal sulfamate ester (Qureshi and Faulkner 1999).

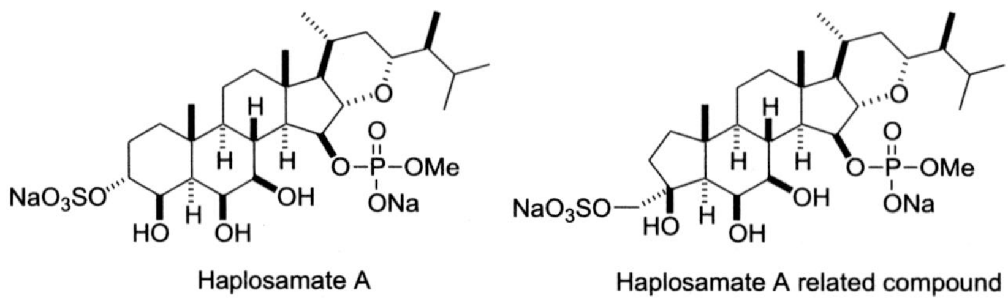

Ancorinosides B-D are terramic glycosides related to ancorinoside A (Ohta *et al*. 1997), isolated from a marine sponge, *Penares sollasi* Thiele. Ancorinosides B, C, and D inhibit MT1-MMP with IC_{50} values of 500, 370, and 180 μg/mL, respectively (Fujita *et al*. 2003a).

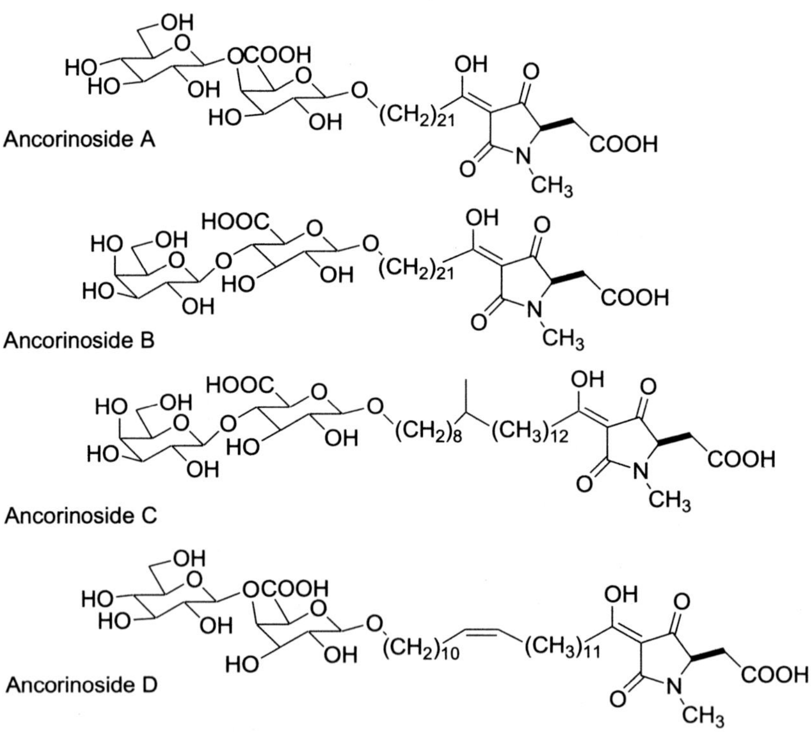

Callysponginol sulfate A is a sulfated C_{24} acetylenic fatty acid, isolated from the marine sponge *Callyspongia truncate*. Callysponginol sulfate A inhibits MT1-MMP with an IC_{50} value of 15.0 μg/mL, although the desulfated derivative does not show any inhibitory activity against MT1-MMP (Fujita *et al.* 2003b).

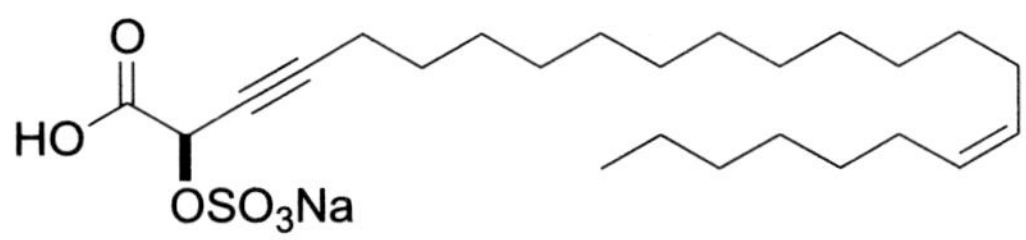

Callysponginol sulfate A

Ageladine A is a fluorescent alkaloid, isolated from the marine sponge *Agelas nakamurai*. Ageladine A inhibits not only MMP-2, but also MMPs-1, -8, -9, -12, and -13 with IC_{50} values of 2.0, 1.2, 0.39, 0.79, 0.33, and 0.47 μg/mL, respectively. The inhibition mechanism of ageladine A is different from those of other MMP-2 inhibitors, because ageladine A was not able to chelate Zn^{2+}. Ageladine A also inhibits cell migration and vascular formation *in vitro* (Fujita *et al.* 2003).

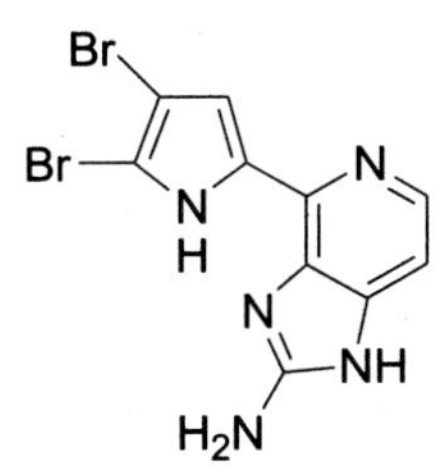

Ageladine A

BRYOZOANS

Bryoanthrathiophene

Bryoanthrathiophene, together with two known compounds, 5, 7-dihydroxy- 1-methoxycarbonyl-6-oxo-6H-anthra [1, 9-bc] thiophene, and 1, 8-dihydroxyanthraquinone (Shindo *et al.* 1993) were isolated from the bryozoan *Watersipora subtorquata* (d'Orbigny 1852) with bioassay-guided fractionation. These three anthraquinones inhibit the bFGF-induced proliferation of BAEC at IC_{50} values of 0.005, 2.2 and 5.7 μM, respectively. Bryoanthrathiophene also inhibits the tube formation of BAEC at 20 nM in a time-dependent manner (Jeong *et al.* 2002).

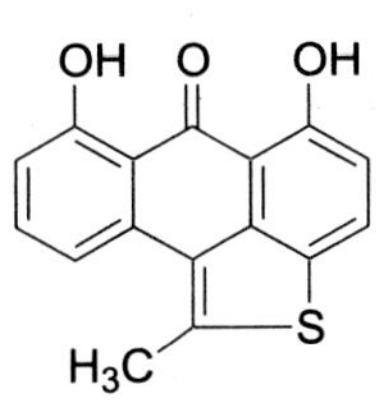

Bryoanthrathiophene

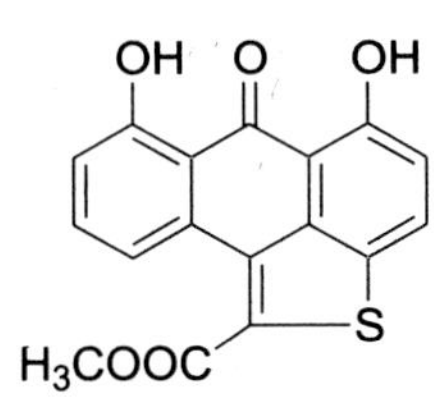

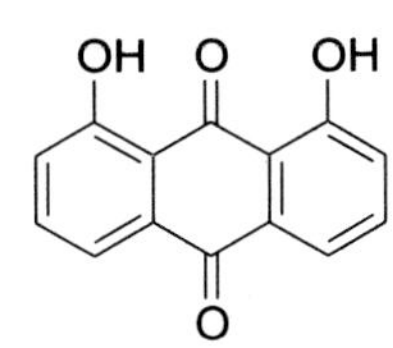

1,8-Dihydroxyanthraquinone

HUVECs (1×10^6 cells/0.3 mL) were suspended in EBM-2 containing 0.5% FBS and seeded in the inner chamber of a Transwell insert (24-well chamber with 1.33 μg/mL fibronectin-coated 8 μM polycarbonate filter). The outer chamber was incubated for 4 hr, and migrated cells were stained with Giemsa regent and counted with the aid of a microscope (Ferrara *et al.* 1997).

CAM Assay *In Vivo*

Fertilized chick eggs were incubated horizontally at 38°C in a humidified incubator, windowed by day 3 of incubation and, processed by day 8. The test samples were added to a 0.7% solution of methylcellulose in water, and 10 μL drops of this solution were allowed to dry on a Teflon-coated surface in a laminar flow hood. Then, the methylcellulose disks were implanted on CAM, and the eggs were sealed with adhesive tape and return to the incubator for 48 hr. After the reincubation, the CAM was examined under a microscope (Ribatti *et al.* 1995).

5-5 Mouse Corneal Neovascularization Assay *In Vivo*

Slow release pellets of Hydron N (Interferon Sciences) containing mouse recombinant VEGF were implanted into corneal pockets prepared by an incision placed 1mm from the limbic vessel. On the day of implantation, animals were treated once daily with either an intraperitoneal or oral

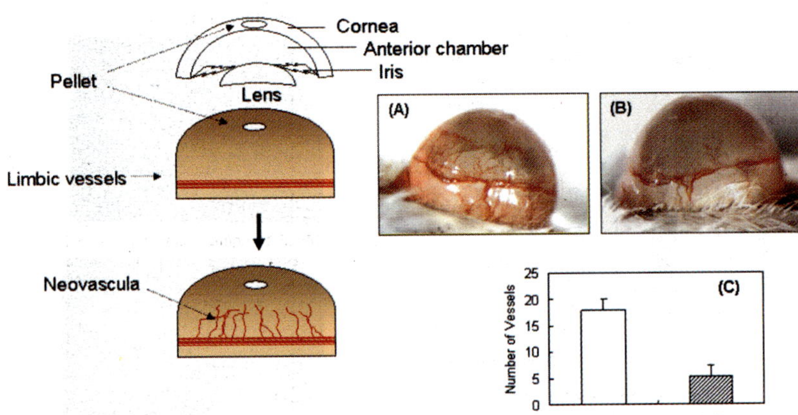

Effect of *cis*-hinokiresinol (CHR) on mouse corneal neovascularization. Corneas of mice ($n = 3$) were implanted with a pellet containing 200 ng VEGF, and the mice were treated once daily with intraperitoneal injection of CHR at 100 mg/kg/day in olive oil or olive oil alone. After 7 days, the animals were anesthetized, and the vessels were photographed. **(A)** VEGF alone, **(B)** VEGF and CHR **(C)** The number of vessels from limbic vessel. Each column represents the mean of three corneas; bars indicate SD.

Fig. 10.5 Mouse corneal neovascularization assay (Jeong *et al.* 2003a,b).

injection of test samples. After 7 days, the animals were anesthetized, and the vessels were photographed. The contiguous circumferential zone of neovascularization was measured (Kenyon *et al.* 1996).

REFERENCES

Aoki, S., Y. Naka, T. Itoh, T. Furukawa, R. Rachmat, S. Akiyama, and M. Kobayashi. 2002. Lembehsterols A and B, novel sulfated sterols inhibiting thymidine phosphorylase, from the marine sponge *Petrosia strongylata*. *Chem. Pharm. Bull.* 50: 827–830.

Barrero, F.A., J.E. Alvarez-Manzaneda, M.M. Herrador, M. Valdivia, and R. Chahboun. 1998. Synthesis of monoterpenic analogues of puupehenone and puupehedione. *Tetrahedron Lett.* 39: 2425–2428.

Barrero, F.A., J.E. Alvarez-Miranda, R. Chahboun, M. Cortes, and V. Armstrong. 1999. Synthesis and antitumor activity of puupehedione and related compounds. *Tetrahedron* 55: 15181–15195.

Brown, J.M. 1999. The hypoxic cell: A target for selective cancer therapy. *Cancer Res.* 59: 5863–5870.

Castro, M.E., M. Gonzalez-Iriarte, A.F. Barrero, N. Salvador-Tormo, R. Munoz-Chapuli, M.A. Medina, and A.R. Quesada. 2004. Study of puupehenone and related compounds as inhibitors of angiogenesis. *Int. J. Cancer* 110: 31–38.

Chehade, C.C., R.L.A. Dias, R.G.S. Berlinck, A.G. Ferreira, L.V. Costa, M.R. Rangel, E.L.A. Malpezzi, J.C. de Freitas, and E. Hajdu. 1997. 1, 3-Dimethylisoguanine, a new purine from the marine sponge *Amphimedon viridis*. *J. Nat. Prod.* 60: 729–731.

Fattorusso, E., L. Minale, and G. Sodano. 1972. Aeroplysinin-1, an antibacterial bromocompound from the sponge *Verongia aerophoba*. *J. Chem. Soc. Perkin 1.* 1: 16-18.

Fenical, W., J.J. Shims, D. Squatrit, R.M. Wing, and P. Radlick. 1973. Zonarol and isozonarol, fungitoxic hydroquinones from brown seaweed *Dictyopteris-zonarioides*. *J. Org. Chem.* 38: 2383–2386.

Ferrara, N., and T. Davis-Smyth. 1997. The biology of vascular endothelial growth factor. *Endocrine Rev.* 18: 4–25.

Folkman, J. 1975. Tumour angiogenesis. *Adv. Cancer. Res.* 43: 175–203.

Folkman, J. 1995. Angiogenesis in cancer, vascular, rheumatoid and other diseases. *Nat. Med.* 1: 27–31.

Fong, T.A.T., L.K. Shawver, L. Sun, C. Tang, H. App, T.J. Powell, Y.H. Kim, R. Schreck, X. Wang, W. Risau, A. Ullrich, K.P. Hirth, and G. McMahon. 1999. SU5416 is a potent and selective inhibitors of the vascular endothelial growth factor receptor (Flk-1/KDR) that inhibits tyrosine kinase catalysis, tumor vascularization, and growth of multiple tumor types. *Cancer Res.* 59: 99–106.

Fujita, M., Y. Nakao, S. Matsunaga, M. Seiki, Y. Itoh, R.W.M. van Soest, M. Heubes, D.J. Faulkner, and N. Fusetani. 2001a. Isolation and structure elucidation of two phosphorylated sterol sulfates, MT1-MMP inhibitors from a marine sponge *Cribrochalina* sp.: revision of the structures of haplosamates A and B. *Tetrahedron* 57: 3885–3890.

Fujita, M., Y. Nakao, S. Matsunaga, M. Seiki, Y. Itoh, R.W.M. van Soest, and N. Fusetani. 2001b. Ancorinosides B-D, inhibitors of membrane type 1 matrix metalloproteinase (MT1-MMP), from the marine sponge *Penares sollasi* Thiele. *Tetrahedron* 57: 1229–1234.

Fujita, M., Y. Nakao, S. Mastunaga, M. Seiki, Y. Itoh, J. Yamashita, R.W.M. van Soest, and N. Fusetani. 2003a. Ageladine A: An antiangiogenic matrixmetalloproteinase inhibitors from the marine sponge *Agelas nakamurai*. *J. Am. Chem. Soc.* 125: 15700–15701.

Fujita, M., Y. Nakao, S. Matsunaga, R.W.M. van Soest, Y. Itoh, M. Seiki, and N. Fusetani. 2003b. Callysponginol sulfate A, an MT1-MMP inhibitor isolated from the marine sponge *Callyspongia truncate*. *J. Nat. Prod.* 66: 569–571.

Gore, M., R. A'Hern, M. Stankiewicz, and M. Slevin. 1996. Tumour marker levels during marimastat therapy. *Lancet* 348: 263–264.

Hamann, T.M., P.J. Scheuer, and M. Kelly-Borges. 1993. Biogenetically diverse, bioactive constituents of a sponge, order Verongida: bromotyramines and sesquterpene-shikimate derived metabolites. *J. Org. Chem.* 58: 6565–6569.

Hoffman, R., D.H. Paper, J. Donaldson, and H. Vogl. 1996. Inhibition of angiogenesis and murine tumour growth by laminarin sulphate. *Brit. J. Cancer* 73: 1183–1186.

Hoffman, R., D.H. Paper, J. Donaldson, S. Alban, and G. Franz. 1995. Characterisation of a laminarin dulphate which inhibits basic fibrobrast growth factor binding and endothelial cell proliferation. *J. Cell Sci.* 108: 3591–3598.

Ingber, D.,T. Fujita, S. Kishimoto, K. Sudo, T. Kanematsu, H. Brem, and J. Folkman. 1990. Synthetic analogues of fumagillin that inhibit angiogenesis and suppress tumour growth. *Nature* 348: 555–557.

Jeong, S.-J.R. Higuchi, T. Miyamoto, M. Ono, M. Kuwano, and S.F. Mawatari. 2002. Bryoanthrathiophene, a new antiangiogenic constituent from the bryozoan *Watersipora subtorquata* (d'Orbigny, 1852). *J. Nat. Prod.* 65: 1344–1345.

Jeong, S.-J., R. Higuchi, M. Ono, M. Kuwano, Y-C. Kim, and T. Miyamoto. 2003a. cis-Hinokiresinol, a norlignan from *Anemarrhena asphodeloides*, inhibits angiogenic response *in vitro* and *in vivo*. *Biol. Pharm. Bull.* 26: 1721–1724.

Jeong, S.-J., M. Inagaki, R. Higuchi, T. Miyamoto, M. Ono, M. Kuwano, and R.W.M. Van Soest. 2003b. 1, 3-Dimethylisoguaninium, an antiangiogenic purine analogue from the sponge *Amphimedon paraviridis*. *Chem. Pharm. Bull.* 51: 731–733.

Jung, J.H., C.J. Sim, and C.O. Lee. 1995. Cytotoxic compounds from a two-sponge association. *J. Nat. Prod.* 58: 1722–1726.

Kenyon, M.B., E.E. Voest, C.C. Chen, E. Flynn, J. Folkman, and R.J.D'Amato. 1996. A model of angiogenesis in the mouse cornea. *Inves. Ophthalmol. Vis. Sci.* 37: 1625–1632.

Matsubara, K., M. Mori, H. Matsumoto, K. Hori, and K. Miyazawa. 2003. Antiangiogenic properties of a sulfated galactan isolated from a marine green alga, *Codium cylindricum*. *J. Appl. Phycol.* 15: 87–90.

Mitchell, S.S., A.B. Whitehill, H.G. Trapido-Rosenthal, and C.M. Ireland. 1997. Isolation and characterization of 1, 3-dimethylisoguanine from the Bermudian sponge *Amphimedon viridis*. *J. Nat. Prod.* 60: 727–728.

Nagle, G.D., Y-D. Zhou, P.U. Park, V.J. Paul, I. Rajbhandari, C.J.G. Duncan, and D.S. Pasco. 2000. A new indanone from the marine cyanobacterium *Lyngbya majuscule* that inhibits hypoxia-induced activation of the VEGF Promoter in Hep3B cells. *J. Nat. Prod.* 63: 1431–1433.

Ohta, S., M. Ohta, S. Ikegami. 1997. Ancorinoside A: A novel tetramic acid glycoside from the marine sponge, *Ancorina* sp. which specifically inhibits blastulation of starfish embryos. *J. Org. Chem.* 62: 6452–6453.

Proteau, P.J., W.H. Gerwick, F. Garcia-Pichel, and R. Castenholtz. 1993. The structure of scytonemin, an ultraviolet sunscreen pigment from the sheaths of cyanobacteria. *Experientia* 49: 825–829.

Qureshi, A., and D.J. Faulkner. 1999. Haplosamates A and B: New steroidal sulfamate esters from two *Haplosclerid* sponges. *Tetrahedron* 55: 8323–8330.

Ribatti, D., C. Urbinati, B. Nico, M. Rusanti, L. Roncali, and M. Presta. 1995. Endogenous basic fibroblast growth factor is implicated in the vascularization of the chick embryo chorioallantoic membrane. *Dev. Biol.* 170: 39–49.

Rodriguez-Nieto, S., M. Gonzalez-Iriarte, R. Carmona, R. Munoz-Chapuli, M.A. Medina, and A.R. Quesada. 2002. Antiangiogenic activity of aeroplysinin-1, a brominated compound isolated from a marine sponge. *FASEB J.* 16: 261–263.

Roskelley, D.C., D.E. Williams, L.M. McHardy, K.G. Leong, A. Troussard, A. Karsan, R.J. Andersen, S. Dedhar, and M. Roberge, 2001. Inhibtion of tumor cell invasion and angiogenesis by motuporamines. *Cancer Res.* 61: 6788–6794.

Sakemi, S., T. Ichiba, S. Kohmoto, G. Saucy, and T. Higa. 1988. Isolation and structure elucidation of onnamide A, a new bioactive metabolite of a marine sponge, *Theonella* sp. *J. Am. Chem. Soc.* 110: 4851–4853.

Shim, S.J., H-Y. Lee, J. Shin, and H.O. Kwon. 2004. Psammaplin A, a marine natural product, inhibits aminopeptidase N and suppresses angiogenesis *in vitro*. *Cancer Lett.* 203: 163–169.

Shindo, T., A. Sato, N. Kasanuki, K. Hasegawa, S. Sato, T. Iwata, and T. Hata. 1993. 6H-anthra [1,9-bc]thiophene derivatives from a bryozoan, *Dakaira subovoidea*. *Experientia* 49: 177–178.

Soeda, S., T. Kozako, K. Iwata, and H. Shimeno. 2000. Oversulfated fucoidan inhibits the basic fibroblast growth factor-induced tube formation by human umbilical vein endothelial cells: its possible mechanism of action. *Biochim. Biophys. Acta* 1497: 127–134.

Soeda, S., Y. Shibata, and H. Shimeno. 1997. Inhibitory effect of oversulfated fucoidan on tube formation by human vascular endothelial cells. *Biol. Pharm. Bull.* 20: 1131–1135.

Stevenson, C.S., E.A. Capper, A.K. Roshak, B. Marquez, K. Grace, W.H. Gerwick, R.S. Jacobs, and L.A. Marshall. 2002. Scytonemin-a marine natural product inhibitor of kinases key in hyperproliferative inflammatory diseases. *Inflamm. Res.* 51: 112–114.

Ushio, S., M. Ono, J. Nakayama, T. Fujiwara, Y. Komatsu, K. Suguyama, and M. Kuwano. 1997. New nortriterpenoid isolated from anti-rheumatoid arthritic plant, *Tripterygium wilfordii*, modulates tumor growth and neovascularization. *Int. J. Cancer* 72: 657–663.

Williams, E.D., K.S. Craig, B. Patrick, L.M. McHardy, R. van Soest, M. Roberge, and R.J. Andersen. 2002. Motuporamines, anti-invasion and anti-angiogenic alkaloids from the marine sponge *Xestospongia exigua* (Kirkpatrick): Isolation, structure elucidation, analogu synthesis, and conformational analysis. *J. Org. Chem.* 67: 245–258.

11

Biologically Active Terpenoids from Sponges

Tatsuo Higa and Masayuki Kuniyoshi

Department of Chemistry, Biology, and Marine Science, University of the Ryukyus, Nishihara, Okinawa 903-0213, Japan

INTRODUCTION

Terpenoids are naturally-occurring compounds, most of which are synthesized through a mevalonic acid pathway. Their carbon backbones can be regarded as consisting of isoprene (C_5) units. Depending on the number of isoprene units, they are classified as monoterpenoids (C_{10}), sesquiterpenoids (C_{15}), diterpenoids (C_{20}), sesterterpenoids (C_{25}), and triterpenoids (C_{30}). Sponges are a major source of terpenoids, ranging from sesquiterpenes to triterpenes as shown in Table 11.1 which gives the approximate numbers of marine terpenoids recorded by the end of 1993 (Higa, unpublished data). A total of some 2000 terpenoids had been reported from various marine organisms by that time. We have no accurate count at present, but the number could well have doubled. The major sources of diterpenoids, which amounted to more than a half of the 1993 count, were octocorals and brown algae. Although sponges contain all types of terpenoids except for monoterpenes, the invertebrates could best be characterized by the wealth of sesterterpenes, *vide infra*. Marine monoterpenes are virtually confined to certain species of red algae.

In Volume 6 of the monograph series "Recent Advances in Marine Biotechnology" edited by Fingerman and Nagabhushanam, we contributed a chapter entitled "Recent Developments on Antimicrobial Metabolites from Marine Sponges" (Kuniyoshi and Higa 2001). In that

Table 11.1 Approximate number of marine terpenoids reported by the end of 1993.

Classification	*Number*	*Percentage*	*Major sources**
Monoterpene	80	4	Red alga
Sesquiterpene	550	28	Red alga, Sponge, Octocoral, Mollusc
Diterpene	1100	55	Octocoral, Sponge, Brown alga, Green alga, Mollusc
Sesterterpene	150	7.5	Sponge, Mollusc
Triterpene	110	5.5	Sponge, Red alga, Green alga, Mollusc
Total	1990	100	

* Most of the terpenoids derived from molluscs are dietary origin.

article we reviewed various classes of antimicrobial sponge compounds published in the period from 1990 to 1999. Nearly 30% of the compounds covered were terpenoids. In the present chapter we review sponge-derived terpenoids reported with biological activity in the period from 1990 to 2003. For antimicrobial compounds, we include only those that appeared in the literature after 2000. A reader is referred to our previous chapter for the earlier works on antimicrobial terpenoids.

In this chapter we also include some bioactive terpenoids derived from nudibranchs, because many of them are believed to have originated in sponges, a dietary source of nudibranchs. Biological activities described for sponge terpenoids vary from antibacterial, antifungal, antiviral, cytotoxic, antitumoral, and antimalarial and to inhibition of various enzymes. We attempt to cover compounds having any of these activities.

The discussion here will be organized by classifying the active compounds into sections dealing with the structural types: sesquiterpenoid, diterpenoid, sesterterpenoid, and triterpenoid. Meroterpenoids, compounds of mixed biogenesis consisting of terpenoid and non-terpenoid skeletons, will be discussed in each relevant section. However, a small group of polyprenylquinones and their derivatives will be treated in a separate section because many of them have been reported to co-occur as a series of compounds having different levels of prenylation.

As before, we focus on the variation of the structures of active compounds and keep the chemical descriptions as brief as possible. No attempt is made to describe structure elucidation. There are abundant papers describing the chemical synthesis of terpenoids. Some of them will briefly be cited without details of synthesis.

SESQUITERPENOIDS

The sesquiterpenoids are C_{15} compounds consisting of three isoprene units. There are more than 100 structural types classified, depending on their carbon skeletons, all of which, however, arise from a common precursor, farnesyl pyrophosphate (**1**). Sesquiterpenoids are known from many organisms, particularly from higher plants. The main sources of marine sesquiterpenes are red algae of the genus *Laurencia*, octocorals, and sponges. In this section we discuss sponge-derived bioactive sesquiterpenoids in several subsections, based on either skeletal classes or functional groups.

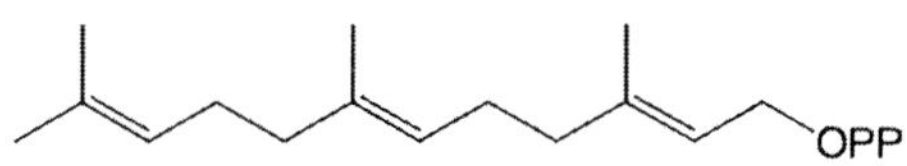

1

Bisabolane Derivatives

Since organotin compounds, widely used synthetic antifouling agents, have become environmental issues, there has been active research to seek naturally-occurring antifouling compounds. Fusetani's group identified a number of isocyanoterpenoids and related compounds of sponge origin as active compounds against barnacle larvae. One of them was 3-isocyanotheonellin (**2**) first discovered from a *Theonella* sponge. It was reisolated from the nudibranch *Phyllidia pustuiosa* for possible use as a potent antifouling substance against larvae of the barnacle *Balanus amphitrite* (Okino *et al*. 1996a, Fusetani *et al*. 1996). The isonitrile **2** and its stereoisomers have been synthesized and their antifouling activity tested (Kitano *et al*. 2002). All synthetic analogues showed similar potency.

An aromatic bisabolene, curcuphenol (**3**), was reported to have a variety of biological properties including cytotoxic and antifungal activity and inhibition of H, K-ATPase. In 1999 several new derivatives of **3**, parahigginols (e.g., parahigginol B: **4**) and parahiggic acid (**5**), were isolated as cytotoxic constituents of a sponge, *Parahigginsia* sp. (Chen *et al*. 1999). These compounds exhibited mild cytotoxicity against P-388 murine leukemia cells and human mouth epidermoid (KB16), lung (A549), and colon (HT-29) cancer cells. Related compounds have also been isolated from an Australian sponge, *Arenochalina* sp. (Butler *et al*. 1991), but no activity was reported.

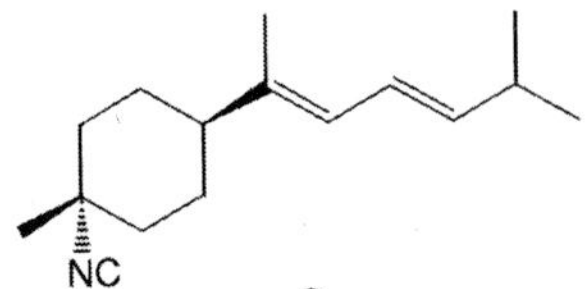

2

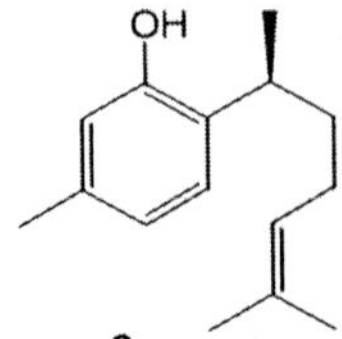

3

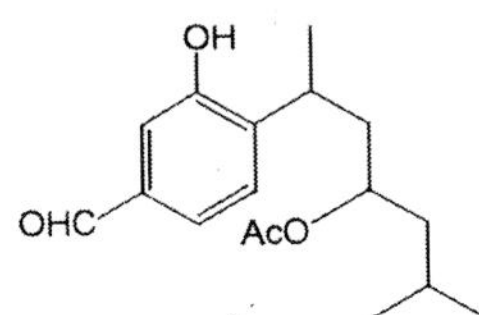

4

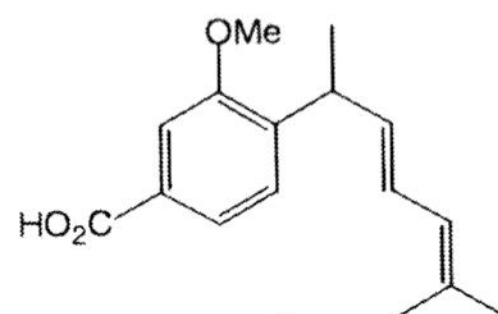

5

Cyclofarnesanes

Two papers describing sesquiterpenes of the cyclofarnesane skeleton enter discussion here. The first of them deals with the metabolites of a nudibranch. Two cyclofarnesoylglycerides, tanyolides A (**6**) and B (**7**), were isolated as feeding deterrent constituents of the Californian nudibranch *Sclerodoris tanya* (Krug *et al.* 1995). The structures of **6** and **7** elucidated with spectroscopic data were confirmed by synthesis. Both compounds showed potent anti-feeding but no ichthyotoxic activity against the ecologically-relevant fish *Gibbonsia elegans*. Tanyolide B (**7**) had no antimicrobial or cytotoxic activity. These tanyolides may not have been derived from a sponge because there are ample biosynthetic studies that reveal *de novo* biosynthesis of terpene glycerides by nudibranchs.

A Thai collection of the sponge *Dysidea arenaria* gave rise to two sesquiterpene cyclic ethers, arenarans A (**8**) and B (**9**), of which **8** was mildly cytotoxic against A549, HT-29, HCT-29 (human colon adenocarcinoma) and P388 cell lines (Horton and Crews 1995). Arenaran B (**9**) was inactive against each of these cell lines.

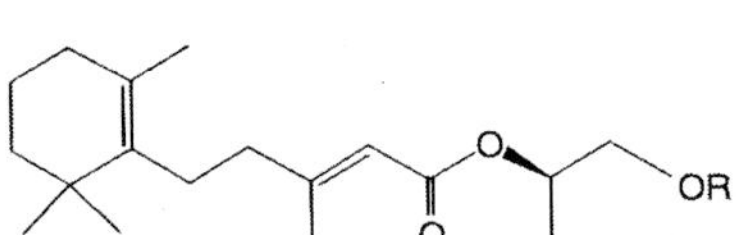

6 R = Ac
7 R = H

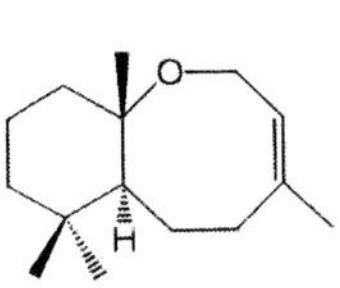

8

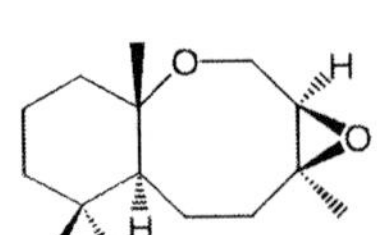

9

Drimane Derivatives

Drimane sesquiterpenes are those having a carbon skeleton (**10**) consisting of two fused six-membered rings. A well-known example is polygodial (**11**), a feeding deterrent constituent of terrestrial plants, marine sponges, and nudibranchs. Polygodial is also known to exhibit antimicrobial and cytotoxic activities. Barrero *et al.* (1995) synthesized polygodial (**11**) and its analogues and tested them for antifeedant, cytotoxic, and antimicrobial properties. A number of compounds related to polygodial have been reported from sponges and nudibranchs.

From an undescribed sponge of the genus *Dysidea* collected in Guam, Paul *et al.* (1997) isolated 10 drimane sesquiterpenes including **11**, 7-deacetoxyolepupuane (**12**), and seven new ones (e.g., **13**, **14**). Compound **12**, from an Australian *Dysidea* sp. was reported to have antimicrobial and cytotoxic properties (Garson *et al.* 1992). In antimicrobial assays, **11** exhibited potent inhibition of the bioluminescence reaction of *Photobacterium leiognathi*, a symbiotic luminous bacterium of tropical fishes, while **14** and others showed weaker activity. Compounds **11**, **13**, and **14** showed moderate inhibition of Na^+/K^+ -ATPase and **14** also weak activity against PLA_2.

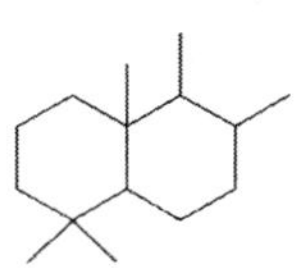

10

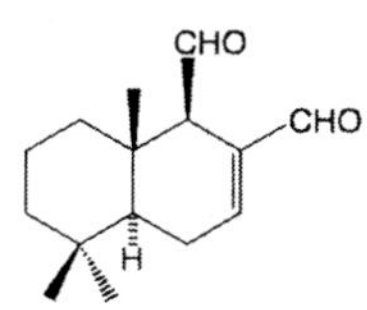

11

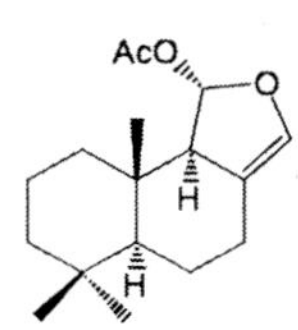

12

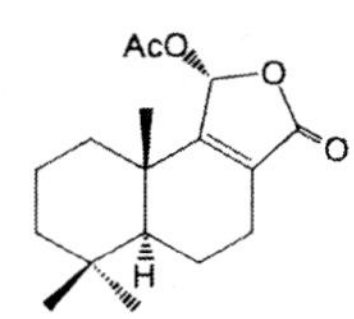

13

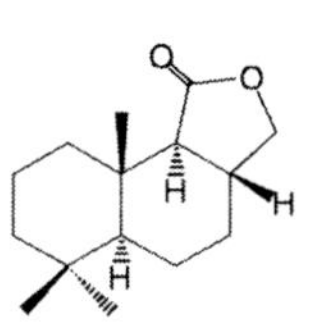

14

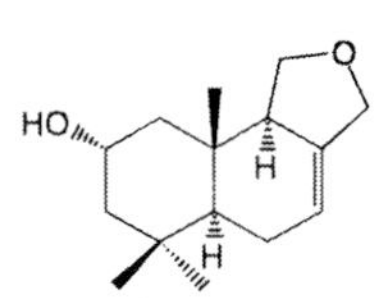

15

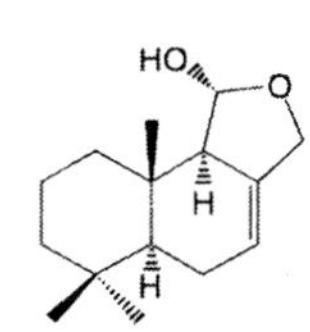

16

Fourteen new (e.g., **15**) and two known related compounds have been reported from the nudibranch *Dendrodoris carbunculosa* collected in Japan (Sakio *et al.* 2001). All of their structures were based on the tricyclic system represented by **15**. Cytotoxicity assays of these compounds were performed using adriamycin- and vincristine-resistant P388 cell lines and the sensitive P388 strain. Some of them exhibited mild activity and others weak activity. The known and major component isodrimeninol (**16**) was shown to be responsible for the peppery taste of the animal.

From another collection of *Dysidea* sp. from temperate Australia several related compounds have been reported without biological activity information (Butler and Capon 1993). Although many of these drimane sesquiterpenes are metabolites of sponges, it is also known that polygodial and some other related compounds are *de novo* biosynthetic products of certain nudibranchs.

Furanosesquiterpenes

The furanosesquiterpene furodysin (**17**) and many related compounds have been reported from *Dysidea* spp. and several species of nudibranchs. These compounds, being antifeedant, are believed to play a defensive role in these organisms. Three compounds (**18-20**) enter discussion here.

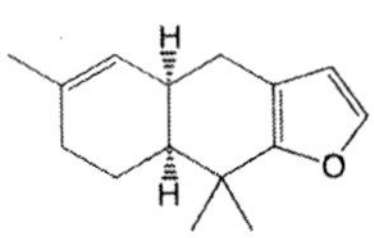

17

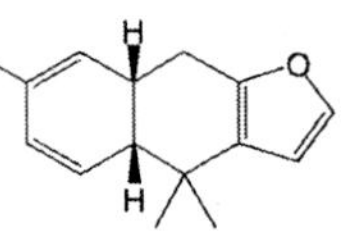

18

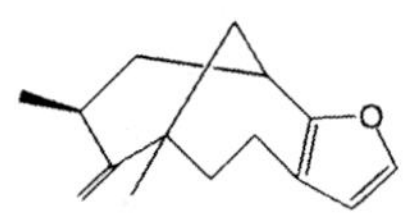

19

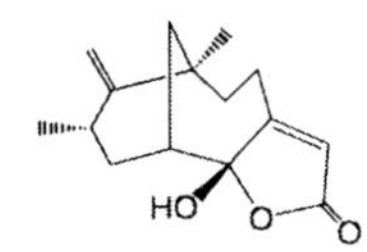

20

9,10-Dehydrofurodysinin (**18**) was isolated from the nudibranch *Hypselodoris webbi* collected off Blanes, Spain (Fontana *et al.* 1994). Compound **18** showed antifeedant activity against the freshwater fish *Carassius auratus* and cytotoxicity (LD_{50} 1.52 ppm) in the brine shrimp assay. It also exhibited ichthyotoxicity against *Gambusia affinis* at 10 ppm.

The Great Barrier Reef sponge *Dysidea* sp. nov. gave rise to two new isonakafurans (Flowers *et al.* 1998). One of them, a hydroperoxide derivative of **19**, was cited as an antimicrobial compound in our previous review (Kuniyoshi and Higa 2001). This and its parent compound, $\Delta^{7,14}$-isonakafuran-9 (**19**), showed growth inhibitory activity against P388 cells.

In search of antifouling substances from marine organisms, Sera *et al.* (1999) found a compound (**20**) in *Dysidea herbacea* that exhibited repellent activity against the blue mussel *Mytilus edulis galloprovincialis*. The compound (**20**) was shown to be a derivative of nakafuran-9 known from *Dysidea* sp.

Sesquiterpenoids with Isocyano and Related Functional Groups

Terpenoids with an isocyano (-NC), isothiocyanato (-NCS), or formamido (-NHCHO) group often co-occur and are characteristic metabolites of a certain group of sponges. Also found, but less abundant, are the metabolites with a cyano (-CN), isocyanato (-NCO), or thiocyanato (-SCN) group. They usually occur as sesquiterpenoids (several classes) and diterpenoids (a few classes). The compounds to be discussed here fall into six skeletal classes and have such biological properties as antimicrobial, antimalarial, antifouling, and antihelminthic activity.

In our earlier review, axisonitrile 3 (**21**) was cited as a compound having antimicrobial (Alvi *et al*. 1991) and antimalarial activities (Angerhofer *et al*. 1992). This compound has also been reported to have anthelminthic activity against *Nippostrongylus brasiliensis* (Alvi *et al*. 1991) and antifouling activity against the barnacle *Balanus amphitrite* (Okino *et al*. 1996a). The isothiocyanato derivative (**22**) of axisonitrile 3 was also reported from the sponges *Acanthella klethra* and *A. cavernosa* (Hirota *et al*. 1996). The antimalarial activity of **22** was 500-fold weaker than that of **21**. Axisonitrile 3 (**21**) showed comparable activity with those of existing drugs (e.g., chloroquine) against the chloroquine-resistant strain of the malaria parasite *Plasmodium falciparum* (Angerhofer *et al*. 1992). An epimer of **21**, 10-*epi*-axisonitrile 3 (**23**) has been isolated as an antifouling constituent of the nudibranch *Phyllidia pustulosa* (Okino *et al*. 1996a). Its hydrated product, exiguamide (**24**), was recently reported from the sponge *Geodia exigua* together with **23** and two other derivatives (Uy *et al*. 2003). Exiguamide (**24**) inhibited cell fate specification in the development of sea urchin embryos and was suggested to be a useful tool to embryogensis studies.

Compounds **25-29** are examples of different carbon skeletons. The isonitriles **25** and **28** are antifouling constituents of the nudibranch *Phyllidia varicosa* (Okino *et al*. 1996a) and **26** that of the sponge *Acanthella cavernosa* (Hirota *et al*. 1996). The isothiocyanate **27**, listed as an example of an eudesmane skeleton, is a constituent of the sponge *A. klethra* (Angerhofer *et al*. 1992). It showed no significant biological activity. Compound **29**, 9-thiocyanatopupukeanane, was isolated as a mixture with its 9-epimer from both the sponge *Axinyssa aculeata* and the nudibranch *Phyllidia varicosa* (Yasman *et al*. 2003). The mixture exhibited toxicity against brine shrimp (LD_{50} 5 ppm) and antimicrobial activity against *Bacillus subtilis* and *Candida albicans*.

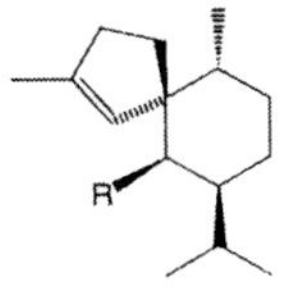

21 R = NC
22 R = NCS

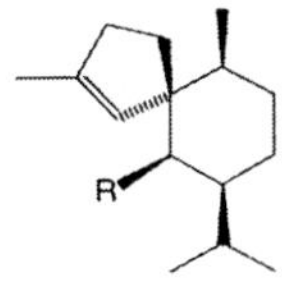

23 R = NC
24 R = NHCHO

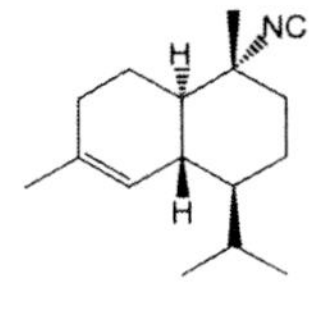

25

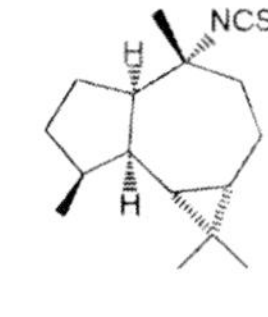

26

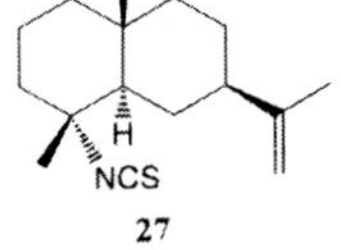

27

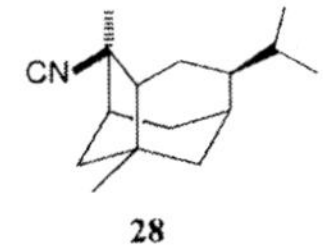

28

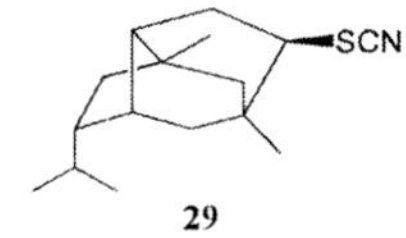

29

Sesquiterpene Carbonimidic Dichlorides

The sesquiterpene carbonimidic dichlorides are a small group of compounds closely related to those discussed in the preceding subsection. Carbonimidic dichloride (= dichloroimine –N=CCl_2) is a rare functionality found so far only with sesquiterpenes of sponge origin. A biosynthetic study by Simpson *et al.* (1997) using ^{14}C-labeled cyanide and thiocyanate suggested that –NC and –NCS groups are involved in the biosynthesis of the carbonimidic dichloride.

Axinyssimide A (**30**) and two related compounds have been reported as the constituents responsible for antifouling activity of the extract of a sponge, *Axinyssa* sp. (Hirota *et al.* 1998). Reticulidin A (**31**) and its isomer, reticulidin B, were isolated together with two known related compounds as cytotoxic constituents of the nudibranch *Reticulidia fungia* collected in Okinawa (Tanaka and Higa 1999). These compounds were moderately active against KB and L1210 cells. More recently Musman *et al.* (2001a) obtained five new (e.g., **32, 33**) and seven known related compounds from the sponge *Stylotella aurantium* also collected in Okinawa. The new compounds showed moderate cytotoxicity against P388, A549, HT29, and MEL28 cell lines with IC_{50} values of 0.1–1 μg/mL. The aldehyde **33** could be either a catabolic product of the corresponding carbonimidic dichloride or its biosynthetic precursor. Related compounds have also been reported from the Australian sponge *Ulosa spongia* (Kehraus *et al.* 2001).

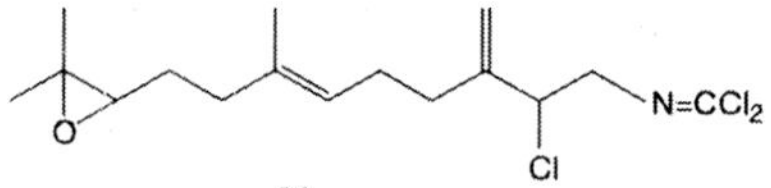

30

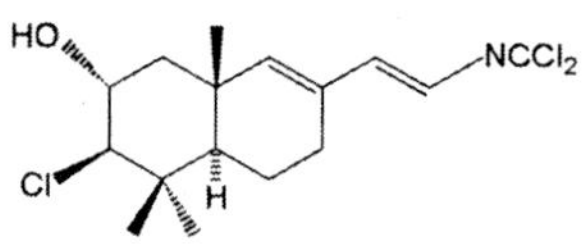

31

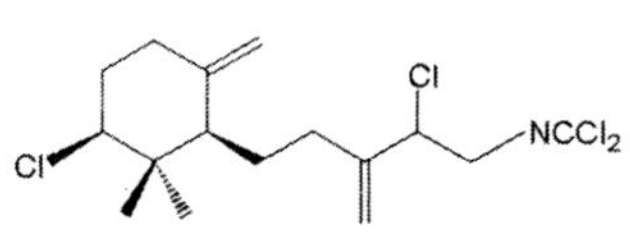

32

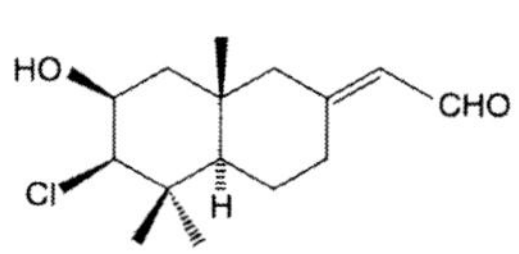

33

Merosesquiterpenoids

Meroterpenoids are of mixed biogenesis, containing a carbon-carbon bond between a terpenoid and a non-terpenoid moiety. A large number of meroterpenoids have been reported from marine organisms, especially sponges. The non-terpenoid portions of marine meroterpenoids are usually quinone/quinol derivatives. The terpenoid moieties of sponge merosesquiterpenoids are based mainly on drimane (**10**) or its rearranged 9,4-friedodrimane skeleton. A few of them are based on cyclofarnesane or its rearranged skeleton.

From the sponge *Hippospongia metachromia* collected in Okinawa, Kobayashi *et al.* (1992) isolated five cytotoxic compounds designated metachromins D-H which exhibited moderate activity against murine lymphoma L1210 cells and human epidermal carcinoma KB cells. Metachromin E (**34**) was most active with IC_{50} values of 0.2 and 0.4 μg/mL against L1210 and KB cells, respectively. The same sponge collected in Taiwan gave rise to two new metabolites, hippochromins A (**35**) and B, in addition to metachromins A, B, and E (**34**). The diacetate of **35** showed moderate cytotoxicity against KB and human colon (COLO-205) cancer cells (Shen *et al.* 2001).

Another related cytotoxic compound, isometachromin (**36**), had earlier been reported from a deep water sponge of the family Spongiidae, collected at a depth of ca. 800 m in the Bahamas (McConnell *et al.* 1992). The Mediterranean sponge *Reniera fulva* afforded two new sesquiterpenes named fulvanins 1 and 2 and four known paniceins (Casapullo *et al.* 1993). Only the paniceins were submitted for cytotoxicity screening at the NCI. Panaceins C and B3 showed activity against NCI-H522 non-small lung cancer cells and CCRF-CEM leukemia cells, while panicein A hydroquinone (**37**) revealed selective cytotoxicity against the latter cells.

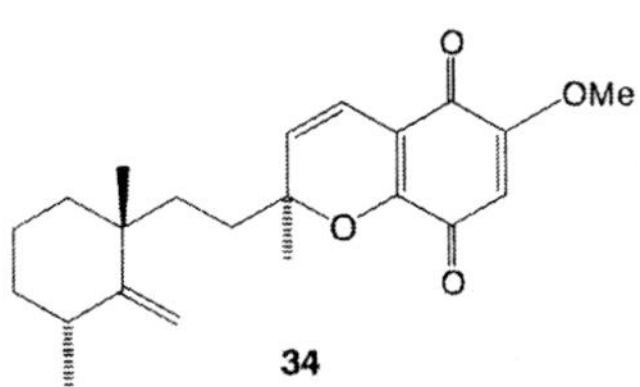

34

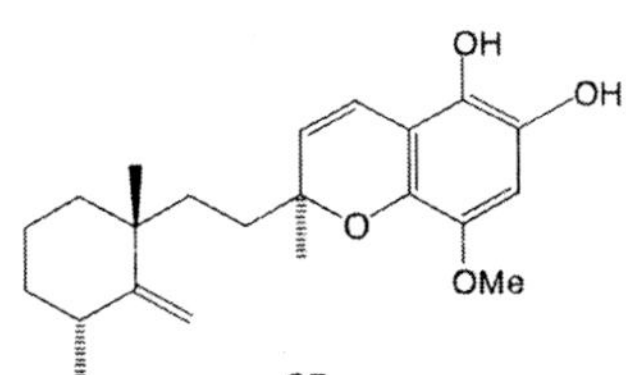

35

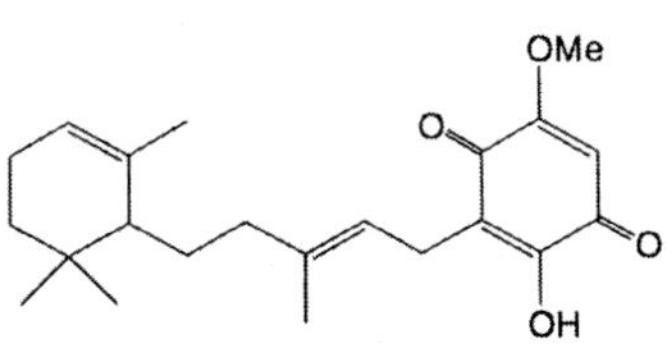

36

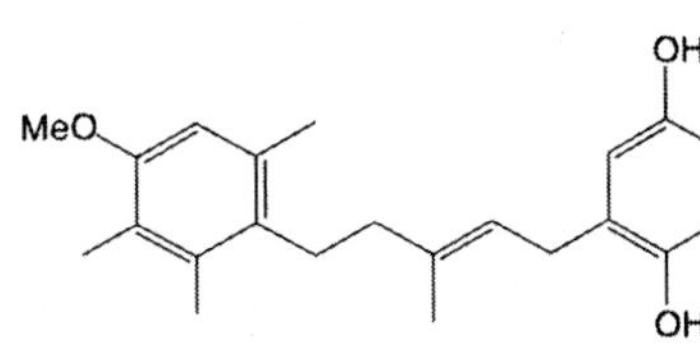

37

Avarol (**38**) and avarone (**39**) are the well-known examples of merosesquiterpenes belonging to the 9,4-friedodrimane class. As noted in our earlier review (Kuniyoshi and Higa 2001), these compounds have received significant attention because of a report of their activity against human immunodeficiency virus. Subsequent studies, however, revealed them to have no therapeutic effect on HIV-infected cells. Avarol, avarone, and five other known and seven new metabolites (e.g., **40**) have been isolated from a New Zealand collection of a sponge, *Dysidea* sp. (Stewart *et al.* 1997). Their biological activities were tested against P388 cells, *Bacillus subtilis*, and *Trichophyton mentagrophytes*. Most of the compounds exhibited moderate cytotoxicity and antimicrobial activity. The known compound avarone A (**41**) was best in the cytotoxicity (IC_{50} 0.6 μg/mL) and a new compound (**42**) in the antimicrobial activity.

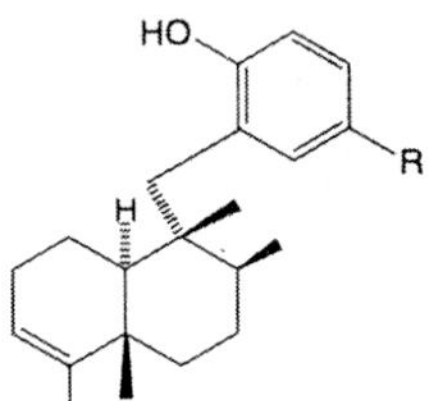

38 R = OH
43 R = CO_2Me

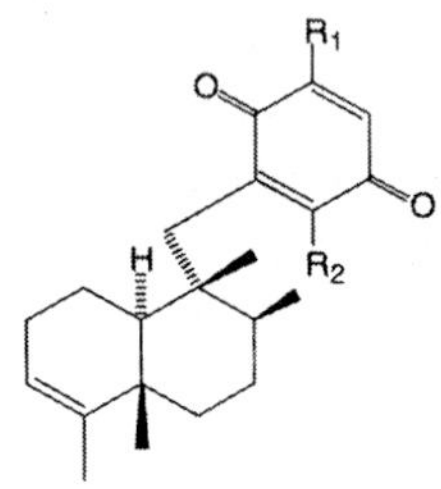

39 $R_1 = R_2 = H$
41 R_1 = OH, R_2 = H
44 R_1 = OMe, R_2 = OH

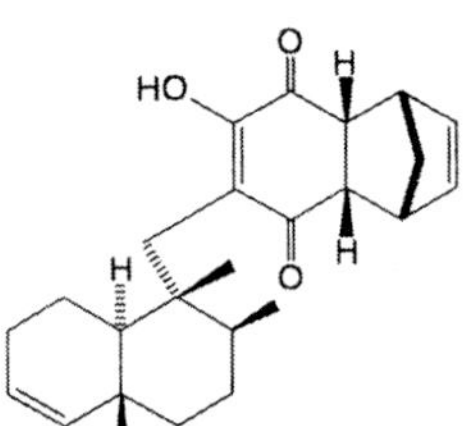

40

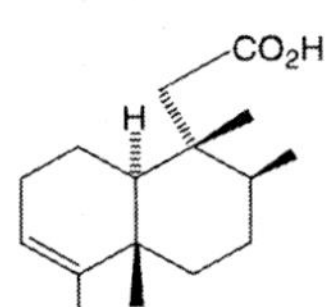

42

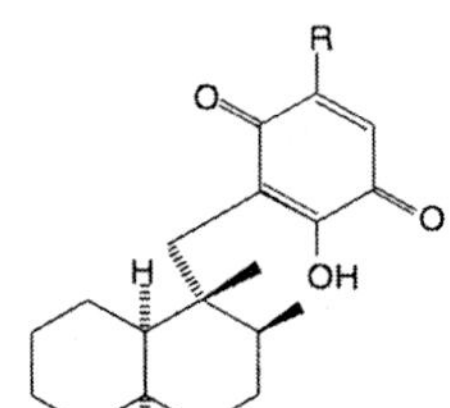

45 R = OMe
46 R = NH-*i* Bu

From two species of the genus *Hyrtios* Salmoun *et al.* (2000) reported five new (e.g., hyrtiophenol **43**) and three known compounds, including isospongiaquinone (**44**) and 5-epiilimaquinone (**45**). The major constituents **44** from *Hyrtios* sp. and **45** from *H. tubulatus* were found to be toxic against *Artemia* larvae with LD_{50} 2 and 4 mg/L, respectively. The authors claimed that this was the first report of the isolation of sesquiterpene quinones of the 9,4-friedodrimane skeleton from the genus *Hyrtios* (Thorectidae, Dictyoceratida). Such compounds had been reported almost exclusively from Dictyoceratid sponges of the genera *Dysidea, Fasciospongia, Smenospongia, Dactylospongia, Fenestraspongia, Hippospongia, Hyatella, Polyfibrospongia, Spongia,* and *Stelospongia.*

Compounds **44** and **45** have also been reported together with 15 other compounds, including **46**, from collections of *Dactylospongia elegans* from three different Indo-Pacific regions, Fiji, Papua New Guinea, and Thailand (Rodriguez *et al.* 1992). Some selected compounds were tested against A549, HT-29, B16/F10, and P388 cell lines. The aminated derivative **46** was most potent in the range of IC_{50} 0.6–0.9 μg/mL. Related to **46** are nakijiquinones A-D (**47-50**), amino acid conjugates of avarone. These compounds are metabolites of a sponge belonging to the family Spongiidae, collected at Nakijin, Okinawa (Kobayashi *et al.* 1995, Shigemori *et al.* 1994). Two of them were quoted as antimicrobial compounds in our previous article. The natural products and several synthetic compounds conjugated with a different L- or D-amino acid were tested for cytotoxicity against L1210 and KB cells and inhibitory activity against two tyrosine kinases, EGF receptor and C-erbB-2 (or Her-2/Neu) kinases, and protein kinase C. Cytotoxicities of these compounds were observed in the range of IC_{50} 1-10 μg/mL. All of nakijiquinones and synthetic derivatives showed relatively potent inhibitory activity against C-erbB-2 kinase rather than EGF receptor kinase or potein kinase C (Kobayashi *et al.* 1995).

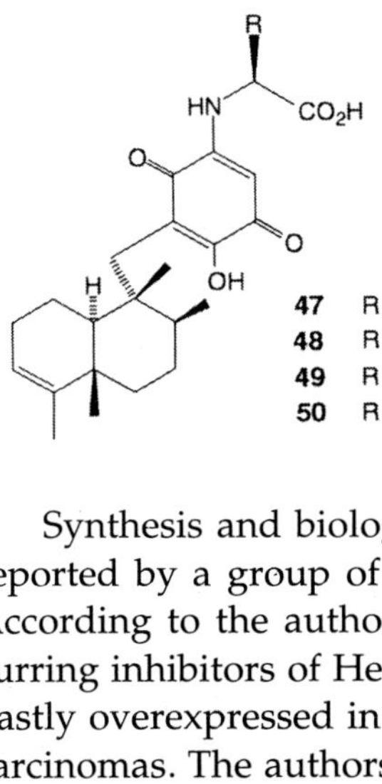

47 R = H
48 R = $CH(CH_3)_2$
49 R = CH_2OH
50 R = $CH(CH_3)OH$

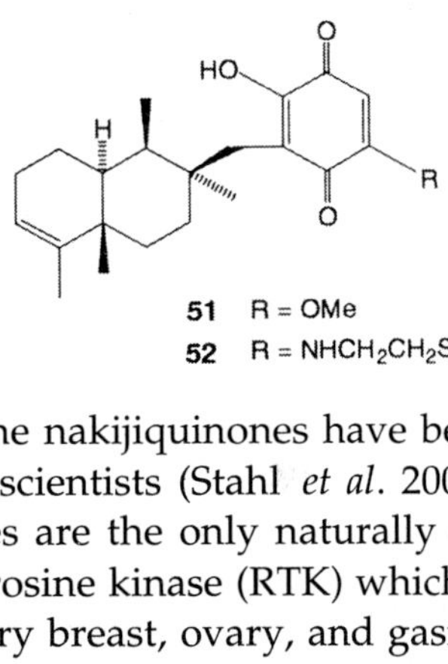

51 R = OMe
52 R = $NHCH_2CH_2SO_3H$

Synthesis and biological evaluation of the nakijiquinones have been reported by a group of German and Swiss scientists (Stahl *et al.* 2001). According to the authors, the nakijiquinones are the only naturally occurring inhibitors of Her-2/Neu receptor tyrosine kinase (RTK) which is vastly overexpressed in about 30% of primary breast, ovary, and gasrtic carcinomas. The authors stated that selective inhibitors of RTK are paramount importance for the development of new anticancer drugs.

Bolinaquinoe (**51**) from a Philippine *Dysidea* sp. was also quoted in our previous review as an antimicrobial merosesquiterpene. It was also reported to have cytotoxic activity against human colon tumor (HCT116) cells, and its action was suggested to be caused by interfering with or damaging DNA (De Guzman *et al.* 1998). From a Vanuatu collection of *Dysidea* sp. Giannini *et al.* (2001) isolated **51** and three new compounds, including dysidine (**52**), an aminoquinone derivative. Both **51** and **52** showed significant inhibition of human synovial phospholipase A_2.

Since its discovery in 1979 puupehenone (**53**) has been found in many species of sponges. It has been reported to have a diverse array of biological properties, including antimicrobial, antimalarial, cytotoxic, antiviral, and immunomodulatory activities. Its absolute stereochemistry was determined by chemical degradation in 1996 (Urban and Capon 1996). Recently, puupehenone was shown to have strong antioxidant activity in a 2′,7′-dichlorodihydrofluorescein diacetate cellular-based assay (Takamatsu *et al.* 2003).

Amagata *et al.* (2003) examined over a dozen marine terpenoids including puupehenone (**53**) and its derivatives for their inhibitory potency and selectivity against 12-human, 15-human, and 15-soybean lipoxygenases. Lipoxygenases are a class of widely occurring, non-heme iron-containing oxygenases of plants and animals. Human 12-lipoxygenase may be involved in psoriasis and controlling cancer cell proliferation and human 15-lipoxygenase in atherosclerosis and tumorigenesis. All five puupehenones tested were potent inhibitors of the lipoxygenases but without much selectivity except for one, dimethoxypuupehenol (**54**), which exhibited notable selectivity.

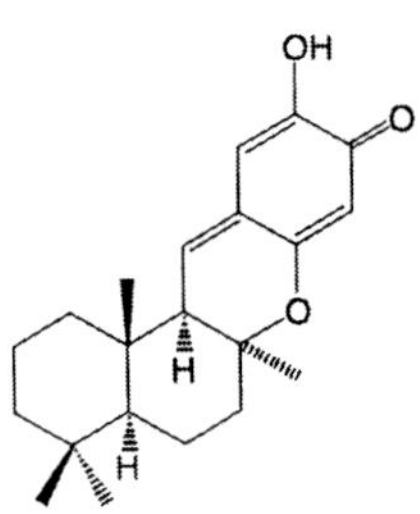

53

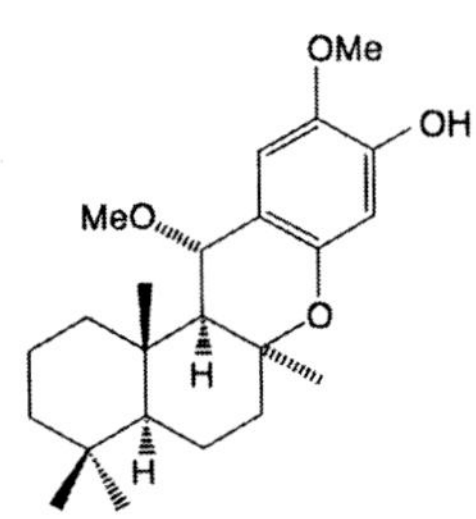

54

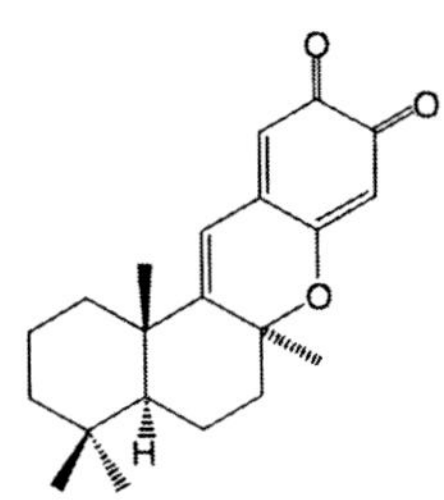

55

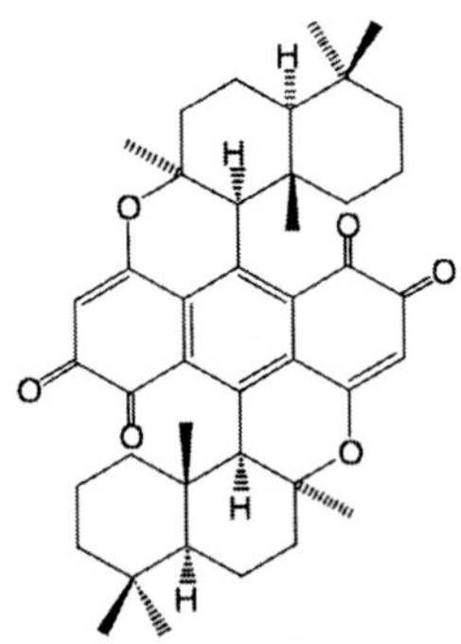

56

Puupehedione (**55**) and related compounds have been synthesized and subjected to cytotoxicity assay against P-388, A-549, HT-29, and MEL-28 (Barrero *et al.* 1999). All showed moderate activity without selectivity. Dipuupehedione (**56**), a dimer of puupehedione (**55**), has been isolated together with puupehenone (**53**) from a New Caledonian collection of *Hyrtios* sp. (Bourguet-Kondracki *et al.* 1996). Dipuupehedinone, a red pigment, showed no antimicrobial activity against *Staphylococcus aureus* or *Candida tropicalis* but cytotoxicity against KB cells (IC_{50} 3 μg/mL).

In puupehenone (**53**) a phenolic hydroxyl oxygen is cyclized to C-8 of the drimane sesquiterpene moiety to form an ether ring. There are some related compounds in which an ether ring is formed by binding the oxygen to another position such as C-10 or C-9. A previous example is smenoqualone (**57**) isolated from *Smenospongia* sp. (Bourguet-Kondracki *et al.* 1992). This compound, however, was shown to be inactive in antimicrobial and cytotoxic assays. Recently Utkina *et al.* (2003a,b) reported a related compound, cyclosmenospongine (**58**), possessing moderate cytotoxic activity from *Spongia* sp. collected in Australia. It also showed hemolytic activity.

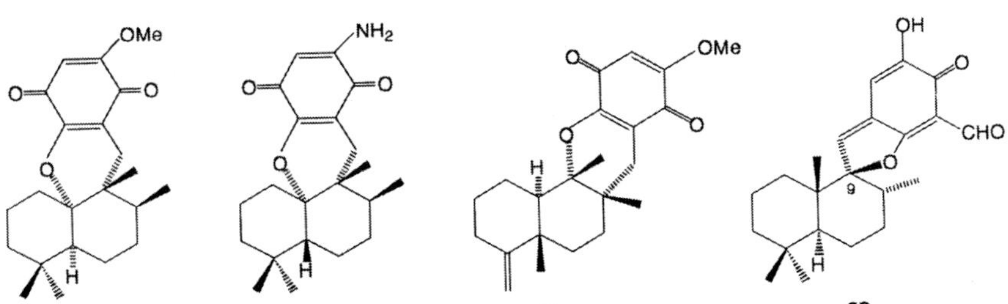

57 58 59 60

Another recent example is a cytotoxic quinone, neodactyloquinone (**59**), isolated together with related compounds from the Okinawan sponge *Dactylospongia elegans* (Mitome *et al.* 2003). Two spirosesquiterpene aldehyde, corallidictyals A (**60**) and B, were isolated as a mixture from the sponge *Aka* (=*Siphonodictyon*) *coralliphagum* (Chan *et al.* 1994). These compounds, diastereomeric at the spiro junction (C-9), exhibited inhibition of protein kinase C.

A tetracarbocyclic sesquiterpene quinol, pelorol (**61**), has been isolated by two independent groups from two different sponges, *Dactylospongia elegans* from Australia (Goclik *et al.* 2000) and *Petrospongia metachromia* from Yap, Federated States of Micronesia (Kwak *et al.* 2000). Goclik *et al.* (2000) reported that pelorol (**61**) and co-occurring ilimaquinone (**62**) showed weak activity against *Trypanosoma rhodesiense* and *Plasmodium falciparum*, while Kwak *et al.* (2000) reported **61** to be toxic to brine shrimp.

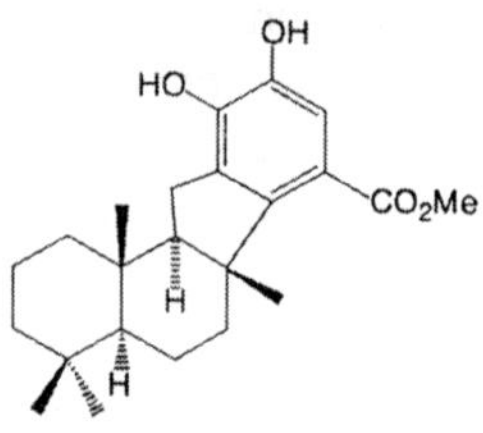

61

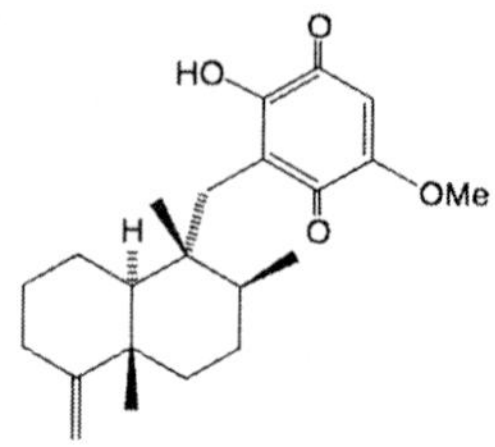

62

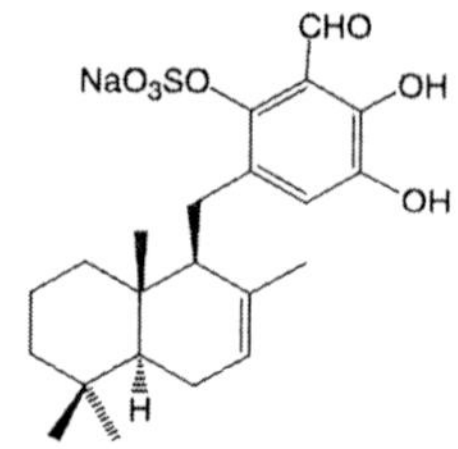

63

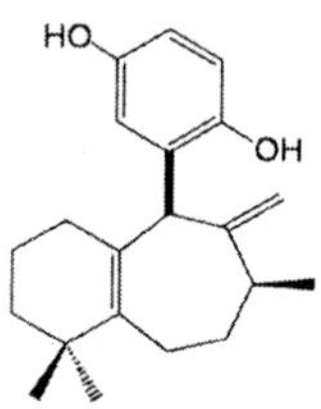

64

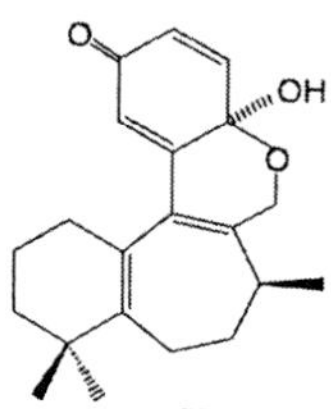

65

Two compounds closely related to **61** and the known siphonodictyal C (**63**) have been isolated from another Yap sponge, *Aka* sp. (Mukku *et al.* 2003). The structure of **63** was revised from the one reported previously. Siphonodictyal C inhibits cyclin-dependent kinase 4/cyclin D1 complexation. The kinase plays an important role in the G1/S transition phase in the cell cycle and needs to be activated by complexation with cyclin D1 for it to function.

A series of unusual sesquiterpene quinones, frondosins A-E (A: **64**, D: **65**), from *Dysidea frondosa* collected in Pohnpei have been reported to be inhibitors of interleukin-8 receptors (Patil *et al.* 1997). In frondosins the B ring of the drimane skeleton is expanded to a seven-membered ring, thereby allowing the quinone/quinol ring to bind directly to the B ring. All frondosins showed inhibition of interleukin-8 receptors (IL-8 Rα and IL-8 Rβ) and protein kinase C (PKC). Frondosin A (**64**) was most potent in the assays against both IL-8 Rα and IL-8 Rβ receptors and also against PKC. The IL-8 receptor antagonist is expected to be a promising lead for the development of a novel anti-inflammatory agent. However, the authors concluded that the non-selective activity of frondosins "may indicate that these compounds are reactive rather than representing a true agonist/antagonist event."

DITERPENOIDS

The diterpenoids, having a C_{20} skeleton, are a large group of metabolites distributed in plants, fungi, insects, and marine organisms. They are biosynthesized through geranylgeranyl pyrophosphate which has an extension of one isoprene unit on the skeleton of farnesyl pyrophosphate (**1**). As shown in Table 11.1 the diterpenoids are the most abundant group among the terpenoids from marine organisms. Their major sources are octocorals, brown algae, and sponges. A variety of skeletal structures have been identified with the sponge diterpenoids. The bioactive ones to be discussed here can be classified into several subsections in which spongian class (**66**) diterpenes are most abundant, followed by bicyclic labdane (**67**), clerodane (**68**), and biflorane (**69**) skeletal classes.

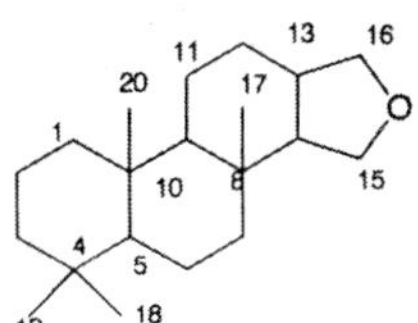

66

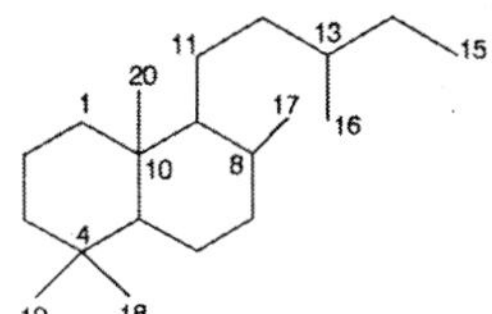

67

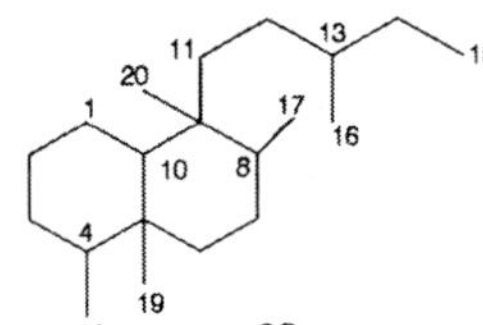

68

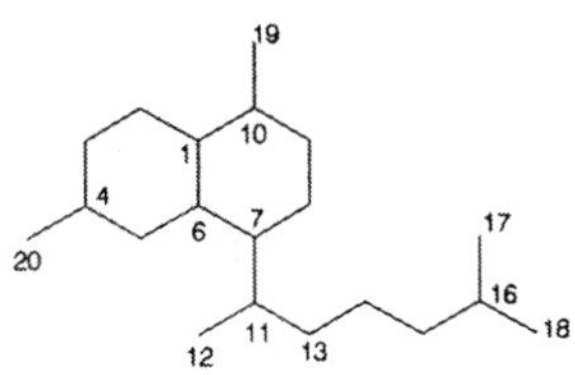

69

Labdane and Clerodane Class Diterpenoids

The clerodane skeleton is formed from the labdane skeleton by consecutive migration of methyl groups, one from C-4 to C-5 and another from C-10 to C-9. In spite of a significant number of these classes of the sponge diterpenoids, only several compounds enter discussion here, because some of them have been covered in our previous review and for many biological activity has not yet been reported.

The first example here is cacofurans A (**70**) and B (**71**), furanoditerpenes of the labdane class, isolated from *Cacospongia* sp. collected in Okinawa (Tanaka *et al.* 2001). Both compounds showed moderate cytotoxicity against P388 and K562 cell lines and disruption of the actin cytoskeleton of NBT-II cells. None of them exhibited antimicrobial activity against *E. coli*, *B. cereus*, or MRSA.

An example of the clerodane skeleton is provided by dactyltronic acids (**72**), an inseparable epimeric mixture, which has been isolated together with ilimaquinone (**62**) and other previously known terpenoids from *Dactylospongia elegans* (López *et al.* 1994). Dactyltronic acids exhibited mild cytotoxic activity against P388 and A549 cell lines.

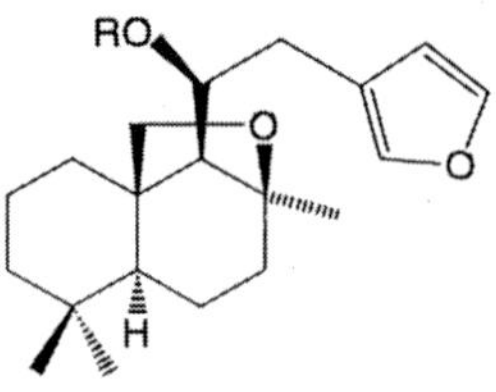

70 R = AC
71 R = H

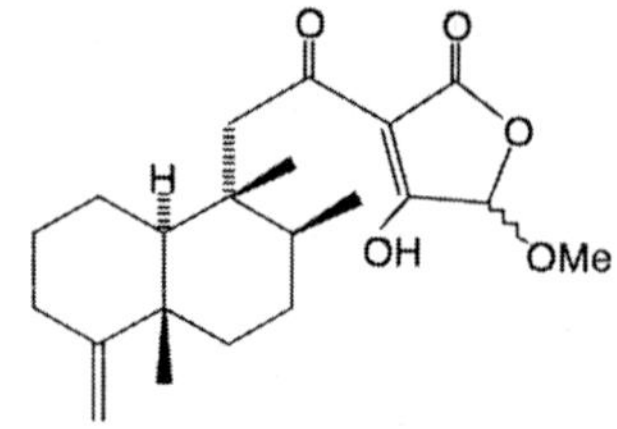

72

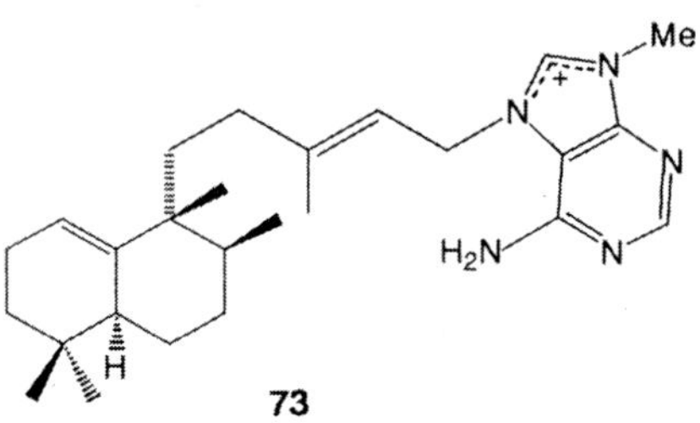

73

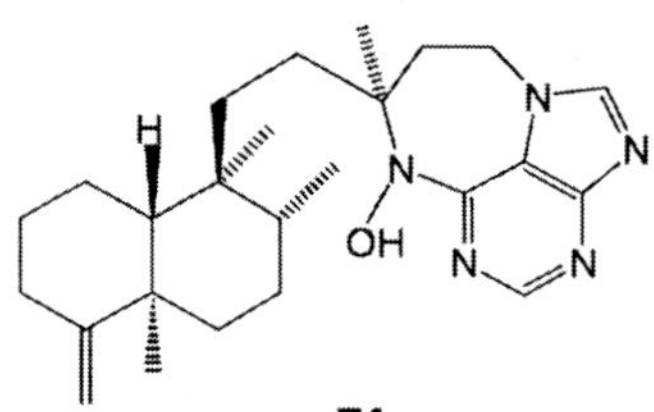

74

Agelasines are 9-methyladenine-containing diterpenes isolated from sponges of the genus *Agelas*. Many related compounds have been described in the literature and some discussed as antimicrobial compounds in our previous article. A new derivative, *epi*-agelasine C (**73**), has been reported as an antifouling compound from the sponge *Agelas mauritiana* (Hattori *et al*. 1997). *epi*-Agelasine C showed lethal activity against a macroalga, *Ulva* fronds, at 50 ppm and the microalgae *Oscillatoria amphibia, Skeletonema costatum, Brachiomonas submaria,* and *Prorocentrum micans* at 1.0-2.5 ppm. No antibacterial activity was observed against *S. aureus, Vibrio costicola, E. coli,* or *B. subtilis*.

A series of novel cytotoxic compounds, asmarines A (**74**)-F, have been reported from a Red Sea sponge, *Raspailia* sp. (Yosief *et al*. 2000). All asmarines have the same terpenoid skeleton and the heterocyclic system shown in **74**. They are closely related to the agelasines. Asmarines A and B (epimeric to A at C-10) exhibited cytotoxic activity against P388, A549, HT29 and MEL28, asmarine B being more active than A (Yosief *et al*. 1998).

Kalihinanes and Related Compounds

The kalihinanes (e.g., kalihinol A: **75**) are diterpenoids based on the biflorane skeleton (**69**) and are usually highly functionalized with –NC, –NCS, –Cl, etc. More than 40 compounds have been reported, mainly from two species of sponges, *Acanthella cavernosa* and *A. klethra*. Since many of these compounds were included in our previous review, discussion in this subsection is limited to those having biological properties other than antimicrobial activity. Also included here are related compounds, the amphilectone class diterpenes.

Kalihinol A (**75**) was first reported by Chang *et al.* (1984) as an antimicrobial substance. Since then, many related metabolites have been discovered, with a variety of biological activities: antifungal, cytotoxic, antihelminthic, and antifouling properties. More recently, kalihinol A (**75**) was shown to have potent antimalarial activity (Miyaoka *et al.* 1998). The absolute stereochemistry of **75** was determined for the first time in 1999 (Shimomura *et al.* 1999).

A number of kalihinanes have been demonstrated to exhibit an antifouling effect against larvae of the barnacle *Balanus amphitrite* (Hirota *et al.* 1996, Okino *et al.* 1995, 1996b). Kalihinene X (**76**) and kalihipyran B (**77**) are just two examples of the active metabolites isolated from *Acanthella cavernosa*. These compounds inhibited attachment and metamorphosis of cyprid larvae of the barnacle at nontoxic low concentration levels. Kalihinol A (**75**) with EC_{50} 0.087 μg/mL was more potent than **76** (EC_{50} 0.49 μg/mL), **77** (EC_{50} 0.85 μg/mL), and $CuSO_4$ (EC_{50} 0.15 μg/mL).

The amphilectanes and cycloamphilectanes are characteristic sponge diterpenoids closely related to the kalihinanes not only in their chemistry but also in their biological activity. A review of marine-derived antimalarial compounds lists a number of amphilectanes (e.g., **78**) and cycloamphilectane (e.g., **79**) (Wright and König 1996). These compounds exhibit significant and selective *in vitro* antimalarial activity. More recently several aromatic analogues of the amphilectanes were synthesized and their *in vitro* antimalarial activity assessed (Schwarz *et al.* 2002). From a Caribbean sponge, *Cribochalina* sp., Ciavatta *et al.* (1999) isolated the known amphiletene **80** as a main constituent along with three new related compounds. Compound **80** reduced the proliferation of lymphocytes T and B under various stimuli.

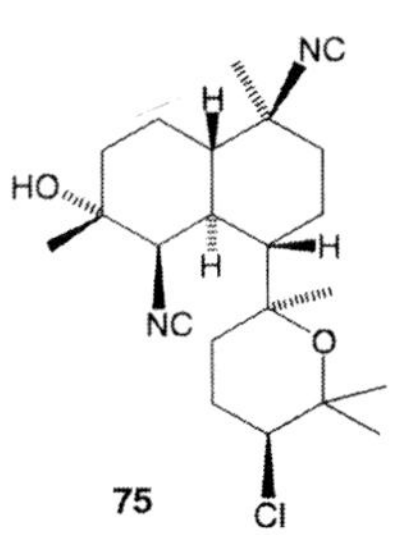

75

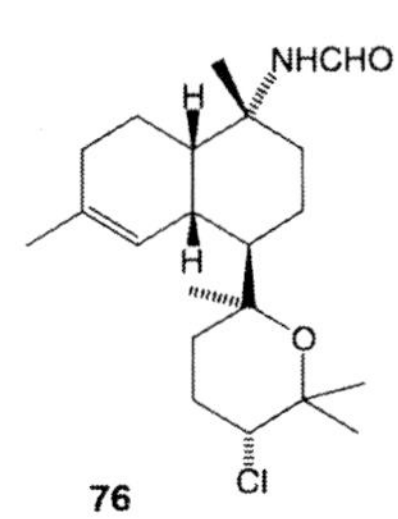

76

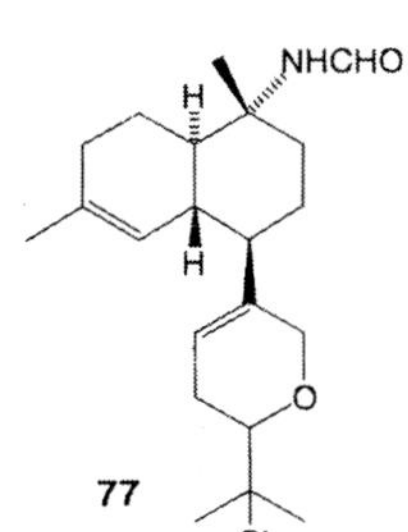

77

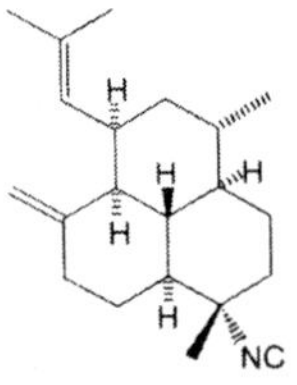

78

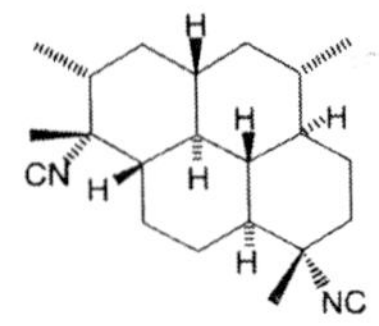

79

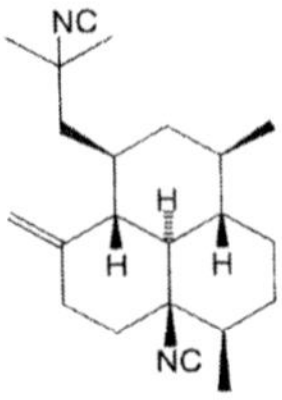

80

Spongian Diterpenoids

A large number of the spongian diterpenes have been reported from sponges of the orders Dictyoceratida and Dendroceratida and also from nudibranchs. A simple example is spongia-13 (16), 14-diene (**81**) isolated from *Spongia officinalis*. Many related compounds have highly oxidized structures (e.g., aplyroseol-2: **82**) and extensively rearranged skeletons (e.g., aplyviolene: **83**).

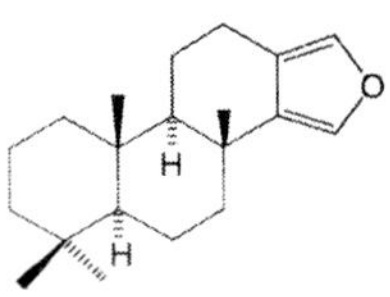

81

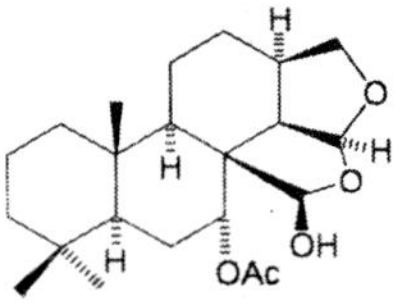

82

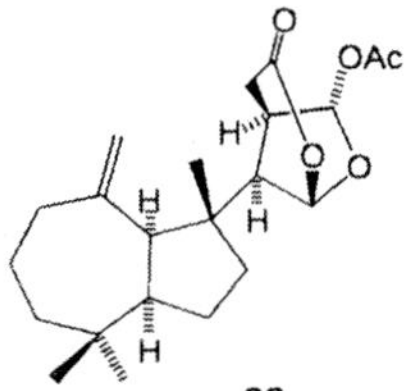

83

In this subsection we discuss a small number of the spongian diterpenes which have been described with biological activity during the review period. A number of the compounds of this class have been reported without biological properties. Gunasekera and Schmitz (1991) isolated four spongian diterpenoids from a Great Barrier Reef sponge, *Spongia* sp. One (**84**) of them, an example of a nor-spongian with a ring-A δ-lactone, exhibited cytotoxicity (EC_{50} 3.5 μg/mL) against P388 cells. A pair of related nor-spongian epimers (**85** and C-4 epimer) have been isolated along with three known related compounds (e.g., **86**) from *Spongia matamata* collected in Yap (Li *et al.* 1998). The epimer **85**, epispongiadiol (**86**), and other known compounds showed mild toxicity in the brine shrimp lethality test. The other epimer of **85** was inactive. Epispongiadiol (**86**) had earlier been reported to have cytotoxic and antiviral activities (Kohmoto *et al.* 1987).

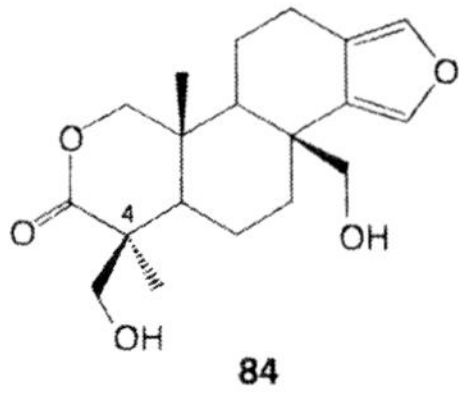

84

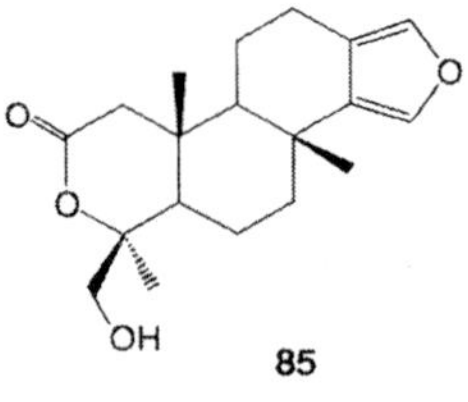

85

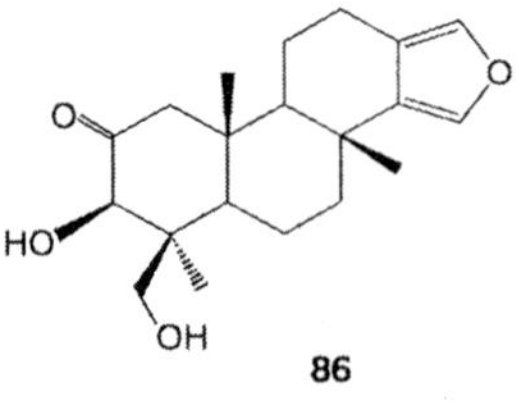

86

Seven new spongians, dorisenones (e.g., **87**, **88**), were isolated together with several known compounds from the nudibranch *Chromodoris obsoleta* collected in Fukuoka, Japan (Miyamoto *et al.* 1996). Cytotoxicity assays revealed that all of the isolated compounds were moderately active against L1210 and KB cell lines. Dorisenone A (**87**) and 15, 16-diacetoxy-11,12-epoxyspongian (**88**) showed more potent activity than did the others with IC_{50} values at the level of 0.2 μg/mL against L1210 and 0.2-1 μg/mL against KB cells. A synthetic effort toward the preparation of oxygenated spongians of the dorisenone type has recently been published (Abad *et al.* 2003).

An international group consisting of Colombian and Spanish scientists has published a paper on the evaluation of cytotoxic and antiviral effects of 14 synthetic spongian diterpenes including six naturally-occurring ones (e.g., **81**) (Betancur-Galvis *et al.* 2002). The antiviral activity of these compounds against herpes simplex virus type 2 was very weak. In the cytotoxicity test most of the spongians were mildly active against HeLa (human cervix epitheloid carcinoma) and HEP-2 (human larynx epidermoid carcinoma) cell lines. The Spanish members of the group also reported revised structures for the spongian diterpenes acetyldendrillol-1 and aplyroseol-14 by synthesis (Arnó *et al.* 2003).

Spongian diterpenoids having a rearranged skeleton like **83** are quite common among the metabolites of sponges and nudibranchs. A bioactive example is norrisolide (**89**), which was isolated along with three related compounds as a cytotoxic constituent of the Red Sea sponge *Dysidea* sp. (Rudi and Kashman 1990). Norisolide (**89**) and four other related spongians (e.g., **90**) have also been reported as ichthyotoxic constituents of the nudibranch *Chromodoris luteorosea* collected in northern Spain (Gavagnin *et al.* 1992).

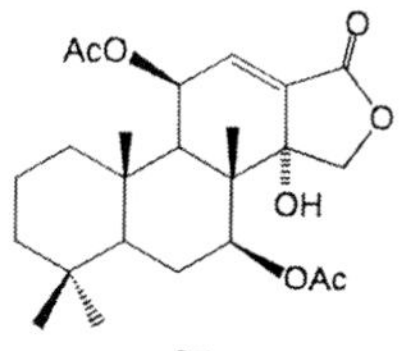

87

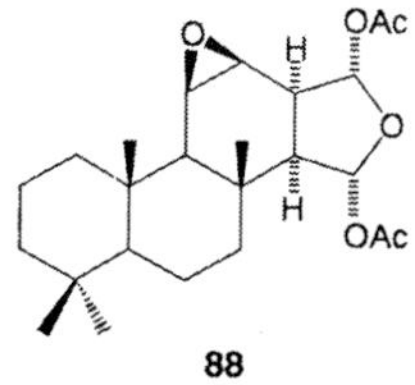

88

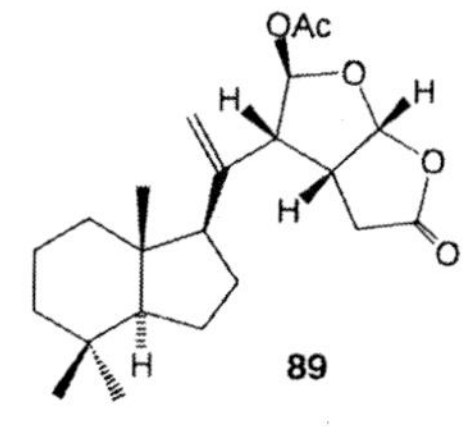

89

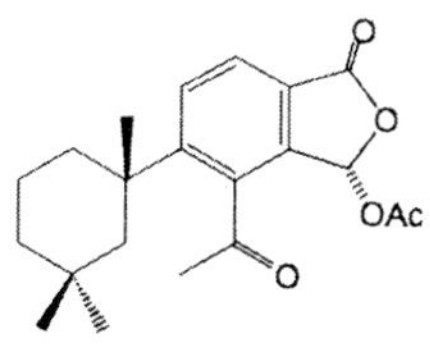

90

Meroditerpenoids

In this subsection we consider only those meroditerpenoids which have a ring structure in the terpenoid moiety. Linear tetraprenyl quinones will be treated with other polyprenyl quinones in a sepatate section.

Three meroditerpenes, cacospongins B (**91**), C, and D (**92**), were isolated together with other terpenoids from the Philippine sponge *Cacospongia* sp. (Tasdemir *et al*. 2000). Of these compounds only **91** and **92** showed inhibition of *Staphylococcus aureus* and methicillin-sensitive *S. aureus*. Cacospongin C, the quinone form of **91** was inactive.

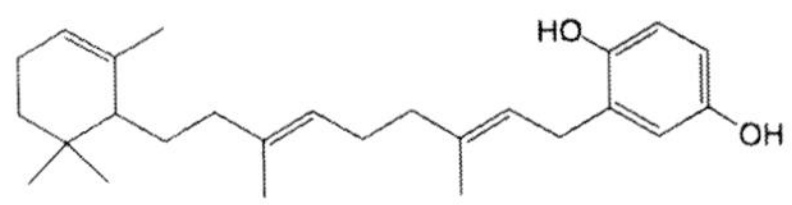

91

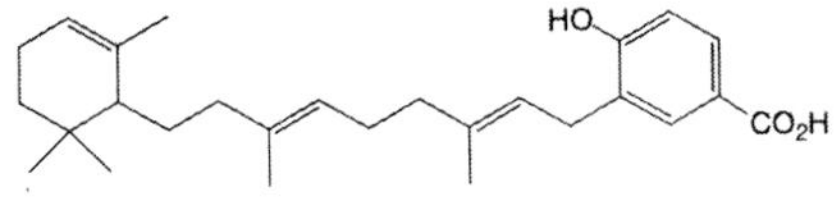

92

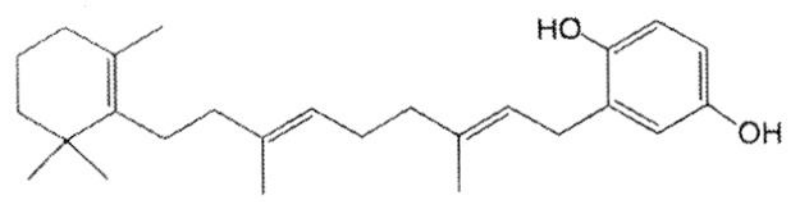

93

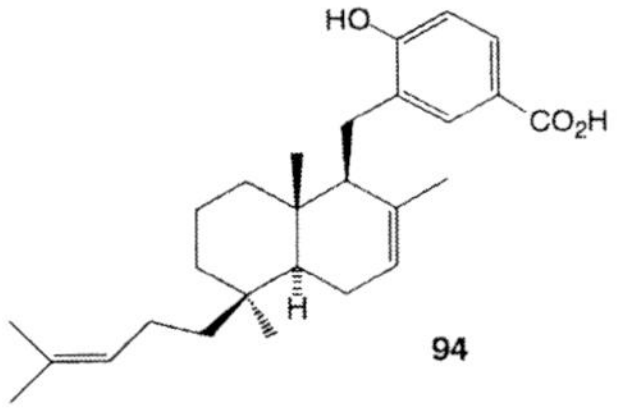

94

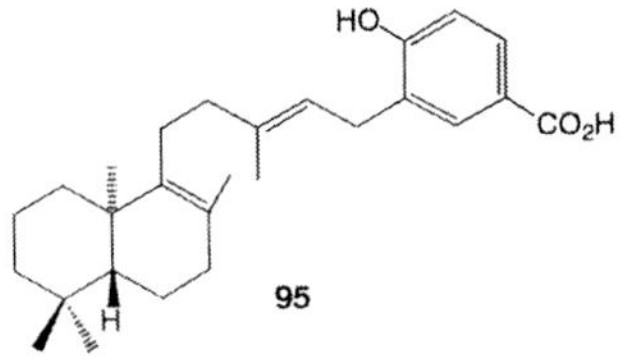

95

Two related compounds, jaspaquinol (**93**) and jaspic acid (**94**), were isolated from the sponge *Jaspis* cf. *johnstoni* from Madang, Papua New Guinea (Murray *et al*. 1997). No biological properties had been reported for these compounds until they were re-isolated together with subersic acid (**95**) from the sponge *Suberea* sp. collected also in Madang (Carroll *et al*. 2001). All three compounds exhibited inhibition of human 15-lipoxygenase, jaspaquinol (**93**) being most active. A short and efficient synthesis of (+)-subersic acid, the enantiomer of **95**, was recently achieved from sclareol and *p*-hydroxybenzoic acid (Basabe *et al*. 2003).

Strongylophorines are meroditerpenoids of pentacyclic chroman structures consisting of an isocopalane skeleton as the diterpene moiety and a hydroquinone. Some of them have an additional lactone ring. Twenty five related compounds have so far been reported. These compounds have been known to possess biological activities such as ichthyotoxicity, antimicrobial property, and cytotoxicity. Recently four new compounds, strongylophorines-22 (**96**), -23, -24 and -25, were isolated along with several known congeners (e.g., strongylophorine-3: **97**) from the Okinawan *Petrosia (Strongylophora) corticata* (Hoshino *et al*. 2003). Some of them exhibited significant cytotoxic activity against HeLa cells. Strongylophorine-3 (**97**) has earlier been described to be active against the phytopathogenic fungus *Cladosporium cucumerinum* and also against the larvae of the polyphagus pest insect *Spodoptera littoralis* (Balbin-Oliveros *et al*. 1998).

A dimer (**98**) of strongylophorine-3 (**97**) has been isolated together with **97** and two other known derivatives from *Strongylophora* sp. from the Philippines (Balbin-Oliveros *et al*. 1998). The dimer **98** was found to be more active than monomeric congeners in the brine shrimp lethality test.

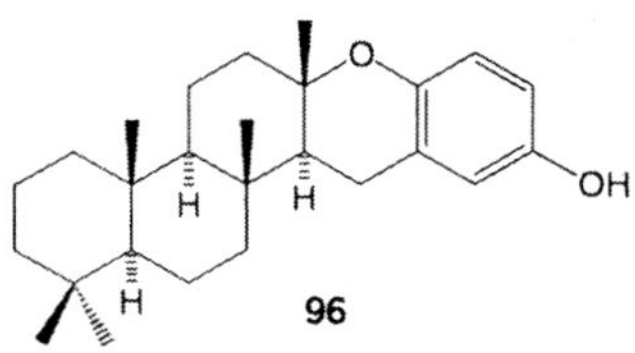

96

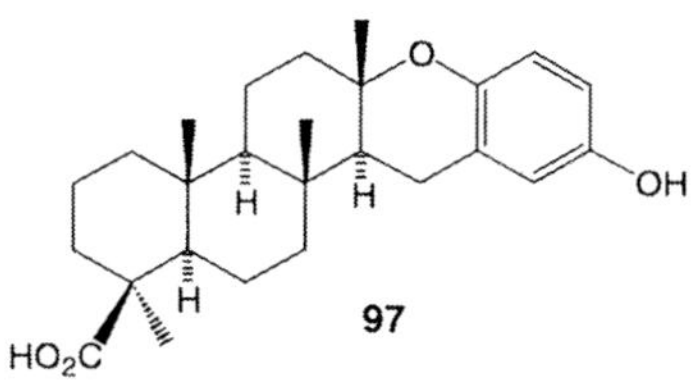

97

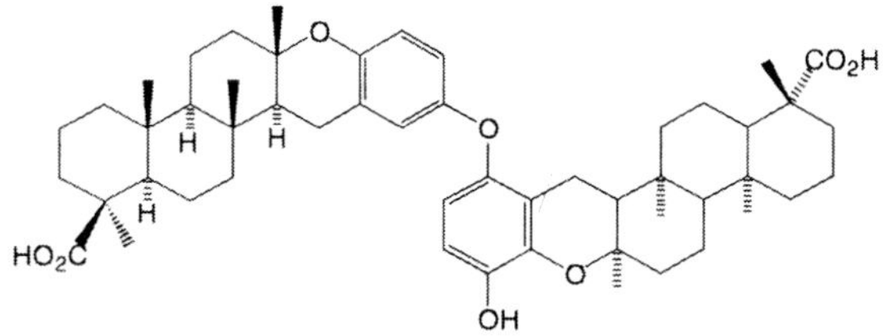

98

Other Diterpenoids

Here we discuss a small number of sponge diterpenes which have not been covered in the preceding subsections. They are diterpenoids belonging to relatively simple skeletal classes or of rare occurrence in sponges. The first example is linear diterpenes (e.g., **99**) isolated from the Caribbean sponge *Myrmekioderma styx* (Albrizio *et al*. 1992). Four compounds have been isolated, and all were lethal to the brine shrimp *Artemia salina*.

The next skeletal class is the cembranes. The cembranes are 14-membered carbocyclic compounds known from terrestrial plants (tobaccos and pine trees) and marine animals. Marine sources of the cembranes are almost exclusively soft corals and gorgonians belonging to the sub-class Octocorallia. A large number of cembrane diterpenes have been known from these invertebrates. It is therefore surprising to find cembranes from a sponge. Isolation of two cembranes, flabellatene A (**100**) and B, from the Antarctic sponge *Lissodendoryx flabellata* has been reported by Fontana *et al.* (1999b). The main product, **100**, exhibited potent cytotoxicity against N18-T62 mouse neuroblastoma cells (IC_{50} 0.16 μM) and significantly reduced cell proliferation of human tumoral cells (DU-145 and MCF-7) in a dose-dependent manner.

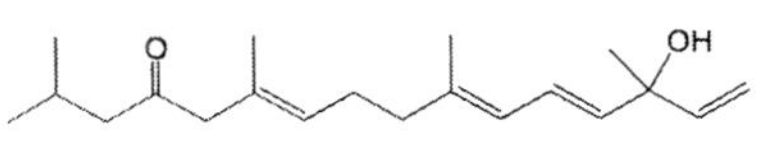

99

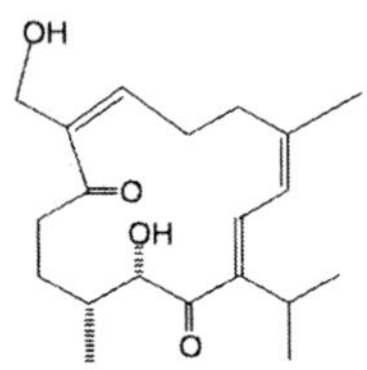

100

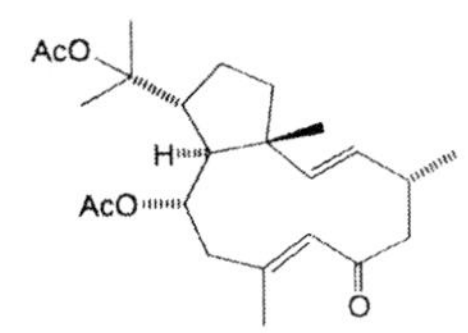

101

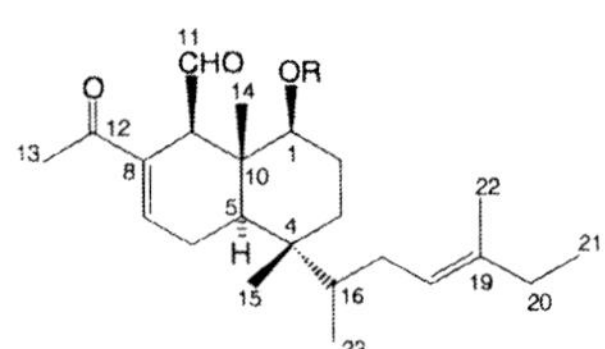

102 R = Ac
103 R = H

Another case of unusual occurrence of diterpenoids is the finding of dolabellans in a sponge. Three dolabellane class diterpenes (e.g., **101**) were isolated from a specimen of *Dysidea* sp. collected in Palau (Lu and Faulkner 1998). These compounds showed low cytotoxicity against HCT116 human colon carcinoma cells and A2780 human ovarian carcinoma cells. The dolabellane diterpenes have previously been known from sea hares of the genus *Dolabella*, brown algae of the genus *Dictyota*, and gorgonians of the genus *Eunicea*, but had never been reported from sponges. The authors suggested that the sponge might have absorbed the terpenoids released into water by organisms in the vicinity.

Cytotoxic diterpenes containing extra methyl groups on the skeleton of sacculatane class have been isolated from the sponge *Psammoclema* sp. from Durban, South Africa (Rudi *et al.* 1995b). All three compounds isolated contain an aldehyde and named durbinals A (**102**), B (**103**), and C. Durbinals A-C were cytotoxic against P388 cells with an IC_{50} value of 1 μg/mL. The skeleton of durbinals A (**102**) and B (**103**) possesses 23 carbons. The authors suggested that biologically it could be formed by alkylation of the sacculatane (prenyldrimane) diterpenoid skeleton at C-12, C-16, and C-20.

SESTERTERPENOIDS

Sesterterpenoids are the next higher isoprenoids to diterpenoids, consisting of five units of isoprene. They have been known as a small group of natural products from insects, fungi, higher plants, and marine organisms. Since sponges and nudibranchs, which sequester toxic metabolites from dietary sponges, have so far been the only marine sources of sesterterpenoids, sponges would be the virtual producers. The number of marine sesterterpenes was some 150 in our 1993 survey (Table 11.1). The number of bioactive sesterterpenoids published after 1994 exceeds 200 by a rough count. Therefore, it would be safe to estimate that the total number of marine sesterterpenoids has now reached over 500. Indeed sesterterpenoids are major metabolites among sponge terpenoids.

Most of the sponge sesterterpenoids can be classified into six structural types: acyclic (**104**), cyclohexane (**105**), bicyclic (**106, 107**), cheilanthane (**108**), and scalarane skeletons (**109**). Furanostertepenes, based on the acyclic skeleton (**104**), and scalaranes (**109**) are perhaps most abundant and characteristic sesterterpenoids of sponges. There are also many nor- and homo-(alkylated) members.

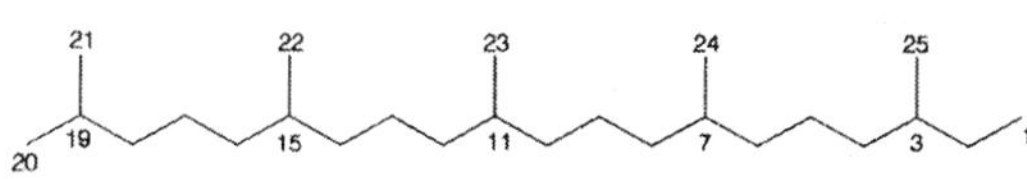

104

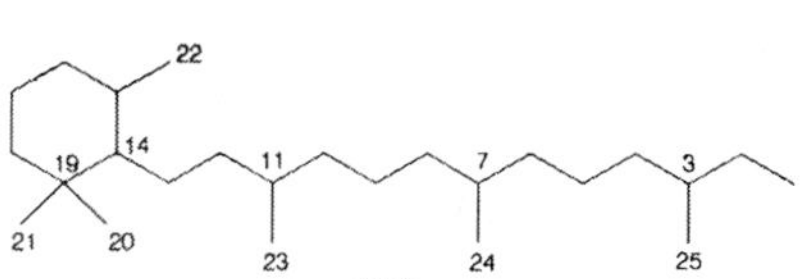

105

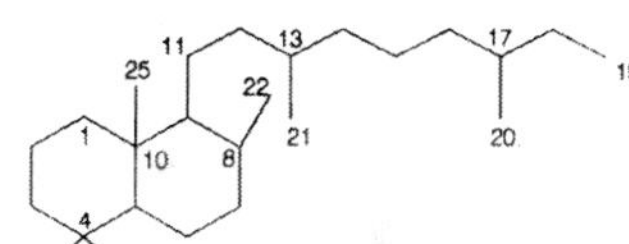

106

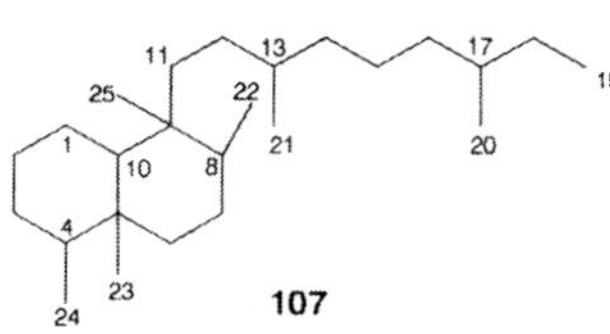

107

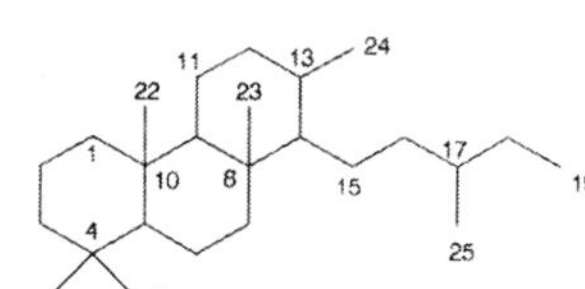

108

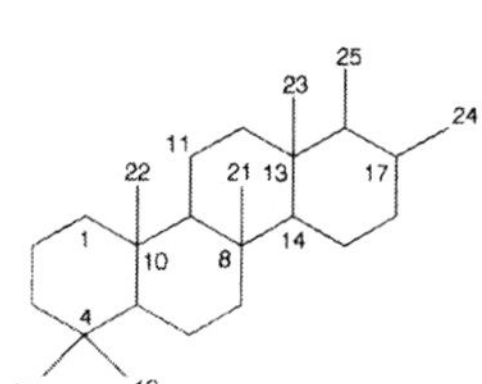

109

Furanosesterterpenoids

Included in this subsection are not only the compounds having the regular C_{25} acyclic skeleton of **104** but also the furanoterpenes having a C_{21} skeleton and other related nor-sesterterpenoids. A typical C_{25} compound has a furan ring at one end and a tetronic acid at the other end of the pentaprenyl chain. Some of these compounds have two or three furan rings. The skeleton of the C_{21} furanoterpenes is believed to be derived from the skeleton of **104** by loss of a four-carbon unit (C_{18}-C_{21}). These compounds usually contain two furan rings at the both ends. The furanoterpenes have been isolated from a number of species belonging to the order Dictyoceratida.

In two recent papers Liu *et al.* (2001, 2002) described the isolation, structure elucidation, and cytotoxic activity of 21 terpenoids from the Korean sponge *Sarcotragus* sp. Of these compounds 10 were furanosesterterpene tetronic acids (e.g., **110, 111**), two trinorsesterterpenes (e.g., **112**), and two diterpenes. The remaining seven were unusual pyrrolosesterterpene tetronic acids (e.g., **113, 114**) which contained an N-substituted pyrrolidone sinstead of a furan. The skeleton of the trinorsesterterpene **112** has 22 carbons. All of these compounds were tested for cytotoxicity against five human cancer cell lines (A549, SK-OV-3, SK-MEL-2, XF498, HCT15) and shown to possess mild to marginal activity.

More recently the same group examined another unidentified species of the genus *Sarcotragus* and obtained additional related compounds (Liu *et al.* 2003). However, none of them showed significant cytotoxicity. In this paper the authors also reported the stereochemistry of **112** and the isolation and cytotoxicity of a C_{21} furanoterpene, kurospongin (**115**). The latter was first described from an Okinawan sponge as an ichthyotoxic and antifeedant compound (Tanaka and Higa 1988).

From the Adriatic sponge *Ircinia oros* De Giulio *et al.* (1990) isolated two C_{24} norsesterterpenes (e.g., **116**) containing a chlorine atom and two previously reported ircinin-1 (**117**) and ircinin-2, all of which were toxic against the brine shrimp *Artemia salina*. The norsesterterpenes were less active than ircinins.

Cytotoxic furanosesterterpenes having a tetrahydrobenzofuran have recently been reported from an Okinawan sponge, *Ircinia* sp. (Issa *et al.* 2003). Two (e.g., **118**) of the isolated compounds were sesterterpene tetronic acids, while three others (e.g., **119**) were norsesterterpenes. The latter were more active (IC_{50} < 1 μg/mL) than the tetronic acid derivatives in a cytotoxicity assay using KB cells. A related C_{21} tetranorsesterterpene, untenic acid (**120**), has earlier been reported as an activator of sarcoplasmic reticulum Ca^{2+}-ATPase from an unidentified sponge of

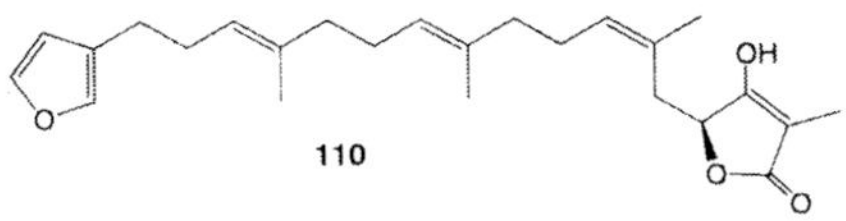

110

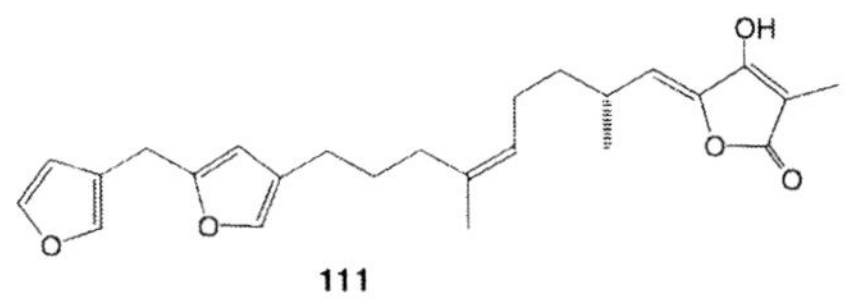

111

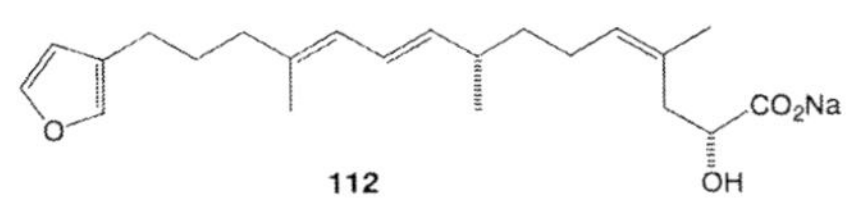

112

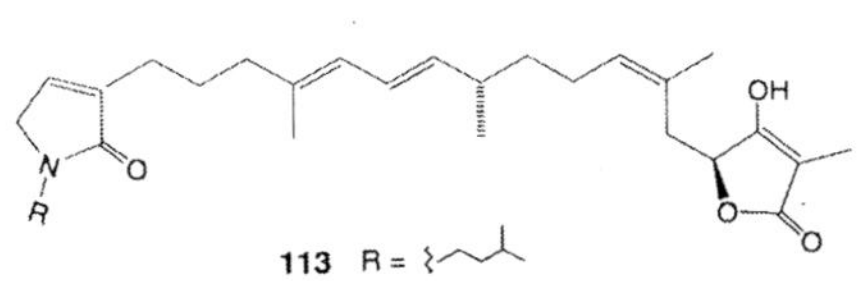

113 R =

114 R = 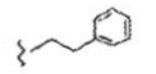

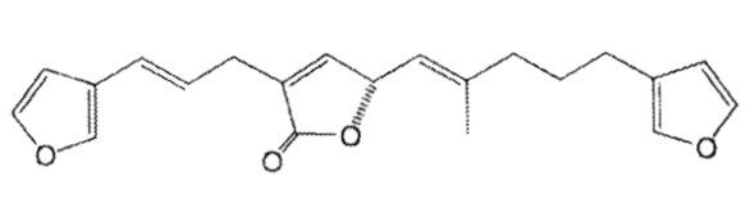

115

the family Spongiidae, collected in Okinawa (Shoji *et al.* 1992a). The Ca^{2+}-ATPase in the sarcoplasmic reticulum membrane plays a key role in muscle relaxation.

Untenospongin C (**121**), a C_{21} furanoterpene, was isolated as a cytotoxic constituent of the Okinawan sponge *Hippospongia* sp. (Kobayashi *et al.* 1993a). Untenospongin C exhibited moderate cytotoxicity against L1210 cells. A specimen of *Spongia officinalis* from Cádiz, Spain furnished a number of known and new furanoterpenes including furospongin-5 (**122**) and also furanosesterterpenes including demethylfurospongin-4 (**123**). These compounds were weakly cytotoxic against P388, A549, HT29, and MEL28 cell lines (Garrido *et al.* 1997).

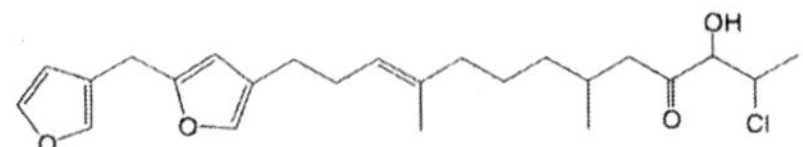

116

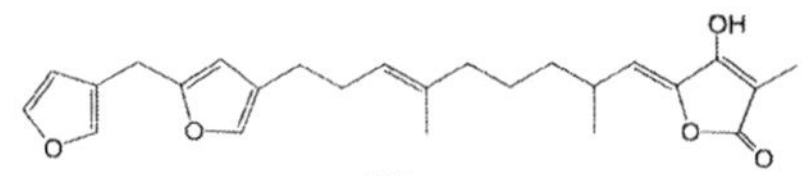

117

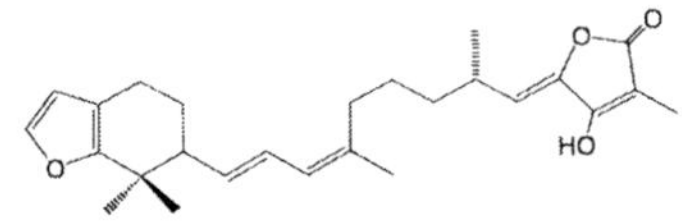

118

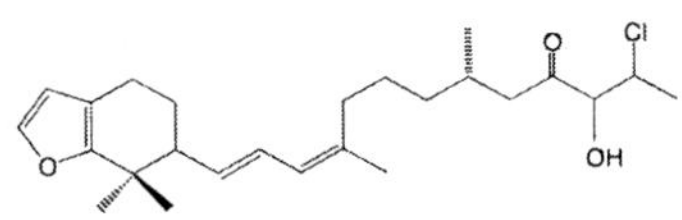

119

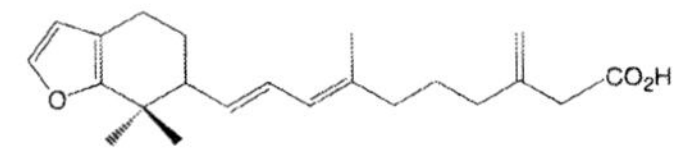

120

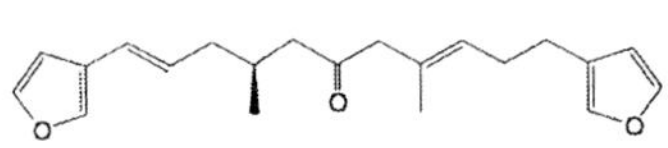

121

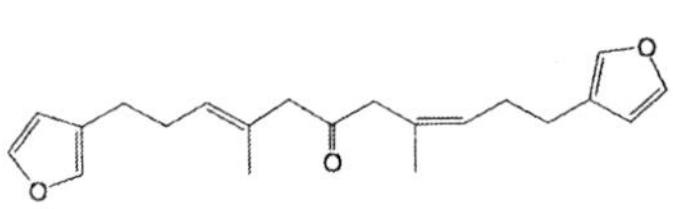

122

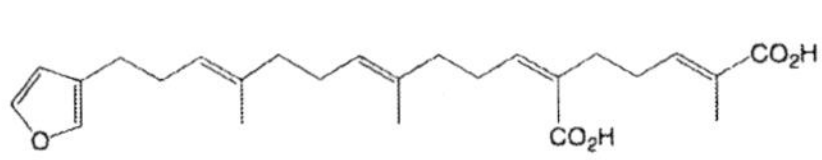

123

Cyclohexane Sesterterpenoids

The best known example of cyclohexane sesterterpenoids is manoalide (**124**), which inhibits phospholipase A_2 and has extensively been used as a reagent in biochemical research. A number of chemically related compounds, including those to be discussed in the next subsection, have been reported from marine sponges. These compounds have been described to have various biological properties including antimicrobial, antiviral, cytotoxic, and anti-inflammatory activities.

From three separate collections of a sponge, *Luffariella* sp. from Okinawa, Kobayashi's group isolated a total of nine compounds, luffariolides A (**125**)-J as cytotoxic constituents (Tsuda *et al.* 1992, 2002, Kobayashi *et al.* 1993b). Most of these compounds exhibited moderate cytotoxic activity against L1210 and KB cells. Manoalide (**124**) was most potent with IC_{50} values of 0.032 (L1210) and 0.31 µg/mL (KB). Luffariolides H and J (**126**) also showed antimicrobial activity against *Staphylococcus aureus*, *Bacillus subtilis*, and *Micrococcus luteus*.

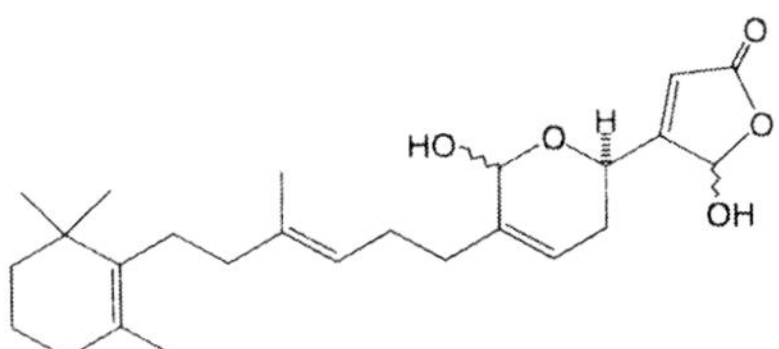

124

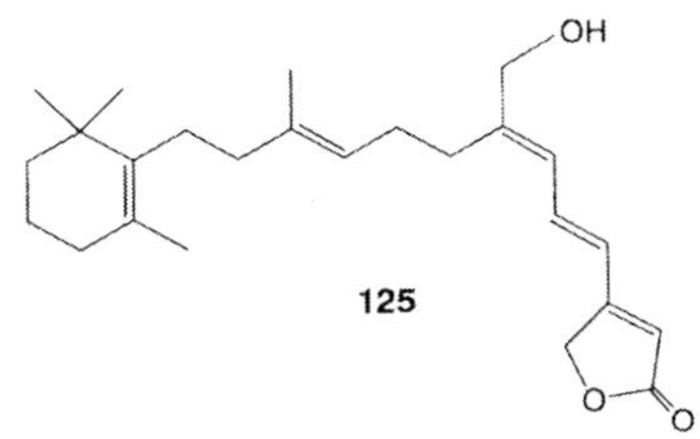

125

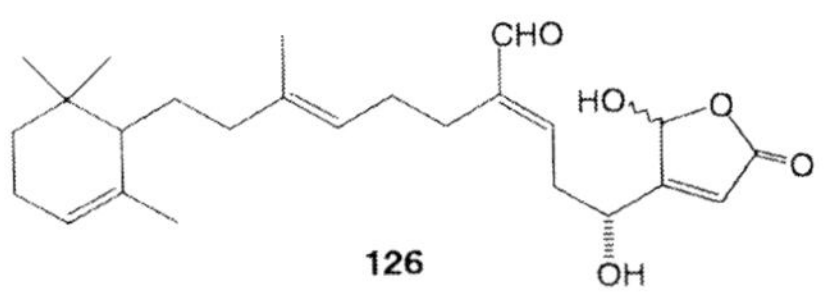

126

Jimênez *et al.* (1991) reported the isolation of three sesterterpenes along with alkaloidal constituents from the sponge *Fascaplysinopsis reticulata* collected in Fiji. One (**127**) of the terpenoids showed inhibition of the HIV reverse transcriptase. Isodehydroluffariellolide (**127**) was a constituent of another sponge, *Hyrtios* cf. *erecta* also from Fiji (Kirsch *et al.* 2000). It was shown to reduce the activity of $p56^{lck}$ tyrosine kinase at 0.5 mM to 45% and to have weak antifungal activity against the fungus *Ustilago violaceae*.

A series of related sesterterpenes, i.e., cyclolinteinone (**128**), cyclolinteinol (**129**), and lintenone (**130**), have been reported from the Caribbean sponge *Cacospongia* cf. *linteiformis* (Conte *et al.* 1994a,b, Carotenuto *et al.* 1995, 1997, Fattorusso *et al.* 1992). It has been suggested that the rearranged skeletons of **128**, **129**, and **130** arose from the same intermediate (Conte *et al.* 1994b). In fact **130** could be derived together with its C-10 epimer, which was also obtained as a natural product, by UV irradiation of a solution of **128** (Carotenuto *et al.* 1995). However, the authors ruled out the possibility that **130** could be an artifact of the isolation procedure by careful processing of the sponge specimen in darkness. These compounds were shown to inhibit the induction of nitric oxide synthase in cultured macrophage J774 stimulated with lipopolysaccharide (Carotenuto *et al.* 1997). Cyclolinteinone (**128**) and lintenone (**130**) were also shown to have icthyotoxic and antifeedant properties. Lintenone was also moderately toxic against brine shrimp (Fattorusso *et al.* 1992).

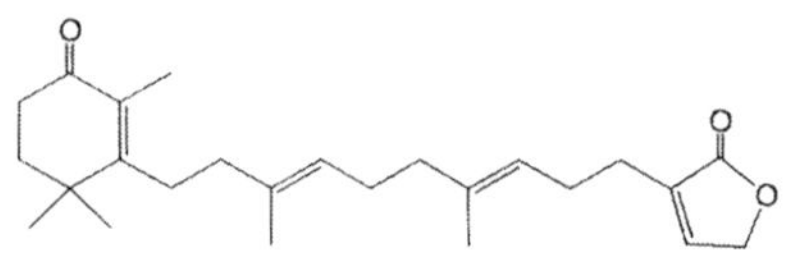

127

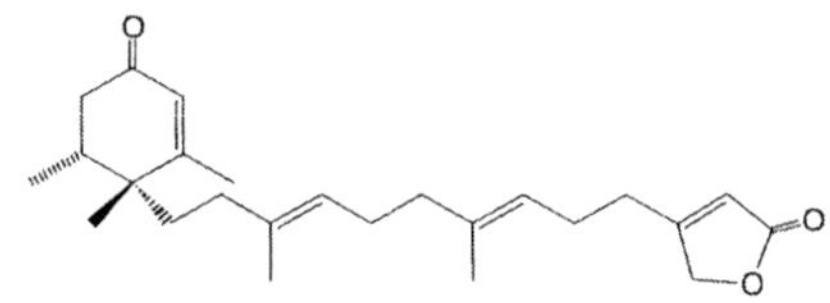

128

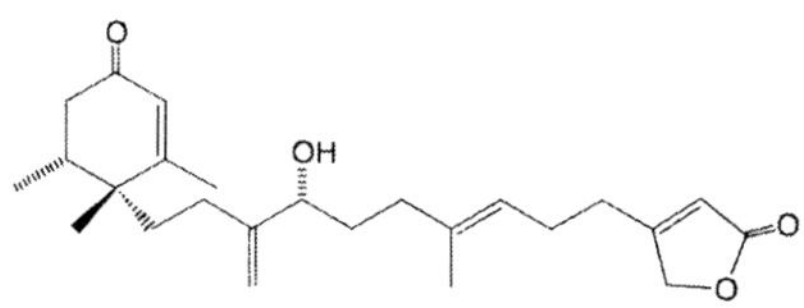

129

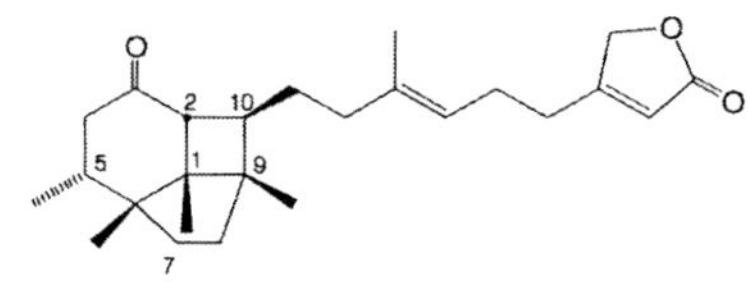

130

Bicyclic Sesterterpenoids

Included in this subsection are sesterterpenoids based on the skeleton of **106** and its rearranged ones, including **107**. Some of these compounds are structurally related to and co-occur with those of the cyclohexane sesterterpenoids and also cheilanthane class tricarbocyclic compound, *vide infra*. Also discussed here are a small group of sesterterpenes having a bicyclic system in the middle of the pentaprenyl chain, represented by dysidiolide (**135**).

Thorectandrols A (**131**) and B have been reported along with palauolol (Schmidt and Faulkner 1996) as cytotoxic constituents of the sponge *Thorectandra* sp. (Charan *et al.* 2001). Both thorectandrols exhibited weak cytotoxic activity against MALME-3M (melanoma) and MCF-7 (breast) cancer cell lines. Palauolol showed stronger activity than the thorectandrols with both cell lines.

25-Deoxycacospongionolide B (**132**), a deoxy derivative of cacospongionolide B known previously, was isolated from the Adriatic sponge *Fasciospongia cavernosa* and shown to be highly active in the brine shrimp assay (De Rosa *et al.* 1995b). The sponge *Luffariella variabilis* is an initial source of manoalide (**124**). A number of compounds related to manoalide have also been discovered from this sponge. One of them is luffalactone (**133**), which showed 52% inhibition of edema in the mouse ear assay at 50 µg/ear (Potts *et al.* 1992).

Halisulfates are a group of sesterterpenes derived from sponges. These compounds have been reported to exhibit antimicrobial activity and inhibitory effects on phospholipase A_2, PMA-induced inflamma-

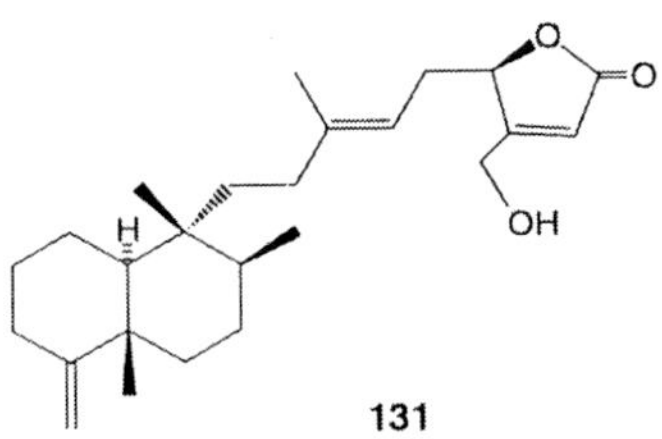

131

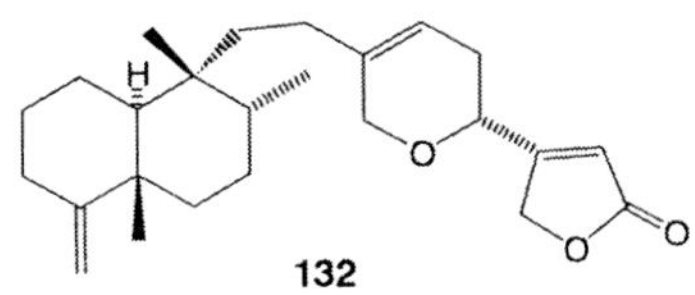

132

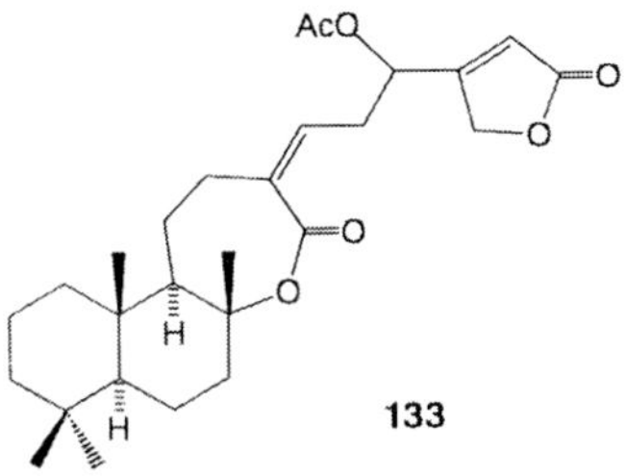

133

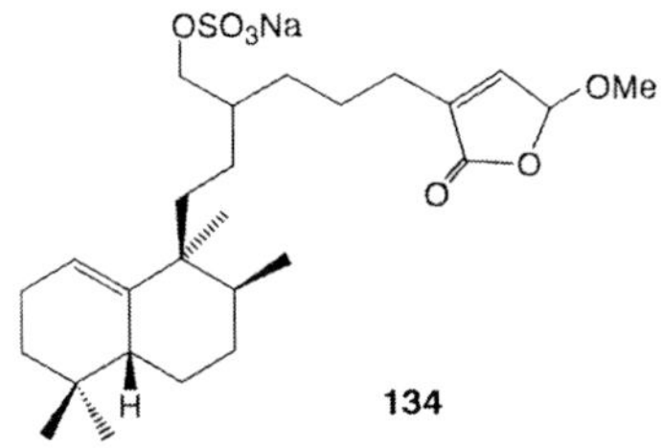

134

tion, and serine protease. Recently Makarieva *et al.* (2003) reported the isolation of three new congeners, halisulfates 8-10, from the Australian sponge *Darwinella australensis*. Halisulfates 9 and 10 (**134**) inhibited cell division of the fertilized eggs of the sea urchin *Strongylocentrotus intermedius*.

Dysidiolide (**135**) is an unusual sesterterpene butenolide isolated from the sponge *Dysidea etheria* collected in the Bahamas (Gunasekera *et al.* 1996). It was shown to inhibit the human cdc25A protein phosphatase (IC_{50} 9.4 μM) and also the growth of A549 and P388 cell lines with IC_{50} values of 4.7 and 1.5 μM, respectively. Three total syntheses of dysidiolide have been reported (Boukouvalas *et al.* 1998, Magnuson *et al.* 1998, Miyaoka *et al.* 2001). Two related compounds, kohamaic acids A and B (**136**), have been isolated from a sponge, *Ircinia* sp., collected on the coast of Kohama Island, Okinawa (Kokubo *et al.* 2001). These compounds exhibited mild cytotoxicity against P388 cells.

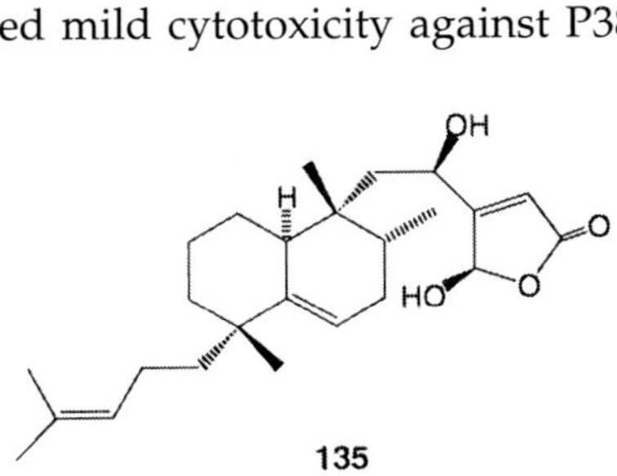

135

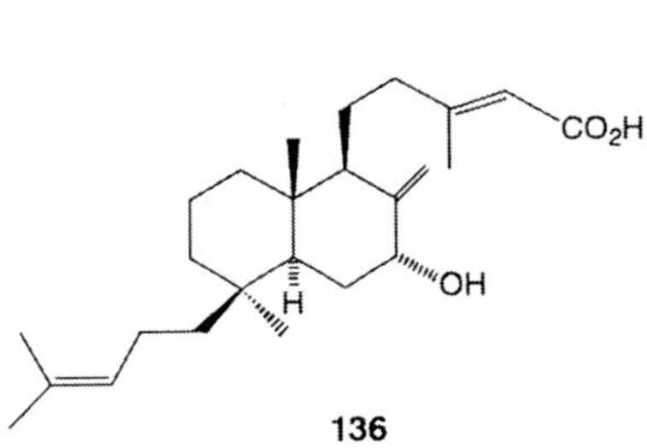

136

Norsesterterpene Cyclic Peroxides

A number of cyclic peroxides composed of a 1,2-dioxane ring have been reported from marine organisms, mainly from sponges. They can be classified into two structural types: fatty acid- and terpenoid-based compounds. The majority of terpenoid cyclic peroxides are C_{24} norsesterterpenes having a 2-substituted propionic acid attached at the C-3 position of a 1,2-dioxane ring. The sesterterpenes can take an acyclic, monocyclic, or bicyclic form, the latter being most abundant. Biological properties of cyclic peroxides include antimicrobial, antiviral, cytotoxic, and antimalarial activities.

Norditerpene cyclic peroxides (e.g., **137**) and norsesterterpene congeners (e.g., **138**) have been isolated from the New Caledonian sponge *Diacarnus levii* (D'Ambrosio *et al.* 1997). Comparison of their cytotoxicity revealed importance of the length of the lipophillic chain for the activity. The norsesterterpenoids showed significant activity (**138:** IC_{50} 0.9 μg/mL) against KB cells, while the norditerpenoids were inactive or marginally active. A related compound, muqubilone (**139**), was isolated along with the known sigmosceptrellin B and muqubilin from the Red Sea sponge *Diacarnus erythraenus* (El Sayed *et al.* 2001). Muqubilone and muqubilin exhibited antiviral activity against herpes simplex virus type 1, while sigmosceptrellin B (**140**) showed significant *in vitro* antimalarial activity against *Plasmodium falciparum*. Muqubilin and **140** also displayed potent *in vitro* activity against *Toxoplasma gondii.*

Compound **139** was also isolated together with a new norditerpene cyclic peroxide by another group (Youssef *et al.* 2001) as cytotoxic constituents of the same sponge collected in the Red Sea. They named **139** aikupikoxide A. This compound could biogenetically be formed by oxidative scission of the cyclohexene double bond of a related monocarbocyclic terpenoid.

From another species of the same genus, *Diacarnus* cf. *spinopoculum*, Sperry *et al.* (1998) reported the isolation and cytotoxic evaluation of a dozen terpenoid cyclic peroxides and related compounds. Eight of them, including epimuqubilin B (**141**), were norsesterterpenes of the muqubilin/sigmosceptrellin classes. Sigmosceptrellin class (e.g., **140**) compounds displayed significant cytotoxicity against several cancer cell lines, while muqubilin class (e.g., **141**) peroxides showed less significant activity.

A Thai sponge, *Mycale* sp., gave rise to two sigmosceptrellin class norsesterterpenes named mycaperoxides A (**142**) and B (**143**), which showed significant cytotoxicity (IC_{50} 0.5-1.0 μg/mL) against P388, A549, and HT29 cell lines (Tanaka *et al.* 1993). These compounds also displayed antiviral activity against herpes simplex virus type 1 and vesicular stomatitis virus. The mycaperoxides were antimicrobial against *Bacillus subtilis* and *Staphylococcus aureus*. Capon *et al.* (1997, 1998) have reported

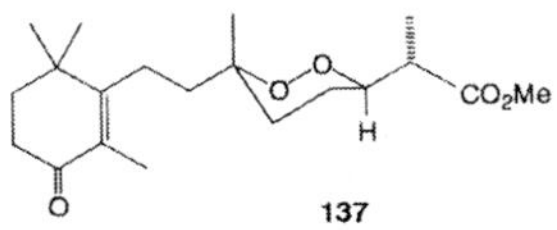

137

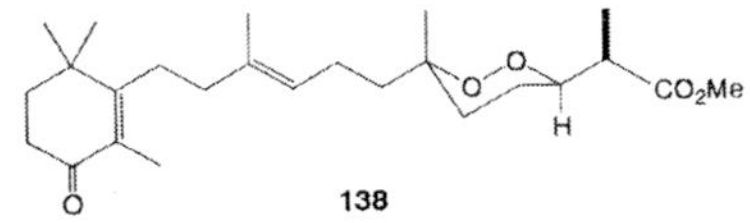

138

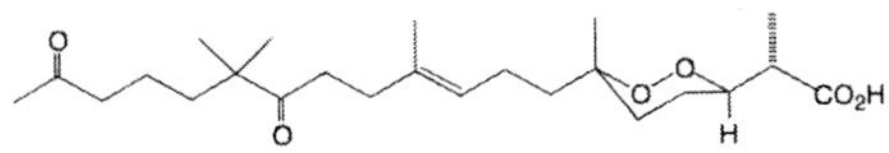

139

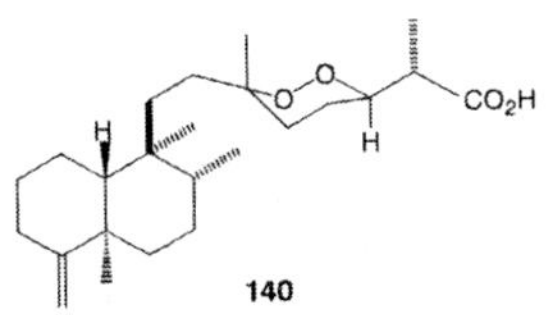

140

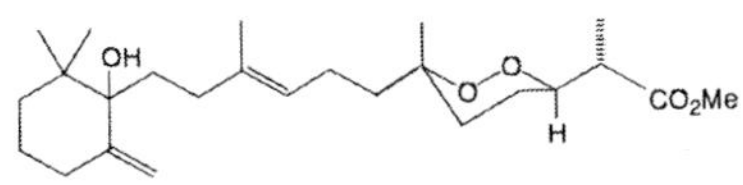

141

several related compounds, mycaperoxides C-G, from Australian *Mycale* spp. No biological activities have been recorded. More recently, mycaperoxide H **(144)** has been isolated from a Thai *Mycale* sp. which was presumably the same species as the earlier specimen which yielded mycaperoxides A and B (Phuwapraisirisan *et al.* 2003). Both specimens were collected at the same locale, Sichang Island, Thailand. Mycaperoxide H showed significant cytotoxicity against HeLa cells (IC_{50} 0.8 μg/mL).

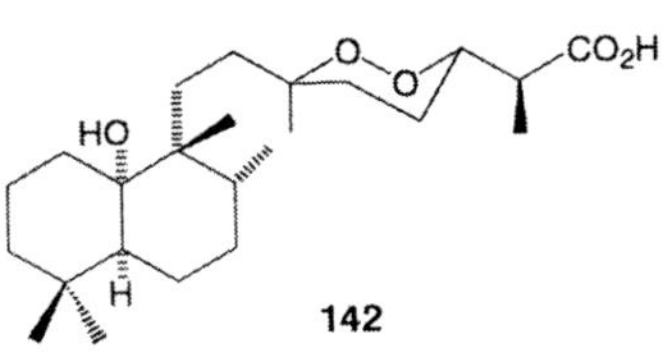

142

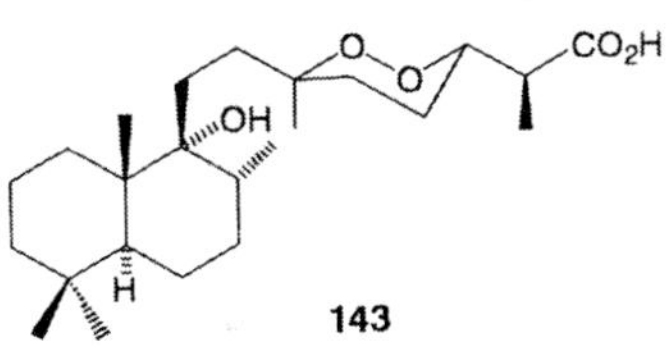

143

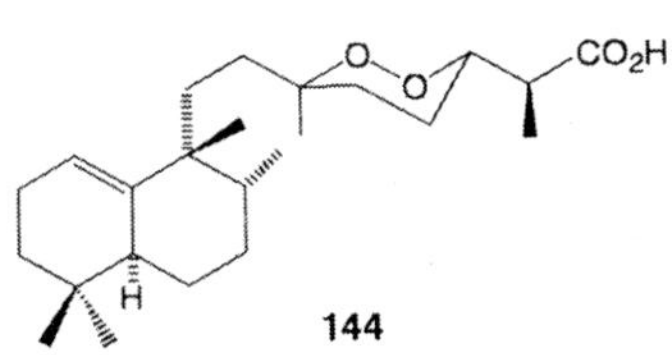

144

Cheilanthane Sesterterpenoids

Cheilanthane sesterterpenoids are those based on the skeleton **108**. Almost all of these compounds from sponges contain a γ-hydroxybutenolide, butanolide, or furan in the side chain portion of the tricyclic skeleton. As stated earlier these compounds from sponges are closely related to some of the compounds discussed in the subsections of "cyclohexane sesterterpenoids" and "bicyclic sesterterpenoids".

Six cheilanthane sesterterpenes, spongianolids A-F (A: **145, C: 146)**, have been isolated as inhibitors of protein kinase C (PKC) from a sponge, *Spongia* sp., collected in Florida (He *et al.* 1994). All of them contained a γ-hydroxybutenolide. These compounds inhibited PKC at IC_{50} 20-30 μM and did not inhibit human 85 kD phospholipase A_2. Spongianolides A-D showed potent activity (IC_{50} 0.5-1.4 μM) against the mammary cancer cell line MCF-7. Spongianolides C **(146)** and D have also been reported by the names lintenolides A and B, respectively, as ichthyotoxic and antifeedant constituents of the Caribbean sponge *Cacospongia* cf. *linteiformis* (Conte *et al.* 1994a). Spongianolide A **(145)** has been synthesized and its absolute stereochemistry determined (Hata *et al.* 1999). A related compound, cavernosolide **(147)**, having a pyrane ring instead of the tetrahydrofuran ring in **146**, has been isolated from the Tyrrhenian sponge *Fasciospongia cavernosa* (De Rosa *et al.* 1997). Cavernosolide showed potent activity (IC_{50} 0.37 μg/mL) in the brine shrimp assay and ichthyotoxicity against the fish *Gambusia affinis*.

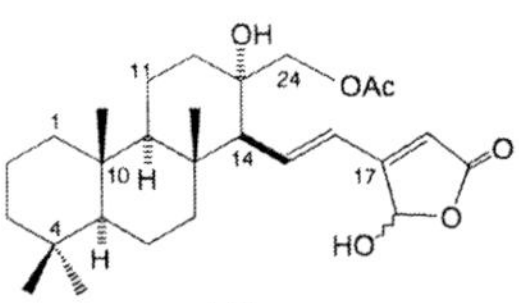

145

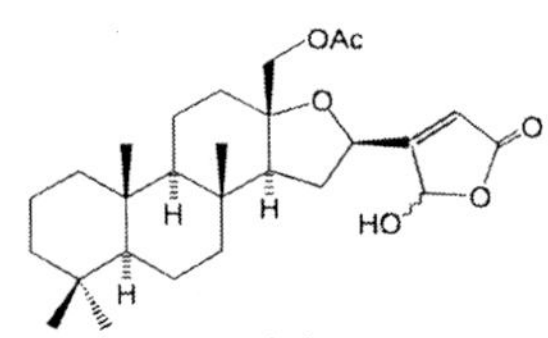

146

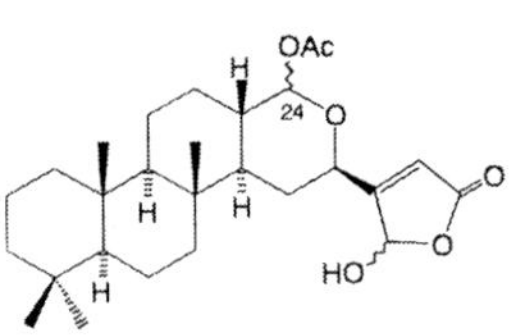

147 24R
148 24S

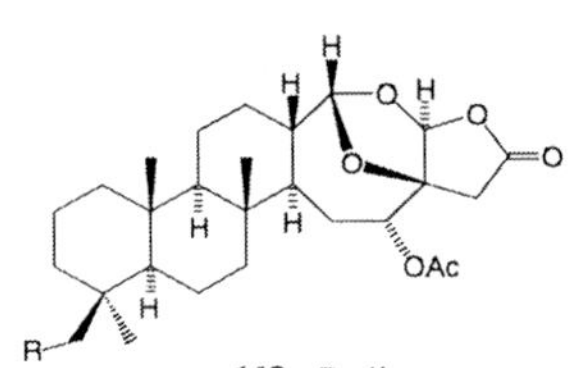

149 R = H
150 R = OAc

An epimer **(148)** at C-24 of **147** was isolated along with four related compounds (petrosaspongiolides N-R) as an inhibitor of phospholipase A_2 (PLA_2) from the New Caledonian sponge *Petrosaspongia nigra* (Randazzo *et al.* 1998). Petrosaspongiolide M **(148)** was most active and slightly more active than manoalide **(124)** against human synovial (IC_{50} 1.6 μM) and bee venom (IC_{50} 0.6 μM) PLA_2 enzymes. Petrosaspongiolides A-L have earlier been reported as cytotoxic compounds by the same group from the same sponge (Gomez-Paloma *et al.* 1997). The majority of them have the same hexacyclic ring system as represented by petrosaspongiolides A (**149**) and C (**150**). These compounds exhibited mild cytotoxicity against the NSCLC-N6 human non-small cell lung cancer cell line. Compound **150** was most active with IC_{50} 0.5 μg/mL. Petrosaspongiolides A and B are the same as the compounds reported earlier from a sponge designated incorrectly as *Dactylospongia* sp. (Lal *et* *al*. 1994). This species was reassigned as *Petrosaspongia nigra* by Bergquist (Gomez-Paloma *et al*. 1997).

Additional cheilanthane sesterterpenes with a γ-hydroxybutenolide have been described as protein kinase inhibitors from *Ircinia* sp. (Buchanan *et al*. 2001). The Vanuatu sponge *Spongia* sp. contained some petrosaspongiolides, their derivatives (e.g., **151**), and several terpene alkaloids (e.g., **152**) possessing a pyridinium moiety (De Marino *et al*. 2000). The pyridinium alkaloids which inhibited human synovial PLA_2 could biosynthetically be derived from a common sesterterpenoid intermediate which could also give rise to the norsesterterpene **151**. Related compounds to **151** and **152** have also been reported along with petrosaspongiolides from the sponge *Petrosaspongia nigra* described above (Gomez-Paloma *et al*. 1997).

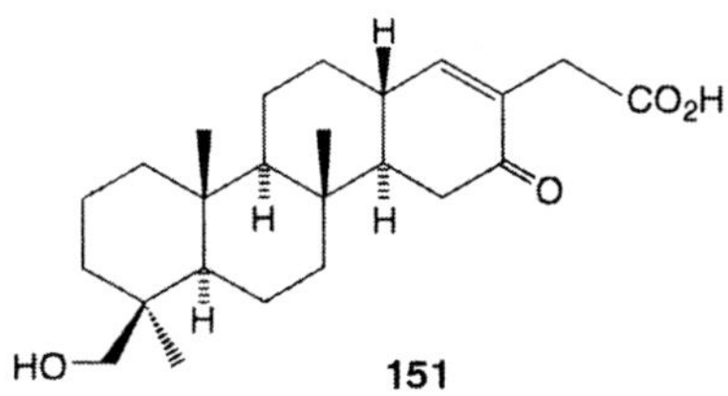

151

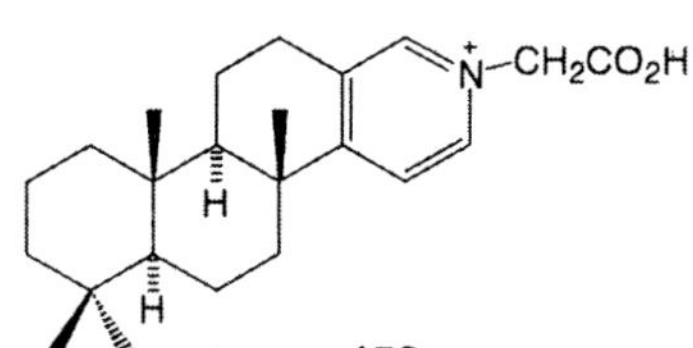

152

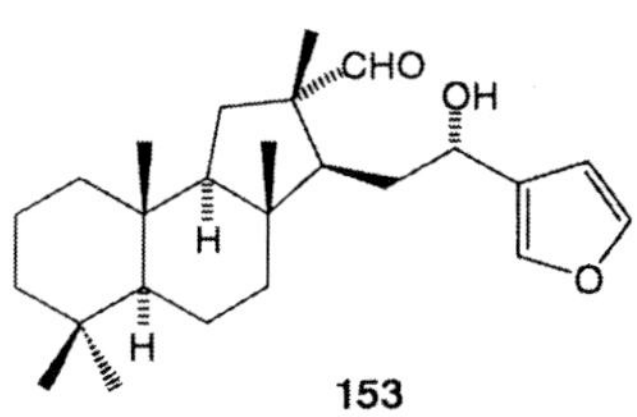

153

A tricarbocyclic sesterterpene having a somewhat different skeleton from cheilanthane has been known as hyrtiosal (**153**) and was isolated from the Okinawan sponge *Hyrtios erecta* (Iguchi *et al*. 1992). The ring C in the cheilanthane skeleton is contracted to a five-membered ring in hyrtiosal (**153**). It showed mild cytotoxicity against KB cells. A total synthesis of hyrtiosal has been described (Lunardi *et al*. 2002).

Scalarane Sesterterpenoids

Scalaranes are characteristic marine sesterterpenoids based on the tetracarbocyclic skeleton **109**. They are known from about a dozen species of sponges belonging to the families Spongiidae and Thorectidae and also from some species of nudibranchs which feed on these sponges (Bergquist and Wells 1983). Many scalaranes also have one or two additional methyl groups on the skeleton **109**. The majority have a five-membered oxygen heterocycle fused on the D-ring as a butenolide, butanolide, tetrahydrofuran, or furan. Biological properties include antimicrobial, cytotoxic, antiinflammatory, and anti-HIV activities and inhibition of enzymes.

A simple member of the scalaranes is 12-deacetoxyscalaradial (**154**), which was isolated along with the known scalaradial (**155**) and furoscalarol (De Rosa *et al.* 1994b). The C/D ring moiety of **154** resembles polygodial (**11**) which is a well-known antifeedant in both terrestrial and marine environments. In the fish antifeedant assay **154** showed similar potency with polygodial, while scalaradial (**155**) required twice the concentration of **11** to elicit the same effect. Like **11, 154** has a hot taste to humans but **155** does not. Both **154** and **155** were toxic (LD_{50} 0.77 and 0.18, respectively) against brine shrimp.

Scalaradial (**155**), scalarin (**156**, the first scalarane class compound described), and 10 other related compounds (e.g., **157**) have been reported as cytotoxic constituents of the Spanish sponge *Cacospongia scalaris* (Rueda *et al.* 1997). Some of these compounds exhibited moderate cytotoxicity against five cancer cell lines. 18-*epi*-Scaladial (**158**) displayed the strongest activity (IC_{50} 0.2 μg/mL: P388, A549, HT29). 12-*epi*-Scalarin (**159**), isolated from the sponge *Hyrtios erecta*, was shown to have potent *in vitro* and *in vivo* antitumor activities (Tsuchiya *et al.* 1998). This compound inhibited *in vitro* cell growth of P388 (IC_{50} 14.5 ng/mL) and MNK-1 (IC_{50} 57.7 ng/mL), MNK-7 (IC_{50} 56.0 ng/mL), and MNK-74 (IC_{50} 36.8 ng/mL) human gastric cancer cell lines. When **159** was administered at a dose 8.0 mg/kg to P388 implanted mice, a 74.4% increase of life span was achieved.

A number of cytotoxic sesterterpenoid derivatives (e.g., **160-162**) of **156** and **157** have been described from the sponge *Hyrtios erecta* (Youssef *et al.* 2002, Miyaoka *et al.* 2000, Pettit *et al.* 1998, Ryu *et al.* 1996b). Salmahyrtisol (**162**) was isolated together with the known hyrtiosal (**153**) from a specimen of *H. erecta* collected in the Red Sea (Youssef *et al.* 2002). This unique ring system could biogenetically be derived from **153**. Related cytotoxic compounds have also been reported from *Spongia* sp. (Tsukamoto *et al.* 2003). Two derivatives (**163, 164**) of scalarin (**156**) and furoscalarol (**157**) from *Hyrtios* sp. were noted to have stimulating effects on the synthesis of nerve growth factor in cultured astroglial cells (Doi *et al.* 1993).

The structures **154-164** cited above were based on the C_{25} scalarane or its rearranged skeleton. A great number of scalarane derivatives can also take a methylated skeleton of C_{26} (*homo*-scalarane) or C_{27} (*dihomo*-scalarane). The methylation occurs at C-19, C-20, C-23, and/or C-24 of the scalarane skeleton **109**. The majority of *homo*- and *dihomo*-scalaranes are metabolites of sponges belonging to two genera, *Phyllospongia* and *Strepsichordaia*.

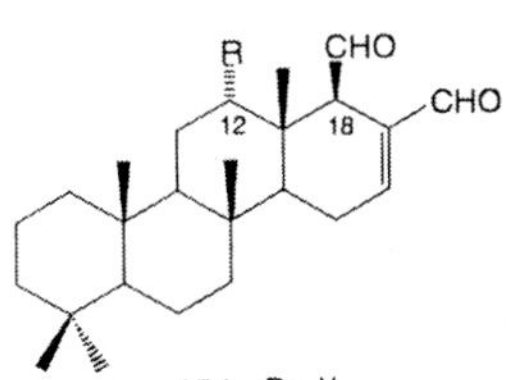

154 R = H
155 R = OAc
158 R = OAc, 18-epi

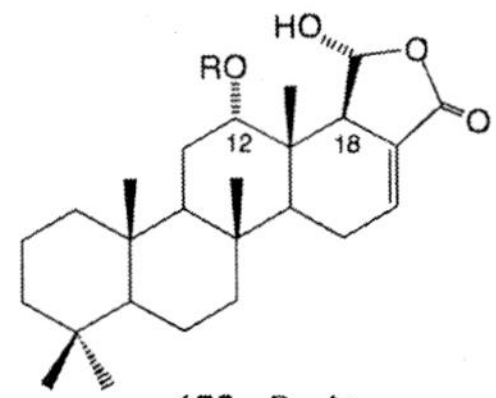

156 R = Ac
159 R = Ac, 12-epi
163 R = H

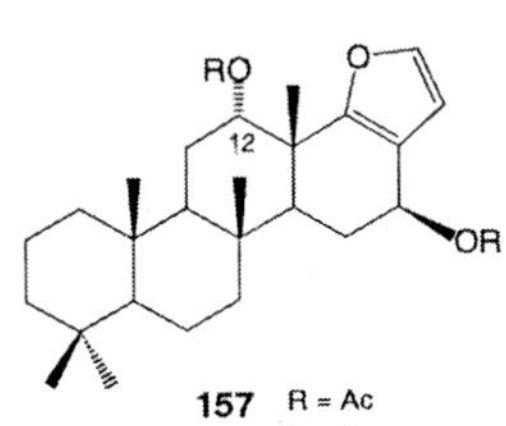

157 R = Ac
164 R = H

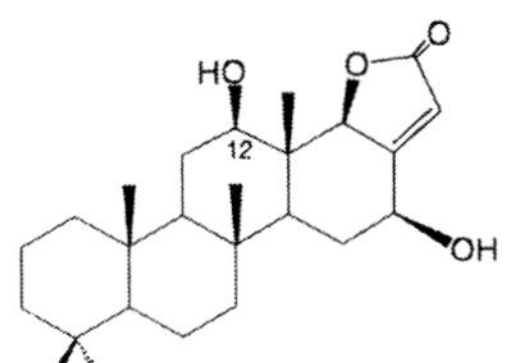

160

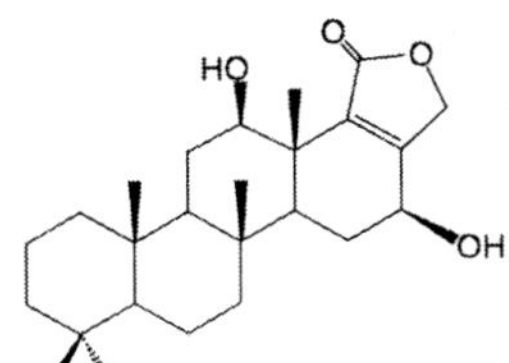

161

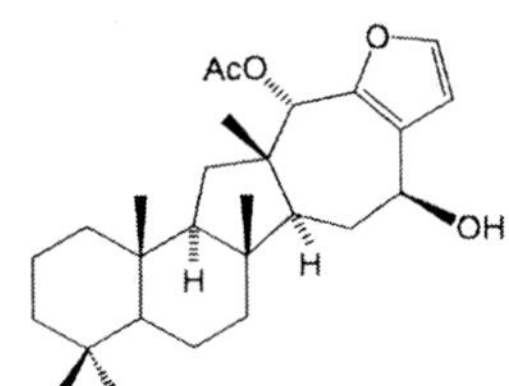

162

Several *homo*-scalaranes having a methyl extention at C-24 on **109** have been reported from the sponge *Lendenfeldia frondosa* (Alvi and Crews 1992). One (**165**) of them, which was isolated after acetylation, exhibited moderate anti-inflammatory activity. An Indonisian specimen of *Phyllospongia* sp. gave rise to a total of eight scalaranes encompassing 24-*homo*- (e.g., **166**), 20,24-*dihomo*- (e.g., **167**) and 24-*homo*-25-*nor*-scalaranes (e.g., **168**). All of them showed mild cytotoxicity against KB cells (Roy *et al.* 2002). More than 10 related compounds (e.g., **169**, **170**) have been reported from *Phyllospongia foliascens* collected in the South China Sea (Zeng *et al.* 1991, Fu *et al.* 1992). The *dihomo*-scalarane phyllofenone B (**169**) possesses a unique pentacarbocyclic skeleton. Some of these compounds including **169** and **170** exhibited cytotoxicity.

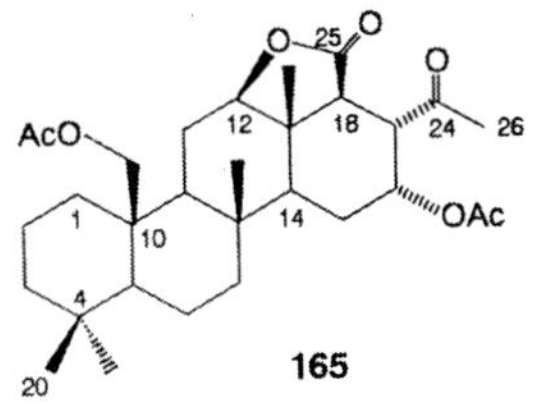

165

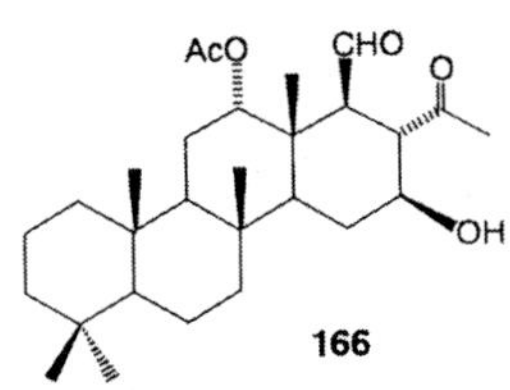

166

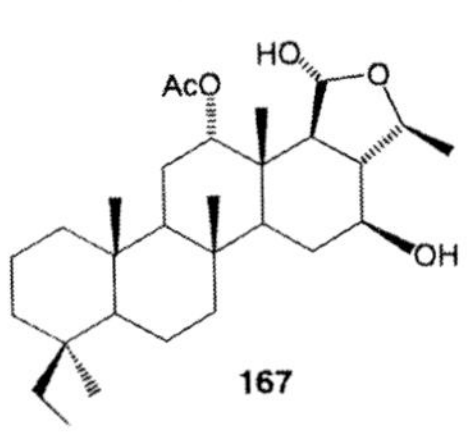

167

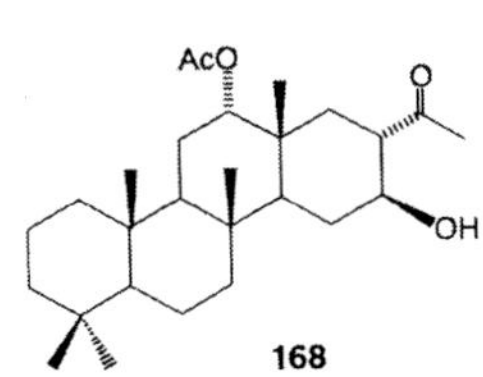

168

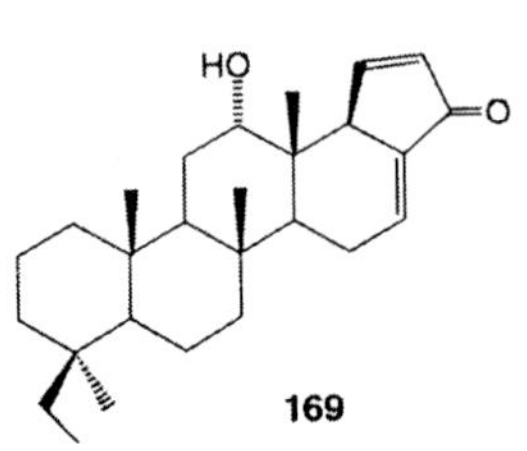

169

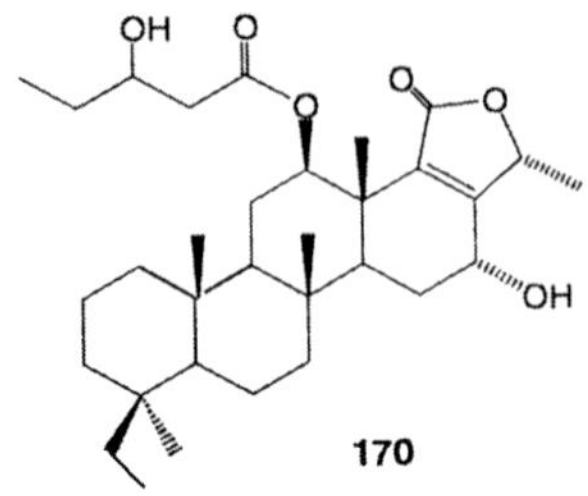

170

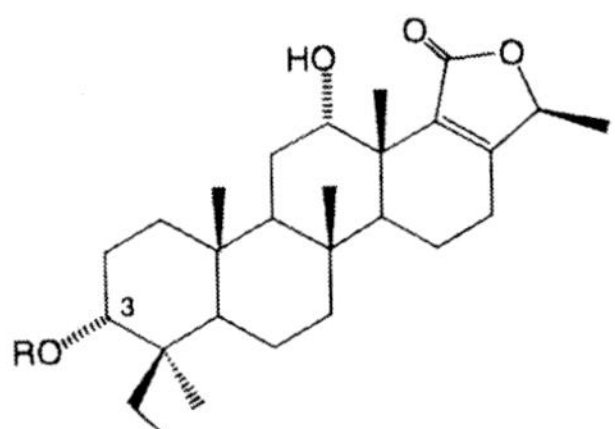

171 R = -COCH$_2$CH$_2$CH$_3$
172 R = -COCH$_3$

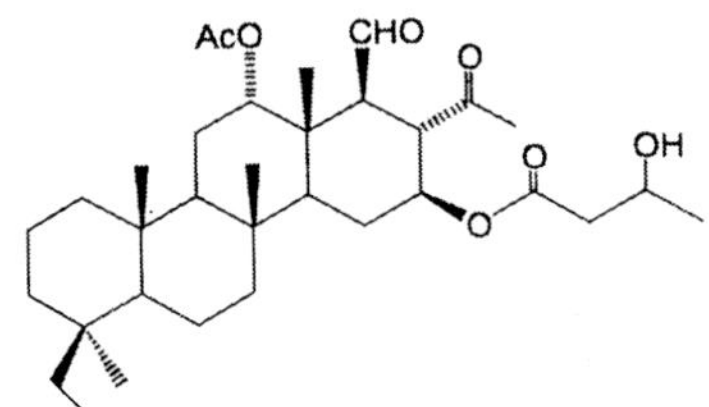

173

A series of 20,24-*dihomo*-scalaranes, phyllolactones A-E (A: **171**, C: **172**), which inhibited HIV-1 envelope-mediated fusion *in vitro*, have been described by an NIH group as metabolites of the sponge *Phyllospongia lamellosa* (Chang *et al*. 2001). Except for Phyllolactone E, all others have an aceloxy function at C-3. As seen in the structures **154-170**, most of the scalaranes are oxygenated at C-12, but oxygenation at C-3 as in the phyllolactones is rare. Phyllolactones A-D modestly inhibited HIV-1 envelope-mediated fusion with IC_{50} values of ~2 μM and showed negligible cytotoxicity toward the cell lines (BS-C-1, NIH3T3) used in the fusion assay.

From the Australian sponge *Strepsichordaia lendenfeldi* Bowden *et al*. (1992) have isolated several cytotoxic *dihomo*-scalaranes with various aceloxy groups attached on C-12 or C-16 as represented by **173**. Additional related compounds from the same sponge have been described with no mention of biological activity (Jahn *et al*. 1999).

20,24-Dihomoscalaranes possessing a cyclopropane ring and related compounds, honulactones A-L (A: **174**, D: **175**), have been reported from the Indonesian sponge *Strepsichordaia aliena* (Jiménez *et al*. 2000a). Honulactones A-D were evaluated for cytotoxicity against P388, A549, HT-29, and MEL-28 and all were active at a level of IC_{50} 1 μg/mL. The Scheuer group also reported additional related metabolites, but no biological properties were described (Jiménez *et al*. 2000b).

Scalaranes and homoscalaranes have also been known from nudibranchs. The Japanese nudibranch *Chromodoris inornata* contains several cytotoxic scalaranes (Miyamoto *et al*. 1999). Two of them, inorolides A (**176**) and B, have a rare skeleton with seven-membered D-ring. One of the constituents was a cheilanthane butenolide (**177**). These compounds showed moderate cytotoxicity against L1210 and KB cell lines. Cimino's group studied constituents of two nudibranchs, *Glossodoris sedna* and *G. dalli,* collected at the same place in Costa Rica (Fontana *et al.* 2000). The nudibranch *G. sedna* contained three 24-homoscalaranes (e.g., **178**) only and *G. dalli* three scalaranes (e.g., **179**) only. The authors suggested different dietary sources of the metabolites. The keto-aldehyde **178** was ichthyotoxic at 0.1 ppm against *Gambusia affinis* and exhibited moderate inhibition (IC_{50} 18 μM) of mammalian phospholipase A_2. New cytotoxic scalaranes have also been reported from the nudibranch *Glossodoris atromarginata* collected off Mandapan, India (Fontana *et al.* 1999a).

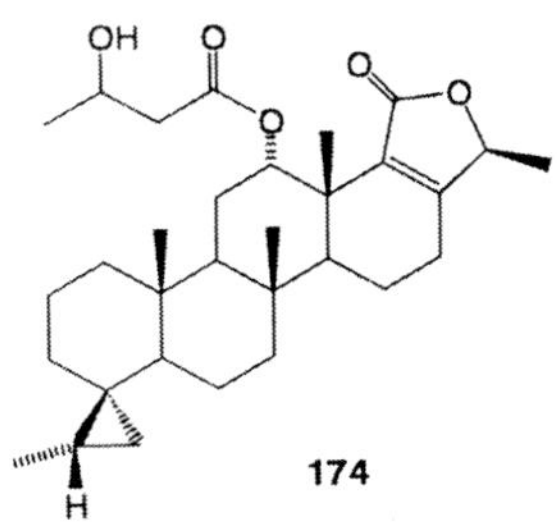

174

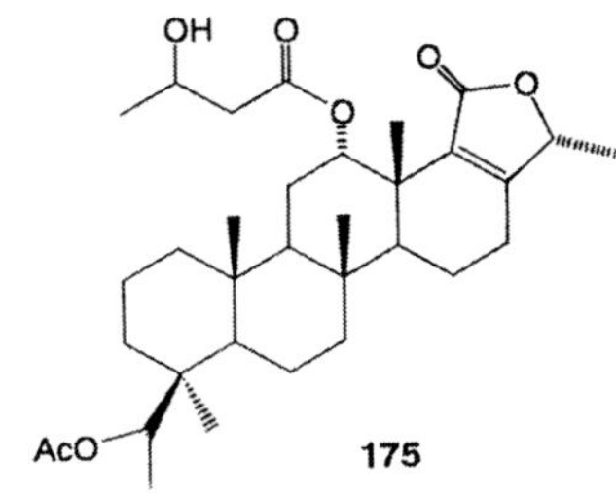

175

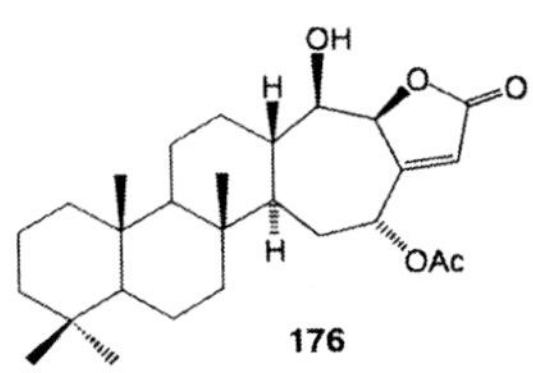

176

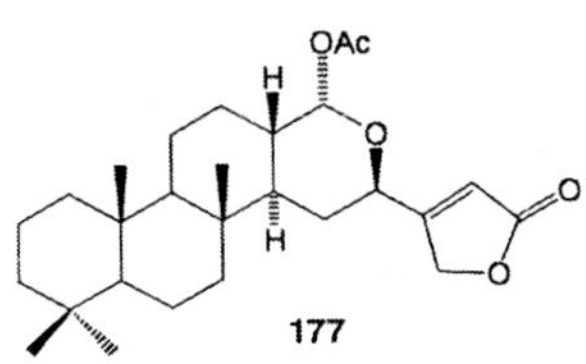

177

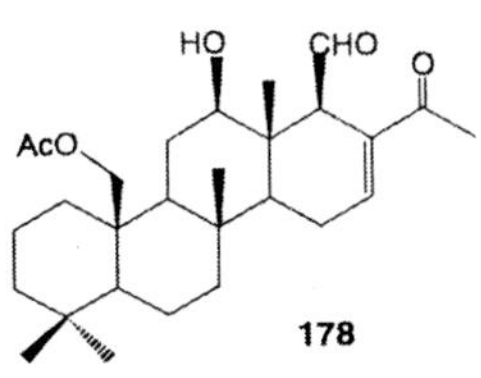

178

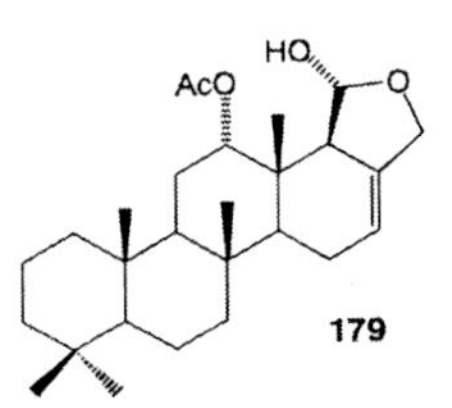

179

Other Sesterterpenoids

Included here are a small number of compounds belonging to three skeletal classes: acyclic sesterterpenoids, suberitanes, and merosesterterpenoids. Other than the furanosesterterpenoids discussed above, only a few compounds with an acyclic skeleton have been described as bioactive sponge metabolites. A norsesterterpene, rhopaloic acid A (**180**), was isolated from the sponge *Rhopaloeides* sp. and found to be an inhibitor of gastrulation of starfish embryos (Ohta *et al.* 1996a). Rhopaloic acid A also exhibited potent cytotoxicity against K-562 human myeloid, MOLT-4 human leukemia, and L1210 murine leukemia cell lines with IC_{50} values of 0.04, 0.05, and 0.10 μM/L, respectively. Additional compounds, rhopaloic acids B (C-6 epimer of **180**) and C (5-dehydro derivative of **180**), have also been reported from the same sponge and were found to act as inhibitors of gastrulation of starfish embryos (Yanai *et al.* 1998).

More recently Craig *et al.* (2002) described the isolation of the related sesterterpenoid barangcadoic acid (**181**) and the norsesterterpenoid rhopaloic acids A-G (D: **182**) from an Indonesian sponge, *Hippospongia* sp. These compounds showed inhibition of RCE-protease activity. RCE (Ras-converting enzyme) is responsible for proteolytic processing of Ras, and inhibitors of these enzymes are potentially important for the development of anticancer drugs. Triterpenoid analogues of **180-182** will be discussed in the next section.

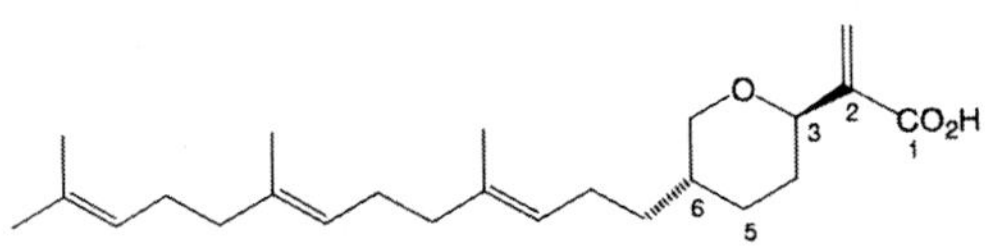

180

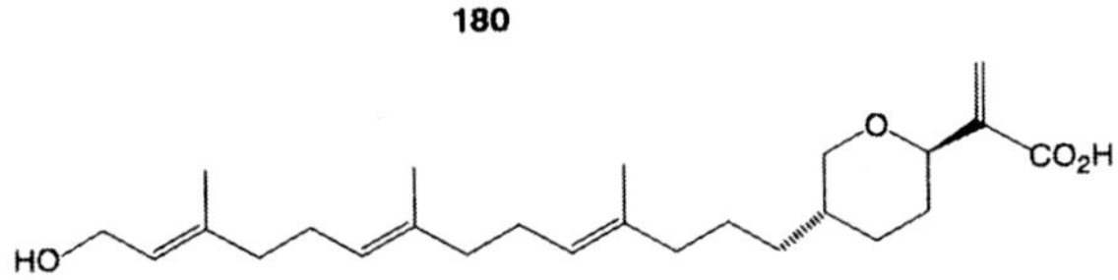

181

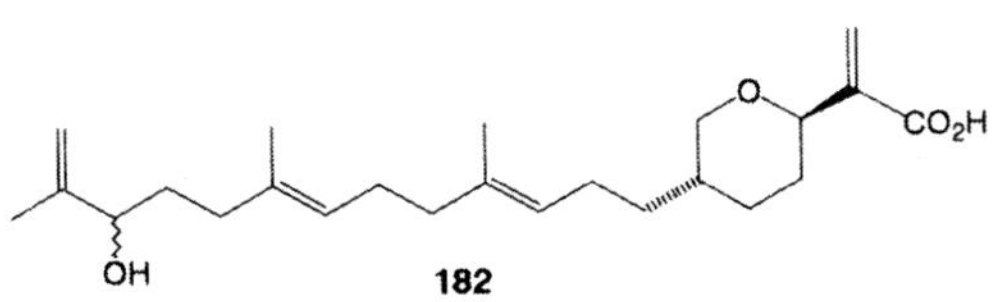

182

Suberitenones A and B (**183**), having a new skeleton, suberitane, have been reported from an Antarctic sponge, *Suberites* sp. (Shin *et al.* 1995). Suberitane B (**183**) showed inhibition of cholesteryl ester transfer protein, which mediates the transfer of cholesteryl ester and triglyceride between high-density lipoproteins and low-density lipoproteins.

Merosesterterpenoids are relatively rare. Only several bicyclic sesterterpene quinols have so far been described as the metabolites of sponges. Hipposulfates A (**184**) and B (16-hydroxyhipposulfate A) were isolated from an Okinawan sponge, *Hippospongia* cf. *metachromia* (Musman *et al.* 2001b). Hipposulfate A was moderately cytotoxic toward P388, A549, HT29, and MEL28 cells.

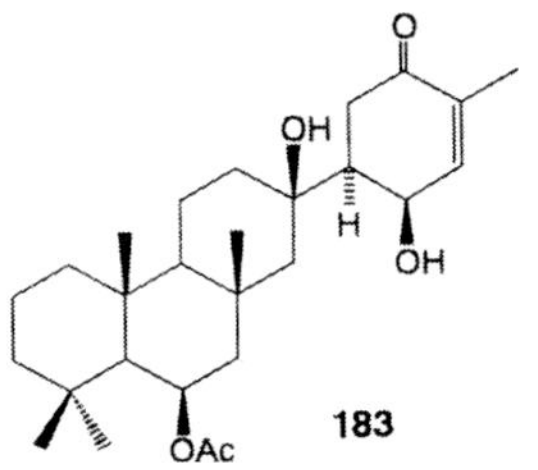

183

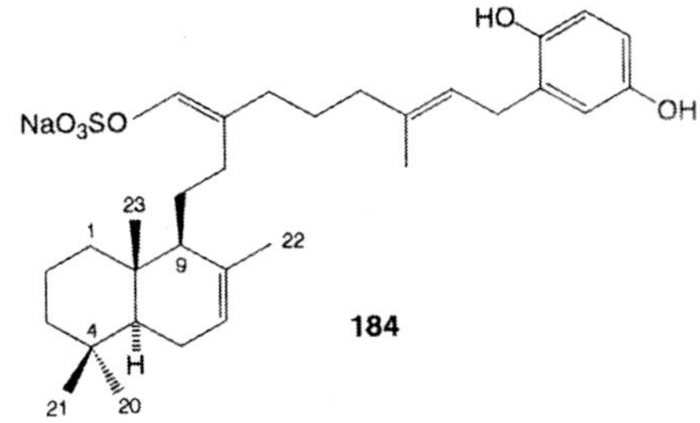

184

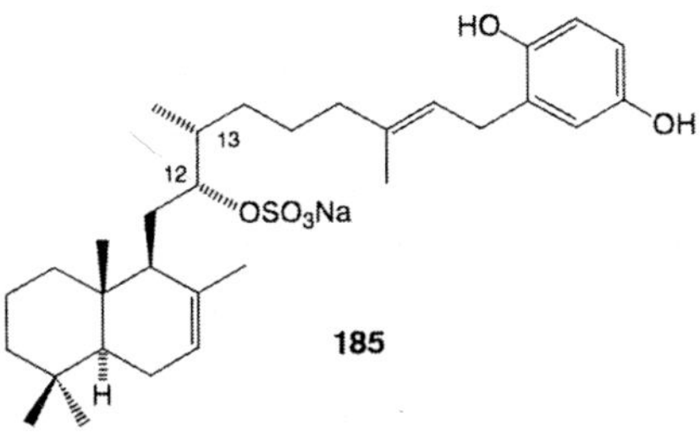

185

Coscinosulfate (**185**) was isolated from the New Caledonian sponge *Coscinoderma mathewsi* as a selective inhibitor of cdc25A protein phosphatase. This enzyme is known to be oncogenic and overexpressed in many cancer cell lines. Thus, its inhibitor can be a potential candidate for a therapeutic agent. The structure with absolute stereochemistry of coscinosulfate (**185**) was unambiguously established by a total synthesis (Poigny *et al.* 2001). The structure **185** turned out to be identical with that reported for halisulfate-1 (Kernan and Faulkner 1988). Therefore, the structure of halisulfate-1 needs to be revised. A salt form of **185** with an organic base, 1-methylherbipoline, instead of sodium, has been reported as an inhibitor of thrombin and trypsin from the same sponge, *Coscinoderma mathewsi,* from Pohnpei (Kimura *et al.* 1998).

TRITERPENOIDS

The triterpenoids are biogenetically derived from squalene (**186**) which is formed by head to head coupling of two molecules of farnesyl pyrophosphate **(1)**. Sponges are a major source of marine triterpenoids. For convenience in our discussion we classify bioactive sponge triterpenoids into several structural groups. Also included here are a small number of hexaprenoids, the carbon backbone of which is formed by head to tail arrangement of six isoprene units.

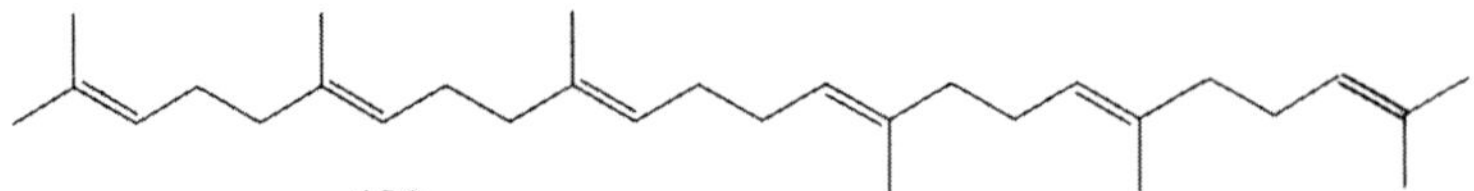

186

Linear Triterpenoids

Only one compound, hippospongic acid A (**187**), enters discussion here. It is a higher prenyl analogue of rhopaloic acids and barangcadoic acid, described above. Hippospongic acid A was isolated from *Hippospongia* sp. (Ohta *et al.* 1996b). Later the structure was revised to **187** and the absolute configuration was determined (Hioki *et al.* 1998). Like rhopaloic acids, it also exhibited selective inhibition of the gastrulation of starfish embryos without affecting fertilization and early embryonic development of the fertilized eggs up to the pre-gastrula stage. Total synthesis of **187** has been described by three groups (Tokumasu *et al.* 1999, Takikawa *et al.* 1999, Hioki *et al.* 2001).

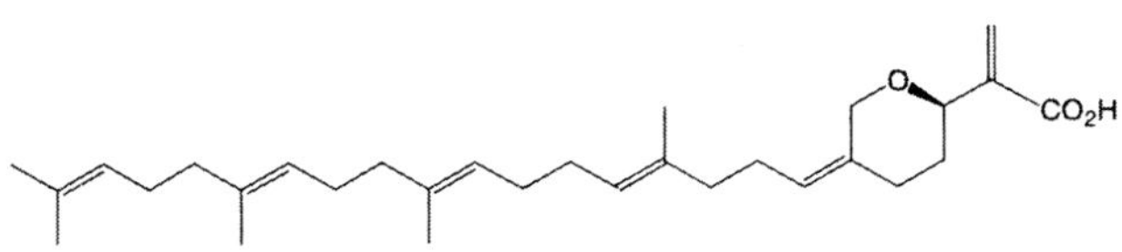

187

Isomalabaricane Triterpenoids

The malabaricanes are yellow pigments, first isolated from the wood of the tree *Alianthus malabarica*. The carbon skeleton (**188**) consists of a 6/6/5 tricarbocyclic nucleus and a side chain. The side chain usually forms a conjugated polyene. Marine counterparts of the malabaricanes are those classified as isomalabaricanes in which the configurations at C-8 and C-9 are opposite to those of malabaricanes. Isomalabaricanes have been reported from several genera of sponges. However, Tasdemir *et al.* (2002) recently pointed out the confusion on the sponge identification in previous papers and stated that many isomalabaricane-producing specimens are misidentified specimens of *Rhabdastrella globostellata*. For details a reader is referred to the paper by Tasdemir *et al.* (2002) where Dr. J.N.A. Hooper, a sponge taxonomist, is a co-author. Nevertheless, here we cite the names of the sponges as they appear in the original papers.

Stellettins H (**189**) and I have been isolated along with several previously known stellettins from the sponge *Rhabdastrella globostellata* collected from the Philippines (Tasdemir *et al.* 2002). These compounds showed significant cytotoxicity against HCT-116 human colorectal cancer cells. Especially, stellettin B (**190**) and (–)-stellettin E displayed selective toxicity toward the p21-deficient HCT cell line. Stellettins A-G have

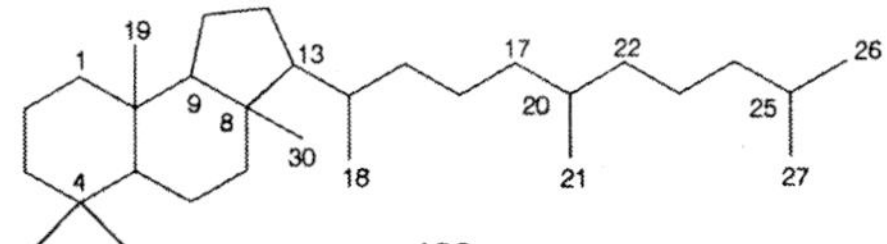

188

earlier been reported as cytotoxic constituents of the Chinese sponge *Stelletta tenuis* (Su *et al.* 1994) and from *Stelletta* sp. from northern Australia (McCormick *et al.* 1996). All stellettins contain either a carboxyl group or a pyrone at the terminus of the side chain as represented by **189** and **190**.

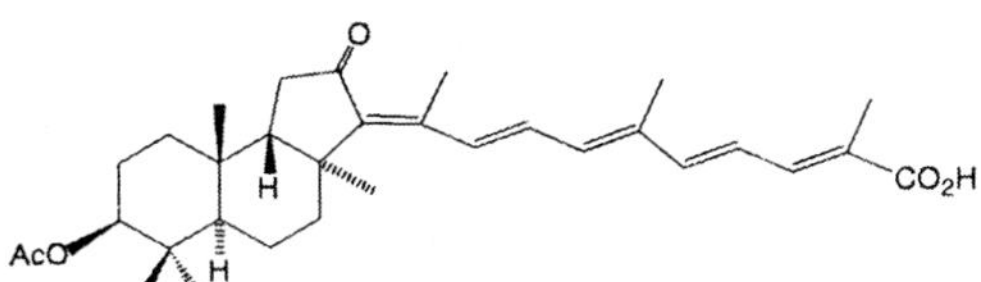

189

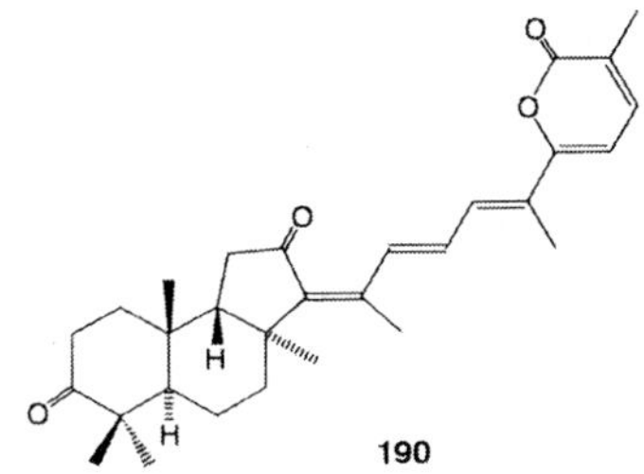

190

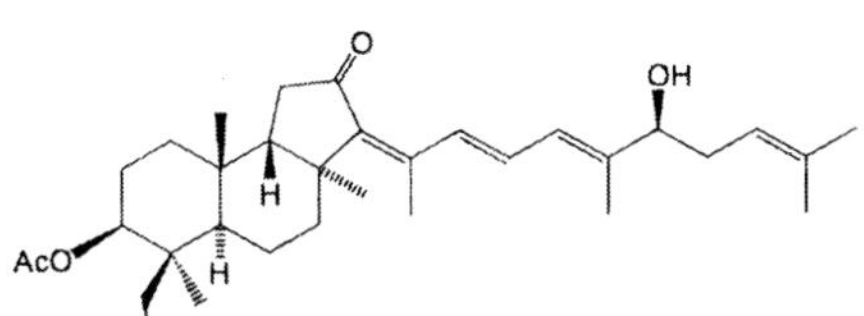

191 R = H
192 R = OH

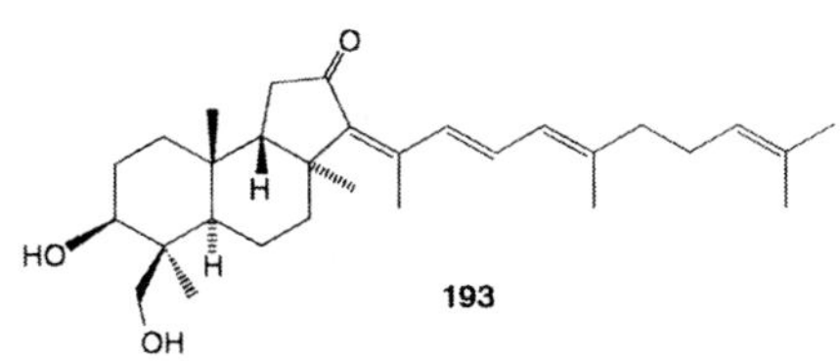

193

Stelliferins A-F (A: **191**) were some of the earliest members of the sponge-derived isomalabaricanes which displayed significant cytotoxicity against L1210 and KB cells (Tsuda *et al.* 1991). Several related compounds having a hydroxyl group at C-29 (29-hydroxystelliferin A: **192**) have recently been isolated as cytotoxic constituents of the sponge *Jaspis* sp. from Tonga (Meragelman *et al.* 2001). A monosaccharide, stelliferin riboside, has been isolated along with stelletins A and B from the Fijian sponge *Geodia globostellifera* (Tabudravu and Jaspars 2001). The riboside was far less active than stelletins in the assay against A2780 ovarian cancer cells.

Some stelliferin derivatives (e.g., **193**) from *Stelletta globostellata* induced characteristic morphological changes in 3Y1 rat fibroblasts at a concentration of 0.2 μM (Oku *et al.* 2000). From the same sponge Fusetani's group has earlier reported cytotoxic isomalabaricanes, globostellatic acids A-D (A: **194**), which contain a carboxyl group at C-29 (Ryu *et al.* 1996a).

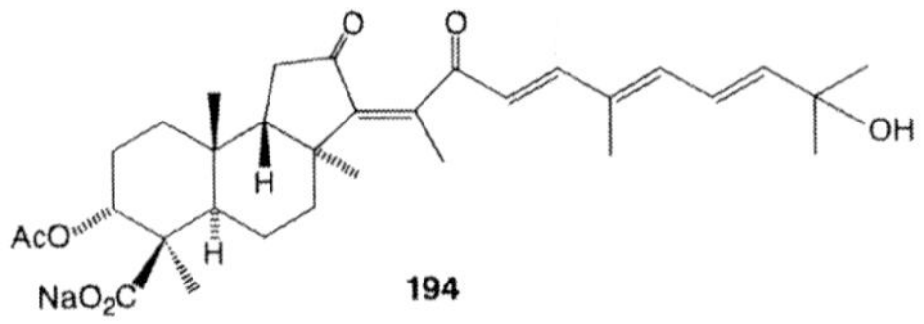

194

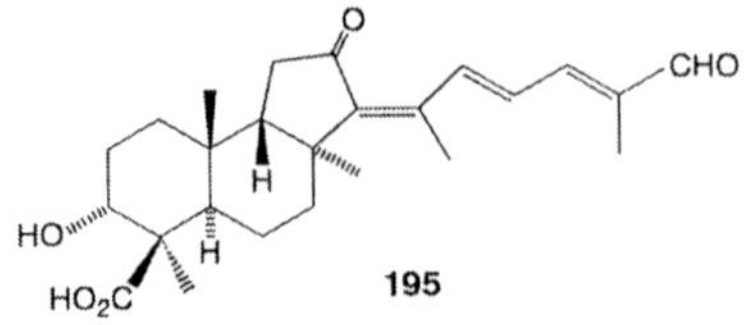

195

Nortriterpenes of the isomalabaricane class, jaspiferals A-G (C: **195**), have also been reported as cytotoxic metabolites from the Okinawan sponge *Jaspis stellifera* (Kobayashi *et al.* 1996). These compounds have a short side chain corresponding to the loss of three to ten carbons from the regular skeleton. Some more derivatives of **194** and **195** having a carboxyl group at C-29 have been isolated from a Vanuatu sponge, *Jaspis* sp. (Zampella *et al.* 2000).

Raspacionins and Sodwanones

Raspacionins and sodwanones are sponge-derived triterpenoids containing a perhydrobenzoxepane (PBO) moiety at both ends of a squalene skeleton or a PBO at one end and a decalin at the other end. The raspacionins are metabolites from the Mediterranean sponge *Raspaciona aculeata* and sodwanones from *Axinella weltneri* from the Indian Ocean. Related sponge metabolites include sipholenols/sipholenones and yardenones.

The isolation and structure determination of raspacionin (**196**) were first reported in 1990 (Cimino *et al.* 1990) and its absolute stereochemistry in 1993 (Cimino *et al.* 1993). Twenty raspacionins have so far been isolated from specimens of *R. aculeata* collected in Blanes, Spain, and Sicily, Italy (Ciavatta *et al.* 2002). All raspacionins have essentially the same skeleton and differ by functionality and stereostructure. These compounds showed antifeedant (*Carassius auratus*) and ichthyotoxic activity (*Gambusia affinis*). Cytotoxicity tests were also performed using the MCF-7 mammary cancer cell line. Raspacionin (**196**) showed an IC_{50} value of 4 μM. All others exhibited inhibition of cell proliferation.

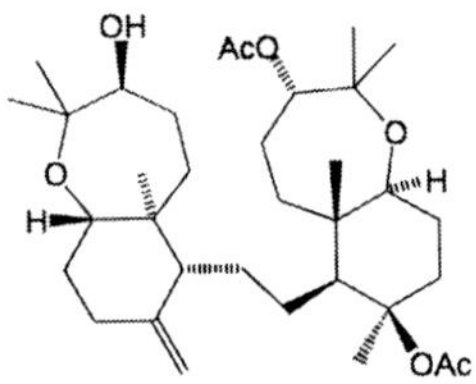

196

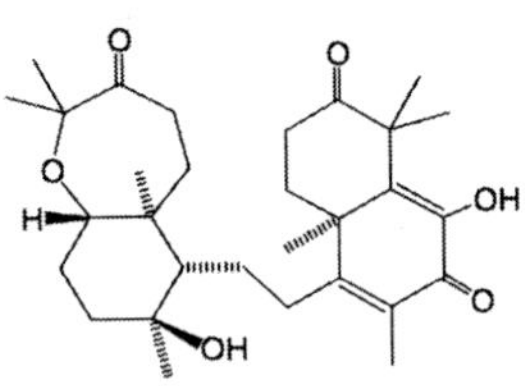

197

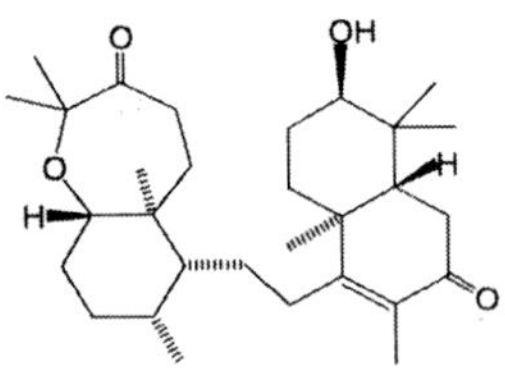

198

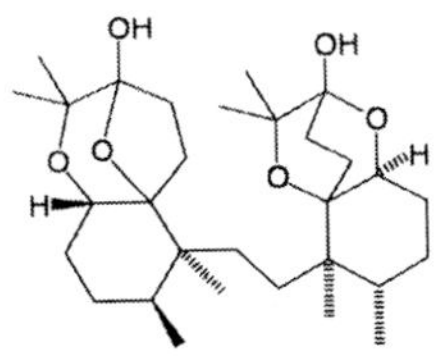

199

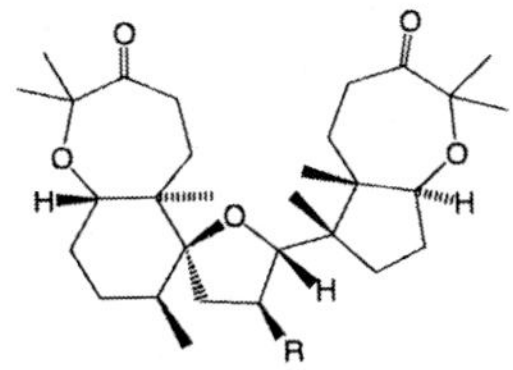

200 R = OH
201 R = H

A total of 18 sodwanones (A: **197**, H: **198**, L: **199**) have been reported by Kashman's group from the sponge *Axinella weltneri* (Rudi *et al*. 1993, 1995a, 1997, 1999). Sodwanones vary in the ring structures, represented by **196-199**. A few of them were cytotoxic against several cancer cell lines. Sodwanone H (198) was most active against A-549 with IC_{50} 0.02 μM (Rudi *et al*. 1995a).

Yardenones A (**200**) and B were isolated along with yardenone (**201**) and sodwanone A (**197**) from *Axinilla* cf. *bidderi* collected in Yemen (Carletti *et al*. 2003). Yardenones A and B were weakly cytotoxic against NSCLC-N6L16 lung carcinoma cells, while yardenone (**201**) was inactive. Yardenone was earlier reported from the Red Sea sponge *Ptilocaulis speculifer* (Rudi *et al*. 1998).

Hexaprenylquinol Sulfates

Some 20 hexaprenylquinol sulfates have been known from sponges. An example is the antifungal compounds shaagrockols (C: **202**) which were covered in our previous review (Kuniyoshi and Higa 2001). In addition to shaagrockols the sponge *Toxiclona toxius* yielded toxiusol (**203**), toxicol A (**204**), and toxicol B (**205**), all of which exhibited inhibition of HIV-1 reverse transcriptase (Loya *et al*. 1993). Toxiusol (**203**) was found to be the most potent inhibitor of the RNA-dependent DNA polymerase.

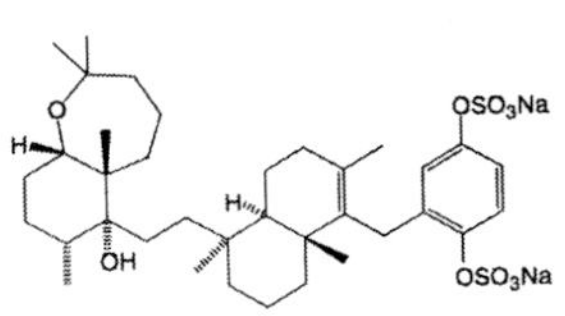

202

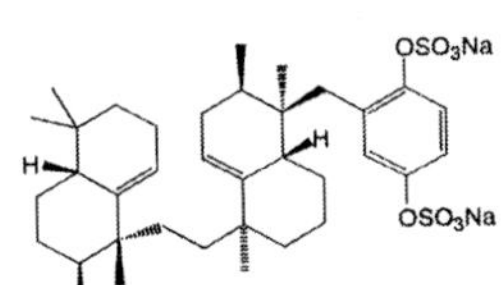

203

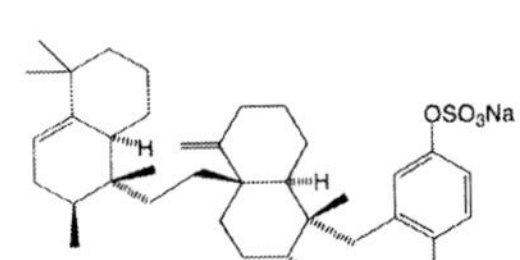

207

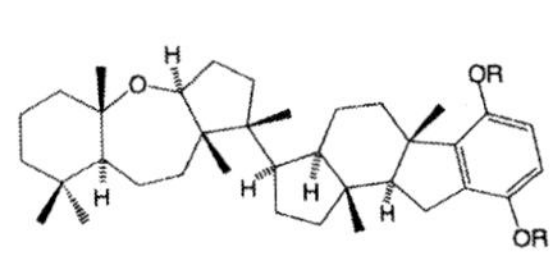

204 R = SO_3Na
205 R = H

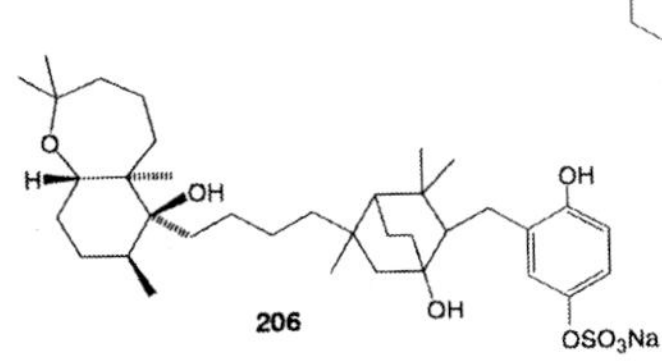

206

Another anti-HIV active compound is phuklona sulfate (**206**) isolated from a sponge, *Haliclona* sp., collected near Phuket, Thailand (Bokesch *et al.* 2002). Phuklona sulfate displayed modest cytoprotection against HIV in the NCI assay. Akaterpin (**207**), possessing a rearranged hexaprenyl skeleton closely related to **203**, has been isolated as an inhibitor of phosphatidylinositol-specific phospholipase C (PI-PLC) from a sponge, *Callyspongia* sp. (Fukami *et al.* 1997). Akaterpin (**207**) showed inhibition of PI-PLC with IC_{50} 0.5 μg/mL and of neutral sphingomyelinase with IC_{50} 30 μg/mL.

A series of hexaprenylhydroquinone sulfates, adociasulfates 1-8, have been described by two independent groups. Adociasulfates 1-6 (1: **208**, 2: **209**, 3: **210**, 6: **211**) were isolated as inhibitors of kinesin motor proteins from the sponge *Haliclona* (aka *Adocia*) sp. collected in Palau (Blackburn *et al.* 1999), while adociasulfates 1 (**208**), 7 (**212**), and 8 (**213**) were obtained as inhibitors of H^+-ATPase proton pump activity from the Australian sponge *Adocia* sp. (Kalaitzis *et al.* 1999). Adociasulfates 1-6 exhibited inhibition of the transport of stabilized microtubules by kinesin motor proteins. Adociasulfates 2 (**209**) and 6 (**211**) were most active. In the proton pump inhibition assays adociasulfate 1 (**208**) reduced the proton pump activity in hen bone-derived membrane vesicles and also in brain-derived vesicles. Adociasulfates 7 (**212**) and 8 (**213**) were less active than 1 (**208**).

Adociasulfate 10 (**214**) was also isolated from the Palauan sponge *Haliclona* (aka *Adocia*) sp. (Blackburn and Faulkner 2000). It contains a glycolic acid residue in the place of one of the sulfate groups. Adociasulfate 10 (**214**) showed similar potency with adociasulfates 2 (**209**) and 6 (**211**) in the inhibition of kinesin motor protein assay.

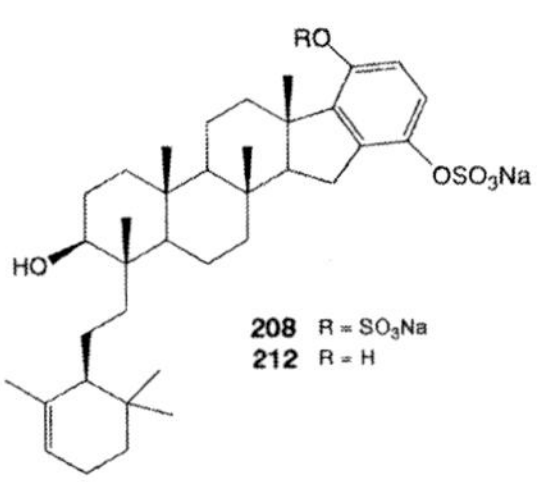

208 R = SO_3Na
212 R = H

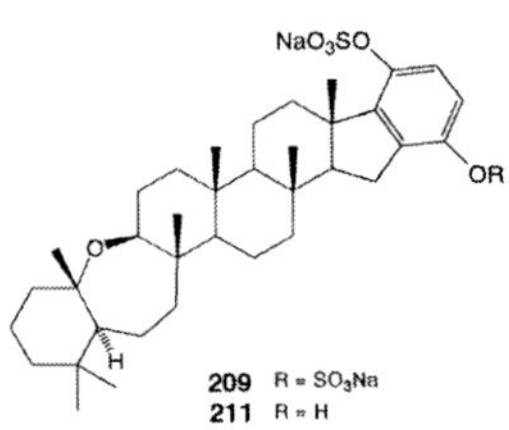

209 R = SO_3Na
211 R = H

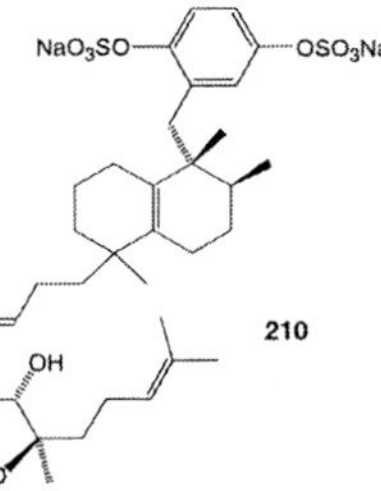

210

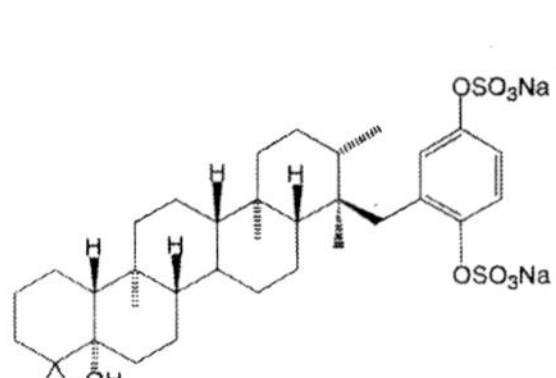

213

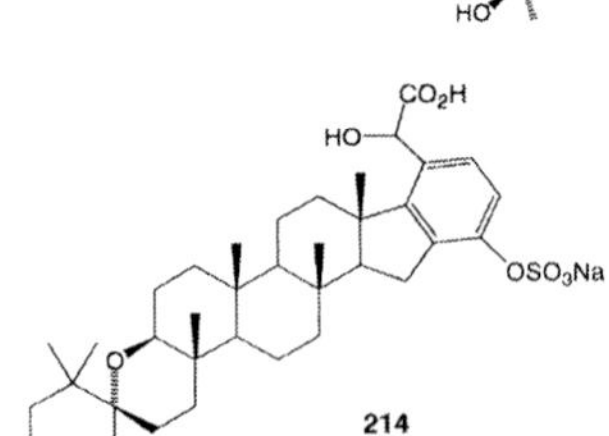

214

Lanostane Class Triterpenoids

Lanostane (**215**) is a key intermediate in steroid biosynthesis. Lanostane and its derivatives constitute common aglycones of a large group of saponins known from echinoderms such as sea cucumbers and sea urchins. Lanostanes and their saponins are relatively rare in sponges. Sponges have so far yielded some 20 bioactive triterpenoid saponins, the majority being cytotoxic compounds.

Two lanostane derivatives, penasterone (**216**) and acetylpenasterol, were isolated from the Okinawan sponge *Penares incrustans* (Shoji *et al.* 1992b). These compounds exhibited potent inhibition of histamine release from rat peritoneal mast cells induced by anti-IgE in a dose-dependent manner. The activity of **216** (IC_{50} 1.5 μM) was about 17 times greater than that of disodium cromoglycate, an antiallergy drug.

A penasterol trisaccharide, nobiloside (**217**), was obtained as an enzyme inhibitor from the sponge *Erylus nobilis* (Takada *et al.* 2002). The saccharide inhibited neuraminidase from the bacterium *Clostridium perfringens* with an IC_{50} value of 0.46 μg/mL.

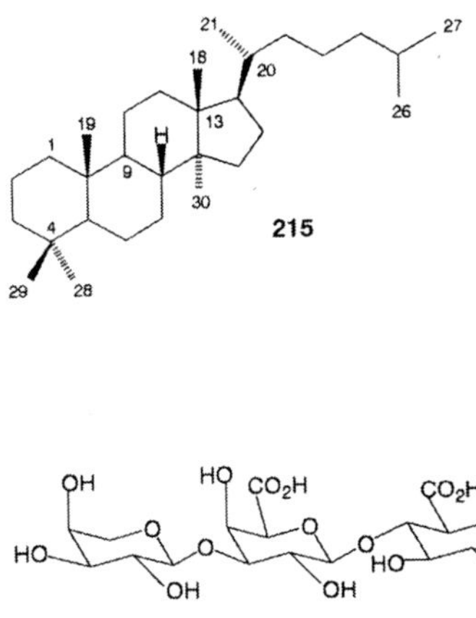

215

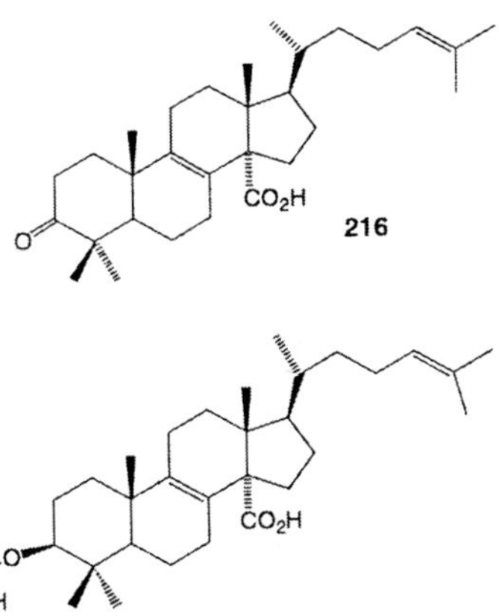

216

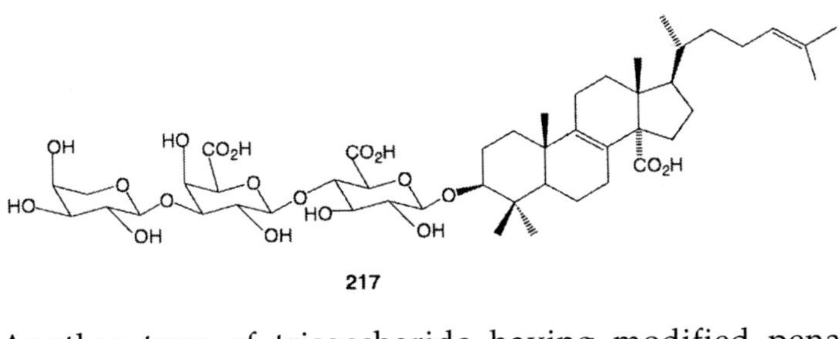

217

Another type of trisaccharide having modified penasterols as aglycones has been reported from a Korean collection of *Erylus nobilis* (Shin *et al.* 2001). The saponins are erylosides G-J (G: **218**, H: **219**) which exhibited mild cytotoxicity against the K562 human leukemia cell line. Cytotoxic saponins with norlanostane skeletons and four to five sugar moieties have been reported from the sponge *Asteropus sarasinosum* (Kobayashi *et al.* 1991, Espada *et al.* 1992), *Melophlus isis* (Lee *et al.* 2000), and *Ectyoplasia ferox* (Campagnuolo *et al.* 2001). None of them were potent cytotoxins.

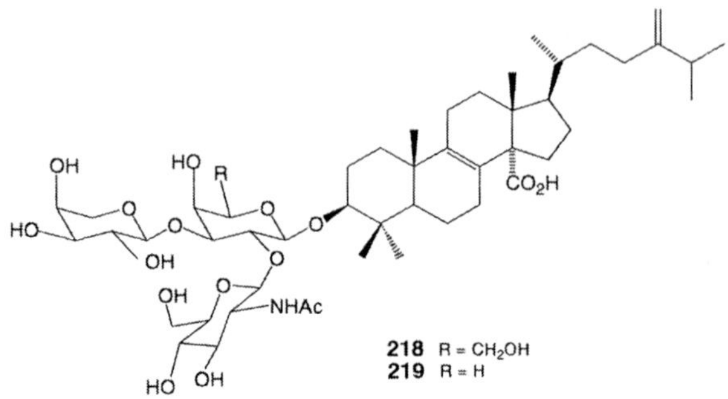

218 R = CH_2OH
219 R = H

POLYPRENYLQUINOLS

Included in this section are a small group of hydroquinones, quinones, and chromenols with a linear polyprenyl side chain. The number of prenyl groups in these compounds is from four to eight. Therefore, some of them could be placed in the preceding sections of meroterpenoids. However, since such compounds having different levels of prenylation often co-occur in sponges, a separate section is assigned for the convenience of our discussion.

Hydroquinone sulfates with a six to eight prenyl chain, sarcohydroquinone sulfates A-C (**220-222**), and the corresponding chromenols, sarcochromenol sulfates A-C (**223-225**), have been isolated as Na^+, K^+-ATPase inhibitors from the Pacific sponge *Sarcotragus spinulosus* (Stonik *et al.* 1992). Compounds **220-223** inhibited Na^+, K^+-ATPase from rat brain with IC_{50} values of 1.3-1.6 μM.

Similar compounds **226-229** and **222** are metabolites of the Croatian sponge *Ircinia spinosula* (De Rosa *et al.* 1995a). These compounds showed toxicity in brine shrimp and fish lethality assays. The sulfated derivatives (**222, 229**) displayed higher toxicity than the corresponding hydroquinones (**226-228**) in both assays.

From *Ircinia* sp. collected in New Caledonia, Bifulco *et al.* (1995) isolated a total of 15 polyprenyl metabolites including **220, 223, 226, 229**, benzoquinones (**230, 231**), a hydroxylated compound (**232**), and a C_{31}

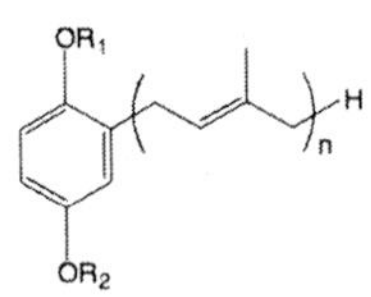

	R_1	R_2	n
220	H	SO_3Na	6
221	SO_3Na	SO_3Na	7
222	H	SO_3Na	8
226	H	H	6
227	H	H	7
228	H	H	8
229	H	SO_3Na	7

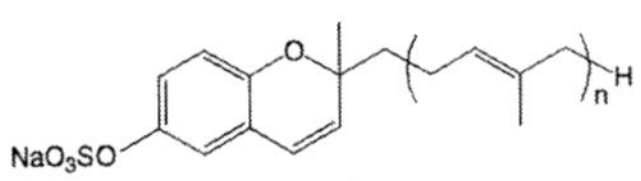

	n
223	5
224	6
225	7

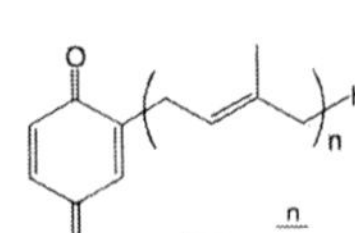

	n
230	4
231	5

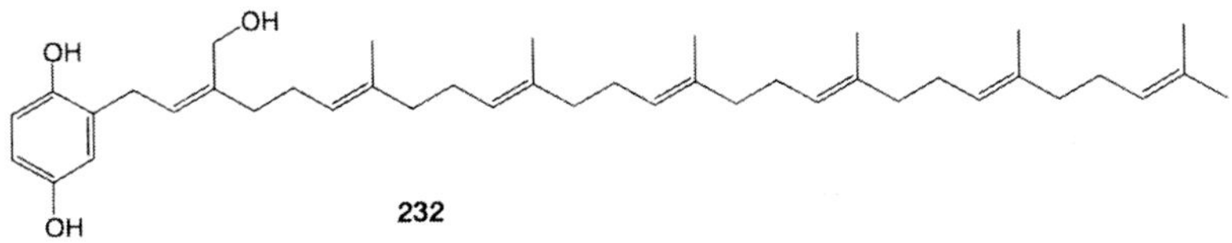

232

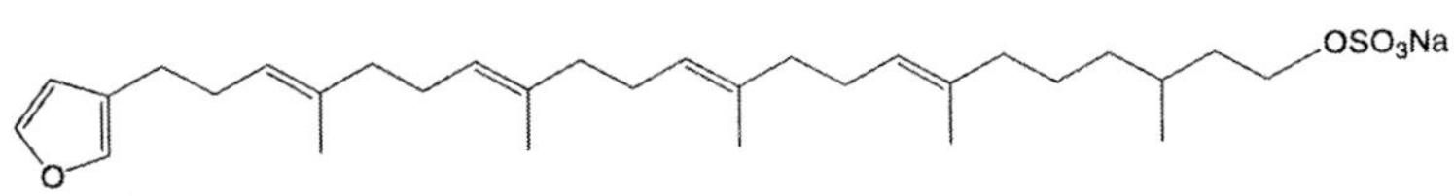

233

furanoterpene (**233**). The sulfated hydroquinones **220**, **229**, and their pentaprenyl congener displayed affinity to the neuropeptide Y receptor and inhibition of the enzymes tyrosine protein kinase (TPK) and HIV-1 integrase. The hydroxylated prenyl compound **232** was also active against TPK and HIV-1 integrase. No activity has been reported for the furanoterpene **233**.

Hydroquinones and sulfated hydroquinones with four to eight prenyl groups and their corresponding chromenols and chromanols have also been reported from the Indian sponge *Ircinia fasciculata* (Venkateswarlu and Reddy 1994). The sulfated derivatives were shown to inhibit H^+, K^+- and Na^+, K^+-ATPases.

De Rosa *et al.* (1994a) reported structure-activity relationship studies of prenylated hydroquinones and benzoquinones having one to eight prenyl groups which have been secured by synthesis or isolation from the sponge *Ircinia spinosula*. The bioassays used were antimicrobial activity against *Bacillus subtilis*, *Micrococcus luteus*, and *Candida albicans* and lethality tests against brine shrimp and the fish *Gambusia affinis*. For antimicrobial activity the optimal length of the side chain was in the range of five to fifteen carbon atoms (1-3 units of prenyl group). The highest lethality was also obtained with the compounds having two to three prenyl groups in the side chain. The quinone derivatives were more toxic than the corresponding hydroquinone.

REFERENCES

Abad, A., C. Agulló, A.C. Cuñat, A.B. García, and C. Gimenez-Saiz. 2003. Synthetic studies on the preparation of oxygenated diterpenes from carvone. *Tetrahedron* 59: 9523–9536.

Albrizio, S., E. Fattorusso, S. Magno, A. Mangoni, and M. Pansini. 1992. Linear diterpenes from the Caribbean sponge *Myrmekioderma styx*. *J. Nat. Prod.* 55: 1287–1293.

Alvi, K.A., and P. Crews. 1992. Homoscalarane sesterterpenes from *Lendenfeldia frondosa*. *J. Nat. Prod.* 55: 859–865.

Alvi, K.A., L. Tenenbaum, and P. Crews. 1991. Anthelmintic polyfunctional nitrogen-containing terpenoids from marine sponges. *J. Nat. Prod.* 54: 71–78.

Amagata, T., S. Whitman, T.A. Johnson, C.C. Stessman, C.P. Loo, E. Lobkovsky, J. Clardy, P. Crews, and T.R. Holman. 2003. Exploring sponge-derived terpenoids for their potency and selectivity against 12-human, 15-human, and 15-soybean lipoxygenases. *J. Nat. Prod.* 66: 230–235.

Angerhofer, C.K., J.M. Pezzuto, G.M. König, A.D. Wright, and O. Sticher. 1992. Antimalarial activity of sesquiterpenes from the marine sponge *Acanthella klethra*. *J. Nat. Prod.* 55: 1787–1789.

Arnó, M., M.A. Gonzalez, and R.J. Zaragoza. 2003. Synthesis of C-17 fanctionalized spongiane diterpenes: diastereoselective synthesis of (–)-spongian-16-oxo-17-al, (–)-acetyldendrillol-l and aplyroseol-14. *J. Org. Chem.* 68: 1242–1251.

Balbin-Oliveros, M., R.A Edrada, P. Proksch, V. Wray, L. Witte, and R.W.M. Van Soest. 1998. A new meroditerpenoid dimer from an undescribed Philippine marine sponge of the genus *Strongylophora*. *J. Nat. Prod.* 61: 948–952.

Barrero, A.F., E.A. Manzaneda, J. Altarejos, S. Salido, J.M. Ramos, M.S.J. Simmonds, and W.M. Blaney. 1995. Synthesis of biologically active drimanes and homodrimanes from (–)-sclareol. *Tetrahedron* 51: 7435–7450.

Barrero, A.F., E.J. Alvarez-Manzaneda, R. Chahboun, M. Cortés, and V. Armstrong. 1999. Synthesis and antitumor activity of puupehedione and related compounds. *Tetrahedron* 55: 15181–15208.

Basabe, P., A. Diego, S. Delgado, D. Diez, I.S. Marcos, and J.G. Urones. 2003. Short and efficient synthesis of (+)-subersic acids. *Tetrahedron* 59: 9173–9177.

Bergquist, P.R., and R.J. Wells. 1983. Chemotaxonomy of the Porifera: The development and current status of the field. Pages 1-50 in *Marine Natural Products, Chemical and Biological Perspectives*, Vol. V., P.J. Scheuer, ed., Academic Press, New York.

Betancur-Galvis, L., C. Zuluaga, M. Arnó, M.A. González, and R.J. Zaragoza. 2002. Cytotoxic effect (on tumor cells) and *in vitro* antiviral activity against herpes simplex virus of synthetic spongian diterpenes. *J. Nat. Prod.* 65: 189–192.

Bifulco, G., I. Bruno, L. Minale, R. Riccio, C. Debitus, G. Bourdy, A. Vassas, and J. Lavayre. 1995. Bioactive prenylhydroquinone sulfates and a novel C_{31} furanoterpene alcohol sulfate from the marine sponge, *Ircinia* sp. *J. Nat. Prod.* 58: 1444–1449.

Blackburn, C.L., and D.J. Faulkner. 2000. Adociasulfate 10, a new merohexaprenoid sulfate from the sponge *Haliclona* (aka *Adocia)* sp. *Tetrahedron* 56: 8429–8432.

Blackburn, C.L., C. Hopmann, R. Sakowicz, M.S. Berdelis, L.S.B. Goldstein, and D.J. Faulkner. 1999. Adociasulfates 1-6, inhibitors of kinesin motor proteins from the sponge *Haliclona* (aka *Adocia*) sp. *J. Org. Chem.* 64: 5565–5570.

Bokesch, H.R., A.C. Stull, L.K. Pannell, T.C. McKee, and M.R. Boyd. 2002. A new pentacyclic sulfated hydroquinone from the marine sponge *Haliclona* sp. *Tetrahedron Lett.* 43: 3079–3081.

Boukouvalas, J., Y.-X. Cheng, and J. Robichaud. 1998. Total synthesis of (+)-dysidiolide. *J. Org. Chem.* 63: 228–229.

Bourguet-Kondracki, M.-L., M.-L. Martin, and M. Guyot. 1992. Smenoqualone, a novel sesquiterpenoid from the marine sponge *Smenospongia* sp. *Tetrahedron Lett.* 33: 8079–8080.

Bourguet-Kondracki, M.-L., C. Debitus, and M. Guyot. 1996. Dipuupehedione, a cytotoxic new red dimer from a New Caledonian marine sponge *Hyrtios* sp. *Tetrahedron Lett.* 37: 3861–3864.

Bowden, B.F., J.C. Coll, H. Li, R.C. Cambie, M.R. Kernan, and P.R. Bergquist. 1992. New cytotoxic scalarane sesterterpenes from the Dictyoceratid sponge *Strepsichordaia lendenfeldi*. *J. Nat. Prod.* 55: 1234–1240.

Buchanan, M.S., A. Edser, G. King, J. Whitmore, and R.J. Quinn. 2001. Cheilanthane sesterterpenes, protein kinase inhibitors, from a marine sponge of the genus *Ircinia*. *J. Nat. Prod.* 64: 300–303.

Butler, M.S., and R.J. Capon. 1993. Beyond polygodial: new drimane sesquiterpenes from a Southern Australian marine sponge, *Dysidea* sp. *Aust. J. Chem.* 46: 1255–1267.

Butler, M.S., R.J. Capon, R. Nadeson, and A.A. Beveridge. 1991. Aromatic bisabolenes from an Australian marine sponge, *Arenochalina* sp. *J. Nat. Prod.* 54: 619–623.

Campagnuolo, C., E. Fattorusso, and O. Taglialatela-Scafati. 2001. Feroxosides A-B, two norlanostane tetraglycosides from the Caribbean sponge *Ectyoplasia ferox*. *Tetrahedron* 57: 4049–4055.

Capon, R.J., S.J. Rochfort, and S.P.B. Ovenden. 1997. Cyclic peroxides and related norterpenes from a southern Australian marine sponge, *Mycale* sp. *J. Nat. Prod.* 60: 1261–1264.

Capon, R.J., S.J. Rochfort, S.P.B. Ovenden, and R.P. Metzger. 1998. Mycaperoxides F and G and related norterpene ketone from southern Australian marine sponge, *Mycale* species. *J. Nat. Prod.* 61: 525–528.

Carletti, I., C. Long, C. Funel, and P. Amade. 2003. Yardenone A and B: new cytotoxic triterpenes from the Indian Ocean sponge *Axinella* cf. *bidderi*. *J. Nat. Prod.* 66: 25–29.

Carotenuto, A., M.R. Conte, E. Fattorusso, V. Lanzotti, and S. Magno. 1995. Stereostructures of new bioactive sesterterpenes isolated from the Caribbean sponge *Cacospongia* cf. *linteiformis*. *Tetrahedron* 51: 10751–10758.

Carotenuto, A., E. Fattorusso, V. Lanzotti, S. Magno, R. Carnuccio, and F. D'Acquisto. 1997. Structure and absolute stereochemistry of the new cyclolinteinol and cyclolinteinol acetate, macrophage activation modulators. *Tetrahedron* 53: 7305–7310.

Carrol, J., E.N. Jonsson, R. Ebel, M.S. Hartman, T.R. Holman, and P. Crews. 2001. Probing sponge-derived terpenoids for human 15-lipoxygenase inhibitors. *J. Org. Chem.* 66: 6847–6851.

Casapullo, A., L. Minale, and F. Zollo. 1993. Paniceins and related sesquiterpenoids from the Mediterranean sponge *Reniera fulva*. *J. Nat. Prod.* 56: 527–533.

Chan, J.A., A.J. Freyer, B.K. Carté, M.E. Hemling, G.A. Hofmann, M.R. Mattern, M.A. Mentzer, and J.W. Westley. 1994. Protein kinase C inhibitors: novel spirosesquiterpene aldehydes from a marine sponge *Aka* (= *Siphonodictyon*) *coralliphagum*. *J. Nat. Prod.* 57: 1543–1548.

Chang, C.W.J., A. Patra, D.M. Roll, P.J. Scheuer, G.K. Matsumoto, and J. Clardy. 1984. Kalihinol A, a highly functionalized diisocyano diterpenoid antibiotic from a sponge. *J. Am. Chem. Soc.* 106: 4644–4646.

Chang, L.C., S. Otero-Quintero, G.M. Nicholas, and C.A. Bewley. 2001. Phyllolactones A-E: new bishomoscalarane sesterterpenes from the marine sponge *Phyllospongia lamellosa*. *Tetrahedron* 57: 5731–5738.

Charan, R.D., T.C. McKee, and M.R. Boyd. 2001. Thorectandrols A and B, new cytotoxic sesterterpenes from the marine sponge *Thorectandra* species. *J. Nat. Prod.* 64: 661–663.

Chen, C.-Y., Y.-C. Shen, Y.-J. Chen, J.-H. Sheu, and C.-Y. Duh. 1999. Bioactive Sesquiterpenes from a Taiwanese marine sponge *Parahigginsia* sp. *J. Nat. Prod.* 62: 573–576.

Ciavatta, M.L., A. Fontana, R. Puliti, G. Scognamiglio, and G. Cimino. 1999. Structures and absolute stereochemistry of isocyanide and isothiocyanate amphilectenes from the Caribbean sponge *Cribochalina* sp. *Tetrahedron* 55: 12629–12636.

Ciavatta, M.L., G. Scognamiglio, E. Trivellone, T. Bisogno, and G. Cimino. 2002. New additional triterpenoids from the Mediterranean sponge *Raspaciona aculeata*. *Tetrahedron* 58: 4943–4948.

Cimino, G., A. Crispino, C.A. Mattia, L. Mazzarella, R. Puliti, E. Trivellone, and M.J. Uriz. 1990. A new triterpenoid skeleton from the Mediterranean sponge *Raspaciona aculeata*: structure of raspacionin. *Tetrahedron Lett.* 31: 6565–6568.

Cimino, G., R.A. Epifanio, A. Madaio, R. Puliti, and E. Trivellone. 1993. Absolute stereochemistry of raspacionin, the main triterpenoid from the marine sponge *Raspaciona aculeata*. *J. Nat. Prod.* 56: 1622–1626.

Conte, M.R., E. Fattorusso, V. Lanzotti, S. Magno, and L. Mayol. 1994a. Lintenolides, new pentacyclic bioactive sesterterpenes from the Caribbean sponge *Cacospongia* cf. *linteiformis*. *Tetrahedron* 50: 849–856.

Conte, M.R., E. Fattorusso, V. Lanzotti, S. Magno, and L. Mayol. 1994b. Structure and absolute stereochemistry of cyclolinteinone a novel monocarbocyclic sesterterpene from *Cacospongia* cf. *linteiformis*. *Tetrahedron* 50: 13469–13476.

Craig, K.S., D.E. Williams, I. Hollander, E. Frommer, R. Mallon, K. Collins, D. Wojciechowicz, A. Tahir, R. Van Soest, and R.J. Andersen. 2002. Novel sesterterpenoid and norsesterterpenoid RCE-protease inhibitors isolated from the marine sponge *Hippospongia* sp. *Tetrahedron Lett.* 43: 4801–4804.

D' Ambrosio, M., A. Guerriero, C. Debitus, J. Waikedre, and F. Pietra. 1997. Relative contributions to antitumoral activity of lipophilic vs. polar moieties in marine terpenoids. *Tetrahedron Lett.* 38: 6285–6288.

De Giulio, A., S. De Rosa, G. Di Vincenzo, G. Strazzullo, and N. Zavodnik. 1990. Norsesterterpenes from the north Adriatic sponge *Ircinia oros*. *J. Nat. Prod.* 53: 1503–1507.

De Guzman, F.S., B.R. Copp, C.L. Mayne, G.P. Consepcion, G.C. Mangalindan, L.R. Barrows, and C.M. Ireland. 1998. Bolinaquinone: a novel cytotoxic sesquiterpene hydroxyquinone from a Philippine *Dysidea* sponge. *J. Org. Chern.* 63: 8042–8044.

De Marino, S., M. Iorizzi, F. Zollo, C. Debitus, J.-L. Menou, L.F. Ospina, M.J. Alcaraz, and M. Paya. 2000. New pyridinium alkaloids from a marine sponge of the genus *Spongia* with human pospholipase A_2 inhibitor profile. *J. Nat. Prod.* 63: 322–326.

De Rosa, S., A. De Giulio, and C. Iodice. 1994a. Biological effects of prenylated hydroquinones: structure-activity relationship studies in antimicrobial, brine shrimp, and fish lethality assays. *J. Nat. Prod.* 57: 1711–1716.

De Rosa, S., R. Puliti, A. Crispino, A. De Giulio, C.A. Mettia, and L. Mazzarella. 1994b. A new scalarane sesterterpenoid from the marine sponge *Cacospongia mollior*. *J. Nat. Prod.* 57: 256–262.

De Rosa, S., A. Crispino, A. De Giulio, C. Iodice, and A. Milone. 1995a. Sulfated polyprenylhydroquinoes from the sponge *Ircinia spinosula*. *J. Nat. Prod.* 58: 1450–1454.

De Rosa, S., R. Puliti, A. Crispino, A. De Giulio, C. De Sena, C. Iodice, and C.A. Mattia. 1995b. 25-Deoxycacospongionolide B and cacospongionolide C, two new terpenoids from the sponge *Fasciospongia cavernosa*. *Tetrahedron* 51: 10731–10736.

De Rosa, S., A. Crispino, A. De Giulio, C. Iodice, and G. Tommonaro. 1997. Cavemosolide, a new sesterterpene from a Tyrrhenian sponge. *J. Nat. Prod.* 60: 844–846.

Doi, Y., H. Shigemori, M. Ishibashi, F. Mizobe, A. Kawashima, S. Nakaike, and J. Kobayashi. 1993. New sesterterpenes with nerve growth factor synthesis-stimulating activity from the Okinawan marine sponge *Hyrtios* sp. *Chem. Pharm. Bull.* 41: 2190–2191.

El Sayed, K.A., M.T. Hamann, N.E. Hashish, W.T. Shier, M. Kelly, and A.A. Khan. 2001. Antimalarial, antiviral, and antitoxoplasmosis norsesterterpene peroxide acids from the Red Sea sponge *Diacarnus erythraeanus*. *J. Nat. Prod.* 64: 522–524.

Espada, A., C. Jiménez, J. Rodríguez, P. Crews, and R. Riguera. 1992. Sarasinosides D-G: four new triterpenoid saponins from the sponge *Asteropus sarasinosurn*. *Tetrahedron* 48: 8685–8696.

Fattorusso, E., V. Lanzotti, S. Magno, L. Mayol, and M. Pansini. 1992. A novel bioactive sesterterpene based on an unprecedented tricyclic skeleton from the Caribbean sponge *Cacospongia* cf. *linteiforrnis*. *J. Org. Chem.* 57: 6921–6924.

Flowers, A.E., M.J. Garson, K.A. Byriel, and C.H.L. Kennard. 1998. Two new isonakafurans from the Great Barrier Reef sponge *Dysidea* sp. nov. *Aust. J. Chem.* 51: 195–200.

Fontana, A., E. Trivellone, E. Mollo, G. Cimino, C. Avila, E. Martinez, and J. Ortea. 1994. Further chemical studies of Mediterranean and Atlantic Hypselodoris nudibranchs: a new furanosesquiterpenoid from *Hypselodoris webbi*. *J. Nat. Prod.* 57: 510–513.

Fontana, A., P. Cavaliere, N. Ungur, L. D'Souza, P.S. Parameswaram, and G. Cimino. 1999a. New scalaranes from the nudibranch *Glossodoris atromarginata* and its sponge prey. *J. Nat. Prod.* 62: 1367–1370.

Fontana, A., M.L. Ciavatta, P. Amodeo, and G. Cimino. 1999b. Single solution phase conformation of new antiproliferative cembranes. *Tetrahedron* 55: 1143–1152.

Fontana, A., E. Mollo, J. Ortea, M. Gavagnin, and G. Cimino. 2000. Scalarane and homoscalarane compounds from the nudibranch *Glossodoris sedna* and *Glossodoris dalli:* chemical and biological properties. *J. Nat. Prod.* 63: 527–530.

Fu, X., L.M. Zeng, J.Y. Su, M. Paris, and P. Potier. 1992. Scalarane-type bishomosesterterpenes from the sponge *Phyllospongia foliascens*. *J. Nat. Prod.* 55: 1607–1613.

Fu, X., L. Zeng, J. Su, and F.J. Schmitz. 1999. Phyllofolactones C and D, two new minor homoscalarane sesterterpenes from the Chinese sponge *Phyllospongia foliascens*. *J. Nat. Prod.* 62: 644–646.

Fukami, A., Y. Ikeda, S. Kondo, H. Naganawa, T. Takeuchi, S. Furuya, Y. Hirabayashi, K. Shimoike, S. Hosaka, Y. Watamabe, and K. Umezawa. 1997. Akaterpin, a novel bioactive triterpene from the marine sponge *Callyspongia* sp. *Tetrahedron Lett.* 38: 1201–1202.

Fusetani, N., H. Hirota, T. Okino, Y. Tomono, and E. Yoshimura. 1996. Antifouling activity of isocyanoterpenoids and related compounds isolated from a marine sponge and nudibranchs. *J. Nat. Toxins* 5: 248–259.

Garrido, L., E. Zubía, M.J. Ortega, and J. Salvá. 1997. New furanoterpenoids from the sponge *Spongia officinalis*. *J. Nat. Prod.* 60: 794–797.

Garson, M.J., A.F. Dexter, L.K. Lambert, and V. Liokas. 1992. Isolation of the bioactive terpene 7-deacetoxyolepupuane from the temperate marine sponge *Dysidea* sp. *J. Nat. Prod.* 55: 364–367.

Gavagnin, M., R.R. Vardaro, C. Avila, G. Cimino, and J. Ortea. 1992. Ichthyotoxic diterpenoids from the Cantaabrian nudibranch *Chromodoris luteorosea*. *J. Nat. Prod.* 55: 368–371.

Giannini, C., C. Debitus, R. Lucas, A. Ubeda, M. Para, J.N.A. Hooper, and M.V. D'Auria. 2001. New sesquiterpene derivatives from the sponge *Dysidea* species with a selective inhibitor profile against human phospholipas A_2 and other leukocyte functions. *J. Nat. Prod.* 64: 612–615.

Goclik, E., G.M. König, A.D. Wright, and R. Kaminsky. 2000. Pelorol from the tropical marine sponge *Dactylospongia elegans*. *J. Nat. Prod.* 63: 1150–1152.

Gomez-Paloma, L., A. Randazzo, L. Minale, C. Debitus, and C. Roussakis. 1997. New cytotoxic sesterterpenes from the New Caledonian marine sponge *Petrosaspongia nigra* (Bergquist). *Tetrahedron* 53: 10451–10458.

Gunasekera, S.P., and F.J. Schmitz. 1991. New spongian diterpenoids from a Great Barrier Reef sponge, *Spongia* sp. *J. Org. Chem.* 56: 1250–1253.

Gunasekera, S.P., P.J. McCarthy, M. Kelly-Borgers, E. Lobkovsky, and J. Clardy. 1996. Dysidiolide: a novel protein phosphatase inhibititor from the Caribbean sponge *Dysidea etheria* de Laubenfels. *J. Am. Chem. Soc.* 118: 8759–8760.

Hata, T., K. Tanaka, and S. Katsumura. 1999. First synthsis of (–)-spongianolide A and determinetion of its absolute structure. *Tetrahedron Lett.* 40: 1731–1734.

Hattori, T., K. Adachi, and Y. Shizuri. 1997. New agelasine compound from the marine sponge *Agelas mauritiana* as an antifouling substance against macroalgae. *J. Nat. Prod.* 60: 411–413.

He, H., P. Kulanthaivel, and B.J. Baker. 1994. New cytotoxic sesterterpenes from the marine sponge *Spongia* sp. *Tetrahedron Lett.* 35: 7189–7192.

Hioki, H., M. Hamano, Y. Miura, M. Kodama, S. Ohta, M. Yanai, and S. Ikegami. 1998. Revised structure, synthesis and absolute configuration of hippospongic acid A. *Tetrahedron Lett.* 39: 7745–7746.

Hioki, H., H. Ooi, M. Hamano, Y. Mimura, S. Yoshio, M. Kodama, S. Ohta, M. Yanai, and S. Ikegami. 2001. Enantioselective total synthesis and absolute stereostructure of hippospongic acid A. *Tetrahadron* 57: 1235–1246.

Hirota, R., Y. Tomono, and N. Fusetani. 1996. Terpenoids with antifouling activity against barnacle larvae from the marine sponge *Acanthella cavernosa*. *Tetrahedron* 52: 2359–2368.

Hirota, H., T. Okino, E. Yoshimura, and N. Fusetani. 1998. Five new antifouling sesquiterpens from two marine sponges of the genus *Axinyssa* and the nudibranch *Phyllidia pustulosa*. *Tetrahedron* 54: 13971–13980.

Horton, P.A., and P. Crews. 1995. The arenarans, sesquiterpene ethers from the marine sponge *Dysidea arenaria*. *J. Nat. Prod.* 58: 44–50.

Hoshino, A., H. Mitome, H. Miyaoka, A. Shintani, Y. Yamada, and R.W.M. Van Soest. 2003. New strongylophorines from the Okinawan marine sponge *Petrosia* (*Strongylophora*) *corticata*. *J. Nat. Prod.* 66: 1600–1605.

Iguchi, K., Y. Shimada, and Y. Yamada. 1992. Hyrtiosal, a new sesterterpenoid with a novel carbon skeleton from the Okinawan sponge *Hyrtios erectua*. *J. Org. Chem.* 57: 522–524.

Issa, R.H., J. Tanaka, and T. Higa. 2003. New cytotoxic furanosesterterpenes from an Okinawan marine sponge, *lrcinia* sp. *J. Nat. Prod.* 66: 251–254.

Jahn, T., G.M. König, and A.D. Wright. 1999. Three new scalarane-based sesterterpenes from the tropical marine sponge *Strepsichordaia lendenfeldi*. *J. Nat. Prod.* 62: 375–377.

Jimênez, C., E. Quiñoã, M. Adamczeski, L.M. Hunter, and P. Crews. 1991. Novel sponge-derived amino acids. 12. Tryptophan-derived pigments and accompanying sesterterpenes from *Fascaplysinopsis reticulata*. *J. Org. Chem.* 56: 3403–3410.

Jiménez, J.I., W.Y. Yoshida, P.J. Scheuer, E. Lobkovsky, J. Clardy, and M. Kelly. 2000a. Honulactones: new bishomoscalarane sesterterpenes from the Indonesian sponge *Strepsichordaia aliena*. *J. Org. Chem.* 65: 6837–6840.

Jiménez, J.I ., W.Y. Yoshida, P.J. Scheuer, and M. Kelly. 2000b. Scalarane-based sesterterpenes from an Indonesian sponge *Strepsichordaia aliena*. *J. Nat. Prod.* 63: 1388–1392.

Kalaitzis, J.A., P.A. Leone, L. Harris, M.S. Butler, A. Ngo, J.N.A. Hooper, and R.J. Quinn. 1999. Adociasulfates 1, 7, and 8: new bioactive hexaprenoid hydroquinones from the marine sponge *Adocia* sp. *J. Org. Chem.* 64: 5571–5574.

Kehraus, S., G.M. König, and A.D. Wright. 2001. New carbonimidic dichlorides from the Australian sponge *Ulosa spongia* and their possible taxonomic significance. *J. Nat. Prod.* 64: 939–941.

Kernan, M.R., and D.J. Faulkner. 1988. Sesterterpene sulfates from a sponge of the family Halichondriidae. *J. Org. Chem.* 53: 4574–4578.

Kimura, J., E. Ishizuka, Y. Nakao, W.Y. Yoshida, P.J. Scheuer, and M. Kelly-Borges. 1998. Isolation of 1-methylherbipoline salts of halisulfate-1 and of suvanine as serine protease inhibitors from a marine sponge, *Coscinoderma mathewsi*. *J. Nat. Prod.* 61: 248–250.

Kirsch, G., G.M. König, A.D. Wright, and R. Kaminsky. 2000. A new bioactive sesterterpene and antiplasmodial alkaloids from the marine sponge *Hyrtios* cf. *erecta*. *J. Nat. Prod.* 63: 825–829.

Kitano, Y., T. Ito, T. Suzuki, Y. Nogata, K. Shinshima, E. Yoshimura, K. Chiba, M. Tada, and I. Sakaguchi. 2002. Synthesis and antifouling activity of 3-isocyanotheonellin and its analogues. *J. Chem. Soc. Perkin Trans. I*, 2251-2255.

Kobayashi, J., K. Naitoh, T. Sasaki, and H. Shigemori. 1992. Metachromins D-H, new cytotoxic sesquiterpenoids from the Okinawan marine sponge *Hippospongia metachromia*. *J. Org. Chem.* 57: 5773–5776.

Kobayashi, J., H. Shinonaga, H. Shigemori, and T. Sasaki. 1993a. Untenospongin C, a new C_{21} furanoterpene from the Okinawan marine sponge *Hippospongia* sp. *Chem. Pharm. Bull.* 41: 381–382.

Kobayashi, J., C.-M. Zeng, M. Ishibashi, and T. Sasaki. 1993b. Luffariolides F and G, new manoalide derivatives from the Okinawan sponge *Luffariella* sp. *J. Nat. Prod.* 56: 436–439.

Kobayashi, J., T. Madono, and H. Shigemori. 1995. Nakijiquinones C and D, new sesquiterpenoid quinones with a hydroxy amino acid residue from a marine sponge inhibiting c-erb B-2 kinase. *Tetrahedron* 51: 10867–10874.

Kobayashi, J., K. Yuasa, T. Kobayashi, T. Sasaki, and M. Tsuda. 1996. Jaspiferals A-G, new cytotoxic isomalabaricane-type nortriterpenoids from Okinawan marine sponge *Jaspis stellifera*. *Tetrahedron* 52: 5745–5750.

Kobayashi, M., Y. Okamoto, and 1. Kitagawa. 1991. Marine natural products XXVIII. The structures of sarasinosides A_1, A_2, A_3, B_1, B_2, C_1, C_2, and C_3, nine new norlanostane-triterpenoidal oligoglycosides from the Palauan marine sponge *Asteropus sarasinosum*. *Chem. Pharm. Bull.* 39: 2867–2877.

Kohmoto, S., O.J. McConnell, A. Wright, and S. Cross. 1987. Isospongiadiol, a cytotoxic and antiviral diterpene from a Caribbean deep water marine sponge, *Spongia* sp. *Chem. Lett.* 1687–1690.

Kokubo, S., K. Yogi, M. J. Uddin, T. Inuzuka, K. Suenaga, K. Ueda, and D. Uemura. 2001. Kohamaic acids A and B, novel cytotoxic sesterterpenic acids, from the marine sponge *Ircinia* sp. *Chem. Lett.* 176-177.

Krug, P.J., K.G. Boyd, and D.J. Faulkner. 1995. Isolation and synthesis of tanyolides A and B, metabolites of the nudibranch *Sclerodoris tanya*. *Tetrahedron* 51: 11063–11074.

Kuniyoshi, M., and T. Higa. 2001. Recent advances on antimicrobial metabolites from marine sponges. Pages 21-83 in *Recent Advances in Marine Biotechnology*, Vol. 6, *Bio-organic Compounds: Chemistry and Biomedical Applications*, M. Fingerman and R. Nagabhushanam, eds., Science Publishers, Enfield, New Hampshire.

Kwak, J.H., F.J. Schmitz, and M. Kelly. 2000. Sesquiterpene quinols/quinones from the Micronesian sponge *Petrospongia metachromia*. *J. Nat. Prod.* 63: 1153–1156.

Lal, A.R., R.C. Cambie, C.E.F. Rickard, and P.R. Bergquist. 1994: Sesterterpene lactones from a sponge species of the genus *Dactylospongia*. *Teterahedron Lett.* 35: 2603–2606.

Lee, H.-S., Y. Seo, K.W. Cho, J.-R. Rho, J. Shin, and V.J. Paul. 2000. New triterpenoid saponins from the sponge *Melophlus isis*. *J. Nat. Prod.* 63: 915–919.

Loya, S., R. Tal, A. Hizi, S. Issacs, Y. Kashman, and Y. Loya. 1993. Hexaprenoid hydroquinones, novel inhibitors of the reverse transcriptase of human immunodeficiency virus type 1. *J. Nat. Prod.* 56: 2120–2125.

Li, C.-J., F.J. Schmitz, and M. Kelly-Borges. 1998. New diterpene lactones from the sponge *Spongia matamata*. *J. Nat. Prod.* 61: 546–547.

Liu, Y., B.H. Bae, N. Alam, J. Hong, C.J. Sim, C.-O. Lee, K.S. Im, and J.H. Jung. 2001. New cytotoxic sesterterpenes from the sponge *Sarcotragus* species. *J. Nat. Prod.* 64: 1301–1304.

Liu, Y., J. Hong, C.-O. Lee, K.S. Im, N.D. Kim, J.S. Choi, and J.H. Jung. 2002. Cytotoxic pyrrolo- and furanoterpenoids from the sponge *Sarcotragus* species. *J. Nat. Prod.* 65: 1307–1314.

Liu, Y., T.A. Mansoor, J. Hong, C.-O. Lee, C.J. Sim, K.S. Im, N.D. Kim, and J.H. Jung. 2003. New cytotoxic sesterterpenoids and norsesterterpenoids from two sponges of the genus *Sarcotragus*. *J. Nat. Prod.* 66: 1451–1456.

López, M.D., E. Quñioá, R. Riguera, and S. Omar. 1994. Dactyltronic acids from the sponge *Dactylospongia elegans*. *J. Nat. Prod.* 57: 992–996.

Lu, Q., and D.J. Faulkner. 1998. Three dolabellanes and a macrolide from the sponge *Dysidea* sp. from Palau. *J. Nat. Prod.* 61: 1096–1100.

Lunardi, I., G.M.P. Santiago, and P.M. Imamura. 2002. Synthsis of (–)- and (+)-hyrtiosal and their C-16 epimers. *Teterahedron Lett.* 43: 3609–3611.

McConnell, O.J., R. Longley, and M. Gunasekera. 1992. Isometachromin, a new cytotoxic sesquiterpenoid from a deep water sponge of the family Spongiidae. *Experientia* 48: 891–892.

McCormick, J.L., T.C. McKee, J.H. Cardellina, M. Leid, and M.R. Boyd. 1996. Cytotoxic triterpenes from a marine sponge, *Stelletta* sp. *J. Nat. Prod.* 59: 1047–1050.

Magnuson, S.R., L. Sepp-Lorenzino, N. Rosen, and S.J. Danishefsky. 1998. A concise total synthesis of dysidiolide through application of a dioxolenium-mediated Diels-Alder reaction. *J. Am. Chem. Soc.* 120: 1615–1616.

Makarieva, T.N., R.-R. Rho, H.-S. Lee, E.A. Santalova, V. Stonik, and J. Shin. 2003. New sesterterpene sulfates from the sponge *Darwinella australensis*. *J. Nat. Prod.* 66: 1010–1012.

Meragelman, K.M., T.C. McKee, and M.R. Boyd. 2001. New cytotoxic isomalabaricane triterpenes from the sponge *Jaspis* species. *J. Nat. Prod.* 64: 389–392.

Mitome, H., T. Nagasawa, H. Miyaoka, Y. Yamada, and R.W.M. Van Soest. 2003. A new sesquiterpenoid quinone and other compounds from the Okinawan marine sponge *Dactylospongia elegans*. *J. Nat. Prod.* 66: 46–50.

Miyamoto, T., K. Sakamoto, K. Arao, T. Komori, R. Higuchi, and T. Sasaki. 1996. Dorisenones, cytotoxic spongian diterpenoids, from the nudibranch *Chromodoris obsoleta*. *Tetrahedron* 52: 8187–8198.

Miyamoto, T., K. Sakamoto, H. Amano, Y. Arakawa, Y. Nagarekawa, T. Komori, R. Higuchi, and T. Sasaki. 1999. New cytotoxic sesterterpenoids from the nudibranch *Chromodoris inornata*. *Tetrahedron* 55: 9133–9142.

Miyaoka, H., M. Shimomura, H. Kimura, Y. Yamada, H.-S. Kim, and Y. Wataya. 1998. Antimalarial activity of kalihinol A and new relative diterpenoids from the Okinawan sponge *Acanthella* sp. *Tetrahedron* 54: 13467–13474.

Miyaoka, H., S. Nishijima, H. Mitome, and Y. Yamada. 2000. Three new scalarane sesterterpenoids from the Okinawan sponge *Hyrtios erectus*. *J. Nat. Prod.* 63: 1369–1372.

Miyaoka, H., Y. Kajiwara, Y. Hara, and Y. Yamada. 2001. Total synthesis of natural dysidiolide. *J. Org. Chem.* 66: 1429–1435.

Mukku, V.J.R.V., R.A. Edrada, F.J. Schmitz, M.K. Shanks, B. Chaudhuri, and D. Fabbro. 2003. New sesquiterpene quinols from Micronesian sponge *Aka* sp. *J. Nat. Prod.* 66: 686–689.

Murray, L.M., A. Johnson, M.C. Diaz, and P. Crews. 1997. Geographic variation in the tropical marine sponge *Jaspis* cf. *Johnstoni:* an unexpected source of new terpene-benzenoids. *J. Org. Chem.* 62: 5638–5641.

Musman, M., J. Tanaka, and T. Higa. 2001a. New sesquiterpene carbonimidic dichlorides and related compounds from the sponge *Stylotella aurantium*. *J. Nat. Prod.* 64: 111–113.

Musman, M., I.I. Ohtani, D. Nagaoka, J. Tanaka, and T. Higa. 2001b. Hipposulfates A and B, new sesterterpene sulfates from an Okinawan sponge, *Hippospongia* cf. *metachromia*. *J. Nat. Prod.* 64: 350–352.

Ohta, S., M. Uno, M. Yoshimura, Y. Hiraga, and S. Ikegami. 1996a. Rhopaloic acid A: a novel norsesterterpene from a marine sponge, *Phopalieides* sp., which inhibits gastrulation of starfish embryos. *Tetrahedron Lett.* 37: 2265–2266.

Ohta, S., M. Uno, M. Tokumasu, Y. Hiraga, and S. Ikegami. 1996b. Hippospongic acid A: an unusual triterpenoic acid from a marine sponge, *Hippospongia* sp., which inhibits gastrulation of starfish embryos. *Tetrahedron Lett.* 37: 7765–7766.

Okino, T., E. Yoshimura, H. Hirota, and N. Fusetani. 1995. Antifouling kalihinenes from the marine sponge *Acanthella cavernosa*. *Tetrahedron Lett.* 36: 8637–8640.

Okino, T., E. Yoshimura, H. Hirota, and N. Fusetani. 1996a. New antifouling sesquiterpenes from four nudibranchs of the family Phyllidiidae. *Tetrahedron* 52: 9447–9454.

Okino, T., E. Yoshimura, H. Hirota, and N. Fusetani. 1996b. New antifouling kalihipyrans from the marine sponge *Acanthella cavernosa*. *J. Nat. Prod.* 59: 1081–1083.

Oku, N., S. Matsunaga, S. Wada, S. Watabe, and N. Fusetani. 2000. New isomalabaricane triterpenes from the marine sponge *Stelletta globostellata* that induce morphological changes in rat fibroblasts. *J. Nat. Prod.* 63: 205–209.

Patil, A.D., A.J. Freyer, L. Killmer, P. Offen, B. Carté, A.J. Jurewicz, and R.K. Johnson. 1997. Frondosins, five new sesquiterpene hydroquinone derivatives with novel skeletons from the sponge *Dysidea frondosa:* inhibitors of interleukin-8 receptors. *Tetrahedron* 53: 5047–5060.

Paul, V.J., Y. Seo, K.W. Cho, J.-R. Rho, J. Shin, and P.R. Bergquist. 1997. Sesquiterpenoids of the drimane class from a sponge of the genus *Dysidea*. *J. Nat. Prod.* 60: 1115–1120.

Pettit, G.R., Z.A. Cichacz, R. Tan, M.S. Hoard, N. Melody, and R.K. Pettit. 1998. Antineoplastic agents. 386. Isolation of sesterstatins 1-3 from the marine sponge *Hyrtios erecta*. *J. Nat. Prod.* 61: 13–16.

Phuwapraisirisan, P., S. Matsuaga, N. Fusetani, N. Chaitanawisuti, S. Kritsanapuntu, and P. Menasveta. 2003. Mycaperoxide H, a new cytotoxic norsesterterpene peroxide from a Thai marine sponge *Mycale* sp. *J. Nat. Prod.* 66: 289–291.

Poigny, S., S. Nouri, A. Chiaroni, M. Guyot, and M. Samadi. 2001. Total synthesis and determination of the absolute configuration of coscinosulfate. A new selective inhibitor of cdc25 protein phosphatase. *J. Org. Chem.* 66: 7263–7269.

Potts, B.C.M., R.J. Capon, and D.J. Falkner. 1992. Luffalactone and (4E, 6E)-dehydromanoalide from the sponge *Luffariella variabilis*. *J. Org. Chem.* 57: 2965-2967.

Randazzo, A., C. Debitus, L. Minale, P.G. Pastor, M.J. Alcaraz, M. Paya, and L. Gomez-Paloma. 1998. Petrosaspongiolides M-R: new potent and selective phospholipase A_2 inhibitors from the New Caledonian marine sponge *Petrosaspongia nigra*. *J. Nat. Prod.* 61: 571–575.

Rodriguez, J., E. Quiñoa, R. Riguera, B.M. Peters, L.M. Abrell, and P. Crews. 1992. The structures and stereochemistry of cytotoxic sesquiterpene quinones from *Dactylospongia elegans*. *Tetrahedron* 48: 6667–6680.

Roy, M.C., J. Tanaka, N. de Voogd, and T. Higa. 2002. New scalarane class sesterterpenes from an Indonesian sponge, *Phyllospongia* sp. *J. Nat. Prod.* 65: 1838–1842.

Rudi, A., and Y. Kashman. 1990. Three new norrisolide related rearranged spongians. *Tetrahedron* 46: 4019–4022.

Rudi, A., I. Goldberg, Z. Stein, Y. Benayahu, M. Schleyer, and Y. Kashman. 1993. Sodwanones A-C, three new triterpenoids from a marine sponge. *Tetrahedron Lett.* 34: 3943–3944.

Rudi, A., I. Goldberg, Z. Stein, Y. Kashman, Y. Benayahu, M. Schleyer, and M.D.G. Gravalos. 1995a. Sodwanones G, H, and I, new cytotoxic triterpenes from a marine sponge. *J. Nat. Prod.* 58: 1702–1712.

Rudi, A., Y. Kashman, Y. Benayahu, and M. Schleyer. 1995b. Durbinals A, B and C: three new cytotoxic sponge metabolites. *Tetrahedron Lett.* 36: 4853–4856.

Rudi, A., M. Aknin, E.M. Gaydou, and Y. Kashman. 1997. Sodwanones K, L, and M: new triterpenes from the marine sponge *Axinella weltneri*. *J. Nat. Prod.* 60: 700–703.

Rudi, A., Z. Stein, I. Goldberg, T. Yosief, Y. Kashman, and M. Schleyer. 1998. Yardenone and abudinol, two new triterpenes from the marine sponge *Ptilocaulis spiculifer*. *Tetrahedron Lett.* 39: 1445–1448.

Rudi, A., T. Yosief, M. Schleyer, and Y. Kashman. 1999. Several new isoprenoids from two marine sponges of the family Axinellidae. *Tetrahedron* 55: 5555–5566.

Rueda, A., E. Zubía, M.J. Ortega, J.L. Carballo, and J. Salvá. 1997. New cytotoxic metabolites from the sponge *Cacospongia scalaris*. *J. Org. Chem.* 62: 1481–1485.

Ryu, G., S. Matsunaga, and N. Fusetani. 1996a. Globostellatic acids A-D, new cytotoxic isomalabaricane triterpenes from the marine sponge *Stelletta globostellata*. *J. Nat. Prod.* 59: 512–514.

Ryu, G., S. Matsunaga, and N. Fusetani. 1996b. Three new cytotoxic sesterterpenes from the marine sponge *Hyrtios* cf. *erectus*. *J. Nat. Prod.* 59: 515–517.

Sakio, Y., Y.J. Hirano, M. Hayashi, K. Komiyama, and M. Ishibashi. 2001. Dendocarbins A-N, new drimane sesquiterpenes from the nudibranch *Dendrodoris carbunculosa*. *J. Nat. Prod.* 64: 726–731.

Salmoun, M., C. Devijver, D. Daloze, J.C. Braekman, R. Gomez, M. de Kluijver, and R.W.M. Van Soest. 2000. New sesquiterpene/quinones from two sponges of the genus *Hyrtios*. *J. Nat. Prod.* 63: 452–456.

Schmidt, E.W., and D.J. Faulkner. 1996. Palauolol, a new anti-inflammatory sesterterpene from the sponge *Fascaplysinopsis* sp. from Palau. *Tetrahedron Lett.* 37: 3951–3954.

Schwarz, O., R. Brun, J.W. Bats, and H.-G. Schmalz. 2002. Synthesis and biological evaluation of new antimalarial isonitriles related to marine-diterpenoids. *Tetrahedron Lett.* 43: 1009–1013.

Sera, Y., K. Adachi, F. Nishida, and Y. Shizuri. 1999. A new sesquiterpene as an antifouling substance from a Palauan marine sponge, *Dysidea herbaceae*. *J. Nat. Prod.* 62: 395–396.

Shen, Y.-C., C.-Y. Chen, and Y.-H. Kuo. 2001. New sesquiterpene hydroquinones from a Taiwanese marine sponge, *Hippospongia metachromia*. *J. Nat. Prod.* 64: 801–803.

Shigemori, H., T. Madono, T. Sasaki, Y. Mikami, and J. Kobayashi. 1994. Nakijiquinones A and B, new antifungal sesquiterpenoid quinones with an amino acid residue from an Okinawan marine sponge. *Tetrahedron* 50: 8347–8354.

Shimomura, M., H. Miyaoka, and Y. Yamada. 1999. Absolute configuration of marine deterpenoid kalihinol A. *Tetrahedron Lett.* 40: 8015–8017.

Shin, J., Y. Seo, J.-R. Rho, E. Baek, B.-M. Kwon, T.-S. Jeong, and S-H. Bok. 1995. Suberitenones A and B : sestertepenoids of an unprecedented skeletal class from the Antarctic sponge *Subertites* sp. *J. Org. Chem.* 60: 7582-7588.

Shin, J., H.-S. Lee, L. Woo, J.-R. Rho, Y. Seo, K.W. Cho, and C.J. Sim. 2001. New triterpenoid saponins from the sponge *Erylus nobilis*. *J. Nat. Prod.* 64: 767–771.

Shoji, N., A. Umeyama, K. Kishi, S. Arihara, Y. Ohizumi, and J. Kobayashi. 1992a. Untenic acid, a novel Ca^{2+}-ATPase activator from an Okinawan marine sponge. *Aust. J. Chem.* 45: 793–795.

Shoji, N., A. Umeyama, S. Motoki, S. Arihara, T. Ishida, K. Nomoto, J. Kobayashi, and M. Takei. 1992b. Potent inhibitors of histamine release, two novel triterpenoids from the Okinawan marine sponge *Penares incrustans*. *J. Nat. Prod.* 55: 1682–1685.

Simpson, J.S., P. Raniga, and M.J. Garson. 1997. Biosynthesis of dichloroimines in the tropical marine sponge *Stylotella aurantium*. *Tetrahedron Lett.* 38: 7947–7950.

Sperry, S., F.A. Valeriote, T.H. Corbett, and P. Crews. 1998. Isolation and cytotoxic evaluation of marine sponge-derived norterpene peroxides. *J. Nat. Prod.* 61: 241–247.

Stahl, P., L. Kissau, R. Mazitschek, A. Huwe, P. Furet, A. Giannis, and H. Waldmann. 2001. Total synthesis and biological evaluation of the nakijiquinones. *J. Am. Chem. Soc.* 123: 11586–11593.

Stewart, M., P.M. Fell, J.W. Blunt, and M.H.G. Munro. 1997. Avarol and related compounds from the New Zealand marine sponge *Dysidea* sp. *Aust. J. Chem.* 50: 341–347.

Stonik, V.A., T.N. Makarieva, and A.S. Dmitrenok. 1992. Sarcochromenol sulfates A-C and sarcohydroquinone sulfates A-C, new natural products from the sponge *Sarcotragus spinulosus*. *J. Nat. Prod.* 55: 1256–1260.

Su, J.Y., H. Meng, L.M. Zeng, X. Fu, and F.J. Schmitz. 1994. Stelletin A, a new triterpenoid pigment from the marine sponge *Stelletta tenuis*. *J. Nat. Prod.* 57: 1450–1451.

Tabudravu, J.N., and M. Jaspars. 2001. Stelliferin riboside, a triterpene monosaccharide isolated from the Fijian sponge *Geodia globostellifera*. *J. Nat. Prod.* 64: 813–815.

Takada, K., Y. Nakao, S. Matsunaga, R.W.M. Van Soest, and N. Fusetani. 2002. Nobiloside, a new neuraminidase inhibitory triterpenoidal saponin from the marine sponge *Erylus nobilis*. *J. Nat. Prod.* 65: 411–413.

Takamatsu, S., T.W. Hodges, I. Rajbhandari, W.H. Gerwick, M.T. Hamann, and D.G. Nagle. 2003. Marine natural products as novel antioxidant prototypes. *J. Nat. Prod.* 66: 605–608.

Takikawa, H., J. Koizumi, Y. Kato, and K. Mori. 1999. Triterpenoid total synthesis. Part 4. Synthesis of (±)-hippospongic acid A, a triterpene isolated from the marine sponge *Hippospongia* sp. *J. Chem. Soc. Perkin Trans.* 1: 2271–2275.

Tanaka, J., and T. Higa. 1988. The absolute configuration of kurospogin, a new furanoterpene from a marine sponge, *Spongia* sp. *Tetrahedron* 44: 2805–2810.

Tanaka, J., and T. Higa. 1999. Two new cytotoxic carbonimidic dichlorides from the nudibranch *Reticulidia fungia*. *J. Nat. Prod.* 62: 1339–1340.

Tanaka, J., T. Higa, K. Suwanborirux, and U. Kokpol, G. Bernardinelli, and C.W. Jefford. 1993. Bioactive norsesterterpene 1,2-dioxanes from a Thai sponge, *Mycale* sp. *J. Org. Chem.* 58: 2999–3002.

Tanaka, J., G. Marriott, T. Higa, and T. Higa. 2001. Cacofurans A and B, new furanoditerpenes from a marine sponge. *J. Nat. Prod.* 64: 1468–1470.

Tasdemir, D., G.P. Concepción, G. C. Mangalindan, M.K. Harper, E. Hajdu, and C. M. Ireland. 2000. New terpenoids from a *Cacospongia* sp. from the Philippines. *Tetrahedron* 56: 9025-9030.

Tasdemir, D., G.C. Mangalindan, G.P. Concepcion, S.M. Verbitski, S. Rabindran, M. Miranda, M. Greenstein, J.N.A. Hooper, M.K. Harper, and C.M. Ireland. 2002. Bioactive isomalabaricane triterpenes from the marine sponge *Rhabdastrella globostellata*. *J. Nat. Prod.* 65: 210–214.

Tokumasu, M., H. Ando, Y. Hiraga, S. Kojima, and K. Ohkata. 1999. Synthesis of rachipposponglc acid A and revision of the structure. *J. Chem. Soc. Perkin Trans. I.* 484–496.

Tsuchiya, N., A. Sato, T. Hata, N. Sato, K. Sasagawa, and T. Kobayashi. 1998. Cytotoxic scalarane sesterterpenes from a sponge, *Hyrtios erecta*. *J. Nat. Prod.* 61: 468–473.

Tsuda, M., M. Ishibashi, K. Agemi, T. Sasaki, and J. Kobayashi. 1991. Stelliferins A-F, new antineoplastic isomalabaricane triterpenes from the Okinawan marine sponge *Jaspis stellifera*. *Tetrahedron* 47: 2181–2194.

Tsuda, M., H. Shigemori, M. Ishibashi, T. Sasaki, and J. Kobayashi. 1992. Luffariolides A-E, new cytotoxic sesterterpenes from the Okinawan marine sponge *Luffariella* sp. *J. Org. Chem.* 57: 3503–3507.

Tsuda, M., T. Endo, Y. Mikami, J. Fromont, and J. Kobayashi. 2002. Luffariolides H and J, new sesterterpenes from a marine sponge *Luffariella* species. *J. Nat. Prod.* 65: 1507–1508.

Tsukamoto, S., S. Miura, R.W.M. Van Soest, and T. Ohta. 2003. Three new cytotoxic sesterterpenes from a marine sponge *Spongia* sp. *J. Nat. Prod.* 66: 438–440.

Urban, S., and R.J. Capon. 1996. Absolute stereochemistry of puupehenone and related metabolites. *J. Nat. Prod.* 59: 900–901.

Utkina, N.K., V.A. Denisenko, O.V. Scholokova, M.V. Virovaya, and N.G. Prokof'eva. 2003a. Cyclosmenospongine, a new sesquiterpenoid aminoquinone from an Australian marine sponge *Spongia* sp. *Tetrahedron Lett.* 44: 101–102.

Utkina, N.K., V.A. Denisenko, O.V. Scholokova, and A.E. Makarchenko. 2003b. Determination of the absolute stereochemistry of cyclosmenospongine. *J. Nat. Prod.* 66: 1263–1265.

Uy, M.M., S. Ohta, M. Yanai, E. Ohta, T. Hirata, and S. Ikegami. 2003. New spirocyclic sesquiterpenes from the marine sponge *Geodia exigua*. *Tetrahedron* 59: 731–736.

Venkateswarlu, Y., and M.V.R. Reddy. 1994. Three new heptaprenylhydroquinone derivatives from the sponge *Ircinia fasciculata*. *J. Nat. Prod.* 57: 1286–1289.

Wright, A.D., and G.M. König. 1996. Antimalarial activity: Search for marine-derived natural products with selective antimalarial activity. *J. Nat. Prod.* 59: 710–716.

Yanai, M., S. Ohta, E. Ohta, and S. Ikegami. 1998. Novel norsesterterpenes, which inhibit gastrulation of the starfish embryo, from the marine sponge *Rhopaloeides* sp. *Tetrahedron* 54: 15607–15612.

Yasman, R.A. Edrada, V. Wray, and P. Proksch. 2003. New 9-thiocyanatopupukeanane sesquiterpenes from the nudibranch *Phyllidia varicosa* and its sponge-prey *Axinyssa aculeata*. *J. Nat. Prod.* 66: 1512–1514.

Yosief, T., A. Rudi, and Y. Kashman. 2000. Asmarines A-F, novel cytotoxic compounds from the marine sponge *Raspailia* species. *J. Nat. Prod.* 63, 299-304.

Yosief, T., A. Rudi, Z. Stein, I. Goldberg, G.M.D. Gravalos, M. Schleyer, and Y. Kashman. 1998. Asmarines A-C: three novel cytotoxic metabolites from the marine sponge *Raspailia* sp. *Tetrahedron Lett.* 39: 3323–3326.

Youssef, D.T.A., W.Y. Yoshida, M. Kelly, and P.J. Scheuer. 2001. Cytotoxic cyclic norterpene peroxides from a Red Sea sponge *Diacarnus erythraenus*. *J. Nat. Prod.* 64: 1332–1335.

Youssef, D.T.A., R.K. Yamaki, M. Kelly, and P.J. Scheuer. 2002. Salmahyrtisol A, a novel cytotoxic sesterterpene from the Red Sea sponge *Hyrtios erecta*. *J. Nat. Prod.* 65: 2–6.

Zampella, A., M.V. D'Auria, C. Debitus, and J.-L. Menou. 2000. New isomalabaricane derivatives from a new species of *Jaspis* sponge collected at the Vanuatu Islands. *J. Nat. Prod.* 63: 943–946.

Zeng, L., X. Fu, J. Su, E.O. Pordesimo, S.C. Treager, and F.J. Schmitz. 1991. Novel bishomoscalarane sesterterpenes from the sponge *Phyllospongia foliascens*. *J. Nat. Prod.* 54: 421–427.

12

Bioactive Polyacetylenic Compounds from Marine Sponges

JEE H. JUNG, KWANG SIK IM, AND BOK HEE BAE

College of Pharmacy, Pusan National University, Busan 609-735, Korea

INTRODUCTION

Polyacetylenic compounds are a diverse group of secondary metabolites from terrestrial plants, most frequently from the family Compositae. These polyacetylenic compounds from the family Compositae and other higher plants are well documented (Bohlmann *et al.* 1973, 1988). As for the physiological roles of these acetylenes in the plants themselves, little is known though there are several indications that these compounds may be linked with special hormonal functions in plant metabolism. Observations have been made showing that the concentrations of special acetylenes vary drastically during the vegetation period. Some natural acetylenes showed powerful biological activity of various kinds, but only a very limited number of these compounds could be investigated since they were available only in small amounts. The acetylenes from higher plants often show antibiotic activity, though their toxicity and instability preclude any practical application of the compounds. When the safflower is infected by the pathogenic fungus *Phytophthora drechsleri*, the concentration of the highly antifungal compound, polyacetylene, increased twenty-fold. These observations throw some light on the role of polyacetylene as antifungals (Allen and Thomas 1971).

Recent investigations of marine secondary metabolites revealed the presence of various polyacetylenic compounds in marine organisms, especially in marine sponges. These substances have also been occasionally reported from nudibranchs, hard corals, and red algae.

Algae are a good source of halogenated acetylenes of terpenoid origin, and mollusks are a source of polyacetylenes that they obtained from their diet

Polyacetylenes in higher plants usually possess short carbon chains ($< C_{18}$) since the biosynthesis of polyacetylenes in higher plants is based on oleic acid ($C_{18}H_{34}O_2$), even though a C_{34} long-chain polyacetylene was recently isolated, as a rare example, from the higher plant *Aciphylla scott-thomsonii* of the family Apiaceae (Perry *et al.* 2001). The long chain polyacetylenes were considered to be characteristic sponge metabolites. Long-chain (C_{46}–C_{48}) acetylenic compounds from sponges may be derived from long-chain fatty acids that are characteristic metabolites of sponges. Fatty acids of sponges are characterized by a long carbon chain (C_{24}–C_{30}), unusual unsaturation pattern ($\Delta^{5,9}$), and methyl branching, which might modulate the function of a cell membrane when they are used as components of these membranes.

In a recent report, a symbiotic-bacterial origin of the acetylenes was suggested by a study of primary cultures of the marine sponge *Xestspongia muta* (Richelle-Maurer *et al.* 2003). The acetylenes were distributed in non-fixed cells (mainly bacteria), and in cultures kept in the dark. This suggested that heterotrophic bacteria were involved in their production. Further evidence was obtained by analyses of Ficoll and Percoll fractions. Significant amounts of the acetylenes were found in the heterotrophic bacterial fractions but not in the sponge cell fractions nor in the cyanobacterial fractions. This study also shows that bioactive metabolites of sponges can be produced in large scale by culture of the sponge cells or symbiotic bacteria.

Though a role of polyacetylenes in chemical defense was speculated, no deterrent activity to fishes was observed. The ecological role of such acetylenic compounds in sponges still remains unknown. However, diverse pharmacological activities such as antitumor, antiviral, antimicrobial, and anti-inflammatory activities have been reported.

Our discussion on polyacetylenes in sponges will be confined to linear polyacetylenes of polyketide origin. For convenience, linear monoacetylenes will be included in this category since they share the same biogenetic origin with linear polyacetylenes. Polyacetylenes of other biogenetic origins, such as those derived from terpenoids, alkaloids, and shikimates, will be left for future discussion.

OCCURRENCE OF POLYACETYLENES IN SPONGES

Most of polyacetylenes from marine organisms have been isolated from sponges. Although the genera *Petrosia* and *Xestospongia* seem to be the most frequently used sources of polyacetylenes, these compounds are in fact distributed in a wide range of taxonomically distinct sponge genera. Quite possibly, polyacetylenes might be ubiquitous sponge metabolites, which are not fully known due to their chemical instability, low content in the organism, and limited chemical investigation. To date, over 270 polyacetylenes from sponges have been reported in the literature, and most of them were isolated during the last two decades. For convenience, these polyacetylenes may be classified into six major types.

1. long-chain (> C_{30}) polyacetylenes
2. C_{30} polyacetylenes
3. short-chain (< C_{30}) polyacetylenes
4. brominated long-chain polyacetylenes
5. brominated short-chain polyacetylenes
6. chlorinated short-chain polyacetylenes

The occurrence of polyacetylenes in sponges and the typical compound types are summarized in Table 12.1.

Petrosia

It was known that the Mediterranean (Bay of Naples, Italy) nudibranch *Peltodoris atromaculata* is able to find its prey, the sponge *Petrosia ficiformis*, by chemotaxis and to store metabolites taken up during the predation (Castiello *et al.* 1979). By chemical study of the organisms, two series of high-molecular-weight polyacetylenes (**1** and **2**) were obtained from both the sponge and its predator (Castiello *et al.* 1980). The structures were only partially defined as polyacetylenes of C_{46}–C_{55} alkyl chains. These polyacetylenes were of much higher-molecular-weight than those from terrestrial plants or other marine plants such as red algae (Paul and Fenical 1980). This Mediterranean sponge usually displays a red-brown color due to the presence of the symbiotic alga *Aphanocapsa feldmanni*. However, the sponge *P. ficiformis* found in dark caves lacks this symbiotic alga and therefore appears white. From this white sponge, similar polyacetylenes (**3-5**) but with an enyne terminus (**3** and **5**) were isolated and the structures were again only partially defined (Cimino *et al.* 1981). These polyacetylenes from the white sponge also provoked a positive food response by the nudibranch. It is worth noting that the same sponge living in different habitats contains similar but different acetylenic compounds. The polyacetylenes isolated from the specimens living in the dark were absent in the sponge living in its usual habitat and *vice versa*. However, among the secondary metabolites of these sponges, this diversity is restricted to the polyacetylenes. All of the specimens examined contained the same sterol pattern, including the unusual sterol petrosterol, which is the major component.

Table 12.1 Occurrence of new acetylenes in marine sponges*.

Order	*Family*	*Genus*	*Typical compound type*	*No. of new compounds*
Haplosclerida	Petrosiidae	*Petrosia*	long-chain acetylenes, C_{30} acetylenes, brominated short-chain acetylenes, C_{31} acetylenes	92
		Xestospongia	brominated short-chain acetylenes, brominated long-chain acetylenes, long-chain acetylenes	42
		Strongylophora	short-chain acetylenes, long-chain acetylene	5
	Niphatidae	*Cribrochalina*	short-chain acetylenes, C_{30} acetylene, long-chain acetylene	13
		Unidentified	C_{30} acetylenes	3
	Callyspongiidae	*Siphonochalina*	short-chain acetylenes	9
		Callyspongia	short-chain acetylenes, long-chain acetylenes	26
	Chalinidae	*Haliclona*	long-chain acetylenes, chlorinated short-chain acetylenes, short-chain acetylenes	24
		Adocia	C_{30} acetylenes	4
		Reniera	brominated short-chain acetylenes, short-chain acetylenes,	6
		Cladocroce	long-chain acetylene, short-chain acetylene	1
	Phloeodictyidae	*Oceanapia*	brominated short-chain acetylenes, short-chain acetylene	3
Halichondrida	Halichondriidae	*Pellina* (=*Halichondria*)	long-chain acetylenes	16
	Axinellidae	*Phakellia*	brominated short-chain acetylenes, short-chain acetylenes	5
Poecilosclerida	Microcionidae	*Clathria*	amides of brominated short-chain acetylenes	3
	Acarnidae	*Acarnus*	short-chain acetylenes	4
Lithistida	Theonellidae	*Theonella*	long-chain acetylenes	4
Dictyoceratida	Spongiidae	*Hippospongia*	short-chain acetylene	1
Astrophorida	Ancorinidae	*Stelletta*	short-chain acetylenes	6
Hadromerida	Spirastrellidae	*Diplastrella*	brominated short-chain acetylenes	8

* 'Systema Porifera, a Guide to the Classification of Sponges', edited by J.N.A. Hooper and R.W.M. Van Soest (2002, Kluwer Academic/Plenum Publishers), was employed for taxonomic classification in this chapter. Note that the genus *Reniera* was changed to the subgenus of *Haliclona* [*Haliclona (Reniera)*]. The genus *Adocia* was changed to the subgenus *Haliclona* [Haliclona (Haliclona)].

An additional two polyacetylenes (**6** and **7**) were isolated from the red variety of *P. ficiformis*, but as in several earlier studies, the structures were only partially elucidated (Cimino *et al.* 1985). Because interpretable mass spectra were not obtained, the lengths of the molecules were determined by hydrogenolysis of the acetyl derivative to the parent hydrocarbons. A defensive role of the acetylenes in the sponge was hypothesized (Fusetani *et al.* 1983). As the acetylenic metabolites of the *P. ficiformis* were found also in the specific predator of the sponge, the nudibranch *Peltodoris atromaculata*, their defensive role was investigated by use of a feeding inhibition bioassay with marine and fresh water fish. However, no deterrent activity of the polyacetylenes was evidenced.

Reinvestigation of the red variety of *P. ficiformis* yielded four new C_{46} polyacetylenes, petroformynes 1–4 (**8–11**), and their planar structures were fully elucidated using spectroscopic methods (Cimino *et al.* 1989). By reduction, all petroformynes yielded the same saturated straight chain hydrocarbon, $C_{46}H_{94}$. The location of one of the double bonds at C-12 of petroformyne 1 (**8**) was determined by ozonolysis, followed by oxidation with H_2O_2, methylation of the bicarboxylic acids with CH_2N_2 and analysis of the esters by GCMS. And it was further confirmed by EIMS analysis of the trimethylsilyl (TMS) derivative. The absolute configurations of the carbinol carbons were determined in the later study (Guo *et al.* 1994). The geometry of the double bonds was defined on the basis of ^{13}C NMR data of the allylic carbons. The petroformynes (**8–11**) were quite toxic to larvae of the brine shrimp, *Artemia salina,* and inhibited development of fertilized sea urchin eggs at concentrations ranging from 1 to 50 μg/mL.

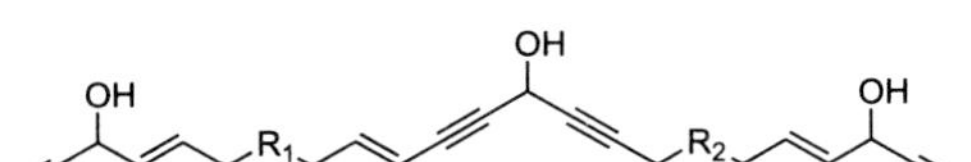

1 $R_1 + R_2 = C_nH_{2n-6}$; n = 25 or 28; ΔZ
2 $R_1 + R_2 = C_nH_{2n-4}$; n = 28, 31, or 34; ΔZ

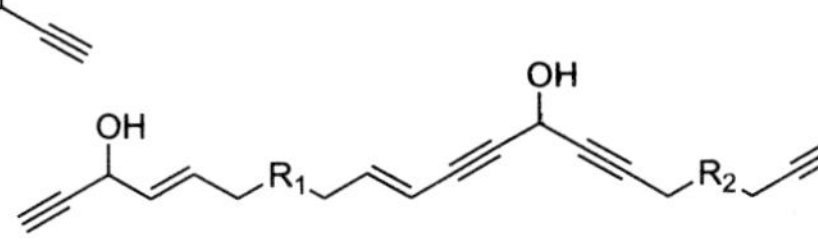

4 $R_1 + R_2 = C_nH_{2n-4}$; n = 28, 31, 34

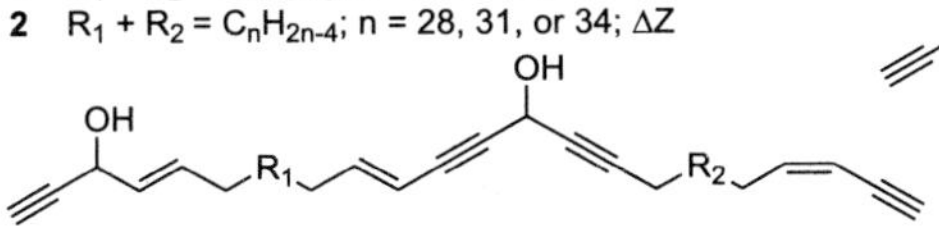

3 $R_1 + R_2 = C_nH_{2n-4}$; n = 26, 29

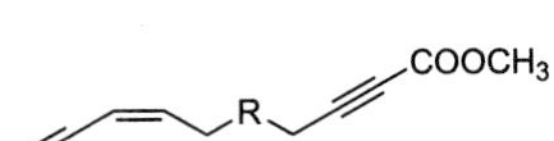

5 $R = C_nH_{2n-2}$; n = 16, 18

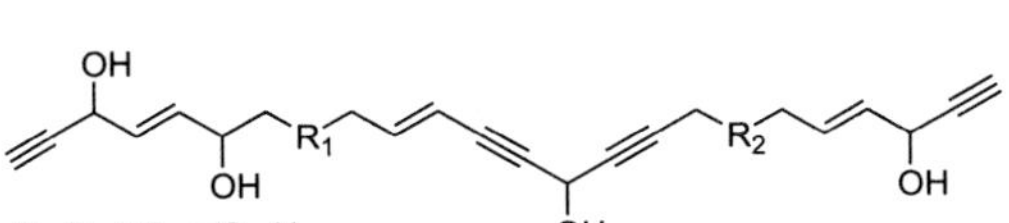

6 $R_1 + R_2 = C_{24}H_{44}$

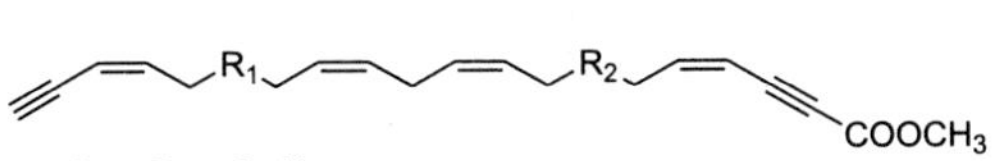

7 $R_1 + R_2 = C_{25}H_{48}$

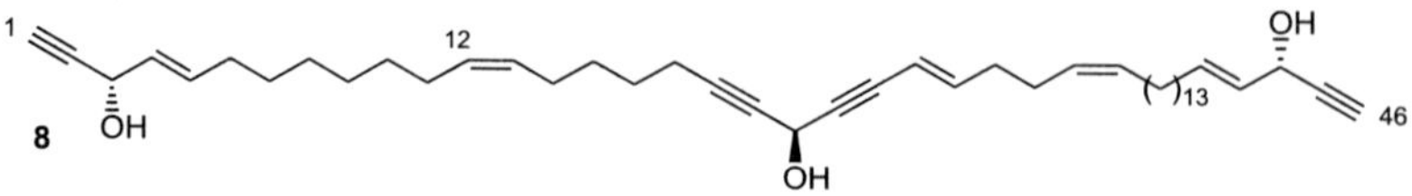

8

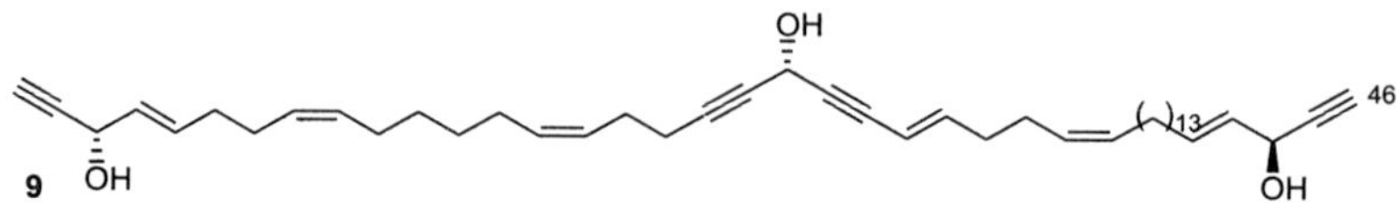

9

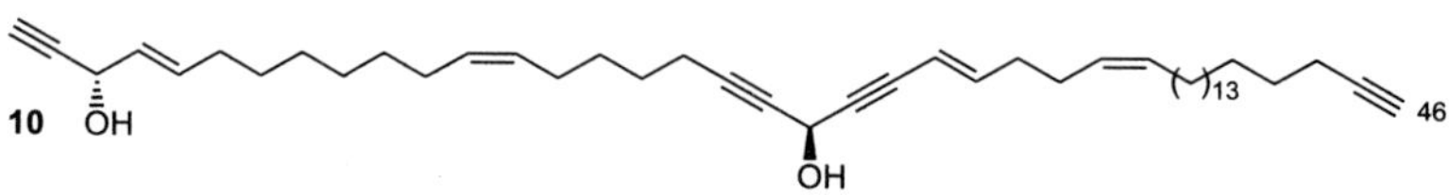

10

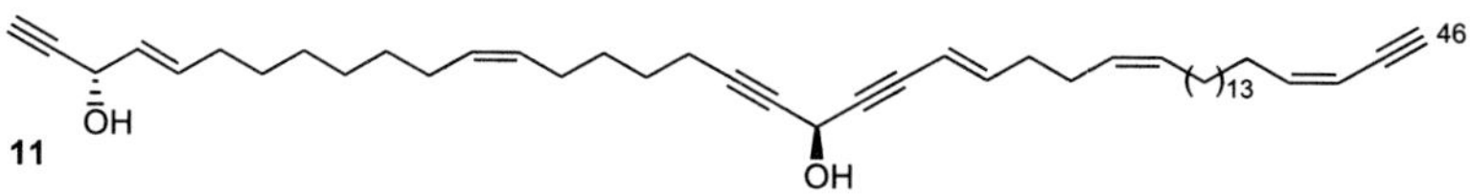

11

Further studies of the bioactive polyacetylenes from the same sponge have yielded minor analogues: petroformynes 5–9 (**12–16**) and petroformynes A and B (**17** and **18**) (Cimino *et al.* 1990). The structures of some of the new petroformynes were not completely elucidated. Both alternative structures, which have a switched terminus, for petroformynes 5 (**12**), 8 (**15**), and 9 (**16**) were proposed. All petroformynes (**8–18**) were very toxic to brine shrimp larvae with LC_{50} values of 0.002 to 0.12 µg/mL. The potency exhibited by the alcohols was usually one order of magnitude higher than that observed for the corresponding ketones. Petroformynes 1–4 (**8–11**) were evaluated by a sea urchin egg assay to show LC_{50} values in the range of 1–50 µg/mL. Petroformynes 3 (**10**) and 4 (**11**) showed antitumor activity in a potato disc assay with 49% and 43% inhibition, respectively, at the concentration of 0.5 µg/mL.

The absolute configurations of petroformynes 1–5 (**8–12**) and petroformynes A and B (**17** and **18**) were determined by a modified Mosher's method (Guo *et al.* 1994) rather than by the exciton chirality method. The validity of determining the absolute stereochemistry of enynols (allylic alcohols with an acetylene adjacent to the alcohol) by the exciton chirality method has been questioned. The application of the exciton chirality method is not recommended when the chiral carbon is flanked by two chromophores (Matthew *et al.* 1994).

Twelve minor C_{46} acetylenic alcohols (**19–30**), closely related to petroformynes 1–5 (**8–12**), were isolated from the same Mediterranean sponge, *P. ficiformis,* and were screened for brine shrimp toxicity (Guo *et al.* 1995). Both alternative structures, which have a switched terminus, were proposed for petroformyne 10 (**28**). Most of these new

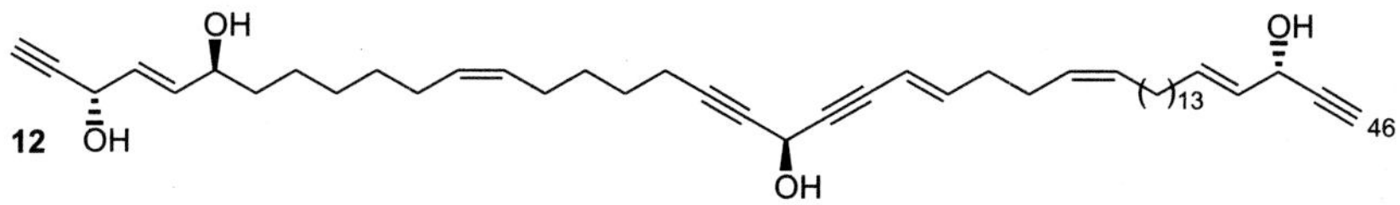

12

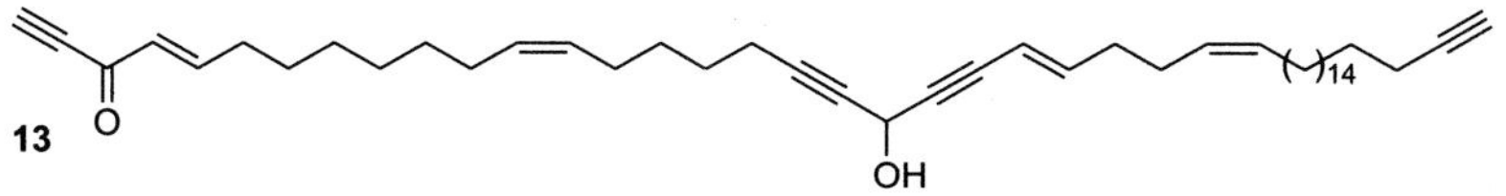

13

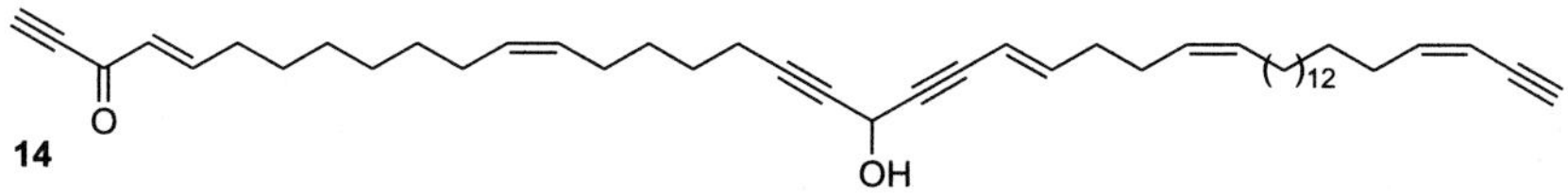

14

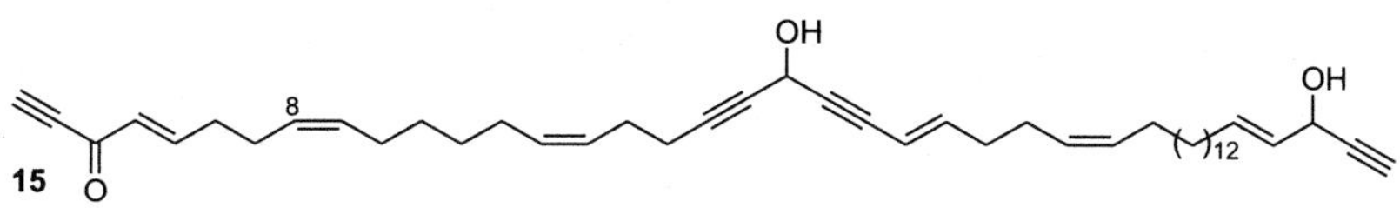

15

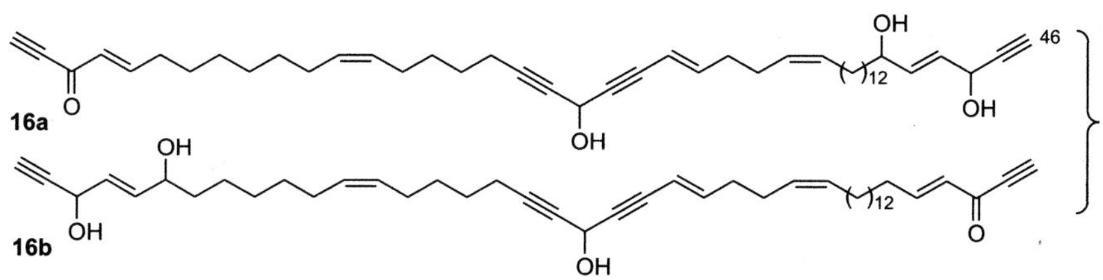

16a

16b

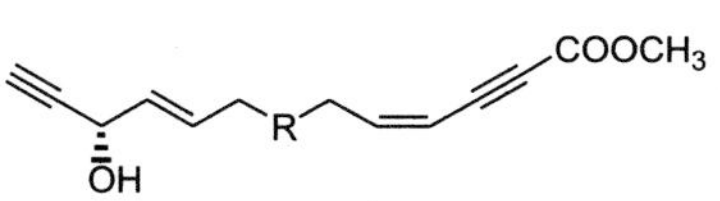

17 R = $C_{31}H_{56}$ (undetermined double bond location)
18 R = $C_{33}H_{60}$ (undetermined double bond location)

polyacetylenes were tested for their toxic potential by a brine shrimp assay and showed LC_{50} values ranging from 0.04 to 6.8 μg/mL.

Five additional polyacetylenes (**31–35**), petroformynic acid (**31**), isopetroformyne 1 (**32**), 23, 24-dihydropetroformyne 4 (**33**), 20-oxoisopetroformyne 4 (**34**), and 20-oxoisopetroformyne 3 (**35**) were isolated from both Atlantic and Mediterranean specimens of *P. ficiformis* (Guo *et al.* 1998).

The C_{30} polyacetylene (**36**) that was isolated from a species of the genus *Tetrosia* (should be corrected to *Petrosia*) collected at Hachijo-jima Island, Japan, inhibited cell division in fertilized eggs of the sea-urchin *Pseudocentrotus depressus* at a concentration of 1 μg/mL (Fusetani *et al.* 1983). The compound was also active against *Penicillium chrysogenum* (data not shown). The ether soluble part of the ethanol extract inhibited

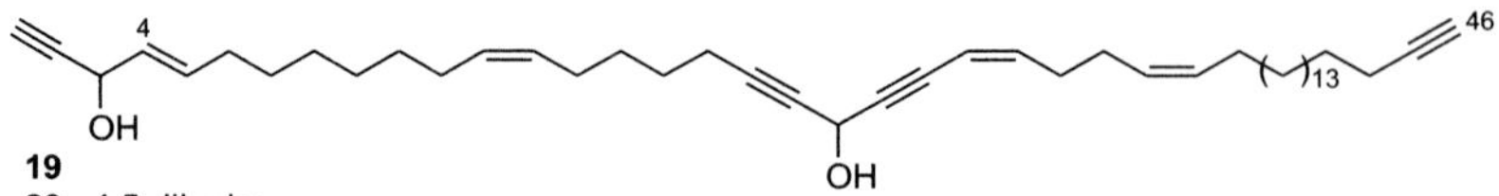

19
20 4,5-dihydro

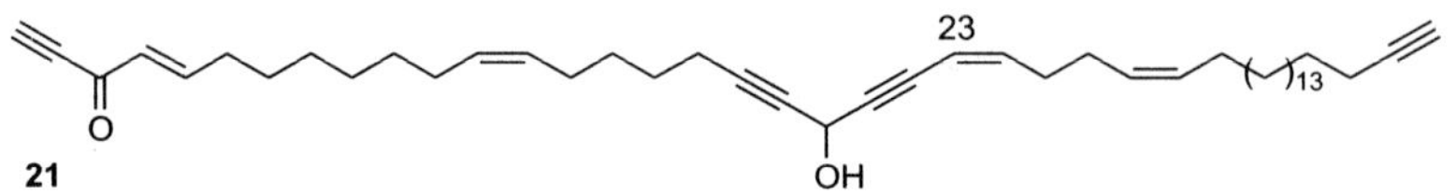

21
22 23,24-dihydro

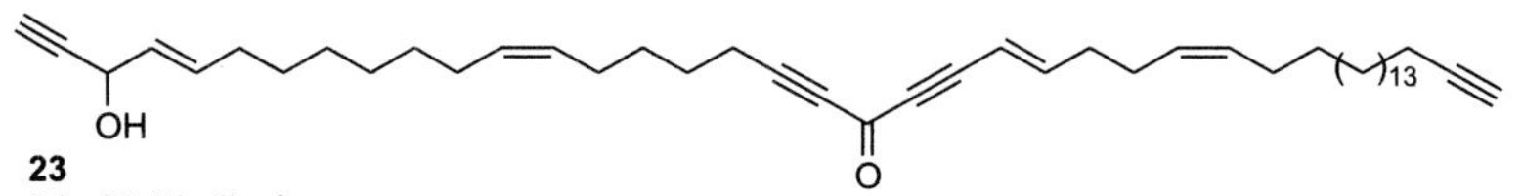

23
24 23,24-dihydro

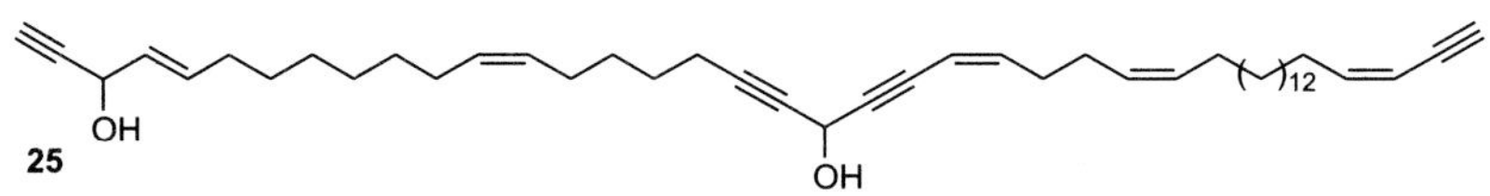

25

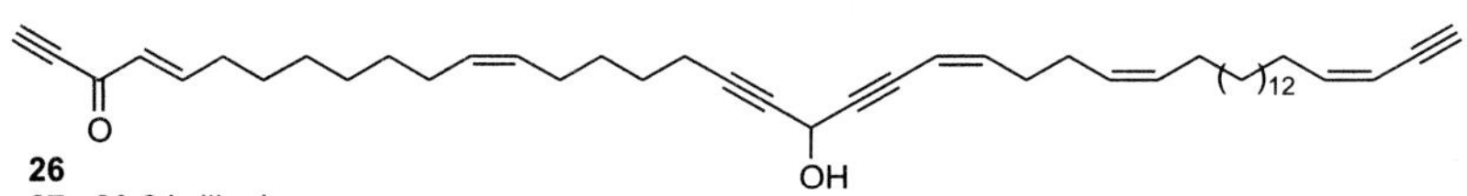

26
27 23,24-dihydro

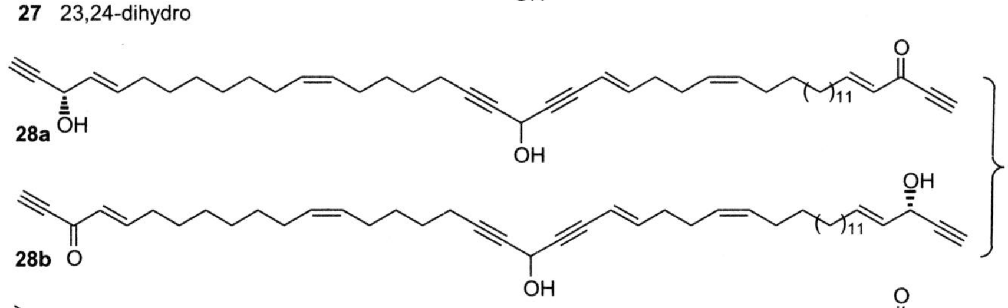

28a
28b

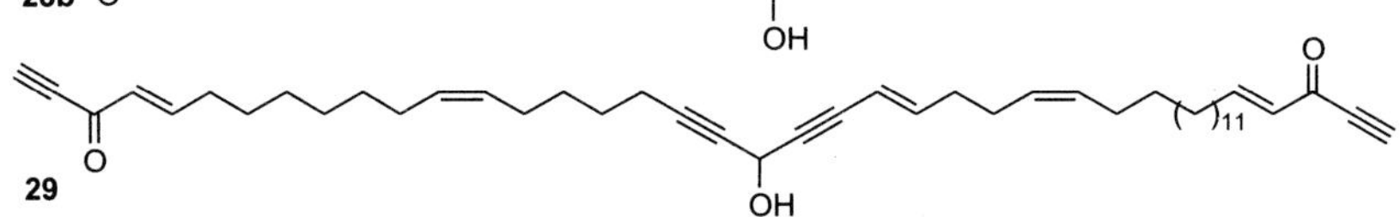

29

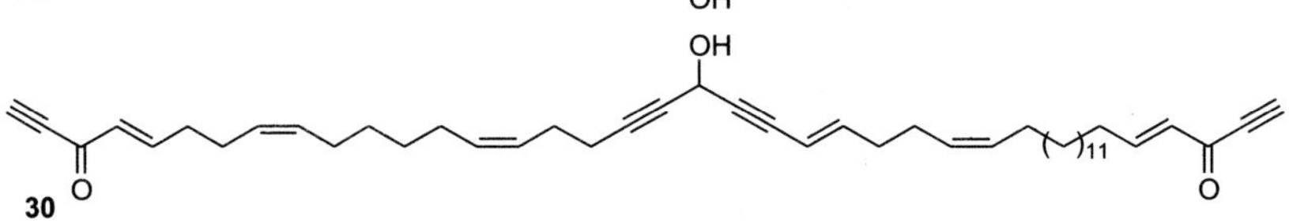

30

the cell division of starfish and sea urchin embryos. The active compound was an unusual polyacetylenic alcohol with a symmetric structure (**36**). All attempts at chemical degradation, including Pd-C hydrogenation, and Jones, DDQ, PDC-DMF, and MnO_2 oxidation, were unsuccessful because of the instability of the molecule. The configuration of the hydroxyls and the geometry of the central double bond were defined in a later study (Fusetani *et al.* 1987) It was speculated that the compound participates in the defense mechanism of the sponge.

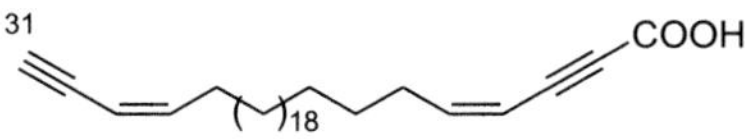

31

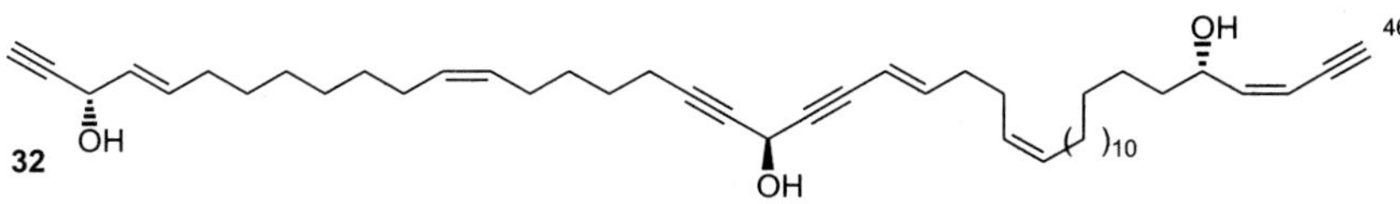

32

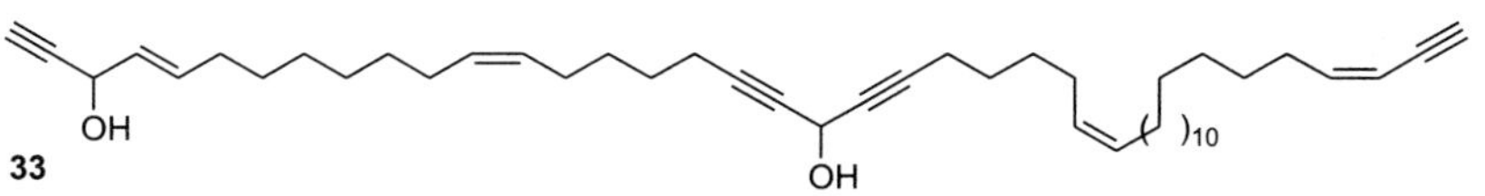

33

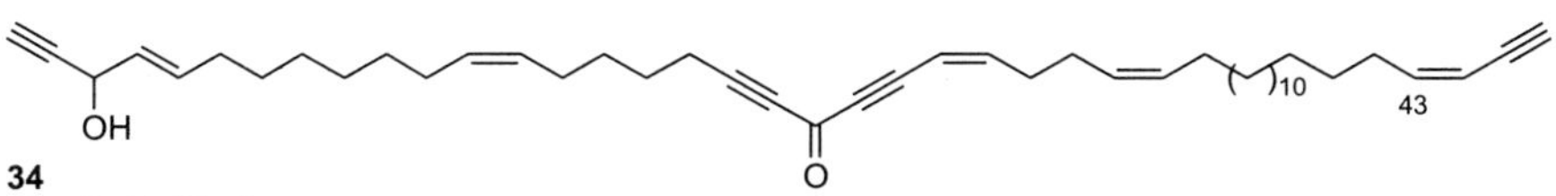

34
35 43,44-dihydro

From another collection of *Petrosia* sp. collected at Hachijo-jima Island, **36** and **37** were isolated and were given the trivial names petrosynol (**36**) and petrosynone (**37**) (Fusetani *et al.* 1987). The absolute configuration of petrosynol (**36**) has now been established by the exciton chirality method. The 3-benzoate and 14-benzoate were prepared by treating the alcohol with 1.2 eq. of benzoyl chloride and catalytic amounts of dimethylaminopyridine (DMAP), followed by HPLC separation. Both benzoates revealed negative first Cotton effects at 228 nm, indicating an *S* configuration for C-3, C-14, C-17, and C-28 (Shim *et al.* 1985). In view of the geometry of the Δ^{15} double bond, introduction of a benzoate at C-14 made it possible to resolve H-15 and H-16 proton signals (J = 10.6 Hz); thus Z geometry was secured for the Δ^{15} double bond. Petrosynol (**36**) was converted to petrosynone (**37**) by oxidation with MnO_2 in CH_2Cl_2. Petrosynol (**36**) showed antifungal activity against *Mortierella ramanniana* and was also active in the starfish egg assay (IC_{99} 1 µg/mL). Petrosynone (**37**) showed antimicrobial activity against *Bacillus subtilis* (data not shown).

Petrosolic acid (**38**) and petrosynol (**36**) were isolated as constituents of a Red Sea sponge, *Petrosia* sp., that inhibits HIV reverse transcriptase (Isaacs *et al.* 1993). The high instability of petrosolic acid (**38**) prevented derivatization of the hydroxyl groups for determination of the absolute configuration. The compounds were evaluated for their inhibitory effect on HIV-1 RT (reverstranscriptase), i.e., RDDP (RNA-dependent DNA polymerase), DDDP (DNA-dependent DNA polymerase), and RNase H (ribonuclease H) activity. In general, the compounds exhibited an inhibitory effect on the DNA polymerase activity of HIV-1 RT but showed almost no inhibition on the RT-associated RNase H. Petrosolic acid (**38**) was found to be the most potent inhibitor of RDDP activity (50% inhibition obtained at 1.2 μM and 95% at 5.9 μM), whereas the DDDP was significantly less sensitive (Table 12.2).

Table 12.2 Inhibition of HIV-1 reverse transcriptase associated RDDP, DDDP and RNase H functions by petrosynol (**36**) and petrosolic acid (**38**)[a].

	RDDP		*DDDP*		*RNase H*	
	IC_{50}	IC_{95}	IC_{50}	IC_{95}	IC_{50}	IC_{95}
36	15.8 ± 0.8	38.5 ± 0.5	36.0 ± 2.0	74.0 ± 3.0	>200	>>200
38	1.2 ± 0.3	5.9 ± 1.5	6.2 ± 0.2	16.5 ± 1.7	39.5 ± 4.5	>>200

[a] The values expressed in μM.

Corticatic acids A–C (**39–41**) are C_{31} polyacetylenic acids from *P. corticata* collected off Hachijo-jima Island, Japan (Li *et al.* 1994). Corticatic acids are structurally related to other C_{31} acetylenic acids such as petroformynic acid (**31**) and aztequynols A and B (**91** and **92**). In order to determine the absolute configuration of the chiral center of the 1-yn-4-en-3-ol group, corticatic acid A was converted to *p*-bromobenzyl ethyl ester by treatment with EtI/DMF, followed by *p*-bromobenzyl chloride/DMAP in pyridine. However, the *p*-bromobenzyl ethyl ester did not exhibit a split CD spectrum. Coincidently, it was reported that the exciton chirality method cannot be applied to secondary alcohols flanked by two chromophores (Matthew *et al.* 1994). Corticatic acids exhibited antifungal activity against *Mortierella ramanniana* (data not shown).

From the same sponge, corticatic acids D and E (**42** and **43**) were isolated and found to be inhibitors of geranylgeranyltransferase type I (GGTase I) (Nishimura *et al.* 2002). Corticatic acids A, D, and E inhibited GGTase I of *Candida albicans* with IC_{50} values of 1.9, 3.3, and 7.3 μM, respectively; corticatic acid A (**39**) also inhibited the growth of *C. albicans* with an MIC value of 54 μM. Since there is little sequence identity between the human and *Candida* GGTase I, they may act as selective antifungal agents.

Four new C_{30} polyacetylenes (**44–47**), along with petrosynol, were isolated from another Japanese *Petrosia* sp. (Ochi *et al.* 1994) collected at Sukumo Bay. The absolute configuration of **44** was determined by the modified Mosher's method. The validity of the method to the polyacetylenes from sponge was confirmed by the application to petrosynol, the absolute configuration of which had been elucidated by the CD allylic benzoate method. The co-occurrence of closely related polyacetylenic alcohols (**45** and **46**) with opposite stereochemistry is of biogenetic interest. Compounds **44–47** were toxic to brine shrimp larvae (LC_{50} 30, 0.1, 0.3, and 5.0 μg/mL, respectively). Compounds **45–47** inhib-

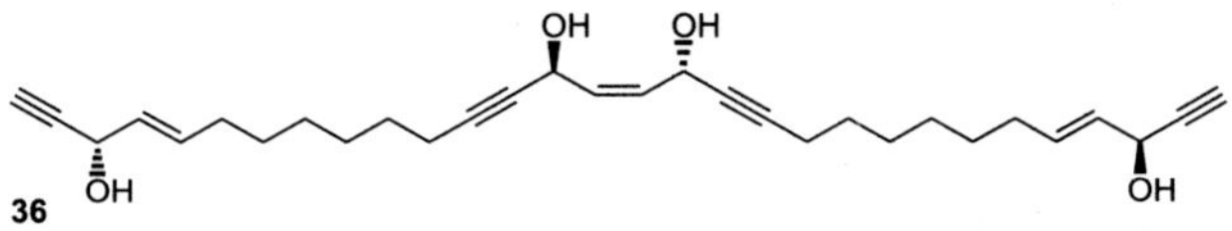

36

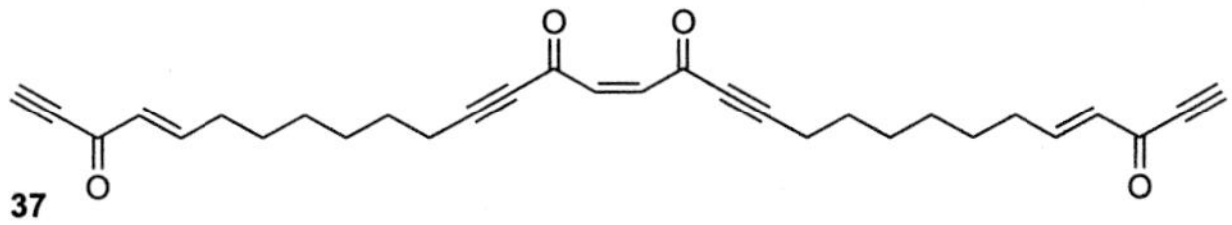

37

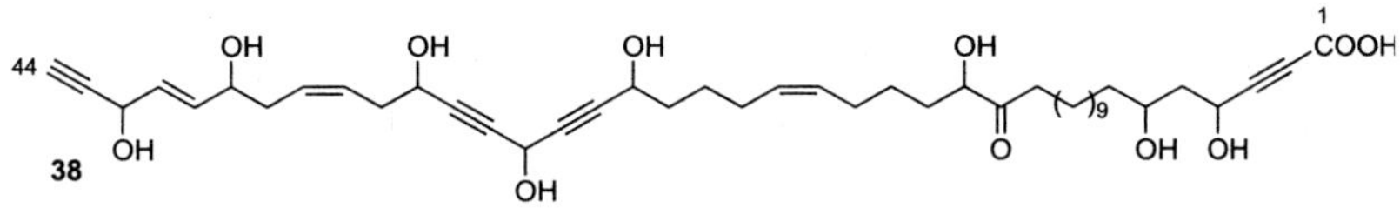

38

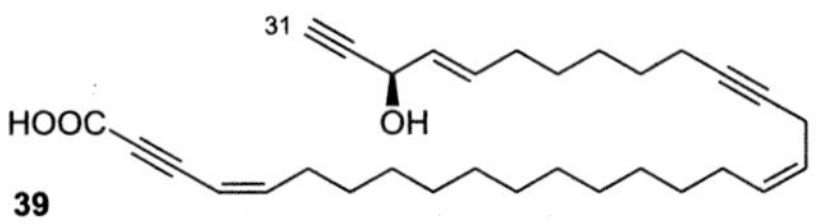

39

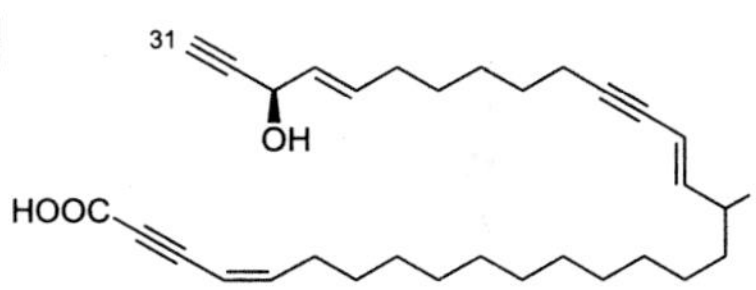

42

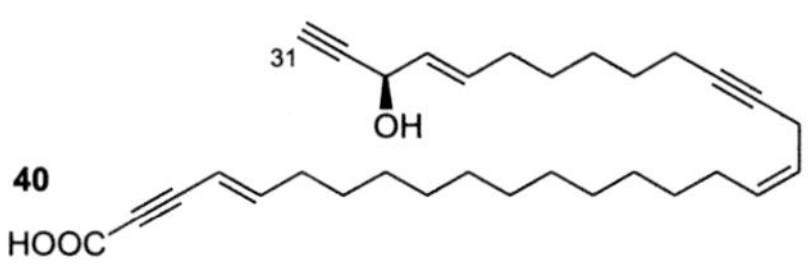

40

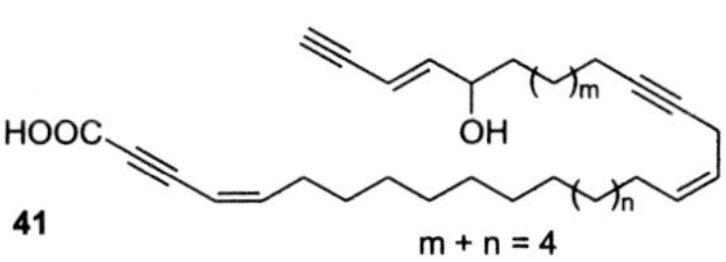

41

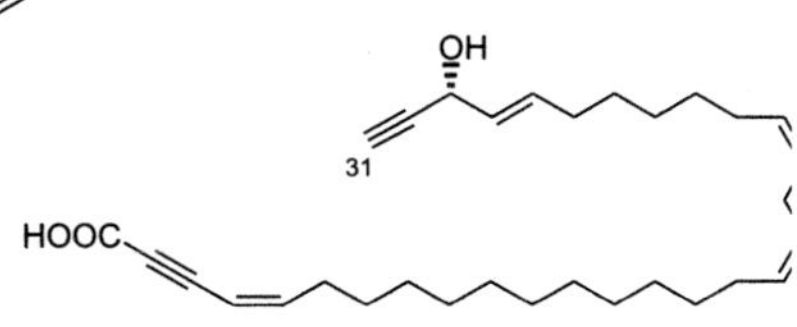

43

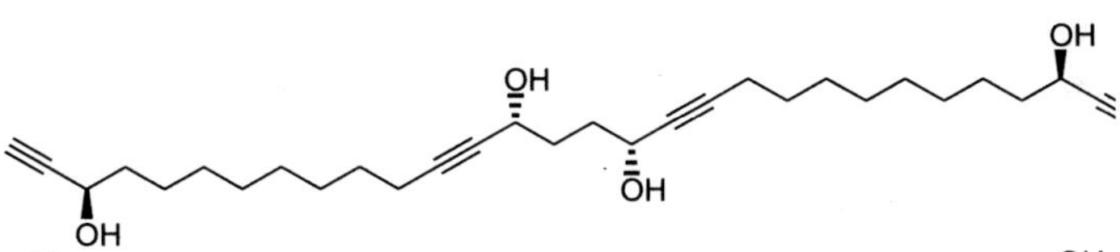

44

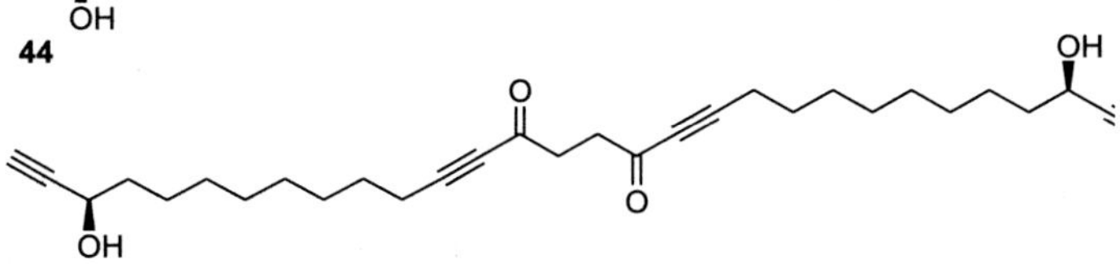

45

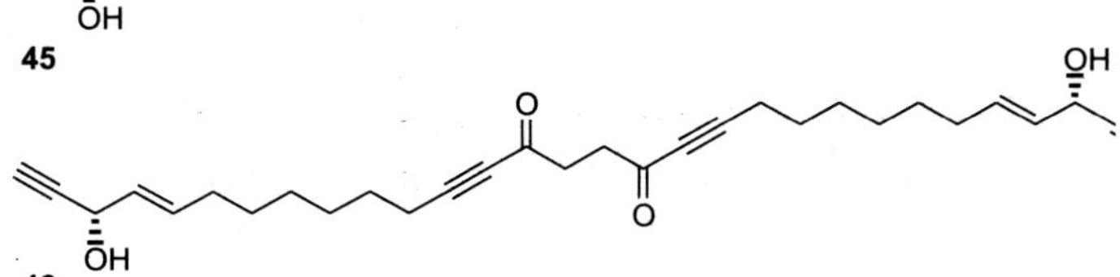

46

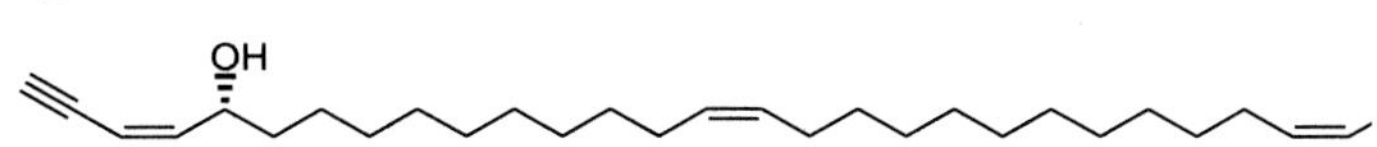

47

ited the cell division of fertilized ascidian (*Styela partite*) eggs with IC_{50} values of 30, 5.0, and 25 μg/mL, respectively.

Two research groups have reported that Korean specimens of *Petrosia* sp. contain petrocortynes A–H (**48–55**), petrosiacetylenes A, B, C, and D (**56–59**), and dideoxypetrosynols A (**56**), B (**58**), C (**57**), and D (**60**), representatives of which inhibit phospholipase A_2 (Seo *et al.* 1998, Shin *et al.* 1998) or show cytotoxicity against several types of human solid tumor cells (Kim *et al.* 1998, 1999a,b). There appears to be some overlap in the metabolites reported, such that petrosiacetylenes A, B, and C have been assigned the same structures as dideoxypetrosynols A, C, and B, respectively, although there still appears to be some confusion regarding the absolute configurations. Petrocortynes (**48–55**) are C_{46} polyacetylenes structurally related to petroformynes isolated from the Mediterranean sponge *P. ficiformis*. However, the structures of these compounds are distinguished from each other on the location of a diacetylenic carbinol functionality as well as the stereochemistry of asymmetric carbon centers. Petrocortyne C (**50**) possesses an unusual γ-pyrone ring possibly formed by an oxidative cyclization of the diacetylenic carbinol group (Fig. 12.1). Petrosiacetylenes (dideoxypetrosynols) are C_{30} polyacetylenes structurally related to petrosynol (**36**). Interestingly, these were isolated as diastereomeric mixtures. The absolute configuration of these chiral 1-yn-4-en-3-ol carbon was unpredictable. In an inhibitory assay against RNA-based reverse transcriptase, compounds **56** and **58** totally cleaved the template 16S rRNA of *Escherichia coli*, at the concentration of 10 μg/20 mL, while other compounds did not cleave it at all. However, none of the compounds cleaved a supercoiled DNA (pUC 119) at the same concentration. Compounds **57** and **58** displayed moderate inhibition of PLA_2 (phospholipase A_2) at the concentration of 50 μg/mL (49 and 36%, respectively), while other petrosiacetylenes were much less active. Among petrocortynes, the γ-pyrone-containing petrocortyne C (**50**) was much more active (42%) than the others. Compounds **50**, **57**, and **58** showed weak inhibition of Na^+/K^+ APTase at the concentration of 20 μg/mL (22, 26, and 25%, respectively). The structures of petrosiacetylenes A–C and dideoxypetrosynols A–C overlap and it looks like the same compounds were isolated, but the stereochemistries of petrosiacetylene C (3*S*) and dideoxypetrosynol B (3*R*) were reported to be opposite. Dideoxypetrosynols A–D showed significant cytotoxicity against five human tumor cell lines, A549 (human lung cancer), SK-OV-3 (human ovarian cancer), SK-MEL-2 (human skin cancer), XF498 (human CNS cancer), and HCT15 (human colon cancer). Dideoxypetrosynol A (**56**) was most potent and the activity was more potent than or comparable to that of doxorubicin (Table 12.3).

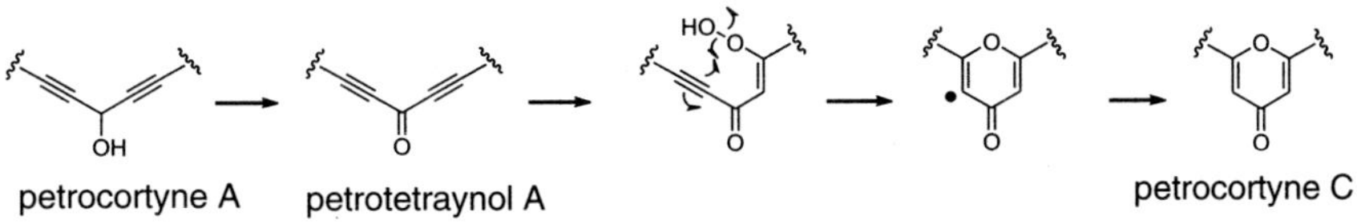

Fig. 12.1 Possible biogenesis of the γ-pyrone ring of petrocortyne C.

Table 12.3 *In vitro* cytotoxicities (EC_{50}, mg/mL) of dideoxypetrosynols against human solid tumors.

Compounds	*A549*	*SK-OV-3*	*SK-MEL-2*	*XF498*	*HCT15*
56	1.43	0.02	0.01	0.16	0.17
57	12.41	1.83	1.27	1.83	1.87
58	1.98	0.21	0.11	1.83	1.56
60	5.78	0.02	0.02	3.02	1.94
doxorubicin	0.09	0.16	0.11	0.13	1.02

A549: human lung cancer; SK-OV-3: human ovarian cancer; SK-MEL-2: human skin cancer; XF498: human CNS cancer; HCT15: human colon cancer.

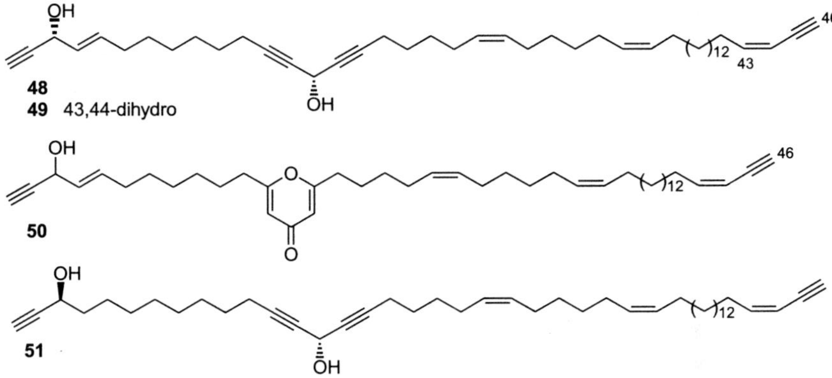

48
49 43,44-dihydro
50
51

From the same sponge which yielded dideoxypetrosynols, five additional cytotoxic polyacetylenes, (3*S*, 14*S*)-petrocortyne A (**61**), petrotetrayndiols A (**62**) and B (**63**), and dideoxypetrosynols E (**64**) and F (**65**) were also isolated (Kim *et al.* 1999b). Some of the absolute configurations of the carbinol groups were reported to be the same as those of petroformynes, but opposite to those of petrocortynes. Compounds **61–65** showed significant cytotoxicity against five human tumor cell lines with the EC_{50} values of 0.1 to 3.7 μg/mL. The potencies were comparable to that of doxorubicin. It is interesting to note that this sponge specimen collected from Korean waters contained polyacetylenic compounds with mixed stereochemistry. The location of double bonds was determined by FAB-CID tandem mass analysis.

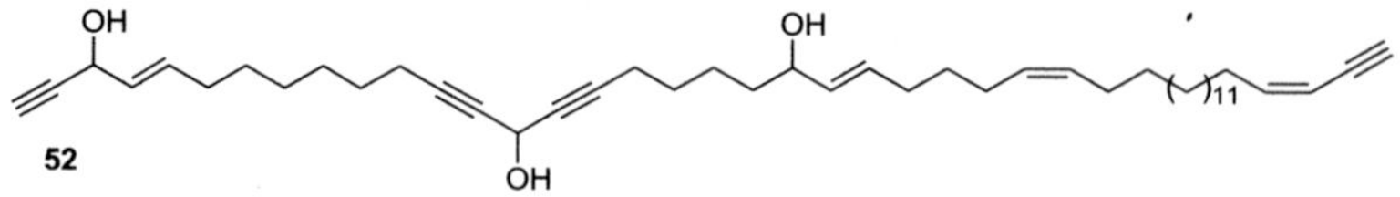

52

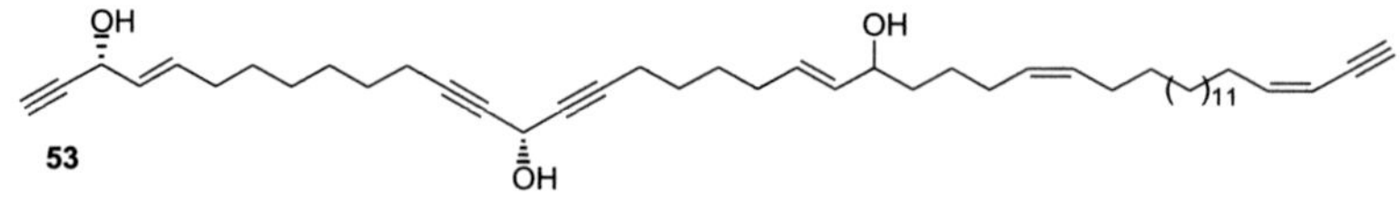

53

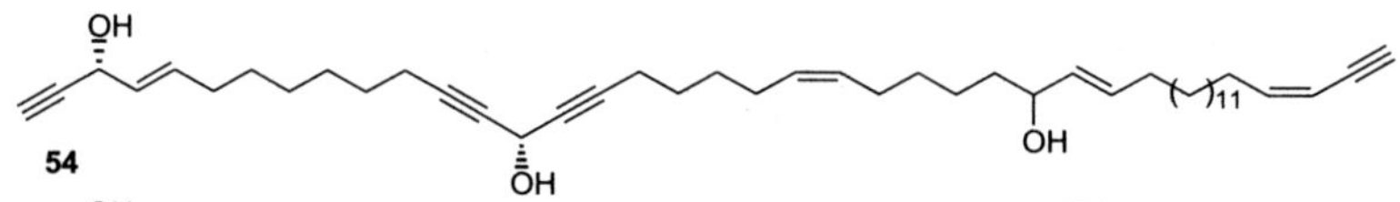

54

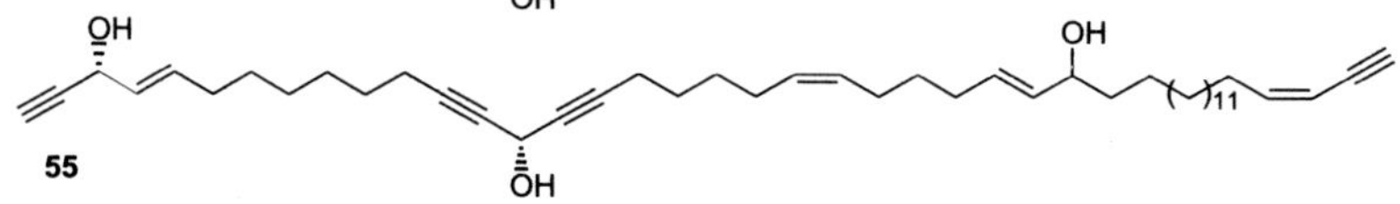

55

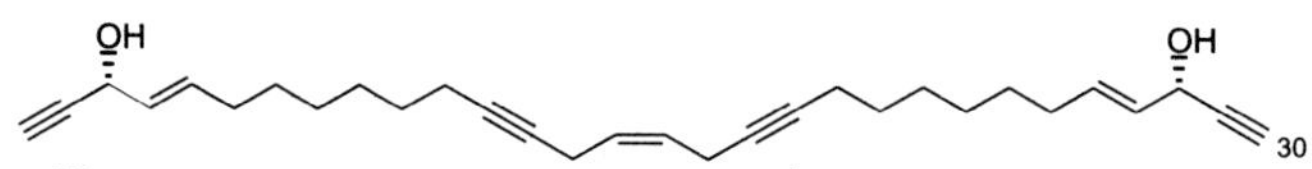

56

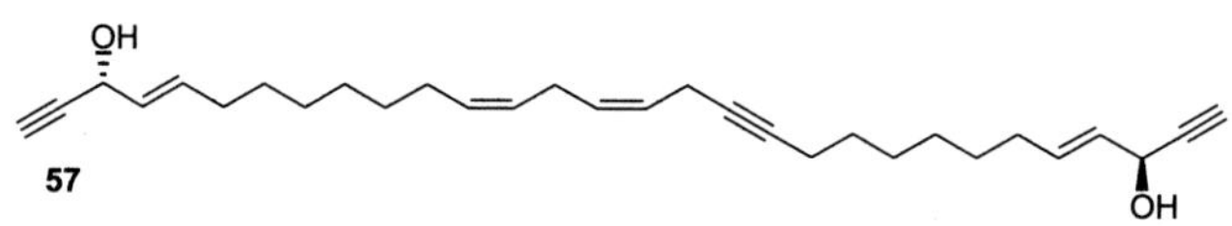

57

58

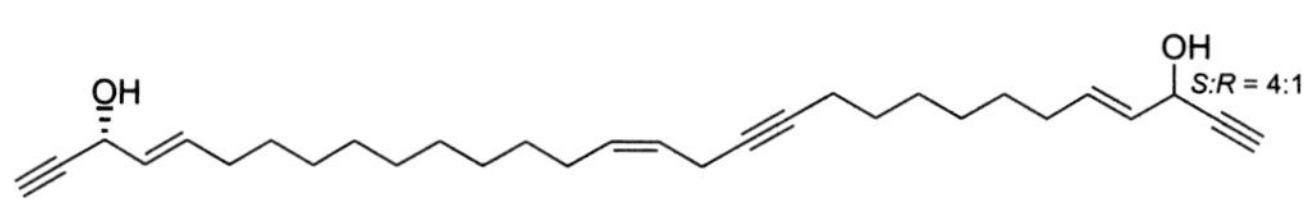

59

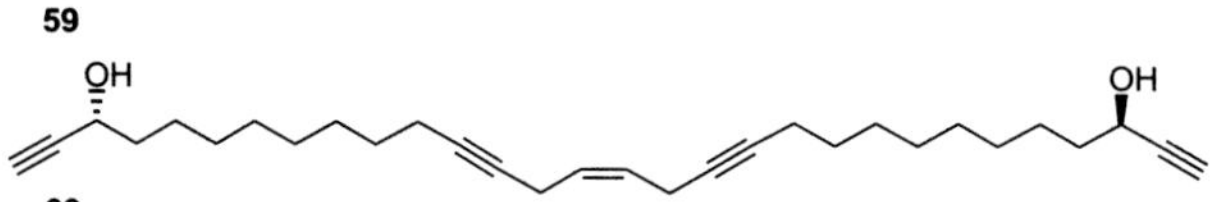

60

A further series of cytotoxic polyacetylenic alcohols (**66–75**) have been isolated along with **50** from the same sponge (Lim *et al.* 1999, 2001). The compounds showed mild to significant cytotoxicity, while the potency (EC_{50} values of 0.6 to 1.8 mg/mL) of petrotriyndiol A (**74**) was comparable to that of cisplatin. These compounds showed significant inhibition on the DNA replication of the simian virus SV40 in a concentration-dependent manner, which could explain their cytotoxicity (Table 12.4).

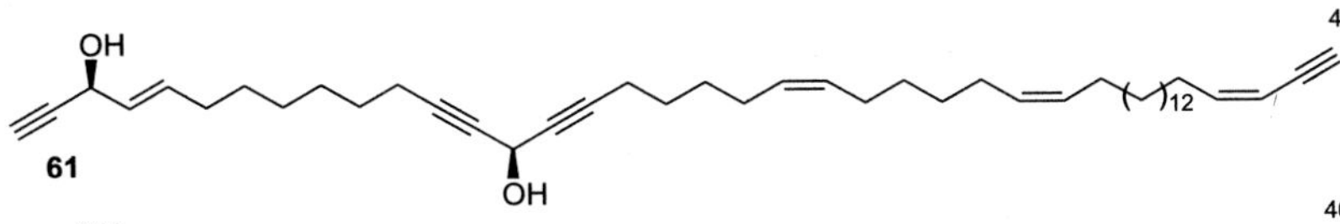

61

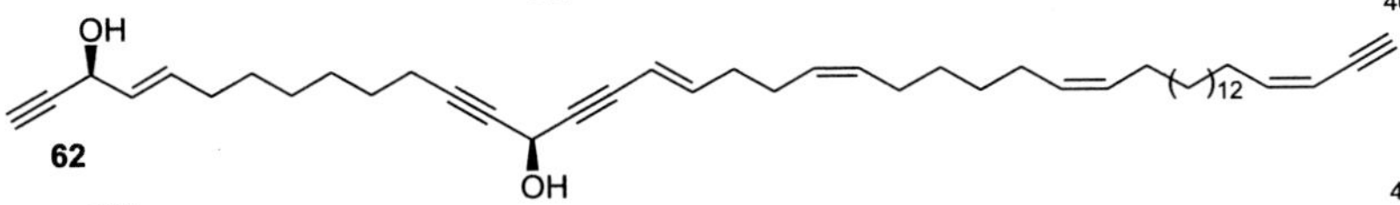

62

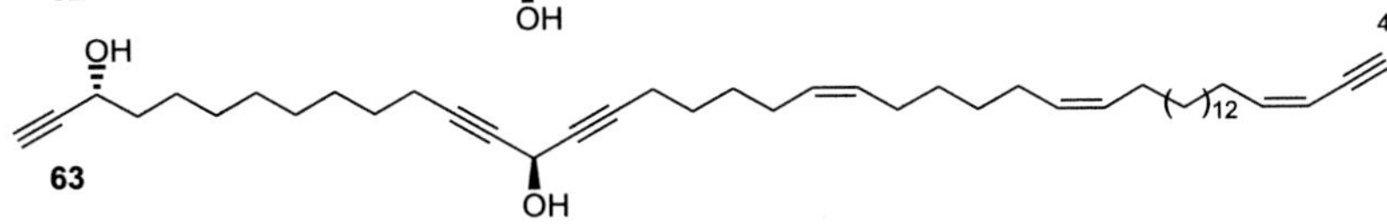

63

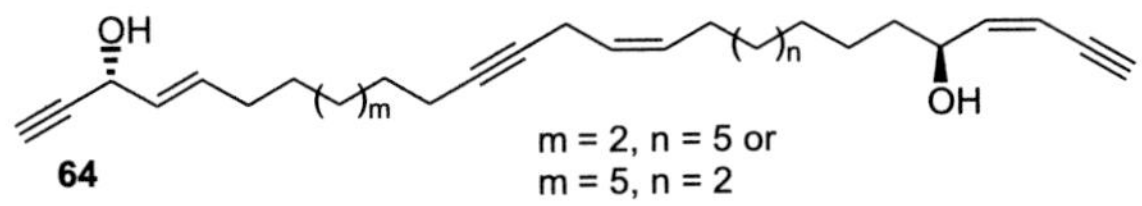

64

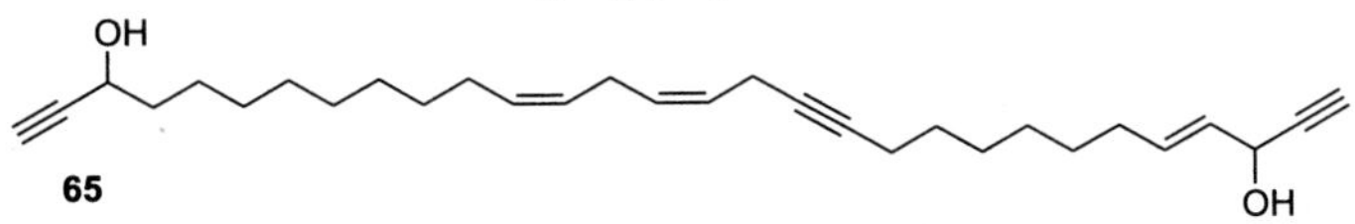

65

Table 12.4 The percent inhibitions of **50** and **66–75** on SV40 DNA replication[a].

conc. (mM)	***66***	***67***	***68***	***69***	***70***	***71***	***50***	***72***	***73***	***74***	***75***
125	45	12	63	79	66	40	89	62	59	62	76
250	46	47	77	94	80	70	100	83	86	90	1
500	81	70	100	100	100	100	100	100	100	100	100

[a]The bioactivity of compounds **66** and **67** was assessed as a separate batch from others.

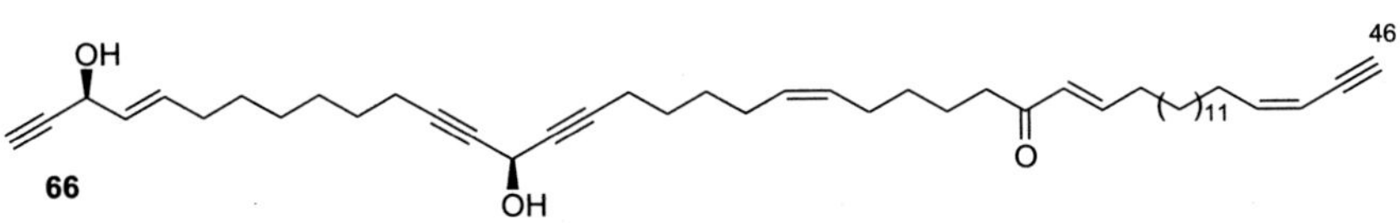

66

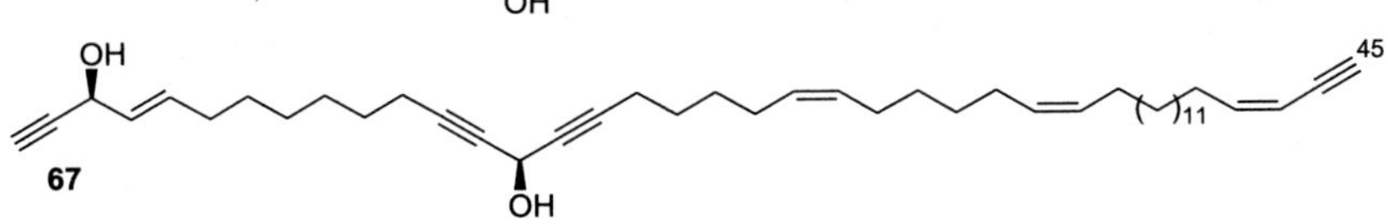

67

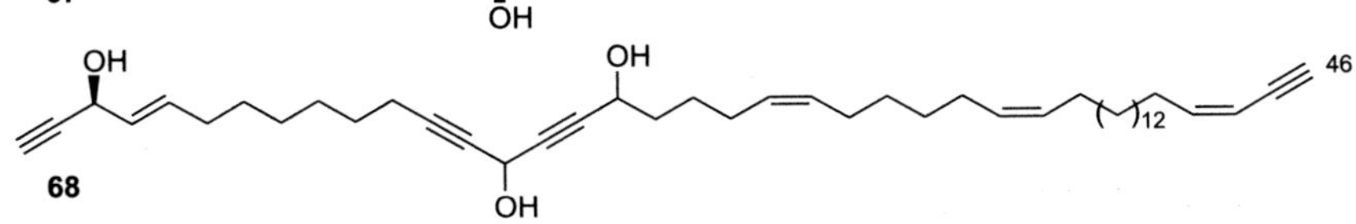

68

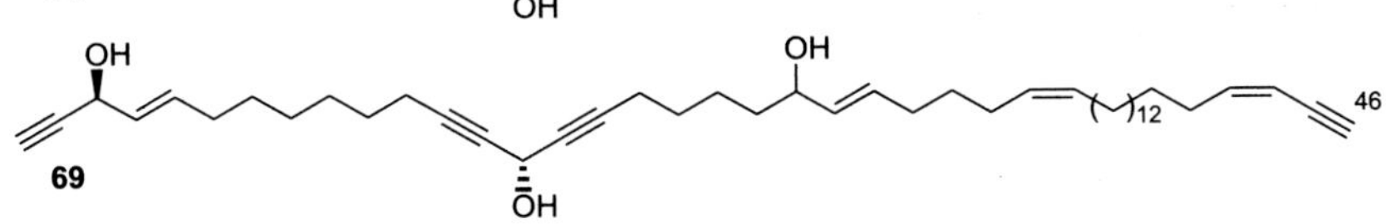

69

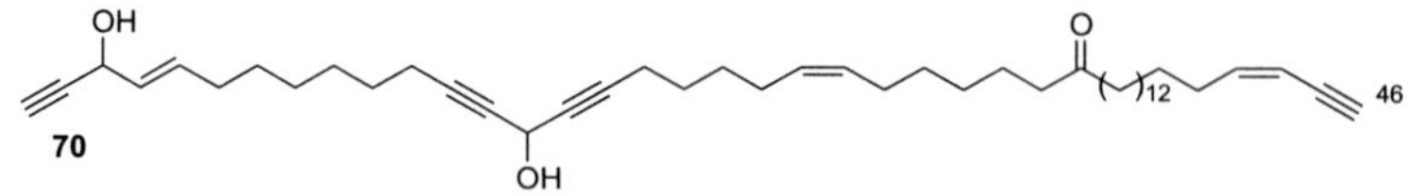

70

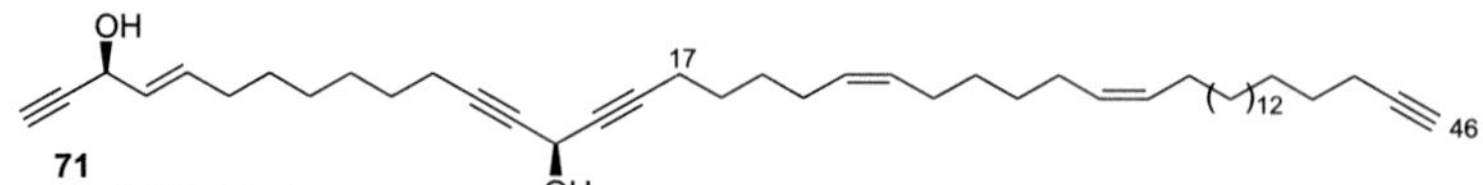

71
72 17,18-dehydro

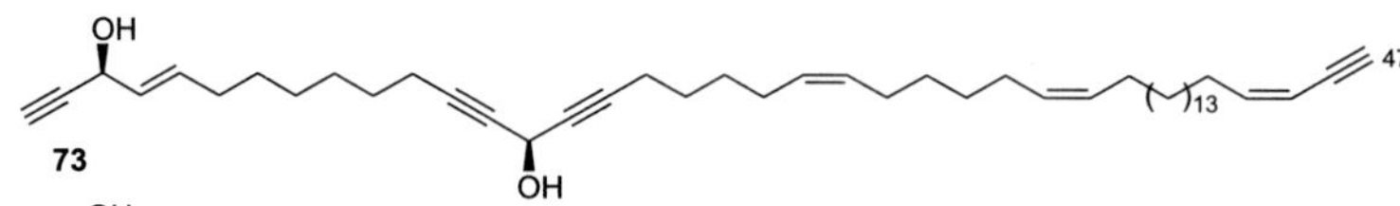

73

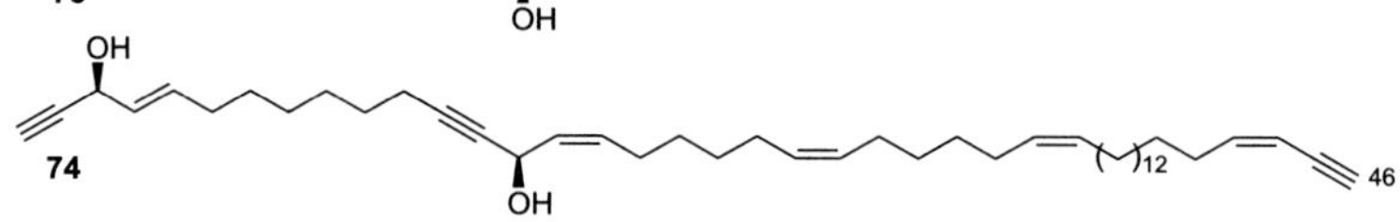

74

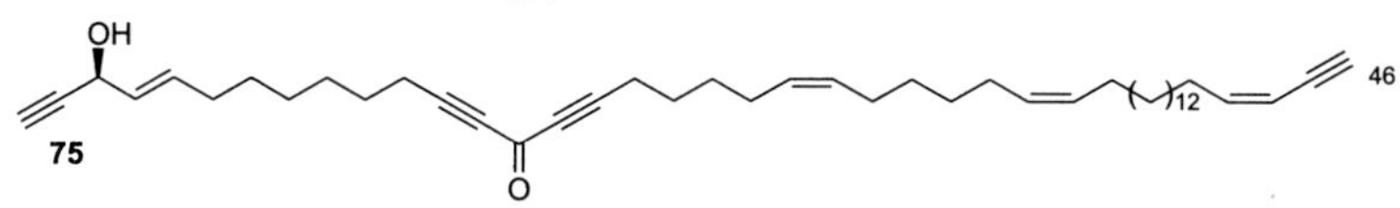

75

From the same sponge, similar derivatives (**76-80**) with various chain lengths (C_{46}-C_{48}) were isolated (Lim *et al.* 2001). Again, the analogous linear polyacetylenes from *Petrosia* sp. revealed mixed stereochemistries of each allylic secondary alcohol (acetylenic carbinol). Nor-(3*S*, 14*S*)-petrocortyne A (**67**), homo-(3*S*, 14*S*)-petrocortyne A (**73**), petrotetrayndiol

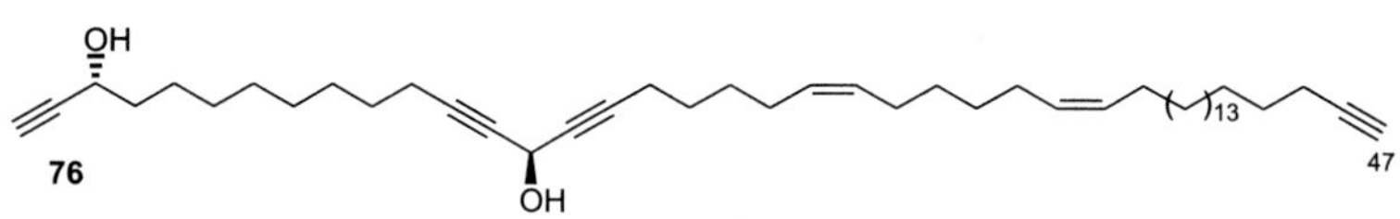

76

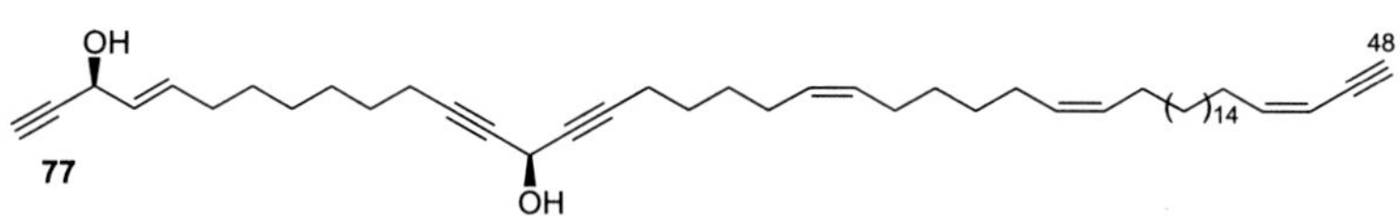

77

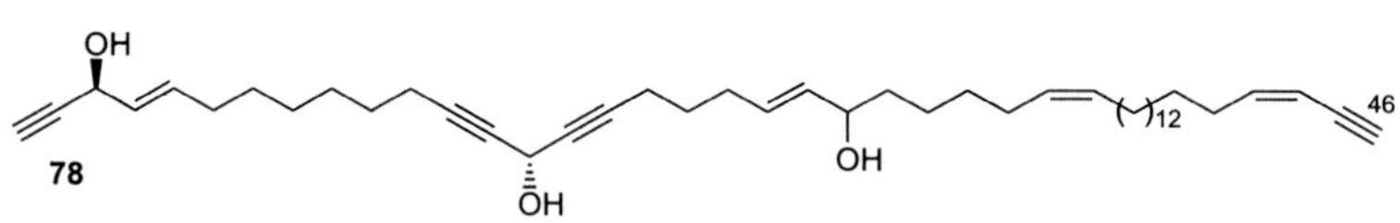

78

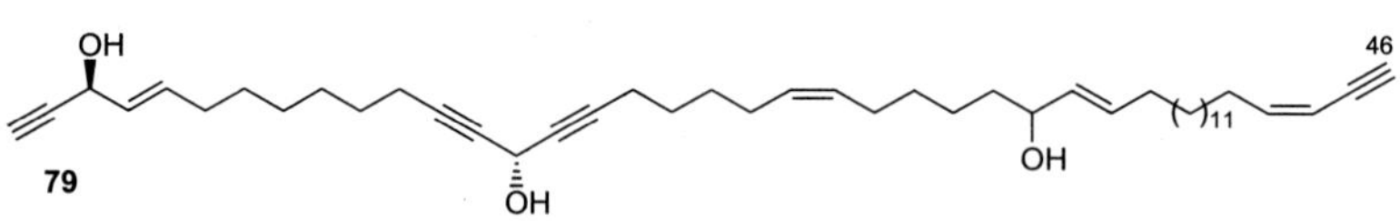

79

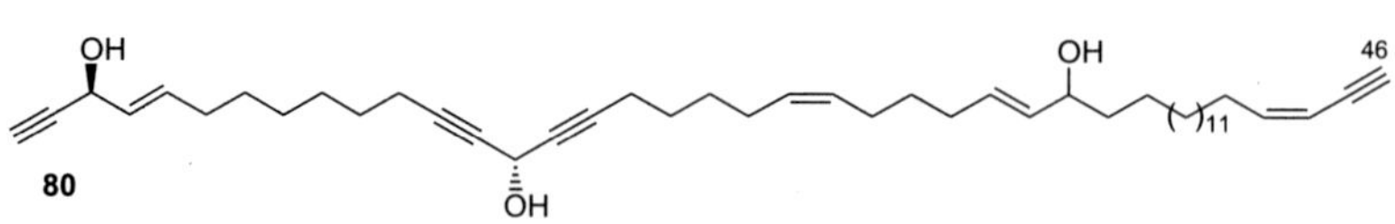

80

F (**76**), and dihomo-(3*S*, 14*S*)-petrocortyne A (**77**) were structurally unique due to their unusual chain lengths. The compounds were significantly cytotoxic to five human tumor cells (EC_{50} values in the range of 0.5 to 10.0 μg/mL) and inhibited SV40 DNA replication (50% inhibition at the concentration of ca. 40 μM).

Seven new brominated C_{18} acetylenic acids (**81–87**) and three methyl esters (**88–90**) were isolated from *P. volcano* from Japan (Fusetani *et al.* 1993). The methyl esters presumably were generated as artifacts from corresponding free acids. They exhibited antifungal activity against *Mortierella ramanniana* (data not shown).

81
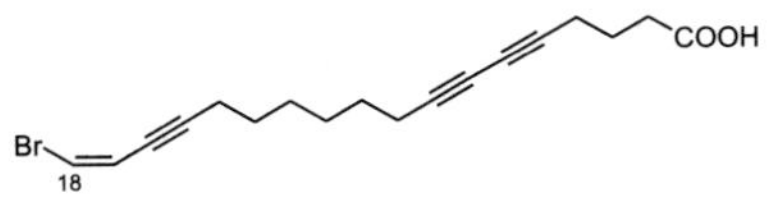

86
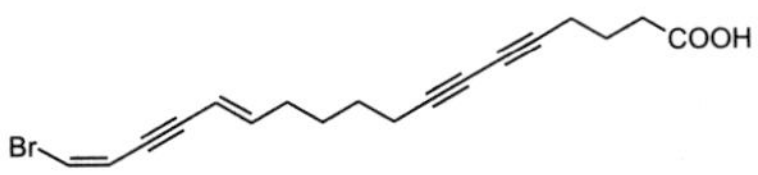

82
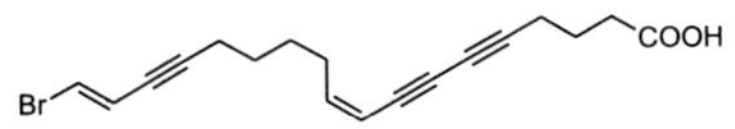

87
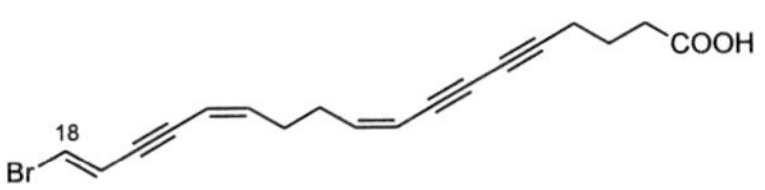

83
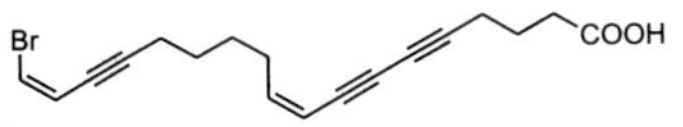

88
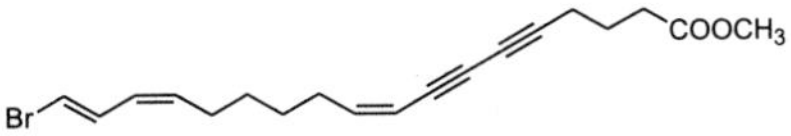

84
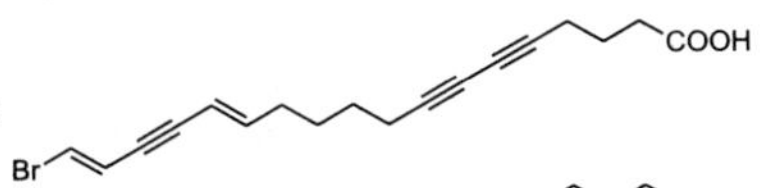

89
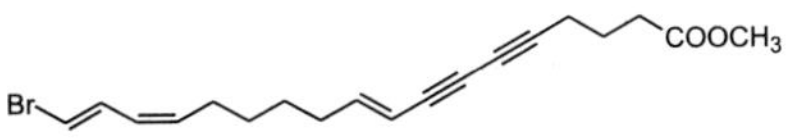

85
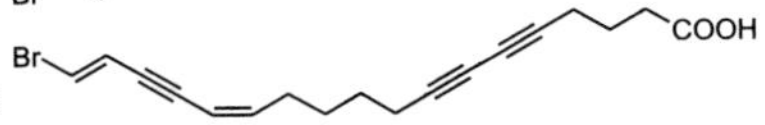

90
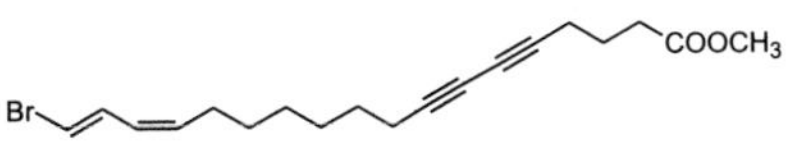

Aztequynols A and B (**91** and **92**) were the first C-branched acetylenes from a *Petrosia* sp. from New Caledonia (Guerriero *et al.* 1998). The location of the methyl branching was defined based on high-energy collisionally–activated decomposition tandem mass spectrometry of lithium adducts which may have wide application in natural product structural analysis. The configurations of the carbinol centers of aztequynols were opposite to those of petroformynes. Neither aztequynol A nor B (< 20% at 10 μg/mL) accounted for the cytotoxicity of the sponge extract on KB (human epidermoid carcinoma) cell line.

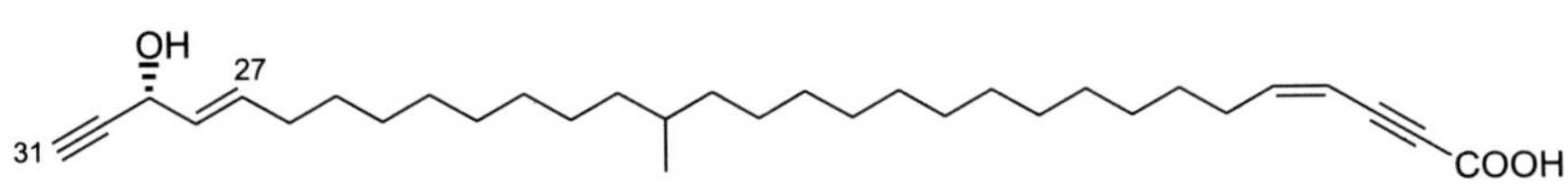

91
92 27,28-dihydro

Xestospongia

A dibrominated C_{16} acetylenic acid (**93**) was first isolated from *Xestospongia muta* collected near Summerland Key, Florida (Schmitz and Gopichand 1978). It has two conjugated vinyl bromide groups and was isolated from the crude extract which showed both antitumor activity and CNS modulating activity *in vivo*. The location and orientation of the conjugated acetylene group was defined by reducing ($LiAlH_4$, -70° C) the methylester of **93** and analyzing the ^1H NMR spectrum of the resulting alcohol in the presence of $Eu(fod)_3$ shift reagent. At 0.21 mole ratio of shift reagent, four methylene group signals were well resolved. The compound and its methylester did not show the pentobarbital effect in mice (Kaul *et al.* 1978). Compound **93** and its methylester showed weak cytotoxicity against PS and L1210 murine leukemia cells with ED_{50} values of **93**, 24, 34 μg/mL; methylester, 29, 34 μg/mL, respectively. They were inactive in the *in vivo* PS antitumor assay.

A similar brominated C_{18} acetylenic acid (**94**) was isolated from the sponge *X. testudinaria* collected near Townsville, Australia (Quinn and Tucker 1985). The location of the central enyne group was defined by COSY-45 spectrum measured in pyridine-d_5. The 8H multiplet was resolved into two 4H multiplets and each showed no correlation, indicating that the partial structures are each joined by two 4H methylene groups. The methylene β to the carboxyl group and a methylene adjacent to an acetylene were coupled to one of the 4H methylenes, while allylic protons and a methylene adjacent to another acetylene were coupled to another 4H methylenes. Two additional brominated C_{18} acetylenic esters (**103** and **104**) have been isolated as minor constituents of the same sponge, and identified by comparison of spectral data with those of known acetylenic acids (Quinn and Tucker 1991). These compounds seem to be artifacts generated from the free acids in the separation process, which involved warming the crude extract with MeOH.

Four brominated C_{16} and C_{18} acetylenic acids (**94–98**) and a brominated C_9 olefinic acid (**99**) were isolated along with **94**, as the corresponding methyl esters from a species of *Xestospongia* that had been collected in the Red Sea (Hirsh *et al.* 1987). The compounds were found to possess moderate antibacterial activity against *Bacillus subtilis* (the best MIC value, 25 μg/mL, being observed for **96**). Isolation of C_9 fatty acid (**99**) is of interest from the biogenetic point of view as this compound might be a naturally oxidative degradation product of the C_{18} fatty acid (**95**).

Three new C_{32}–C_{34} polyacetylenic alcohols, melynes A–C (**100–102**), have been isolated from a species of *Xestospongia* from Vanatu (Quinoa and Crews 1988). The structural features overlap part of those existing in duryne (**140**) (Wright *et al.* 1987) and 18-hydroxyrenierin-2 (**211**) (Cimino and De Stefano 1977). The location of central double bond or triple bond was defined based on ozonolysis and HREIMS. Melyne A (**100**) was active against the intestinal protozoan *Giardia* (data not shown).

X. testudinaria from Mayotte contained two antimicrobial brominated polyacetylenes, xestospongic acid (**105**) and the corresponding ethyl ester (**106**), which also inhibited Na^+/K^+-ATPase with an IC_{50} value of 10^{-4}–10^{-3} M, measured by the method described by Hosie (Hosie 1965). Xestospongic acid (**105**) and its ethyl ester (**106**) showed weak antimicrobial activity against *Staphylococcus aureus*: **106**, ϕ inhibition 15 mm at 500 μg/disc; **105**, ϕ inhibition 12 mm at 100 μg/disc (Bourguet-Kondracki *et al.* 1992).

Five brominated polyacetylenic acids (**107–111**), along with **96** and **104**, were obtained from *X. muta* from the Bahamas (Patil *et al.* 1992). These compounds inhibited HIV-1 protease with an IC_{50} of 6 μM to 12 μM. The minimal differences in the inhibitory potency within this series of compounds suggest a nonspecific mode of action. To test for nonspecific inhibition of enzymes, these compounds were assayed for inhibition of rat ATP citrate lyase and porcine malic dehydrogenase. No inhibition of either enzyme was observed when these compounds were tested at 100 μg/mL. However, cytotoxicity was observed when MOLT-4 (human T lymphocyte leukemia) cells were exposed to these compounds at a concentration near their IC_{50} values.

Spectroscopic analysis has resulted in a tentative structure for nepheliosyne A (**112**) from an Okinawan sponge, *Xestospongia* sp. (Kobayashi *et al.* 1994a). Nepheliosyne A (**112**) possesses a closely related structure to petrosolic acid (**38**) (Isaacs *et al.* 1993) and was the first example of a C_{47} polyacetylene. The locations of the ketone and hydroxyl group were not determined and the position of each group may be reversed. Nepheliosyne A (**112**) showed weak cytotoxicity against murine lymphoma L1210 (40% inhibition) and human epidermoid carcinoma KB cells (14% inhibition) at 20 μg/mL.

An Indian Ocean specimen of *Xestospongia* sp. that contained high proportions (ca. 50%) of eubacteria in its tissues yielded two new brominated acetylenic acids (**113** and **114**), and a dibrominated olefinic acid (**115**), which was characterized as its methyl ester (Brantley *et al.* 1995). Compounds **113, 114** (C_{20}), and **115** (C_{14}) are brominated fatty acids with an unusual carbon number that have not been encountered previously in marine invertebrates. The known brominated acetylenic acids are usually based on C_{16} and C_{18} chain lengths, with the exception of one C_9 acid (**99**).

Seventeen brominated acetylenic acids, of which 14 (**116–129**) were new and three (**130–132**) had previously been found as components of sponge phospholipids, were reported as metabolites of an Okinawan speci-

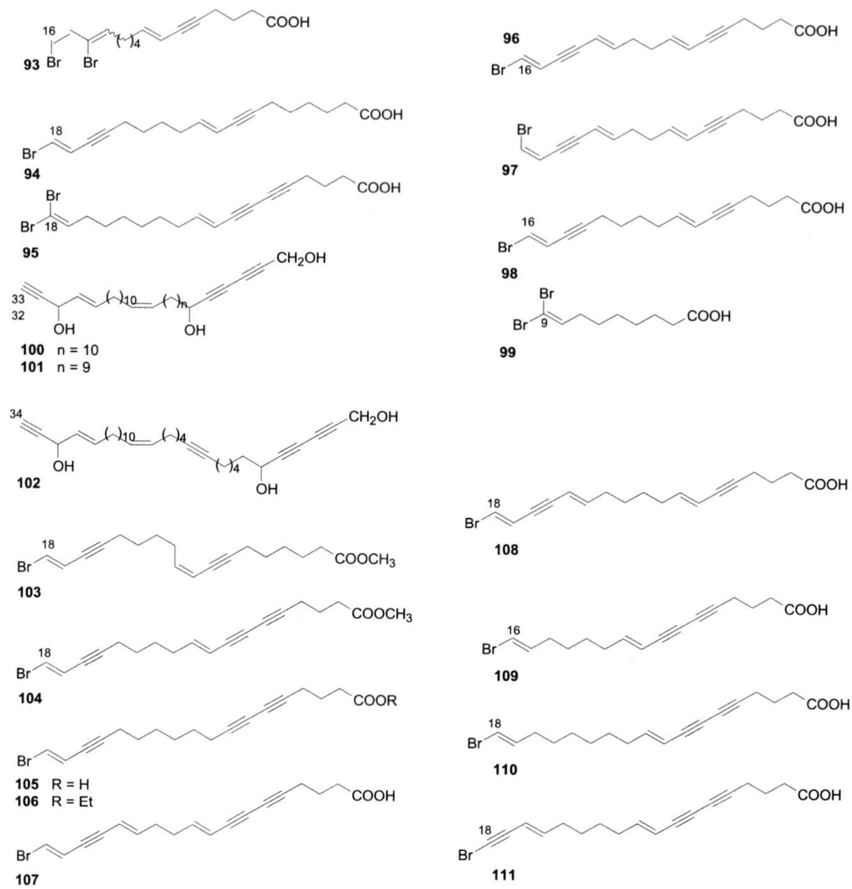

men of *Xestospongia* sp. (Li *et al.* 1995). Brominated C_{27}–C_{29} fatty acids (**126–132**) were unprecedented. Compounds **116–119**, **126**, and **131** exhibited cytotoxicities against L1210 murine leukemia cells with IC_{50} values of 9.2, 1.3, 2.5, 2.1, 8.8, and 11 μg/mL, and KB human epidermoid carcinoma cells with those of >20, 6.8, 7.2, 2.0, 15.5, and 18.5 μg/mL, respectively. Other compounds did not show cytotoxicity (IC_{50} >20 μg/mL).

The xestosterol esters of two brominated acetylenic acids (**133** and **134**), along with the free acid of **133,** were obtained from an Australian specimen of *X. testudinaria* (Pham *et al.* 1999). The free brominated acid inhibited [^{3}H]DPCPX (8-cyclopentyl-1, 3-dipropylxanthine) binding to rat-brain adenosine A_1 receptor with an IC_{50} of 106 μM, while the esters were tested as a mixture and had no effect up to the concentration of 10^{-2} M. Activation of adenosine A_1 receptor results in bradycardia and brochoconstriction. Therefore, antagonist of adenosine A_1 receptor could be useful in the treatment of arrhythmia and CNS depressant action.

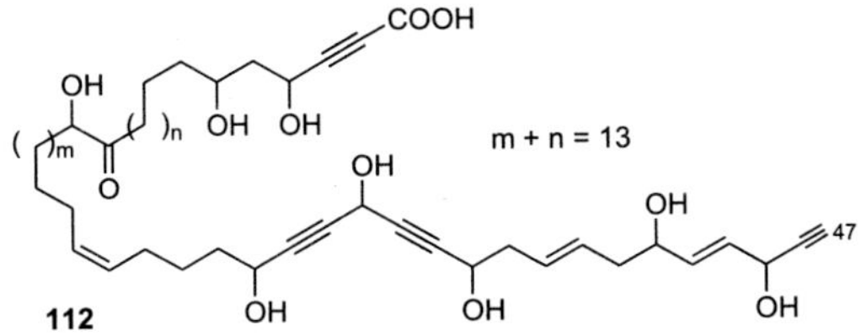

112

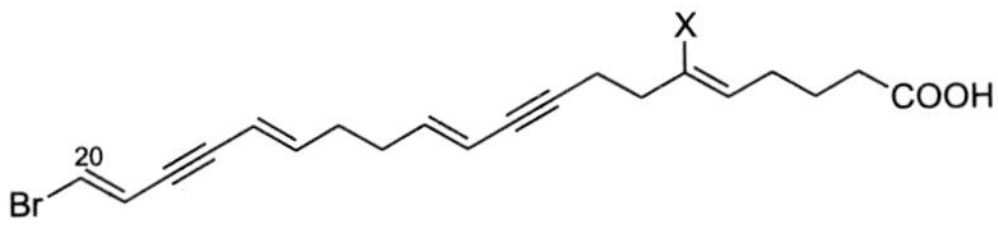

113 X = H
114 X = Br

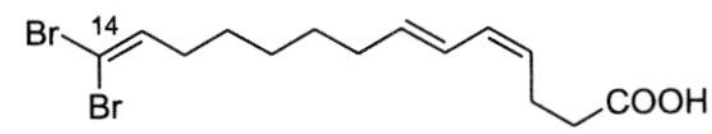

115

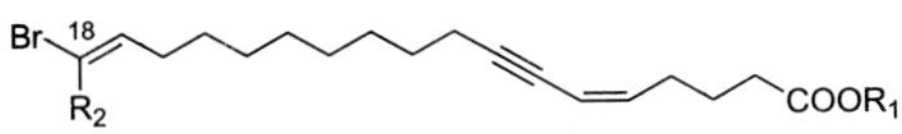

116 R1 = R2 = H
117 R1 = H, R2 = Br
118 R1 = Me, R2 = H
119 R1 = Me, R2 = Br

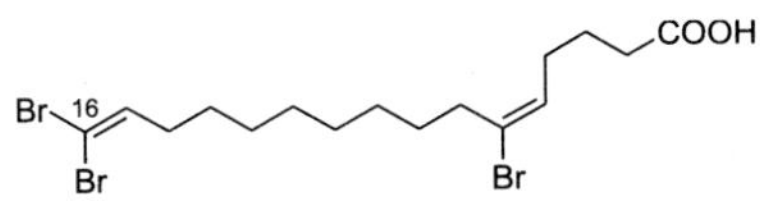

124

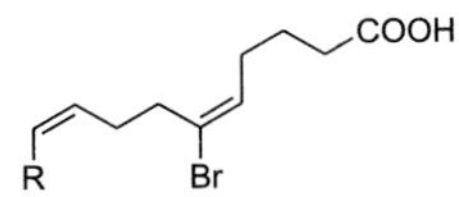

125 R = -C_6H_{13}
126 R = -$(CH_2)_{13}CH{=}CHC_3H_7$ (*Z*)
127 R = -$(CH_2)_{13}CH{=}CHC_2H_5$ (*Z*)
128 R = -$(CH_2)_{15}CH(CH_3)_2$
129 R = -$(CH_2)_{16}CH(CH_3)_2$
130 R = -$(CH2)_{14}CH(CH_3)_2$
131 R = -$C_{17}H_{35}$
132 R = -$C_{18}H_{37}$

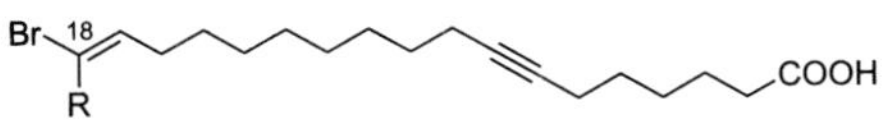

120 R = H
121 R = Br

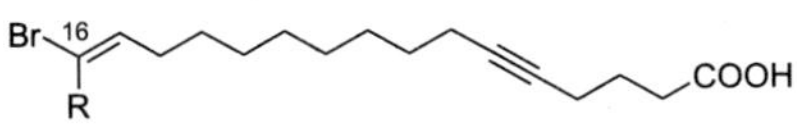

122 R = H
123 R = Br

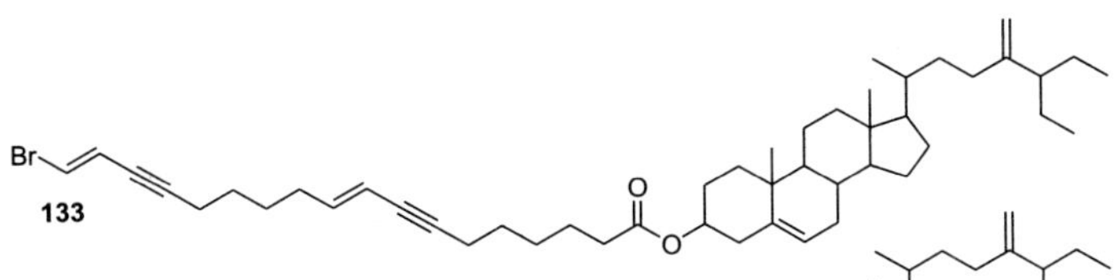

133

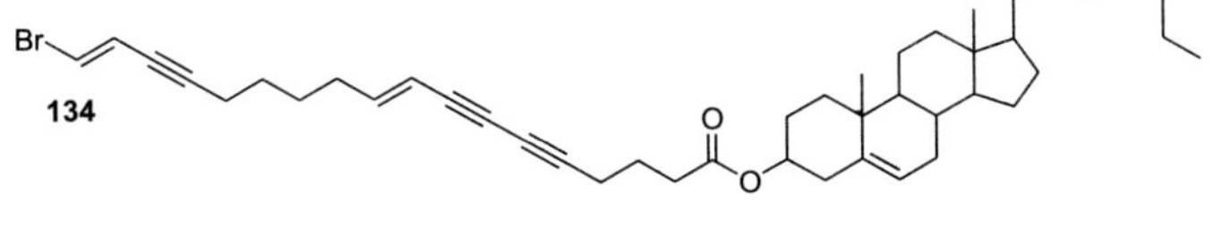

134

Strongylophora

Strongylodiols A–C (**135–137**), which occurred as enantiomeric mixtures of different ratios (A; 91% *R*, B: 97% *R*, C: 84% *R*), are cytotoxic acetylenic alcohols from a *Strongylophora* sp. from Okinawa (Watanabe *et al.* 2000). The absolute stereochemistry at C-6 was determined by the modified Mosher's method. MTPA esters were used instead of 2NMA esters since there is no possibility of epimerization at the chiral center of the reagent during the preparation of esters. These compounds were found to be enantiomeric mixtures with the 6*R* isomers as the major components. Cytotoxicities of the compounds against DLD-1 (human colorectal adenocarcinoma), MOLT-4 (human T lymphocyte leukemia), and IMR-90 (normal human diploid lung fibroblast) were examined. Significant activities of strongylodiols A, B, and C were observed against MOLT-4 cells with IC_{50} values of 0.35, 0.8, and 0.85 μg/mL, respectively. Strongylodiols are the first linear acetylenic compounds isolated from a sponge of the genus *Strongylphora*. Thus far, only meroditerpenoids such as strongylophorines have been recognized as chemical markers for this genus. Isolation of acetylenic compounds from *Strongylophora* with structural similarity to compounds isolated from some *Petrosia* species confirms the close relationship between the two genera. Strongylodiol A (**135**) has been synthesized confirming the (*R*) stereochemistry (Yadav and Mishra 2002). The key step of this process involves the selective introduction of the *R* configuration to the acetylenic carbinol at C-6, which was achieved by *β*-elimination of epoxychlorides.

A Taiwanese specimen of *S. durissima* contained two additional acetylenes, durissimols A and B (**138** and **139**), as well as a new meroditerpenoid, in addition to several known acetylenic compounds, renierin-2 (**210**), 18-hydroxyrenierin-2 (**211**), melyne-C (**102**), duryne (**140**), and siphonochalinol (**157**) (Shen and Prakash 2000). Preliminary biological study revealed that durissimol B (**139**) and duryne (**140**) exhibit potent cytotoxicity against human gastric (NUGC) cells at 10 μM.

135
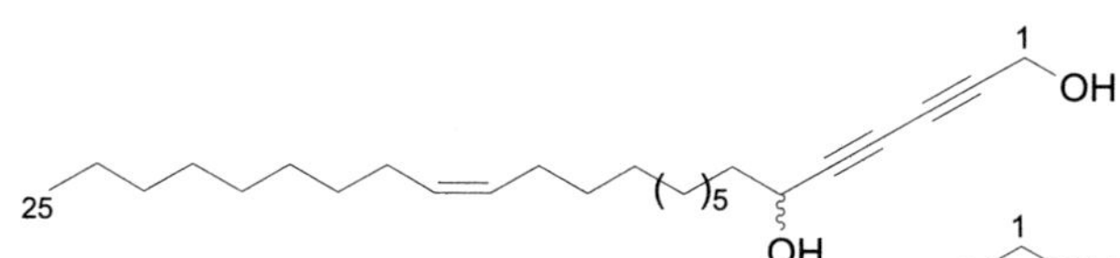

136
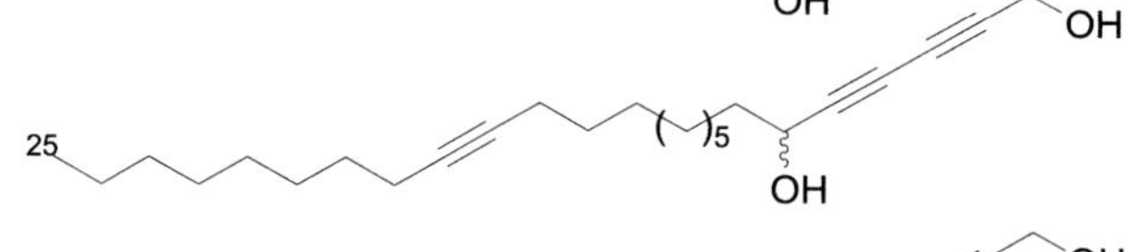

137
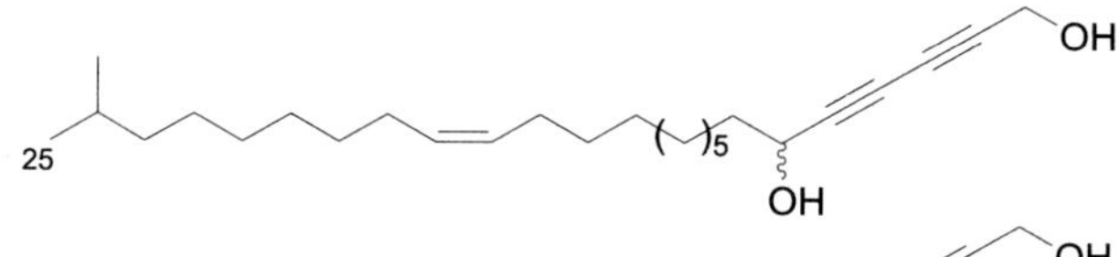

138
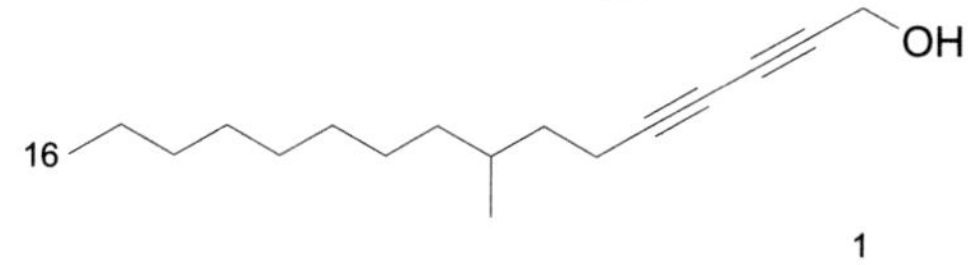

139
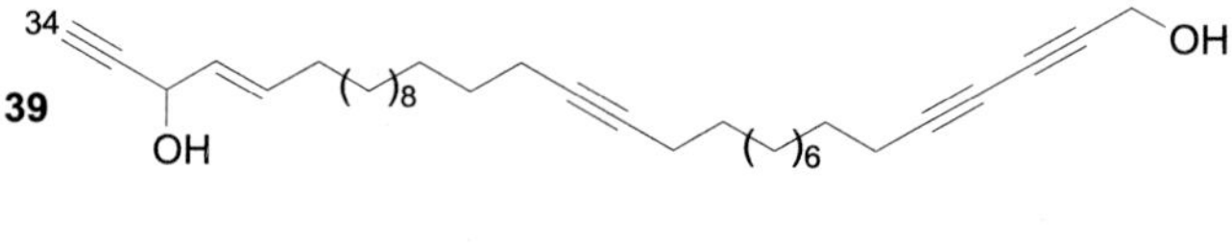

Cribrochalina

A symmetrical C_{30} acetylene, duryne (**140**), was isolated as a cytotoxic agent from specimens of *Cribrochalina dura* that had been collected in the Bahamas (Wright *et al.* 1987). The isolation of duryne (**140**) was directed by *in vitro* cytotoxicity against P388 murine leukemia. Duryne inhibited the growth of P388 murine leukemia (IC_{50} 0.07 μg/mL) and of colon (HCT-8), lung (A549), and mammary (MCF7) human tumor cell lines (MIC 0.1 μg/mL). Duryne (**140**) has been synthesized using a strategy that takes advantage of the symmetry of the compound (Deshpande *et al.* 1989). Although both central double bond isomers were synthesized, it was not possible to unambiguously determine the stereochemistry of duryne by comparison of the published spectral data with those of the synthetic isomers.

Five new acetylenic alcohols (**141–145**) from *C. vasculum* collected in Belize showed *in vitro* immunosuppressive and antitumor activities (Gunasekera and Faircloth 1990). The position of the isolated double bond in **142** was confirmed by the mass spectrum of the hydroxy-dimethylamino derivative obtained by peroxidation of the double bond with *m*-chlorobenzoic acid followed by heating the epoxide solution in dimethylamine in a sealed tube. The strongest fragment obtained in the mass spectrum, the ammonium ion $CH_3(CH_2)_5CH=N^+(CH_3)_2$ (*m/z* 142), determined unequivocally the position of the isolated double bond. The assignment of the trans configuration of the isolated double bond was based on the absence of any IR absorptions between 730 and 675 cm^{-1}. Compounds **141–145** had *in vitro* antitumor activity against murine leukemia P388 with IC_{50} values of 1.0, 1.3, 1.1, 0.2, and 0.1 μg/mL, respectively. Compounds **141–145** showed 100% immunosuppressive activity on mixed lymphocyte reaction (MLR) assays (Gunasekera *et al.* 1989).

In addition to three major metabolites (**142–144**) described previously, a Bahamian specimen of *C. vasculum* contained four minor acetylenic compounds (**146–149**) that were toxic in the brine shrimp assay (Aiello *et al.* 1992). The stereochemistry at C-3 of the known compounds **142–144** was determined by the circular dichroic excitation method (Gonnella *et al.* 1982). The *p*-bromobenzoates, prepared by the usual procedure, showed a positive first Cotton effect at 238 nm, which indicated the *R* configuration for these compounds. The position of branching at C-14 of **146** was deduced by the mass spectral fragmentation pattern. The spectrum showed two relatively intense peaks at *m/z* 221.1901 ($C_{15}H_{25}O$, 221.1906) and 193.1587 ($C_{13}H_{21}O$, 193.1593), separated by 28 amu, corresponding to the ions $[HC{\equiv}C\text{-}CH(OH)CH{=}CH(CH_2)_8\text{-}CH(CH3)]^+$ and $[HC{\equiv}C\text{-}CH(OH)CH{=}CH(CH_2)_8]^+$, and absence of a peak at *m/z* 207. The stereochemistry at C-3 was assigned as *R* due to the positive first Cotton effect observed for its *p*-bromobenzoate. The location of the isolated double bond at C-15 of **149** was determined by periodate-permanganate oxidation. This gave a mixture of products, which was analyzed by GC-MS to reveal heptanoic acid as the major product. The stereochemistry of the isolated double bond was assigned as *Z* based on the ^{13}C NMR chemical shift of the allylic methylene carbons which resonated at δ 27.33, a value in good agreement with a *Z* configuration. The compounds were toxic to brine shrimp larvae with LC_{50} values in the range of 0.03–1.2 μg/mL.

Three bioactive acetylenic alcohols (**142**, **144**, and **145**) from *C. vasculum* (Gunasekera and Faircloth 1990) have been synthesized (Kulkarni *et al.* 1993). The synthesis of (+)-(4*E*, 15*E*)-docosa-4, 15-dien-1-yn-3-ol (**142**) has been accomplished using a chemo-enzymatic strategy (Ohtani *et al.* 1996). It was synthesized in enantiomerically pure form by lipase-catalyzed transesterification with Novozym 435 (*Candida antarctica*). A subsequent chemoenzymatic synthesis of both the (4*E*, 15*Z*) and (4*E*, 15*E*) isomers resulted in assignment (Morishita *et al.* 1999) of the (3*S*, 4*E*, 15*Z*) stereochemistry to the natural (+)-docosa-4, 15-dien-1-yn-3-ol (**142**). The absolute configuration of C-3 was assigned as *S* on the basis of the conversion of (+)-docosa-4, 15-dien-1-yn-3-ol (**142**) into (*R*)-(-)-docosan-3-ol, which is also prepared from oct-7-en-3-ol via biotransformation with lipase Novozym. The total synthesis of (*S*, 4*E*)-eicos-4-en-1-yn-3-ol (**141**) employed an enzyme-catalyzed reaction to prepare a key intermediate with high enantiomeric purity (Sharma and Chattopadhyay 1998). The proposed stereochemistry at C-3 for both (*S*, 4*E*)-eicos-4-en-1-yn-3-ol (**141**) and (*S*, 4*E*)-19-methyleicos-4-en-1-yn-3-ol (**144**) has been confirmed by two independent total syntheses (Gung *et al.* 2002). The key step was an enzymatic resolution of a racemic mixture of the compound.

Two new acetylenic alcohols, (3*S*, 4*E*)-3-hydroxyheneicos-4-en-1-yne (**150**) and (5*S*, 3Z)-5-hydroxy-16-methyleicos-3-en-1-yne (**151**), were reported as a part of the discussion of the antitumor activity and absolute stereochemistry of a series of acetylenic alcohols from *C. vasculum* (Hallock *et al.* 1995). Other acetylenes with the same planar structure, but with opposite stereochemistry from known compounds, were isolated. The compounds were defined as enantiomers (3*S* or 5*S*) of the compounds (**141–146**) reported from *C. vasculum* by Aiello *et al.* (1992). It is somewhat surprising to find enantiomers of the same compounds in different collections of the same species, although the collections were made at different depths. Scalemic mixtures are known from nature but are rare (Cardellina *et al.* 1995). The geometries of **142** and **145** were defined as cis, which are different from those reported by Gunasekera and Faircloth (1990), based on the ^{13}C chemical shifts of the allylic methylene carbons. The acetylenic alcohols were evaluated in the NCI's 60-cell-line human tumor screen. A non-small cell lung cancer cell line (H-522) and an ovarian cancer cell line (IGROV-1) showed an approximately ten-fold to one hundred-fold greater sensitivity than the panel average. The ovarian IGROV-1 cell line was used in a preliminary *in vivo* study of racemic (*S*, 4*E*)-eicos-4-en-1-yn-3-ol (**141**), synthesized by a published procedure. Intraperitoneal treatment (once daily doses of 50 mg/kg beginning on the day following tumor cell implantation) of the compound resulted in a small, but significant increase in lifespan (130%).

The same sponge, *C. vasculum,* collected at a different depth (174 m) yielded a longer-chain (C_{43}) polyacetylene, vasculyne (**152**) (Dai *et al.* 1996). Vasculyne yielded mean average GI_{50} (50% growth inhibitory concentration), TGI (total growth inhibitory concentaration), and LC_{50} values of 0.2, 0.7, and 6.7 μg/mL, respectively, and modest differential cytotoxicity toward the melanoma and colon tumor cell line subpanels when tested in the NCI's 60 cell line panel. Structurally, vasculyne (**152**) is closely related to the C_{30} compound duryne (**140**).

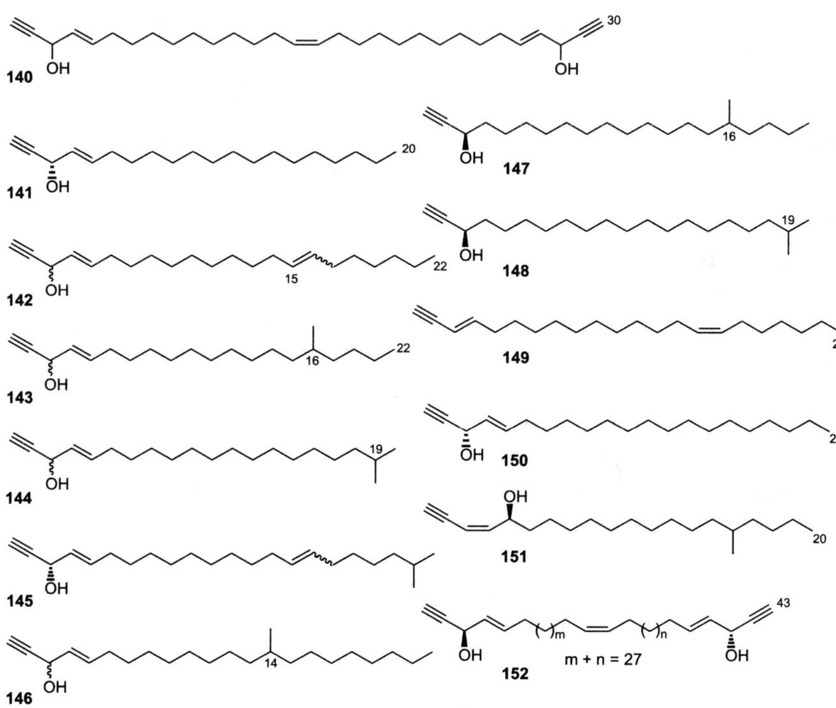

Siphonochalina

Six C_{22} acetylenes, Δ^{15}-docos-1-yne (**153**), siphonochalyne (**155**), octahydrosiphonochalyne (**154**), dihydrosiphonochalyne (**156**), siphonochalynol (**157**), and dehydrosiphonochalynol (**158**) have been isolated from a species of the genus *Siphonchalina* from the Red Sea (Rotem and Kashman 1979). Four of the acetylenes have been fully characterized but the structures of two additional acetylenes, octahydrosiphonochaline (**154**) and dihydrosiphonochaline (**156**), remain to be fully elucidated. All compounds, except for one (**153**), possess an enyne terminus on one side of the molecule and either a terminal acetylene or a propargyl alcohol on the other end. The location of the double bond in **153** was determined by MS analysis of its hydroxy dimethylamino derivative which was prepared by mono peroxidation of the double bond, followed by reaction with Me_2NH. Assignment of a cis configuration to the isolated double bond was based on the absence of IR absorption at 970 cm^{-1}. The highly characteristic UV spectrum of **155** with a very sharp vibrational fine structure (λ_{max} 310, 284, 268, 253 nm) confirmed unequivocally the triyne unit (Bohlman *et al.* 1973). The location of the triyne unit was determined by LIS (lanthanide induced shift) experiment on the monoepoxide of **155** which prepared by epoxidation of the double bond with *m*-chlorobenzoic acid. Interesting is the finding of the same chain length (C_{22}) in the compounds as in the acetylenes of *Reniera fulva* (Cimino and Stefano 1977). Accordingly, a C_{23} fatty acid precursor could also be suggested for the compounds, however, the fact that compounds **157** and **158** bearing the propargyl terminus (first stage in a possible oxidative degradation) are C_{22} and not C_{23} casts doubt on such a biogenesis in this case.

A very similar C_{23} polyacetylene, siphonodiol (**159**), has been obtained from *Siphonochalina truncata* (Tada and Yasuda 1984). The location of the internal double bond was determined by an LIS experiment. Siphonodiol showed weak activity against *Trichophyton asteroids* (MIC 25.0 µg/mL) and medium activity against *Staphylococcus aureus* (MIC 12.5 µg/mL) and *Streptococcus pyogenes* C-203 (MIC 6.2 µg/mL).

Siphonodiol (**159**) and two related metabolites, dihydrosiphonodiol (**160**) and tetrahydrosiphonodiol (**161**), were isolated from *S. truncata* collected in the Gulf of Suruga (Fusetani *et al.* 1987). They strongly inhibited gastric H^+/K^+-transporting ATPase with IC_{50} of 1.0-10^{-5} M. The absolute configuration at C-2 of siphonodiol (**159**) was determined by the exciton chirality method. The *p*-dimethylaminobenzoate prepared by reacting the diol with *p*-dimethylbenzoic acid in the presence of DCC gave a negative first and a positive second Cotton effects ($\Delta\varepsilon_{324}$–25.3°, $\Delta\varepsilon_{303}$+ 24.2°), which indicated the 2*R* configuration.

153 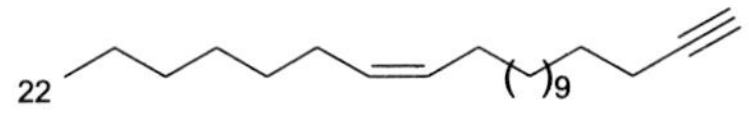

155 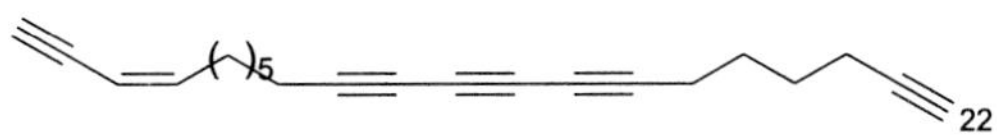

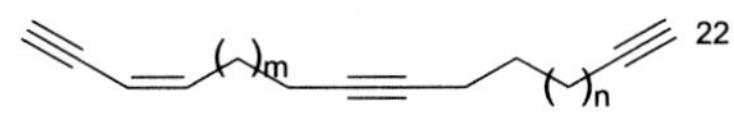

154 m + n = 11

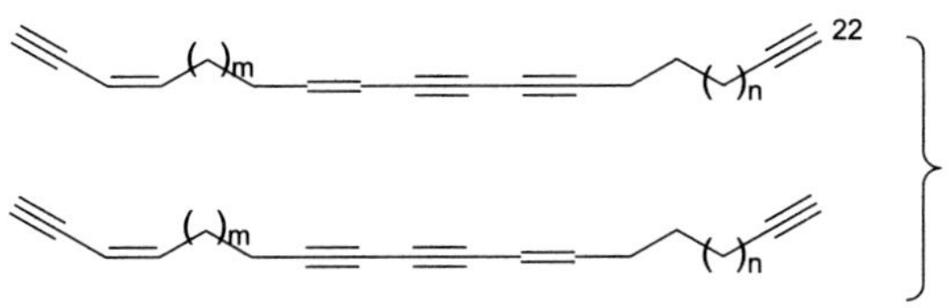

156 m + n = 7

157 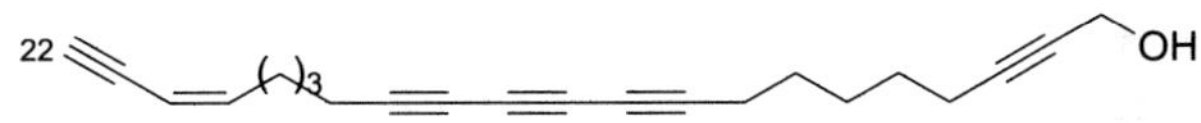

158 

159 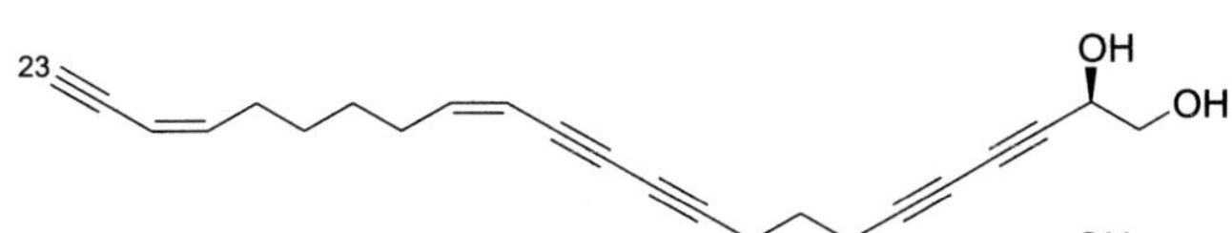

160 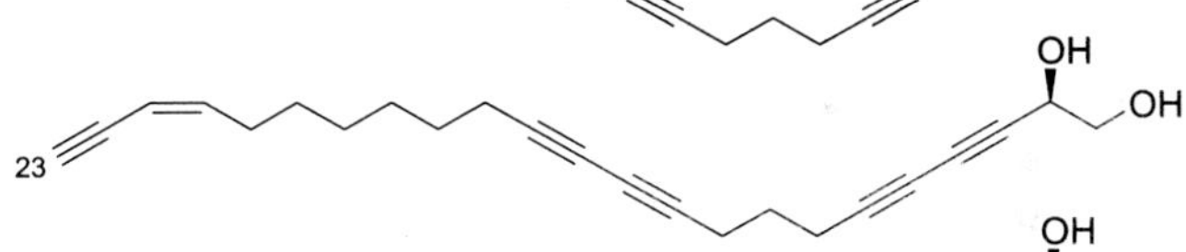

161 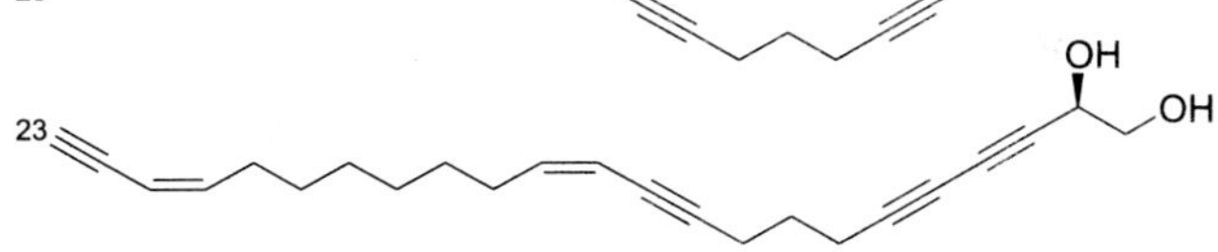

Callyspongia

Callydiyne (**162**) is a symmetrical bis-enyne from *Callyspongia flammea* from Papua New Guinea that was identified by spectral methods (Miao and Andersen 1991).

Callyspongins A and B (**163** and **164**) were isolated from a Japanese specimen of *C. truncata* and they were the first sulfates of polyacetylene hydrocarbons (Uno *et al.* 1996). Callyspongin A (**163**) was hydrolyzed with *Helix pomatia* sulfatase (50 °C, 12h) to provide the diol derivative. Its ^{1}H and ^{13}C NMR data were identical to those of siphonodiol (**159**), which is an antifungal acetylene from the marine sponge *Siphonochalina* *truncata*. Callyspongins A and B inhibited fertilization of starfish (*Asterias amurensis*) gametes. Minimum inhibitory concentrations of callyspongins A and B on sperm fertilization of starfish gametes were 6.3 and 50 μM, respectively. While the minimum inhibitory concentrations of callyspongins A and B on sperm motility were 50 and 100 μM, respectively. The diol type (siphonodiol, **159**) exerted non-selective toxicity against gametes and embryo.

Callyspongia truncata from Japan contained the acetylenic hydrocarbons callytetrayne (**165**) and callypentayne (**166**), together with five triols, callytriols A–E (**167–171**) (Tsukamoto *et al.* 1997). Siphonodiol (**159**) and its mono- and disulfates, callyspongins A and B (**163** and **164**) were also isolated. All seven acetylenes induced metamorphosis in larvae of the ascidian *Halocynthia roretzi* (ED_{100} 0.13–1.3 μg/mL) and the triols also showed antifouling activity against the barnacle *Balanus amphitrite* (ED_{50} 0.24–4.5 μg/mL).

The same acetylenic hydrocarbons, callytetrayne (**165**) and callypentayne (**166**), together with a third hydrocarbon callyberyne (**172**) and siphonodiol (**159**) were isolated from a *Callyspongia* sp. from Japan (Umeyama *et al.* 1997). Biogenetically, these compounds (**165** and **166**) are considered most likely to be produced by decarboxylation of a C_{22} fatty acid precursor in the sponges. Reports of polyacetylenes with no oxygen functional groups from marine organisms are relatively rare.

Callyspongyns A and B (**173** and **174**, different from callyspongins A and B) are two poorly characterized metabolites of a *Callyspongia* sp. from the southern Australian sea (Rooney and Capon 1998). Although callyspongynes A and B incorporate terminal acetylenic functionalities common to other polyacetylenic lipids from marine sponges, the isolated central acetylenic functionality is uncommon.

Six polyacetylenes, aikupikanynes A–F (**175–180**) were obtained together with octahydrosiphonochalyne (**154**), from a *Callyspongia* sp. from the Red Sea (Youssef *et al.* 2000). Aikupikanyne F (**180**) is the first example of a polyacetylenic α-hydroxy carboxylic acid.

A Japanese specimen of *Callyspongia truncata* yielded the *α*-glucosidase inhibitor callyspongynic acid (**181**) (Nakao *et al.* 2002). Callyspongynic acid inhibited *α*-glucosidase with an IC_{50} value of 0.25 μg/mL, but was inactive against *β*-glucosidase, *β*-galactosidase, thrombin, and trypsin at 100 μg/mL. Since corticatic acid A (**39**) and petrosynol (**36**) were also inhibitory against α-glucosidase with IC_{50} values of 0.16 and 4.08 μg/mL, respectively, but the polyacetylenic hydrocarbon callytetrayne (**165**) and methyl callyspongynate were inactive even at 100 μg/mL, perhaps the carboxylic acid and the allylic alcohol linked to an acetylene are

162
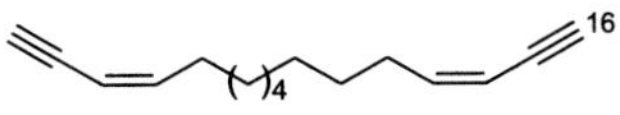

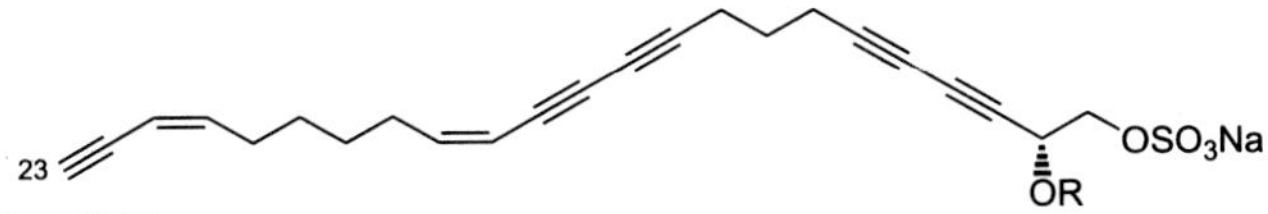

163 R = SO_3Na
164 R = H

165
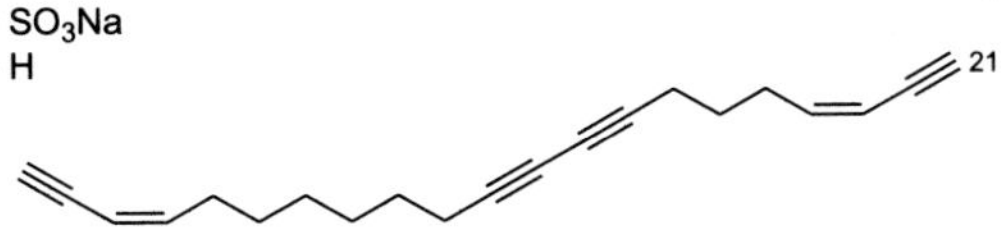

166
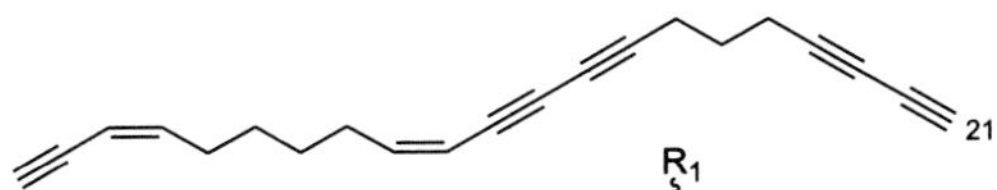

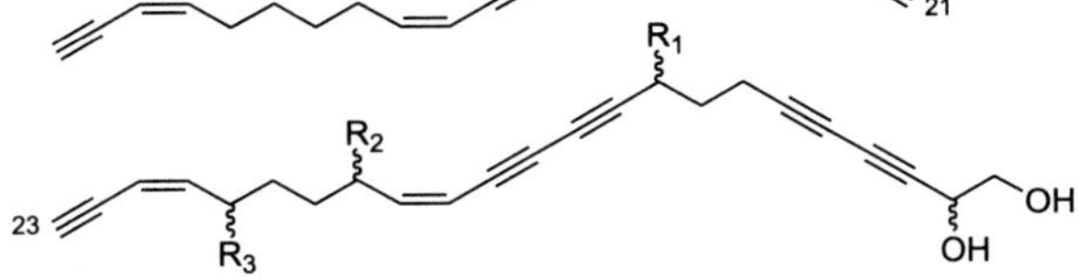

167 R_1 = OH, R_2 = R_3 = H
168 R_2 = OH, R_1 = R_3 = H
169 R_3 = OH, R_1 = R_2 = H

170

171
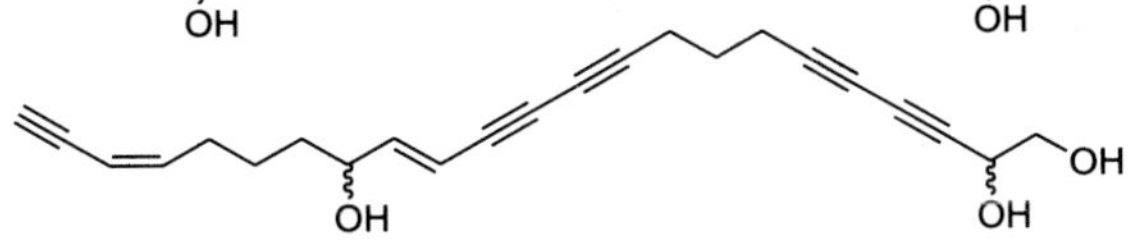

172
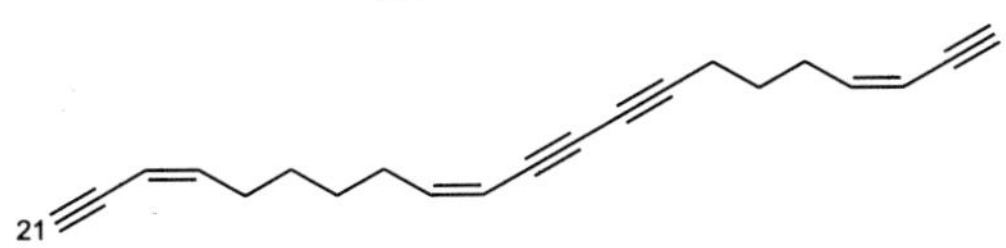

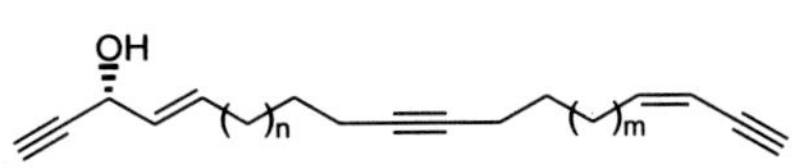

173 n + m = 18
174 n + m = 20

important for activity. α-Glucosidase inhibitors may delay the digestion of oligosaccharides and disaccharides to monosaccharides on the small intestinal brush-border, and reduce the rate of glucose absorption. Therefore, inhibition of α-glucosidase is considered important in managing non-insulin dependent diabetes (Bischo, 1994). The C_{32} polyacetylenes were also reported from marine sponges of the genera *Adocia, Pellina, Petrosia,* and *Xestospongia,* which all belong to the order Haplosclerida. Interestingly, a C_{32} polyacetylene (**245**) was also isolated from a *Theonella* sponge (order Lithistida).

Callysponginol sulfate A (**182**) was isolated along with halistanol sulfates from *Callyspongia truncata* as an inhibitor membrane type 1 matrix metalloproteinase (MT1-MMP) (Fujita *et al.* 2003). Callysponginol sulfate A (**182**) is the first example of an acetylenic acid containing a sulfate group from marine organisms. Callysponginol sulfate A (**182**) inhibited MT1-MMP with an IC_{50} value of 15.0 μg/mL, while halistanol sulfate showed an IC_{50} of 19.0 μg/mL. The desulfated form of **182** did not show any inhibitory activity. Considering this result as well as similar activity of structurally unrelated sulfated compounds, the MT1-MMP inhibition probably results from the presence of the sulfate group. Membrane type 1 matrix metalloproteinase (MT1-MMP) is one of the key enzymes involved in tumor growth, migration, angiogenesis, invasion, and metastasis. Therefore, MT1-MMP inhibitors are considered as potential antimetastatic and antiangiogenic agents.

Three cytotoxic C_{22} polyacetylenic alcohols, callyspongenols A–C (**183–185**), together with dehydrosiphonochalynol (**158**), were isolated from a Red Sea sponge, *Callyspongia* sp. (Youssef *et al.* 2003a). The compounds were tested for cytotoxicity against P388 and HeLa cells and significant activities were observed (Table 12.5).

Table 12.5 Cytotoxicity of callyspongenols A–C (**183–185**) and dehydrosiphonochalynol (**158**) (IC_{50}: μg/mL).

	183	*184*	*185*	*158*	adriamycin
P388	2.2	10.0	2.2	2.2	0.04
HeLa	4.5	10.0	3.9	5.1	0.066

Callyspongamide A (**186**), a new cytotoxic polyacetylenic amide, has been isolated from *Callyspongia fistularis* collected in the Red Sea (Youssef *et al.* 2003b). Callyspongamide A (**186**) represents a novel polyacetylenic amide within the family Callyspongidae. It showed structural similarity with hermitamide A, which was isolated recently from a marine cyanobacterium, *Lyngbya majuscula* (Tan *et al.* 2000). Such structural similarity supported the microbial origin of callyspongamide A. Callyspongamide A (**186**) showed moderate cytotoxicity against HeLa cells with an IC_{50} value of 4.1 μg/mL.

A new C_{20} polyacetylene (**187**) was isolated from the Indonesian sponge *Callyspongia pseudoreticulata* (Braekman *et al.* 2003). NMR spectra of the (*R*)- and (*S*)-MTPA esters revealed the presence of traces of the diastereomer (ca. 10%).

Haliclona

A *Haliclona* sp. from Palau yielded three relatively simple acetylenic metabolites (**188–190**) whose structural assignments were complicated by

175

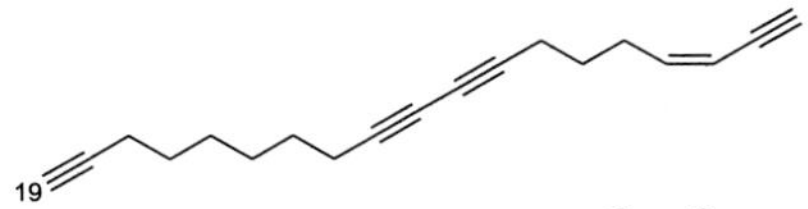

176

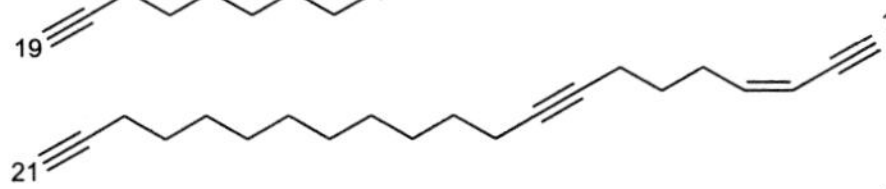

177

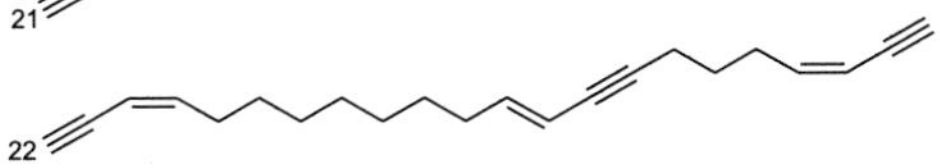

178

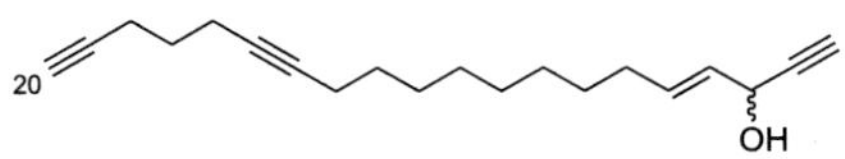

179

180

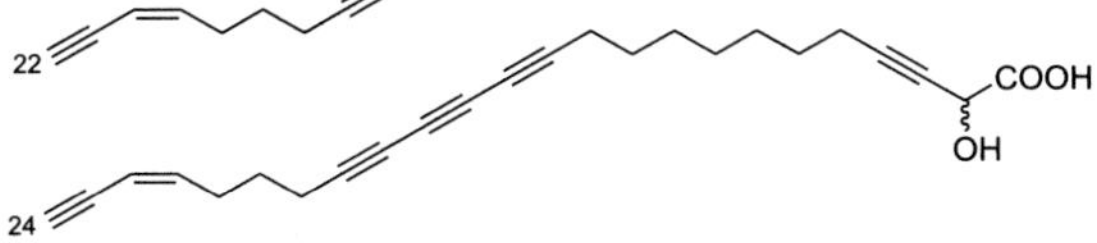

181

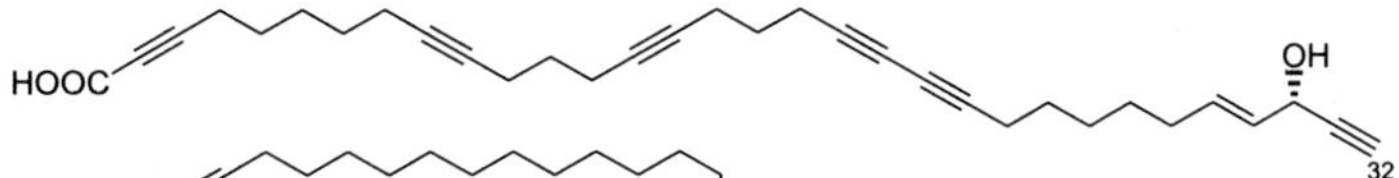

182

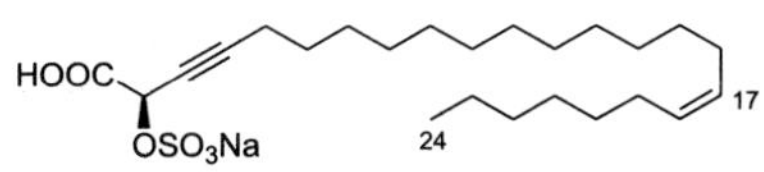

183

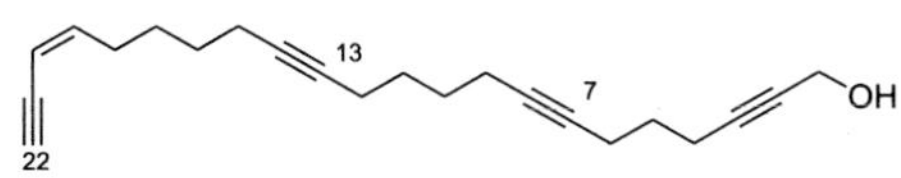

184

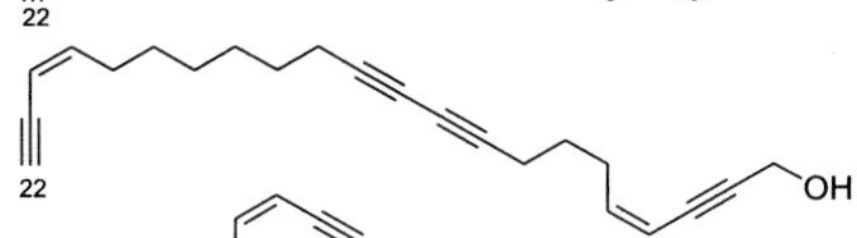

185

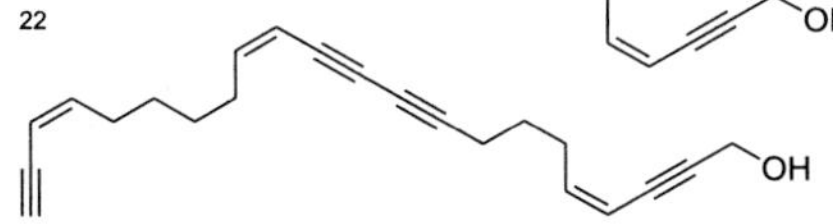

186

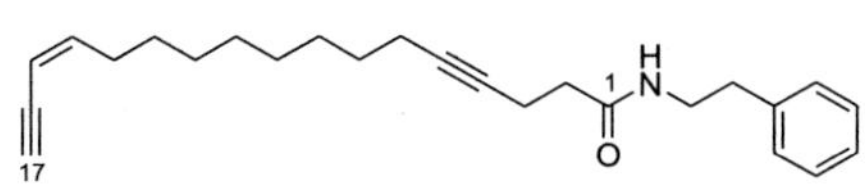

187

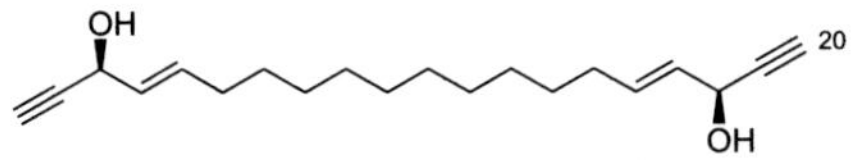

having a methyl substituent on the alkyl chain (Williams and Faulkner 1996). The location of the methyl branching of **188** was tentatively assigned on the basis of the fragmentation pattern of the 11-methylhexacosanol which was derived by hydrogenation over Adam's catalyst.

A more complex series of highly oxygenated C_{47} polyacetylenic acids, osirisynes A–F (**191–196**), were isolated as cytotoxins from a specimen of *H. osiris* collected from Guam (Shin *et al.* 1998). These compounds possess a diacetylenic carbinol and an α-acetylenic carboxyl group as common structural features that made osirisynes structurally comparable to petrosolic acid (**38**) and nepheliosyne A (**112**), previously isolated from the sponges *Petrosia* sp. and *Xestospongia* sp., respectively. These compounds exhibited moderate cytotoxicity and inhibitory activities against Na^+/K^+-ATPase and reverse transcriptase. Osirisynes A–F (**191–196**) exhibited moderate cytotoxicity against a human leukemia cell line (K562) with LC_{50} values of 25, 48, 52, 25, 20, and 22 μM, respectively. Osirisynes B (**193**), E (**195**), and F (**196**) exhibited inhibitory activities against Na^+/K^+-ATPase and reverse transcriptase at a concentration of 1 μg/10 mL.

An Indonesian *Haliclona* sp. contains lembehyne A (**197**), which induced neuritogenesis in rat pheochromocytoma (PC12) and neuroblastoma (Neuro 2A) cells (Aoki *et al.* 2000). This compound induced neurite outgrowth of PC12 cells and Neuro 2A cells at 2 and 0.1 μg/mL, respectively. Lembehyne A (**197**) also induced neuritogenesis in a mouse at 0.1 μg/mL. Treatment with cytochalasin B, an inhibitor of actin polymerization, or cycloheximide, an inhibitor of protein synthesis in eukaryotes, blocked neurite outgrowth induced by lembehyne A. This finding indicates that the morphological change induced by lembehyne A is dependent upon actin polymerization and *de novo* protein synthesis. Lembehyne A also induced neuronal cell differentiation in a Neuro 2A cell line (Aoki *et al.* 2001). The treatment predominantly resulted in a morphological change with bipolar neurites. The acetylcholinesterase activity of Neuro 2A was also increased by treatment with lembehyne A. Furthermore, the cell cycle of Neuro 2A cells was found to be specifically blocked at the G1 phase by lembehyne A. The structure-activity relationship study using the lembehyne A analogues revealed the importance of the terminal 1-yn-3-ol and unsaturated long-chain alkyl moieties for the neuronal differentiation activity. The lembehynes B and C (**198** and **199**) were isolated from the same Indonesian sponge, and were found to possess neuritogenic activity against neuroblastoma cells (Aoki *et al.* 2002). The configuration of the hydroxyl group at C-3, and the length of the alkyl chain were important for neuritogenic activity.

The polyacetylene carboxylic acid haliclonyne (**200**) was obtained from a *Haliclona* sp. from the Red Sea (Chill *et al.* 2000). Partial structures were connected by TOCSY and mass analysis. Similarities were found between haliclonyne (**200**) and osirisyne F (**196**).

Two new C_{33} polyacetylenic compounds, halicynones A and B (**201** and **202**) were isolated from the Micronesian sponge *Haliclona* sp. along with known analogs, such as triangulynes A (**217**) and E (**221**), pellynols A–D (**227–230**), and pellynol I (**232**) (Zhou and Molinski 2003). Halicynones A and B (**201** and **202**) are related to the highly cytotoxic polyacetylenes, pellynols (**227–230**) (Fu *et al.* 1997) and triangulynes (**217–224**) (Dai *et al.*

1996) from *Pellina triangulata*. The known compound pellynol A (**227**), possessing a 1-yn-3-ol terminus, exhibited strong antitumor activity against HCT-116 (human colon tumor cell) with an IC_{50} value of 0.026 μg/mL. However, the corresponding 1-yn-3-one, halicynone A (**201**), was inactive, which suggests an important role of the terminal 1-yn-3-ol functional group in the cytotoxic activity. Cultured HCT-116 cells were very sensitive to pellynols A (**227**) (IC_{50} 0.026 μg/mL), B (**228**) (0.12 μg/mL), C (**229**) (0.127 μg/mL), D (**230**) (0.103 μg/mL), and I (**232**) (<0.008 μg/mL), however, they are unaffected by halicynones A and B (**201** and **202**) (IC_{50} >78 μg/mL). The importance of the 1-yn-3-ol has been also demonstrated in other bioactivity relationships.

Three new chlorinated acetylenes (**hc–3**, **hc–8**, and **hc–9**) were isolated from the sponge *Haliclona lunisimilis* from California (Jesus and Faulkner 2003). Six co-occurring chlorinated acetylenes (**hc-1**, **2**, and **hc-4–hc-7**) were previously isolated from the dorid nudibranch *Diaulua sandiegenesis*. Brominated acetylenes are frequently isolated from sponges, but isolation of chlorinated acetylenes is unusual, and these three compounds comprise the only example.

188
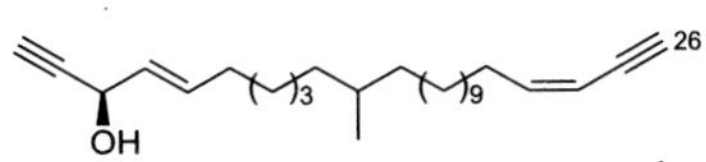

189
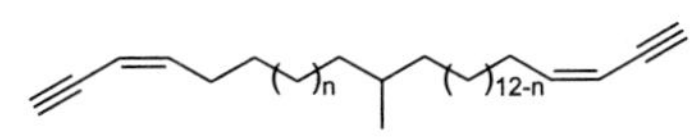

190
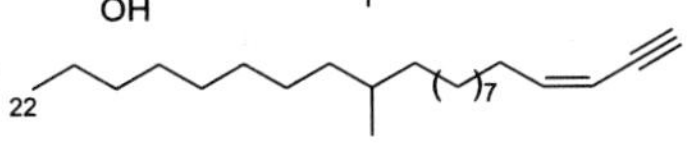

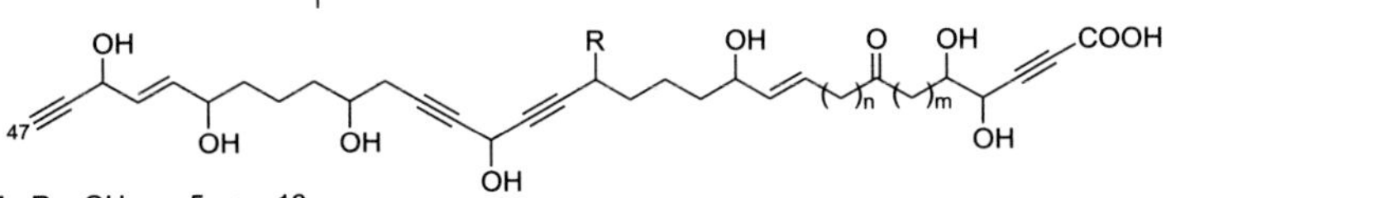

191 R = OH, n = 5, m = 13
192 R = H, n = 5, m = 13
193 R = OH, n = 3, m = 15
194 R = H, n = 3. m = 15

195
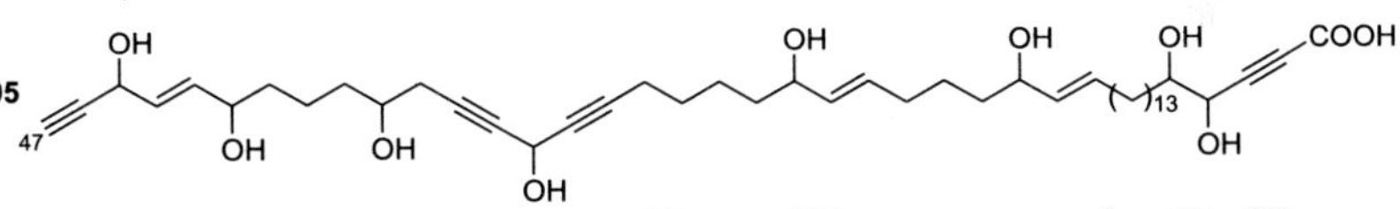

196
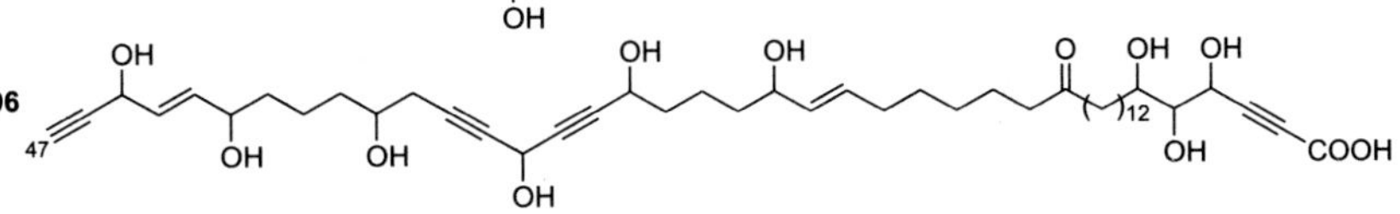

197
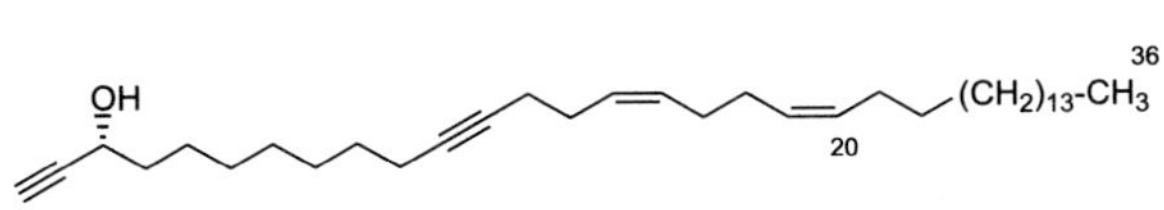

198
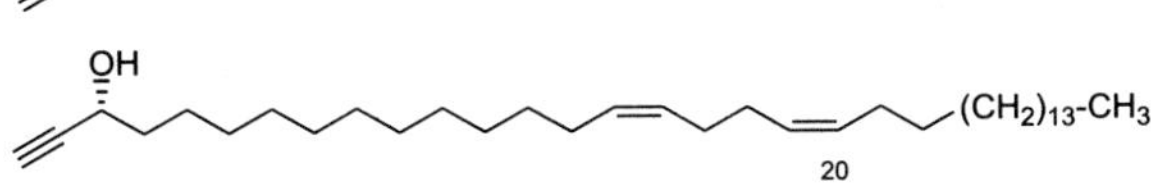

199
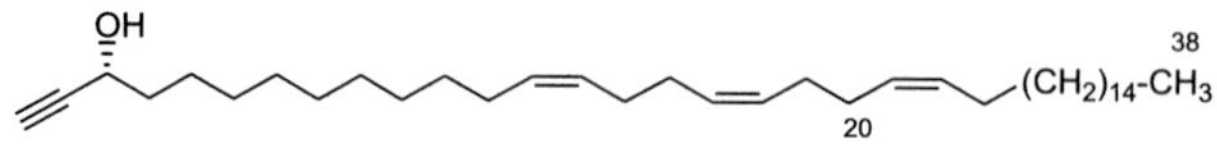

200
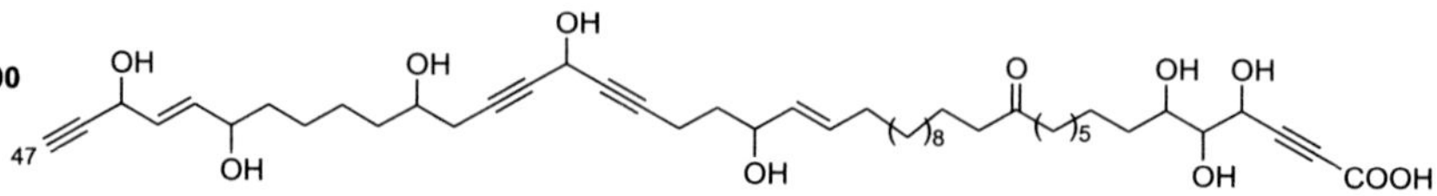

201
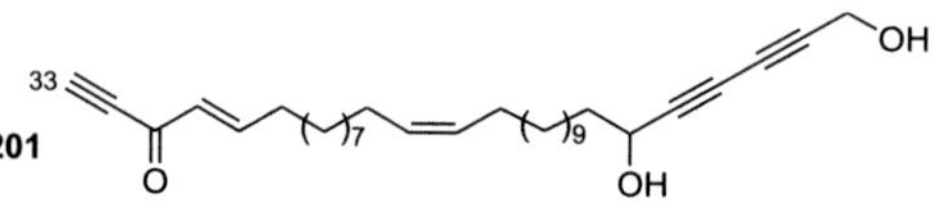

202
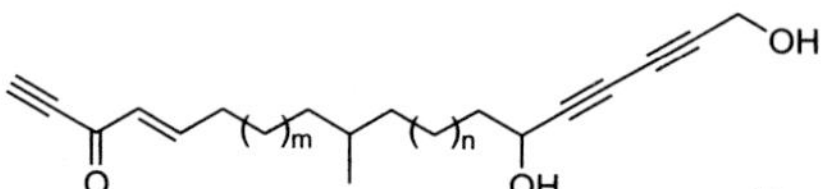

m + n = 15

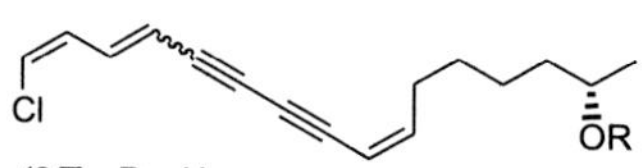

hc-1 (3*E*), R = H
hc-2 (3*Z*), R = H
hc-3 (3*E*), R = Ac

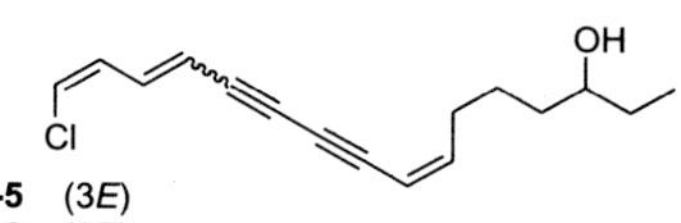

hc-5 (3*E*)
hc-6 (3*Z*)

hc-4
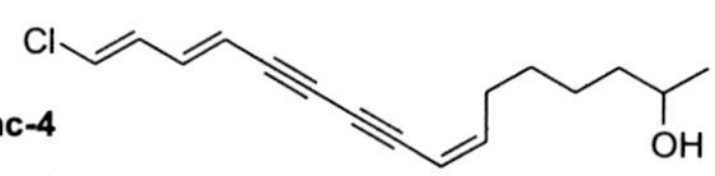

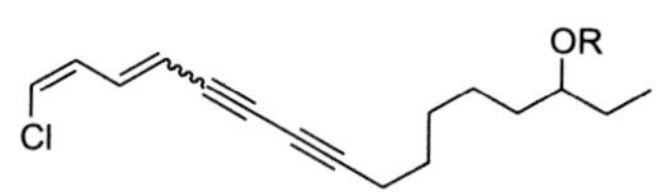

hc-7 (3*E*), R = H
hc-8 (3*Z*), R = H
hc-9 (3*E*), R = Ac

***Adocia* [→*Haliclona* (*Haliclona*)]**

The genus *Adocia* was changed to the subgenus *Haliclona* of the genus *Haliclona* in a recent classification (De Weerdt 2002). Adociacetylenes A–D (**203**–**206**) are cytotoxic constituents of an Okinawan *Adocia* sp. that exhibit activity in an *in vitro* endothelial cell-neutrophil leukocyte adhesion assay (Kobayashi *et al.* 1996). Adociacetylenes A–D were isolated along with petrosynol (**36**). Adociacetylenes A and B were previously isolated from an unidentified sponge of the Niphatidae, and given the trivial names 3-ketopetrosynol and petrofuran, respectively (Higa *et al.* 1994). It was noticed that petrosynol in $CDCl_3$ was gradually converted (presumably via air-oxidation) to adociacetylenes A and B on standing at room temperature for a long time (ca. 40 days, Fig. 12.2). Petrosynol (**36**) and adociacetylene A (**203**) showed weak antibiotic activity [10 and 11 mm diameter growth inhibitions for *E. coli* and *B. subtilis* at 50 μg/disk (i.d. = 8 mm)]. Adociacetylene A showed moderate cytotoxicity against KB (human epidermoid carcinoma) cells (IC_{50} 0.8 μg/mL). Furthermore, it was found that adociacetylenes A, B, and D inhibited neutrophil leukocyte

203 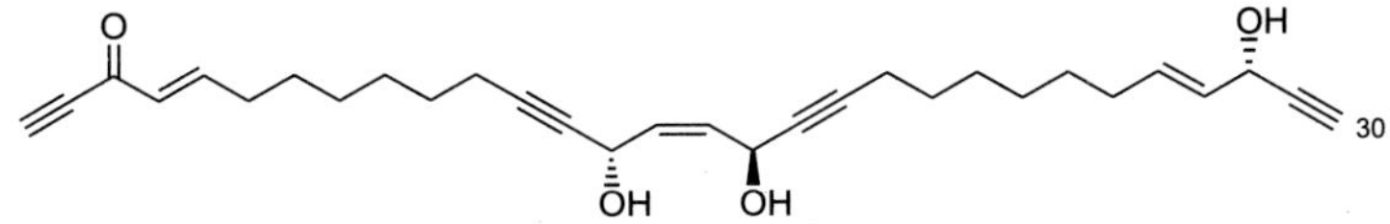

204 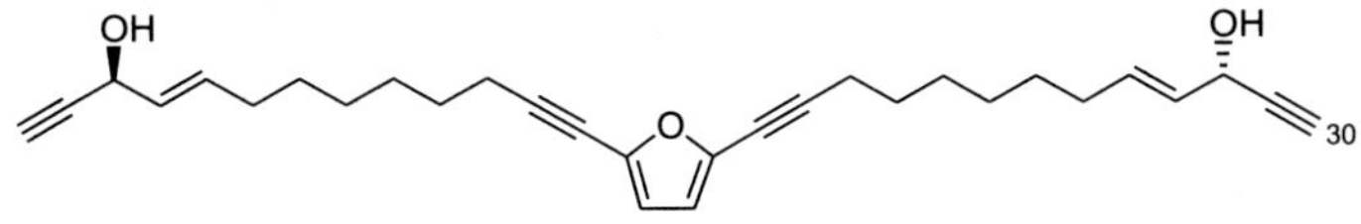

205 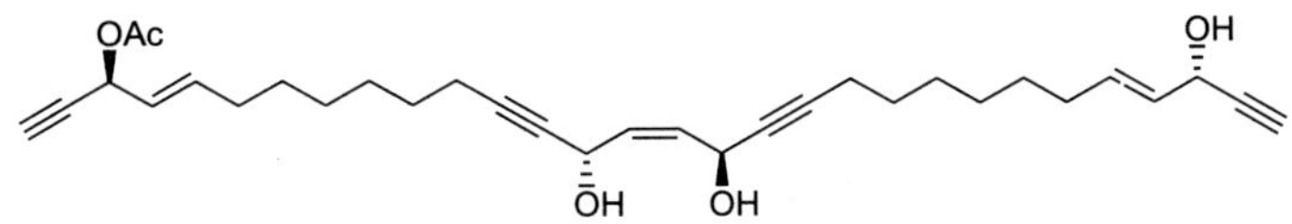

206 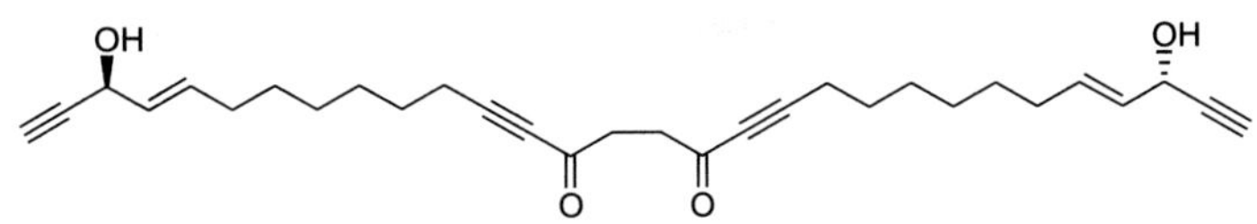

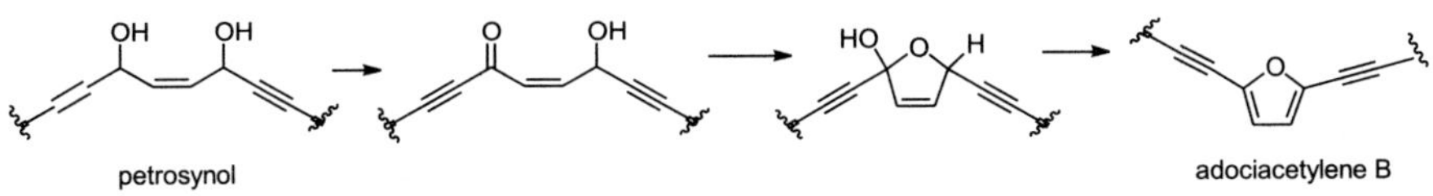

Fig. 12.2 Chemical conversion of petrosynol into adociacetylene B.

adhesion to tumor necrosis factor-α (5JRU/mL)-stimulated endothelial cells at 1 μg/mL concentration.

Petrofuran (adociacetylene B, **204**) has been synthesized using an enantioselective reduction to produce the two chiral alcohols (Garcia *et al.* 1999). Highly enantioenriched 1-yn-4-en-3-ol moiety has been efficiently obtained by reduction of the parent 1-trimethyl-silyl-4-alken-1-yn-3-one with Alpine-Borane or with $BH_3 . SMe_2$ in the presence of chiral oxazaborolidines, followed by desilylation of the resulting alcohol. This strategy has been applied to the first stereoselective synthesis of petrofuran (**204**). The absolute stereochemistry of (-)-adociacetylene B was confirmed by a synthesis of both (+)- and (-)-isomers employing enzymatic resolution (Gung *et al.* 2001). The synthesis of (-)-adociacetylene B was completed in six steps in a yield of 9.3%, while the (+)-enantiomer was synthesized in seven steps and 7.5%. An enzymatic resolution of racemate using the lipase from *Pseudomonas* sp. was employed to obtain (+)- and (-)-adociacetylene B.

Reniera [→Haliclona (Reniera)]

The genus *Reniera* was changed to the subgenus *Reniera* of the genus *Haliclona* in the recent classification (De Weerdt 2002). From a sponge *Reniera fulva* collected in the bay of Naples, a series of C_{22}–C_{23}

polyacetylenic compounds, debromorenierin-1 (**207**), renierin-1 (**208**), 18-dihydrorenierin-1 (**209**), renierin-2 (**210**), and 18-hydroxyrenierin-2 (**211**), were isolated (Cimino and Stefano 1977). They were the first isolated acetylenes from a sponge. They were defined as unsaturated and/or brominated diacetylenes, two of which also incorporated a vinyl bromide moiety. Assignment of the cis configuration of the isolated double bond was based on the absence of IR absorption at 970 cm^{-1}. The location of the double bond in the chain was established by ozonolysis of debromorenierin-1 (**207**) in methanol. The geometry of the double bond in renierin-1 (**208**) was suggested based on the chemical shift of the olefinic proton (δ 5.8). The chirality of 18-dihydrorenierin-1 (**209**) was assigned as *R* on the basis of a gas-chromatographic modification of Horeau's method. It is interesting that C_{22} (**207–209**) and C_{23} (**210** and **211**) polyacetylenes coexist in the same specimen. Three polyacetylenes (**207, 210**, and **211**) from *R. fulva* have been synthesized (Dorta de Marquez and Thaller 1985). With the exception of electron-impact mass spectra, the spectral data were compatible with those of the natural compounds, but the optical rotations of one compound (**211**), although of the same sign, do not agree in magnitude.

Fulvinol (**212**) is a slightly more complex cytotoxin from a Spanish specimen of *R. fulva* (Ortega *et al.* 1996). It was isolated along with 18-hydroxyrenierin-2 (**211**). Fulvinol showed *in vitro* cytotoxicity against P-388 mouse lymphoma, A-549 human carcinoma, HT-29 human colon carcinoma, and MEL-28 human melanoma (EC_{50} = 1 μg/mL). These EC_{50} values are in the range of other acetylenic compounds isolated from marine sources. The presence of polyacetylenes in *R. fulva* represents the only example of this kind of compound, in the genus *Reniera*. Phylogenetic studies using ribosomal RNA analyses have suggested that *R. fulva* is more closely related to the *Petrosia* genus than it is to *Reniera mucosa*. Interestingly, fulvinol (**212**) closely resembled the long-chain acetylenes isolated from some *Petrosia* sponges.

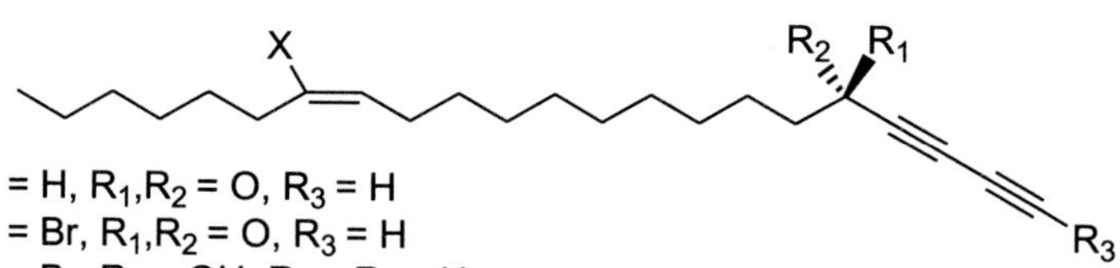

207 X = H, R_1,R_2 = O, R_3 = H
208 X = Br, R_1,R_2 = O, R_3 = H
209 X = Br, R_1 = OH, R_2 = R_3 = H
210 X = H, R_1 = R_2 = H, R_3 = CH_2OH
211 X = H, R_1 = OH, R_2 = H, R_3 = CH_2OH

212

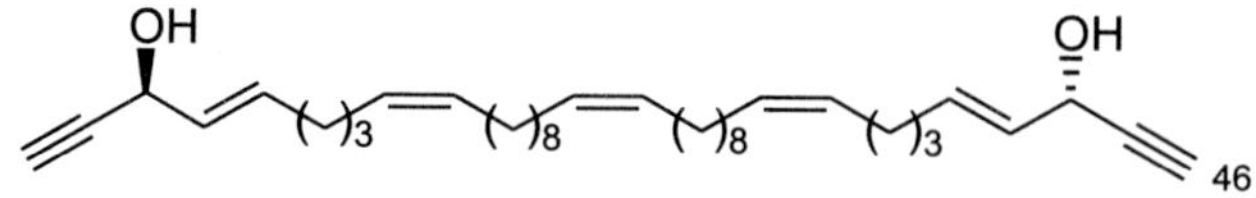

Cladocroce

Cladocroic acid (**213**) is an unusual cyclopropane-containing acetylenic acid from a deep-water New Caledonian sponge, *Cladocroce incurvata* (D'Auria *et al.* 1993). The stereochemistry of the cyclopropane ring was derived after the synthesis of cis- and trans-2, 3-methanohexanoic acid models and comparison of nmr spectral data. Generally downfield-shifted ^{13}C NMR signals of the cyclopropane moiety were observed for the trans isomer.

213

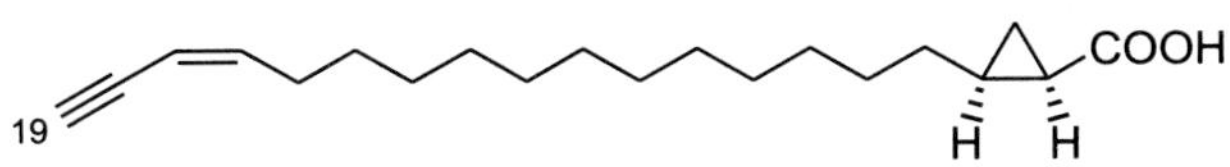

Oceanapia

An Indonesian specimen of *Oceanapia* sp. contained two new brominated acetylenic acids (**214** and **215**) along with a known one (**93**) (Ichiba *et al.* 1993). The mixture of three compounds showed only weak cytotoxicity against KB cells (++ at 10 µg/mL). Compound **215** showed mild antimicrobial activity against Gram-positive bacteria (data not shown).

Another antimicrobial constituent of a Japanese *Oceanapia* sp. was identified as the bis-acetylene (**216**) (Matsunaga *et al.* 2000). Compound **216** exhibited a rather selective antimicrobial activity. It was moderately active against four mutants of *S. cerevisiae* and *Candida albicans*, but was inactive against *Penicillium chrysogenum* and *Mortierella ramanniana*. The compound was also inhibitory against both Gram-positive and Gram-negative bacteria (Table 12.6). Previously reported C_{16}, C_{18}, and C_{20} acetylenic acids from marine sponges of the genera *Xestospongia* and *Oceanapia* possess one or two bromine atoms attached to sp^2 carbons. Therefore compound **216** is the first example of a short-chain acetylenic acid without a bromine atom.

214

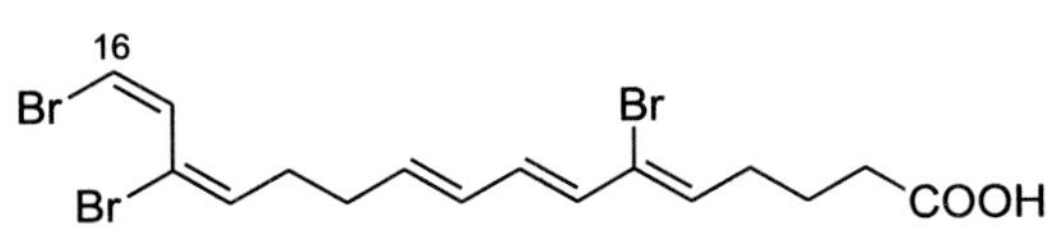

215

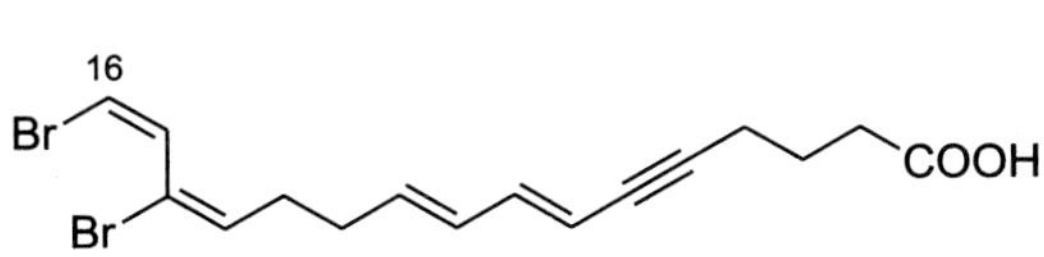

216

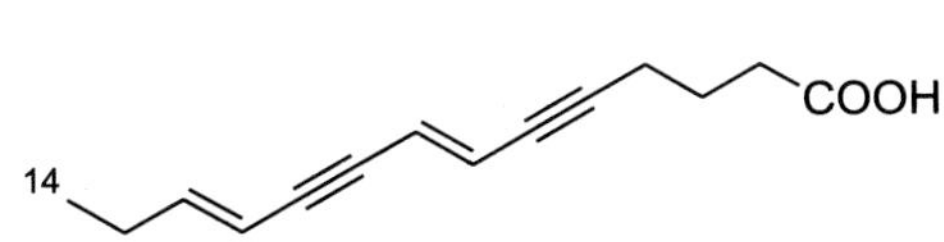

Table 12.6 Antimicrobial activities of **216.**

Species/strain	*50 μg*[a]	*100 μg*
Yeast		
S. cerevisiae	n.a.	n.a.
GT160-45C[c]	n.a.	6.5[b]
cdc5[c]	7.0	7.5
act1-1[c]	9.0	10.0
YAT2296[c]	6.5	7.0
Fungi		
P. chrysogenum	n.a.	n.a.
M. ramanniana	n.a.	n.a.
C. albicans	n.a.	8.5
Bacteria		
E. coli	8.5	12.0
P. aeruginosa	8.5	13.0
B. subtilis	n.a.	11.0
S. aureus	9.5	13.5

[a] Amount loaded on a 6 mm ϕ disk. [b] Diameter of inhibitory zone in mm. [c] Mutant of *S. cerevisiae*.

Pellina (=Halichondria)

The genus *Pellina* was changed to *Halichondria* in a recent classification publication (De Weerdt 2002). *Pellina triangulata* from Truk in Micronesia contained C_{31}–C_{41} acetylenes, triangulynes A–H (**217–224**) and triangulynic acid (**225**), which were a series of cytotoxic polyacetylenes that were most active against leukemia and colon tumor cell lines (Dai *et al.* 1996). The absolute configurations of the compounds were determined by application of Mosher's method. Triangulynic acid (**225**) showed structural similarities to the corticatic acids (**39–41**), reported from the sponge of the genus *Petrosia*. Triangulynes A–H (**217–224**) demonstrated similar *in vitro* cytotoxicity profiles and generally comparable overall potency against the NCI human tumor cell line panels (data not shown). In general, leukemia, colon, and melanoma tumor cell lines showed greater sensitivity to triangulynes A–H (**217–224**). It is interesting to note that triangulynic acid (**225**) was less potent than triangulynes A–H (**217–224**) and did not exhibit differential cytotoxicity. In addition, the differential cytotoxicity patterns produced by the triangulynes did not resemble those produced by the simple C_{20}–C_{23} enynols from *Cribrochalina vasculum*, but they did closely match that exhibited by vasculyne (**152**), a C_{43} bis-enynol from a different collection of *C. vasculum*. Repetitive testing of triangulyne A (**217**), as representative of the series, yielded mean panel GI_{50}, TGI, and LC_{50} concentrations of 0.5, 2.0, and 12 μM, respectively.

Pellynic acid (**226**), pellynols A–D (**227–230**), and pellynone (**231**) were obtained from *P. triangulata* from Chuuk (Truk) Atoll. The position of the ketone group in pellynone (**231**) was tentatively assigned (Fu *et al.* 1997). Collisional activation decomposition (CAD) tandem mass spectrometry of the lithium adducts of the acetylenic compounds was applied to ascertain if charge remote fragmentation would yield definitive information regarding the site of internal unsaturation in these compounds. Pellynic acid (**226**) inhibited IMPDH (inosine monophosphate dehydrogenase), which is one of anticancer target, *in vitro* with an IC_{50} of 1.03 μM.

An additional polyacetylene, pellynol I (**232**) was isolated from a South African collection of *Pellina* sp., along with pellynols A–D and F (Rashid *et al.* 2000). The six pellynols were cytotoxic to human tumor cell lines. The IC_{50} values (μg/mL) for pellynols against the LOX (melanoma) and OVCAR-3 (ovarian) human tumor cell lines were: pellynol A, LOX (0.39), OVCAR-3 (2.23); pellynol B, LOX (0.15), OVCAR-3 (1.54); pellynol C, LOX (0.14), OVCAR-3 (1.0); pellynol D, LOX (0.12), OVCAR-3 (1.75); pellynol F, LOX (0.08), OVCAR-3 (1.7); pellynol I, LOX (0.41), OVCAR-3 (2.0).

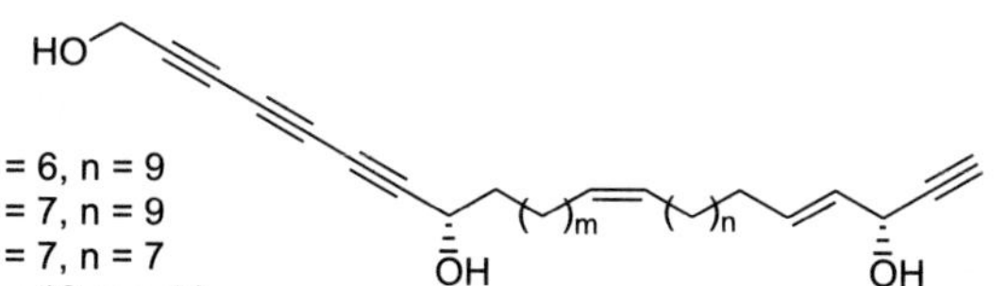

217 m = 6, n = 9
218 m = 7, n = 9
219 m = 7, n = 7
220 m = 13, n = 11

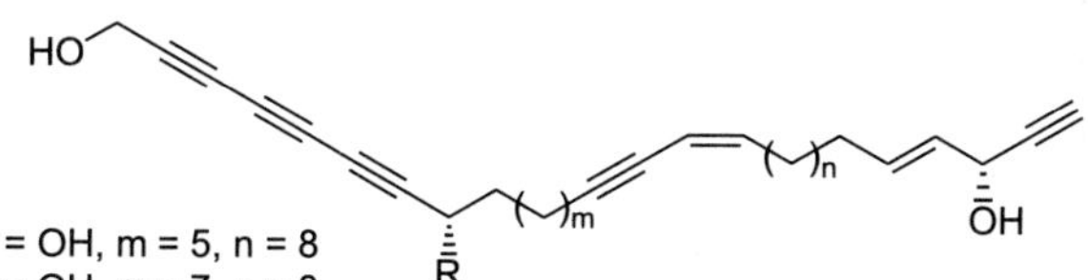

221 R = OH, m = 5, n = 8
222 R = OH, m = 7, n = 8
223 R = H, m = 7, n = 8

224

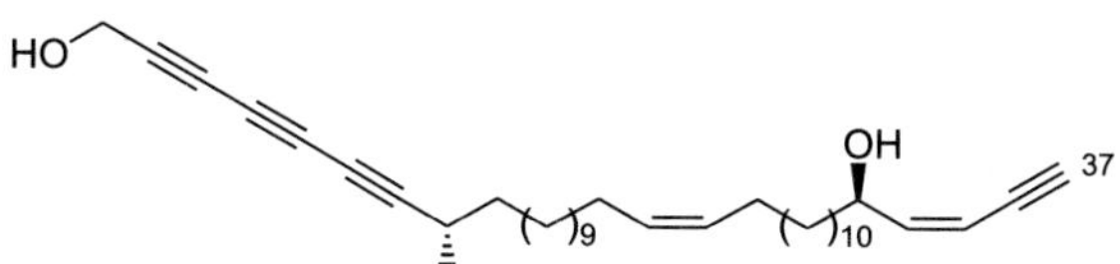

225

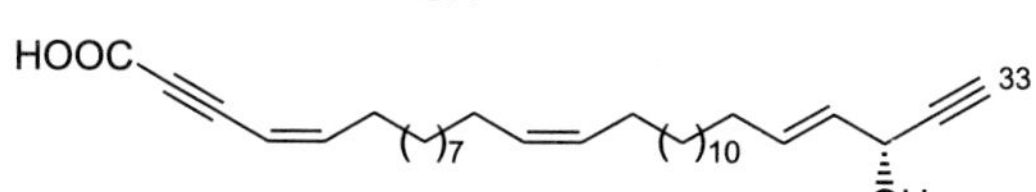

226

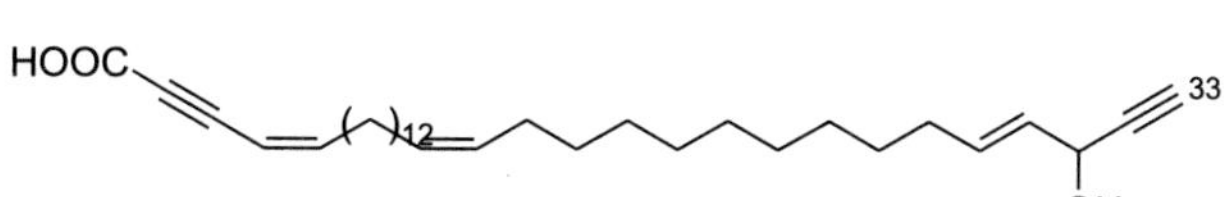

227 n = 11
228 n = 10

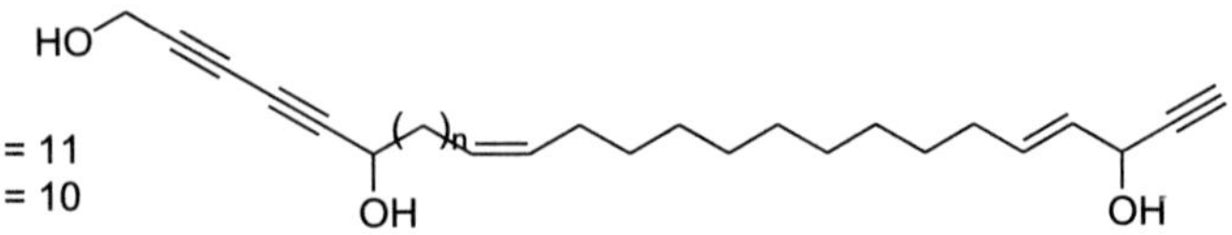

229 n = 11
230 n = 13

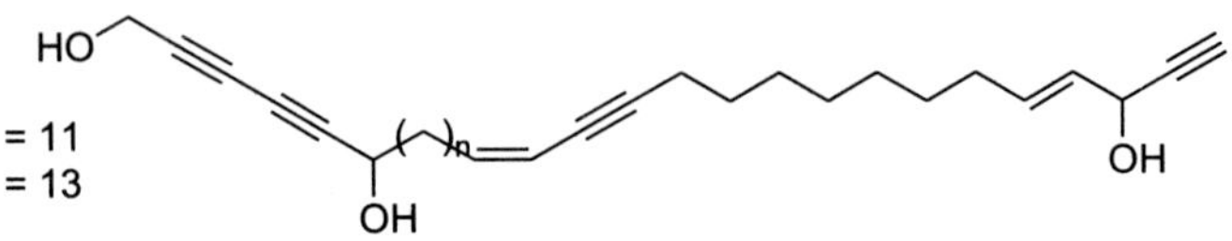

231

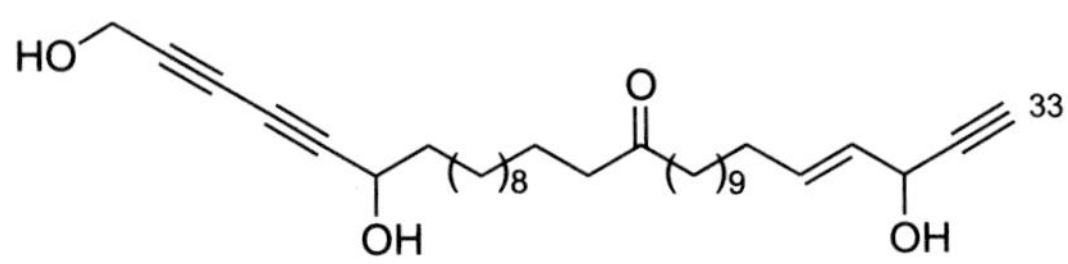

232

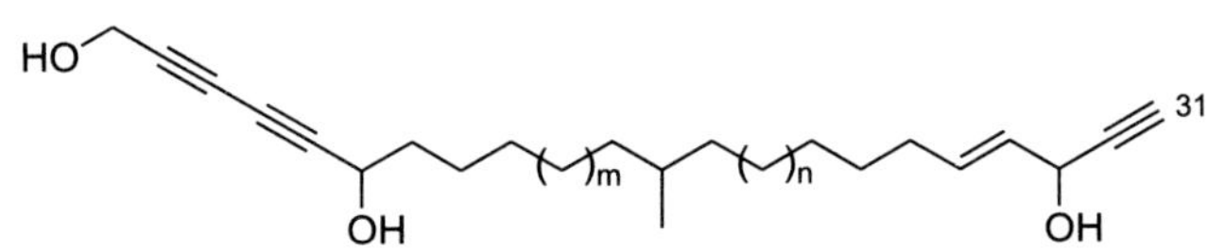

m + n = 11

Phakellia

Five acetylenic acids (**233–237**), identified as the corresponding ethyl esters, were obtained from a specimen of *Phakellia carduus* obtained from a depth of 350 m by trawling (Barrow and Capon 1994). Carduusynes C (**235**) and E (**236**) existed as mixtures of epimers in a ratio of 83:17. There was no evidence to suggest that the epimerization was a consequence of the isolation process. The antimicrobial activity observed in the CH_2Cl_2 extract was attributed to the metabolites isolated, although further quantitative biological evaluation was limited due to lack of material. The structure of carduusyne A (**233**) has been confirmed by a stereocontrolled synthesis of conjugated dienynes (Charoenying *et al.* 1996).

Clathria

Clathrynamides A, B, and C (**238–240**) are brominated acetylenic amides from a Japanese *Clathria* sp. that inhibit cell division in fertilized starfish eggs (Ohta *et al.* 1993). Clathrynamide A (**238**) inhibited cell division of starfish eggs at an extraordinarily low concentration (IC_{50} 6 ng/mL). However, clathrynamides B and C were not very active, with the IC_{50} values of 0.2 and 1 μg/mL, respectively. It is noteworthy that the difference of the structure of the amide moiety affects profoundly biological activities. Clathrynamide A (**238**) also inhibited the growth of human myeloid K-562 cells *in vitro* with an IC_{50} value of 0.2 μg/mL.

233

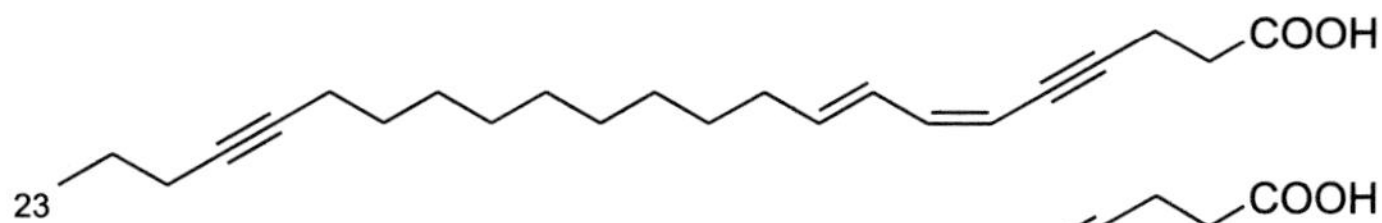

234

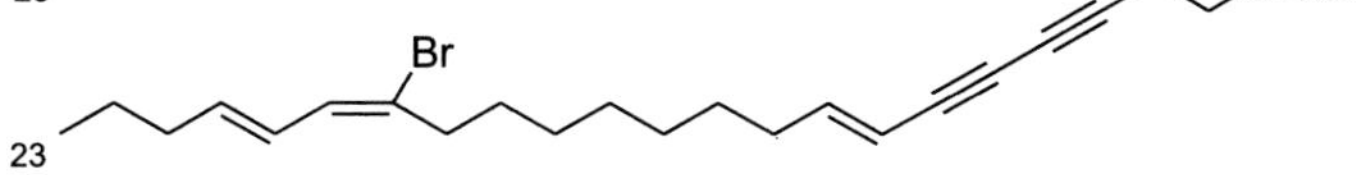

235 R = OH
236 R = OCH3

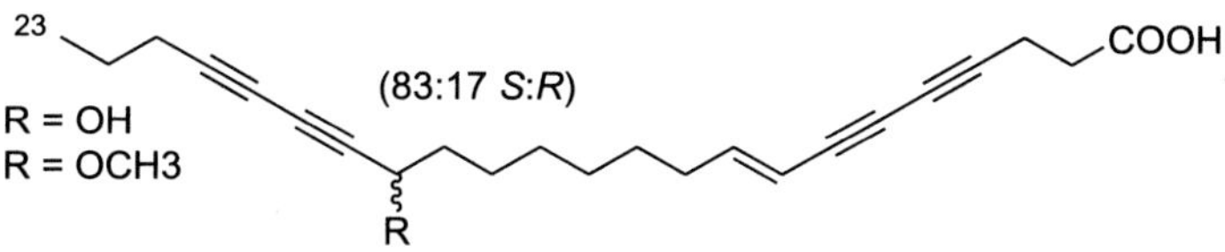

237

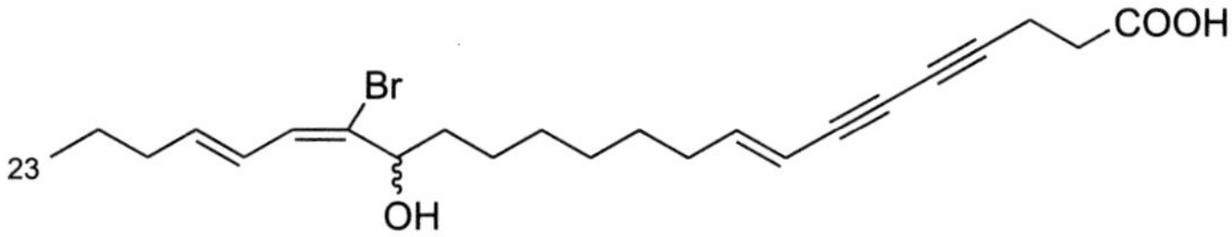

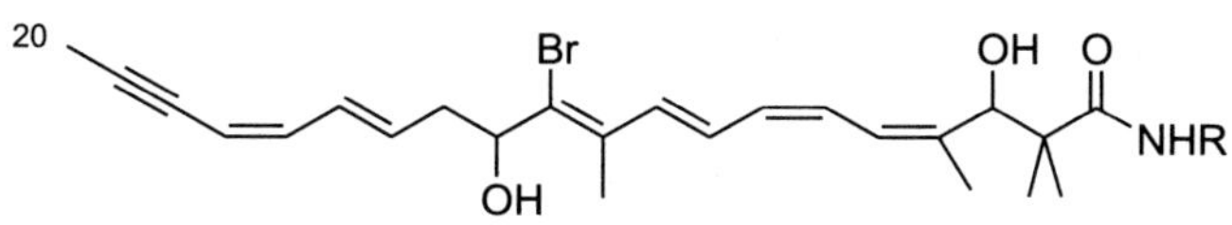

238 R = H
239 R = $-CH(CH_3)CH_2CH_2CH(OH)CH_3$
240 R = $-CH(CH_3)CH_2CH_2COCH_3$

Acarnus

Two very unusual cytotoxic polyacetylenes peroxyacarnoic acid methylesters A and B (**241** and **242**), which also contain a cyclic peroxide ring, were isolated from *Acarnus* cf. *berquistae* from the Red Sea (Yosief *et al.* 1998).

Acarnus bicladotylota collected from the South West Coast of India yielded the acetylenic cycloperoxides, peroxyacranoic acids C and D (**243** and **244**) which were isolated as their methyl esters (Fontana *et al.* 2001).

Theonella

Pellynols E–H (**245–248**), together with known polyacetylenes, melynes A and B (**100** and **101**), pellynols C (**229**), D (**230**), and macrolides (swinholides), were isolated from an undescribed *Theonella* sp. from Chuuk Atoll, Micronesia (Fu *et al.* 1999). This was the first report of the isolation of acetylenic compounds from a *Theonella* sponge or other Lithistida sponges, and is interesting from a chemotaxonomic point of view.

241

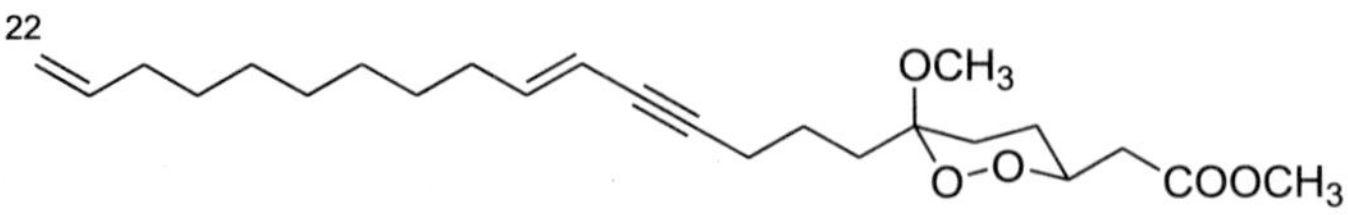

242

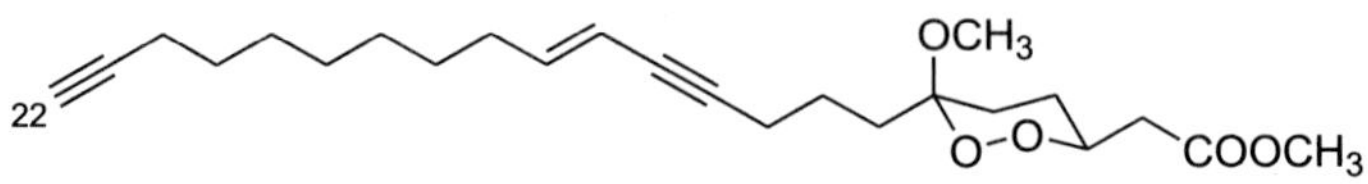

243

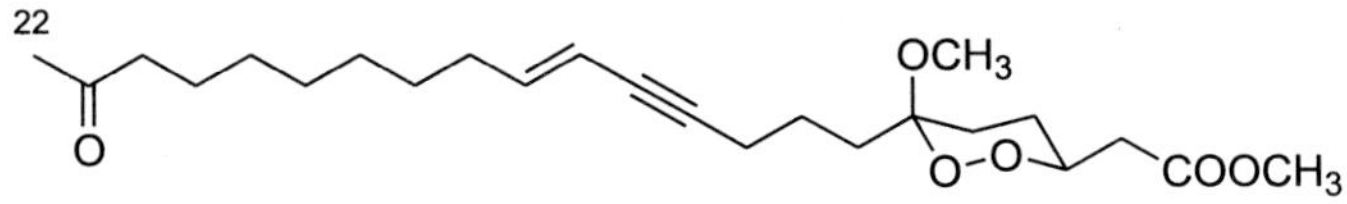

244

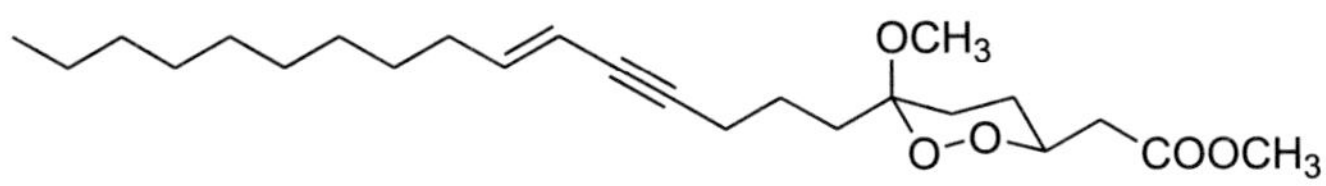

245

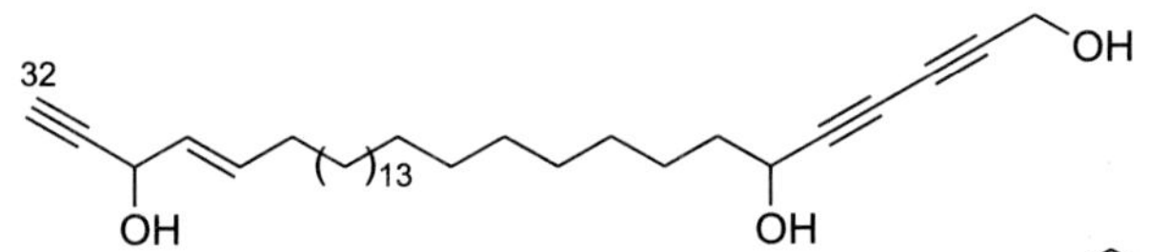

246

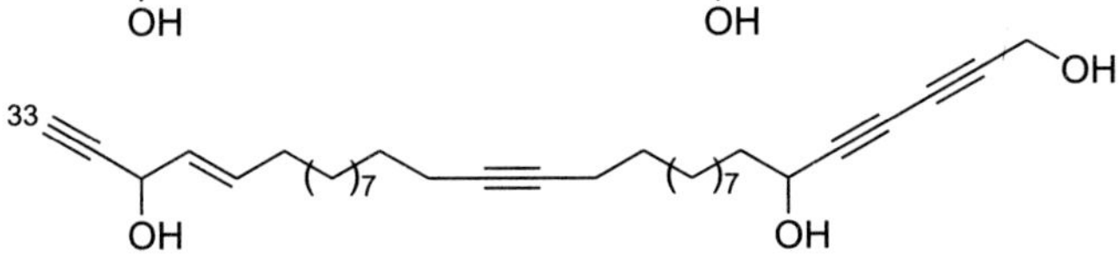

247

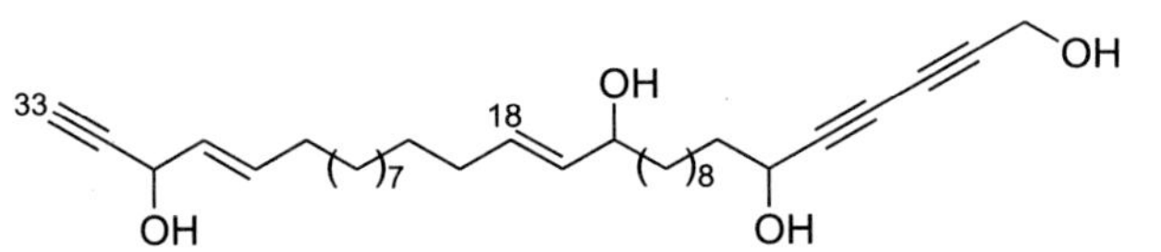

248

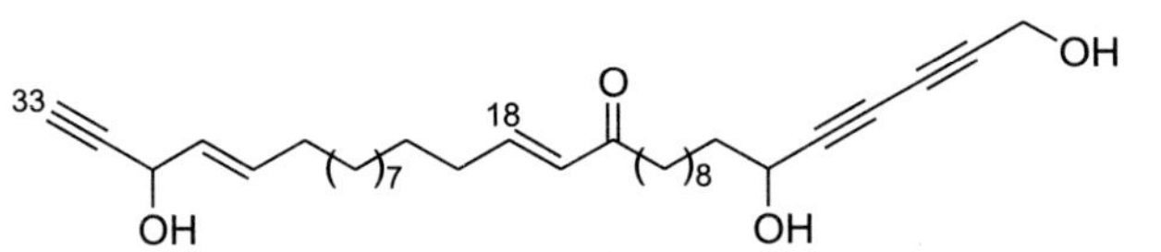

Hippospongia

Taurospongin A (**249**) is an unusual acetylenic fatty acid derivative from an Okinawan *Hippospongia* sp. that inhibits both DNA polymerase β and HIV reverse transcriptase (Ishiyama *et al.* 1997). The absolute configurations of taurospongin A were established as 3*R*, 7*S*, and 9*R* by synthesis of the MTPA ester of a degradation product of taurospongin A. The C_{25} fatty acid contained in taurospongin A is new, while a C_{23} acetylenic acid with more unsaturation was previously isolated from the Australian sponge *Phakellia carduus* (**233–237**). Taurospongin A (**249**) exhibited potent inhibitory activity against DNA polymerase β and HIV reverse transcriptase with IC_{50} values of 7.0 and 6.5 μM, respectively. This inhibitory activity has been frequently associated with the sulfonic acid function. Taurospongin A also showed weak inhibition of c-erbB-2 kinase (IC_{50} 28 μg/mL), while there was no cytotoxicity (IC_{50} > 10 μg/mL) against murine lymphoma L1210 and human epidermoid carcinoma KB cells. Taurospongin was synthesized in 14 steps, in 6% overall yield, including the kinetic resolution (Lebel and Jacobsen 1998).

249

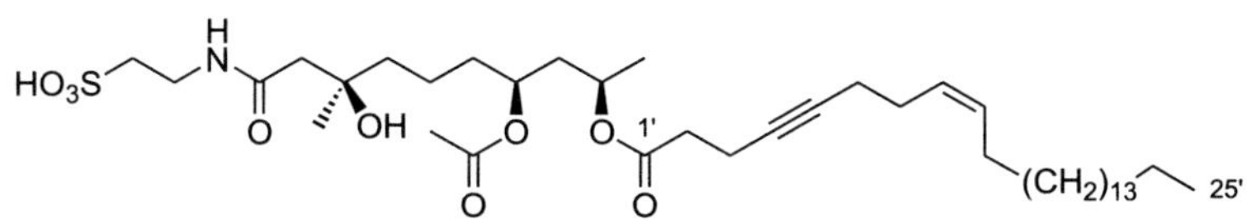

Stelletta

Four new cytotoxic acetylenic acids, stellettic acids A–D (**250–253**), were isolated from a marine sponge *Stelletta* sp. collected from Korean waters (Zhao *et al.* 2003). Stellettic acids A–C (**250–252**) are the first example of 2-methoxy fatty acids with 2*S* configuration, while most of the 2-methoxy fatty acids of sponge phospholipids were reported to possess 2*R* configuration. Stellettic acid D (**253**) showed significant cytotoxicity against five human tumor cell lines, A549 (human lung cancer), SK-OV-3 (human ovarian cancer), SK-MEL-2 (human skin cancer), XF498 (human CNS cancer), and HCT15 (human colon cancer) with EC_{50} values in the range of 3.4-4.8 μg/mL.

Similar acetylenic acid (**254**) and anhydride (**255**) were isolated along with stellettic acid A (**250**) from the sponge, *Stelletta* sp., from Korea (Lee *et al.* 2003). These compounds exhibited no significant antimicrobial activity and displayed only weak cytotoxicity against the human leukemia cell line K562 with LC_{50} values of 43.5, 51.3, and 62.5 μg/mL, respectively. The configuration of the α-methoxyl group was not determined.

Diplastrella

Brominated acetylenes, diplynes A–E (**256, 257, 259, 261**, and **263**), diplyne A 1-sulfate (**258**), diplyne C 1-sulfate (**260**), and 2-deoxydiplyne D sulfate (**262**) were isolated from the Philippines sponge *Diplastrella* sp. by employing the HIV-1 integrase inhibition assay (Lerch *et al.* 2003). Diplynes A-E did not show significant inhibition on integrase (IC_{50} >50 μg/mL), while the sulfated analogues exhibited mild inhibition (~30 to 90 μg/mL), which is consistent with the general activity observed for sulfated compounds in this assay.

250
251 Δ^{7Z}
252 Δ^{7E}

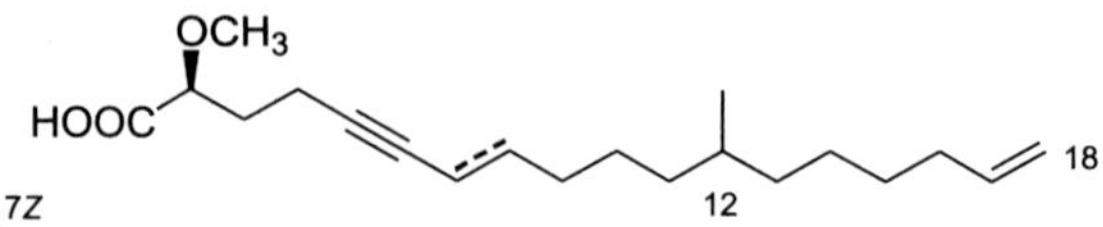

253

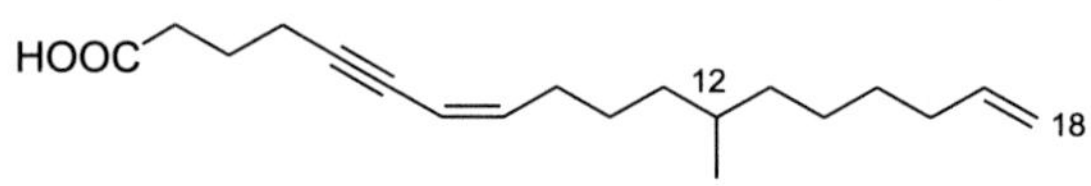

254

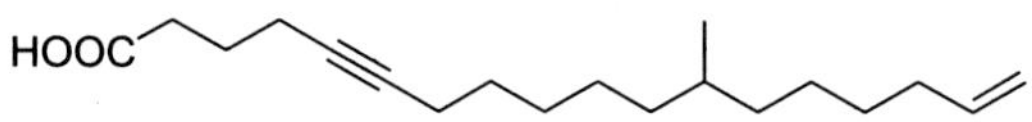

255

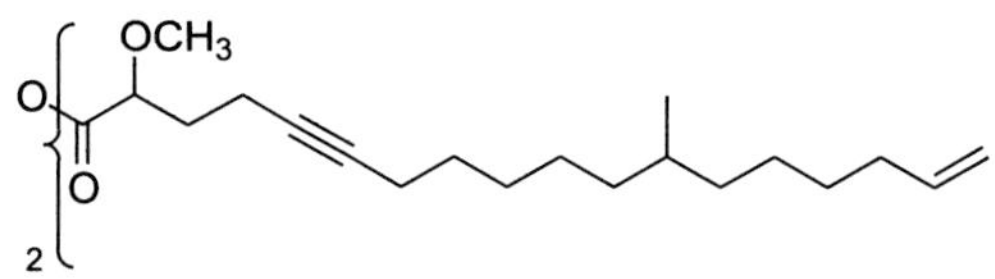

256 R_1 = H, R_2 = Br, R_3 = H
257 R_1 = Br, R_2 = H, R_3 = H
258 R_1 = H, R_2 = Br, R_3 = SO_3H

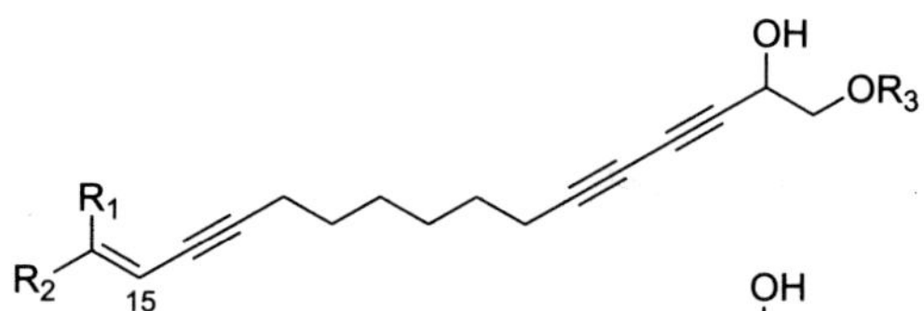

259 R = H
260 R = SO_3H

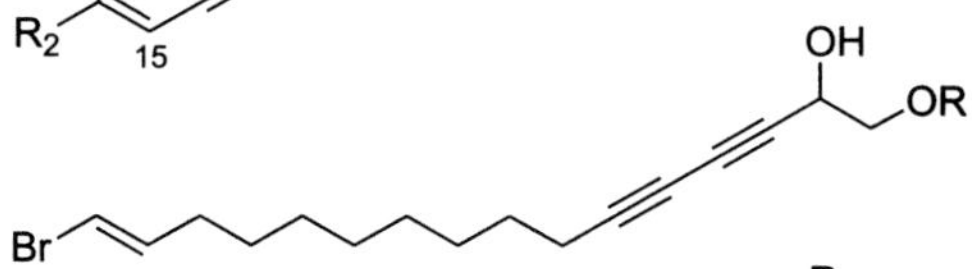

261 R_1 = OH, R_2 = H
262 R_1 = H, R_2 = SO_3H

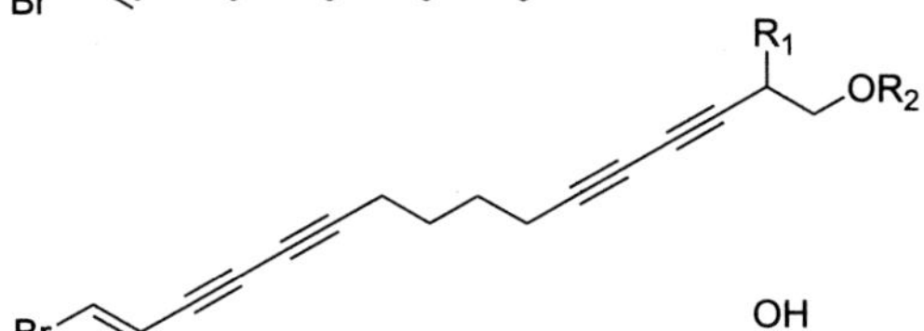

263

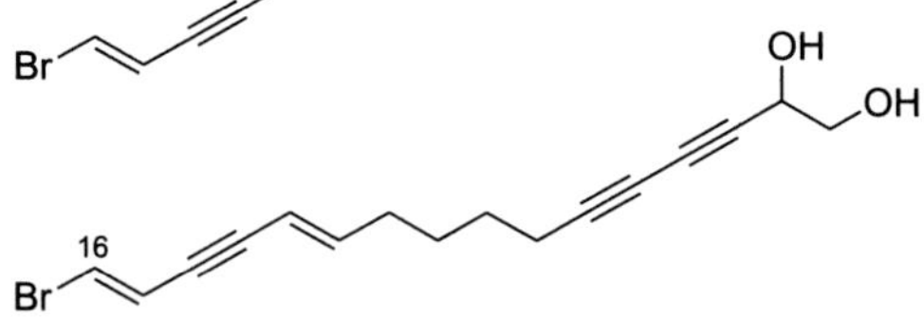

Unidentified Sponge

From an unidentified sponge of the family Niphatidae, petrosynol (**36**), petrosynone (**37**), and five new C_{30} acetylenes, petrofuran (**204**), 15, 16-dihydropetrosynone (**264**), 15*E*-petrosynone (**265**), 3, 28-diketopetrosynol (**266**), 3-ketopetrosynol (**203**), were isolated (Higa *et al.* 1994). Petrofuran (**204**) and 3-ketopetrosynol (**203**) were later isolated again from an *Adocia* sp. and given the trivial names of adociacetylene B and A, respectively (Kobayashi *et al.* 1996). These compounds, except petrofuran, which was not tested, were highly cytotoxic against P388, A-549, HT-29, and MEL-28 tumor cells (data not shown).

264

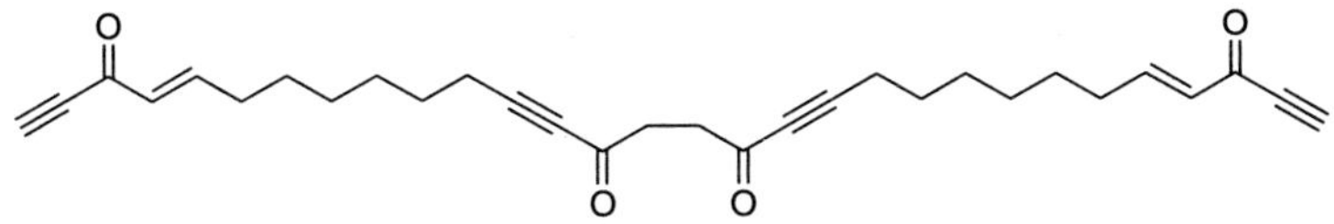

265

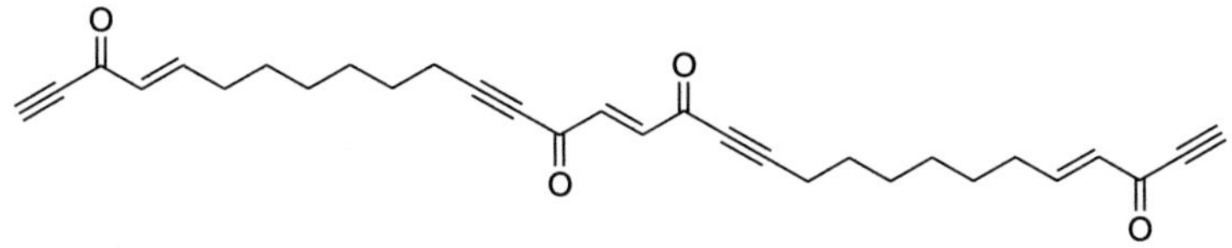

266

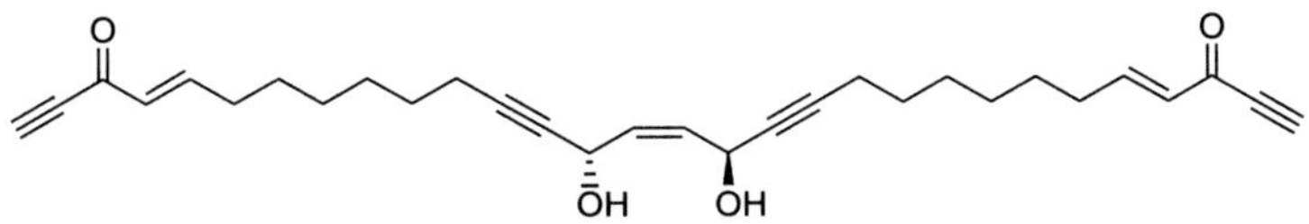

Biological Activities

The linear acetylenes found in sponges exhibited diverse biological activities, such as antitumor, antimicrobial, antiviral, and anti-inflammatory activities. Most often they possess remarkable cytotoxicity against tumor cells. Their cytotoxicity might be attributed to their inherent non-specific toxicity because of their simple functionality that adorns the structures of these molecules, or a result of non-specific lipid aggregation under assay conditions. However, a recent report demonstrating that analogues of a petrosynol, dideoxypetrosynol A (**56**), directly inhibit DNA replication at the level of initiation supports a suggestion that the compounds interfere with proteins required for initiation forks (Kim *et al.* 2002). Indeed, earlier studies show that petrosynol (**36**) inhibits both RNA- and DNA-dependent DNA polymerase function of HIV reverse transcriptase (Isaacs *et al.* 1993). Petroformyne 2 (**9**) was also found to be a general potent inhibitor of DNA polymerases (Loya *et al.* 2002). It inhibits equally well RNA- and DNA-dependent DNA polymerase activities of retroviral reverse transcriptases (i.e., of HIV, murine leukemia virus and mouse mammary tumor virus) as well as cellular DNA polymerases (i.e., DNA polymerases α and β, and *E. coli* polymerase I). Petroformyne 2 was shown to be a reversible non-competitive inhibitor. Nevertheless, a unifying structural model to explain these unusually high activities should be advanced by further study.

Antitumor Related Activities

Brine Shrimp, Sea Urchin, Starfish, Ascidian, and Potato Disc Assays

Lethality to brine shrimp larvae, inhibition of *Agrobacterium tumefaciens*-induced plant tumors (potato disc assay), and inhibition of cell division of sea urchin, starfish, or ascidian eggs were often employed as indicators of antitumor activity of natural products. Many polyacetylenes from *Petrosia* sp. were toxic to brine shrimp larvae. The C_{46} acetylenes (**8–30**) and C_{30} acetylenes (**44–47**) were very toxic to brine shrimp larvae (Guo *et al.* 1995, Ochi *et al.* 1994). Other C_{46}, C_{30}, and related acetylenes (**56–58, 60**, and **61–80**) from *Petrosia* sp. were isolated from the fractions that were toxic to brine shrimp larvae (Kim *et al.* 1998, 1999a,b). Short-chain acetylenes (**146–149**) were also toxic to brine shrimp larvae (Aiello *et al.* 1992). It appears that linear acetylenic compounds exert general toxicity to brine shrimp larvae.

The C_{46} acetylenes (**8–11**) and C_{30} petrosynol (**36**) inhibited cell division in fertilized eggs of the sea urchin (Cimino *et al.* 1990, Fusetani *et al.* 1983). The C_{30} acetylenes (**45–47**) inhibited the cell division of fertilized ascidian (*Styela partite*) eggs (Ochi *et al.* 1994). Callyspongins A and B (**163** and **164**) inhibited fertilization of starfish (*Asterias amurensis*) gametes (Uno *et al.* 1996). Clathrynamide A (**238**) inhibited cell division of starfish eggs (Ohta *et al.* 1993).

The C_{46} acetylenes, petroformynes 3 (**10**) and 4 (**11**) showed inhibition of plant tumors in the potato disc assay (Cimino *et al.* 1990).

Cytotoxicity

Dideoxypetrosynols A–D (**56–58, 60**) and other C_{46} and C_{30} acetylenes (**61–65**) showed significant cytotoxicity against five human tumor cell lines, A549 (human lung cancer), SK-OV-3 (human ovarian cancer), SK-MEL-2 (human skin cancer), XF498, (human CNS cancer), and HCT15 (human colon cancer) (Kim *et al.* 1998, 1999a,b). Dideoxypetrosynol A (**56**) was most potent and the activity was more potent than or comparable to that of doxorubicin. A further series of long-chain polyacetylenic alcohols (**66–85**) showed mild to significant cytotoxicity against five human tumor cell lines, while the potency of petrotriyndiol A (**74**) was comparable to that of doxorubicin. (Lim *et al.* 1999, 2001).

Other C_{30} or long-chain acetylenes were frequently reported to be cytotoxic. Nepheliosyne A (**112**) showed weak cytotoxicity against murine lymphoma L1210 and human epidermoid carcinoma KB cells (Kobayashi *et al.* 1994). Durissimol B (**139**) and duryne (**140**) exhibit potent cytotoxicity against human gastric (NUGC) cells (Shen and Prakash, 2000). Duryne (**140**) inhibited the growth of P388 murine leukemia and of colon (HCT-8), lung (A549), and mammary (MCF7) human tumor cell lines (Wright *et* *al.* 1987). Vasculyne (**152**) showed significant cytotoxicity and modest differential cytotoxicity toward the melanoma and colon tumor cell-line subpanels when tested in the NCI's 60-cell-line antitumor screening panel (Dai *et al.* 1996). Osirisynes A–F (**191–196**) exhibited moderate cytotoxicity against a human leukemia cell line (K562) (Shin *et al.* 1998). Fulvinol (**212**) showed cytotoxicity against P-388 mouse lymphoma, A-549 human carcinoma, HT-29 human colon carcinoma, and MEL-28 human melanoma (Ortega *et al.* 1996). Pellynol A (**227**) exhibited strong antitumor activity against HCT-116 (human colon tumor cell). Triangulynes A-H (**217–224**) demonstrated overall potency against the NCI human tumor cell line panels. In general, leukemia, colon, and melanoma tumor lines showed greater sensitivity to these compounds (Dai *et al.* 1996). Pellynols (**227–230**, **232**, **246**) were cytotoxic to human tumor cell lines (Rashid *et al.* 2000).

Various short-chain acetylenes showed cytotoxicity against cancer cells. Strongylodiols A–C (**135–137**) were cytotoxic against MOLT-4 (human T lymphocyte leukemia) cells (Watanabe *et al.* 2000). Compounds **141–145** had *in vitro* antitumor activity against murine leukemia P388 (Gunasekera and Faircloth 1990). The acetylenic alcohols (**150** and **151**) were evaluated in the NCI's 60-cell-line human tumor screen. A lung cancer cell line (H-522) and an ovarian cancer cell line (IGROV-1) showed an approximately tenfold to one hundred-fold greater sensitivity than the panel average. Callyspongenols A–C (**183–185**) and dehydrosiphonochalynol (**158**) (Youssef *et al.* 2003a) were cytotoxic against P388 and HeLa cells. Callyspongamide A (**186**) showed moderate cytotoxicity against HeLa cells (Youssef *et al.* 2003b). Clathrynamide A (**238**) inhibited the growth of human myeloid K-562 cells (Ohta *et al.* 1993). Stellettic acid D (**253**) showed significant cytotoxicity against five human tumor cell lines, A549 (human lung cancer), SK-OV-3 (human ovarian cancer), SK-MEL-2 (human skin cancer), XF498 (human CNS cancer), and HCT15 (human colon cancer) (Zhao *et al.* 2003).

The brominated short-chain acetylene, compound **93** and its methylester showed weak cytotoxicity against PS and L1210 murine leukemia cells (Schmitz and Gopichand 1978). Brominated short-chain acetylenes, compounds **116–119**, **126**, and **131** exhibited cytotoxicities against L1210 murine leukemia and KB human epidermoid carcinoma cell (Li *et al.* 1995).

Mechanism Based Assays

Cytotoxic polyacetylenes, dideoxypetrosynol A (**56**) and (3*S*, 14*S*)-petrocortyne A (**61**), inhibited DNA replication of simian virus (SV40)(Kim *et al.* 2002). In vitro SV40 DNA replication system was extensively used as a model to understand eukaryotic DNA replication because it uses the host replication machinery together with the virally encoded SV40 T antigen (T-Ag) for its own DNA replication. The SV40 T-Ag is a multifunctional protein that binds to specific DNA sites at the replication origin to initiate replication. All other replication functions are supplied by the extract of HeLa cells. Replication protein A (RPA) mediates unwinding of SV40 origin-containing DNA in the presence of T-Ag and topoisomerase. It interacts with T-Ag and the DNA polymerase α-primase complex (pol α-primase), which is necessary for the initiation of SV40 DNA replication. Dideoxypetrosynol A (**56**) and (3*S*, 14*S*)-petrocortyne A (**61**) were found to inhibit SV40 DNA replication at the level of initiation. The replication activity was strongly inhibited if the compounds were added to reaction mixture before the incubation, whereas weakly inhibited if added after the incubation started. Dideoxypetrosynol A (**56**) also significantly inhibited topoisomerase I. However, the ssDNA binding activity of RPA was little affected by these acetylenes. The activity of pol α-primase was inhibited by these compounds. Especially, dideoxypetrosynol A significantly inhibited pol α-primase activity at the concentration of 125 μM. Therefore, inhibition of SV40 DNA replication by these polyacetylenes may be mostly due to the inhibition of pol α-primase activity. Similar long-chain acetylenes, compounds (**66–75**) also showed significant inhibition on the SV40 DNA replication in a concentration-dependent manner, which could explain their cytotoxicity (Lim *et al.* 1999, 2001).

Dideoxypetrosynol A (**56**) inhibited the growth of human skin melanoma SK-MEL-2 cells with an apoptosis induction (Choi *et al.* 2004). The apoptotic events were mediated by an increase in Bax expression and an activation of caspases. Dideoxypetrosynol A increased the expression of pro-apoptotic modulator, Bax, and it decreased the expression of anti-apoptotic modulator, Bcl-2. The apoptotic effects of dideoxypetrosynol A were also associated with specific inhibition of cIAP-1 expression. The anti-apoptotic factors IAP (inhibitors of apoptosis protein) family proteins bind to caspases, which lead to inactivation of caspases in eukaryotic cells. The levels of cIAP-1 protein and mRNA were slightly decreased in response to dideoxypetrosynol A treatment. Dideoxypetrosynol A stimulated both caspase-3 and caspase-9 activities in a dose-dependent manner. Caspases are known as important mediators of apoptosis induced by various apoptotic stimuli. These results suggest that dideoxypetrosynol A-induced apoptosis in SK-MEL-2 cells is mediated, at least in part, by the mitochondrial-signaling pathway.

The short-chain acetylene, callysponginol sulfate A (**182**), inhibited MT1-MMP (Fujita *et al.* 2003). The desulfated form of **182** did not show any inhibitory activity. Considering this result as well as similar activity of structurally unrelated sulfated compounds, the activity of MT1-MMP inhibition is probably a consequence of the sulfate. MT1-MMP is one of the key enzymes involved in tumor angiogenesis and metastasis. Therefore, MT1-MMP inhibitors are considered as potential antineoplastic agents.

The short-chain acetylene, pellynic acid (**226**), inhibited IMPDH (inosine monophosphate dehydrogenase), which is one of anticancer targets (Fu *et al.* 1997).

Taurospongin A (**249**) exhibited potent inhibitory activity against a DNA replication enzyme, DNA polymerase β (Ishiyama *et al.* 1997). This inhibitory activity has been frequently associated with the sulfonic acid function. Taurospongin A also showed weak inhibition of the oncogene, c-erbB-2 kinase, while there was no cytotoxicity against murine lymphoma L1210 and human epidermoid carcinoma KB cells.

In Vivo Antitumor Activity

The recemic mixture of 3-hydroxyeicos-(4*E*)-en-1-yne (**141**) showed *in vivo* antitumor activity against the human ovarian cancer (IGROV-1) implanted to athymic mouse. Intraperitoneal treatment (once daily doses of 50 mg/kg beginning on the day following tumor cell implantation) of the compound resulted in a small, but significant increase (130%) in lifespan (Hallock *et al.* 1995).

Antiviral Activities

Petrosolic acid (**38**) and petrosynol (**36**) exhibited inhibitory activity of the RNA-dependent and DNA-dependent DNA polymerase functions of HIV-1 RT but almost showed no inhibition of the reverse transcriptase-associated RNase H (ribonuclease H) (Isaacs *et al.* 1993).

In an inhibitory assay against RNA-based reverse transcriptase, petrosiacetylene A (**56**) and C (**58**) totally cleaved the template 16S rRNA of *E. coli*, at the concentration of 10 μg/20 μL, while other compounds did not cleave it at all. However, all of the compounds did not cleave a supercoiled DNA (pUC 119) at the same concentration (Shin *et al.* 1998).

Brominated polyacetylenic acids (**107–111**), **96**, and **104** inhibited HIV-1 protease. The minimal differences in the inhibitory potency within this series of compounds might imply a nonspecific mode of action (Patil *et al.* 1992).

Taurospongin A (**249**) exhibited potent inhibitory activity against HIV reverse transcriptase (Ishiyama *et al.* 1997).

Brominated acetylenes, diplynes A–E (**256**, **257**, **259**, **261**, and **263**), diplyne A 1-sulfate (**258**), diplyne C 1-sulfate (**260**), and 2-deoxydiplyne D sulfate (**262**), were isolated by employing the HIV-1 integrase inhibition assay (Lerch *et al.* 2003). Diplynes A-E did not show significant inhibition of integrase, while the sulfated analogues exhibited mild inhibition, which is consistent with the general activity observed for sulfated compounds in this assay (Lerch *et al.* 2003).

Petroformyne 2 (**9**) was found to be a potent inhibitor of DNA polymerases (Loya *et al.* 2002). It inhibits equally well the RNA- and DNA-dependent DNA polymerase activities of retroviral reverse transcriptases (i.e. of HIV, murine leukemia virus and mouse mammary tumor virus) as well as cellular DNA polymerases (i.e., DNA polymerases α and β, and *E. coli* polymerase I). A study on the mode of the polymerase inhibition by petroformyne 2 was conducted with HIV-1 RT. Petroformyne 2 was shown to be a reversible non-competitive inhibitor. Petroformyne 2 binds RT independently and at a site different from that of the primer-template and dNTP substrates with high affinity. Blocking the polar hydroxyl groups of petroformyne 2 has only a marginal effect on the inhibitory capacity, thus hydrophobic interactions of RT with primer-template substrate prior to the interaction with petroformyne 2 reduces substantially the inhibition capacity, probably by preventing these contacts. Petroformyne 2 does not interfere with the first step of polymerization, the binding of RT to DNA, nor does the inhibitor interfere with the binding of dNTP to RT/DNA complex, as evident from the steady-state kinetics, whereby K_m remains unchanged. Therefore, it was assumed that petroformyne 2 interferes with subsequent catalytic steps of DNA polymerization. The inhibitor may alter the optimal stereochemistry of the polymerase active site relative to the primer terminus, bound dNTP and the metal ions that are crucial for efficient catalysis or, alternatively, may interfere with the thumb sub-domain movement and thus, with the translocation of the primer-template following nucleotide incorporation.

Antimicrobial Activities

The C_{30} petrosynol (**36**) was active against *Penicillium chrysogenum* (data not shown) (Fusetani *et al.* 1983). Petrosynol (**36**) also showed antifungal activity against *Mortierella ramannianus* and was also active in the starfish egg assay (Fusetani *et al.* 1987). Petrosynol (**36**) and adociacetylene A (**203**) showed weak antibacterial activity against *E. coli* and *B. subtilis* (Kobayashi *et al.* 1996).

Petrosynone (**37**) showed antibacterial activity against *Bacillus subtilis* (data not shown) (Fusetani *et al.* 1987).

Corticatic acids (**39–41**) exhibited antifungal activity against *Mortierella ramannianus* (data not shown) (Li *et al.* 1994). Corticatic acids A, D, and E (**39, 42, 43**) inhibited GGTase I (geranylgeranyl transferase type I) of *Candida albicans*; corticatic acid A (**39**) also inhibited the growth of *C. albicans*. GGTase I is the key player in cell wall biosynthesis and is essential for viability of *Saccharomyces cerevisiae*. Since there is little sequence homology between the human and *Candida* GGTase I, selective inhibitors of *Candida* GGTase I could be good antifungal agents (Nishimura *et al.* 2002).

Brominated C_{18} acetylenic acids (**81–87**) and three methyl esters (**88–90**) exhibited antifungal activity against *Mortierella ramannianus* (data not shown) (Fusetani *et al.* 1993).

Brominated short-chain acetylenes (**94–98**) were found to possess moderate antibacterial activity against *Bacillus subtilis* (Hirsh *et al.* 1987).

Xestospongic acid (**105**) and its ethyl ester (**106**) showed antibacterial activity against *Staphylococcus aureus* (Bourguet-Kondracki *et al.* 1992).

Siphonodiol (**159**) showed weak activity against *Trichophyton asteroids* and medium activity against *Staphylococcus aureus* and *Streptococcus pyogenes* C-203 (Tada and Yasuda 1984).

The brominated short-chain acetylene, compound **215**, showed mild antibacterial activity against Gram-positive bacteria (data not shown) (Ichiba *et al.* 1993).

The short-chain acetylene, compound **216** exhibited a rather selective antimicrobial activity. It was moderately active against four mutants of *S. cerevisiae* and *Candida albicans*, but was inactive against *Penicillium chrysogenum* and *Mortierella ramannianna*. The compound was also inhibitory against both Gram-positive and Gram-negative bacteria (Matsunaga *et al.* 2000).

Anti-inflammatory Activities

(3*S*, 14*S*)-Petrocortyne A (**61**) was investigated for *in vitro* anti-inflammatory and pro-aggregative effects at non-cytotoxic concentrations on various cellular inflammatory phenomena using the murine macrophage (RAW264.7) and human monocytic cells (U937) (Hong *et al.* 2003). (3*S*, 14*S*)-Petrocortyne A blocked tumor necrosis factor-α (TNF-α) production strongly and concentration-dependently in lipopolysaccharide (LPS)-activated RAW264.7 cells and phorbol 12-myristate 13-acetate (PMA)/LPS-treated cells. It also blocked NO production concentration-dependently in LPS or interferon (IFN)-γ-treated RAW264.7 cells. Among the tested factors stimulating cell migration, the compound selectively blocked the expression of hepatocyte growth factor/scatter factor (HGF/SF). On the other hand, as assessed by a cell-cell adhesion assay, (3*S*, 14*S*)-petrocortyne A did not block the activation of adhesion molecules, but suppressed PMA-induced cell-cell adhesion significantly. Intriguingly, (3*S*, 14*S*)-petrocortyne A induced U937 homotypic aggregation following a long-time exposure (2 or 3 days), accompanied by weak induction of pro-aggregative signals such as tyrosine phosphorylation of p132 and phosphorylation of extracellular signal-related kinase 1 and 2 (ERK 1/2). (3*S*, 14*S*)-Petrocortyne A may thus inhibit cellular inflammatory processes by inhibition of both pro-inflammatory signals and leukocyte migration to the inflammed tissues. Inflammation is a multi-component process mediated by activated inflammatory or immune cells. Of these cells, macrophages play a central role in managing many different immunopathological phenomena, such as over production of cytokines (e.g., TNF-α and interleukin-1), migratory chemokines (e.g. HFG/SF), and inflammatory mediators (e.g., NO and prostaglandin E_2). Under inflammatory conditions, immune cells are also stimulated by adhesion molecule-activating signals to enhance migration capacity to the inflammed tissues. As soon as surface adhesion molecules are activated, cells move to the target tissues and show heterotypic cell clustering between immune cells and inflammed cells. Modulation of molecular events in cellular inflammatory phenomena caused by those stimuli and immune cell migration to the inflammed tissues may thus be a therapeutic target for the amelioration of chronic inflammatory disease.

Adociacetylene A (**203**) showed moderate cytotoxicity against KB (human epidermoid carcinoma) cells. Furthermore, it was found that adociacetylenes A (**203**), B (**204**), and D (**206**) inhibited neutrophil leukocyte adhesion to TNF-α (5JRU/mL)-stimulated endothelial cells at a 1 μg/mL concentration (Kobayashi *et al.* 1996).

Miscellaneous Bioactivities

Long-chain acetylenes from *Petrosia ficiformis* (**1-5**) provoked a positive food response by the nudibranch (Castiello *et al.* 1979). The nudibranch, *Peltodoris atromaculata*, is able to find its prey, the sponge, *Petrosia ficiformis*, by chemotaxis.

C_{30} acetylenes, petrosiacetylene B (**57**) and C (**58**), displayed moderate inhibition of phospholipase A_2 (PLA_2), while other petrosiacetylenes were much less active. Phospholipase A_2 has a crucial role in the inflammation process. Among petrocortynes, the γ-pyrone-containing petrocortyne C (**50**) was significantly more active than others. Compounds **50**, **57**, and **58** showed weak inhibition of Na^+/K^+ ATPase (Seo *et al.* 1998, Shin *et al.* 1998). Na^+/K^+ ATPase inhibitors such as digoxin have been used as cardiotonics.

Brominated polyacetylenes, xestospongic acid (**105**) and the corresponding ethyl ester (**106**) inhibited Na^+/K^+ATPase (Bourguet-Kondracki *et al.* 1992).

Osirisynes B (**193**), E (**195**), and F (**196**) exhibited inhibitory activities against Na^+/K^+ATPase (Shin *et al.* 1998).

Siphonodiol (**159**) and two related metabolites, dihydrosiphonodiol (**160**) and tetrahydrosiphonodiol (**161**) strongly inhibited gastric H^+/K^+-transporting ATPase (Fusetani et al. 1987). Gastric H^+/K^+-transporting ATPase plays a role in gastric HCl secretion and development of gastric ulcer.

Melyne A (**100**) was active against the intestinal protozoan *Giardia* (data not shown) (Quinoa and Crews 1988).

The xestosterol esters of two brominated acetylenic acids (**133** and **134**), along with the free acid of **133,** were isolated from an Australian specimen of *X. testudinaria* (Pham *et al.* 1999). The free acid inhibited [^{3}H]DPCPX binding to the rat-brain adenosine A_1 receptor. Stimulation of the adenosine A_1 receptor induces bronchoconstriction, block of cardiac AV conduction, and reduction of contraction force, leading to bradycardia. New adenosine antagonists are in development, mainly for treatment of ischaemic heart disease and stroke.

Compounds **141–145** showed 100% immunosuppressive activity in a mixed lymphocyte reaction (MLR) assay (Gunasekera *et al.* 1989). The immunomodulatory effects were determined using murine splenocytes in the two-way mixed lymphocyte reaction and the mitogenic response to Con A.

Callytetrayne (**165**), callypentayne (**166**), and five triols, callytriols A–E (**167–171**), induced metamorphosis in larvae of the ascidian *Halocynthia roretzi*. The triols (**167–171**) also showed antifouling activity against the barnacle *Balanus amphitrite* (Tsukamoto *et al.* 1997).

Callyspongynic acid (**181**) inhibited α-glucosidase but was inactive against β-glucosidase, β-galactosidase, thrombin, and trypsin. Corticatic acid A (**39**) and petrosynol (**36**) were also inhibitory against α-glucosidase (Nakao *et al.* 2002). α-Glucosidase is involved in the development of diabetes and obesity. α-Glucosidase inhibitors may delay the digestion of oligosaccharides and disaccharides to monosaccharides on the small intestinal brush-border, and reduce the rate of glucose absorption. Therefore, inhibition of α-glucosidase is considered important in managing non-insulin dependent diabetes (Bischo 1994).

Lembehynes A-C (**197–199**) induced neuritogenesis in pheochromocytoma PC12 and neuroblastoma Neuro 2A cells (Aoki *et al.* 2000, 2001, 2002). These compounds induced neurite outgrowth of rat pheochromocytoma (PC12) cells and neuroblastoma (Neuro 2A) cells. Treatment with cytochalasin B, an inhibitor of actin polymerization, or cycloheximide, an inhibitor of protein synthesis in eukaryotes, blocked neurite outgrowth induced by lembehyne A. This finding indicates that the morphological change induced by lembehyne A is dependent upon actin polymerization and *de novo* protein synthesis. Lembehyne A also induced neuronal cell differentiation in a Neuro 2A cell line (Aoki *et al.* 2001). The treatment predominantly resulted in a morphological change with bipolar neurites. The acetylcholinesterase activity of Neuro 2A was also increased by the treatment of lembehyne A. Furthermore, the cell cycle of Neuro 2A cells was found to be specifically blocked at the G1 phase by lembehyne A. The structure-activity relationship study using the lembehyne A analogues revealed the importance of the terminal 1-yn-3-ol and unsaturated long-chain alkyl moieties for the neuronal differentiation activity. To identify the target protein for lembehyne A (**197**), a radioactive photoaffinity probe was synthesized (Aoki *et al.* 2003). As a result of *in situ* labeling experiments against Neuro 2A cells, a protein of *Mr* 30 kDa was photolabeled specifically. This labeling was inhibited in the presence of lembehyne A or the active analogues of lembehyne A, whereas the inactive analogues showed no inhibitory effect on this labeling. These results suggest that this protein of 30 kDa is the target protein for lembehyne A and may play an important role in the neuronal differentiation in neuroblastoma.

CONCLUDING REMARKS

Most of the polyacetylenes from marine organisms have been isolated from sponges that are distributed in a wide range of taxonomically distinct sponge genera. Quite possibly, polyacetylenes might be ubiquitous sponge metabolites, which are not yet fully understood due to their chemical instability, low content in the organisms, and limited chemical investigation.

Polyacetylenes and related compounds occurring in nature generally comprise biologically active compounds. However, their instability and the mammalian toxicity of this type of molecule are two problems that have hampered application of these compounds. Nevertheless, some biological effects of the marine polyacetylenes are observed at non-cytotoxic concentrations (Hong *et al.* 2003), encouraging further pharmacological investigations.

ACKNOWLEDGEMENT

The authors thank Prof. D.S. Im, College of Pharmacy, Pusan National University, for the critical reading of the manuscript and helpful discussions. Thanks are also due to Prof. C.J. Sim and Dr. K.J. Lee of Hannam University for the valuable comments on sponge taxonomy.

REFERENCES

Aiello, A., E. Fattorusso, M. Menna, and M. Pansini. 1992. Further bioactive acetylenic compounds from the Caribbean sponge *Cribrochalina vasculum*. *J. Nat. Prod.* 55: 1275–1280.

Allen, E.H., and C.A. Thomas. 1971. Trans-trans-3, 11-tridecadiene-5, 7, 9-triyne-1, 2-diol, an antifungal polyacetylene from diseased safflower (*Carthamus tinctorius*). *Phytochemistry* 10: 1579–1582.

Aoki, S., K. Matsui, K. Tanaka, R. Satari, and M. Kobayashi. 2000. Lembehyne A, a novel neuritogenic polyacetylene from a marine sponge of *Haliclona* sp. *Tetrahedron* 56: 9945–9948.

Aoki, S., K. Matsui, T. Takata, W. Hong, and M. Kobayashi. 2001. Lembehyne A, a spongean polyacetylene, induces neuronal differentiation in neuroblastoma cell. *Biochem. Biophys. Res. Commun.* 289: 558–563.

Aoki, S., K. Matsui, H. Wei, N. Murakami, and M. Kobayashi. 2002. Structure-activity relationship of neuritogenic spongean acetylene alcohols, lembehynes. *Tetrahedron* 58: 5417–5422.

Aoki, S., K. Matsui, T. Takata, and M. Kobayashi. 2003. In situ photoaffinity labeling of the target protein for lembehyne A, a neuronal differentiation inducer. *FEBS Lett.* 544: 223–227.

Barrow, R.A., and R.J. Capon. 1994. Carduusynes (A–E): acetylenic acids from a Great Australian Bight marine sponge *Phakellia carduus*. *Aust. J. Chem.* 47: 1901–1918.

Bischo, H. 1994. Pharmacology of α-glucosidase inhibition. *Eur. J. Clin. Invest.* 24: 3-10.

Bohlmann, F., J. Lam, H. Breteler, T. Arnason, and L. Hansen. 1988. *Chemistry and Biology of Naturally-Occurring Acetylenes and Related Compounds*. Elsevier, Amsterdam.

Bohlmann, F., T. Burkhardt, and C. Zdero. 1973. *Naturally Occurring Acetylenes*. Academic Press, London and New York.

Bourguet-Kondracki, M.L., M.T. Rakotoarisoa, M.T. Martin, and M. Guyot. 1992. Bioactive bromopolyacetylenes from the marine sponge *Xestospongia testudinaria*. *Tetrahedron Lett.* 33: 225–226.

Brantley, S.E., T.F. Molinsky, C.M. Preston, and E.F. DeLong. 1995. Brominated acetylenic fatty acids from *Xestospongia* sp., a marine sponge-bacteria association. *Tetrahedron* 52: 7667–7672.

Braekman, J.C., D. Daloze, C. Devijver, D. Dubut, and R.W.M. van Soest. 2003. A new C-20 polyacetylene from the sponge *Callyspongia pseudoreticulata*. *J. Nat. Prod.* 66: 871–872.

Castiello, D., G. Cimino, S. De Rosa, S. De Stefano, G. Izzo, and G. Sodano. 1979. *Biologie des Spongiaires*. C.N.R.S. Paris.

Castiello, D., G. Cimino, S. De Rosa, S. De Stefano, and G. Sodano. 1980. High molecular weight polyacetylenes from the nudibranch *Peltodoris atromaculata* and the sponge *Petrosia ficiformis Tetrahedron Lett.* 21: 5047–5050.

Cardellina II, J.H., H.R. Bokesch, T.C. McKee, and M.R, Boyd, 1995. Resolution and comparative anti-HIV evaluation of the enantiomers of calanolides A and B. *BioMed. Chem. Lett.* 5: 1011–1014.

Charoenying, P., D.H. Davies, D. McKerrecher, and R.J.K. Taylor. 1996. A new stereocontrolled, pyrylium-based route to conjugated dienynes: the first synthesis of carduusyne A. *Tetrahedron Lett.* 37: 1913–1916.

Chill, L., A. Miroz, and Y. Kashman. 2000. Haliclonyne, a new highly oxygenated polyacetylene from the marine sponge *Haliclona* species. *J. Nat. Prod.* 63: 523–526.

Choi, H.J., S.-J. Bae, N.D. Kim, and Y.H. Choi. 2004. Induction of apoptosis by dideoxypetrosynol A, a polyacetylene from the sponge *Petrosia* sp., in human skin melanoma cells. *Int. J. Mol. Med.* 14: 1091-1096.

Cimino, G., A. Crispino, S. De Rosa, S. De Stefano, and G. Sodano. 1981. Polyacetylenes from the sponge *Petrosia ficiformis* found in dark caves. *Experientia* 37: 924–926.

Cimino, G., A. de Giulio, S. de Rosa, S. de Stefano, and G. Sodano. 1985. Further high molecular weight polyacetylenes from the sponge *Petrosia ficiformis*. *J. Nat. Prod.* 48: 22–27.

Cimino, G., A. De Giulio, S. De Rosa, and V. Di Marzo. 1989. High molecular weight polyacetylenes from *Petrosia ficiformis*: Further structural analysis and biological activity. *Tetrahedron Lett.* 30: 3563–3566.

Cimino, G., A. De Guilio, S. De Rosa, and V. Di Marzo, 1990. Minor bioactive polyacetylenes from *Petrosia ficiformis*. *J. Nat. Prod.* 53: 345–353.

Cimino. G., and De Stefano. 1977. New acetylenic compounds from the sponge *Reniera fulva*. *Tetrahedron Lett.* 15: 1325–1328.

D'Auria, M.V., L.G. Paloma, L. Minale, R. Riccio, A. Zampella, and C. Debitus. 1993. Metabolites of the New Caledonian sponge *Cladocroce incurvata*. *J. Nat. Prod.* 56: 418–423.

Dai, J.R., Y.F. Hallock, J.H. Cardellina II, and M.R. Boyd. 1996. Vasculyne, a new cytotoxic acetylenic alcohol from the marine sponge *Cribrochalina vasculum*. *J. Nat. Prod.* 59: 88–89.

Dai, J.R., Y.F. Hallock, J.H. Cardellina II, G.N. Gray, and M.R. Boyd. 1996. Triangulynes A–H and triangulynic acid, new cytotoxic polyacetylenes from the marine sponge *Pellina triangulata*. *J. Nat. Prod.* 59: 860–865.

De Weerdt, W.H. 2002. Family Chalinidae Gary. 1867. Pages 852-873 in *Systema Porifera, a Guide to the Classification of Sponges*, J.N.A. Hooper, and R.W.M. Van Soest, eds., Kluwer Academic/Plenum Publishers, New York.

Deshpande, V.H., B.K. Upadhye, and R.D. Wakharkar. 1989. Synthesis of (±)duryne. *Tetrahedron Lett.* 30: 1991–1992.

Dorta de Marquez, M., and V. Thaller. 1985. Natural acetylenes. Part 60. Synthesis of C_{22} and C_{23} diacetylenes, constituents of the sponge *Reniera fulva*. *J. Chem. Res.(S)*. 104–105.

Fontana, A., G. d'Ippolito, L. D'Souza, E. Mollo, P.S., Parameswaram, and G. Cimino. 2001. New acetogenin peroxides from the Indian sponge *Acarnus bicladotylota*. *J. Nat. Prod.* 64: 131–133.

Fu, X., S.A. Abbas, F.J. Schmitz, I. Vidavsky, M.L. Gross, M. Laney, R.C. Schatzman, and R.D. Cabuslay. 1997. New acetylenic metabolites from the marine sponge *Pellina triangulata*. *Tetrahedron* 53: 799–814.

Fu, X., F.J. Schmitz, and M. Kelly. 1999. Swinholides and new acetylenic compounds from an undescribed species of *Theonella* sponge. *J. Nat. Prod.* 62: 1336–1338.

Fujita, M., Y. Nakao, S. Matsunaga, R. van Soest, Y. Itoh, M. Seiki, and N. Fusetani. 2003. Callysponginol sulfate A, an MT1-MMP inhibitor isolated from the marine sponge *Callyspongia truncata*. *J. Nat. Prod.* 66: 569–571.

Fusetani, N., Y. Kato, S. Matsunaga, and K. Hashimoto. 1983. Bioactive marine metabolites III. A novel polyacetylene alcohol, inhibitor of cell division in fertilized sea urchin eggs, from the marine sponge *Petrosia* sp. *Tetrahedron Lett.* 24: 2771–2774.

Fusetani, N., H-y. Li, K. Tamura, and S. Matsunaga. 1993. Antifungal brominated C_{18} acetylenic acids from the marine sponge, *Petrosia volcano* Hoshino. *Tetrahedron* 49: 1203–1210.

Fusetani, N., T. Shiragaki, S. Matsunaga, and K. Hashimoto. 1987. Bioactive marine metabolites XX. Petrosynol and petrosynone, antimicrobial C_{30} polyacetylenes from the marine sponge *Petrosia* sp.: Determination of the absolute configuration. *Tetrahedron Lett.* 28: 4313–4314.

Garcia, J., M. Lopez, and J. Romeu, 1999. A synthesis of petrofuran based on the enantioselective reduction of 1-trimethylsily-4-alken-1-yn-3-ones. *Synlett* 429–431.

Gonnella, N.C., K. Nakanishi. V.S. Martin, and K.B. Sharpless. 1982. General method for determining absolute configurations of acyclic allylic alcohols. *J. Am. Chem. Soc.* 104: 3775–3776.

Guerriero, A., C. Debitus, D. Laurent, M. D'Ambrosio, and F. Pietra. 1998. Aztequynol A, the first clearly defined, C-branched polyacetylene and the analogue aztequynol B. Isolation from the tropical marine sponge *Petrosia* sp. *Tetrahedron Lett.* 39: 6395–6398.

Gunasekera, S.P., and G.T. Faircloth, 1990. New acetylenic alcohols from the sponge *Cribrochalina vasculum*. *J. Org. Chem.* 55: 6223–6225.

Gunasekera, S.P., S. Cranick, and R.E. Longley. 1989. Immunosuppressive compounds from a deep water marine sponge, *Agelas flabelliformis*. *J. Nat. Prod.* 52: 757–761.

Gung, B.W., H. Dickson, and S. Shockley. 2001. A concise synthesis of (+)- and (-)-adociacetylene B. *Tetrahedron Lett.* 42: 4761–4763.

Gung, B.W., H.D. Dickson, S. Seggerson, and K. Bluhm. 2002. A short synthesis of an acetylenic alcohol from the sponge *Cribrochalina vasculum*. *Synth. Commun.* 32: 2733–2740.

Guo, Y., M. Gavagnin, E. Trivellone, and G. Cimino. 1994. Absolute stereochemistry of petroformynes, high molecular polyacetylenes from the marine sponge *Petrosia ficiformis*. *Tetrahedron* 50: 13261–13268.

Guo, Y., E. Gavagnin, E. Trivellone, and G. Cimino. 1995. Further structural studies on the petroformynes. *J. Nat. Prod.* 58: 712–722.

Guo, Y., M. Gavagnin, C. Salierno, and G. Cimino. 1998. Further petroformynes from both Atlantic and Mediterranean populations of the sponge *Petrosia ficiformis*. *J. Nat. Prod.* 61: 333–337.

Hallock, Y.F., J.H. Cardellina II, M.S. Balaschak, M.R. Alexander, T.R. Prather, R.H. Shoemaker, and M.R. Boyd. 1995. Antitumor activity and stereochemistry of acetylenic alcohols from the sponge *Cribrochalina vasculum*, *J. Nat. Prod.* 58: 1801–1807.

Higa, T., J. Tanaka, A. Kitamura, T. Koyama, M. Takahashi, and T. Uchida. 1994. Bioactive compounds from marine sponges, *Pure Appl. Chem.* 66: 2227–2230.

Hirsh, S., S. Carmely, and Y. Kashman. 1987. Brominated unsaturated acids from the marine sponge *Xestospongia* sp. *Tetrahedron* 43: 3257–3261.

Hong, S., S.H. Kim, M.H. Rhee, A.R. Kim, J.H. Jung, T. Chun, E.S. Yoo, and J.Y. Cho. 2003. In vitro anti-inflammatory and pro-aggregative effects of a lipid compound, petrocortyne A, from marine sponges. *Naunyn-Schmeideberg's Arch. Pharmacol.* 368: 448–456.

Hosie, R.J.A. 1965. The localization of adenosine triphosphatases in morphologically characterized subcellular fractions in guinea pig brain. *Biochem. J.* 96: 404–412.

Ichiba, T., P.J. Scheuer, and M. Kelly-Borges. 1993. Sponge-derived polyunsaturated C_{16} di- and tribromocarboxylic acids. *Helv. Chim. Acta* 7: 2814–2816.

Isaacs, S., Y. Kashman, S. Loya, A. Hizi, and Y. Loya. 1993. Petrosynol and petrosolic acid, two novel natural inhibitors of the reverse transcriptase of human immunodeficiency virus from *Petrosia* sp. *Tetrahedron* 49: 10435–10438.

Ishiyama, H., M. Ishibashi, A. Ogawa, S. Yoshida, and J. Kobayashi. 1997. Taurospongin A, a novel acetylenic fatty acid derivative inhibiting DNA polymerase β and HIV reverse transcriptase from sponge *Hippospongia* sp. *J. Org. Chem.* 62: 3831–3836.

Jesus, R.P., and D.J. Faulkner. 2003. Chlorinated acetylenes from the San Diego sponge *Haliclona lunisimilis*. *J. Nat. Prod.* 66: 671–674.

Kaul, P.N., S.K. Kulkarni, F.J. Schmitz, and K.H. Hollenbeak. 1978. *Drugs and Food from the Sea, Myth or Reality*. Univ. of Oklahoma Press, Norman, Oklahoma.

Kim, J.S., K.S. Im., J.H. Jung, Y.-L. Kim, J. Kim, C.J. Sim, and C.-O. Lee. 1998. New bioactive polyacetylenes from the marine sponge *Petrosia* sp. *Tetrahedron* 54: 3151–3158.

Kim, J.S., K.S. Im., J.H. Jung, Y.-L. Kim, J. Kim, C.J. Sim, and C.-O. Lee. 1999a. Corrigendum to "New bioactive polyacetylenes from the marine sponge *Petrosia* sp." *Tetrahedron* 55: 2113.

Kim, J.S., Y.J. Lim, K.S. Im., J.H. Jung, C.J. Sim, C.O. Lee, J. Hong, and H.K. Lee. 1999b. Cytotoxic polyacetylenes from the marine sponge *Petrosia* sp. *J. Nat. Prod.* 62: 554–559.

Kim, D.-K., M.-Y. Lee, H.S. Lee, D.S. Lee., J.-R. Lee, B.-J. Lee, and J.H Jung. 2002. Polyacetylenes from a marine sponge *Petrosia* sp. inhibit DNA replication at the level of initiation. *Cancer Lett.* 185: 95–101.

Kobayashi, J., K. Naitoh, K. Ishida, H. Shigemori, and M. Ishibashi, M. 1994a. Nepheliosyne A, new C_{47} acetylenic acid from the Okinawan marine sponge *Xestospongia* sp. *J. Nat. Prod.* 57: 1300–1303.

Kobayashi, M., K. Kawazoe, T. Okamoto, T. Sasaki, and I. Kitagawa. 1994b. Marine natural products. XXXI. Structure-activity correlation of a potent cytotoxic dimeric macrolide swinholide A, from the Okinawan marine sponge *Theonella swinhoei*, and its isomers. *Chem. Pharm. Bull.* 42: 19–26.

Kobayashi, M., T. Mahmud, H. Tajima, W. Wang, S. Aoki, S. Nakagawa, T. Mayumi, and I. Kitagawa. 1996. Marine natural products. XXXVI. Biologically active polyacetylenes, adociacetylenes A, B, C, and D, from an Okinawan marine sponge, *Adocia* sp. *Chem. Pharm. Bull.* 44: 720–724.

Kulkarni, B.A., S. Chattopadhyay, A. Chatoopadhyay, and V.R. Mamdapur. 1993. Synthesis of some bioactive acetylenic alcohols; components of the marine sponge *Cribrochalina vasculum*. *J. Org. Chem.* 58: 5964–5966.

Lebel, H., and E.N. Jacobsen. 1998. Enantioselective total synthesis of taurospongin A. *J. Org. Chem.* 63: 9624–9625.

Lee, H.-S., J.-R. Rho, C.J. Sim, and J. Shin. 2003. New acetylenic acids from a sponge of the genus *Stelletta*. *J. Nat. Prod.* 66: 566–568.

Lerch, M.L., M.K. Harper, and D.J. Faulkner. 2003. Brominated polyacetylenes from the Philippines sponge *Diplastrella* sp. *J. Nat. Prod.* 66: 667–670.

Li, H.-Y., S. Matsunaga, and N. Fusetani. 1994. Cotricatic acids A-C, antifungal acetylenic acids from the marine sponge, *Petrosia corticata*. *J. Nat. Prod.* 57: 1464–1467.

Li, Y., M. Ishibashi, T. Sasaki, and J. Kobayashi. 1995. New bromine-containing unsaturated fatty acid derivatives from the Okinawan marine sponge *Xestospongia* sp. *J. Chem. Res. (S)*. 126–127; see also *J. Chem. Res. (M)* . 0901–0909.

Lim, Y.J., J.S. Kim, K.S. Im, J.H. Jung, C.-O. Lee, J. Hong, and D.-K. Kim. 1999. New cytotoxic polyacetylenes from the marine sponge *Petrosia*. *J. Nat. Prod.* 62: 1215–1217.

Lim, Y.J., H.S. Park, K.S., Im, C.-O. Lee, J. Hong, M.-Y. Lee, D.-K. Kim, and J.H. Jung. 2001. Additional cytotoxic polyacetylenes from the marine sponge *Petrosia* species. *J. Nat. Prod.* 64: 46–53.

Lim, Y.J., C.-O. Lee, J. Hong, D.-K. Kim, K.S. Im, and J.H. Jung. 2001. Cytotoxic polyacetylenic alcohols from the marine sponge *Petrosia* species. *J. Nat. Prod.* 64: 1565–1567.

Loya, S., A. Rudi, Y. Kashman, and A. Hizi. 2002. Mode of inhibition of HIV-1 reverse transcriptase by polyacetylenetriol, a novel inhibitor of RNA- and DNA-directed DNA polymerases. *Biochem. J.* 362: 685-692.

Matsunaga, S., Y. Okada, N. Fusetani, and R.W.M. van Soest. 2000. An antimicrobial C_{14} acetylenic acid from a marine sponge *Oceanapia* species. *J. Nat. Prod.* 63: 690–691.

Matthew, W.B., F.H. Yali, H.C. Join, and R.B. Michael. 1994. Stereochemistry of enynols- a caveat on the exciton chirality method. *Tetrahedron Lett.* 35: 993–994.

Miao, S., and R.J. Andersen. 1991. Callydiyne, a new diacetylenic hydrocarbon from the sponge *Callyspongia flammea*. *J. Nat. Prod.* 54: 1433–1434.

Morishita, K., M. Kamezawa, T. Ohtani, H. Tachibana, M. Kawase, M. Kishimoto, and Y. Naoshima. 1999. Chemoenzymatic synthesis of (+)-docosa-4, 15-dien-1-yn-3-ol, a component of the marine sponge *Cribrochalina vasculum*, and confirmation of the structure and absolute configuration of the acetylenic alcohol, by lipase-catalysed biotransformations. *J. Chem. Soc. Perkin Trans.* I. 513–518.

Nishimura, S., S. Matsunaga, M. Shibazaki, K. Suzuki, N. Harada, H. Naoki, and N. Fusetani. 2002. Corticatic acids D and E, polyacetylenic geranylgeranyltransferase type I inhibitors, from the marine sponge *Petrosia corticata. J. Nat. Prod.* 65: 1353–1356.

Nakao, Y., T. Uehara, S. Matsunaga, N. Fusetani, and R. van Soest, R. 2002. Callyspongynic acid, a polyacetylenic acid which inhibits α-glucosidase, from the marine sponge *Callyspongia truncata. J. Nat. Prod.* 65: 922–924.

Ochi, M., S. Arika, A. Tatsukawa, H. Kotsuki, Y. Fukuyama, and K. Shibata. 1994. Bioactive polyacetylenes from the marine sponge *Petrosia* sp. *Chem. Lett.* 89–92.

Ohta, S., H. Okada, H. Kobayashi, J.M. Oclarit, and S. Ikegami. 1993. Clathrynamides A, B, and C: Novel amides from a marine sponge *Clathria* sp. that inhibit cell division of fertilized starfish eggs. *Tetrahedron Lett.* 34: 5935–5938.

Ohtani, T., K. Kikuchi, M. Kamezawa, H. Hamatani, H. Tachibana, T. Totani, and Y. Naoshima. 1996. Chemoenzymatic synthesis of (+)-(*4E, 15E*)-docosa-4, 15-dien-1-yn-3-ol, a component of the marine sponge *Cribrochalina vasculum. J. Chem. Soc., Perkin Trans.* 1: 961–962.

Ortega, M.J., E. Zubia, J.L. Carballo, and J. Salva. 1996. Fulvinol, a new long-chain diacetylenic metabolite from the sponge *Reniera fulva. J. Nat. Prod.* 59: 1069–1071.

Patil, A.D., W.C. Kokke, S. Cochran, T.A. Francis, T. Tomszek, and J.W. Westley. 1992. Brominated polyacetylenic acids from the marine sponge *Xestospongia muta*: Inhibitors of HIV protease. *J. Nat. Prod.* 55: 1170–1177.

Paul, V.J., and W. Fenical. 1980. Toxic acetylene-containing lipids from the red marine alga *Liagora farinosa* Lamouroux. *Tetrahedron Lett.* 21: 3327–3330.

Perry, N.B., E.M. Span, and C. Zidorn, 2001. Aciphyllal–a C_{34}-polyacetylene from *Aciphylla scott-thomsonii* (Apiaceae). *Tetrahedron Lett.* 42: 4325–4328.

Pham, N.B., M.S. Butler, J.N. Hooper, R.W. Moni, and R.J. Quinn. 1999. Isolation of xestosterol esters of brominated acetylenic fatty acids from the marine sponge *Xestospongia testudinaria. J. Nat. Prod.* 62: 1439–1442.

Quinn, R.J., and D.J. Tucker. 1985. A brominated bisacetylenic acid from the marine sponge *Xestospongia testudinaria. Tetrahedron Lett.* 26: 1671–1672.

Quinn, R.J., and D.J. Tucker. 1991. Further acetylenic acids from the marine sponge *Xestospongia testudinaria. J. Nat. Prod.* 54: 290–294.

Quinoa, E., and P. Crews. 1988. Melynes, polyacetylene constituents from a Vanuatu marine sponge. *Tetrahedron Lett.* 29: 2037–2040.

Rashid, M.A., K.R. Gustafson, and M.R. Boyd. 2000. Pellynol I, a new cytotoxic polyacetylene from the sponge *Pellina* sp. *Nat. Prod. Lett.* 14: 387–392.

Richelle-Maurer, E., R. Gomez, J-C. Braekman, G. Van de Vyver, R.W.M. Van Soest, and C. Devijver. 2003. Primary cultures from the marine sponge *Xestospongia muta. J. Biotechnol.* 100: 169–176.

Rooney, F., and R.J. Capon. 1998. Callyspongynes A and B: New polyacetylenic lipids from a Southern Australian marine sponge, *Callyspongia* sp. *Lipids* 33: 639–642.

Rotem, M., and Y. Kashman. 1979. New polyacetylenes from the sponge *Siphonochalina* sp. *Tetrahedron Lett.* 3193–3196.

Schmitz, F.J., and Y. Gopichand. 1978. (*7E, 13E,* 15*Z*)-14, 16-dibromo-7, 13, 15-hexadecatrien-5-ynoic acid. A novel dibromo acetylenic acid from the marine sponge *Xestospongia muta. Tetrahedron Lett.* 39: 3637–3640.

Seo, Y., K.W. Cho, J.-R. Rho, J. Shin, and C.J. Sim. 1998. Petrocortynes and petrosiacetylenes, novel polyacetylenes from a sponge of the genus *Petrosia*. *Tetrahedron* 54: 447–462.

Sharma, A., and S. Chattopadhyay. 1998. Chemoenzymatic synthesis of (*S*)-eicos-(4*E*)-en-1-yn-3-ol, the cytotoxic principle of the marine sponge, *Cribrochalina vasculum*. *Tetrahedron: Asymmetry* 9: 2635–2639.

Shen, Y.C., and C.V. Prakash. 2000. Two new acetylenic derivatives and a new meroditerpenoid from a Taiwanese marine sponge *Strongylophora durissima*. *J. Nat. Prod.* 63: 1686–1688.

Shim, S.C., H.Y. Koh, and S. Chang. 1985. Determination of absolute stereochemistry of panaxynol. *Tetrahedron Lett.* 26: 5775–5776.

Shin, J., Y. Seo, K.W. Cho, J.-R., Rho, and V.J. Paul. 1998. Osirisynes A-F, highly oxygenated polyacetylenes from the sponge *Haliclona osiris*. *Tetrahedron* 54: 8711–8720.

Shin, J., Y. Seo, and K.W. Cho. 1998. Five new polyacetylenes from a sponge of the genus *Petrosia*. *J. Nat. Prod.* 61: 1268–1273.

Tada, H., and F. Yasuda. 1984. Siphonodiol, a new polyacetylenic metabolite from the sponge *Siphonochalina truncata*. *Chem. Lett.* 779–780.

Tan, L.T., T. Okino, and W.H. Gerwick. 2000. Hermitamides A and B, toxic malyngamide-type natural products from the marine cyanobacterium *Lyngbya majuscula*. *J. Nat. Prod.* 63: 952–955.

Tsukamoto, S., H. Kato, H. Hirota, and N. Fusetani. 1997. Seven new polyacetylenes derivatives, showing both potent metamorphosis-inducing activity in ascidian larvae and antifouling activity against barnacle larvae, from the marine sponge *Callyspongia truncata*. *J. Nat. Prod.* 60: 126–130.

Umeyama, A., C. Nagano, and S. Arihara. 1997. Three novel C_{21} polyacetylenes from the marine sponge *Callyspongia* sp. *J. Nat. Prod.* 60: 131–133.

Uno, M., S. Ohta, E. Ohta, and S. Ikegami. 1996. Callyspongins A and B: novel polyacetylenes sulfates from marine sponge *Callyspongia truncata* that inhibit fertilization of starfish gametes. *J. Nat. Prod.* 59: 1146–1148.

Watanabe, K., Y. Tsuda, Y. Yamane, H. Takahashi, K. Iguchi, H. Naoki, T. Fujita, and R. Van Soest. 2000. Strongylodiols A, B and C, new cytotoxic acetylenic alcohols isolated from the Okinawan marine sponge of the genus *Strongylophora* as each enantiomeric mixture with a different ratio. *Tetrahedron Lett.* 41: 9271–9276.

Williams, D.H., and D.J. Faulkner. 1996. Three new acetylenes from the Palauan sponge *Haliclona* sp. *J. Nat. Prod.* 59: 1099–1101.

Wright, A.E., O.J. McConnell, S. Kohmoto, M.S. Lui, W. Thompson, and K.M. Snader. 1987. Duryne, a new cytotoxic agent from the marine sponge *Cribrochalina dura*. *Tetrahedron Lett.* 28: 1377–1380.

Yadav, J.S., and R.K. Mishra. 2002. First total synthesis of strongylodiol A. *Tetrahedron Lett.* 43: 1739–1741.

Yosief, T., A. Rudi, Y. Wolde-ab, and Y. Kashman. 1998. Two new C_{22} 1, 2-dioxane polyketides from the marine sponge *Acarnus* cf. *bergquistae*. *J. Nat. Prod.* 61: 491–493.

Youssef, D.T.A., W.Y. Yoshida, M. Kelly, and P.J. Scheuer. 2000. Polyacetylenes from a Red Sea sponge *Callyspongia* species. *J. Nat. Prod.* 63: 1406–1410.

Youssef, D.T.A., R.W.M. van Soest, and N. Fusetani. 2003a. Callyspongenols A–C, new cytotoxic C_{22}-polyacetylenic alcohols from a Red Sea sponge, *Callyspongia* species. *J. Nat. Prod.* 66: 679–681.

Youssef, D.T.A., R.W.M. van Soest, and N. Fusetani. 2003b. Callyspongamide A, a new cytotoxic polyacetylenic amide from the red sea sponge *Callyspongia fistularis*. *J. Nat. Prod.* 66: 861–862.

Zhao, Q., S.-Y. Lee, J. Hong, C.-O. Lee, K.S. Im, C.J. Sim, D.S. Lee, and J.H. Jung. 2003. New acetyelenic acids from the marine sponge *Stelletta* species. *J. Nat. Prod.* 66: 408–411.

Zhou, G.-X., and T.F. Molinski. 2003. Long-chain acetylenic ketones from the Micronesian sponge *Haliclona* sp. Importance of the 1-yn-3-ol group for antitumor activity. *Mar. Drugs* 1: 46–53.

13

Bioactive Natural Products from Nudibranchs

Masami Ishibashi,[1] Yoshihiro Yamaguchi[1] and Yoshiaki J. Hirano[2]

[1]Graduate School of Pharmaceutical Sciences, Chiba University, Chiba 263-8522, Japan
[2]Marine Biosystems Research Center, Chiba University, Amatsu-Kominato, Chiba 299-5502, Japan

INTRODUCTION

Opisthobranchs, in which nudibranchs are included, represent a diverse group of gastropod molluscs. They express all grades of body organization from fully shelled to completely shell-less, and include the most colorful and structurally diversified forms of gastropods. The evolutionary loss or reduction of shell, the most characteristic molluscan organ of mechanical defense, brings with it several problems, but opisthobranchs have few known predators. It is known that other defense adaptations including behavioral, morphological and biochemical strategies may compensate for their shell loss (Thompson 1960, Ros 1976, Faulkner and Ghiselin 1983, Avila 1995).

Within opisthobranchs, the ability of nudibranchs to sequester either morphological or chemical defenses from their prey has been well documented (reviewed in Todd 1981, Avila 1995). Nudibranchs are carnivorous feeding primarily upon sessile animals such as sponges, bryozoans, acidians and cnidarians; most nudibranch species are specialist predators showing a preference for a single prey species, or group of species (Todd 1981, Rudman and Willan 1998). Aeolid nudibranchs which are specialists on cnidarians are known to translocate nematocysts into cerata to use them for their own defense (Thompson 1960, Harris 1973, Edmunds 1974). However, sequestration of distasteful secondary metabolites is much more widely distributed (see Avila 1995), and perhaps more important as a defense strategy. Some nudibranchs may even biosynthesize defensive compounds *de novo* (e.g., Cimino *et al.* 1983, Avila *et al.* 2000, Iken *et al.* 2002).

In relation to chemical defense, warning coloration has been suggested for some nudibranchs (Pawlik *et al.* 1988, Johnson 1989, Brunckhorst 1991, Rudman 1986, 1991). Nudibranchs may provide some critical evidence relevant to the question of the evolutionary origin of aposematism and related topics, as suggested by Faulkner and Ghiselin (1983). It has been also suggested that chemical defense may be the catalyst for the evolution of opisthobranch gastropods, collection of which and analysis may provide opportunities to examine predator-prey coevolution (Fahey and Garson 2002). In order to develop these studies co-operative, multidisciplinary research is indispensable, and experiments and activity tests with ecological meaning are urgently needed.

Opisthobranch natural products have also proved to be a useful source for medical applications (Avila 1995). Thus, studies of opisthobranch natural products are still much required and these animals are most promising subjects for multiple research fields such as chemistry, pharmacology and biology.

CHEMICAL STUDIES

Since the 1995 review article "Natural products of opisthobranch mollusks: a biological review" was published by C. Avila (Avila 1995), a number of new natural products have been isolated from nudibranchs. Avila reported that most of those natural products possess unique biological activities, such as ichthyotoxicity, feeding deterrent and antifeedant, toxicity against *Artemia salina*, alarm pheromone, sexual pheromone, antibacterial and antifungal activities and others. This section describes recent studies on isolation and structure determination of natural products from nudibranchs together with their unique biological activities, which were published after the review by Avila. Because many natural products have been isolated from nudibranchs in these recent years, the selection of compounds here may seem arbitrary but it depends on the interest of the reviewer to some extent. This section consists mainly of two parts, which deal with (1) isolation studies of nudibranch natural products; and (2) our recent studies on natural products from nudibranchs collected in the area of Chiba, Japan.

Isolation Studies of Nudibranch Natural Products

Sesquiterpenes

Nudibranchs are well known to sequester secondary metabolites from their diets, such as from sponges, to protect their soft bodies from predators, and their chemical defense could be used for development of an antifouling strategy (Kobayashi and Ishibashi 1999). Fusetani and Hirota's group investigated antifouling substances from Japanese marine invertebrates, and they found that extracts of four nudibranchs of the family Phyllidiidae inhibited larval settlement and metamorphosis of the barnacle *Balanus amphitrite*. Three new sesquiterpene isocyanides, 10-*epi*-axisonitrile-3 (**1**), 10-isocyano-4-cadinene (**2**), and 2-isocyanotrachyopsane (**3**), and a new sesquiterpene peroxide, 1,7-epidioxy-5-cadinene (**4**) were isolated from *Phyllidia pustulosa*, *P. varicosa*, and *P. krempfi*, along with six known sesquiterpenes, axisnitrile-3 (**5**), (-)-10-isothiocyanato-4-amorphene (**6**), 2-thiocyanato*neo*pupukeanane (**7**), and 4-thiocyanato*neo*pupukeanane (**8**), 3-isocyanotheonellin (**9**), and 10-isocyano-4-amorphene (**10**). These compounds with isocyano, isothiocyanato, and thiocyanato functionalities (**1-3** and **5-10**) showed potent antifouling activity against larvae of the barnacle *Balanus amphitrite* (IC_{50} values, 0.13-10 µg/mL), while compound **4** with a peroxide moiety showed no activity at 50 µg/mL (Okino *et al.* 1996).

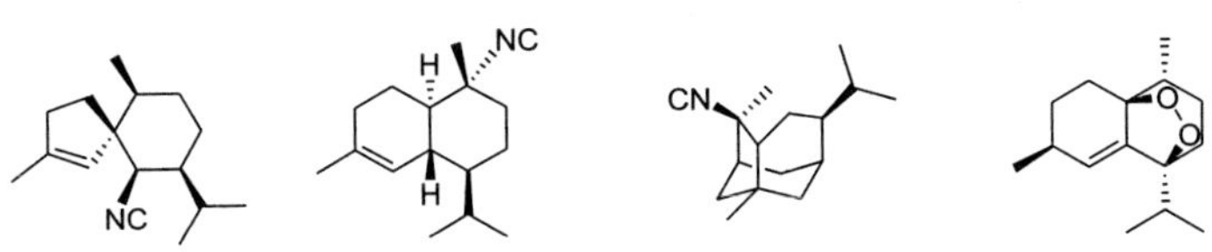

1 **2** **3** **4**

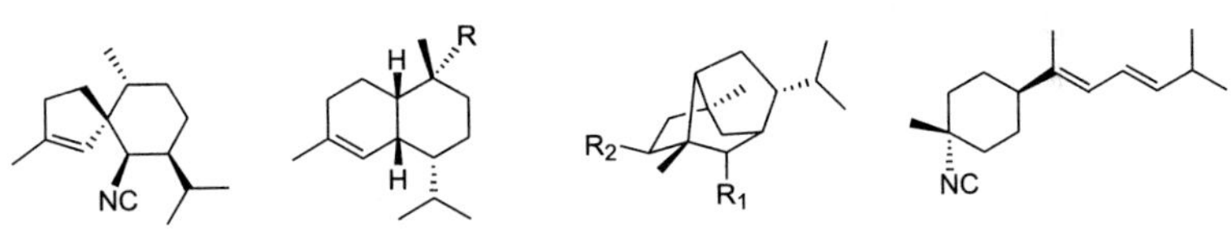

5 **6** R = NCS **10** R = NC **7** R_1 = SCN, R_2 = H **8** R_1 = H, R_2 = SCN **9**

Another new isocyanosesquiterpene alcohol (**11**) was obtained from the nudibranch *Phyllidia pustulosa* and showed antifouling activity against cypris larvae of the barnacle *Balanus amphitrite* with an IC_{50} value of 0.17 µg/mL. This compound (**11**) had almost the same potent antifouling activity as $CuSO_4$ with very weak lethality and this is the first case of an antifouling isocyanide with a hydroxyl moiety from nudibranchs (Hirota *et al.* 1998).

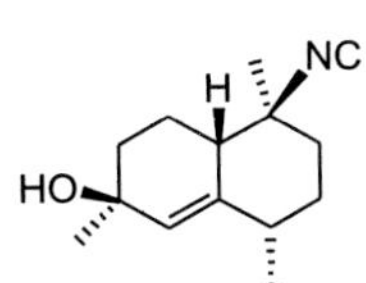

11

From the mantle extract of the Antarctic opisthobranch mollusk *Bathydoris hodgsoni*, a new drimane sesquiterpene, hodgsonal (**12**), was isolated, and this compound was the first example of a 2-substituted drimane sesquiterpene from marine organisms. Levels of **12** in individual animals are very similar and independent of both locality and depth, and the localization of hodgsonal (**12**) in the mantle and the analogies with related drimane sesquiterpenes may indicate a defensive role of **12** against predators (Iken *et al.* 1998).

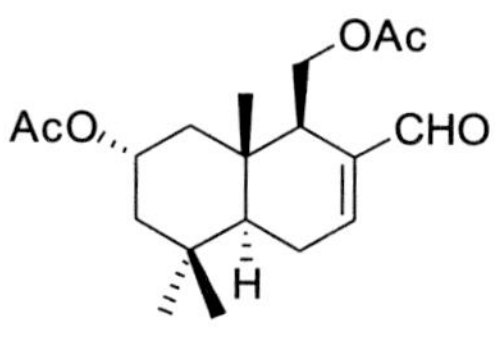

12

The Patagonian nudibranch *Tyrinna nobilis* contained a number of terpenoids, a new seco-11,12-spongiane tyrinnal (**13**) and the known sesquiterpenoids dendrolasin (**14**), pallescenisn A (**15**), and dehydropallescensin-2 (**16**). These metabolites were probably derived from dietary sponges (Fontana *et al.* 1998).

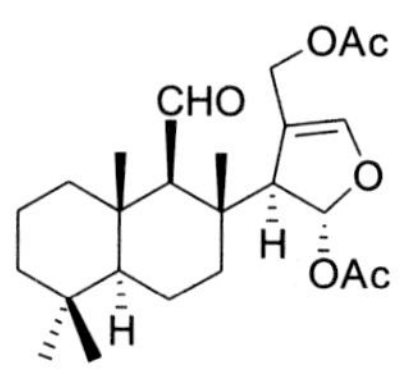

13

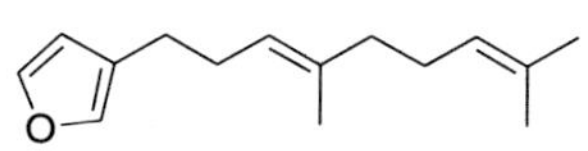

14

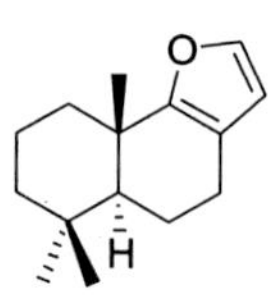

15

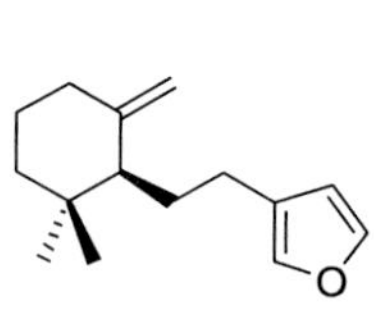

16

Two new sesquiterpenes, reticulidins A (**17**) and B (**18**), containing a rare functional group, $N{=}CCl_2$, were isolated together with two known related carbonimidic dichlorides (**19** and **20**) from the nudibranch *Reticulidia fungia* (family Phyllidiidae) collected from Irabu Island, Okinawa. Both of these unique chlorine-containing compounds (**17** and **18**) exhibited moderate cytotoxicity against KB and L1210 cells (IC_{50} values 0.11-0.59 μg/mL) (Tanaka and Higa 1999).

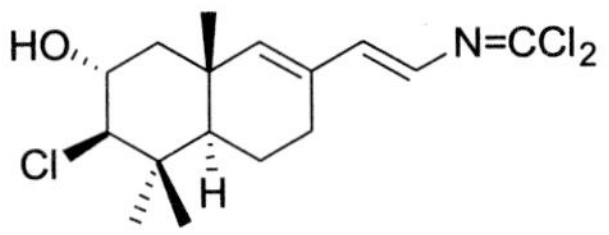

17

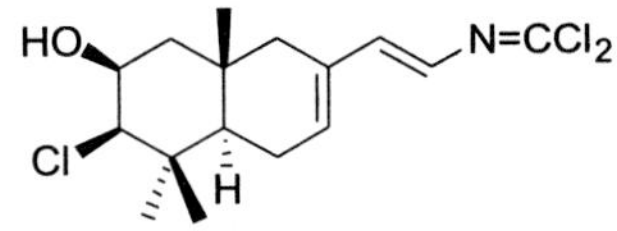

18

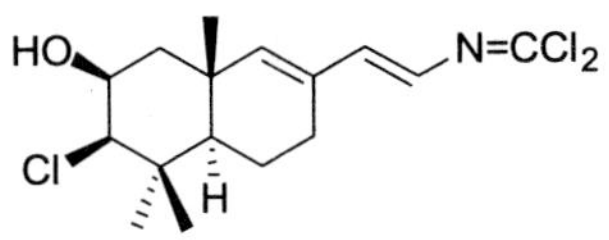

19

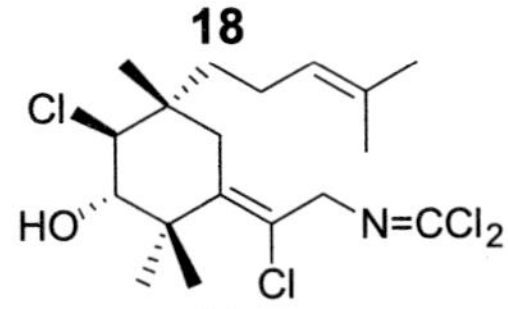

20

From the Arminacean nudibranch *Leminda millecra* collected in Algoa Bay, South Africa, nine new compounds, two sesquiterpenes (**21** and **22**) and seven triprenylquinones (**23-29**), were isolated together with known sesquiterpenes such as **30** and **31**. The GC-MS survey of extracts of *L. millecra*'s possible octocoral prey species in Algoa Bay identified the soft coral *Alcyonium fauri* as the source of **30** and the gorgonian *Leptogorgia palma* as the species producing **22** and **31** (McPhail *et al*. 2001).

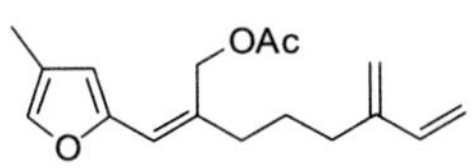

21

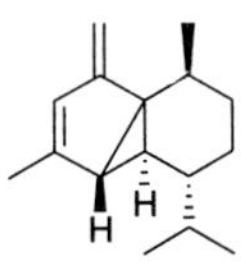

22

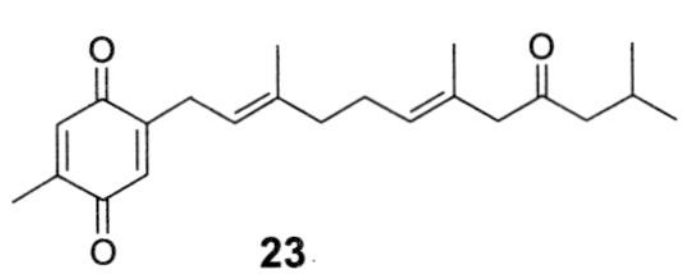

23

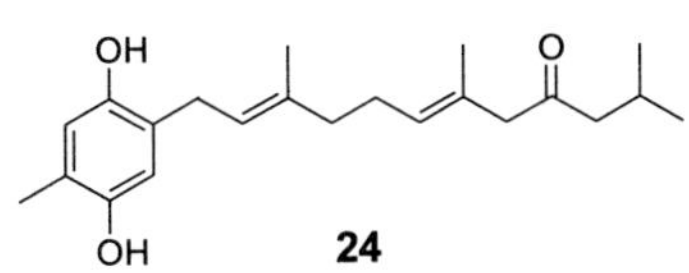

24

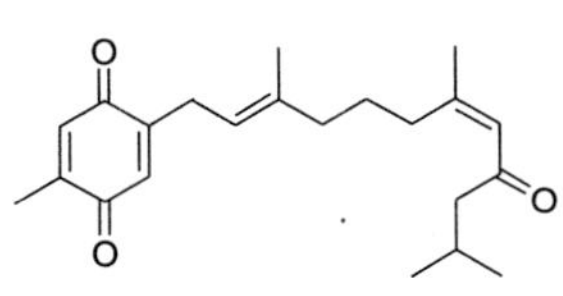

25

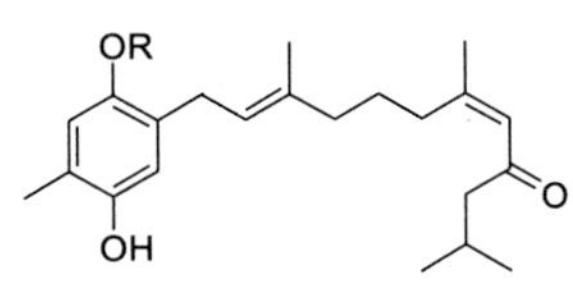

26 R = H
27 R = Ac

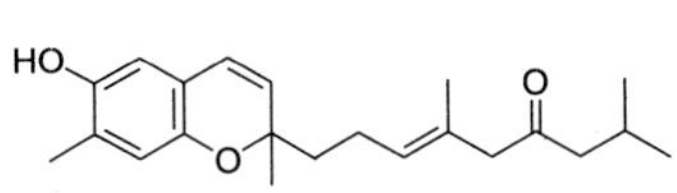

28

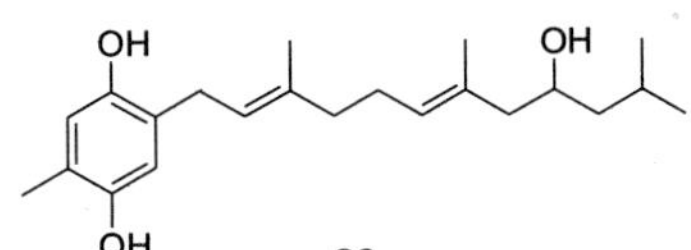

29

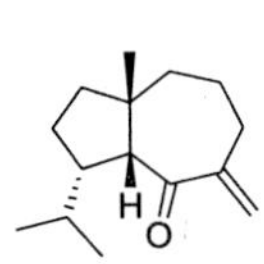

30

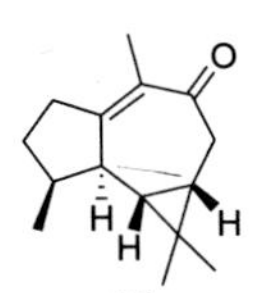

31

Two known furanosesquiterpenes, nakafuran-8 (**32**) and nakafuran-9 (**33**) were obtained as a 3:1 mixture from the chromodorid nudibranch *Hypselodoris infucata,* collected from Kaneohe Bay, Oahu, Hawaii, and the rather uncommon nudibranch, *Chromodoris petechialis,* collected in Koko Head, Oahu, contained a diterpene, spongian-16-one (**34**). Extraction of the nudibranch *Nembrotha kubaryana,* collected from the lagoon of Pohnpei in the Federated State of Micronesia, yielded the known though unusual blue pigment **35**. This pigment was previously isolated from blue ascidians and a blue bryozoan. Since nudibranchs of the family Nembrothiadae are known to feed on either ascidians or bryozoans, the metabolite is presumably diet-derived (Karuso and Scheuer 2002).

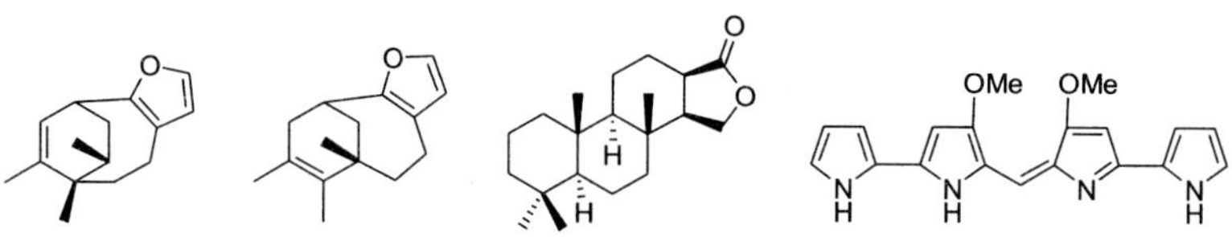

32 **33** **34** **35**

Two *nor*-sesquiterpenes with a new carbon skeleton, austrodoral (**36**) and its oxidized derivative austrodoric acid (**37**), were isolated from the skin of the marine dorid *Austrodoris kerguelenesis,* collected in Antarctica. It may be interesting to note that the level of austrodoral (**36**) detected in the animal was significantly more abundant in those individuals kept in an aquarium for 15 days, suggesting for austrodoral (**36**) a probable role of a stress-metabolite (Gavagnin *et al.* 2003).

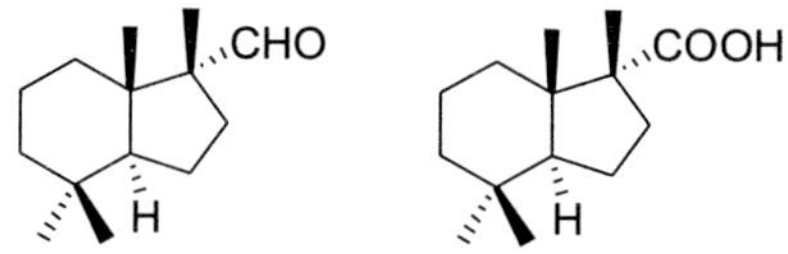

36 **37**

Gas-chromatographic and mass-spectrometric and NMR analyses of the lipophilic extract of the nudibranch *Phyllidiella pustulosa* indicated that the nudibranch accumulated some of the sponge metabolites. The main components of the extracts were identified as the new sesquiterpene, 10-isothiocyano-4-cadinene (**38**) and axisonitrile-3 (**39**). The new compound (**38**) was shown to have moderate antiplasmodial activity (IC_{50} 1.5 μg/mL towards *Plasmodium falciparum*) (Wright 2003).

Two new 9-thiocyanatopupukeanane sesquiterpene isomers (**40** and **41**) were isolated as major metabolites from the MeOH extract of the nudibranch *Phyllidia varicosa,* and they were also isolated from the diet sponge *Axinyssa aculeata.* The isolated sesquiterpenes were found to be

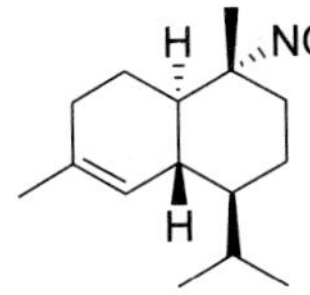

38

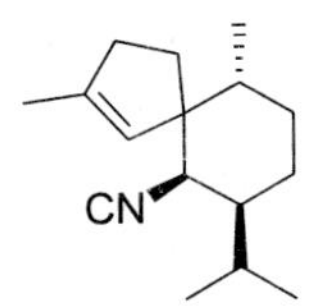

39

toxic toward brine shrimp at an LC_{50} of 5 ppm. At a dose level of 20 μg, they were found to be weakly and moderately active against *Bacillus subtilis* and *Candida albicans* (Yasman *et al.* 2003).

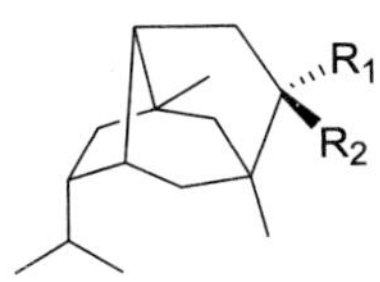

40 R_1 = SCN, R_2 = H
41 R_1 = H, R_2 = SCN

Diterpenes

Seven new cytotoxic spongian diterpenoids, dorisenones A (**42**), B (**43**), C (**44**), D (**45**), 7α-hydroxyspongian-16-one (**46**), 15α,16α-diacetoxy-11,12β-epoxyspongian (**47**), and 7α-acetoxydendrillol-3 (**48**) were isolated from a Japanese marine mollusk, *Chromodoris obsoleta* (Chromodorididae), collected from the rocky coast of Koinoura, Fukuoka Prefecture, Japan. The structure of **46** was elucidated by single crystal X-ray analysis. Compound **42** and **47** showed relatively strong cytotoxicity against L1210 and KB cells (IC_{50} values, 0.18-0.98 μg/mL), and the antitumor activity of the related known diterpenes (**49** and **50**), which were concurrently isolated from this nudibranch, was examined against P388 *in vivo*; however, they showed no significant activity (Miyamoto *et al.* 1996).

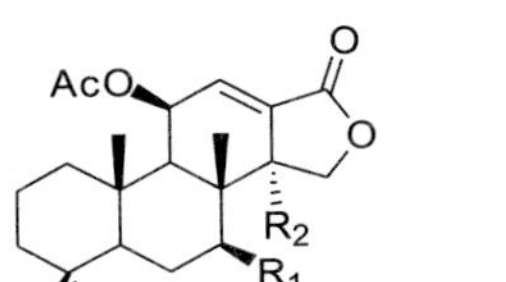

42 R_1 = OAc, R_2 = OH
43 R_1 = H, R_2 = OH
44 R_1 = OAc, R_2 = H

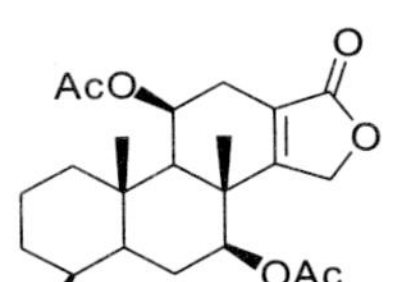

45

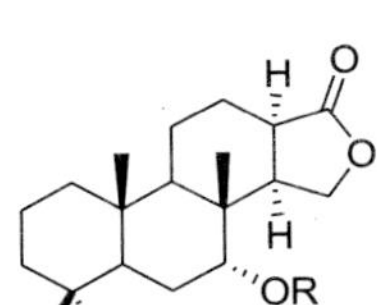

46 R = H
49 R = Ac

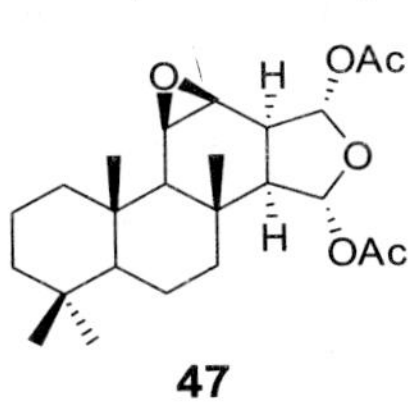

47

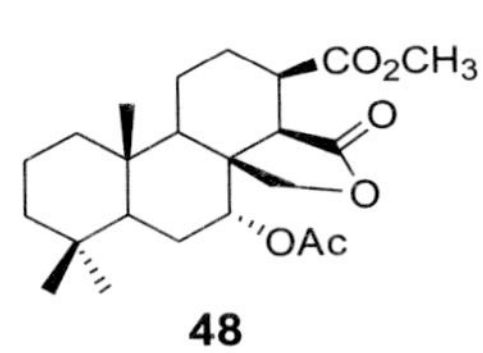

48

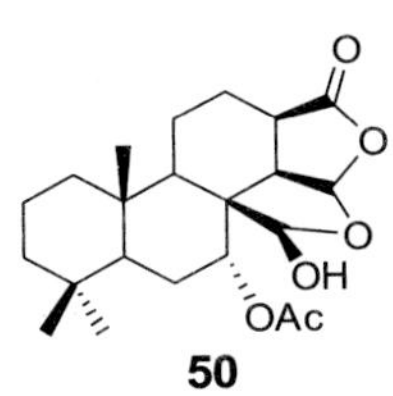

50

The brilliantly colored nudibranch *Chromodoris hamiltoni*, collected from Aliwal Shoal, near Durban, South Africa, was found to contain four unusual chlorinated homo-diterpenes, hamiltonins A – D (**51** – **54**) and a new sesterterpene hamiltonin E (**55**). Hamiltonins A (**51**) and B (**52**), however, showed no significant antimicrobial or cytotoxic activity (Pika and Faulkner 1995).

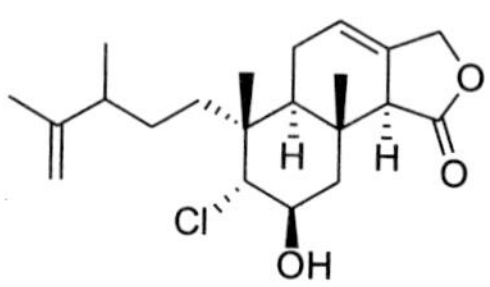

51

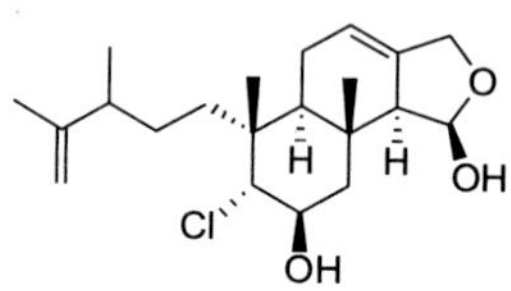

52

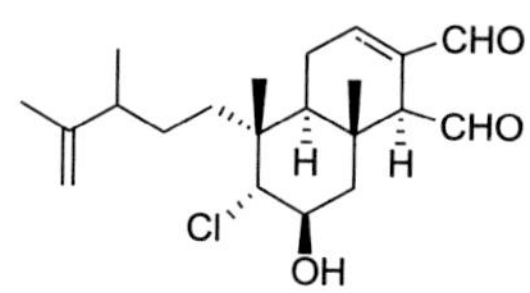

53

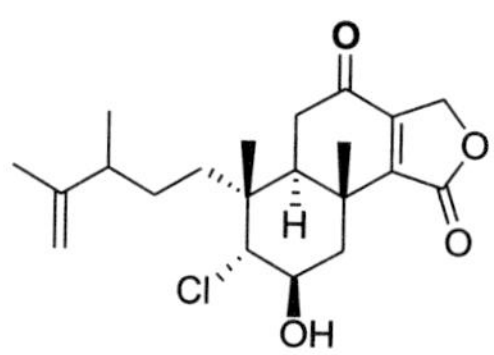

54

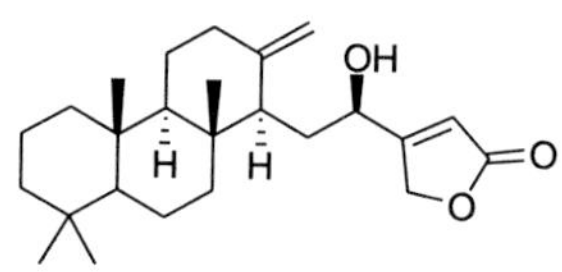

55

From the same nudibranch, *Chromodoris hamiltoni*, collected from Malangaan Reef off southern Mozambique (east African coast), two new spongian diterpene lactones, 7*β*,11*β*-diacetoxy-16-oxospongian-17-al (**56**) and 7*β*,11*β*-diacetoxy-16-oxospongi-12-en-17-al (**57**) were isolated (McPhail and Davies-Coleman 1997).

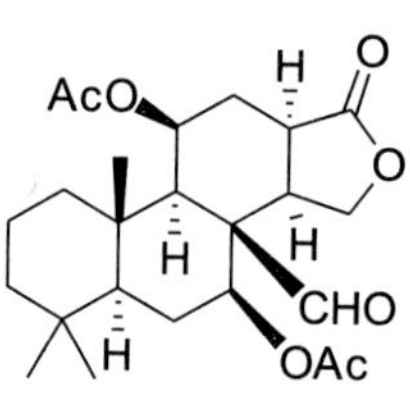

56

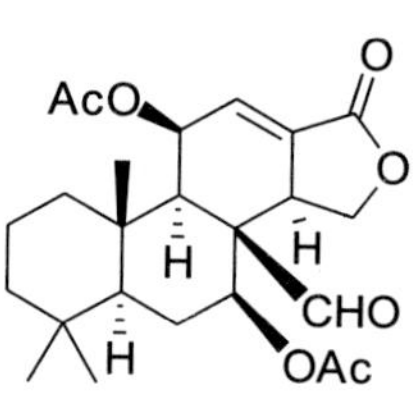

57

Terpenoid Glycerides

A relatively uncommon nudibranch, *Sclerodoris tanya*, collected from the intertidal zone in Southern California was found to contain two sesquiterpene glyceride esters, tanyolides A (**58**) and B (**59**). The structures including absolute configurations of these compounds were determined by synthesis. Tanyolides A (**58**) and B (**59**), which were located primarily in the dorsal mantle, deterred feeding by two ecologically-relevant predatory fish (Krug *et al*. 1995).

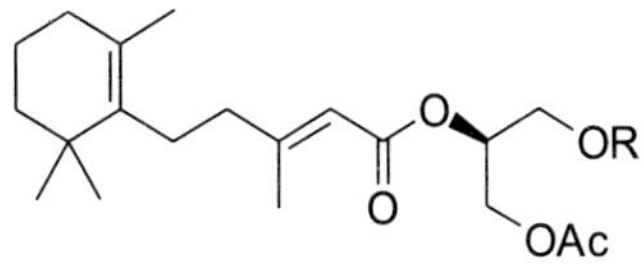

58 R = Ac
59 R = H

Diterpenoic acids linked to glycerine may be a chemical marker of dorid nudibranchs. The biological role of glyceryl esters, only present in the mantle tissue, should be linked to the protection of these shell-less molluscs. The monoglyceryl ester of a new halimane diterpenoid acid, austrodorin (**60**), was contained in the mantle tissue of the Antarctic nudibranch *Austrodoris kerguelenensis* (Gavagnin *et al.* 1995).

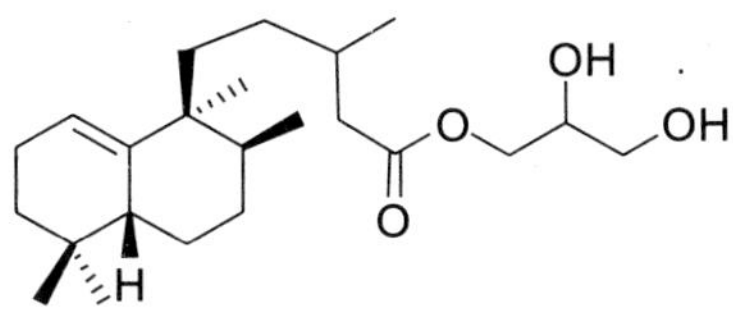

60

Nine new diterpenoid acid glycerides (**61** - **69**) were isolated from the skin extract of the Mediterranean nudibranch *Doris verrucosa*. Some of these diacylglycerols were shown to exhibit *in vivo* PKC-mediated morphogenetic properties in the *Hydra* tentacle regeneration assay (Gavagnin *et al.* 1997).

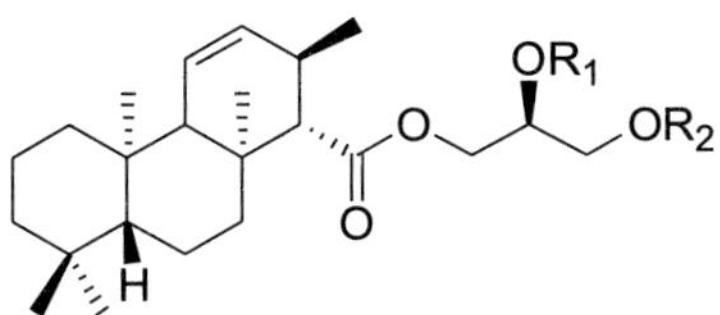

61 R_1 = H, R_2 = Ac
62 R_1 = Ac, R_2 = H

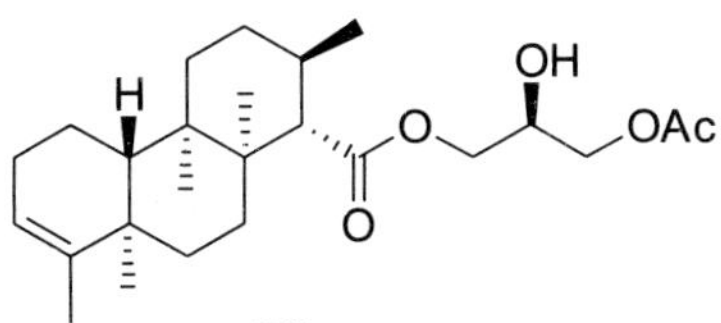

63

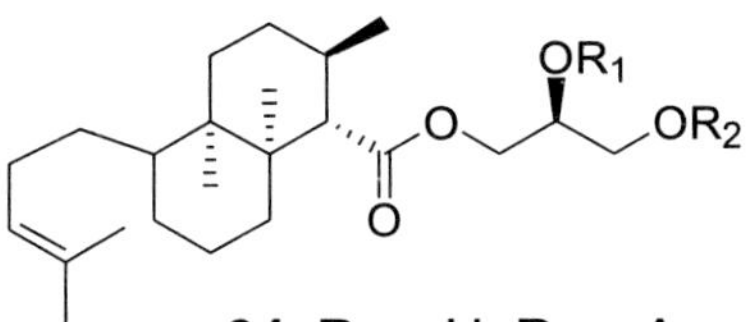

64 R_1 = H, R_2 = Ac
65 R_1 = Ac, R_2 = H

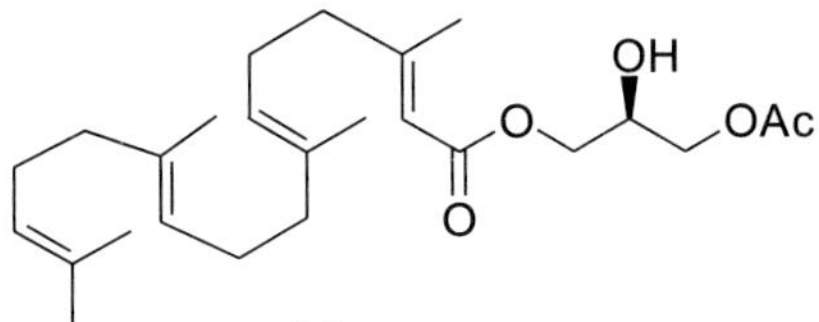

66

0.35 and 3.1, respectively]. Compound **90** exhibited neuritogenic activity against the rat phochromocytoma PC-12 cell line at a concentration above 10 μg/mL (Miyamoto *et al.* 1999).

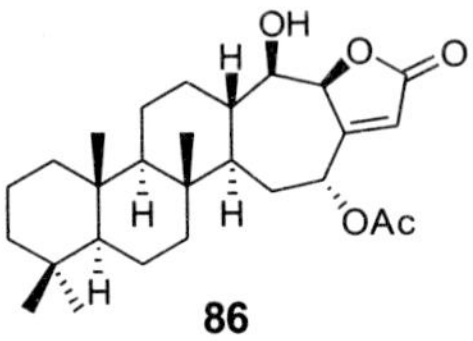

86

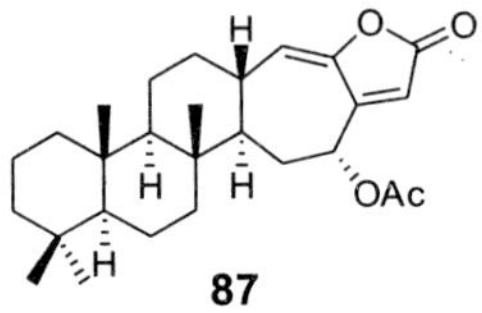

87

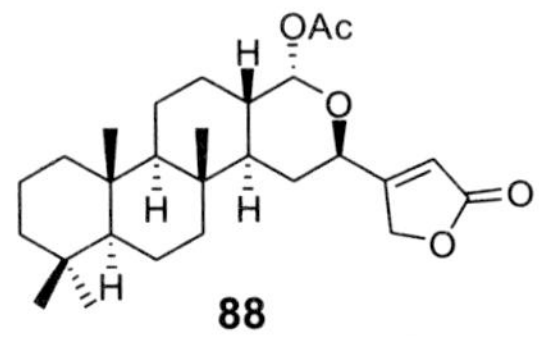

88

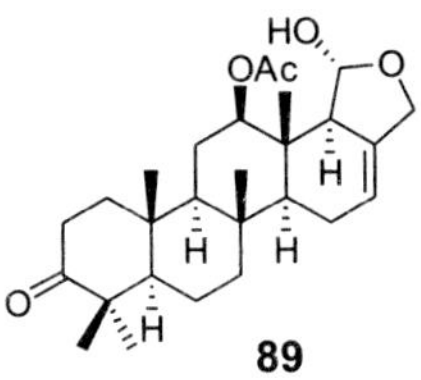

89

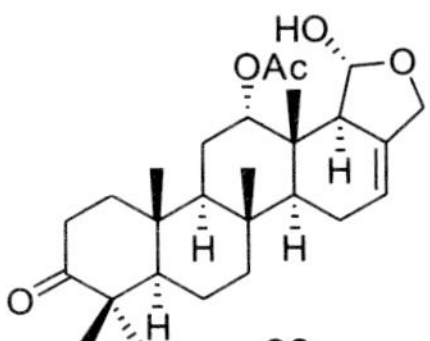

90

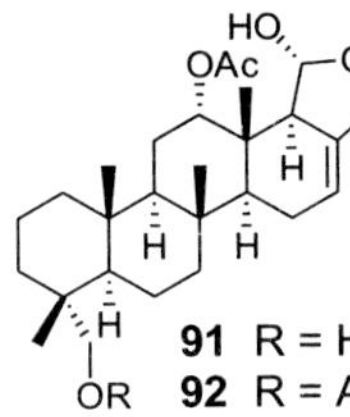

91 R = H
92 R = Ac

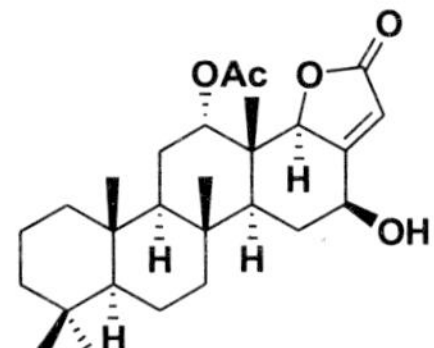

93

Two new scalarane sesterterpenes, 12-deacetoxy-12-oxo-deoxoscalarin (**94**) and 12-deacetyl-12-*epi*-deoxoscalarin (**95**), were isolated together with heteronemin (**96**) from the Indo-Pacific dorid nudibranch *Glossodoris atromarginata* (Chromodorididae) and its prey, an uncharacterized sponge probably belonging to the genus *Spongia*. Compounds **94** and **96** showed specific cytotoxicity against human thyroid carcinoma (25 and 30% of mortality, respectively), but compound **95** was inactive in this test (Fontana *et al.* 1999).

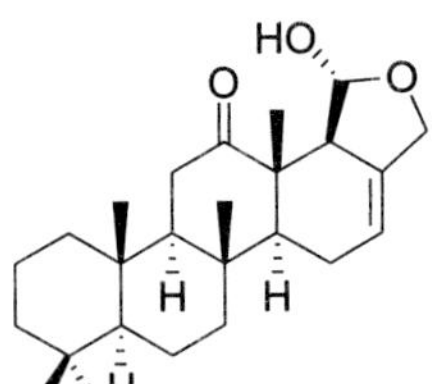

94

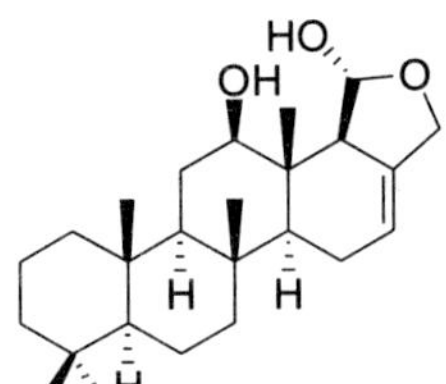

95

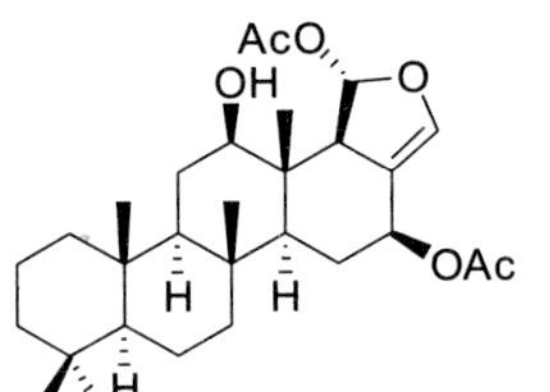

96

From two species of Pacific nudibranchs, *Glossodoris sedna* and *Glossodoris dalli*, a series of homoscalarane and scalarane sesterterpenoids (**97** - **102**) were isolated. The keto-aldehydo (**97**) was ichthyotoxic at 0.1 ppm against *Gambusia affinis* and showed moderate activity to inhibit mammalian phospholipase A_2 (Fontana *et al.* 2000a).

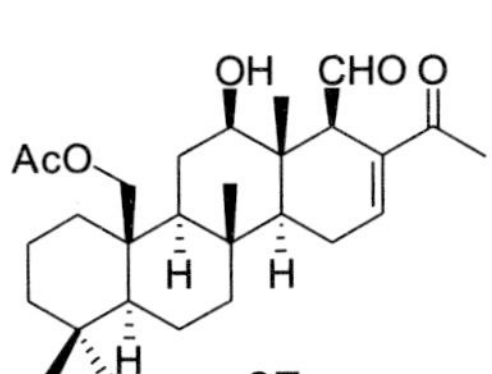
97

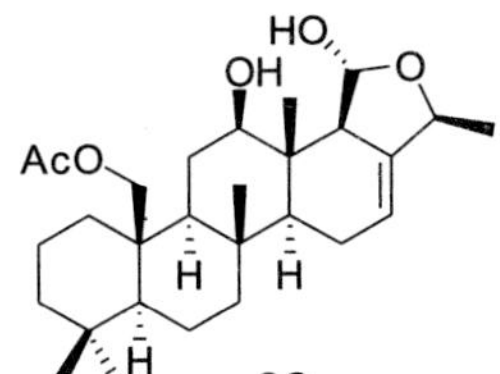
98

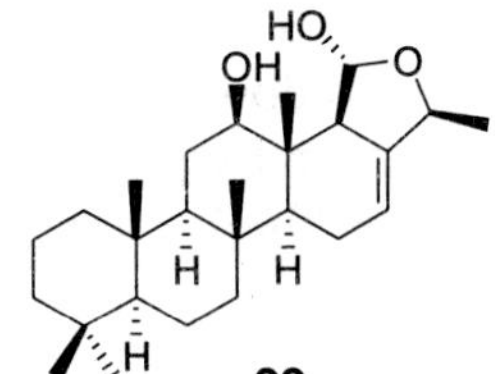
99

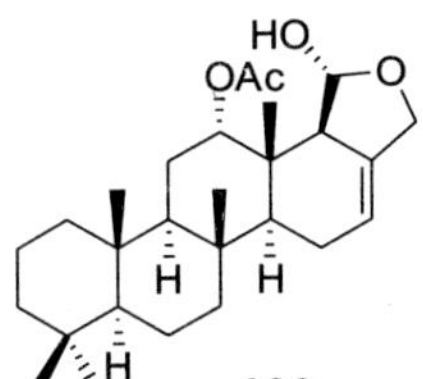
100

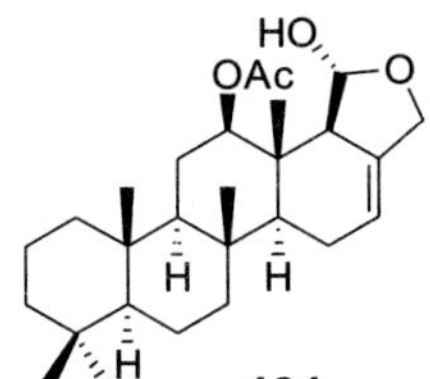
101

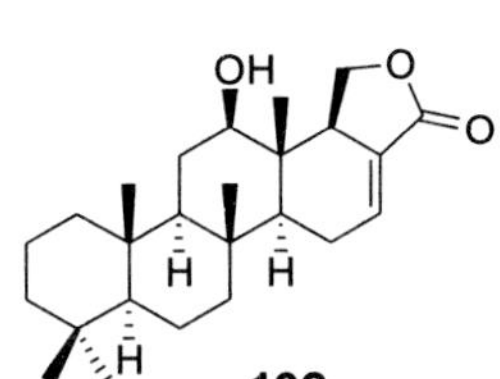
102

A new linear sesterterpene, 22-deoxy-23-hydroxymethylvariabilin (**103**), was isolated from the endemic South African nudibranch *Hypselodoris capensis* together with related known sesterterpenes (**104** - **106**). These sesterterpenes were also yielded by one of *H. capensis*' dietary sponges, a *Fasciospongia* sp. collected simultaneously with the nudibranch (McPhail *et al.* 1998).

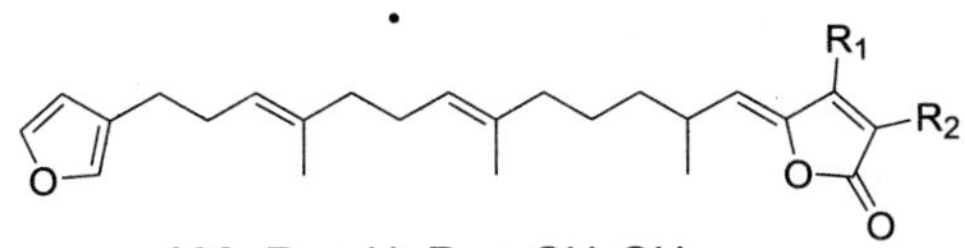

103 R_1 = H, R_2 = CH_2OH
104 R_1 = OH, R_2 = CH_3
105 R_1 = H, R_2 = CH_3
106 R_1 = H, R_2 = CH_2OAc

Others

From the Red Sea nudibranch *Hexabranchus sanguineus*, which is a large brilliantly colored shell-less sea slug, a deep-red pigment, hurghadin (**107**), was isolated from the mantle Et_2O extract of the animal (Guo *et al.* 1998).

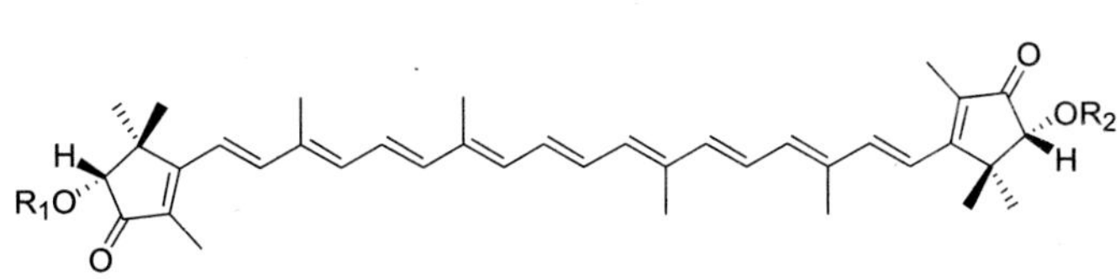

107 R_1 = 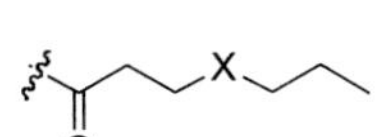 R_2 = 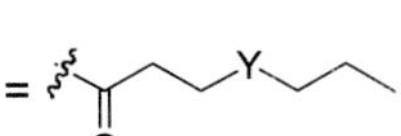

$R_1 + R_2 = C_{27:2}+C_{20:1}$ or $C_{20:2}+C_{18:0}$ or $C_{20:1}+C_{18:1}$ or $C_{19:0}+C_{18:1}$ or $C_{18:1}+C_{18:0}$

A degraded triterpenoid, lovenone (**108**), was isolated from the white or pale yellow dorid nudibranch *Adalaria loveni*, collected off Flatory Island near Bergen, Norway. The ring A degraded carbon skeleton of lovenone (**108**) appeared to be without precedent, and this compound showed modest *in vitro* cytotoxicity against human ovarian carcinoma HEY (ED_{50} 11 µg/mL) and human glioblastoma/astrocytoma U373 (ED_{50} 11 µg/mL) cell lines (Graziani *et al.* 1995).

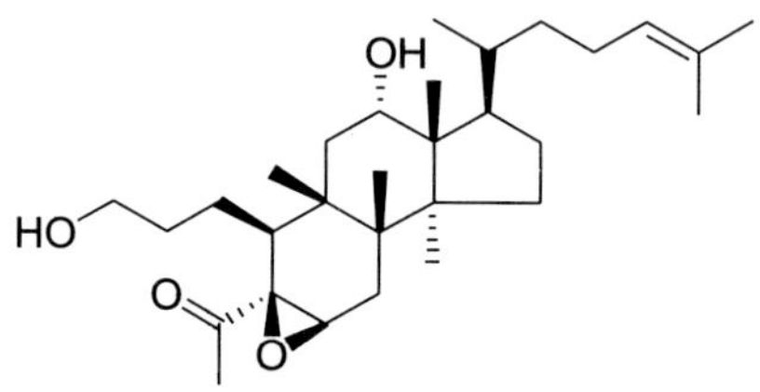

108

The short-side-chain steroid, 24-norchol-4-ene-3,22-dione (**109**), was isolated as a main metabolite from both the skin and mucus of the doridacean nudibranch *Aldisa smaragdina*, collected off Cabo Cope (Murcia, SE Spain). This unusual steroid (**109**) was previously obtained by chemical synthesis and by microbial conversion of cholesterol, but was never before reported as a natural product. The absolute stereochemistry of **109** was determined by synthesis from commercial stigmasterol (Gavagnin *et al.* 2002).

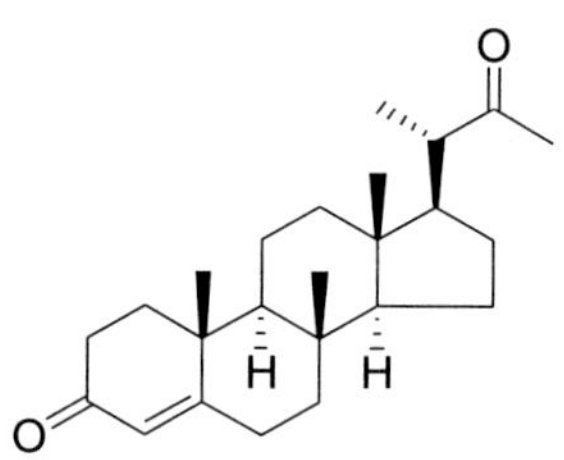

109

A new dimeric isoquinoline alkaloid, jorumycin (**110**), was isolated from the skin and the mucus of the Pacific nudibranch *Joruuna funebris*, collected off Mandapam, India. Compound **110** was structurally related to saframycin, ecteinascidins and other dimeric isoquinoline alkaloids, independently isolated from bacteria, marine sponges and tunicates and these metabolites often revealed interesting antimicrobial and antitumor activities. Jorumycin (**110**) was cytotoxic against NIH3T3 tumor cells (100% of inhibition at 50 ng/mL), and was also active against other cell lines such as P388, A549, HT29, and MEL28, at low concentrations (with IC_{50} values of 12.5 ng/mL). Jorumycin (**110**) also inhibited the growth of various Gram-positive bacteria (e.g., *Bacillus subtilis, Staphylococcus aureus*) at a concentration lower than 50 ng/mL, while no inhibition was observed in *Escherichia coli* at the same concentration (Fontana *et al.* 2000b).

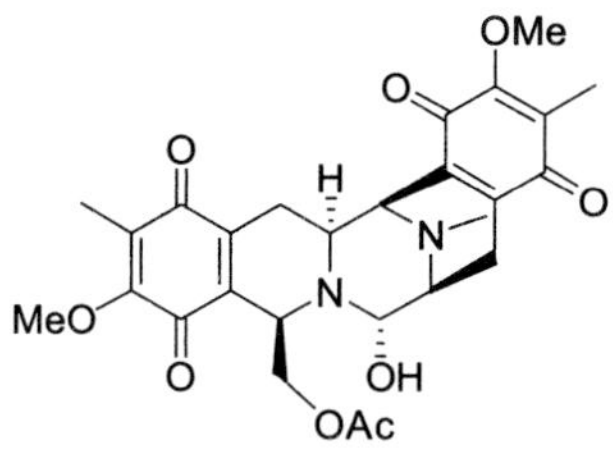

110

Our Recent Studies on Natural Products from Nudibranchs Collected in Japan

Dendrodoris carbunculosa

During our studies in a search for bioactive substances from marine and terrestrial natural resources, we examined chemical constituents of several nudibranchs collected in Japan, mainly from the Boso Peninsula, Chiba Prefecture. In 2001, we investigated a Japanese nudibranch *Dendrodoris carbunculosa* (family Dendrodorididae). This animal specimen was exceptionally large; it was 35 cm long in the relaxed condition in a tide pool and could be flattened to a maximum length of 40 cm on the laboratory sink. The wet weight was 2 kg. This species of *D. carbunculosa* usually lives in a tropical area and it was rare to be found in Japan. This nudibranch specimen was extracted with EtOH, and the EtOAc-soluble fraction of the EtOH extract was subjected to silica gel column chromatography with a gradient elution of EtOAc in hexane, followed by further purification with HPLC on ODS to give 16 sesquiterpenes as colorless oils, 14 new sesquiterpenes (**111** – **124**) and two known compounds, isodrimeninol (**125**) and 11-epivaldiviolide (**126**). Most of these sesquiterpenes exhibited cytotoxicity against murine leukemia P388 cell lines, and isodrimeninol (**125**), the major sesquiterpene of this animal, proved to have sharp peppery taste (Sakio *et al.* 2001).

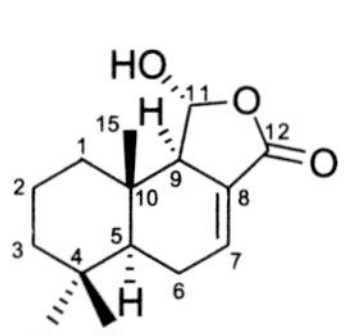

111

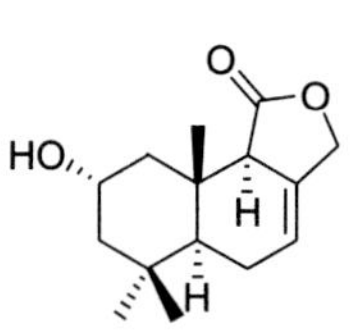

112

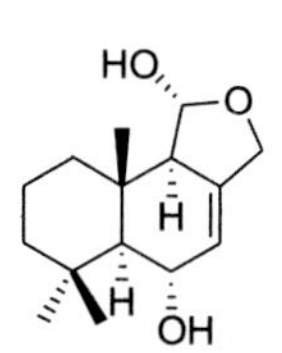

113

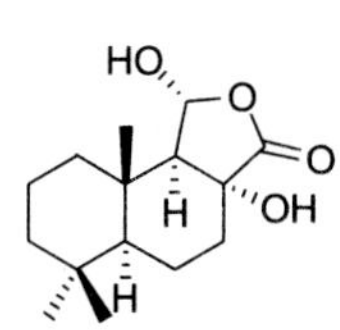

114

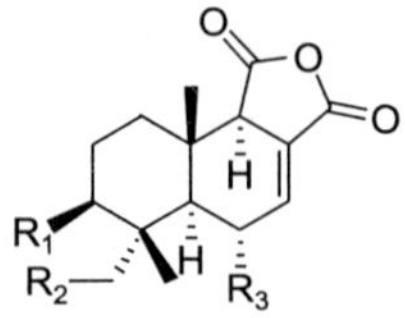

115 R1= OH, R2= R3= H
116 R1= H, R2=OH, R3= H
117 R1= R2=H, R3= OH
118 R1= R2=R3= H

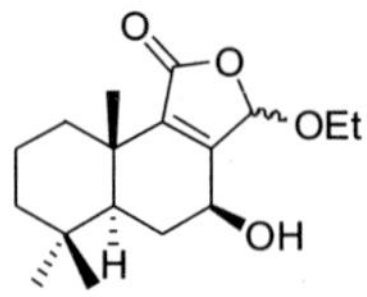

119

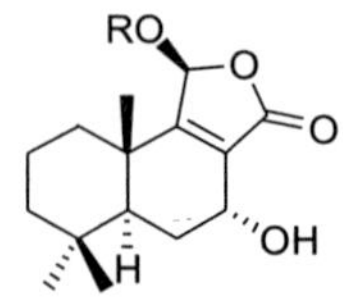

120 R = Et
122 R = H

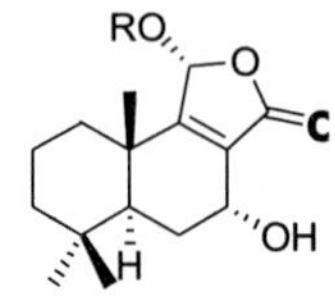

121 R = Et
123 R = H

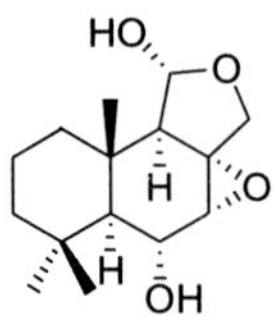

124

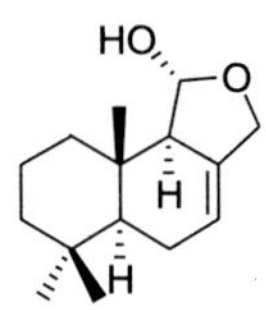

125

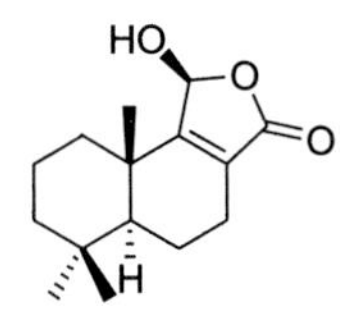

126

Armina babai

The nudibranch *Armina babai* (30 specimens, 1.12 kg wet weight) was collected off the Chita Peninsula, Aichi Prefecture, in 2002, by Dr. K. Iwasaki (Aichi Fisheries Research Institute). These nudibranch specimens were extracted with MeOH, and the EtOAc-soluble fraction of the MeOH extract was subjected to silica gel column chromatography with a gradient elution of MeOH in $CHCl_3$, followed by further purification with Sephadex LH-20 (eluted with MeOH) and silicagel column chromatography (eluted with EtOAc in hexane) to give a ceramide (**127**, 2.4 mg). This ceramide showed characteristic peaks at *m/z* 535 (M^+) and 517 $(M\text{-}H_2O)^+$ in the EI mass spectrum, and by comparison of its ^1H and ^{13}C NMR spectral data with literature data this ceramide was identified as **127**, which was previously isolated from the gorgonian *Acabaria undulata* (Shin and Seo 1995).

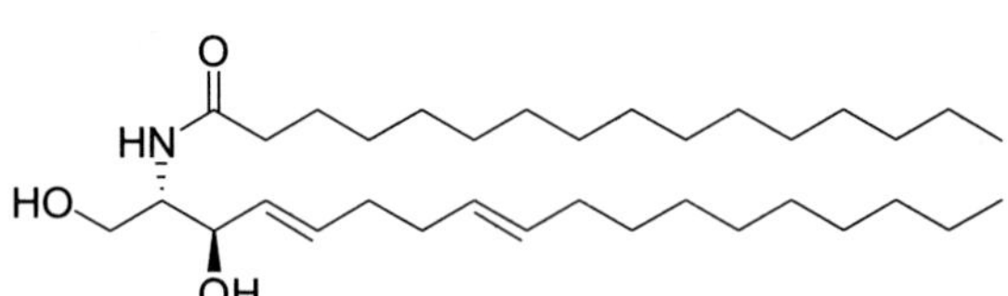

127

Dermatobranchus otome

The nudibranch *Dermatobranchus otome* (77 specimens, 15.5 g wet weight) was collected at Amatsu-kominato, Boso peninsula, Chiba Prefecture, in 2002-2003. These nudibranch specimens were extracted with MeOH,

REFERENCES

Avila, C. 1995. Natural products of opisthobranch molluscs: a biological review. *Oceanogr. Mar. Biol.* 33: 487–559.

Avila, C., K. Iken, A. Fontana, and G. Cimino. 2000. Chemical ecology of the Antarctic nudibranch *Bathydoris hodgsoni* Eliot, 1907: defensive role and origin of its natural products. *J. Exp. Mar. Biol. Ecol.* 252: 27–44.

Brunckhorst, D.J. 1991. Do phyllidiid nudibranchs demonstrate behaviour consistent with their apparent warning coloration? Some field observations. *J. Molluscan Stud.* 57: 481–483.

Cimino, G., S. De Rosa, S. De Stefano, G. Sodano, and G. Villani. 1983. Dorid nudibranch elaborates its own chemical defense. *Science* 219: 1237–1238.

Edmunds, M. 1974. *Defence in Animals*, Longman, London.

Fahey, S.J., and M.J. Garson. 2002. Geographic variation of natural products of tropical nudibranch *Asteronotus cespitosus*. *J. Chem. Ecol.* 28: 1773–1785.

Faulkner, D.J., and M.T. Ghiselin. 1983. Chemical defense and evolutionary ecology of dorid nudibranchs and some other opisthobranch gastropods. *Mar. Ecol. Prog. Ser.* 13: 295–301.

Fontana, A., C. Muniaín, and G. Cimino. 1998. First chemical study of Patagonian nudibranchs: a new seco-11,12-spongiane, tyrinnal, from the defensive organs of *Tyrinna nobilis*. *J. Nat. Prod.* 61: 1027–1029.

Fontana, A., P. Cavaliere, N. Ungur, L. D'Souza, P.S. Parameswaram, and G. Cimino. 1999. New scalaranes from the nudibranch *Glossodoris atromarginata* and its sponge prey. *J. Nat. Prod.* 62: 1367–1370.

Fontana, A., E. Mollo, J. Ortea, M. Gavagnin, and G. Cimino. 2000a. Scalarane and homoscalarane compounds from the nudibranchs *Glossodoris sedna* and *Glossodoris dalli*: chemical and biological properties. *J. Nat. Prod.* 63: 527–530.

Fontana, A., P. Cavaliere, S. Wahidulla, C.G. Naik, and G. Cimino. 2000b. A new antitumor isoquinoline alkaloid from the marine nudibranch *Jorunna funebris*. *Tetrahedron* 56: 7305–7308.

Gavagnin, M., E. Trivellone, F. Castelluccio, G. Cimino, and R. Cattaneo-Vietti. 1995. Glyceryl ester of a new halimane diterpenoic acid from the skin of the Antarctic nudibranch *Austrodoris kerguelensis*. *Tetrahedron Lett.* 36: 7319–7322.

Gavagnin, M., N. Ungur, F. Castelluccio, and G. Cimino. 1997. Novel verrucosins from the skin of the Mediterranean nudibranch *Doris verrucosa*. *Tetrahedron* 53: 1491–1504.

Gavagnin, M., N. Ungur, F. Castelluccio, C. Muniain, and G. Cimino. 1999a. New minor diterpenoid diacylglycerols from the skin of the nudibranch *Anisodoris fontaini*. *J. Nat. Prod.* 62: 269–274.

Gavagnin, M., A. De Napoli, G. Cimino, K. Iken, C. Avila, and F.J. Garcia. 1999b. Absolute configuration of diterpenoid diacylglycerols from the Antarctic nudibranch *Austrodoris kerguelenensis*. *Tetrahedron: Asymmetry* 10: 2647–2650.

Gavagnin, M., A. De Napoli, F. Castelluccio, and G. Cimino. 1999c. Austrodorion-A and –B: first tricyclic diterpenoid 2′-monoglyceryl esters from an Antarctic nudibranch. *Tetrahedron Lett.* 40: 8471–8475.

Gavagnin, M., N. Ungur, E. Mollo, J. Templado, and G. Cimino. 2002. Structure and synthesis of a progesterone homologue from the skin of the dorid nudibranch *Aldisa smaragdina*. *Eur. J. Org. Chem.* 1500–1504.

Gavagnin, M., M. Carbone, E. Mollo, and G. Cimino. 2003. Further chemical studies on the Antarctic nudibranch *Austrodoris kerguelenensis*: new terpenoid acylglycerols and revision of the previous stereochemistry. *Tetrahedron* 59: 5579–5583.

Graziani, E.I., T.M. Allen, and R.J. Andersen. 1995. Lovenone, a cytotoxic degraded triterpenoid isolated from skin extracts of the North Sea dorid nudibranch *Adalaria loveni*. *Tetrahedron Lett*. 36: 1763–1766.

Guo, Y., M. Gavagnin, E. Mollo, E. Trivellone, and G. Cimino. 1998. Structure of the pigment of the Red Sea nudibranch *Hexabranchus sanguineus*. *Tetrahedron Lett*. 39: 2635–2638.

Harris, L.G. 1973. Nudibranch associations. Pages 213–215 in *Current Topics in Comparative Pathobiology* (2), T.C. Cheng, ed., Academic Press, New York.

Hirota, H., T. Okino, E. Yoshimura, and N. Fusetani. 1998. Five new antifouling sesquiterpenes from two marine sponges of the genus *Axinyssa* and the nudibranch *Phyllidia pustulosa*. *Tetrahedron* 54: 13971–13980.

Iken, K., C. Avila, M.L. Ciavatta, A. Fontana, and G. Cimino. 1998. Hodgsonal, a new drimane sesquiterpene from the mantle of the Antarctic nudibranch *Bathydoris hodgsoni*. *Tetrahedron Lett*. 39: 5635–5638.

Iken, K., C. Avila, A. Fontana, and M. Gavagnin. 2002. Chemical ecology and origin of defensive compounds in the Antarctic nudibranch *Austrodoris kerguelenensis* (Opisthobranchia: Gastropoda). *Mar. Biol*. 141: 101–109.

Johnson, S. 1989. Temporal patterns of nudibranch mollusk activity on a subtidal Hawaiian reef. *Veliger* 32: 1–7.

Karuso, P., and P.J. Scheuer. 2002. Natural products from three nudibranchs: *Nembrotha kubaryana, Hypselodoris infucata* and *Chromodoris petechialis*. *Molecules* 7: 1–6.

Kobayashi, J., and M. Ishibashi. 1999. Marine natural products and marine chemical ecology. Pages 415–649 in *Comprehensive Natural Products Chemistry*, Vol. 8, K. Mori, ed., Elsevier Science, Oxford.

Krug, P.J., K.G. Boyd, and D.J. Faulkner. 1995. Isolation and synthesis of tanyolides A and B, metabolites of the nudibranch *Sclerodoris Tanya*. *Tetrahedron* 51: 11063–11074.

McPhail, K.L., and M.T. Davies-Coleman. 1997. New spongian diterpenes from the east African nudibranch *Chromodoris hamiltoni*. *Tetrahedron* 53: 4655–4660.

McPhail, K.L., M.T. Davies-Coleman, and P. Coetzee. 1998. A new furanosesterterpene from the South African nudibranch *Hypselodoris capensis* and a Dictyoceratida sponge. *J. Nat. Prod*. 61: 961–964.

McPhail, K.L., M.T. Davies-Coleman, and J. Starmer. 2001. Sequestered chemistry of the arminacean nudibranch *Leminda millecra* in Algoa Bay, South Africa. *J. Nat. Prod*. 64: 1183–1190.

Miyamoto, T., K. Sakamoto, K. Arao, T. Komori, R. Higuchi, and T. Sasaki. 1996. Dorisenones, cytotoxic spongian diterpenoids, from the nudibranch *Chromodoris obsolete*. *Tetrahedron* 52: 8187–8198.

Miyamoto, T., K. Sakamoto, H. Amano, Y. Arakawa, Y. Nagarekawa, T. Komori, R. Higuchi, and T. Sasaki. 1999. New cytotoxic sesterterpenoids from the nudibranch *Chromodoris inornata*. *Tetrahedron* 55: 9133–9142.

Okino, T., E. Yoshimura, H. Hirota, and N. Fusetani. 1996. New antifouling sesquiterpenes from four nudibranchs of the family Phyllidiidae. *Tetrahedron* 52: 9447–9454.

Pawlik, J.R., M.R. Kernan, T.F. Molinski, M.K. Harper, and D.J. Faulkner. 1988. Defensive chemicals of the Spanish dancer nudibranch *Hexabranchus sanguineus* and its egg ribbons: macrolides derived from a sponge diet. *J. Exp. Mar. Biol. Ecol*. 119: 99–109.

Pika, J., and D.J. Faulkner. 1995. Unusual chlorinated homo-diterpenes from the South African nudibranch *Chromodoris hamiltoni*. *Tetrahedron* 51: 8189–8198.

Ros, J. 1976. Sistemas de defensa en los opisthobranquios. *Oecologia Aquatica* 2: 41–77.

Rudman, W.B. 1986. The Chromodorididae (Opisthobranchia: Mollusca) of the Indo-West Pacific: *Noumea purpurea* and *Chromodoris decora* colour groups. *Zool. J. Linn. Soc*. 86: 309–353.

Rudman, W.B. 1991. Purpose in pattern: the evolution of colour in chromodorid nudibranchs. *J. Molluscan Stud*. 57: 5–21.

Rudman, W.B., and R.C. Willan. 1998. Opisthobranchia Introduction. Peges 915–942 in *Mollusca: The Southern Synthesis. Fauna of Australia*, Vol. 5, Part B, P.L. Beesley, G.J.B. Ross, and A. Wells, eds., CSIRO Publishing, Melbourne.

Shin, J., and Y. Seo. 1995. Isolation of new ceramides from the gorgonian *Acabaria undulata*. *J. Nat. Prod.* 58: 948–953.

Sakio, Y., Y.J. Hirano, M. Hayashi, K. Komiyama, and M. Ishibashi. 2001. Dendocarbins A-N, new drimane sesquiterpenes from the nudibranch *Dendrodoris carbunculosa*. *J. Nat. Prod.* 64: 726–731.

Tanaka, J., and T. Higa. 1999. Two new cytotoxic carbonimidic dichlorides from the nudibranch *Reticulidia fungia*. *J. Nat. Prod.* 62: 1339–1340.

Thompson, T.E. 1960. Defensive adaptations in opisthobranchs. *J. Mar. Biol. Assoc. U.K.* 39: 123–134.

Todd, C.D. 1981. The ecology of nudibranch molluscs. *Ocean. Mar. Biol. Annu. Rev.* 19: 141–234.

Ungur, N., M. Gavagnin, E. Mollo, and G. Cimino. 1999. Absolute stereochemistry of anisodorin 5. *Tetrahedron: Asymmetry* 10: 1635–1636.

Wright, A.D. 2003. GC-MS and NMR analysis of *Phyllidiella pustulosa* and one of its dietary sources, the sponge *Phakellia carduus*. *Comp. Biochem. Physiol.* 134A: 307–313.

Yasman, R., A. Edrada, V. Wray, and P. Proksch. 2003. New 9-thiocyanatopupukeanane sesquiterpenes from the nudibranch *Phyllidia varicosa* and its sponge-prey *Axinyssa aculeate*. *J. Nat. Prod.* 66: 1512–1514.

14

Barnacle Underwater Adhesive: Complexity from Multi-Functionality in a Multi-Protein Complex

Kei Kamino

Marine Biotechnology Institute, 3-75-1 Heita, Kamaishi, Iwate 026-0001, Japan

A huge number of organisms, at least during part of their life cycle, benefit from habitation at interfaces and have developed particular means to hold themselves at the interface. Sessile organisms, examples of which are seen among crustaceans, mollusks, actinaria, porifera, algae and microorganisms, flourish in an aquatic environment and are supported by their unique holdfasts, although examples of biotic holdfasts are limited. This review article is intended to provide an overview of recent advances in the study of barnacle underwater adhesives.

The barnacle is the only sessile crustacean, and permanently attaches to several material surfaces in the water. The adhesion is remarkably strong, and an accurate survey of the adhesive strength on to a natural substratum is usually difficult because the body structure is destroyed by any effort to dislodge it. Research history shows that the barnacle has long been noted for its economic damage as a fouling organism, although Charles Darwin, the first taxonomist of cirripedes, was attracted by their remarkable individuality (Darwin 1854). Although there are a large number of reports on the larval preference for attachment in relation to antifouling technology, this is beyond the scope of the present article. Cirripedia are, in general, grouped into stalked and acorn types by their morphology. Acorn barnacles have either a calcareous or membranous base. This present review is mainly concerned with the calcareous base acorn barnacles, unless otherwise noted.

LARVAL ATTACHMENT

In the life cycle of the barnacle, the hatched nauplius larva repeatedly molts and metamorphoses into the cypris larva, this being the final larval stage in its development (Yule and Walker 1987). The cypris larva is free-swimming, and the antennular attachment organ holds the cypris on the substratum to explore favorable conditions to inhabit. This exploration is followed by the release from the cypris cement gland of a proteinaceous secretion called larval cement (Okano *et al.* 1996) in which the antennular attachment organ embeds itself as a part of metamorphosis to the juvenile barnacle. Although the biochemical nature of cypris cement is not well understood, the results of our northern blotting analysis suggested no expression of the gene for the adult cement protein, cp-100k, as described in a later section, by either nauplius or cypris larvae. Larval cement and adult cement therefore seem to be distinct.

ADULT BARNACLE UNDERWATER ADHESIVE (CEMENT)

The adult barnacle attaches itself to the substratum by secreting an underwater adhesive, historically called "cement" (Fig. 14.1). In *Megabalanus rosa,* the base can reach more than 5 cm in diameter. Histological studies (Walker *et al.* 1970, Lacombe 1970, Saroyan *et al.* 1970, Fyhn and Costlow 1976) have indicated the presence of giant cells (200-300 μm in diameter) which were laid on connective tissue in close association with ovarian tissue and joined together by ducts. These ducts eventually lead to the base of the animal. The results of these histological studies lead to the view that the cement is biosynthesized in a group of cells called the "cement gland", and is secreted into the seawater-filled space through the calcareous base to attach to the substratum.

This cement has historically been characterized into two types, primary and secondary, based on the mode of formation (Saroyan *et al.* 1970). Primary cement is a material found in the narrow gap at the joint between the calcareous base and foreign substratum when the animal is naturally attached in the water (Fig. 14.1). Secondary cement is an underwater secretion by the intact barnacle through the calcareous base after it has been dislodged from the substratum (Fig. 14.2), and is usually hardened and results in a white opaque jelly-like material. Both types of cement have been found to be proteinaceous substances (Walker 1972, Kamino *et al.* 1996). The primary cement hampers the collection of samples for a detailed analysis, because the joint is usually less than 5 μm in thickness (Saroyan *et al.* 1970). On the other hand, the secondary cement makes it easier to collect samples, so correct identification of the two cements is essential for their study. Reattachment of a dislodged

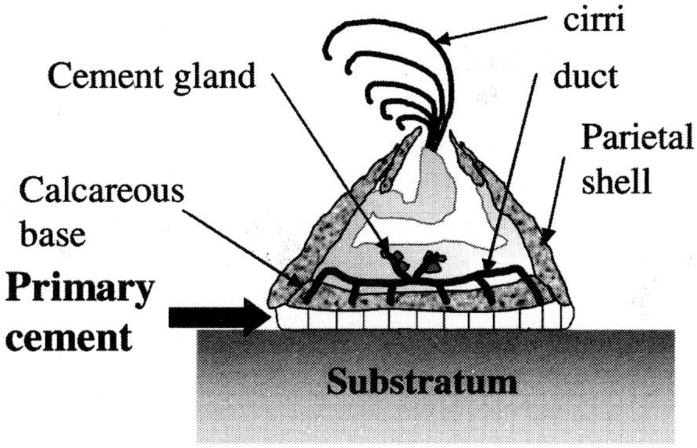

Fig. 14.1 Schematic diagram of a barnacle. The barnacle attached to the substratum is illustrated in cross-section. The underwater adhesive, cement, is probably biosynthesized in the cement glands which are linked together by ducts, and eventually secreted into the space between the calcareous base and substratum. Primary cement is the substance secreted into the space when the animal naturally attaches.

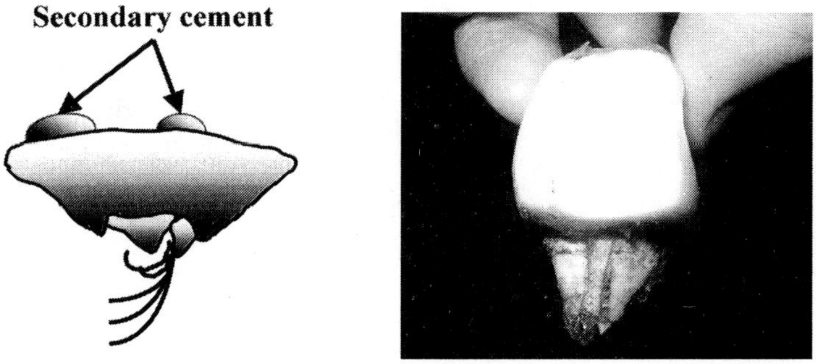

Fig. 14.2 Concept of secondary cement. *Left:* Secondary cement is an underwater secretion from the intact barnacle through the calcareous base after the animal has been dislodged from the substratum. It usually hardens and results in a white opaque jelly-like material. *Right:* A barnacle dislodged from the substratum reattaches itself to a new substratum by the secondary cement.

barnacle is accomplished by the secondary cement (Fig. 14.2) (Saroyan *et al.* 1970, Dougherty 1990), and the similarity in amino acid composition (Naldrett 1993), identity of peptide maps by cyanogen bromide (CNBr)-protein fragmentation (Kamino *et al.* 1996), and cross-reactivity of an antibody have allowed us to identify the secondary cement as being similar to the primary cement in their protein composition, whereby further biochemical analyses became practical.

SEM observation by Wiegemann and Watermann (2003) indicated differences in the microscopic structure of the cement according to the characteristics of the substratum, roughly equating to a "non-stick" surface or "easy-to-attach" surface. The cement secreted by a barnacle attached to a non-stick surface such as a poly(dimethylsiloxane)-coated one is a swollen substance, and is formed by a loose network of hydrated adhesive threads. The cement from a barnacle on an easy-to-attach surface such as aluminium foil or polycarbonate is a dense sheet formed by a closely woven network of threads. The former has a much lower adhesive strength than the latter. The former might correspond to secondary cement in the previous terminology, and can also be found where a barnacle needs to fill a wide gap for attachment even to an easy-to-attach surface. It is obvious that current cement study has not yet reached the point of being able to describe differences in cement microstructures. Future studies on the topological details of cement proteins and the curing process would help analyses of the differences in cement microstructures.

Biochemical studies on cement are hampered by two major obstacles: the insoluble nature and the slippery underwater adhesive activity. Self-assembly of cement proteins is one of essential sub-functions to join two materials, so the sample to be analyzed is usually insoluble. The insolubility of cement, which is a multi-protein complex as described later, limits its analysis and presents a major obstacle to its study.

The expression "underwater attachment" means the sticking together of two different substances in water. It is actually a rapid multi-step process, and can only be accomplished by fulfilling all the requirements in the correct order with proper timing (Waite 1987). An underwater adhesive, therefore, needs the ability to prevent random aggregation at the cement gland and during transport *via* the duct, to displace sufficient seawater to prime and spread on the surface of the foreign substratum without being dispersed in the water, to strongly couple with a variety of material surfaces, and to self-organize itself to join the calcareous base and foreign substratum. After the initial attachment, it is then necessary to cure the cement so that the holdfast remains stiff and tough, and to protect it from continuous water erosion and microbial degradation. The underwater adhesive can orchestrate all these requirements, thus allowing the barnacle to inhabit the liquid-solid interface. The particular ability to handle the water on the foreign substratum without becoming dispersed in it is unique to an underwater adhesive, and remains to be elucidated. Since underwater attachment is only possible by accomplishing all the sub-functions in the correct order with proper timing and the cement is a multi-protein complex as described in the following section, proper micro-analytical methods for each sub-function need to be developed before there can be any research advances. These two current obstacles are common to biotic underwater adhesive studies.

Although a similar amino acid composition has been found among crude cement samples from several barnacle species (Walker 1972, Naldrett 1993, Kamino *et al.* 1996), further progress in identifying the cement proteins awaits the development of a method to render the cement soluble. The quest to identify the cement proteins is summarized in the next section.

IDENTIFICATION OF THE CEMENT PROTEINS

Cement is evidently a proteinaceous substance. Quantification by an amino acid analysis is a suitable approach, especially for an insoluble proteinaceous substance, and has revealed that more than 90% of the cement is protein (Walker 1972, Yan and Tang 1981, Kamino *et al.* 1996). Several studies have been focused on rendering the cement soluble. Approaches to render soluble an insoluble proteinaceous substance would mainly be classified into partial proteolytic and non-proteolytic methods. Although it still remains out of reach, some researchers have nearly accomplished it. Barnes and Blackstock (1976), Yan and Pan (1981), Naldrett (1993) and Naldrett and Kaplan (1997) approached the problem by using the non-proteolytic method with a 2-mercaptoethanol reduction treatment, and have reported the appearance of several protein bands by SDS-PAGE. Kamino *et al.* (1996) applied the CNBr treatment as a partial proteolytic approach, and found that *M. rosa* cement could be rendered soluble. Naldrett and Kaplan (1997) also found several peptide fragments from *Balanus eburneus* and *B. crenatus* cement by using the CNBr treatment. Kamino *et al.* (2000) persevered with development of the non-proteolytic method, whereby more than 90% of *M. rosa* cement could be rendered soluble without any hydrolysis of the proteins (Fig. 14.3). It has now been confirmed that this treatment is also applicable to *B. amphitrite* cement. We have not attempted to use the method on cement from other species or on so-called primary cement, but there seems no reason why it should not be possible. The method involves a protein-denaturation treatment with an excess amount of the reductant, dithiothreitol (DTT), and guanidine hydrochloride (GdnHCl) at 60°C. The reason why this treatment is effective is discussed in the next section.

This last method revealed for the first time that underwater cement is a multi-protein complex. The treatment has so far enabled three major cement proteins (cps) and three minor cps to be identified (Kamino and Shizuri 1998), while a further four cps await identification. The six identified cps each have distinct characteristics from each other, five of them being novel in their primary structure. These results have confirmed the cement to be a unique protein complex.

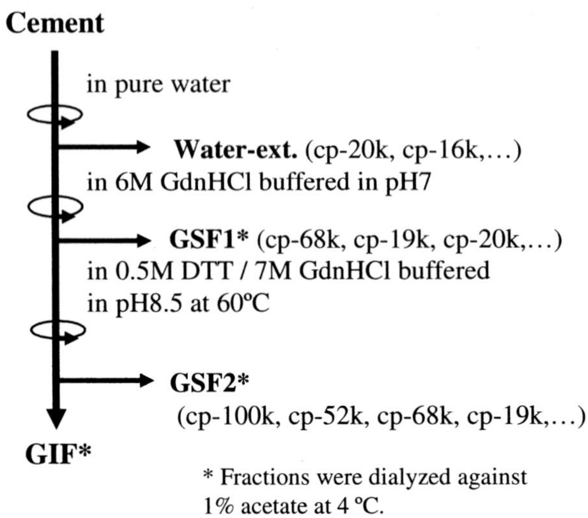

Fig. 14.3 Sequence for solubilizing non-hydrolytic cement and representative cement proteins rendered soluble. GSF1, guanidine-HCl-soluble fraction 1; GSF2, guanidine-HCl-soluble fraction 2; and GIF, guanidine-HCl-insoluble fraction.

As already described, underwater attachment requires several sub-functions to be accomplished. Each cement protein has unique characteristics as described in the next section. The individual cement proteins are presumed to cooperate for underwater attachment, each protein specializing in at least one of the sub-functions required to achieve underwater attachment. The cement is therefore a self-organizing, multi-protein and multi-functional complex.

Several LM, SEM and TEM observations of underwater cement have been reported (Wiegemann and Watermann 2003). Barnacles attached to a "non-stick" surface such as an anti-fouling coating were used in these studies because they had low adhesive strength on this surface and could be easily dislodged for observation. This is not the case with barnacles attached to naturally occurring "easy-to-attach" surfaces, including rocks, metals and such polymers as polycarbonate and polyethylene. Dislodging a barnacle from an "easy-to-attach" surface is extremely difficult and generally breaks the parietal shell, leaving the calcareous base on the substratum. Microscopic observations indicated that the cement on the "non-stick" surface had a fibrous structure formed by a loosely woven network. In contrast, the cement on the "easy-to-attach" surface seemed to have a more dense structure, although detailed observation of the cement layer was limited due to the difficulties in preparing a sample. The loosely woven network of the cement on the "non-stick" surface is presumed to be responsible for the adhesive strength being very low, although it is not evident whether the microscopic structure was the cause or effect. Observation of the cement on a "non-stick" surface might give useful information for developing antifouling surfaces. However, observation of the cement on "an easy-to-attach" surface is more important to elucidate the mechanism for underwater attachment of the cement. Barnacles have long adapted to naturally occurring surfaces for attachment, while they have a much shorter history adapting to "non-stick" man-made surfaces. It is likely, therefore, that they actually fail to attach to such man-made surfaces, the loosely woven network possibly being an artifact of the complex structure. Further efforts to observe the cement on "easy-to-attach" surfaces are needed to elucidate the adhesive mechanism. Antibodies corresponding to the six identified cps have been prepared, and we are attempting to link the microscopic structure with the localization of the cps.

UNIQUE CHARACTERISTICS OF THE CEMENT PROTEINS

The underwater cement is composed of more than 10 cps, of which four have not yet been identified. Five of the six identified cps are novel in respect of their primary structure, and the six cps can be classified into four categories (Table 14.1). The cement proteins are identified by their apparent molecular weight, as estimated by SDS-PAGE, and the unique characteristics of the six cps are summarized next.

Table 14.1 Properties of the cement proteins. The cement proteins are identified by cement protein (cp) and their apparent molecular weight as estimated by SDS-PAGE.

cp	*Characters*	*Possible function*
cp-100k	Hydrophobic, high-content in cement	Bulk adhesive
cp-52k	Hydrophobic, high-content in cement, repetitive sequences	do.
cp-68k	Bias in 4 aa, high-content in cement	Primer, spreading, coupling
cp-19k	Bias in 6 aa, low-content in cement	Primer, spreading, coupling
cp-20k	Charged aa- and Cys-rich, low-content in cement, repetitive sequences	Coupling
cp-16k	Lysozyme homologue, low-content in cement	Protection and/or removal of biofilm

Cement proteins cp-100k and cp-52k are the major ones (Kamino *et al.* 2000), with similar contents. Both proteins have similar behavior in the solubilization process. The DTT-denaturing treatment is the only non-hydrolytic method known to render these proteins soluble. Both cement proteins were first isolated from *M. rosa* cement. The isolation of cp-100k in *B. amphitrite* cement allowed universal PCR primers to be designed, whereby several cp-100k homologous genes were isolated from species including *B. albicostatus, B. trigonus, B. rostratus* and *Semibalanus cariosus*. cDNA cloning indicated both cp-100k and cp-52k to be novel from database searches, and to have low homology with each other. The primary structure of cp-52k contains an approximately 120 amino acid-long repetitive sequence, while there is no repetitive sequence in cp-100k. Both proteins probably play a similar role in underwater attachment.

Cp-100k and cp-52k are both highly hydrophobic, and composed of a relatively low number of Cys residues (1.4% and 1.1%, respectively). The requirement for an excess amount of reductant to render these proteins soluble may, therefore, be questioned, although it is possible to explain. Hydrophobic interaction at an intermolecular level is essential for their insoluble nature, and results in disulfide bonds being buried, such bonds being essential for the conformational stability of each cement protein. A denaturation treatment by using GdnHCl and heating would loosen the intermolecular hydrophobic barrier, whereby an excess amount of reductant could cleave the disulfide bond, and further loosen the intermolecular hydrophobic interaction, finally rendering these proteins soluble. There are no data to indicate whether the disulfide bonds in these cps are intermolecular or intramolecular. Some researchers have proposed intermolecular disulfide involvement in the insoluble nature of cement (Wiegemann and Watermann 2003). However, this proposal needs to be carefully verified, because inter-molecular non-covalent interaction is unusually strong when the molecular conformation is optimized for mutual interaction. This has particularly been shown in the assembly of beta-sheet protein (Otzen *et al.* 2000).

The process to render the cement soluble results in such other cement proteins as cp-68k and cp-19k, which are partially rendered soluble in the GdnHCl solution before the DTT treatment, being completely solubilized due to destruction of the insoluble framework by the DTT-denaturation treatment. Thus, cp-100k and cp-52k are the bulk materials in the cement, and are thought to tie up other cement proteins and give a framework that provides adhesive strength. Although the involvement of cp-100k and cp-52k in the curing process is likely, there have only been limited studies on the process. There has been speculation about the involvement of DOPA-mediated cross-linking (Lindner and Dooley 1973) similar to that in mussel byssus (Deming 1999) but no DOPA is evident in barnacle cement (Kamino *et al.* 1996, Naldrett 1993). Elucidation of the curing process, including unknown cross-linking and participation of bio-mineralization, should be continued.

Another major cement protein is cp-68k (Kamino *et al.*, submitted). As already briefly mentioned, cp-68k is not extractable from the cement in pure water, but is partially soluble in a GdmHCl solution, and rendered completely soluble by the DTT-GdnHCl treatment which rendered cp-100k and cp-52k soluble for the first time. Therefore, cp-68k is presumed not to be covalently connected to cp-100k and cp-52k, but embedded in their framework. Cp-68k is known to be glycosylated, and is characterized by an amino acid composition of Ser, Thr, Ala, and Gly evenly comprising 60%. Cp-68k was isolated from both *M. rosa* and *B. amphitrite*, each with a similar amino acid composition (Fig. 14.4), although the sequence homology was low (47% identical, and 65% similar). An attempt to design a universal PCR primer failed, probably due to the high amino acid exchange rate in the protein. It is likely that the functional significance of cp-68k is based on the content of the four amino acids. Naldrett and Kaplan (1997) found a 58-kDa protein in *B. eburneus* cement, this protein having a similar amino acid composition, probably being a cp-68k homologue.

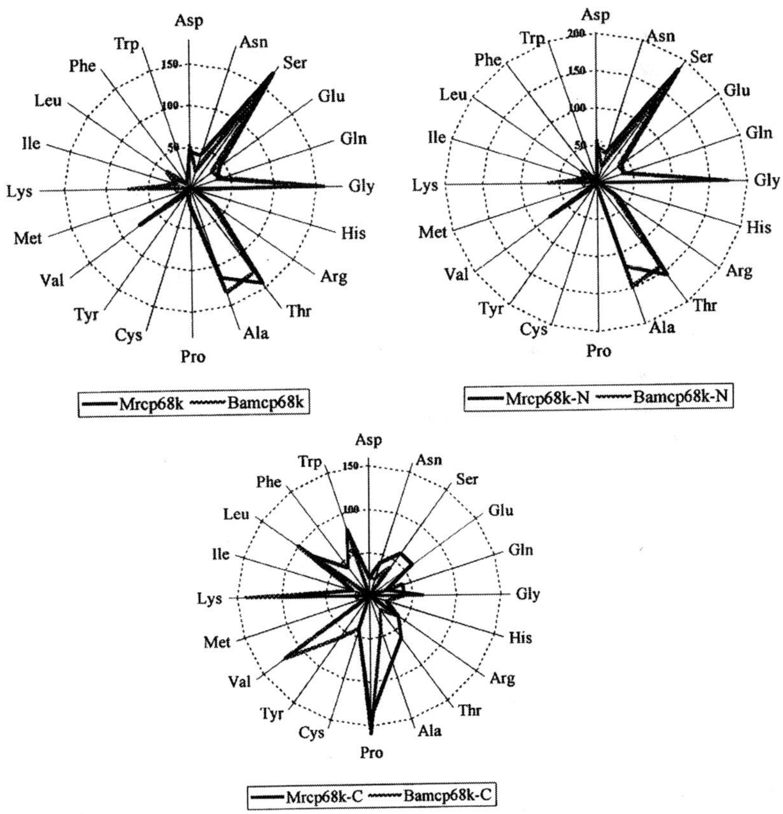

Fig. 14.4 Comparison of the amino acid composition of cp-68k materials. *Upper left:* Mrcp-68k and Bamcp-68k. *Upper right:* Possible N-terminal domains in both Mrcp-68k and Bamcp-68k. *Lower:* Possible C-terminal domains in both Mrcp-68k and Bamcp-68k. The amino acid compositions of the N-terminal regions are inherited from those of the entire cp-68ks, while the C-terminal regions are not inherited at all.

The alignment of the two cp-68ks indicates that the primary structure is divided into two regions: a Ser, Thr, Ala, and Gly-rich N-terminal region (from the mature N-terminal end to residue 364 in cp-68k), and a C-terminal region (residue 365 to the C-terminal end) composed of much less of the four dominant amino acids in the N-terminal region and of such amino acids as Lys, Pro, Trp and hydrophobic amino acids (Fig. 14.4). The C-terminal region seems to have a compact conformation, because Pro, Trp and the hydrophobic amino acids are generally advantageous for packing the structure by their backbone fixation, pi-pi stacking and hydrophobic core formation. If we accept the hypothesis that those amino acids contribute to structural stabilization, the molecular surface would be covered with a strong charge due to the Lys residues, and this charge might be essential to the sub-function of the underwater adhesive. The N-terminal region has a thin polypeptide chain whose surface is probably covered with such amino acids as dominant Ser and Thr residues and the glycosyl moiety. An essential element in the cp-68k protein for its sub-function might therefore be the dominant hydroxyl group on the molecular surface of the N-terminal region and the charge on that of the C-terminal region.

An underwater adhesive is first required to approach the liquid-solid interface without being dispersed in the environmental seawater. Sub-functions called "priming", "spreading" and "condensation" are the most elusive in underwater attachment, and we have not yet been able to elucidate the mechanism for these functions. Waite (1987) has attempted to explain the molecular event with the primer by classifying the targeted solid surfaces as non-polar and polar. Water would weakly bind to an underwater non-polar surface, and a macromolecule with many hydroxyl groups may be a good primer to displace this water. Such a molecule on an underwater polar surface, however, may not have sufficient ability to displace any water that is strongly bound to the surface. A subsequent molecular event such as support of the surface coupling ability would be required. In the case of mussel foot proteins (fps), the catechol moiety of DOPA residues as the common ingredient in all fps is a possible supporter for priming the polar surface by the strong surface-coupling ability. If we assume the requirement of a primer, cp-68k may fulfill this. The possible dominant hydroxyl group on the molecular surface of the N-terminal region may be useful for priming activity, and the strong charge on the molecular surface of the C-terminal region may contribute to surface coupling with the polar surface. The most probable natural surface would be a silicate which is known to be negatively charged. Recombinant cp-68k in *Escherichia coli* had good adsorption ability to a glass surface in seawater, while truncation of the C-terminal region showed lower adsorption ability. The possible sub-function of the cp-68k protein is therefore thought to be priming and surface coupling.

Condensation of the adhesive after its secretion from the duct into the seawater is also an essential sub-function of underwater attachment. Cp-68k is glycosylated. Although there is no further information about the glycosyl moiety in cp-68k, glycosylated cp-68k may be sticky, and this would probably be useful for retaining the condensation of cps. The speculative sub-functions of cp-68k therefore also include condensation. A cp-68k recombinant preparation in an insect system is now being studied by our group, and the results of this analysis will contribute to identifying the significance of the glycosyl moiety in cp-68k to its sub-function.

Several other proteins in small amounts are also present in the underwater cement, although their content does not match any functional significance in underwater attachment. Cp-20k was identified as a simple 20-kDa protein and a minor cement protein (Kamino 2001). Cp-20k was completely recovered from the insoluble cement in the GdnHCl-soluble fraction designated as GSF1 and was also soluble in pure water. After extracting GSF1, cp-20k could not be found in the DTT/GdnHCl-treated soluble fraction designated as GSF2. This protein is characterized by the abundance of charged amino acids and Cys residues. Of these charged amino acids, Asp, Glu and His comprised 32% and Cys comprised 17% of the protein. ToF-MS showed the protein to have mass as a monomer, thus precluding intermolecular disulfide bond formation in cp-20k. The fact that this protein was not recovered in GSF2 also indicates no formation of intermolecular disulfide bonds among the other cement proteins. If anything, the lack of reactivity with alkylating agents suggests that all Cys in the protein formed intra-molecular disulfide bonds. Although polymerization by cp-20k based on intermolecular disulfide bonds has been discussed for cement curing in elsewhere (Wiegemann and Watermann 2003), experimental data do not support this hypothesis for the sub-function of this protein.

Our recent NMR analysis of the recombinant in *E. coli* indicated that an intra-molecular disulfide bond in the protein was essential for stabilizing the conformation. The abundant Cys in cp-20k was more essential for the structure than for the sub-function. The protein was composed of a much lower number of hydrophobic amino acids than usual proteins. Many intra-molecular disulfide bonds may therefore be introduced instead of the poor hydrophobic core formation.

Abundant charged amino acids such as Glu, Asp, His, Lys and Arg would be positioned on the molecular surface and would thus be essential for the sub-function in underwater attachment. The isolation of a homologous gene from *B. albicostatus* revealed high contents of charged amino acids and Cys to be of conserved chemical character (Fig. 14.5), although the homology between Mrcp-20k and Bal-homologous protein was remarkably low. These results confirm the functional significance of

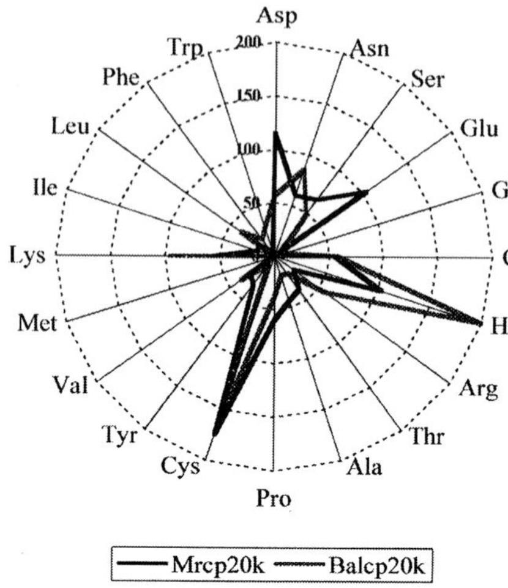

Fig. 14.5 Amino acid compositions of the Mrcp-20k and Balcp-20k. Cp-20k is a charged amino acid and Cys-rich protein (Kamino 2001).

the charged amino acids in cp-20k. Another possible sub-function of cp-20k in underwater attachment involves surface chemistry, although there is no data to support this yet.

When the Cys residues were appropriately aligned, the primary structure of cp-20k was found to comprise six repetitions. However, the amino acids, except for Cys and Pro, were not at all conserved among the repetitive modules. Strangely, there were four repetitions in the primary structure of the homologous gene in *B. albicostatus*, indicating that the molecular weight of the *Bal*-homologous protein was 12 kDa. This suggests that a single or double repetitive unit would be the minimum structural and/or functional unit in the protein. Our recent NMR study seems to support the notion of a single unit being the minimum structural unit (Suzuki *et al*., submitted).

The isoelectric points in the two homologous proteins are opposite, being 4.7 in cp-20k and 8.3 in the *Bal*-homologous protein. In particular, the charge distribution in the space between the third and fourth Cys residues in the repetitive units is completely reversed. *M. rosa* and *B. albicostatus* usually inhabit different marine environments, the former being offshore and the latter coastal, so the surfaces to be frequently attached are necessarily different. The difference in charge of the two proteins may have evolved from the requirements of the most frequent material surfaces to which the two species attach themselves.

Cp-19k is a minor cement protein, and has been newly identified (Urushida *et al*., manuscript in preparation). Cp-19k is characterized by the similar amino acid bias to that of cp-68k (Fig. 14.6). Six amino acids,

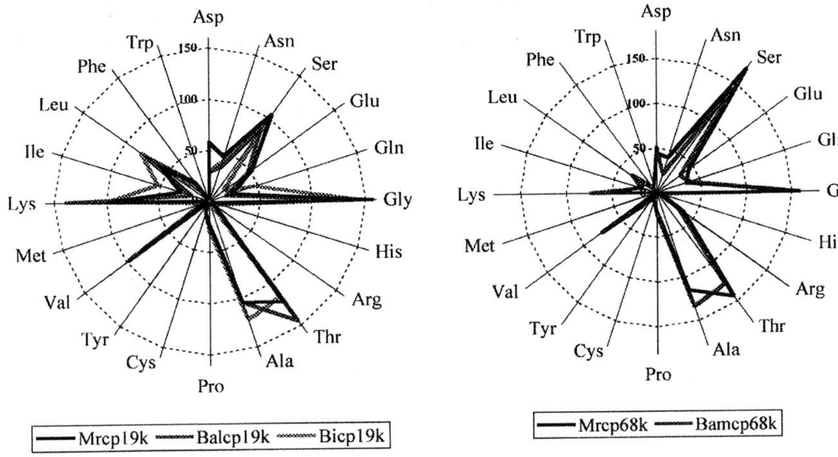

Fig. 14.6 Comparison of the amino acid compositions among the cp-68k and cp-19k materials. Both proteins have distinct primary structures, but the bias of the amino acid content is similar, possibly leading to a similar sub-function in underwater attachment.

Ser, Thr, Gly, Ala, Lys and Val, comprise more than 67% of the total. The development of a universal PCR primer allowed us to amplify homologous cDNA from other species, whereby the primary structures from *M. rosa*, *B. albicostatus* and *B. improvisus* have so far been understood. The homology among the three cp-19ks is as follows: 51% for the *Mr vs. Bal* proteins, 54% for the *Mr vs. Bi* proteins, and 60% for the *Bal vs. Bi* proteins. Although the sequence homology is not high, it is basically similar. A simple repetitive sequence could not be found in the primary structures. This bias in the contents of the 6 amino acids in cp-19k is probably of functional significance. There was no homology between cp-19k and cp-68k in their primary structures. Ser, Thr, Gly and Ala were predominant in cp-68k, and represent the outstanding components. The next most dominant amino acids in cp-68k were actually Lys and Val. The high-content amino acids were the same in both the cp-68k and cp-19k proteins, indicating that these proteins probably contribute to a similar sub-function.

Cp-19k has been successfully expressed in the soluble fraction of *E. coli*. SDS-PAGE of the recombinants and protein in GSF1 indicated almost the same mobility. Purification of the recombinant allowed it to be characterized, and strong adsorption to a glass surface in seawater was demonstrated. The possible sub-function of cp-19k is priming, spreading and surface coupling. Although cp-68k and cp-19k are different proteins, they should strictly provide similar but distinct roles in underwater attachment. Finding the difference among these proteins awaits a detailed analysis.

Cp-16k is a minor cement protein, and can be extracted from the cement in either GSF1 or pure water. The primary structure is homologous to that of lysozyme-P in *Drosophila melanogaster*, being 47%, and all catalytic amino acids and Cys are conserved. The crude cement shows lytic activity to the *Micrococcus luteus* membrane fraction, so cp-16k is probably a lysozyme. Its possible sub-functions include removing biofilm from the substratum and/or protecting the cement from microbial degradation. Dougherty (1997) has reported exo-carboxypeptidase activity in the cement of *Chthamalus fragilis*. The contribution of enzymatic proteins to the curing process is also attracting interest, although the specific role in underwater attachment needs to be thoroughly investigated.

Further cement proteins remain to be characterized, including cp-168k, cp-135k, cp-86k, and cp-9k. These cement proteins are commonly found in the aqueous extract or GSF1 of *M. rosa* cement, and have a hydrophilic amino acid composition. Although the DTT-GdnHCl treatment renders soluble almost 90% of the secondary cement, a portion always remains as an insoluble fraction and has been named GIF. The amino acid composition of GIF seems to be similar to that of the crude cement, so the protein components may be the same. Repetitive treatment does not completely render GIF soluble, and it may be covalently cross-linked, except for disulfide bonds. We have not tested the effectiveness of the DTT-GdnHCl treatment on primary cement. The process of curing may provide mechanical strength, and secondary cement may not be perfect in this respect. Further characterization of GIF and primary cement is awaited.

A northern blotting analysis by using isotope-labeled DNA probes has indicated that the cement protein genes are only expressed in the basal portion where the cement gland is located, and not in the upper portion where almost all of the viscera are located. Both portions include body fluids, so expression by the body fluid cells is negligible.

Barnacle cement is a multi-protein complex whose subunits are unique. It is speculated that six cement proteins, respectively, handle the functions of priming, spreading, surface coupling, organization, curing, and removing the biofilm and/or protecting from microbial degradation. All the requirements for underwater attachment would be fulfilled by cooperation of these cement proteins with as yet unidentified ones.

COMPARISON WITH HOLDFAST PROTEINS IN OTHER ORGANISMS

The counterpart to consider in a biotic underwater adhesive study is the mussel foot protein. The mussel is a bivalve that attaches to an underwater foreign substratum by generating 30 or more byssal threads or simply byssus (Taylor and Waite 1997). Byssus is a proteinaceous holdfast basically composed of plaque and thread. Adhesive plaque of about 2-3 mm in diameter is distally attached to the substratum, while threads of about 0.1 mm in diameter are proximally connected to the mussel's byssus retractor muscles *via* a stem. The special organ for forming byssus is the foot, by which each byssus is generated within 5 min. The discovery of peptydyl-DOPA in byssus in early 1980 by Waite and Tanzer (1981) has enabled several DOPA-containing proteins to be isolated from the foot by monitoring DOPA and applying cold shock: five DOPA-containing proteins, named foot proteins (fp)-1 to -5, and two thread-forming proteins, named preCol-D and -P, have so far been isolated (Waite 1999, Waite and Qin 2001). Determination of the location of each fp in the byssus and foot has made it possible to identify the likely sub-function of each fp. Considerable work has been done to investigate DOPA chemistry and DOPA-mediated cross-linking (Ooka and Garrell 2000, McDowell *et al.* 1999, Sever *et al.* 2004, Deming 1999, Holl *et al.* 1993), and to measure the adsorption behavior to a material surface (Fant *et al.* 2002, Hansen *et al.* 1998, Suci and Greesey 2001), although almost all studies have been focused on only fp-1 and a few on fp-2. No homology at all has been found in the primary structures between mussel fps and barnacle cps.

The mussel holdfast, byssus, is composed of macroscopically or microscopically separate modular structures. Adhesive plaque distally couples with the foreign surface, the bulk of the plaque being a solid foam-like structure, the thread is composed of a stiff distal zone and elastic proximal zone, and the surface of the byssus is coated with a thin layer (5-10 μm). A mussel holds itself against strong and repeated shock by waves and currents. On the other hand, a barnacle attaches directly to the substratum by its whole base *via* cement, this cement seeming not to have a macroscopically modular structure. This difference is also apparent from the distribution of the secretory glands. In the mussel foot, several glands specializing in the biosynthesis of individual fps are appropriately aligned along the longitudinal axis of the ventral groove in the foot (Waite 1992), making possible a macroscopically modular structure. In the barnacle, however, only the cement gland has been found so far, and all cement proteins seem to be secreted together *via* the duct system, although biochemical analyses remain to be conducted. The macroscopic structures of the holdfasts are therefore basically different between the mussel and barnacle.

Fp-3 and fp-5 of the mussel foot proteins are known to be located at the interface between the plaque and substratum (Floriolli *et al.* 2000), so those two proteins are probably involved in priming, spreading and/or surface coupling. Fp-3 is the smallest of the proteins, being about 6kDa, and undergoes post-translational modifications to DOPA and 4-hydroxyarginine (Inoue *et al.* 1996). Fp-5 is also a small protein of about 9.5kDa, and is post-translationally modified to DOPA and *O*-phosphoserine, whose primary structure includes the possible motif for mineral binding (Waite and Qin 2001). Many studies have shown the catechol group in the DOPA residue to be versatile, and it is speculated to be involved in such sub-functions as curing, surface coupling and fiber elasticity. Fp-3 and fp-5 have the highest DOPA contents among the foot proteins, and the residue would be one of the most significant ingredients for the sub-functions. Such remarkable and attractive individuality as DOPA in the mussel is not apparent in barnacle cement proteins. However, any charged amino acid residues in cp-20k and the C-terminal region of cp-68k may be optimized to couple with the material surface by molecular conformation. The hydroxyl group of 4-hydroxyarginine in fp-3 and *O*-phosphoserine in fp-5 could also be concerned with those sub-functions involving competition with water such as priming and spreading. The antifreeze protein is well known in aquatic organisms, and protects their tissue from damage by freezing. This protein binds to the ice crystal nucleus to inhibit nucleation. The 3-D structure of the protein clearly indicates that the site for binding to the ice nucleus is composed of six Thr residues, and that hydroxyl groups in the protein are essential for interacting with a certain form of the water molecule (Liou *et al.* 2000). The abundance of hydroxyl groups in underwater adhesive proteins is thought to be useful for competition with water on the material surface. Glycosylation and the abundant Ser and Thr contents in cp-68k and cp-19k may also be useful for priming and spreading on to the surface. These two cps are suggested to be the counterparts of fp-3 and fp-5. *O*-phosphoserine in fp-5 might have the special function of coupling to calcareous material such as the shell surface of other mussels and barnacles (Waite and Qin 2001). Further investigation of the post-translational modification to cps would be valuable.

Cp-100k and cp-52k of the barnacle cement proteins are hydrophobic, resulting in the formation of an insoluble framework, and are probably involved in the organization of cement proteins and curing. The bulk of the adhesive plaque is formed by fp-2, whose abundant Cys content seems to be involved in plaque organization and curing by intermolecular disulfide formation, although there is no evidence indicating whether disulfide bonds are intra-molecular or inter-molecular. No hydrophobic protein has so far been found in mussel foot proteins, so cross-linking has been suggested as the major contributor to the organization of fps. Barnacle cement studies should pay further attention to the curing process, while mussel byssus studies would benefit from more attention to the non-covalent protein organization process.

Protection from bacterial degradation is a possible sub-function required by extracellular proteinaceous structures. Mussel byssus is evidently protected by being varnished with fp-1 which is cross-linked *via* the DOPA residue. Barnacle cement provides lytic activity originating from cp-16k, a lysozyme homologue, and this is probably useful for protecting the cement. These two typical sessile organisms, mussel and barnacle, have evidently evolved individual molecular systems.

Reef-building Polychaeta build a tube for residence, this tube being constructed by binding particulate material with proteinaceous cement. Jensen *et al.* (1988) have constructed an aquaculture system to successfully simulate this, whereby the amino acid composition of the Polychaeta cement was apparent, with Ser, Gly and Lys comprising about 65% of the total residues. DOPA was also a component of the residues, and a DOPA-containing protein has been isolated from the thorax of the animal where the cement gland is located (Waite *et al.* 1992). This protein had a Gly- and Lys-rich amino acid composition, and several partial amino acid sequences were determined, although the entire primary structure has not yet been reported. The cement is also presumed to be a multi-protein complex, because the Ser-rich amino acid composition of the cement does not match that of the DOPA-containing protein that had been isolated. Polychaeta have a unique proteinaceous cement system, and are a candidate for the third phylum to be biochemically investigated. Comprehensive research is thus anticipated.

FUTURE APPLICATIONS OF PEPTIDE-BASED MATERIALS

Molecular self-assembly is one of most attractive events in supramolecular chemistry (Whitesides and Boncheva 2002), nano-scale technology (Zhao and Zhang 2004) and tissue engineering. Biological molecular systems are specialists in self-assembly, so learning from them is the first step towards the mimetic design of an artificial self-assembly system. Peptides offer great advantages in interdisciplinary biotechnological material science due to their huge molecular diversity, simple design from a protein-based biological molecule, easy insertion of a biological motif, and freedom from contamination of viral and prion origins. Although several peptide-based materials have so far been designed, amphiphilicity and a beta-sheet structure have been favored. The cement is a self-organized multi-functional complex, and an individual cement protein is a specialist for the respective sub-function involved in underwater attachment. Each subfunction has individual attraction for material science, and mimetic molecular design from individual cement proteins is therefore likely. A biotic underwater adhesive protein complex provides a novel molecular template with which to design novel peptide-based materials.

The primary structures of cp-20k from two barnacle species are respectively composed of 6 and 4 repetitions by the alignment of Cys residues, while the amino acid residues, except for Cys and Pro, in each repetitive unit are exchanged with high frequency. The difference of two repetitive units in the polypeptide length between two cps suggests that a double or single repetitive unit would be the minimum structural and/or functional unit. Peptides whose primary structures were designed from the repetitive sequences in cp-20k have been chemically synthesized, and their ability for self-assembly with several salt concentrations was evaluated by using dynamic light scattering (DLS) measurement and atomic force microscopic (AFM) imaging (Nakano *et al.*, manuscript in preparation). The DLS measurements suggested an increment in the molecular size of the designed peptides with the addition of several kinds of salt. These peptides were soluble in water, and self-assembly was suddenly triggered by more than 0.6 M of NaCl. AFM imaging indicated that the peptide assembly had a consistent rod-shape several 30-70 nm in width. Similar molecular assembly was also confirmed by SEM observation. Peptides designed from the cp-20k underwater adhesive protein had self-assembly activity with a threshold near 0.6 M of NaCl concentration, which is close to that of seawater. The consistent diameter of the peptide suggests its ordered self-assembly. Images indicating extension of the long axis were also captured, and it is suggested that improving the peptide sequence and self-assembly conditions might lead to the development of various peptide-based materials. DTT addition during the self-assembly process did not affect the result, suggesting that self-assembly upon the addition of salt was built on non-covalent molecular interaction. A peptide in which all Cys residues were altered to Ser residues, indicating similar self-assembly, also supports the mechanism. Peptide-based materials are generally a class of simple amphiphilic peptides. Those examined in this study, however, contained numerous charged amino acids, few hydrophobic amino acids, had no simple amphiphilicity and had no beta-sheet structure; therefore, self-assembly of these peptides may lead to novel principles for peptide self-assembly. Peptide-based design of materials from underwater adhesive proteins with individual underwater attachment subfunctions is a promising future application.

Studies on the application of mussel fps have been focused on the DOPA moiety due to its remarkable functionality, while efforts have been more limited on the protein/peptide conformation of these mussel fps. DOPA is certainly intriguing and versatile, and is easy to incorporate in a polymer. A recent successful application of DOPA functionality is its utilization as a metal surface binder. In the healthcare arena as well, the functions of therapeutic and diagnostic devices can be compromised by non-specific adsorption of proteins and cells to surfaces during their long-term exposure to physiological fluids. A polyethylene glycol conjugate to DOPA overlaid on a material surface resulted in a successful cellular non-fouling surface for up to 2 weeks (Dalsin *et al.* 2003). Although there have been several approaches to applying DOPA functionality (Deming 1999), mussel foot proteins would also provide individuality in molecular conformation. The incorporation of conformational information is to be encouraged, because it may be optimized to offer a unique application.

SUMMARY

Sessile attachment at a liquid-solid interface is common in an aquatic environment. Such attachment is only possible with unique holdfasts, although molecular illustrations of biotic holdfasts are limited. The barnacle is a unique sessile marine arthropod, and its underwater adhesive is a multi-protein and multi-functional complex called cement. Marked advances have recently been made in elucidating this cement, and it has become a joint objective with the byssus of the mollusk, mussel, in underwater adhesive studies. This paper has reviewed the history of barnacle cement studies, summarized cement proteins, interpreted the molecular events in underwater attachment, and suggested peptide-based material development as an application. Biotic underwater adhesives provide several functions that mankind has so far been unable to achieve. Studies on underwater adhesives are interdisciplinary among several research areas, including biotechnology and material science, have great attraction for the future.

ACKNOWLEDGEMENT

Part of this work was performed under The Industrial Science and Technology Project, Technological Development for Biomaterials Design based on Self-organizing Proteins that was supported by the New Energy and Industrial Technology Development Organization (NEDO).

REFERENCES

Barnes, H., and J. Blackstock. 1976. Further observations on the biochemical composition of the cement of *Lepas Fascicularis Ellis & Solander*; Electrophoretic examination of the protein moieties under various conditions. *J. Exp. Mar. Biol. Ecol.* 25: 263–271.

Dalsin, J.L., B.H. Hu, B.P. Lee, and P.B. Messersmith. 2003. Mussel adhesive protein mimetic polymers for the preparation of nonfouling surfaces. *J. Am. Chem. Soc.* 125: 4253–4258.

Darwin, C. 1854. A monograph on the sub-class Cirripedia. *London: Ray Soc.* 2: 1–684.

Deming, T.J. 1999. Mussel byssus and biomolecular matrials. *Curr. Opin. Chem. Biol.* 3: 100–105.

Dougherty, W.J. 1990. Barnacle adhesion: reattachment of the adult barnacle *Chthamalus fragilis* Darwin to polystyrene surfaces followed by centrifugational shearing. *J. Crust. Biol.* 10: 469–478.

Dougherty, W.J. 1997. Carboxypeptidase activity of the zinc metalloprotease in the cement precursor secretion of the barnacle, *Chthamalus fragilis* Darwin. *Comp. Biochem. Physiol.* 117B: 565–570.

Fant, C., H. Elwing, and F. Höök. 2002. The influence of cross-linking on protein-protein interactions in a marine adhesive: the case of two byssus plaque proteins from the blue mussel. *Biomacromolecules* 3: 732–741.

Floriolli, R.F., J. von Langen, and J.H. Waite. 2000. Marine surfaces and the expression of specific byssal adhesive protein variants in *Mytilus. Mar. Biotech.* 2: 352–363.

Fyhn, U.E., and J.D. Costlow. 1976. A hitochemical study of cement secretion during the intermolt cycle in barnacles. *Biol. Bull.* 150: 47–56.

Hansen, D.C., S.G. Corcoran, and J.H. Waite. 1998. Enzymatic tempering of a mussel adhesive protein film. *Langmuir* 14: 1139–1147.

Holl, S.M., D. Hansen, J.H. Waite, and J. Scaefer. 1993. Solid-state NMR analysis of cross-linking in mussel protein glue. *Arch. Biochem. Biophys.* 302: 255–258.

Inoue, K., Y. Takeuchi, D. Miki, S. Odo, S. Harayama, and J.H. Waite. 1996. Cloning, sequencing and sites of expression of genes for the hydroxyarginine-containing adhesive-plaque protein of the mussel *Mytilus galloprovincialis. Eur. J. Biochem.* 239: 172–176.

Jensen, R.A., and D.E. Morse. 1988. The bioadhesive of *Phragmatopoma californica* tubes: a silk-like cement containing L-DOPA. *J. Comp. Physiol.* 158B: 317–324.

Kamino, K., S. Odo, and T. Maruyama. 1996. Cement proteins of the acorn barnacle, *Megabalanus rosa. Biol. Bull.* 190: 403–409.

Kamino, K., and Y. Shizuri. 1998. Structure and function of barnacle cement proteins. Pages 77-80 in New Developments in Marine Biotechnology, Y. Le Gal, and Halvorson, H. eds., Plenum Press, New York.

Kamino, K., K. Inoue, T. Maruyama, N. Takamatsu, S. Harayama, and Y. Shizuri. 2000. Barnacle cement proteins. *J. Biol. Chem.* 275: 27360–27365.

Kamino, K. 2001. Novel barnacle underwater adhesive protein is a charged amino acid-rich protein constituted by a Cys-rich repetitive sequence. *Biochem. J.* 356: 503–507.

Lacombe, D. 1970. A comparative study of the cement glands in some balanid barnacles (Cirripedia, Balanidae). *Biol. Bull.* 139: 164–179.

Lindner, E., and C.A. Dooley. 1973. Chemical bonding in cirripede adhesive. In Proceedings of the Third International Congress on Marine Corrosion and Fouling, Gaithersburg R.F. Acker *et al.*, eds., Gaithersburg, Maryland, pp. 653–673.

Liou, Y.C., A. Tocilj, P.L. Davies, and Z. Jia. 2000. Mimicry of ice structure by surface hydroxyls and water of beta-helix antifreeze protein. *Nature* 406: 322–324.

McDowell, L.M., L.A. Bursio, J.H. Waite, and J. Schaefer. 1999. Rotational echo double resonance detection of cross-links formed in mussel byssus under high-flow stress. *J. Biol. Chem.* 274: 20293–20295.

Naldrett, M.J. 1993. The importance of sulphur cross-links and hydrophobic interactions in the polymerization of barnacle cement. *J. Mar. Biol. Assoc. UK.* 73: 689–702.

Naldrett, M.J., and D.L. Kaplan. 1997. Characterization of barnacle (*Balanus eburneus* and *B. crenatus)* adhesive proteins. *Mar. Biol.* 127: 629–635.

Okano, K., K. Shimizu, C.G. Satuito, and N. Fusetani. 1996. Visualization of cement exocytosis in the cypris cement gland of the barnacle *Megabalanus rosa*. *J. Exp. Biol.* 199: 2131–2137.

Ooka, A.A., and R.L. Garrell. 2000. Surface-enhanced raman spectroscopy of DOPA-containing peptides related to adhesive protein of marine mussel, *Mytilus edulis*. *Biopolymers* 57: 92–102.

Otzen, D.E., O. Kristensen, and M. Oliveberg. 2000. Designed protein tetramer zipped together with a hydrophobic Alzheimer homology: A structural clue to amyloid assembly. *Proc. Nat. Acad. Sci. USA*. 97: 9907–9912.

Saroyan, J.R., E. Lindner, and C.A. Dooley. 1970. Repair and reattachment in the Balanidae as related to their cementing mechanism. *Biol. Bull.* 139: 333–350.

Sever, M.J., J.T. Weisser, J. Monahan, S. Srinivasan, and J.J. Wilker. 2004. Metal-mediated cross-linking in the generation of a marine-mussel adhesive. *Angew. Chem. Int. Ed.* 43: 448–450.

Suci, P.A., and G.G. Geesey. 2001. Use of attenuated total internal reflection fourier transform infrared spectroscopy to investigate interactions between *Mytilus edulis* foot proteins at a surface. *Langmuir* 17: 2538–2540.

Taylor, S.W., and J.H. Waite. 1997. Marine adhesives from molecular dissection to application. Pages 217-248 in *Protein Based Materials*, K. McGrath, and D. Kaplan, eds., Birkhäuser, Boston.

Waite, J.H., 1987. Nature's underwater adhesive specialist. *Int. J. Adhesives* 7: 9–14.

Waite, J.H. 1992. In Results and Problems in Cell Differentiation. 19 Biopolymers, S.T. Case, ed., Springer-Verlag, Berlin. pp. 27–53.

Waite, J.H. 1999. Reverse engineering of bioadhesion in marine mussels. *Ann. N.Y. Acad. Sci.* 875: 301–309.

Waite, J.H., R.A. Jensen, and D.E. Morse. 1992. Cement precursor proteins of the reef-building polychaete *Phragmatopoma californica* (Fewkes). *Biochemistry* 31: 5733–5738.

Waite, J.H., and X.X. Qin. 2001. Polyphosphoprotein from the adhesive pads of *Mytilus edulis*. *Biochemistry* 40: 2887–2893.

Waite, J.H., and M.L. Tanzer. 1981. Polyphenolic substance of *Mytilus edulis:* Novel adhesive containing L-Dopa and hydroxyproline. *Science* 212: 352–354.

Walker, G. 1970. The histology, histochemistry and ultrastructure of the cement apparatus of three adult sessile barnacles, *Elminius modestus, Balanus balanoides* and *Balanus haemri. Mar. Biol.* 7: 239–248.

Walker, G. 1972. The biochemical composition of the cement of two barnacle species, *Balanus haemri* and *Balanus crenatus*. *J. Mar. Biol.* Assoc. *UK*. 52: 429–435.

Wiegemann, M., and B. Watermann. 2003. Peculiarities of barnacle adhesive cured on non-stick surfaces. *J. Adhesive Sci. Tech.* 17: 1957–1977.

Whitesides, G.M., and M. Boncheva. 2002. Beyond molecules: Self-assembly of mesoscopic and macroscopic components. *Proc. Nat.Acad. Sci. USA*. 99: 4769–4774.

Yan, W., and S. Pan. 1981. The solubilizing effect of denaturation chemicals on the cement of *Balanus reticulates* Utinomi. *Oceanologica Limnologia Sinica* 12: 125–132.

Yan, W., and Y. Tang. 1981. The biochemical composition of the secondary cement of *Balanus reticulates* Utinomi and *Balanus amaryllis* Darwin. *Nanhai Studia Marina Sinica* 2: 145–152.

Yule, A.B., and G. Walker. 1987. Adhesion in barnacles. Pages 389-402 in "Barnacle Biology", A.J. Southward, ed., Balkema, Rotterdam.

Zhao, X., and S. Zhang. 2004. Fabrication of molecular materials using peptide construction motifs. *Trends Biotech.* 22: 470–476.

15

Secondary Metabolites of Biological Significance from Echinoderms

Marta S. Maier[1] and Ana P. Murray[2]

[1]Departamento de Química Orgánica, Facultad de Ciencias Exactas y Naturales, Universidad de Buenos Aires, Pabellón 2, Ciudad Universitaria (1428), Buenos Aires, Argentina
[2]Instituto de Investigaciones en Química Orgánica, Departamento de Química, Universidad Nacional del Sur, Avenida Alem 1253, (8000) Bahía Blanca, Buenos Aires, Argentina

INTRODUCTION

The extraordinary development of marine natural products chemistry during the last three decades has been prompted by the fact that marine organisms are sources of new natural products with unusual structural features that have no precedent among structures of terrestrial origin (Blunt *et al.* 2003). The toxicological and pharmacological properties of many of these secondary metabolites (Mayer and Hamann 2002) as well as the growing interest in their biological role in the natural environment (Paul and Puglisi 2004) have led to the discovery of thousands of new molecules. Some of these have shown interesting biological activities and are actually candidates for the development of new drugs (Natori *et al.* 2000, Hart *et al.* 2000, Kijjoa and Sawangwong 2004). Among marine organisms, echinoderms produce complex mixtures of secondary metabolites that may play an important role as defensive molecules (Komori 1997).

The phylum Echinodermata (Greek, *echinos*, spiny; *derma*, skin) comprises some of the most familiar seashore animals, widely distributed in all oceans and depths. The phylum is divided into five classes: Asteroidea (starfishes or sea stars), Holothuroidea (sea cucumbers or holothurians), Ophiuroidea (ophiuroids or brittle stars), Crinoidea (sea lilies and feather stars) and Echinoidea (sea urchins). The presence of polar saponins is constant in starfishes and sea cucumbers, while they have been found only occasionally in sponges, gorgonians, alcyonarians, green alga, and in fishes of the genus *Pardachirus* (Minale *et al.* 1996). Ophiuroids contain sulfated polyhydroxylated steroids and only two sulfated steroidal monoglycosides have been reported in the ophiuroid *Ophioderma longicaudum* (Riccio *et al.* 1986a). On the contrary, there is no report of saponins in the classes Echinoidea and Crinoidea. Starfishes and sea cucumbers also are rich in glycosphingolipids (cerebrosides and gangliosides) and several bioactive novel structures have been isolated of both classes of echinoderms (Chludil *et al.* 2002a).

The purpose of the present communication is to offer a general view of the structural characteristics and biological activities of the secondary metabolites of the classes Asteroidea, Holothuroidea and Ophiuroidea, focusing on recent examples, some of these from our laboratory.

STEROIDAL GLYCOSIDES FROM ASTEROIDEA

Sulfated steroidal oligoglycosides (asterosaponins) have been recognized as predominant and characteristic metabolites of starfishes. These molecules, responsible for the toxicity of these marine organisms, are accompanied by a variety of steroidal mono- and diglycosides and by free and sulfated polyhydroxysteroids leading to very complex mixtures of highly oxygenated compounds (Iorizzi and Riccio 1999). Starfish extracts containing asterosaponins have shown a wide spectrum of biological effects: cytotoxic, hemolytic, antibacterial, antiviral, and antifungal activities (Verbist 1993). However, few pharmacological studies have been carried out on the pure glycosides and only recently the trend has changed and the new structures published are accompanied by biological studies. Asterosaponins contain a $\Delta^{9(11)}$-3β, 6α-dihydroxysteroidal nucleus with a sulfate group at C-3 and an oligosaccharide chain, usually made up of five or six units, glycosidically linked to C-6. Steroidal side chains usually contain 20-α-hydroxy and 23-oxo functionalities and 18 different native aglycones with C_7 to C_{10} steroidal chains have been identified in *ca.* 70 asterosaponins described so far. The more common monosaccharide units are D-xylose, D-quinovose, D-fucose, and D-galactose, always in their pyranose form with β-anomeric configuration. An L-arabinose unit with α-anomeric configuration has been found only in a limited number of asterosaponins (Iorizzi *et al.* 2001). The first glycosyl residue directly linked to the aglycon (usually D-glucosyl or D-quinovosyl) is always glycosylated at C-3 by a branching glycosyl group (usually D-xylosyl or D-quinovosyl) and a terminal quinovose unit is always linked at C-2 of the branching sugar. The latter is glycosylated at C-4 by the remaining part of the oligosaccharide chain, that may contain two or three additional glycosyl units. Recently we have isolated and characterized labidiasteroside A (**1**), the first saponin containing an unusual glucose unit as a branching point in the oligosaccharide chain, from the starfish *Labidiaster annulatus* (Díaz de Vivar *et al.* 1999), together with the known ovarian asterosaponin 1 (**2**) (Fig. 15.1). The latter oligoglycoside is one of the few examples of asterosaponins with a D-6-deoxy-*xylo*-hex-4-ulosyl residue linked to C-3 of the aglycone.

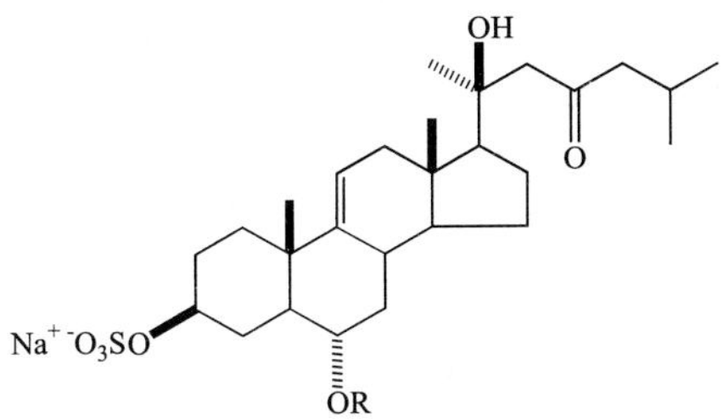

1 Labidiasteroside A R = Fuc —(1-2)→ Qui —(1-4)→ Glc —(1-2)→ Glc —
Glc ← (1-2) Qui

2 Ovarianasterosaponin 1
R = Fuc —(1.-2)→ Fuc —(1-4)→ Qui —(1-3)→ D-6-deoxy-xylo-hex-4-ulosyl —
Qui ← (1-2) Qui

Fig. 15.1 Steroidal oligoglycosides from the starfish *Labidiaster annulatus.*

Labidiasteroise A and the recently isolated luidiaquinoside (De Marino *et al.* 2003), differing from **1** in the oligosaccharide chain, are the only two examples reported so far containing a branched glucose in the oligosaccharide chain. Both asterosaponins contain the typical sodium 3, β, 6α, 20*S*-trihydroxy-5α-cholest-9(11)-en-23-one 3β-yl sulfate aglycone, characteristic of thornasteroside A, the most widely distributed oligoglycoside in starfishes (Minale *et al.* 1993). The latter was the first asterosaponin characterized completely by a combination of spectroscopic analysis and chemical transformations (Kitagawa and Kobayashi 1978). Prior work in this field was based on the analysis of aglycones obtained after acid hydrolysis of saponin mixtures. These procedures led to the formation of artifacts derived from degradation reactions, such as asterone (5α-pregn-9(11)-en-3β, 6α-diol-20-one) easily derived by retro-aldol cleavage of the common 3β, 6α, 20*S*-trihydroxy-5α-cholest-9(11)-en-23-one 3β-yl sulfate aglycone.

Although the majority of asterosaponins are pentaglycosides, two groups of hexaglycosides have been isolated. Versicoside A (**3**) was the first example of the group of hexaglycosides containing a single branching sugar unit (Itakura *et al.* 1983) while pectinioside E was the first hexaglycoside with two branching points in the oligosaccharide chain (Dubois *et al.* 1988). Bioactivity-guided fractionation of the ethanolic extract of the Patagonian starfish *Anasterias minuta* (Chludil *et al.* 2002b) using the brine shrimp (*Artemia salina* L.) larvae mortality assay led us to the isolation of two new hexaglycosides, anasterosides A (**4**) and B (**5**), together with the known versicoside A (**3**) (Fig. 15.2). The three asterosaponins contained the same hexasaccharide chain, but differed in the side chain of the aglycone. Saponin **4** contained (20*R*)-5α-cholest-9(11)-en-23-one-3β, 6α-diol 3-sulfate, the aglycone of asterosaponins isolated from the starfishes *Marthasterias glacialis* (Dini *et al.* 1983), *Coscinasterias tenuispina* (Riccio *et al.* 1986b), *Luidia maculata* (Krebs *et al.* 1984), and *Neosmilaster georgianus* (Vásquez *et al.* 1992) while asterone was the aglicone present in **5**. Although asterone is an artifact easily obtained by retroaldol cleavage of glycosides containing aglycones with 20-hydroxy-23-oxo side chains, the fact that a number of asterosaponins containing this aglycone have been reported in the literature (Iorizzi and Riccio 1999) and the mild extraction and processing conditions employed in the isolation of **5**, suggested that anasteroside B was a naturally occurring asterosaponin and not an artifact.

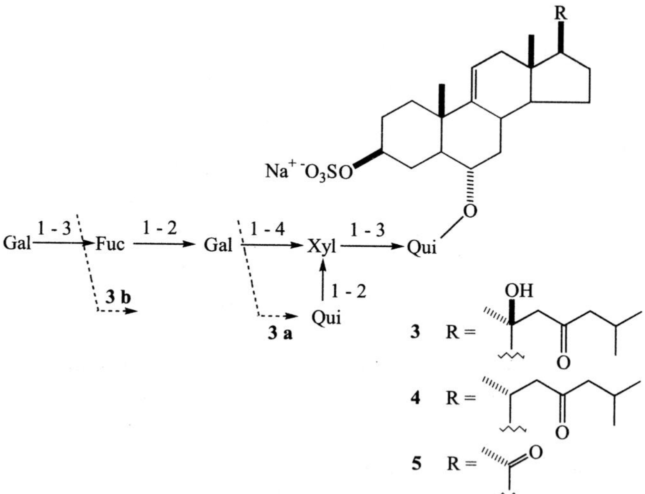

Fig. 15.2 Steroidal oligoglycosides from the starfish *Anasterias minuta*.

Asterosaponins are complex molecules and the structural elucidation of the native steroidal oligoglycosides requires a combination of spectroscopic analysis (NMR and FABMS), chemical transformations, enzymatic degradations and analysis of shortened glycosides.

Characteristic signals in the highfield region of the ^{1}H NMR spectrum gives information about methyl groups belonging to the aglycone and to 6-desoxy sugars, typical in these saponins. These data are confirmed by the analysis of the ^{13}C NMR spectrum. For example, ^{1}H, ^{13}C NMR and DEPT spectra of **4** were consistent with the presence of a steroidal aglycone with a 9(11)-double bond [δ_C 145.3 (s, C-9) and 116.2 (d, C-11); δ_H 5.19 (1H, m, H-11)], one sulfated oxomethine [δ_C 78.0 (d, C-3); δ_H 4.84 (1H, m, H-3)], one oxomethine [δ_C 79.5 (d, C-6); δ_H 3.82 (1H, m, H-6)], and one carbonyl group (δ_C 211.2, C-23). Glycosidation at C-6 was supported by the downfield shift of C-6 in the ^{13}C NMR spectrum with respect to the corresponding value in synthetic asterosaponin aglycones containing the 3β, 6α-dihydroxy oxidation pattern (Honda and Komori 1986). The presence of a methyl singlet at δ 2.17 ppm in the ^{1}H NMR spectrum of **5** and the signals at δ 210.0 and 31.0 ppm in the ^{13}C NMR spectrum confirmed the presence of asterone as the aglycone of glycoside **5**. In addition to aglycone signals, ^{1}H and ^{13}C NMR spectra of both asterosaponins showed six anomeric protons as doublets at δ 4.82, 4.86, 4.90, 5.06, 5.09, and 5.18 ppm and carbons at δ 101.7, 103.9, 104.2, 105.9, 106.1, and 106.2 ppm. The β-stereochemistries at the anomeric carbons were deduced from the coupling constant values (J = 7.3–7.7 Hz). Methyl carbon signals at δ 16.8, 18.0, and 18.3 ppm and the doublet methyl proton signals at δ 1.39, 1.58, and 1.83 ppm indicated the presence of three 6-deoxy sugars. The presence of galactose, fucose, xylose and quinovose in a 2:1:1:2 ratio was confirmed by acid hydrolysis with aqueous 2N trifluoroacetic acid followed by GC analysis of the corresponding alditol peracetates. Due to the overlapping of signals corresponding to the sugar residues in the ^{1}H NMR spectrum, the position of the interglycosidic attachments was determined using a combination of ^{1}H-^{1}H COSY, relayed COSY, and HETCOR experiments in combination with analysis of the ^{13}C NMR data of the carbons involved in the interglycosidic linkages. We confirmed these data by methylation of **4** followed by acid hydrolysis and GC-MS analysis of the partially methylated alditol acetates derived of each sugar unit.

Since some asterosaponins have been reported to show antifungal properties (Verbist 1993), hexaglycosides **3-5** were evaluated by a bioautographic technique (Homans and Fuchs 1970) for their antifungal activity against the plant pathogenic fungus *Cladosporium cucumerinum*. To evaluate the influence of the oligosaccharide moiety on the antifungal activity, we prepared the triglycoside forbeside H (**3a**) and the pentaglycoside thornasteroside A (**3b**) by enzymatic hydrolysis of **3** with *Charonia lampas* glycosidase mixture (Fig. 15.2). Hexaglycosides **3** and **4** and pentaglycoside **3b** were active in a concentration-dependent manner. Anasteroside A (**4**), lacking the hydroxyl group at C-20 was the most active compound while anasteroside B (**5**), containing the same hexasaccharide moiety as **3** and **4** but with a shorter side chain was inactive at all the tested concentrations. On the other hand, while the pentaglycoside **3b** was active, the triglycoside **3a** showed no activity in all concentration ranges. Desulfation of versicoside A (**3**) by solvolysis in dioxane/pyridine (1:1) rendered a totally inactive saponin. These results suggest that the side chain in the steroidal aglycone moiety, together with the structure of the sugar portion and the presence of the sulfate group at C-3 may play an important role in the antifungal activity of these steroidal saponins.

Few examples of asterosaponins containing tetra-and trisaccharide chains have been described in the literature. Cosmasterosides C (**6**) and D (**7**) isolated from the Patagonic starfish *Cosmasterias lurida* (Roccatagliata *et al.* 1994) (Fig. 15.3) together with santiagoside from the Antarctic starfish *Neosmilaster georgianus* (Vásquez *et al.* 1992) and forbeside F from *Asterias forbesi* (Findlay *et al.* 1990) are the only examples of tetrasaccharides in starfishes. Forbesides H and G (Findlay *et al.* 1990) are two of the four examples of triglycosides isolated from starfishes so far.

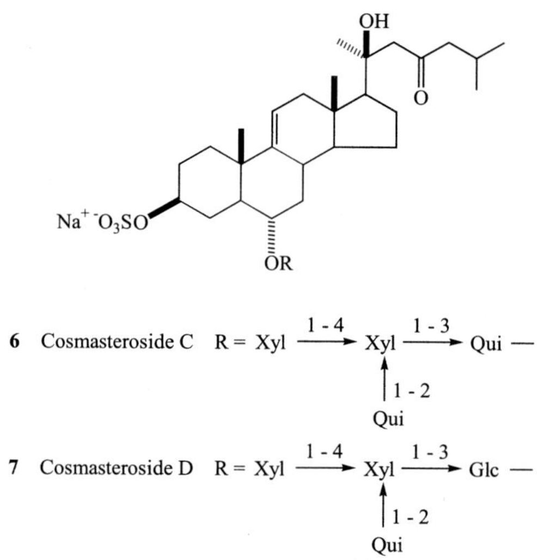

Fig. 15.3 Tetraglycosides from the starfish *Cosmasterias lurida.*

Two starfishes of the genus *Echinaster*, *E. sepositus* (De Simone *et al.* 1981) and *E. luzonicus* (Riccio *et al.* 1982) are characterized by the presence of cyclic steroidal glycosides instead of the typical asterosaponins distributed in starfishes. Among these saponins, sepositoside A (**8**) is the major glycoside in *E. sepositus* (Fig. 15.4). These glycosides are characterized by a Δ^7, 3β, 6β-dihydroxysteroidal nucleus and a trisaccharide chain cyclized between C-3 and C-6 of the aglycone. They lack a sulfate group and the anionic charge is due to a glucuronic acid unit attached to C-3. Initially, they were considered to be chemotaxonomic markers for the genus *Echinaster* but the absence of these glycosides in the starfish *E. brasiliensis* (Iorizzi *et al.* 1993), which contained typical asterosaponins, contradicted this view.

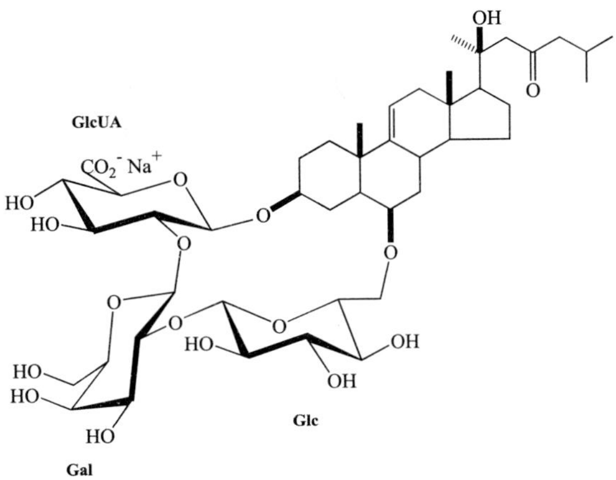

8 Sepositoside A

Fig. 15.4 Cyclic steroidal glycoside sepositoside A from the starfish *Echinaster sepositus*.

Besides asterosaponins, steroidal glycosides composed of a polyhydroxylated steroidal aglycone and a carbohydrate portion of one or two monosaccharide units are widespread in starfishes. They show a large degree of structural variability due to the hydroxylation pattern of the steroidal nucleus, the identity and location of the sugar residue and the presence of a sulfate or phosphate residue (Iorizzi and Riccio 1999). Several examples include in their structure a 5α-cholestane-3β, 6α, 8, 15β, 24*S*-pentaol aglycone moiety and a sugar portion that is glycosidically attached at C-24 of the steroid. Sulfate residues have been found attached to the aglycone, as in the diglycoside antarcticoside P (**9**), isolated from an Antarctic starfish (Iorizzi *et al.* 1996) or at the sugar residue, as in Luridoside A (**10**) from the Patagonic starfish *Cosmasterias lurida* (Maier *et al.* 1993) (Fig. 15.5).

Several excellent reviews on polyhydroxylated glycosides from starfishes with descriptions of their structures, distribution and spectroscopic characteristics have been published (Minale *et al.* 1993, D'Auria *et al.* 1993a, Iorizzi and Riccio 1999, Iorizzi *et al.* 2001). We will discuss now some recent examples of this type of steroidal glycosides and their biological activities.

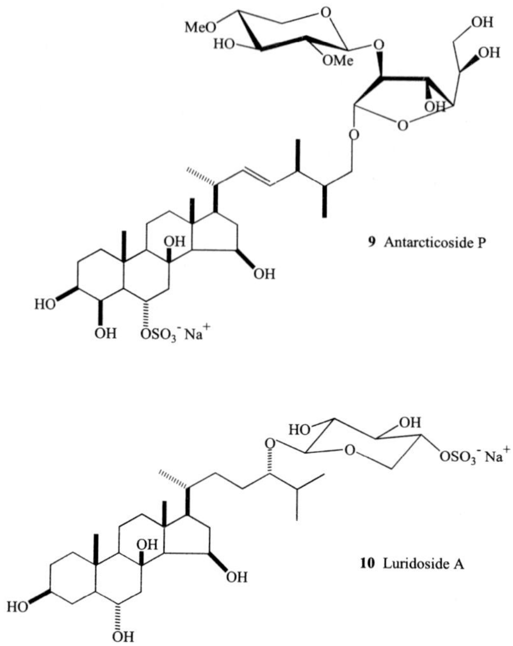

9 Antarcticoside P

10 Luridoside A

Fig. 15.5 Sulfated polyhydroxylated steroidal glycosides from starfishes.

Two new monoglycosides, rathbuniosides R_1 (**11**) and R_2 (**12**) have been isolated from the starfish *Asterias rathbuni* (Fig. 15.6) (Ivanchina *et al.* 2001). The 3β, 6α, 15β, 24-tetrahydroxy substitution pattern in **11** has been found for the first time in steroids from starfish. Both glycosides contain a D-xylopyranose unit attached to C-24, the most common glycosidation position in this type of compounds (Iorizzi and Riccio, 1999).

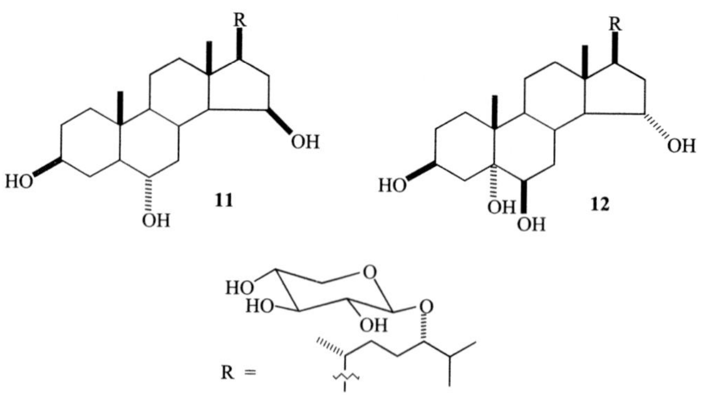

Fig. 15.6 New polyhydroxylated steroidal glycosides from the starfish *Asterias rathbuni*.

Rathbuniosides R_1 and R_2 and six related diglycosides isolated from the starfishes *Mediaster murrayi* (Kicha *et al.* 1999), *Ceramaster patagonicus* (Kicha *et al.* 1997) and *Culcita novaeguineae* (Kicha *et al.* 1986) showed inhibition of cell division of the fertilized eggs of the sea urchin *Strongylocentrotus intermedius* as well as hemolytic activity. Differences in the activity of the eight compounds were correlated with the number of hydroxyl groups in the aglycones. Cytotoxic effects of the compounds were the result of cell membrane damage due to their ability to bind to membrane cholesterol (Prokof'eva *et al.* 2003).

Recently, two new monoglycosides of polyhydroxysteroids, sanguinosides A and B, were isolated from two starfishes of the genus *Henricia*, *H. sanguinolenta* and *H. leviuscula leviuscula* (Kalinovsky *et al.* 2004). Sanguinoside A has a 24-propylcholest-4-ene- 3β, 6β, 8, 15α, 16β, 29-hexaol aglycone with an unusual structure of its side chain, while sanguinoside B has a unique 2, 3, 4-tri-*O*-methylxylopyranosyl unit attached to C-3.

Two new neuritogenic steroid diglycosides, linckosides A (**13**) and B (**14**) were isolated from the Okinawan starfish *Linckia laevigata* (Qi *et al.* 2002). These compounds (Fig. 15.7) are examples of a group of rare steroidal diglycosides in which the monosaccharide residues are located in two different positions of the steroidal nucleus (Iorizzi *et al.* 2001). Both glycosides differ in the sugar unit attached to C-29, an α-arabinofuranose in **13** and a 2-*O*-methylxylopyranose unit in **14**. Linckoside B showed higher neuritogenic activity than A, suggesting that the sugar moiety at C-29 may play an important role in the activity of both glycosides.

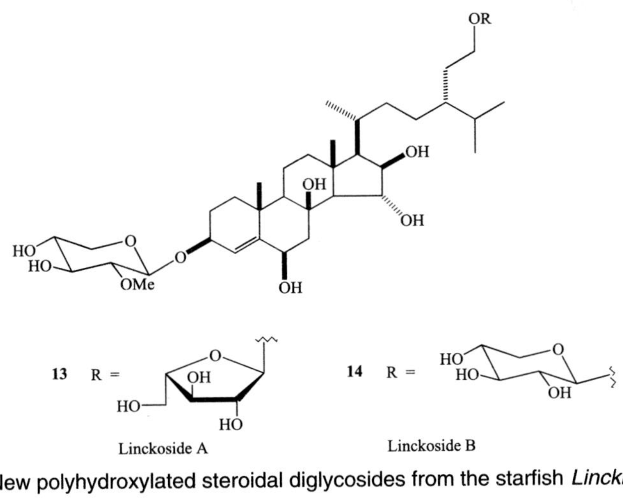

Fig. 15.7 New polyhydroxylated steroidal diglycosides from the starfish *Linckia laevigata*.

The starfish *Certonardoa semiregularis* has shown to be an extremely rich source of a complex mixture of bioactive sterols and glycosides of polyhydroxylated steroids. Purification of the brine shrimp active fraction of the methanolic extract of *C. semiregularis* led to the isolation of eight new polyhydroxylated steroidal diglycosides (**15-22**), together with 10 new monoglycosides and eleven new polyhydroxysteroids (Wang *et al.* 2002, 2003, 2004) (Fig. 15.8). Diglycosides **15-19** are sulfated at C-3′ of the xylopyranose unit attached to C-26 of the steroidal side chain, while compounds **20** and **21** contain a sulfate group at C-6 (α) from the steroid nucleus. The structures of the compounds were elucidated with the aid of 1D and 2D NMR experiments and chemical derivatization. Evaluation of the antiviral activity of the diglycosides showed that the compounds were not active within the range of noncytotoxic concentrations (Wang *et al.* 2002). The monoglycosides of polyhydroxylated steroids and their corresponding aglycones displayed considerable cytotoxicity against a small panel of human solid tumor cells with the free sterols as the more potent compounds (Wang *et al.* 2004).

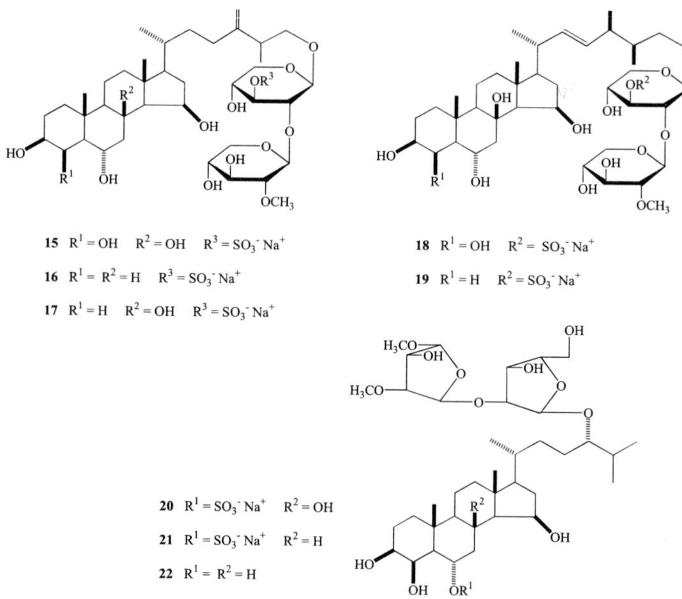

Fig. 15.8 New polyhydroxylated steroidal diglycosides from the starfish *Certonardoa semiregularis*0

Finally, we wish to mention some uncommon secondary metabolites isolated from starfishes. Unusual salts of sulfated steroids and 1-methyl-6, 7-dihydroxy-1, 2, 3, 4-tetrahydroisoquinoline (salsolinol) (**23-25**), were obtained from the aqueous extract of the starfish *Lethasterias nanimensis chelifera* (Kicha *et al.* 2003) (Fig. 15.9).

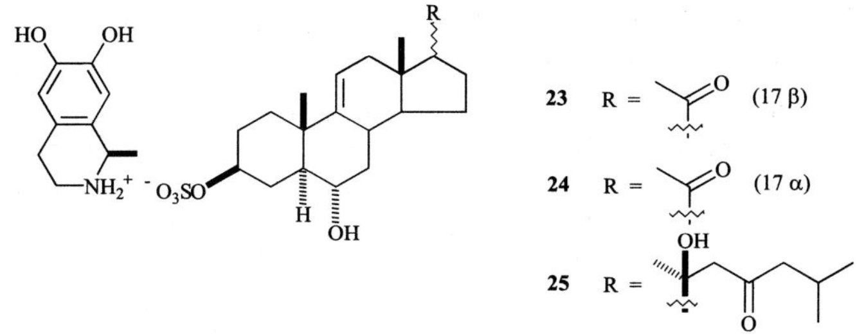

Fig. 15.9 Alkaloidosteroids from the starfish *Lethasterias nanimensis chelifera.*

Sulfated steroids **23** and **24** differ from each other in the stereochemistry at C-17 in the steroidal nucleus. The anionic moiety of **25** was assigned as 3-*O*-sulfothornasterol. The sodium salt of this steroidal sulfate has been previously isolated from *Asterias rathbuni* (Ivanchina *et al.* 2001). The alkaloidosteroids **23-25** are the first examples of ionic hybrids of marine sulfated steroids and an isoquinoline alkaloid reported so far from a natural source.

Some starfishes belonging to the genus *Pteraster* contain ophiuroid-type 3, 21-disulfates instead of typical polyhydroxylated steroids and their glycosides, characteristic for the Asteroidea class (Levina *et al.* 1998). Recently, two further new examples of this type of polar steroids (**26** and **27**) have been isolated from the starfish *Pteraster pulvillus* (Ivanchina *et al.* 2003) (Fig. 15.10).

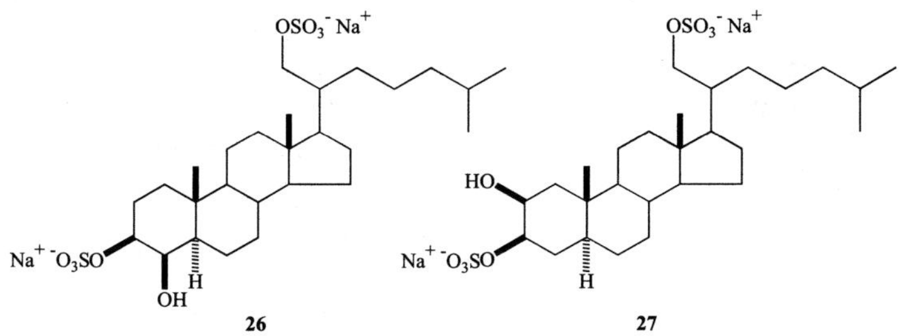

Fig. 15.10 Sulfated trihydroxysteroids from the starfish *Peraster pulvillus.*

Compound **26** possesses a new 3β, 4β-dihydroxy-5α-cholestane skeleton, not previously described in steroids from starfishes or ophiuroids. This polar steroid showed hemolytic activity to mouse erythrocytes.

TRITERPENOIDAL OLIGOGLYCOSIDES FROM HOLOTHUROIDEA

Triterpenoid saponins are typical metabolites of plant origin, but extensive investigation on toxins from marine organisms in the last three decades has shown that triterpenoid glycosides are widely distributed in sea cucumbers (Stonik *et al.* 1999, Chludil *et al.* 2003a). It has been suggested that these saponins have a defensive role due to their ability to induce hemolysis by formation of complexes with erythrocyte membrane Δ^5 sterols and formation of channels that result in the loss of K^+ ions followed by cytolysis (Kalinin *et al.* 1996a). This membranotropic action is responsible for their wide spectrum of biological effects, such as antifungal, cytotoxic, cytostatic and immunomodulatory activities (Kalinin *et al.* 1996b, Agafonova *et al.* 2003). Sea cucumbers are resistant to their own toxins due to the presence of Δ^7, 14α-methyl- and 14α-dimethyl-$\Delta^{9,11}$ sterols as well as their conjugated forms such as steryl sulfates and steryl xylosides (Ponomarenko *et al.* 2001). Some triterpene glycosides are specific for different taxonomic groups of sea cucumbers and their distribution has been successfully applied in the reclassification of *Stichopus mollis* into the new genus *Australostichopus* (Moraes *et al.* 2004) and in the taxonomy of sea cucumbers belonging to the genus *Cucumaria* (Avilov *et al.* 2004).

Most of the triterpenoid glycosides isolated from holothurians present a sugar chain of two to six monosaccharide units linked to the C-3 (β) of an aglycone based on an "holostanol" skeleton (Fig. 15.11). The carbohydrate moieties of these glycosides may include quinovose, glucose, 3-*O*-methylglucose, xylose and 3-*O*-methylxylose. The first monosaccharide unit is always xylose and in several glycosides it is sulfated at C-4. 3-*O*-methylglucose and 3-*O*-methylxylose are always terminal sugars. While starfish saponins contain a sulfate group attached to C-3 of the aglycone, nearly sixty percent of the triterpenoid glycosides isolated so far from sea cucumbers are sulfated at the oligosaccharide moiety. Although most of them are monosulfated, several di-and tri sulfated glycosides have been isolated, mainly from the order Dendrochirotida.

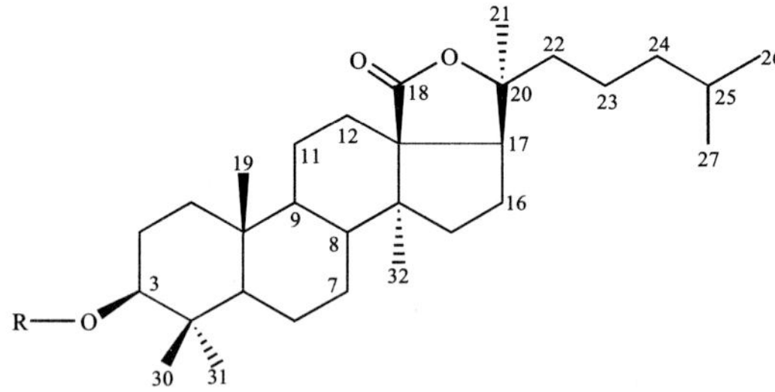

Fig. 15.11 Structure of hypothetical holostanol.

Recently, we have reviewed the chemical structures and the methods applied in the structural elucidation of these complex molecules (Chludil *et al.* 2003a). Two main series of aglycones can be distinguished: those containing a 3β-hydroxyholost-9(11)-ene skeleton and those based on a

3β-hydroxyholost-7-ene aglycone. Bivittoside C (**28**) (Fig. 15.12), isolated from the sea cucumber *Bohadschia bivittata* (Kitagawa *et al.* 1989) is the simplest triterpene glycoside with a $\Delta^{9,11}$ double bond.

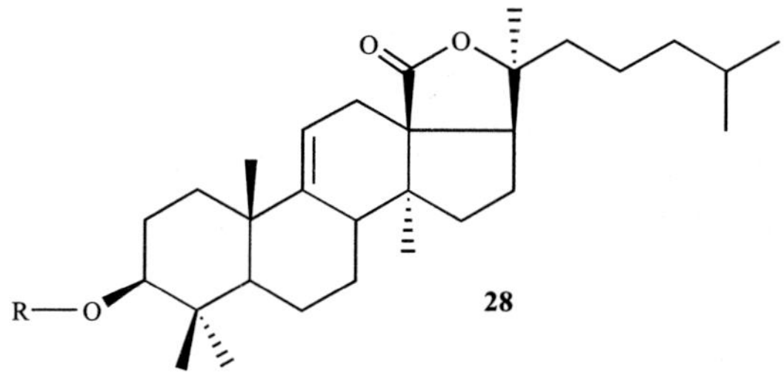

28

R = [3-*O*-Me-Glc-(1→3)-Glc-(1→4)]-[3-*O*-Me-Glc-(1→3)-Glc-(1→4)-Qui-(1→2)]-Xyl

Fig. 15.12 Structure of bivittoside C.

Further glycosides containing aglycones of this series show additional oxygenated functions, such as a carbonyl group at C-16 and 12α- and 17α-hydroxyl groups as well as additional acetoxy or hydroxy groups in the side chain. Recently, we have isolated and characterized two cytotoxic and antifungal triterpene glycosides from the Patagonian sea cucumber *Hemoiedema spectabilis* (Chludil *et al.* 2002c). Compounds **29** and **30** (Fig. 15.13) contain the same aglycone, 3β-hydroxyholosta-9(11), 25-dien-16-one, and differ in the degree of sulfation of the tetrasaccharide chain.

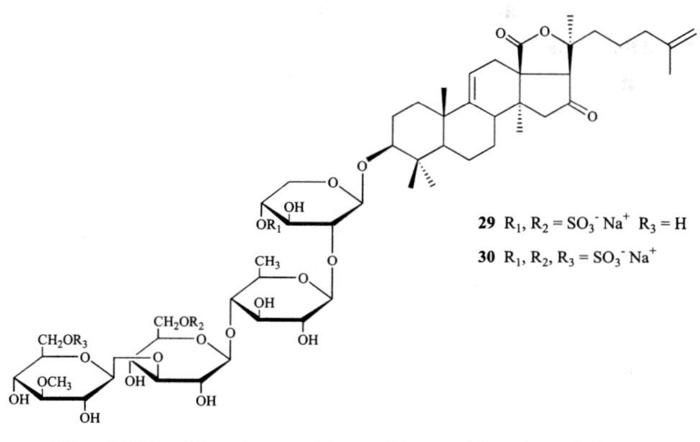

29 $R_1, R_2 = SO_3^- Na^+$ $R_3 = H$

30 $R_1, R_2, R_3 = SO_3^- Na^+$

Fig. 15.13 Structures of hemoidemosides A and B.

Hemoiedemosides A and B showed antifungal activity against the phytopathogenic fungus *Cladosporium cucumerinum*, while the semisynthetic desulfated analog was less active. The zootoxicities of **29** and **30** were evaluated using the brine shrimp (*Artemia salina* L.) larvae mortality bioassay. Hemoiedemoside A (**29**) showed a noteworthy toxicity in this assay, while the trisulfated glycoside **30** was two times less active than **29** and nearly 10 times more active than the desulfated derivative.

Only a few examples of holothurins containing the side chain of the triterpenoid moiety in a furan form were isolated from the sea cucumbers *Holothuria leucospilota* (Kitagawa *et al.* 1981) and *H. forskalii* (Rodríguez *et al.* 1991). Recently, another bioactive glycoside with this structural feature has been isolated from the sea cucumber *Telenata ananas* (Hedge *et al.* 2002). Triglycoside **31** (Fig. 15.14) has been isolated together with the known bivittoside D (Kitagawa *et al.* 1989).

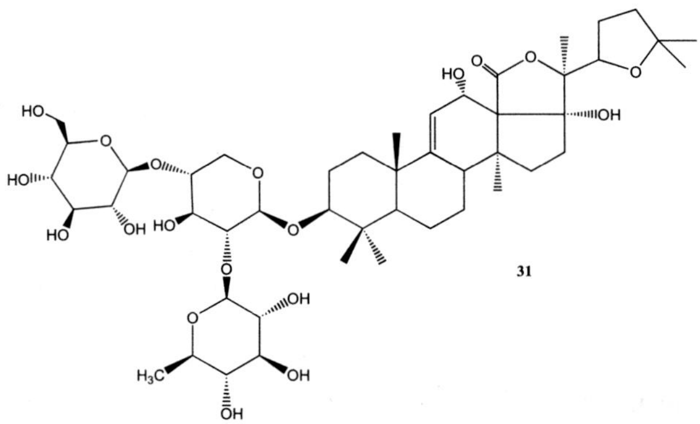

31

Fig. 15.14 A new triglycoside with a furan moiety in the side chain isolated from *Telenata ananas*.

The second series of holothurins comprises several glycosides containing the simple 3β-hydroxyholost-7-ene aglycone, such as frondoside B (**32**) (Fig. 15.15), isolated from the sea cucumber *Cucumaria frondosa* (Findlay *et al.* 1992).

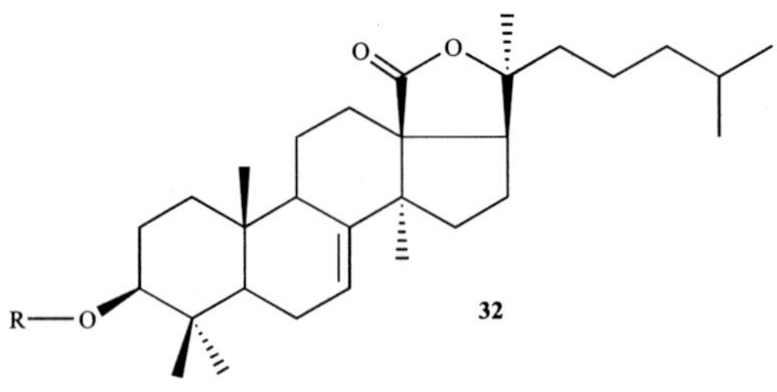

32

R = [3-*O*-Me-Glc-(1→3)-6-OSO_3Na-Glc-(1→4)]-[Xyl-(1→2)]-Qui-(1→2)-4-OSO_3Na-Xyl; Δ^{24}

Fig. 15.15 Structure of frondoside B.

One structural feature that has been found only in this series of aglycones is the presence of an acetoxyl group at C-16 (α or β). Recently, we have isolated two new cytotoxic and virucidal trisulfated triterpene glycosides from the Antarctic sea cucumber *Staurocucumis liouvillei* (Maier *et al.* 2001). Liouvillosides A (**33**) and B (**34**) (Fig. 15.16) contain the same

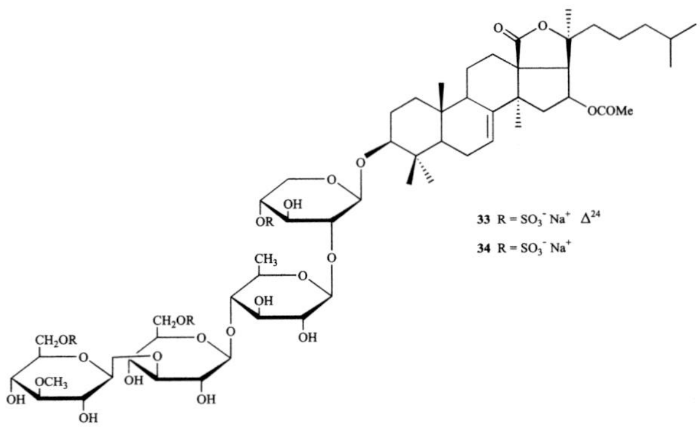

Fig. 15.16 Structures of liouvillosides A and B.

trisulfated tetrasaccharide chain and differ in the side chain of the 16β-acetoxyholost-7-en-3β-ol nucleus. Both saponins exerted an irreversible virucidal effect on HSV-1, but glycoside **33** with a Δ^{24} double bond was less active.

Recently a triterpene glycoside containing the same triterpenoid skeleton as **33** and **34** but with a carbonyl group at C-23 in the side chain has been isolated from the sea cucumber *Cucumaria coniscopermium* (Avilov *et al*. 2003). This is the second example of a triterpene glycoside containing this aglycone, the first one, cucumarioside A_0-1 has also been isolated from a sea cucumber of the genus *Cucumaria* (Drozdova *et al*. 1993).

Further examples of triterpene glycosides containing a 16β-acetoxyholost-7-en-3β-ol nucleus are thyonosides A and B, two new holothurins isolated from the sea cucumber *Thyone aurea* (Bonnard and Reinhart, 2004) as well as intercedensides A-C, three cytotoxic sulfated triterpene glycosides present in *Mensamaria intercedens* (Zou *et al*. 2003). Intercedenside B has the same aglycone as liouvilloside A (**33**) but differs in the oligosaccharide chain.

Two sea cucumbers belonging to the order Aspidochirota, *Stichopus chloronotus* and *Thelenota ananas*, are characterized by their content in triterpenoid glycosides with a 3β-hydroxyholost-7-ene aglycone with an acetoxyl group at C-23 in the side chain (Chludil *et al*. 2003a). Recently, five non sulfated holothurins containing the aglycone 23(*S*)-acetoxyholsta-7, 25-dien-3β-ol have been isolated from *Synallactes nozawai* (Silchenko *et al*. 2002)

Purification of the glycosidic fraction of the sea cucumber *Psolus patagonicus*, led us to isolate patagonicoside A (**35**) (Fig. 15.17), the main triterpene glycoside in the mixture (Murray *et al*. 2001). This holothurin is the first example of a 3β-hydroxyholost-7-ene aglycone substituted with 12α- and 17α-hydroxy groups. All proton and carbon resonances of the aglycone and the oligosaccharide chain were successfully assigned

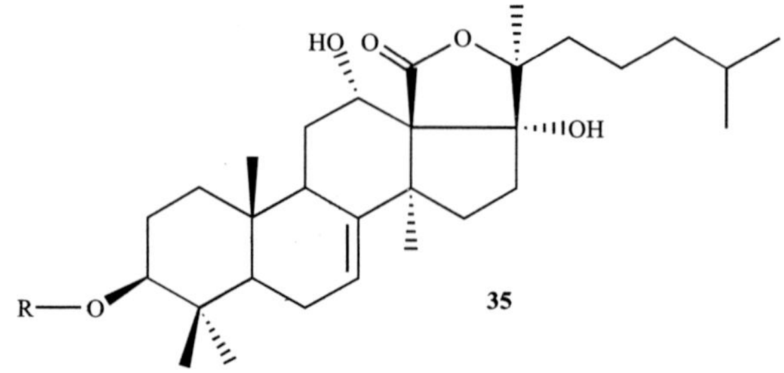

35

R = 3-*O*-Me-Glc-(1→3)-6-OSO_3Na-Glc-(1→4)-Qui-(1→2)-4-OSO_3Na-Xyl

Fig. 15.17 Structure of patagonicoside A, an antifungal oligoglycoside isolated from the sea cucumber *Psolus patagonicus*.

by a combination of ^1H-^1H COSY, COLOC, HETCOR and NOESY experiments. Noteworthy is the *β*-configuration of H-9 in 3*β*-hydroxyholost-7-ene aglycones instead of the characteristic 9*α*-configuration in natural steroids and triterpenoids.

Patagonicoside A and its desulfated analog were evaluated for their antifungal activity against *Cladosporium cucumerinum*. Glycoside **35** was considerably active in a concentration dependent manner, while the desulfated derivative was inactive at the lowest concentrations (1.5 and 3 µg/spot) and weakly active at the highest tested concentrations (6-50 µg/spot). As both compounds differ only in the substitutions at C-4 of the xylose and at C-6 of the glucose units of the tetrasaccharide chain, these results suggest that the presence of sulfate groups in the oligosaccharide chain plays an important role in the antifungal activity of these triterpene glycosides.

Besides the two series of aglycones mentioned before, some examples of triterpenoid glycosides having uncommon non-holostane aglycones have appeared in the literature (Chludil *et al.* 2003a). All are sulfated compounds. Five glycosides contain aglycones with a 18(16) lactone and a Δ^7-unsaturation and a few, such as koreoside A (**36**) (Fig. 15.18) are devoid of a lactone function and have a shortened side chain (Avilov *et al.* 1997).

Recently, three new sulfated glycosides containing the same side chain as koreoside A, together with this last compound, have been isolated from the sea cucumber *Cucumaria conicospermium* (Avilov *et al.* 2003). All the non-holostane glycosides characterized so far were isolated from sea cucumbers belonging to the order Dendrochirota.

As mentioned above, several triterpene glycosides isolated from sea cucumbers show antifungal activity. Nevertheless, it has been shown that fungi can be isolated from all organs of holothurians (Pivkin 2000). Recently, eight new diterpene monoglycosides and one monoglycoside, virescenosides M-U, have been isolated from a marine strain of

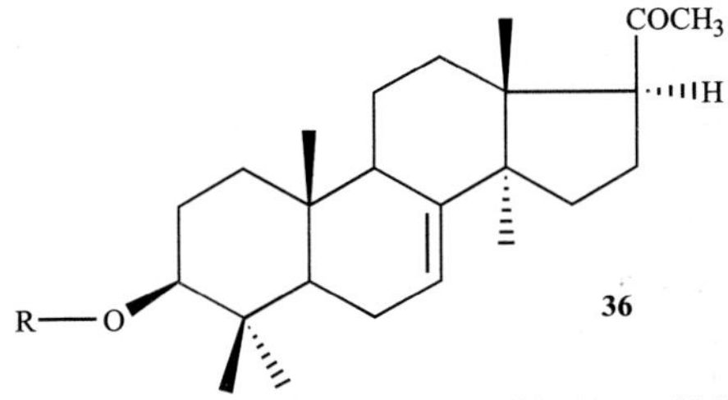

36

R = [6-OSO$_3$ Na-3-*O*-Me-Glc-(1 → 3)-6-OSO$_3$ Na-Glc-(1 → 4)]-[Xyl-(1 → 2)]-Qui-(1 → 2)-4-OSO$_3$ Na-Xyl

Fig. 15.18 Koreoside A isolated from the sea cucumber *Cucumaria koraiensis.*

Acremonium striatisporum associated with the sea cucumber *Eupentacta fraudatrix* (Afiyatullov *et al.* 2000, 2002, 2004). The compounds exhibited cytotoxic action against tumor cells of Ehrlich carcinoma *in vitro* and cytotoxic effects on developing eggs of the sea urchin *Strongylocentrotus intermedius*. These are the first reports on bioactive secondary metabolites from fungi isolated from echinoderms. As marine-derived fungi have become a potential source of new pharmacologically active compounds (Bugni and Ireland 2004), the ongoing exploration of microorganisms associated with echinoderms, in particular those containing antifungal metabolites, will help to understand the importance of fungi in marine ecosystems.

SULFATED POLYHYDROXYSTEROIDS FROM OPHIUROIDEA

Sulfated polyhydroxysterols are naturally occurring metabolites in sponges and echinoderms (D'Auria *et al.* 1993a). They are sometimes very complex mixtures of highly functionalized compounds with additional oxygenation of both the nucleus and the side chains (Aiello *et al.* 1999). These compounds have attracted considerable attention as a growing group of marine sterols with interesting biological and pharmacological activities (Kornprobst *et al.* 1998, Mc Kee *et al.* 1994). Among echinoderms, ophiuroids are characterized by their content of sulfated steroids with sulfate groups located at C-3 and C-21 and additional hydroxy groups located in rings A and C. In comparison to starfishes, rich in 3β-sulfated steroidal glycosides, only two examples of 3α-sulfated steroidal monoglycosides have been reported in the ophiuroid *Ophioderma longicaudum* (Riccio *et al.* 1986a). On the other hand, the content of sulfated polyhydroxysteroids proved to be very low in the Patagonic ophiuroids *Ophiactis asperula*, *Ophiacantha vivipara* and *Gorgonocephalus chilensis*, where mixtures of nucleosides and nucleobases were the major metabolites present in these organisms (Murray *et al.*

2002). Recently, a sulfated alkene has been isolated from the ophiuroid *Ophiocoma echinata* (Roccatagliata *et al.* 1997), being this the first report of this type of compounds in the class Ophiuroidea. It has been suggested that evolution of sulfated polyhydroxysteroids in ophiuroids may go through the diol derivatives and that some polyhydroxysteroids may be used as taxonomic markers (Shubina *et al.* 1998).

Nearly 40 different chemical structures of these compounds have been published in the last 20 years. Two main series of sulfated polyhydroxysteroids can be distinguished in ophiuroids.

The first one is that of 3α, 21-disulfoxy-4α-hydroxy-5β-sterols (Fig. 15.19). Most of these sterols contain a cholestane side chain and additional hydroxy groups located in ring C. Sterol **37**, the most widespread compound of this series, has been isolated as the major polar steroid in the ophiuroids *Ophiocoma dentata, Ophiarthrum elegans, Ophiarachna incrassata* (D'Auria *et al.* 1987), *Ophiosparte gigas* (D'Auria *et al.* 1993b), *Ophioplocus januarii* (Roccatagliata *et al.* 1996), *Ophiomastix annulosa* (D'Auria *et al.* 1992), *Ophionereis reticulata, Ophiozona impressa, Ophiocoma wendti* (D'Auria *et al.* 1995), *Ophiocoma picta* and *Ophiocoma erinaceus* (Shubina *et al.* 1998), and as a minor component in *Ophiocoma echinata* (Roccatagliata *et al.* 1997). Its structure has been determined from NMR and FAB-MS data. The low field chemical shift of the 19-methyl protons (δ_H 1.15 ppm) and the 19-methyl carbon (δ_H 27.4 ppm) are characteristic of the *cis* A/B ring junction.

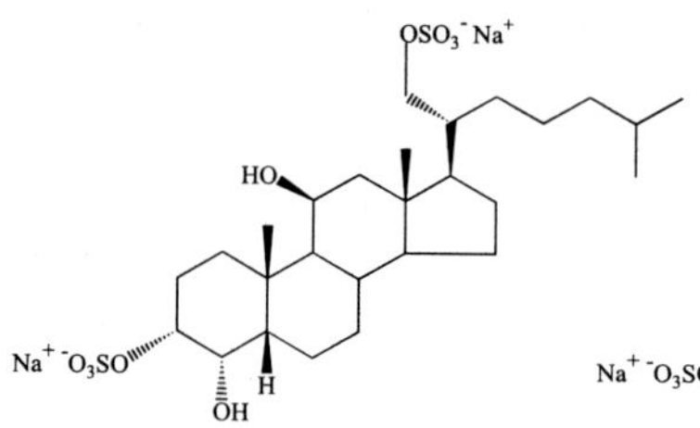

37

39 Δ^{22E}

40 24-Me, $\Delta^{24(28)}$

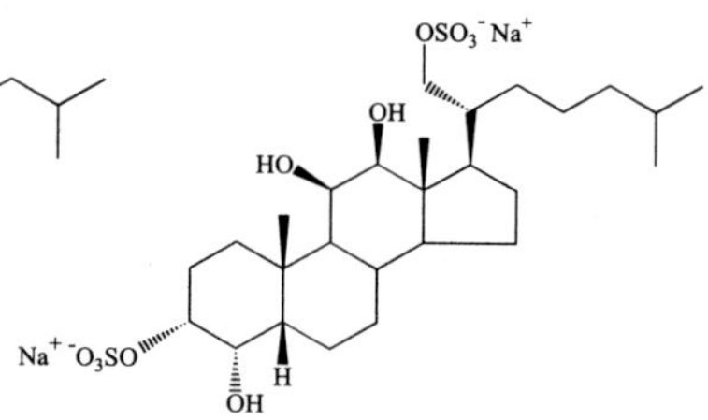

38

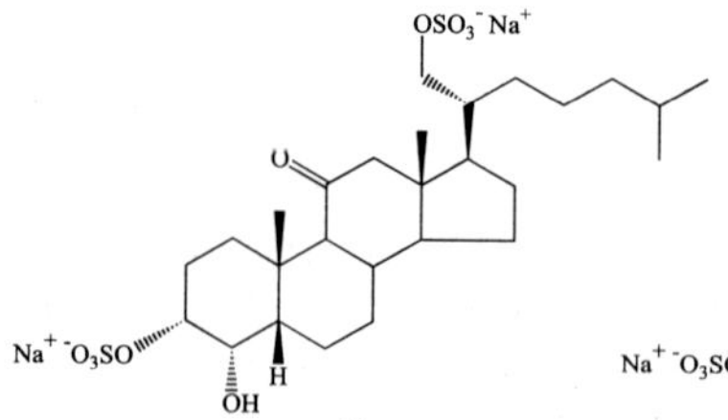

41

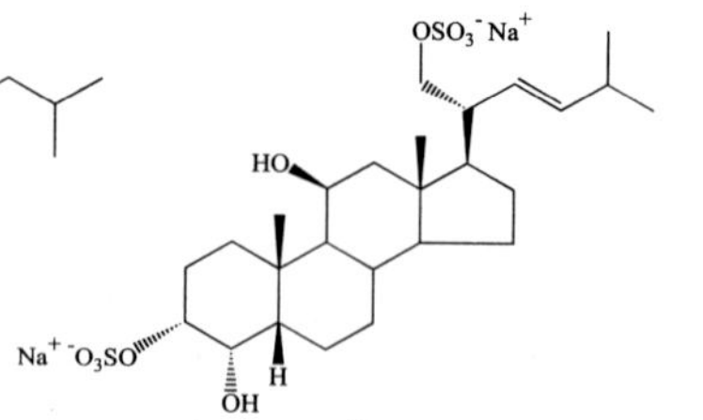

42

Fig. 15.19 Typical 3α, 21-disulfoxy-4α-hydroxy-5β-sterols in ophiuroids.

Disulfated sterol **38**, related to **37** by the presence of an additional hydroxy group at C-12 (β), has been isolated from *Ophiocoma echinata* (Roccatagliata *et al.* 1997), *Ophiozona impressa, Ophionereis reticulata* and *Ophiocoma wendti* (D'Auria *et al.* 1995), and as the major compound in *Ophiocoma scolopendrina* (D'Auria *et al.* 1995) and *Ophioderma longicaudum* (Riccio *et al.* 1985). *Ophiocoma dentata* and *Ophioplocus januarii* also contained two minor compounds (**39** and **40**) that possessed the same nucleus as **37** and differed in the side chain, while the 11-keto derivative (**41**) of **37** was found in *Ophiarthrum elegans* (D'Auria *et al.* 1987) and *Ophiomastix annulosa* (D'Auria *et al.* 1992). Recently we characterized a new disulfated sterol (**42**) in *Ophioplocus januarii* (Roccatagliata *et al.* 1996) that contained the same tetracyclic nucleus found in compounds **37**, **39**, and **40**, but with a shorter unsaturated chain. We tested these compounds for their inhibitory effect on the replication of RNA (polio virus (PV), Junin virus (JV) and respiratory syncytial virus (RSV)) and DNA (herpes simplex virus type 1 (HSV-1)) viruses and found that the viral replication was inhibited in a concentration-dependent manner. As compounds **37**, **39**, **40** and **42** differed only in their side chain, results obtained in the antiviral tests indicated the importance of the Δ^{22} double bond (compounds **39** and **42**) in the inhibitory effect on the replication of RSV as well as the $\Delta^{24(28)}$ double bond (compound **40**) on JV replication. Sulfated sterol **37** has previously shown modest antiviral activity against HIV-1 and HIV-2 (Mc Kee *et al.* 1994) and moderate inhibition of protein tyrosine kinase (PTK) pp60$^{v\text{-}src}$ (Fu *et al.* 1994).

A novel 5β-steroid disulfate (**43**), with a unique 4α, 9α-ether bridge has been isolated from the ophiuroid *Ophiomastix annulosa* together with the minor related compound **44** (D'Auria *et al.* 1992) (Fig. 15.20).

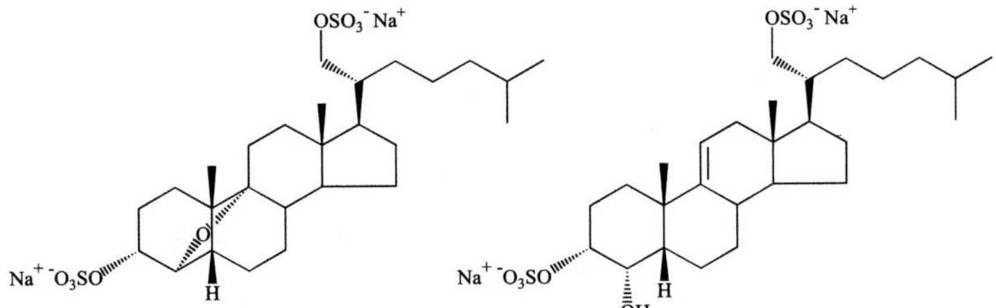

43 44

Fig. 15.20 Sulfated polyhydroxysteroids from *Ophiomastix annulosa*.

One sulfated polyhydroxysterol with a Δ^{25} double bond in the side chain (**45**) and three novel compounds containing a 24-methyl-26-hydroxy (**46**), 24-ethyl-26-hydroxy (**47**) and 24-hydroxyethyl 24(28)-en (**48**) structures were isolated from the Pacific ophiuroid *Ophiolepis superba* (D'Auria *et al.* 1989) (Fig. 15.21). This ophiuroid also contained two unprecedent

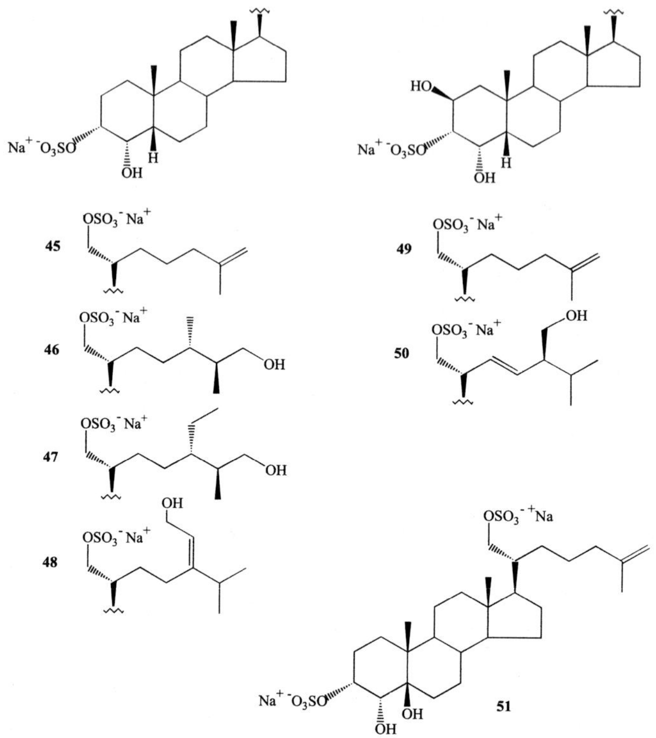

Fig. 15.21 Sulfated polyhydroxysteroids from *Ophiolepsis superba.*

sulfated polyhydroxysterols (**49** and **50**) with an additional hydroxy group at C-2 (β) and compound **51** with a 5β-cholest-25-ene-3α, 4α, 5, 21-tetrol 3, 21-disulfate structure. Compounds **45** and **51** were modestly active against HIV-1 and HIV-2 viruses (Mc Kee *et al.* 1994).

The second series of sulfated polyhydroxysteroids comprises 3, 21-disulfoxy-Δ^5-sterols that present additional hydroxy groups at C-1 (β), C-2 (α or β), C-4 (β), and C-11 (β). Sterol **52** (Fig. 15.22) is the most simple example of this series and has been isolated for the first time from the ophiuroid *Ophioderma longicaudum* together with the $\Delta^{5, 22E}$ (**53**) and 24-methyl-$\Delta^{5, 24(28)}$ (**54**) analogs (Riccio *et al.* 1985). Compound **52** has also been isolated from *Ophiocomina nigra, Ophiotrix fragilis, Ophiura texturata, Ophiozona impressa* and *Ophionotus victoriae* (D'Auria *et al.* 1995) together with the saturated analog **55**.

Four sterols related to **52** with an additional hydroxy group at C-4 (β) have been characterized (Fig. 15.23).

Sterol **56** is the most abundant disulfated steroid of this type in the ophiuroids *Ophiura sarsi* (Levina *et al.* 1990), *Ophiopholis aculeata* (Fedorov *et al.* 1994), *Ophiotrix fragilis, Ophiura texturata* and *Ophionotus victoriae*

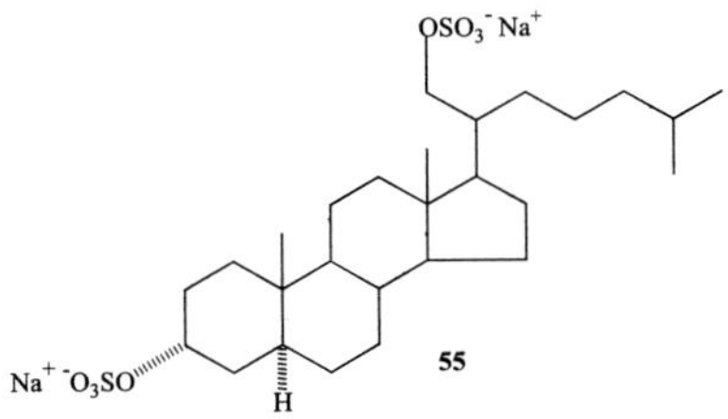

55

52 Δ^{5}

53 $\Delta^{5,\,22E}$

54 24-Me, $\Delta^{5,\,24(28)}$

Fig. 15.22 Simple 3, 21-disulfated polyhydroxysteroids in ophiuroids.

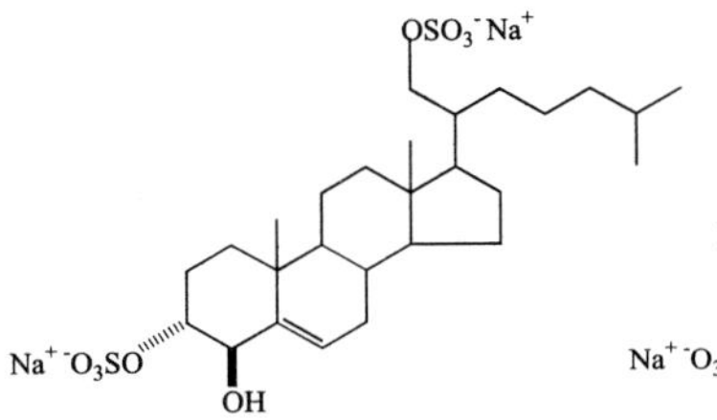

56

57 Δ^{22}

58 Δ^{25}

59 24-Me, $\Delta^{24(28)}$

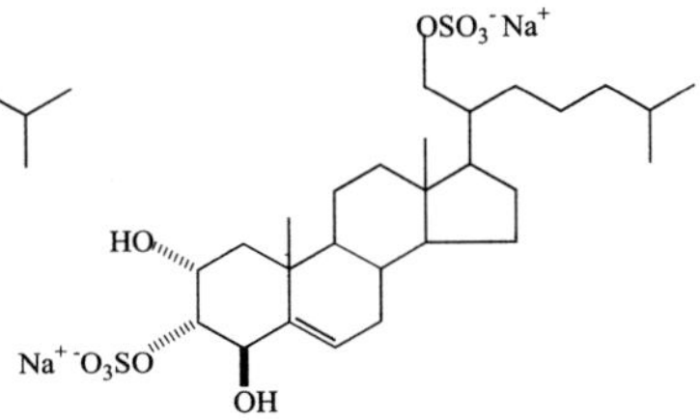

60

61 24-Me, $\Delta^{24(28)}$

Fig. 15.23 3, 21-Disulfated polyhydroxysteroids with additional hydroxy groups in ring A.

(D'Auria *et al.* 1995). Compounds **57** and **58** are Δ^{22} and Δ^{25} analogs of **56** and have been isolated as minor components from *O. aculeata* (Fedorov *et al.* 1994), while **57** was the principal disulfated sterol in *Ophiotrix fragilis* and *Ophiura texturata* (D'Auria *et al.* 1995). The latter two ophiuroids also contained the 24-methyl, $\Delta^{24(28)}$ derivative **59**. Recently, we have characterized two 3, 21-disulfoxy-Δ^{5}-sterols (compounds **60** and **61**) containing two additional hydroxy groups at C-2 (α) and C-4 (β) in the Antarctic ophiuroid *Astrotoma agassizi* (Roccatagliata *et al.* 1998). Compound **60** has been also isolated from two other ophiuroids from cold waters, *Ophiosparte gigas* (D'Auria *et al.* 1993b) and *Amphiophiura ponderosa* (Shubina *et al.* 1998).

Sulfated polyhydroxysteroids **56** and **60** have shown modest antiviral activity against HIV-1 and HIV-2 (Mc Kee *et al.* 1994). On the other hand it has been found that steroids **56** and **58** stimulated the influx of Ca^{2+} into different cell types from the extracellular medium and that sterol **56** was less active than sterol **58** despite the slight differences in structure (Agafonova *et al.* 2002). Both compounds may act as Ca^{2+}-agonists and increase Ca^{2+}-transport across cell membranes.

Two examples of 3, 21-disulfoxy-Δ^5-sterols with additional hydroxy groups at C-1(β) and C-11 (β) (compounds **62** and **63**) (Fig. 15.24) were isolated from the ophiuroids *Ophiarachna incrassata* (Fu *et al.* 1994) and *Ophionereis reticulata* (D'Auria *et al.* 1995). Sterol **62** showed moderate inhibition of protein tyrosine kinase (PTK) $pp60^{v\text{-}src}$ (Fu *et al.* 1994).

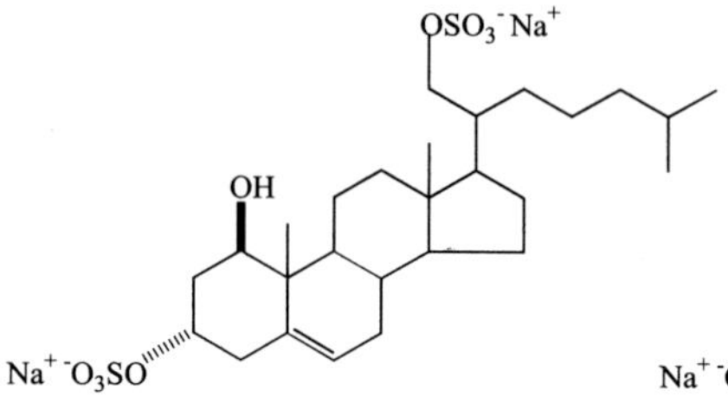

62

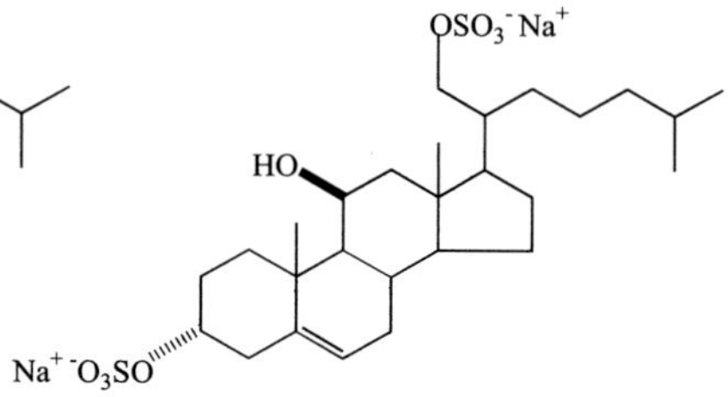

63

Fig. 15.24 Sulfated polyhydroxysteroids from *Ophiarachna incrassata* and *Ophionereis reticulata*.

Several disulfated 2β, 3α, 21-trihydroxy-Δ^5-sterols have been isolated from different ophiuroid species. All compounds have a sulfoxy group located at C-21 and an additional sulfate group at C-2 or C-3 (Fig. 15.25). Sterol **64** with a sulfate group at C-3 (α) has been isolated from the ophiuroids *Astrocladus exiguus* (Shubina *et al.* 1998), *Ophiotrix fragilis* and *Ophiura texturata* (D'Auria *et al.* 1995). This last ophiuroid also contained the Δ^{22} analogs **65** and **66**. Sulfated steroids **67** and **68** were isolated from the ophiuroid *Ophiopholis aculeata* (Fedorov *et al.* 1994) as an inseparable mixture of epimers at C-24. Recently, we were able to isolate from the Antarctic ophiuroid *Astrotoma agassizii* compound **69**, a new 2β-hydroxy-3α, 21-disulfoxisteroid with a terminal isopropenyl group in the side chain and *trans* A/B ring fusion (Roccatagliata *et al.* 1998). *A. agassizii* also contained the known sterol **70**, previously isolated from *Astrocladus exiguus* and *Asteronix loveni* (Shubina *et al.* 1998) and from *Ophiosparte gigas* (D'Auria *et al.* 1993b), and the novel Δ^{24} analog **71**. Compounds **70** and **71** are the only two examples of sulfated sterols isolated from ophiuroids with a sulfoxy group at the 2β-position instead of the common 3α-position. The relative stereochemistry of the protons on C-2 and C-3 was defined by the coupling constants for H-2 (br d, J = 2.5 Hz) and H-3 (br d, J = 2.5 Hz) and assigned as equatorial due to the absence of any large vicinal couplings. The sulfate group was located at C-2 on the basis of the chemical shifts of C-1 (39.0 ppm) and C-4 (35.8 ppm) and by the chemical shift value of H-2 (4.42 ppm). The axial orientation of the hydroxyl group at C-2 was also supported by a pyridine-induced solvent shift of the 19-methyl protons (Δ +0.58) of the desulfated derivative obtained by solvolysis (Roccatagliata *et al.* 1998).

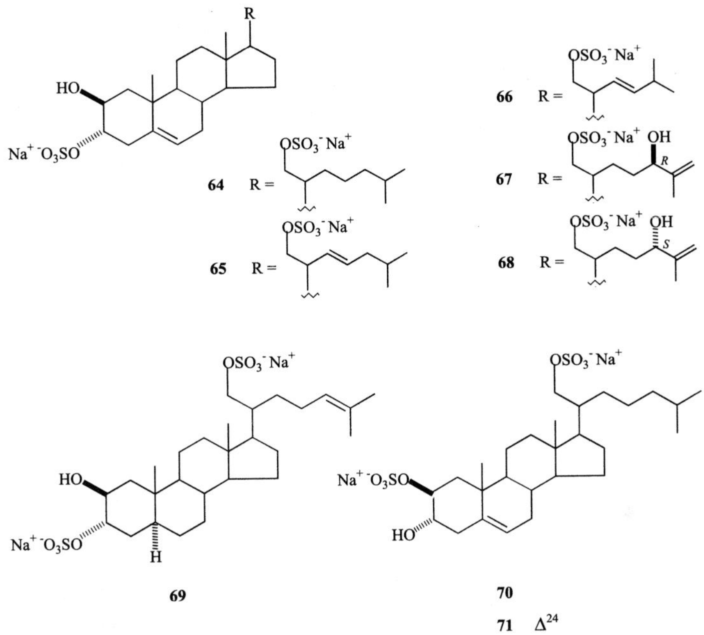

Fig. 15.25 Disulfated 2β, 3α, 21-trihydroxy-Δ^5-steroids.

We tested the antiviral activity of sterols **69**, **70** and **71** against three pathogenic viruses of humans: herpes simplex virus type 2 (HSV-2), Junin virus (JV) and polio virus type 3 (PV-3). The three compounds elicited a marked HSV-2 inhibitory activity. Sterol **70** also proved to be active against PV-3 at a concentration of 80 μg/ml, while **71** was inactive at this concentration and **69** was weakly active (Comin *et al.* 1999). These results prompted us to synthesize disodium 2β, 3α-dihydroxy-5α-cholestane (**72**) as well as monosulfated and acetylated derivatives (**73**-**76**) (Fig. 15.26) in order to evaluate the antiviral activity of these compounds and gain insight into structure-activity correlations (Garrido Santos *et al.* 2003).

The antiviral activity of the five compounds was tested against HSV-2. Compounds **72** and **74** were the most active with IC_{50} values of 19.3 and 23.9 μg/ml, respectively. Monosulfated steroid **73**, differing from **74** in the position of the sulfate group, as well as acetylated and sulfated compounds **75** and **76** were not active against HSV-2 at a concentration of 25 μg/ml. Several sulfated polyhydroxysterols have been recently isolated from sponges (Kornprobst *et al.* 1998). Most of these natural products are characterized by the 2β, 3α, 6α-tri-*O*-sulfate functions but several bioactive sterols sulfated at C-2 (β) and C-3 (α) exhibited antiviral activity. In order to determine the role of the sulfate groups in the

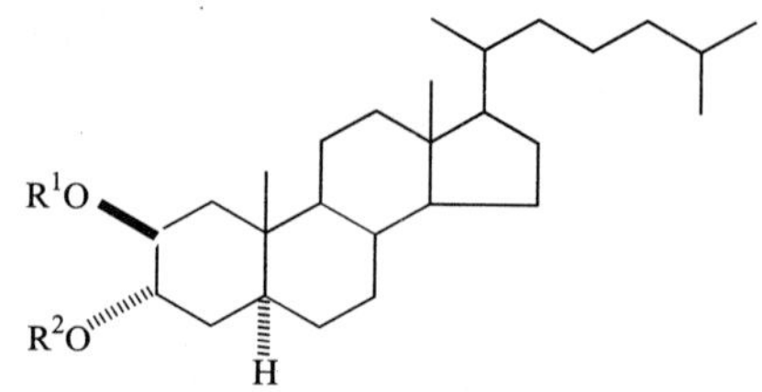

72 $R^1 = SO_3^-Na^+$; $R^2 = SO_3^-Na^+$

73 $R^1 = H$; $R^2 = SO_3^-Na^+$

74 $R^1 = SO_3^-Na^+$; $R^2 = H$

75 $R^1 = SO_3^-Na^+$; $R^2 = Ac$

76 $R^1 = Ac$; $R^2 = SO_3^-Na^+$

77 $R^1 = R^2 = OH$

Fig. 15.26 Synthetic sulfated 2*β*, 3*α*-dihydroxy-5*α*-cholestane derivatives.

antiviral activity of **72**, we evaluated the *in vitro* activity of diol **77** against HSV-2. Although compound **77** showed no cytotoxic effect on Vero cells it was not active aginst HSV-2 at a concentration of up to 50 μg/ml. These results are indicative of the role of the sulfate groups in the antiviral activity of **72** and this compound may be a good model for further introducing additional functional groups at the steroidal skeleton or its side chain in order to evaluate structure-activity correlations in the inhibitory activity against pathogenic viruses in humans.

The trisulfated sterol **78** and its Δ^5 analog **79** (Fig. 15.27) have been isolated from the ophiuroid *Ophiarachna incrassata* (D'Auria *et al.* 1987) as a mixture of 25*R* and 25*S* isomers. These compounds are the first examples of trisulfated polyhydroxysteroids isolated so far from ophiuroids. Although hydroxylation at C-26 is a common feature among the polyhydroxylated sterols isolated from starfishes (Roccatagliata *et al.* 1995) it is uncommon in sulfated sterols from ophiuroids.

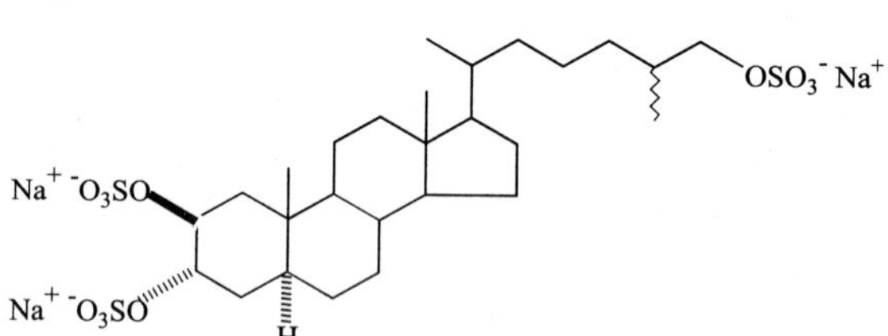

78

79 Δ^5

Fig. 15.27 Trisulfated polyhydroxysteroids from the ophiuroid *Ophiarachna incrassata*.

SPHINGOLIPIDS (CEREBROSIDES AND GANGLIOSIDES)

Glycosphingolipids are a large group of amphipathic molecules that contribute to the structural integrity of a cell and act as mediators of biological events such as cell recognition, intracellular communication and cell development (Spiegel 1989). Chemically, glycosphingolipids are composed of a lipophilic membrane anchor, the ceramide portion, linked at position 1 by a β-glycosidic bond to an hydrophilic carbohydrate moiety. The hydrophobic ceramide portion involves a sphingoid base and an amide-linked fatty acyl chain (Van Echten-Deckert 2000, Wiegandt, 1989). Cerebrosides, the simplest glycosphingolipids, consist of ceramides with a galactose or glucose unit as the sugar residue. Galactocerebrosides are present in the white matter of the brain, whereas glucocerebrosides are ubiquitous membrane components in plant and animal tissues. Gangliosides, the most complex group of glycosphingolipids, contain at least one sialic acid (*N*-acetylneuraminic acid or its derivative) residue and are especially enriched in the brain and nervous tissues.

In the marine environment, a number of glycosphingolipids have been isolated from algae, sponges, echinoderms, fungi and tunicates. The monoglucosylceramides are the most widespread glycosphingolipids in these organisms, although ceramide lactosides have been reported from the starfishes *Acanthaster planci* (Kawano *et al*. 1988) and *Luidia maculata* (Inagaki *et al*. 2003). Some of these compounds have shown biological activities. An example are the agelasphins, potent antitumor monoglucosylceramides isolated from the sponge *Agelas mauritiana* (Natori *et al*. 2000). Among echinoderms, glucosylceramides, gangliosides and sphingosine derivatives have been isolated from a sea urchin (Babu *et al*. 1997), ophiuroids (Inagaki *et al*. 2001), starfishes and sea cucumbers (Tan and Chen 2003). These compounds occur as complex mixtures, whose separation into individual components requires a combination of various extraction and chromatographic procedures. The application of reversed-phase HPLC to the purification of mixtures of sphingolipids allowed the separation of compounds displaying very similar structures (Christie 1987). Recently, we have reviewed the spectral characteristics of cerebrosides and gangliosides isolated from starfishes and sea cucumbers (Chludil *et al*. 2002a). Starfish and sea cucumber glucosylceramides contain several novel long-chain bases (LCB) derived from sphingosine ((2*S*, 3*R*, 4*E*)-2-aminooctadecene-1, 3-diol) as well as from the saturated 4-hydroxysphingosine (Fig. 15.28). Novel glucosylceramides have been isolated from the sea cucumbers *Cucumaria echinata, Pentacta australis* (Chludil *et al*. 2002a) and *Holothuria pervicax* (Yamada *et al*. 2002).

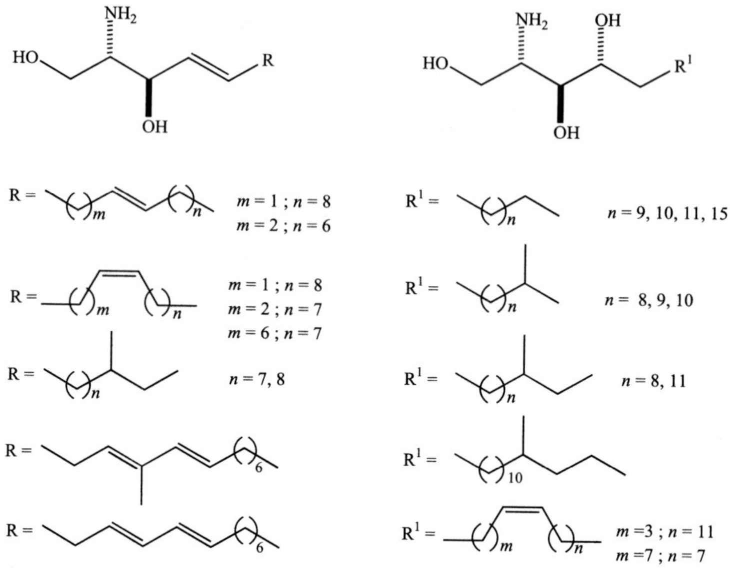

Fig. 15.28 LCB present in starfish and sea cucumber glucosylceramides.

A recent investigation on the water-insoluble lipid fraction from the chloroform/methanol/water extract of the gonads and body walls of the Patagonian starfish *Allostichaster inaequalis* led to the isolation and characterization of two new monoglucosylceramides, **80** and **81** (Díaz de Vivar *et al.* 2002) (Fig. 15.29).

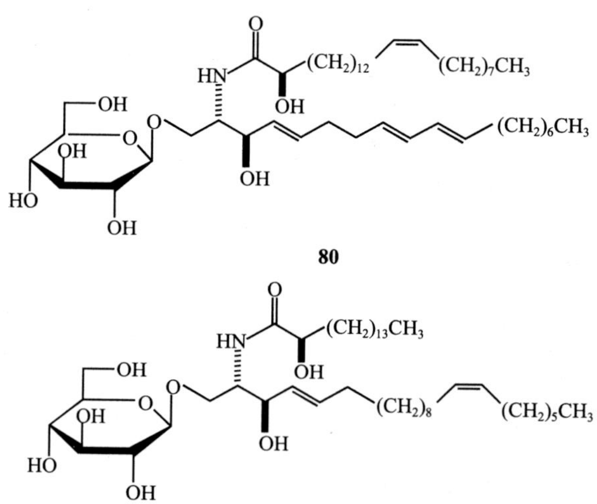

80

81

Fig. 15.29 New glucosylceramides isolated from the starfish *Allostichaster inaequalis*.

Cerebroside **80** containing (2*S*, 3*R*, 4*E*, 8*E*, 10*E*)-2-amino-4, 8, 10-octadecatriene-1, 3-diol, an unusual triunsaturated LCB with a conjugated diene, was also isolated from the Patagonian starfish *Anasterias minuta*, together with the new Anasterocerebroside A (**82**) (Chludil *el al*. 2003b) (Fig. 15.30).

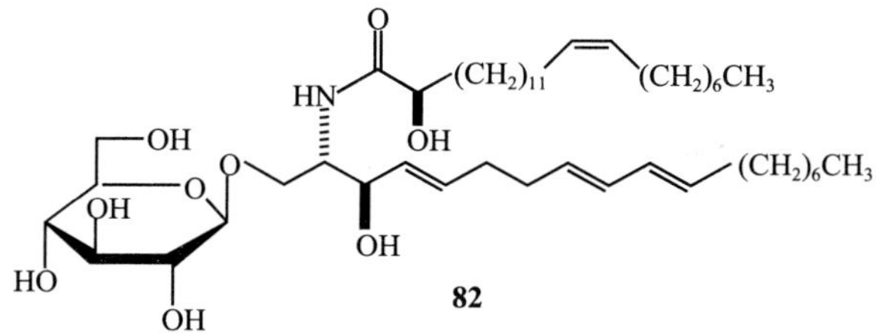

82

Fig. 15.30 Anasterocerebroside A isolated from the starfish *Anasterias minuta*.

The structures of the new glucosylceramides were established by spectroscopic and chemical methods. ^{1}H and ^{13}C-NMR spectra showed characteristic signals due to C-1-C-11, C-1′, C-2′, C-1″-C-6″ of a 1-*O*-*β*-glucopyranoside containing a sphingosine-type LCB moiety and a 2-hydroxy fatty acid (Chludil *et al*. 2002a). Application of homo- and hetero-nuclear NMR techniques (^{1}H-^{1}H COSY, HETCOR) allowed us to establish the olefinic pattern of the LCB unit of **80** and anasterocerebroside A (**82**) as a 4, 8, 10-triene. Negative FAB-MS confirmed the presence of a single molecular ion peak for each compound displaying a single peak on reversed-phase HPLC. Additional information that confirmed the structures proposed was obtained by acid methanolysis of each glucosylceramide yielding a fatty acid methyl ester (FAME), a methylglucoside and a LCB. GC-MS analysis of each FAME showed a single peak. The double-bond position in the FAME derived from glucosylceramide **80** and anasterocerebroside A (**82**) was determined by GC-MS analysis of the dimethyl disulfide derivatives (Vincentini *et al*. 1987). The absolute stereochemistry of glucose was determined as D by hydrolysis of each glucosylceramide with 2 N trifluoroacetic acid, reaction of the monosaccharide residue with L-cysteine methyl ester hydrochloride, acetylation and analysis and comparison with D- and L-glucose derivatives (Hara *et al*. 1987).

Gangliosides present in starfishes and sea cucumbers possess unique structures and biological activities. They are derived from (2*S*, 3*S*, 4*R*)-phytosphingosine-type ceramides possessing *α*-hydroxy and nonhydroxy fatty acids with D-glucose as the first sugar attached by a *β*-glycosidic bond to the ceramide moiety (Chludil *et al*. 2002a). A ganglioside molecular species isolated from the starfish *Leptychaster anomalus* contained sphingosines (predominantly C-20 and C-22) in their ceramide portion

(Smirnova and Chekareva 1997). The most simple gangliosides are disaccharides containing one unit of *N*-acetyl- or *N*-glycolylneuraminic acid and D-glucose. The presence of a lactosylceramide moiety has been established in gangliosides from the starfishes *Aphelasterias japonica, Luidia maculata, Linckia laevigata, Acanthaster planci, Astropecten latespinosus* and *Asterias amurensis* (Chludil *et al.* 2002a). An interesting characteristic is the presence of 8-*O*-methyl-*N*-acetyl or glycolylneuraminic acid in gangliosides isolated from the starfishes *Linckia laevigata* (Smirnova 2000a), *Evasterias echinosoma* (Smirnova, 2000b) and *Evasterias retifera* (Smirnova 2003).

Several gangliosides isolated from sea cucumbers have exhibited interesting biological activities (Yamada *et al.* 2001, Yamada 2002). In particular, gangliosides isolated from *Holothuria pervicax* (Yamada *et al.* 2000), *Cucumaria echinata* (Yamada *et al.* 1998), *Stichopus chloronotus* (Yamada *et al.* 2003) and *Stichopus japonicus* (Kaneko *et al.* 2003) have exhibited neuritogenic activity toward the rat pheochromocytoma cell line. One of these glycosphingolipids exhibited a unique carbohydrate moiety containing a branched sugar chain and an *N*-acetylgalactosamine residue (Kaneko *et al.* 2003) (Fig. 15.31).

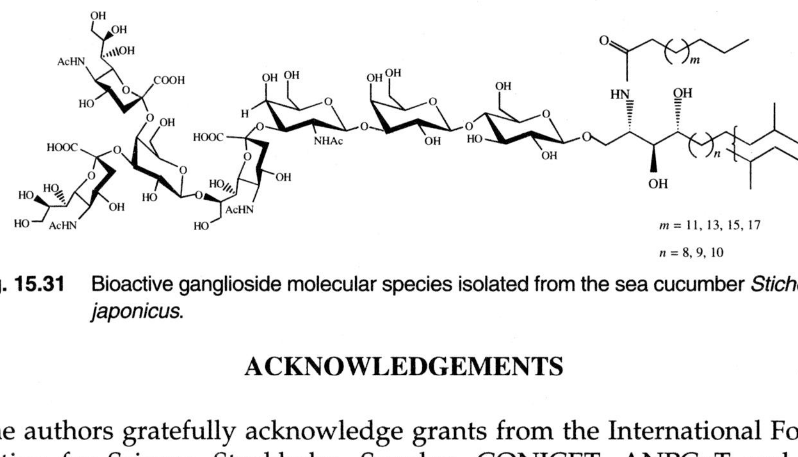

Fig. 15.31 Bioactive ganglioside molecular species isolated from the sea cucumber *Stichopus japonicus*.

ACKNOWLEDGEMENTS

The authors gratefully acknowledge grants from the International Foundation for Science, Stockholm, Sweden, CONICET, ANPCyT and the Universidad de Buenos Aires. M.S. Maier and A.P. Murray are Research Members of the National Research Council of Argentina (CONICET).

REFERENCES

Afiyatullov, Sh. Sh., T.A. Kuznetsova, V.V. Isakov, M.V. Pivkin, N.G. Prokof'eva, and G.B. Elyakov. 2000. New diterpenic altrosides of the fungus *Acremonium striatisporum* isolated from a sea cucumber. *J. Nat. Prod.* 63: 848–850.

Afiyatullov, Sh. Sh., A.I. Kalinovsky, T.A. Kuznetsova, V.V. Isakov, M.V. Pivkin, P.S. Dmitrenok, and G.B. Elyakov. 2002. New diterpene glycosides of the fungus *Acremonium striatisporum* isolated from a sea cucumber. *J. Nat. Prod.* 65: 641–644.

Afiyatullov, Sh. Sh., A.I. Kalinovsky, T.A. Kuznetsova, M.V. Pivkin, N.G. Prokof'eva, P.S. Dmitrenok, and G.B. Elyakov. 2004. New glycosides of the fungus *Acremonium striatisporum* isolated from a sea cucumber. *J. Nat. Prod.* 67: 1047–1051.

Agafonova, I.G., D.L. Aminin, L.K. Shubina, and S.N. Fedorov. 2002. Influence of polyhydroxysteroids on $[Ca^{2+}]_i$. *Steroids* 67: 695–701.

Agafonova, I.G., D.L. Aminin, S.A. Avilov, and V.A. Stonik. 2003. Influence of cucumariosides upon intracellular $[Ca^{2+}]_I$ and lysosomal activity of macrophages. *J. Agric. Food Chem.* 51: 6982–6986.

Aiello, A., E. Fattorusso, and M. Menna. 1999. Steroids from sponges: recent reports. *Steroids* 64: 687–714.

Avilov, S.A., A.I. Kalinovsky, V.I. Kalinin, V.A. Stonik, R. Riguera, and C. Jimenez. 1997. Koreoside A, a new nonholostane triterpene glycoside from the sea cucumber *Cucumaria koraiensis*. *J. Nat. Prod.* 60: 808–810.

Avilov, S.A. A.S. Antonov, A.S. Silchenko, V.I. Kalinin, A.I. Kalinovsky, P.S. Dmitrenok, V.A. Stonik, R. Riguera, and C. Jimenez. 2003. Triterpene glycoside from the Far Eastern sea cucumber *Cucumaria conicospermium*. *J. Nat. Prod.* 66: 910–916.

Avilov, S.A., V.I. Kalinin, and A.V. Smirnov. 2004. Use of triterpene glycosides for resolving taxonomic problems in the sea cucumber genus *Cucumaria* (Holothuroidea, Echinodermata). *Biochem. Syst. Ecol.* 32: 715–733.

Babu, U.V., S.P.S. Bhandari, and H.S. Garg. 1997. Temnosides A and B, two new glycosphingolipids from the sea urchin *Temnopleurus toreumaticus* of the Indian coast. *J. Nat. Prod.* 60: 732–734.

Blunt, J.W., B.R. Copp, M.H.G. Munro, P.T. Northcote, and M.R. Prinsep. 2003. Marine natural products. *Nat. Prod. Rep.* 20: 1–48.

Bonnard, I., and K. Reinhart. 2004. Thyonosides A and B, two new saponins isolated from the holothurian *Thyone aurea*. *Tetrahedron* 60: 2987–2992.

Bugni, T.S., and Ch.M. Ireland. 2004. Marine-derived fungi: a chemically and biologically diverse group of microorganisms. *Nat. Prod. Rep.* 21: 143–163.

Chludil, H.D., A.M. Seldes, and M.S. Maier. 2002a. Glycosphingolipids from starfishes and sea cucumbers. *Res. Adv. Lipids* 2: 9–18.

Chludil, H.D., A.M. Seldes, and M.S. Maier. 2002b. Antifungal steroidal glycosides from the Patagonian starfish *Anasterias minuta*: structure-activity correlations. *J. Nat. Prod.* 65: 153–157.

Chludil, H.D., C.C. Muniain, A.M. Seldes, and M.S. Maier. 2002c. Cytotoxic and antifungal triterpene glycosides from the Patagonian sea cucumber *Hemoiedema spectabilis*. *J. Nat. Prod.* 65: 860–865.

Chludil, H.D., A.P. Murray, A.M. Seldes, and M.S. Maier. 2003a. Biologically active triterpene glycosides from sea cucumbers (Holothuroidea, Echinodermata). Pages 587-615 in *Studies in Natural Products Chemistry*, Atta-ur-Rahman, ed., Elsevier Science B. V., Amsterdam.

Chludil, H.D., A.M. Seldes, and M.S. Maier. 2003b. Anasterocerebroside A, a new glucosylceramide from the Patagonian starfish *Anasterias minuta*. *Z. Naturforsch.* 58c: 433–440.

Christie, W.W. 1987. *High Performance Liquid Chromatography and Lipids*, Pergamon Press, Oxford.

Comin, M.J., M.S. Maier, A.J. Roccatagliata, C.A. Pujol, and E.B. Damonte. 1999. Evaluation of the antiviral activity of natural sulfated polyhydroxysteroids and their synthetic derivatives and analogs. *Steroids* 64: 335–340.

D'Auria, M.V., R. Riccio, E. Uriarte, L. Minale, S. La Barre, and J. Pusset. 1987. Novel marine steroid sulfates from Pacific ophiuroids. *J. Org. Chem.* 52: 3947–3952.

D'Auria, M.V., R. Riccio, E. Uriarte, L. Minale, J. Tanaka, and T. Higa. 1989. Isolation and structure elucidation of seven new polyhydroxylated sulfated sterols from the ophiuroid *Ophiolepis superba*. *J. Org. Chem.* 54: 234–239.

D'Auria, M.V., L. Gómez Paloma, L. Minale, R. Riccio, A. Zampella, J. Tanaka, and T. Higa. 1992. Isolation, structure characterization and conformational analysis of a unique 4α, 9α-epoxysteroid sulphate from the Okinawan ophiuroid *Ophiomastix annulosa*. *Tetrahedron Lett.* 33: 4641–4644.

D'Auria, M.V., L. Minale, and R. Riccio. 1993a. Polyoxygenated steroids of marine origin. *Chem. Rev.* 93: 1839–1895.

D'Auria, M.V., L. Gómez Paloma, L. Minale, R. Riccio, and A. Zampella. 1993b. Isolation and structure characterization of two novel bioactive sulphated polyhydroxysteroids from the Antarctic ophiuroid *Ophiosparte gigas*. *Nat. Prod. Lett.* 3: 197–201.

D'Auria, M.V., L. Gómez Paloma, L. Minale, R. Riccio, and A. Zampella. 1995. On the composition of sulfated polyhydroxysteroids in some ophiuroids and the structure determination of six new constituents. *J. Nat. Prod.* 58: 189–196.

De Marino, S., N. Borbone, M. Iorizzi, G. Esposito, J.B. McClintock, and F. Zollo. 2003. Bioactive asterosaponins from the starfish *Luidia quinaria* and *Psilaster cassiope*. Isolation and structural characterization by two-dimensional NMR spectroscopy. *J. Nat. Prod.* 66: 515–519.

De Simone, F., A. Dini, E. Finamore, L. Minale, C. Pizza, R. Riccio, and F. Zollo. 1981. Starfish saponins. Part 5. Structure of sepositoside A, a novel steroidal cyclic glycoside from the starfish Echinaster sepositus. *J. Chem. Soc. Perkin Trans.* I 1855–1862.

Díaz de Vivar, M.E., M.S. Maier, and A.M. Seldes. 1999. Polar metabolites from the Antarctic starfish *Labidiaster annulatus*. *Anales Asoc. Quím. Argent.* 87: 247–253.

Díaz de Vivar, M.E., A.M. Seldes, and M.S. Maier. 2002. Two novel glucosylceramides from gonads and body walls of the Patagonian starfish *Allostichaster inaequalis*. *Lipids* 37: 597–603.

Dini, A., F.A. Mellon, L. Minale, C. Pizza, R. Riccio, R. Self, and F. Zollo. 1983. Starfish saponins-XI. Isolation and partial characterization of the saponins from the starfish *Marthasterias glacialis*. *Comp. Biochem. Physiol.* 76B: 839–844.

Drozdova, O.A., S.A. Avilov, A.I. Kalinovsky, V.A. Stonik, Y.M. Milgrom, and Y.W. Rashkes. 1993. New glycosides from the holothurian *Cucumaria japonica*. *Khim. Prirod. Soedin* 2: 242–248.

Dubois, M.A., R. Higuchi, T. Komori, and T. Sasaki. 1988. Structures of two new oligoglycoside sulfates, pectiniosides E and F, and biological activities of the six new pectiniosides. *Liebigs Ann. Chem.* 845–850.

Fedorov, S.N., E.V. Levina, A.I. Kalinovsky, P.S. Dmitrenok, and V.A. Stonik. 1994. Sulfated steroids from Pacific brittle stars *Ophiopholis aculeata, Ophiura sarsi*, and *Stegophiura brachiactis*. *J. Nat. Prod.* 57: 1631–1637.

Findlay, J.A., Z.-Q. He, and B. Blackwell. 1990. Minor saponins from the starfish *Asterias forbesi*. *Can. J. Chem.* 68: 1215–1217.

Findlay, J.A., N. Yayli, and L. Radics. 1992. Novel sulfated oligosaccharides from the sea cucumber *Cucumaria frondosa*. *J. Nat. Prod.* 55: 93–101.

Fu, X., F.J. Schmitz, R.H. Lee, J.S. Papkoff, and D.L. Slate. 1994. Inhibitors of protein tyrosine kinase $pp60^{v\text{-}src}$: sterol sulfates from the brittle star *Ophiarachna incrassata*. *J. Nat. Prod.* 57: 1591–1594.

Garrido Santos, G.A., A.P. Murray, C.A. Pujol, E.B. Damonte, and M.S. Maier. 2003. Synthesis and antiviral activity of sulfated and acetylated derivatives of 2β, 3α-dihydroxy-5α-cholestane. *Steroids* 68: 125–132.

Hara, Sh., H. Okabe, and K. Mihashi. 1987. Gas-liquid chromatography separation of aldose enatiomers as trimethylsilyl ethers of methyl 2-(polyhydroxyalkyl)-thiazolidine-4(*R*)-carboxylates. *Chem. Pharm. Bull.* 35: 501–506.

Hart, J.B., R.E. Lill, S.J.H. Hickford, J.W. Blunt, and H.G. Murray. 2000. The halichondrins: chemistry, biology, supply and delivery. Pages 134-153 in *Drugs from the Sea*, N. Fusetani, ed., Karger, Basel.

Hedge, V.R., T.-M. Chan, H. Pu, V.P. Gullo, M.G. Patel, P. Das, N. Wagner, P.S. Parameswaran, and C.G. Naik. 2002. Two selective novel triterpene glycosides from the sea cucumber *Telenata ananas*: inhibitors of chemokine receptor-5. *Bioorg. Med. Chem. Lett.* 12: 3203–3205.

Homans, A.L., and A. Fuchs. 1970. Direct bioautography on thin-layer chromatograms as a method for detecting fungitoxic substances. *J. Chromatog.* 51: 327–329.

Honda, M., and T. Komori. 1986. Structures of thornasterols A and B (biologically active glycosides from Asteroidia, XI). *Tetrahedron Lett.* 27: 3369–3372.

Inagaki, M., M. Shibai, R. Isobe, and R. Higuchi. 2001. Constituents of Ophiuroidea.1.Isolation and structure of three ganglioside molecular species from the brittle star *Ophiocoma scolopendrina*. *Chem. Pharm. Bull.* 49: 1521–1525.

Inagaki, M., K. Hakamura, S. Kawatake, and R. Higuchi. 2003. Isolation and structural determination of four new ceramide lactosides from the starfish *Luidia maculata*. *Eur. J. Org. Chem.* 325–331.

Iorizzi, M., F. De Riccardis, L. Minale, and R. Riccio. 1993. Starfish saponins, Part 52. Chemical constituents from the starfish *Echinaster brasiliensis*. *J. Nat. Prod.* 56: 2149–2162.

Iorizzi, M., S. De Marino, L. Minale, F. Zollo, V. Le Bert, and C. Roussakis. 1996. Investigation of the polar steroids from an Antarctic starfish of the family *Echinasteridae*: isolation of twenty seven polyhydroxysteroids and steroidal oligoglycosides, structures and biological activities. *Tetrahedron* 52: 10997–11012.

Iorizzi, M., and R. Riccio. 1999. Steroidal oligoglycosides from marine sources. Pages 345-397 in *Naturally Occurring Glycosides*, R. Ikan, ed., Wiley & Sons Ltd., New York.

Iorizzi, M., S. De Marino, and F. Zollo. 2001. Steroidal glycosides from Asteroidea. *Curr. Org. Chem.* 5: 951–973.

Itakura, Y., T. Komori, and T. Kawasaki, 1983. Steroid oligoglycosides from the starfish *Asterias amurensis* [cf.] *versicolor* Sladen, 1 Structural elucidation of a new oligoglycoside sulfate. *Liebigs Ann. Chem.* 2079–2091.

Ivanchina, N.V., A.A. Kicha, A.I. Kalinovsky, P.V. Dmitrenok, N.G. Prokofeva, and V.A. Stonik. 2001. New steroid glycosides from the starfish *Asterias rathbuni*. *J. Nat. Prod.* 64: 945–947.

Ivanchina, N.V., A.A. Kicha, A.I. Kalinovsky, P.V. Dmitrenok, and V.A. Stonik. 2003. Hemolytic steroid disulfates from the Far Eastern starfish *Pteraster pulvillus*. *J. Nat. Prod.* 66: 298–301.

Kalinin, V.I., N.G. Prokofieva, G.N. Likhatskaya, E.B. Shentsova, I.G. Agafonova, S.A. Avilov, and O.A. Drozdova. 1996a. Hemolytic activity of triterpene glycosides from the dendrochirotida order holothurian. Pages 557-564 in *Saponins Used in Traditional and Modern Medicine*, G.R. Waller, and K. Yamasaki, eds., Plenum Press, New York.

Kalinin, V.I., M.M. Anisimov, N.G. Prokofieva, S.A. Avilov, Sh. Sh. Afiyatullov, and V.A. Stonik. 1996b. Biological activity and biological role of triterpene glycosides from holothuroids (Echinodermata). Pages 139-181 in *Echinoderm Studies* 5, M. Jangoux, J.M. Lawrence, eds., A.A. Balkema, Rotterdam.

Kalinovsky, A.I., E.V. Levina, V.A. Stonik, P.S. Dmitrenok, and P.V. Andrijashenko. 2004. Steroid polyols from the Far Eastern starfish *Henricia sanguinolenta* and *H. leviuscula leviuscula*. *Russian J. Bioorg. Chem.* 30: 191–195.

Kaneko, M., F. Kisa, K. Yamada, T. Miyamoto, and R. Higuchi. 2003. Structure of a new neuritogenic-active ganglioside from the sea cucumber *Stichopus japonicus*. *Eur. J. Org. Chem.* 1004–1008.

Kawano, Y., R. Higuchi, R. Isobe, and T. Komori. 1988. Isolation and structure of two new ceramide lactosides. *Liebigs Ann. Chem.* 1181–1183.

Kicha, A.A., A.I. Kalinovsky, E.V. Levina, and P.V. Andrijashenko. 1986. Culcitosides C_2 and C_3 from the starfish *Culcita novaeguineae*. *Khim. Prir. Soedin 46*: 592–596.

Kicha, A.A., A.I. Kalinovsky, and V.A. Stonik. 1997. New polyhydroxysteroids and steroid glycosides from the Far East starfish *Ceramaster patagonicus*. *Izv. Acad. Nauk, Ser. Khim*. 190–195.

Kicha, A.A., A.I. Kalinovsky, N.V. Ivanchina, and V.A. Stonik. 1999. New steroid glycosides from the deep-water starfish *Mediaster murrayi*. *J. Nat. Prod*. 62: 279–282.

Kicha, A.A., N.V. Ivanchina, A.I. Kalinovsky, P.S. Dmitrenok, and V.A. Stonik 2003. Alkaloidosteroids from the starfish *Lethasterias nanimensis chelifera*. *Tetrahedron Lett*. 44: 1935–1937.

Kijjoa, A., and P. Sawangwong. 2004. Drugs and cosmetics from the sea. *Mar. Drugs* 2: 73–82.

Kitagawa, I., and M. Kobayashi. 1978. Saponin and sapogenol. XXVI. Steroidal saponins from the starfish *Acanthaster planci* L. (Crown of thorns). (2). Structure of the major saponin Thornasteroside A. *Chem. Pharm. Bull*. 26: 1864–1873.

Kitagawa, I., T. Nishino, M. Kobayashi, and Y. Kyogoku. 1981. Marine natural products. VIII. Bioactive triterpene-oligoglycosides from the sea cucumber *Holothuria leucospilota* Brandt (2). Structure of holothurin A. *Chem. Pharm. Bull*. 29: 1951–1956.

Kitagawa, I., M. Kobayashi, M. Hori, and Y. Kyogoku. 1989. Marine natural products. XVIII. Four lanostane-type triterpene oligoglycosides, bivittosides A, B, C, and D, from the Okinawan sea cucumber *Bohadschia bivittata* Mitsukuri, *Chem. Pharm. Bull*. 37: 61–67.

Komori, T. 1997. Toxins from the starfish *Acanthaster planci* and *Asterina pectinifera*. *Toxicon* 35: 1537–1548.

Kornprobst, J.-M., C. Sallenave, and G. Barnathan. 1998. Sulfated compounds from marine organisms. *Comp. Biochem. Physiol*. 119B: 1–51.

Krebs, H. Chr., T. Komori, and T. Kawasaki. 1984. Steroid-oligoglycoside aus dem Seestern *Luidia maculata* Müller et Troschel, 2. Die Strukturen von zwi neuen Oligoglycosidsulfaten. *Liebigs Ann. Chem*. 296–305.

Levina, E.V., S.N. Fedorov, V.A. Stonik, P.V. Andriyashchenko, A.I. Kalinovskii, and V.V. Isakov. 1990. Steroid compounds from Ophiuroids. II. Sulfated steroids from *Ophiura sarsi* and *Ophiura leptoctenia*. *Khim. Prirod. Soedin* 4: 483–487.

Levina, E.V., P.V. Andriyashchenko, A.I. Kalinovsky, and V.A. Stonik. 1998. New ophiuroid-type steroids from the starfish *Pteraster tesselatus*. *J. Nat. Prod*. 61: 1423–1426.

Maier, M.S., A. Roccatagliata, and A.M. Seldes. 1993. Two novel steroidal glycoside sulfates from the starfish *Cosmasterias lurida*. *J. Nat. Prod*. 56: 939–942.

Maier, M.S., A. Roccatagliata, A. Kuriss, H. Chludil, A.M. Seldes, C.A. Pujol, and E.B. Damonte. 2001. Two new cytotoxic and virucidal trisulfated triterpene glycosides from the Antarctic sea cucumber *Staurocucumis liouvillei*. *J. Nat. Prod*. 64: 732–736.

Mayer, A.M.S., and M.T. Hamann. 2002. Marine pharmacology in 1999: compounds with antibacterial, anticoagulant, antifungal, anthelmintic, anti-inflammatory, antiplatelet, antiprotozoal and antiviral activities affecting the cardiovascular, endocrine, immune and nervous systems, and other miscellaneous mechanisms of action. *Comp. Biochem. Physiol*. 132C: 315–339.

Mc Kee, T.C., J.H. Cardellina, II, R. Riccio, M.V. D' Auria, M. Iorizzi, L., Minale, R.A. Moran, R.J. Gulakowski, J.B. McMahon, R.W. Buckheit, Jr., K.M. Snader, and M.R. Boyd. 1994. HIV-Inhibitory Natural Products. 11. Comparative studies of sulfated sterols from marine invertebrates. *J. Med. Chem*. 37: 793–797.

Minale, l., R. Riccio, and F. Zollo. 1993. Steroidal oligoglycosides and polyhydroxylated steroids from echinoderms. Pages 75-308 in *Progress in the Chemistry of Organic Natural Products*, W. Herz, G.W. Kirby, R.E. Moore, W. Steglich, and Ch. Tamm, eds., Springer-Verlag, New York.

Minale, l., M. Iorizzi, E. Palagiano, and R. Riccio. 1996. Steroid and triterpenoid oligoglycosides of marine origin. Pages 335-356 in *Saponins Used in Traditional and Modern Medicine,* G.R. Waller, and K. Yamasaki, eds., Plenum Press, New York.

Moraes, G., P.C. Norhcote, V.I. Kalinin, S.A. Avilov, A.S. Silchenko, P.S. Dmitrenok, V.A. Stonik, and V.S. Levin. 2004. Structure of the major triterpene glycoside from the sea cucumber *Stichopus mollis* and evidence to reclassify this species into the new genus *Australostichopus*. *Biochem. Syst. Ecol.* 32: 637–650.

Murray, A.P., C. Muniain, A.M. Seldes, and M.S. Maier. 2001. Patagonicoside A: a novel antifungal disulfated triterpene glycoside from the sea cucumber *Psolus patagonicus*. *Tetrahedron* 57: 9563–9568.

Murray, A.P., E. Araya, M.S. Maier, and A.M. Seldes. 2002. Nucleosides and nucleobases from *Ophiactis asperula, Ophiacantha vivipara* and *Gorgonocephalus chilensis*. *Biochem. Syst. Ecol.* 30: 259–262.

Natori, T., K. Motoki, T. Higa, and Y. Koezuka. 2000. KRN7000 as a new type of antitumor and immunostimulatory drug. Pages 86-97 in *Drugs from the Sea,* N. Fusetani, ed., Karger, Basel.

Paul, V.J., and M.P. Puglisi. 2004. Chemical mediation of interactions among marine organisms. *Nat. Prod. Rep.* 21: 189–209.

Pivkin, M.V. 2000. Filamentous fungi associated with holothurians from the sea of Japan, off the primorye coast of Russia. *Biol. Bull.* 198: 101–109.

Ponomarenko, L.P., A.I. Kalinovsky, O.P. Moiseenko, and V.A. Stonik. 2001. Free sterols from the holothurians *Synapta maculata, Cladolabes bifurcates* and *Cucumaria* sp. *Comp. Biochem. Physiol.* 128B: 53–62.

Prokof'eva, N.G., E.L. Chaikina, A.A. Kicha, and N.V. Ivanchina, 2003. Biological activities of steroid glycosides from starfish. *Comp. Biochem. Physiol.* 134B: 695–701.

Qi, J., M. Ojika, and Y. Sakagami. 2002. Linckosides A and B, two new neuritogenic steroid glycosides from the Okinawan starfish *Linckia laevigata*. *Bioorg. Med. Chem.* 10: 1961–1966.

Riccio, R., A. Dini, L. Minale, C. Pizza, F. Zollo, and T. Sevenet. 1982. Starfish saponins VII: Structure of luzonicoside, a further steroidal cyclic glycoside from the Pacific starfish *Echinaster luzonicus*. *Experientia* 38: 68–70.

Riccio, R., M.V. D'Auria, and L. Minale. 1985. Unusual sulfated marine steroids from the ophiuroid *Ophioderma longicaudum*. *Tetrahedron* 41: 6041–6046.

Riccio, R., M.V. D'Auria, and L. Minale. 1986a. Two new steroidal glycoside sulfates, Longicaudoside-A and –B, from from the Mediterranean ophiuroid *Ophioderma longicaudum*. *J. Org. Chem.* 51: 533–536.

Riccio, R., M. Iorizzi, and L. Minale. 1986b. Starfish saponins XXX. Isolation of sixteen steroidal glycosides and three polyhydroxysteroids from the Mediterranean starfish *Coscinasterias tenuispina*. *Bull. Soc. Chim. Belg.* 95: 869–893.

Roccatagliata, A.J., M.S. Maier, A. M. Seldes, M. Iorixxi, and L. Minale. 1994. Starfish saponins, Part 2. Steroidal oligoglycosides from the starfish *Cosmasterias lurida*. *J. Nat. Prod.* 57: 747–754.

Roccatagliata, A.J., M.S. Maier, and A.M. Seldes. 1995. A novel marine steroidal sulfate from the starfish *Luidia ludwigi*. *J. Nat. Prod.* 58: 1941–1944.

Roccatagliata, A.J., M.S. Maier, A.M. Seldes, C.A. Pujol, and E.B. Damonte. 1996. Antiviral sulfated steroids from the ophiuroid *Ophioplocus januarii*. *J. Nat. Prod.* 59: 887–889.

Roccatagliata, A.J., M.S. Maier, A.M. Seldes, S. Zea, and C. Duque. 1997. A new sulfated alkene from the ophiuroid *Ophiocoma echinata*. *J. Nat. Prod.* 60: 285–286.

Roccatagliata, A.J., M.S. Maier, and A.M. Seldes. 1998. New sulfated polyhydroxysteroids from the Antarctic ophiuroid *Astrotoma agassizii*. *J. Nat. Prod.* 61: 370–374.

Rodríguez, J., R. Castro, and R. Riguera. 1991. Holothurinosides: new antitumor nonsulphated triterpenoid glycosides from the sea cucumber *Holothuria forskalii*. *Tetrahedron* 47: 4753–4762.

Shubina, L.K., S.N. Fedorov, E.V. Levina, P.V. Andriyaschenko, A.I. Kalinovsky, A.I. Stonik, and I.S. Smirnov. 1998. Comparative study on polyhydroxylated steroids from echinoderms. *Comp. Biochem. Physiol.* 119 B: 505–511.

Silchenko, A.S., S.A. Avilov, A.S. Antonov, V.I. Kalinin, A.I. Kalinovsky, A.V. Smirnov, R. Riguera, and C. Jimenez. 2002. Triterpene glycosides from the deep-water north-Pacific sea cucumber *Synallactes nozawai* Mitsukuri. *J. Nat. Prod.* 65: 1802–1808.

Smirnova, G.P., and N.V. Chekareva. 1997. A ganglioside from starfish *Leptychaster anomalus* with subterminal N-acetylneuraminic acid residue. *Bioorganicheskaya Khimiya* 23: 586–590.

Smirnova, G.P. 2000a. Hematoside with 8-*O*-methyl-N-glycolylneuraminic acid from the starfish *Linckia laevigata. Russ. Chem. Bull.* 49: 165–168.

Smirnova, G.P. 2000b. Gangliosides from the starfish *Evasterias echinosoma*. Identification of a disialoganglioside containing 8-*O*-methyl-N-acetylneuraminic acid and N-formylgalactosamine. *Russ. Chem. Bull.* 49: 159–164.

Smirnova, G.P. 2003. Structure of gangliosides from gonads of the starfish *Evasterias retifera. Russ. Chem. Bull.* 52: 2270–2275.

Spiegel, S. 1989. Bimodal regulation of cell growth by endogenous gangliosides. Pages 17-30 in *Gangliosides and Cancer*, H.F. Oettgen, ed., VCH, New York.

Stonik, V.A., V.I. Kalinin and S.A. Avilov. 1999. Toxins from sea cucumbers (Holothuroids): chemical structures, properties, taxonomic distribution, biosynthesis and evolution. *J. Nat. Toxins* 8: 235–248.

Tan, R.X., and J.H. Chen. 2003. The cerebrosides. *Nat. Prod. Rep.* 20: 509–534.

Van Echten-Deckert, G. 2000. Sphingolipid extraction and analysis by thin-layer chromatography. *Methods Enzymol.* 312: 64–79.

Vásquez, M.J., E. Quiñoá, R. Riguera, A. San Martín, and J. Dárias. 1992. Santiagoside, the first asterosaponin from an Antarctic starfish (*Neosmilaster georgianus*). *Tetrahedron* 48: 6739–6746.

Verbist, J.F. 1993. Pharmacological effects of compounds from echinoderms. Pages 111-186 in *Echinoderm Studies* 4, M. Jangoux, and J.M. Lawrence, eds., A.A. Balkem, Amsterdam.

Vincentini, M., G. Guglielminetti, G. Cassani, and C. Tonini. 1987. Determination of double bond position in diunsaturated compounds by mass spectrometry of dimethyl disulfide derivatives. *Anal. Chem.* 59: 694–699.

Wang, W., F. Li, N. Alam, Y. Liu, J. Hong, Ch.-K. Lee, K.S. Im, and J.H. Jung. 2002. New saponins from the starfish *Certonardoa semiregularis. J. Nat. Prod.* 65: 1649–1656.

Wang, W., F. Li, Y. Park, J. Hong, Ch.-O. Lee, J.Y. Kong, S. Shin, K.S. Im, and J.H. Jung. 2003. Bioactive sterols from the starfish *Certonardoa semiregularis. J. Nat. Prod.* 66: 384–391.

Wang, W., J. Hong, Ch.-O. Lee, K.S. Im, J.S. Choi, and J.H. Jung. 2004. Cytotoxic sterols and saponins from the starfish *Certonardoa semiregularis. J. Nat. Prod.* 67: 584–591.

Wiegandt, H. 1989. The chemical structures of gangliosides. Pages 5-16 in *Gangliosides and Cancer*, H.F. Oettgen, ed., VCH, New York.

Yamada, K. 2002. Chemo-pharmaceutical studies on the glycosphingolipid constituents from echinoderm, sea cucumbers, as the medicinal materials. *Yakugaku Zasshi* 122: 1133–1143.

Yamada, K., E. Hara, T. Miyamoto, R. Higuchi, R. Isobe, and S. Honda. 1998. Constituents of Holothuroidea, 6. Isolation and structure of biologically active glycosphingolipids from the sea cucumber *Cucumaria echinata. Eur. J. Org. Chem.* 371–378.

Yamada, K., Y. Harada, T. Miyamoto, R. Isobe, and R. Higuchi. 2000. Constituents of Holothuroidea, 9. Isolation and structure of a new ganglioside molecular species from the sea cucumber *Holothuria pervicax. Chem. Pharm. Bull.* (Tokyo) 48: 157–159.

Yamada, K., R. Matsubara, M. Kaneko, T. Miyamoto, and R. Higuchi. 2001. Constituents of Holothuroidea, 10. Isolation and structure of a biologically active ganglioside molecular species from the sea cucumber *Holothuria leucospilota. Chem. Pharm. Bull.* (Tokyo) 49: 447–452.

Yamada, K., K. Sasaki, Y. Harada, R. Isobe, and R. Higuchi. 2002. Constituents of Holothuroidea, 12. Isolation and structure of glucocerebrosides from the sea cucumber *Holothuria pervicax*. *Chem. Pharm. Bull.* (Tokyo) 50: 1467–1470.

Yamada, K., A. Hamada, F. Kisa, T. Miyamoto, and R. Higuchi. 2003. Constituents of Holothuroidea, 13. Structure of neuritogenic active ganglioside molecular species from the sea cucumber *Stichopus chloronotus*. *Chem. Pharm. Bull.* (Tokyo) 51: 46–52.

Zou, Z.-R., Y.-H. Yi, H.-M. Wu, J.-H. Wu, Ch.-Ch. Liaw, and K.-H. Lee. 2003. Intercedensides A-C, three new cytotoxic triterpene glycosides from the sea cucumber *Mensamaria intercedens* Lampert. *J. Nat. Prod.* 66: 1055–1060.

Index